D0778792

OPTICAL RADIATION MEASUREMENTS

Volume 5

VISUAL MEASUREMENTS

OPTICAL RADIATION MEASUREMENTS

A Treatise

Edited by FRANC GRUM and C. JAMES BARTLESON

Volume 1 **Radiometry**
Franc Grum and Richard J. Becherer

Volume 2 **Color Measurement**
Edited by Franc Grum and C. James Bartleson

Volume 3 **Measurement of Photoluminescence**
Edited by Klaus D. Mielenz

Volume 4 **Physical Detectors of Optical Radiation**
W. Budde

Volume 5 **Visual Measurements**
Edited by C. James Bartleson and Franc Grum

OPTICAL RADIATION MEASUREMENTS

Volume 5

Visual Measurements

Edited by

C. JAMES BARTLESON

Research Laboratories
Eastman Kodak Company
Rochester, New York

and

FRANC GRUM

Rochester Institute of Technology
Rochester, New York

1984

ACADEMIC PRESS, INC.
A Subsidiary of Harcourt Brace Jovanovich, Publishers
Orlando San Diego San Francisco New York London
Toronto Montreal Sydney Tokyo São Paulo

ACADEMIC PRESS, INC.
Orlando, Florida 32887

United Kingdom Edition published by
ACADEMIC PRESS, INC. (LONDON) LTD.
24/28 Oval Road, London NW1 7DX

Library of Congress Cataloging in Publication Data
Main entry under title:

Optical radiation measurements.

Includes bibliographies and index.
Contents: v. 1. Grum, F. C., Becherer, R. Radiometry.
--v. 2. Grum, F., Bartleson, C. J. Color measurement.--
[etc.].--v. 5. Bartleson, C. J., Grum F. Visual measurement.
1. Radiation--measurement. 2. Optical measurements.
3. Colorimetry. I. Grum, Franc C.
QC475.067 539.2'028'7 78-31412
ISBN 0-12-304905-9 (v. 5)

PRINTED IN THE UNITED STATES OF AMERICA

84 85 86 87 9 8 7 6 5 4 3 2 1

Contents

Part III Advanced Methods of Photometry and Colorimetry

Contributors

Numbers in parentheses indicate the pages on which the authors' contributions begin.

C. J. BARTLESON (1, 225, 367, 441, 491, 510, 615), Research Laboratories, Eastman Kodak Company, Rochester, New York 14650

ROBERT M. BOYNTON (183, 335), Department of Psychology, University of California, San Diego, La Jolla, California 92093

G. A. FRY (11, 131), College of Optometry, The Ohio State University, Columbus, Ohio 43210

PETER K. KAISER (563), Department of Psychology, York University, Downsview, Ontario M3J 1P3, Canada

Preface

This volume in the treatise "Optical Radiation Measurements" is devoted to visual measurement of light. It differs from the other volumes in the treatise in that it deals with the attempt to derive scales that are isomorphic with the perception of light rather than physical measurements of that light or conditions for determining characteristics of light that match some stimulus. Basically, this book is an introductory text on visual psychophysics as applied to the measurement of light. The emphasis is on applications, and considerable effort has been made to provide the reader with information that is necessary to the successful practice of light measurement by visual means. In most applications of optical radiation measurements, the purpose of the measurement is to express what is seen in quantitative terms. Consequently, this treatise would not be complete without a volume on visual measurement of optical radiations.

The book is divided into three major sections. They deal with characteristics of the visual mechanism, summarized in simplified expositions that should provide a basic understanding of the detection and signal-processing stages of that mechanism as a means for measuring light; with the methodology of psychophysical measurement applied to light measurement; and with advanced methods of photometric and colorimetric application of those methods.

We hope that by treating the subject in this way it will be possible for the reader to learn enough about the procedures for measuring light by psychophysical methods that he will be able to conduct experiments and make measurements in such a way as to avoid major pitfalls and enjoy the satisfaction of success.

Franc Grum
C. James Bartleson

1

Introduction

C. J. BARTLESON
Research Laboratories, Eastman Kodak Company
Rochester, New York

A. THE PURPOSE OF THIS VOLUME

The first four volumes in this treatise deal with radiometry, color measurement, measurement of photoluminescence, and photodetectors. This volume addresses the measurement of light by methods involving the use of the human visual mechanism. It differs from the previous volumes in two fundamental respects. First, the others concern the means for measuring and specifying light energy or power for the characterization of that which exists in the physical world; this volume deals with attempts to characterize what a human observer sees when stimulated by the physical world. Second, physical measurements of light—the subject of the first four volumes in the treatise—are best effected by linear devices in which a doubling of light power is registered as a doubling of the output of the device and the addition of light power is reflected additively in the measurement indication. Visual measurement methods, by contrast, necessarily involve the use of a complex, nonlinear, light-sensing device: the visual mechanism.

ISBN-0-12-304905-9

Although the initial stage of the operation of the visual mechanism, in which photons of light are absorbed, may have linear properties, the subsequent amplification and processing of signals generated by photon absorption do not. As a consequence of this complex signal processing and integration we form the conscious responses to stimulating energy that cause us to "see" what exists in the physical world. The general purpose of visual measurement is to characterize what we see, that is, to measure the amount and kind of response to light that we experience rather than the light itself.

Classical methods for light measurement that invoke the visual mechanism, such as photometry (a neologism derived from the Greek roots *φοτοσ*- and *μετρον* meaning light measurement) and colorimetry (a cognate neologism), avoid the complexities of the nonlinear response of the visual mechanism by using it merely as a null indicator. Photometry and colorimetry measure light properties, not what we see: they determine the amount and kind of light that is required to evoke a criterion response when the photodetector is the visual mechanism. Neither photometry nor colorimetry provide indications of what is seen.

Visual measurement, in the general sense, attempts to measure what is seen. To do this it is necessary to come to grips with the properties of the nonlinear detector system that we call the visual mechanism. Just as we must understand and specify the properties of linear devices for collection, imaging, detection, and amplification of light in radiometry, so must we also understanding those kinds of properties of the visual mechanism if we are to make and interpret visual measurements usefully.

The purpose of this volume is to provide introductory information about the visual mechanism and the ways in which it can be used to make visual measurements. We shall not attempt a definitive description of the visual mechanism or the methodology of visual measurement which would require much more certain knowledge than is available at present. Instead we present a summary of information that we believe has some engineering utility for people who wish to make measurements that require an indication of the kinds of responses to be expected from the sensing human observer who is stimulated in various ways by light of various kinds. We ask, "What should we expect an observer to see?" rather than "What is the amount and kind of light?" Therefore we have structured this volume into three basic parts. The first part, which describes the visual mechanism, includes chapters about the eye as an optical system and as a detector, the visual pigments that initially absorb photons, the sensitivity of detection, and a summary of what is known or generally agreed upon about the mechanism by which the absorptions of photons are processed and integrated to form a visual response in the cortex of the brain. This part is analogous to a description of the characteristics of radiometric measuring instruments that aids in understanding what an instrument is measuring.

The second part concerns the use of methods that make it possible to measure responses to light stimulation. First there is a general discussion of psychophysics, the science of measuring relationships between stimuli and responses, followed by chapters on the methodology of psychophysical measurement of thresholds, matches, differences, and ratios or magnitudes, and on dealing with multidimensional psychophysical situations. The emphasis is on application rather than theory. We attempt to provide a body of basic information that will enable the reader to conduct and interpret psychophysical experiments that will yield useful visual measurements.

The third part concerns the extension of photometric and colorimetric measurement techniques from null-instrument technologies to advanced methods for specifying the characteristics of perceptions that arise from visual stimulation by light. Each chapter provides a part of the total information that is required to understand techniques for making effective visual measurements and interpreting them correctly.

B. THE EYE AS AN OPTICAL SYSTEM

Some knowledge of the anatomy and physiology of the eye is necessary to understand how it performs as an optical device. Light gathered from an entire visual field is projected onto the rear two-thirds of the eyeball forming an image on a conglomerate layer of tissues and nerves called the retina. The image has many optical imperfections. Some of the light is absorbed by ocular media before it reaches the retina. That which does reach and penetrate into the retina is diffused, reflected, transmitted, and absorbed. The portion absorbed by specialized nerve-endings called photoreceptors is of primary interest. These absorptions provide the basis for seeing by setting off a train of neural events that proceed to the occipital cortex of the brain where integration with other signals in the brain forms the conscious response that we call seeing. Only a small portion of the light that forms the initial, optical image is absorbed by the photoreceptors. The influence of the eye's magnification, distortion, aberration, transmission, diffusion, fluorescence, bioluminescence, electroluminescence, and polarization on that part of the retinal image that yields photons for photoreceptor absorption must be considered if we are to understand the differences between physically measured stimulation energies and effective light-stimuli. Chapter 2 discusses these factors and attempts to provide the reader with an elementary understanding of how the light-stimulus for vision is formed in the orbit of the eye.

C. THE EYE AS A DETECTOR

Chapter 3 is concerned with the detection of light in the retina of the eye. It discusses thresholds, the conditions for and characteristics of adequate stimulation for seeing. Light-stimuli that are above threshold, that is, adequate stimuli, may be visible in various ways. Their visibility is related in complex ways to the sensitivities of retinal receptors. These matters are discussed and ways are described for measuring the characteristics of the eye as a detector in the initial stage of the process of seeing.

D. VISUAL PIGMENTS AND SENSITIVITY

A general discussion of the sensitivity of the eye to light points up the fact that photons of light are absorbed by certain photolabile pigments located in the outer segments of the photoreceptors in the retina of the eye. There are four kinds of receptors, categorized by the kinds of photopigments they appear to contain, that selectively absorb light of different spectral frequencies. One kind of receptor, called rods, contains a photopigment named rhodopsin and is effective principally at low levels of light. Scotopic sensitivity (from the Greek *σκοτοσ*, dark, and *οπτοσ*, visibility) is the term used to describe our sensitivity as a function of wavelength to low levels of illumination that primarily involve rod receptors. Its spectral selectivity differs from that of photopic sensitivity (from the root *φοτοσ*, meaning light) which applies to higher levels of illumination such as those encountered in daylight. Three kinds of receptors form a second class of photoreceptors called cones. Each cone contains one of three different spectrally selective pigments: erythrolabe, cyanolabe, and chlorolabe. At least two of these cone-type receptors combine signals to form photopic sensitivity; color discrimination is possible because of their different spectral selectivities. The distribution of rods and cones is not uniform throughout the retina, however, so sensitivity depends on the location of the optical image on the retina and on a variety of other factors. These matters are discussed in Chapter 4.

E. MECHANISMS OF VISION

The formation of an optical image on the retina of the eye and the light absorption by the photoreceptors of the retina are only the first steps in seeing. Absorption of light energy causes the formation of an electrophysiological signal that is transmitted in a variety of ways along neural pathways that terminate in certain cortical columns of the brain. It is here that "seeing" takes place. Chapter 5 provides a description of this mechanism of vision, including a simplified discussion of neurophysiology and a simple but specific treatment of the neurophysiology of vision. These matters are related to the ways in which we see colors, different levels of brightness, and the spatial-frequency characteristics of vision. Although our knowledge of the visual mechanism is far from complete, we attempt in this chapter to bring together and simplify what we do know (or at least accept as reasonable speculation) as a basis for understanding what can be expected to be accomplished with precision in visual measurement and what constraints should be imposed on the interpretation of such measurements.

F. PSYCHOPHYSICS

The second part of the book starts with an overview of visual measurement. The kinds of visual measurement of interest here are those that build a descriptive bridge between stimulus and response. That is, methods are described that can be used to describe quantitatively the relationships between stimulation and response for specified conditions of viewing. These methods form the basis of the science and technology of psychophysics (*ψυχο-*, meaning mind or mental, and *φυσικοσ*, physical, which together represent the relationship between physical and mental events). Visual measurement is a form of psychophysics as discussed here: the elucidation and description of relationships between stimulation and response. The classical form of visual psychophysics involves the visual mechanism only as a null instrument, a method for assessing whether two stimuli are matched or not. Most psychophysical methodology centers around such techniques. Some, such as the theory of signal detection, have undergone elaborate development. These are reviewed in Chapter 6. That chapter also describes methods for estimating sensory magnitudes, methods in which the visual mechanism is used as a variable response device, not merely a null instrument. These form the basis for what has been called the new psychophysics. A number of miscellaneous topics about the practice of psychophysics are also discussed, together with some helpful suggestions about ways in which to ensure the reliability of psychophysical measurements.

G. THRESHOLDS AND MATCHING

Classical methods of visual psychophysics (the null-instrument methods) are the techniques usually used for determining thresholds. Absolute and difference thresholds both represent stimulus conditions in which a test field or a difference is barely perceptible. Matching, by contrast, is a stimulus condition in which a difference is not perceptible. For this reason thresholds and matching methods are treated together in Chapter 7. A general discussion of these psychophysical methods shows that thresholds and matching are probabilistic events; therefore a brief summary of probability is given. The discussion indicates how measurement data can be represented by probability distributions of various kinds. Finally there are examples of psychophysical methods for determining thresholds and match-points. These are illustrated by working out problems for experimental data.

H. MEASURING DIFFERENCES

Chapter 8 provides a general discussion of the difference between determining conditions of equality or threshold differences from equality and attempts to scale suprathreshold differences in visual appearance. Most of the experimental methods used to scale suprathreshold differences are derived as extensions of the threshold methods; they both rely on distributions of uncertainty as a basis for determining scale units. Worked examples are given for the most common methods of scaling differences or intervals. Rank-order, paired comparisons, rating-scale, and category methods are illustrated. Finally some reflections on interval scaling techniques and suggestions for increased precision and enhanced utility of such methods are made.

I. DIRECT RATIO SCALING

The threshold, matching, and suprathreshold methods that rely on discriminal dispersions to form scales of differences or intervals have been called confusibilty scaling, because they require a degree of confusion in the responses to different stimuli. In general this means that the stimuli should be similar in the appearance of the attribute being measured. This may not always be possible. It is often the case that stimuli are obviously different, either because their number is fixed or because it may be impractical to include as many stimuli as would be required for appearances to be similar, and differences among the stimuli are not confused. In such cases confusibilty scaling cannot be applied satisfactorily. Other approaches that do not require confusion in the data use direct estimates of the relationships among the stimuli. These are called the direct methods, and they usually involve estimating the ratios of the magnitudes of differences in appearance among stimuli. Chapter 9 discusses these direct methods and examines three of them in some detail: magnitude estimation, magnitude production, and cross-modality matching. Worked examples are provided to illustrate how scales can be derived by such direct methods, and a discussion of the meanings and limitations of these scales is given.

J. MULTIDIMENSIONAL SCALING

All of the methods previously discussed were developed for use in situations in which only one attribute of appearance varies. It is often the case, however, that differences among stimuli give rise to variations in two or more attributes of appearance. Sometimes the experimenter does not know ahead of time whether this will be the case. Methods of characterizing the relationships among the appearances of stimuli in such cases have been developed and are loosely grouped here under the single rubric *multidimensional scaling*. As used, the term multidimensional implies the alternative condition to unidimensional. The techniques discussed are multiple regression methods, multivariate analysis, and newer techniques called multidimensional scaling. Worked examples are again provided to illustrate the application of these methods.

K. PHOTOMETRIC MEASUREMENT

The second part of this volume concerns methods for scaling responses to light-stimuli to derive measurement scales or relational maps that characterize expected visual responses. The third part of the volume addresses the two major classes of experiments for gathering data: photometry and colorimetry. They are discussed mainly in terms of current methodologies and techniques that, although not fully developed, represent expected directions for future uses of photometry and colorimetry. It is generally accepted that the goal of photometry should be to provide psychophysical specifications for the perceived brightness appearances elicited by stimuli and that the goal for colorimetry should be to specify the color appearances perceived from stimulation by light arrays. Chapter 11 discusses photometry. The topics are physical photometry, visual photometry and visual photometers, research photometric methods, the determination of brightness and lightness (perceptual attributes) from photometric measurements, a review of certain theoretical considerations, and a section on how to select the most appropriate photometric method for a problem.

L. COLORIMETRIC MEASUREMENT

Chapter 12 completes this volume with a discussion of colorimetric methods. After a brief review of conventional colorimetry, this chapter provides sections that illustrate the most up-to-date, state-of-the-art methods for attempting to specify color appearances. These sections use mathematical models that are selected to be representative of the variety of models currently being tested, in order to show the present direction of developing methodologies for dealing with threshold color differences, suprathreshold color differences, and chromatic adaptation. Finally there is some speculation on the future of color measurement, including a consideration of the kinds of problems that must be overcome before universally useful methods can be developed.

PART

I

Characteristics of the Visual Mechanism

2

The Eye as an Optical System

G. A. FRY

College of Optometry
The Ohio State University
Columbus, Ohio

ISBN-0-12-304905-9

A. INTRODUCTION*

Analysis of the eye as an optical system can be divided into two parts. One is the formation of the optical image on the retina, which we will deal with in this chapter. Our discussion will be greatly simplified if we describe the optical image as a distribution of flux on a smooth hypothetical surface that coincides with the external limiting membrane. This hypothetical surface is usually referred to as the retina.

The second part of the analysis deals with transforming the distribution of flux on the hypothetical retina into a distribution of responses in the photoreceptors. It is necessary for the flux to be directed into the photoreceptors where some of it can be absorbed by the photopigment. This second part, which we can identify as *retinal optics*, will be deferred to Chapter 3.

The present chapter includes a discussion of the anatomy and physiology of the eye to aid in visualizing the image-forming mechanism and in understanding how the eye is mounted in its orbit and how it turns in different directions. We consider the quality, magnification, and distortion of the image formed on the retina, devices for correcting the image, and the absorption and scattering of the light that enters the eye. Entoptic phenomena are not systematically covered. Reviews of this topic can be found elsewhere (Southall, 1924, pp. 204–206; Tscherning, 1924, pp. 176–191; Duke-Elder, 1944, pp. 806–818).

* Appreciation is expressed to the authors and publishers who have given permission to duplicate their figures. The source of each figure is indicated at the end of the figure legend.

B. ANATOMY AND PHYSIOLOGY OF THE EYE

To understand how the eye functions as an optical instrument, it is helpful to know how its parts are held together, how it is maintained as a living organ, and how it is suspended in its orbit and rotated in different directions by the extraocular muscles.

1. The Coats of the Eyeball

The eyeball is a bulb composed of three coats, or tunics. It is filled with aqueous and vitreous humors which maintain pressure from within. Figure 1 shows the coats and internal spaces.

a. OUTER COAT

The outer tunic consists of two parts, the *sclera* and the *cornea*. Each is a segment of a sphere, and the two are joined at the sclerocorneal margin (limbus). The line connecting the pole of the cornea and the pole of the sclera is the anatomical axis of the eye. The outside radius of curvature of the cornea

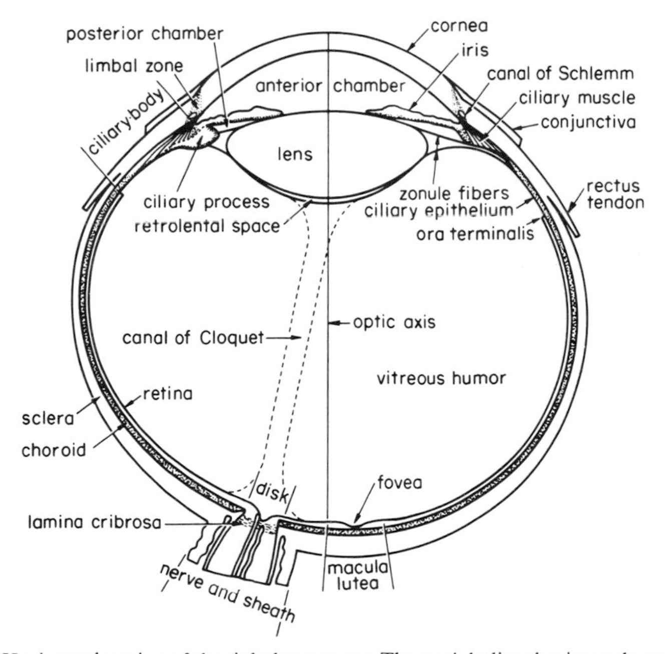

Fig. 1. Horizontal section of the right human eye. The straight line that is nearly normal to the front surface of the cornea and to the two surfaces of the lens is called the optic axis. It may be assumed that it coincides with the anatomical axis which connects the pole of the cornea with the pole of the sclera. (Walls, 1942.) Reproduced with permission of Cranbrooks Institute of Science.

is about 8 mm, and that of the sclera is about 12 mm. The outer tunic, which is fibrous and inelastic, determines the size and shape of the eyeball. The sclera is translucent; the cornea is transparent.

The Sclera At the back of the eye the sclera is continuous with the sheath of the optic nerve. The opening into the optic nerve is covered with the lamina cribosa, which transmits the central artery and vein of the retina and has many minute pores that transmit the optic nerve fibers. The outer surface of the sclera is smooth and white. To it are attached the tendons of the extraocular muscles which control the direction in which the eye is pointing. It is penetrated by numerous nerves and blood vessels. Its inner surface is brown and is grooved to provide channels for nerves and blood vessels.

The Cornea The cornea is made up of five layers as shown in Fig. 2.

(1) The epithelium, the outer layer. This layer is continuous with the conjunctiva, which covers the front of the eye and lines the insides of the upper and lower lids.
(2) Bowman's membrane
(3) The substantia propia, the thickest layer, which involves a criss-crossing of fibers
(4) Descemet's membrane
(5) The endothelial layer, the inner layer, which continues over the front surface of the iris

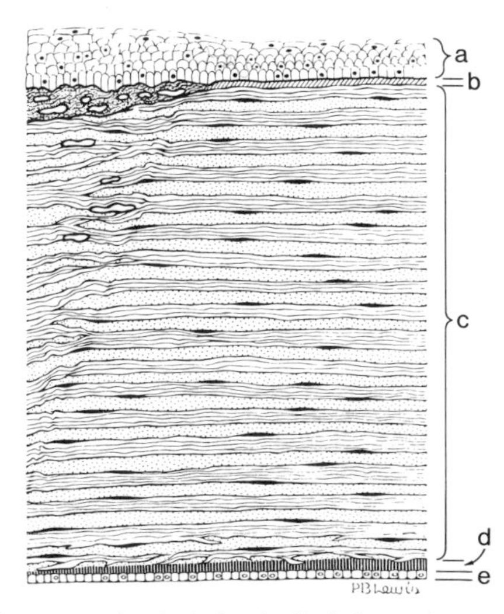

Fig. 2. Section of the cornea that includes the limbal zone between the cornea and sclera. (a) Epithelial layer. (b) Bowman's membrane. (c) Stroma. (d) Descemet's membrane. (e) Endothelium. (courtesy J. E. King.)

The cornea is normally transparent and uniformly thick. It is nearly avascular but is richly supplied with pain nerve fibers which make it sensitive to foreign bodies, cold air, and chemical irritation. The cornea obtains its nutrition from the aqueous humor and the blood vessels at the margin of the cornea. The margin is nearly circular about 12 mm in diameter. The cornea is slightly wider than high.

b. The Intermediate Coat

The intermediate vascular tunic is known also as the *uveal tract*. It has three parts: the *choroid*, the *ciliary body*, and the *iris*.

The Choroid The choroid lines the posterior part of the sclera. It is penetrated by the optic nerve, but adjacent to this it is firmly attached to the sclera. The network of blood vessels in the choroid provides the major source of nutrition for the eye.

The Ciliary Body The ciliary body is joined to the choroid and is continuous with it. The forward margin of the ciliary body is attached to the scleral spur at the sclerocorneal margin. The ciliary body contains the intraocular muscle that controls accommodation. This is made possible by the fact that the crystalline lens is connected to the ciliary body by the zonular fibers. The ciliary body also continuously generates aqueous humor.

The Iris The iris is connected by its root to the scleral spur and extends into the interior of the eye, dividing the space between the lens and cornea into anterior and posterior chambers. The hole at the center of the iris is the pupil, which admits light into the eye. The iris is heavily pigmented and contains the intraocular muscles that constrict and dilate the pupil. The iris has five layers.

(1) Endothelial layer. This layer is a single stratum of mesothelial cells and is an extension of the endothelial layer of the cornea.
(2) Stroma. A meshwork of cells, fibers, nerves, and blood vessels.
(3) Muscular fibers. Circular fibers at the margin form a sphincter which constricts the pupil; radiating fibers dilate it.
(4) Pars iridica. This is the part of the nervous tunic that covers the back of the iris. It consists of two layers of pigmented columnar epithelium. The color of the iris depends on the distribution of pigment. In blue eyes the pigment is confined to the pars iridica. In brown and black eyes the pigment is also found in the cells of the stroma and the endothelium. In albino eyes the pigment is absent.

The pupil is displaced to a slight extent toward the nasal side of the iris.

c. The Inner Coat

The inner coat forms a lining for the vascular tunic, is coextensive with it, and is divided into three parts. The retina, which is the nervous part, lines the choroid. The optic nerve fibers from the different parts of the retina run across the retina to their point of exit through the lamina cribosa into the optic nerve. At this point they form the *optic nerve head*, which is also called the *optic disk* because of its appearance through an ophthalmoscope. The retina and choroid both terminate at the optic nerve head. The outer margin of the retina is called the *ora serrata*. Beyond the ora serrata the inner tunic extends as a thin membrane over the ciliary body (pars ciliaris) and the back of the iris (pars iridica).

2. The Humors of the Eye

The spaces enclosed by the coats of the eye contain the *vitreous* body, the *lens*, and the *aqueous* humor.

a. The Aqueous Humor

The aqueous humor fills the anterior and posterior chambers, and the zonular spaces that separate the vitreous body from the posterior chamber. It has a specific viscosity only slightly greater than 1.0 and a volume of about 2 ml. It is generated from blood plasma flowing through the ciliary body and is then secreted from the ciliary processes. It enters the *posterior chamber*, passes through the pupil and leaves the eye by way of *Schlemm's canal* at the angle between the iris and the cornea at the *sclerocorneal margin*. The complete renewal of the aqueous humor requires about an hour. The rates of generation and outflow control the intraocular pressure, which is 15 to 18 mm of mercury. This internal pressure maintains the shape of the eye and the spacing of elements of the optical system.

b. The Crystalline Lens

The crystalline lens (Fig. 3) is a transparent body enclosed in an elastic capsule. The lens has a firm nucleus that is surrounded by a soft cortex. The lens is built of fibers, each of which begins near the axis and arches outward toward the equator and back toward the axis on the other side. The path of each fiber is nearly parallel to the surface of the lens, which leads to laminations that are parallel to the surface. A young person can change the shape of the lenses by contracting the ciliary muscle; this changes the tension of the zonular fibers on the capsule which changes the shape of the lens. This ability to change the shape of the lens decreases with age until it ceases at an age of about 54 years. The *index of refraction* increases from the surface to the center. The *isoindicial* surfaces are nearly parallel to the surface and roughly

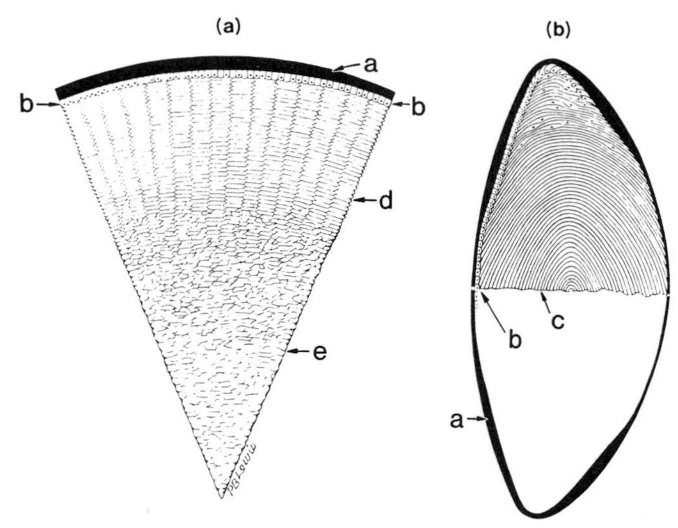

Fig. 3. (a) Sector of an equatorial section of the lens showing the radial arrangement of Rabl's lamellae d which break up into an irregular array at the center of the lens e. (b) Sagittal section showing the arching course of each fiber of the lens from a point near the axis in front of the center to a second point near the axis on the other side of the center. The lens fibers cannot all terminate at the axis. They terminate at surfaces that radiate out from the axis and are called sutures. Each section shows the capsule a and the epithelial layer b from which the lens fibers c are derived by proliferation at the equator.

parallel to the fibers running from pole to pole. This arrangement of fibers controls the internal structure of the lens during changes in accommodation.

The lens obtains its nutrition from the aqueous humor.

c. The Vitreous Humor

The vitreous body fills the space between the lens and the retina. It is a transparent gelatinous body (with a specific viscosity of about 1.8 to 2.0) which obtains its nutrition from the blood vessels on the surface of the retina and ciliary body and from the aqueous humor.

3. The Retina

The retina (Fig. 4) is made up of layers that may be identified as follows starting with the pigment epithelium, which contacts the choroid, and ending with the layer adjacent to the vitreous:

(1) Pigment epithelium
(2) Rods and cones
(3) Outer limiting membrane
(4) Outer nuclear layer, containing the cell bodies of the rods and cones

(5) Outer plexiform layer
(6) Inner nuclear layer (bipolar cells)
(7) Inner plexiform layer
(8) Ganglion cell layer
(9) Optic nerve fibers, which are the axons of the ganglion cells running across the retina to the optic nerve head
(10) External limiting membrane

The gross structure of the rods and the cones is indicated in Fig. 5. The cones differ in size and shape at the different parts of the retina. As indicated in Fig. 6,

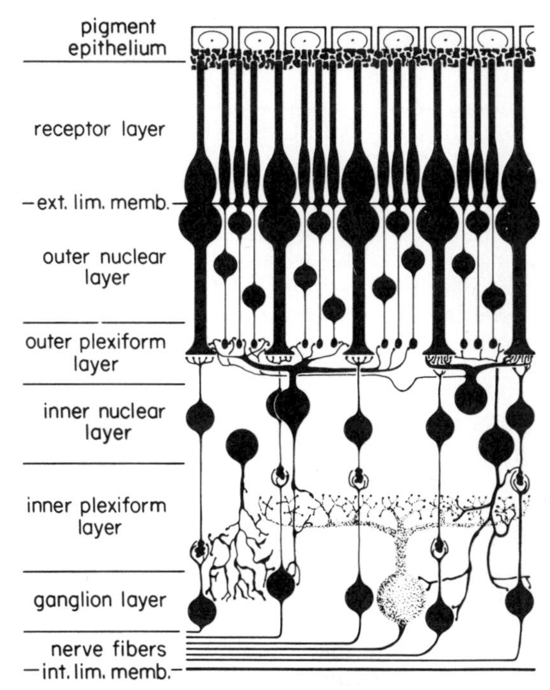

Fig. 4. Cross section of the retina showing the different layers from the pigment epithelium to the internal limiting membrane, which is in contact with the vitreous. Light entering the pupil of the eye passes through the internal limiting membrane toward the photoreceptors. The receptor layer shows the inner and outer segments of the rods and cones. The cell bodies of the rods and cones lie in the outer nuclear layer on the internal side of the external limiting membrane. In the outer plexiform layer the feet of the rods and cones contact the dendrites of the bipolar cells, which have their cell bodies in the inner nuclear layer. The bipolar cells and ganglion cell processes come together in the inner plexiform layer. The axons of the ganglion cells run across the retina and become the fibers of the optic nerve. The horizontal cells in the outer plexiform layer, the amacrine cells in the inner nuclear layer, and the branching dendrites of the bipolar and ganglion cells provide for interaction across the retina (Walls, 1942). Reproduced with permission of Cranbrooks Institute of Science.

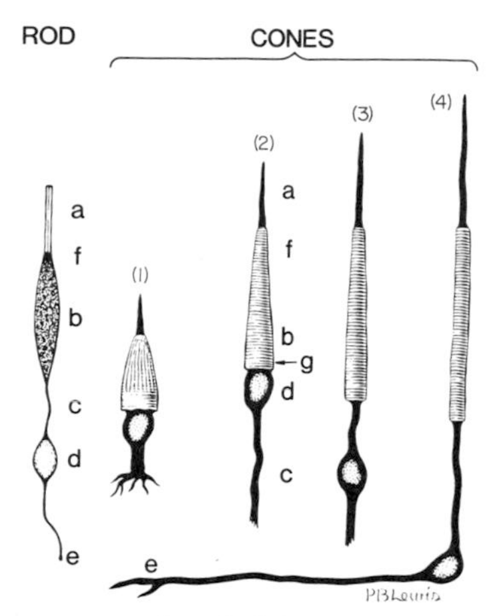

Fig. 5. Types of photoreceptors: rod on the left and four types of cones on the right. (1) Near the ora serrata. (2) Near the equator. (3) Outside the fovea. (4) At the center of the fovea. (a) Outer segment. (b) Inner segment. (c) Fiber. (d) Nucleus. (e) Foot. (f) Ellipsoid. (g) Myoid. (courtesy of J. E. King.)

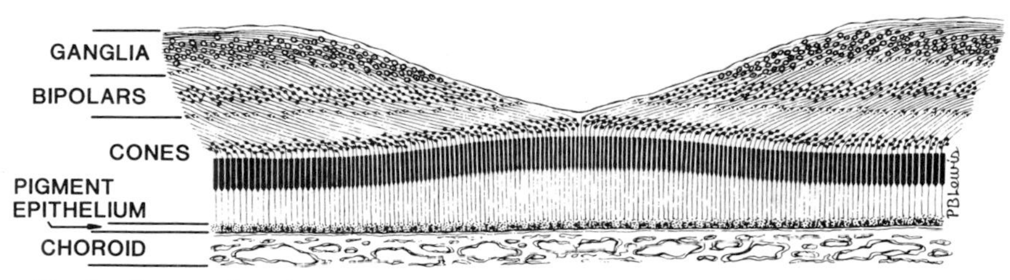

Fig. 6. Cross section of the central fovea. Bipolar and ganglion cells are absent at the center. No attempt has been made to include horizontal cells and amacrine cells, and no provision is made for interaction across the center of the fovea. The macular pigment lies in the layers internal to the outer limiting membrane. The cone fibers that run horizontally across the retina are known as Henle fibers. (courtesy J. E. King.)

only cones are found in the center of the fovea. The bipolar and ganglion cells to which they are connected are pushed toward the rim of the *foveal pit*. The cone fibers connecting the cell bodies and the cone feet run horizontally across the retina and constitute the Henle layer.

Figure 7, based on Østerberg's data (Østerberg, 1935), shows the concentration of rods and cones at different parts of the retina. The distribution of cones is most dense at the center of the fovea. There is a rod-free area centered at the fovea that is about 1° in diameter. The rod density increases out to about 20° and then decreases.

Vilter (1949) used a sagittal section of a human eye 2 μm thick to assess the distribution of ganglion and bipolar cells relative to rods and cones. The

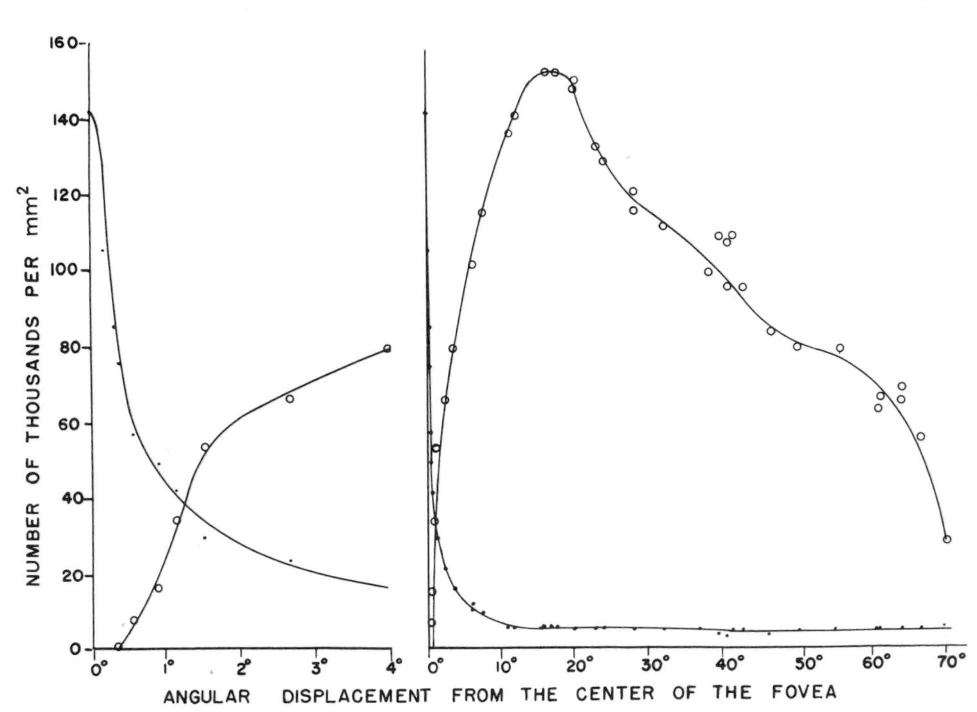

Fig. 7. Concentration of rods (circle) and cones (dots) at different distances in a temporal direction from the center of the fovea [based on the data of Østerberg (1935)] (Fry, 1970).

section was divided into intervals 100 μm wide. The number of neurons of each type in each interval was counted. At the center of the fovea the bipolar and ganglion cells are pushed toward the edge of the fovea, so that the Henle fibers that connect the cell bodies with the feet of the cones radiate across the retina. Hence, one count was made of the external segments, and another was made of the cone feet. The curves in Fig. 8 show the distributions of rods, external segments of the cones, cone feet, and bipolar and ganglion cells. The total distance from the center of the fovea to the ora serrata includes 144 intervals 100 μm wide. The ganglion cells are absent at the center of the fovea but increase in density out to about 0.8 mm from the center. The same is true for cone feet. It is possible to determine that the ratio of ganglion cells to cone feet is about 2:1 out to this point, gradually decreases to about 1:1 at 3 mm and from this point the ratio continues to decrease to about 1:10.

The data show that the concentration of rods drops to zero only at the very center of the fovea; in this connection the data do not support the concept of a rod-free area of finite size. The concentration does increase out to about 3.6 mm and after this decreases.

According to Vilter the ratio of cones to cone bipolar cells is about 1:3 in the

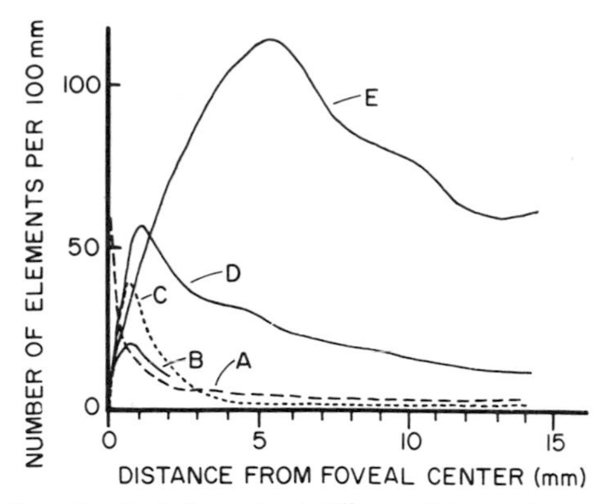

Fig. 8. Concentration of retinal elements at different distances in a temporal direction from the center of the fovea (based on data from Vilter, 1949). (A) cone outer segments, (B) cone feet, (C) ganglion cells, (D) bipolar cells, and (E) rods.

central retina out to 20°. Between 20 and 30° it decreases to 1:1 and remains constant from that point on.

The rods and cones obtain their nutrition from the choroid. The inner layers are nourished from the arteries and veins that branch out from the central retinal artery and vein, which enter the eye through the optic nerve head and run across the internal surface of the retina.

It is important to identify the rods and cones as constituting the layer of elements that responds to the light transmitted through the pupil to various parts of the retina. The cell bodies lie internal to the external limiting membrane, and the inner and outer segments lie between the external limiting membrane and the *pigment epithelium.* The total distance from the external limiting membrane to the tip of an outer segment can be as much as 70 μm.

4. The Primary Line of Sight

A person told to look at a given point (*fixation point*) directs each eye so that the image of that point falls on a specific point on the retina (*anchor point*). The line connecting the fixation point and the center of the entrance pupil is the *primary line of sight* (Fig. 9). The primary line of sight is relatively fixed with respect to the eyeball, so it can be used to specify the direction in which the eye is pointing (Fry *et al.*, 1945).

5. Movements of the Eye with Respect to the Head

The eyeball is imbedded in a pocket of fat, and its movements are those of a ball-and-socket joint. Rotations from right to left are accompanied by a small amount of translation in the lateral direction, but translations in the fore and

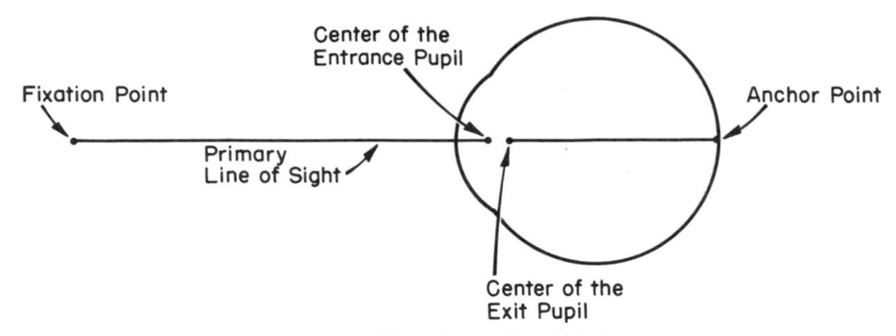

Fig. 9. The primary line of sight.

aft and up and down directions associated with rotation are negligible. When the eyes move up and down, they undergo a lateral translation called *screw movement* (Fry and Hill, 1962, 1963).

In accordance with the concept of a ball-and-socket joint it may be assumed that the center of curvature of the scleral sphere is the *center of rotation* of the eye. But a more important center is the *sighting center*. The line of sight does not pass through the center of the scleral sphere, and as the eye rotates around a vertical axis, the different positions of the line of sight do not intersect at a common point. The point at which they nearly intersect (the sighting center, Fig. 10) lies in a temporal direction from the center of rotation on the straight-ahead position of the primary line of sight and about 13 mm from the cornea.

The sighting center is the point used as the center of rotation in the design of a spectacle lens or an eyepiece for a mobile eye. It is also the center of rotation for specifying the direction of regard and cyclorotation and for specifying convergence in binocular vision. The sighting centers for the two eyes are assumed to be fixed relative to the head.

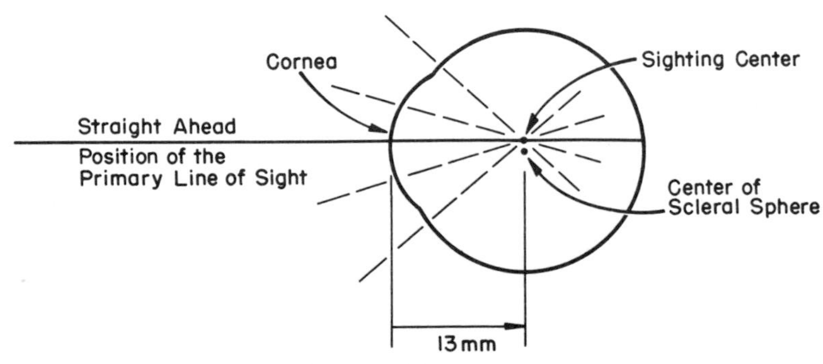

Fig. 10. Sighting center.

The plane that is tangent to the chin and the two superciliary ridges constitutes the *face plane.* The plane that contains the occipital protuberance on the back of the head, the glabella between the two eyebrows, and the nasal depression below the nose is the *median plane* of the head.

The erect position of the head is one in which the face plane is vertical and the *base line* connecting the sighting centers of the two eyes is horizontal.

In binocular vision the primary lines of sight of the two eyes converge at a common fixation point. The maximum level of convergence is about 60°. The plane that contains the fixation point and the base line is the *plane of regard.* This plane rotates around the base line to produce a change in *elevation,* and the primary line of sight rotates around an axis perpendicular to this plane to produce changes in *azimuth.* The angles of elevation and azimuth specify the *direction of regard.* The straightforward position of the primary line of sight is defined as the position in which the head is erect and the primary line of sight is horizontal and perpendicular to the base line. It represents zero elevation and azimuth.

a. Field of Fixation

The extent to which an observer can rotate an eye in a given direction away from the straight-ahead position is the *limit of the field of rotation* in that direction. When the limits are tested for one eye at a time, it is found that the typical eye can turn inward 50°, outward 45°, up 40°, and down 60°.

b. Scanning and Following Movements

Scanning movements can best be studied by covering one eye of an observer and immobilizing the head with a bite board. As the observer scans an array of objects in front of him, a series of saccadic eye movements is made, with the eye coming to rest between each movement. The problems encountered during these periods of rest are the same as those encountered during an extended period of steady fixation on a given object. The observer can follow a moving target in the sense that the primary line of sight can be kept centered on some part of the target. The observer also can scan a line, but this involves a series of saccadic movements as is involved in scanning a row of letters.

c. Steady Fixation with One or Both Eyes

During attempted steady fixation the eye is not stationary but undergoes tremor, drifts to and fro, and occasionally flicks (Riggs, 1965). These movements affect the visibility of fine details in various types of displays. There are several ways to immobilize the image on the retina so that the effects of the micromovements are eliminated. When both eyes attempt steady fixation, the problem is made more complex by the reflex fusional movements and voluntary changes in convergence that are involved in maintaining single vision. In the discussion that follows the effects of these micromovements during steady fixation have been ignored.

6. The Lacrymal System

The lacrymal gland is mounted above the eye, inside the bony orbit in which the eye is suspended and secretes tears that pour into the space between the eye and the upper lid. Tears are swept across the cornea by the blinking of the lids and drain through the lacrymal duct and canals into the nasal sinus. The normal rate of secretion by each gland is about $\frac{1}{24}$ of a gram per hour. The film of tears covering the cornea improves its optical performance.

7. The Eyelids

The upper lid is held open by the *levator muscle* and moves up when the eyeball is turned upward, so that normally the pupil is not occluded by the lids. The lids are closed by the *orbicularis muscles*. They close during sleep and blink periodically during the waking hours. The time cycle of blinking is important to seeing. There is also reflex blinking, identified as corneal reflex, dazzle reflex, and menace reflex. Winking is a voluntary closure of the lids; the *blepharospasm* is an involuntary closure.

One can hold the lids partially closed to produce the visual effect of looking through a narrow slit. In an earlier era when many myopes were not provided with glasses, they used this method of improving their vision; the name *myopia* is derived from this practice.

a. The Eyelashes

The eyelashes of the upper lid help to prevent foreign bodies from falling on the cornea and sclera and screen out part of the light from the upper portion of the field of view. They have to be considered when assessing the distribution of luminance in the field of view.

C. THE EYE AS AN OPTICAL DEVICE

1. The Image-Forming Mechanism

a. The Gullstrand Schematic Eye

In tracing rays through the eye to demonstrate how an image of an external object is formed on the retina, it is customary to make certain approximations about the media and the refracting surfaces. For example the structure of the cornea is complex, and its index is not uniform, but one can visualize the cornea as having a uniform medium bounded in front and behind by smooth spherical surfaces. This makes it a simple lens.

In the case of the lens the problems are more serious, because the index varies considerably from the center to the surface. Gullstrand (Southall, 1924, Vol. I, p. 351) proposed a model in which the lens is considered to be a triplet with a nucleus at the center and with cortical layers in front and behind. He computed that the nucleus would have a uniform index of 1.406 and the cortical layers an index of 1.386. Gullstrand used such approximations to create a schematic eye that can be used to demonstrate image formation.

Gullstrand's schematic eye is illustrated in Fig. 11. The front and back surfaces of the cornea, the front surface of the lens, the front and back surfaces of the cortex of the lens, and the back surface of the lens are the refracting

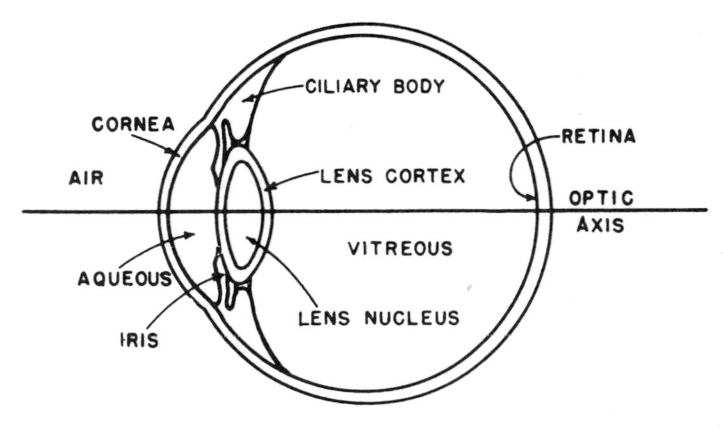

Thickness of cornea	0.5 mm
Displacement of front surface of lens behind front surface of cornea	3.6 mm
Displacement of nucleus from front surface of lens	0.546 mm
Thickness of nucleus	2.419 mm
Thickness of lens	3.6 mm
Index of refraction of cornea	1.376
Index of aqueous and vitreous	1.336
Index of lens cortex	1.386
Index of lens nucleus	1.406
Radius of front surface of cornea	7.7 mm
Radius of back surface of cornea	6.8 mm
Radius of front surface of lens	10.0 mm
Radius of front surface of nucleus	7.911 mm
Radius of back surface of nucleus	−5.76 mm
Radius of back surface of lens	−6.0 mm

Fig. 11. Gullstrand's schematic eye and its dimensions and indices (Fry, 1959).

surfaces. Each of the spaces between the surfaces is filled with a medium, the index of which for a given wavelength is uniform at all points and for light polarized in any direction and travelling in any direction. The surfaces are assumed to be spherical and centered on a common optic axis that coincides with the anatomical axis of the eye. The *pupil* is assumed to be in a plane tangent to the front surface of the lens and normal to the axis and to be centered on the axis.

b. Image Formation

In a homogeneous isotropic medium a ray of light propagates along a straight path (law of rectilinear propagation of light). When it strikes a smooth refracting surface at an angle α from the normal at the point of incidence, its direction is changed to an angle α' from the normal. It still lies in the same plane as the incident ray and the normal. The size of the angle α' is given by the formula:

$$n \sin \alpha = n' \sin \alpha', \tag{1}$$

where n and n' are the indices of the first and second media, respectively. This is known as Snell's Law (Southall, 1933, p. 73). These two laws can be used to trace any ray through any sequence of refracting surfaces.

Figure 12 illustrates how a system of refracting surfaces such as that shown in Fig. 11 can form an image Q′ of a point source Q of monochromatic light, which for the purpose at hand can be called an *object point*. The pupil limits the size of the bundle of rays that can be transmitted through the system. Because in this case all the rays lie close to the optic axis after passing through the system, they will verge at a common *image point* Q′.

c. Cardinal Points and Planes

In the case of a centered optical system it is possible to establish the cardinal points and planes of the system and use them to trace rays through the system and assess various features of the system. The cardinal points and planes are the *nodal*, *principal*, and *focal points* and the *focal* and *principal planes* that are normal to the axis. These are illustrated in Fig. 13. The procedure for locating the cardinal points of a centered system is described elsewhere (Fry, 1969).

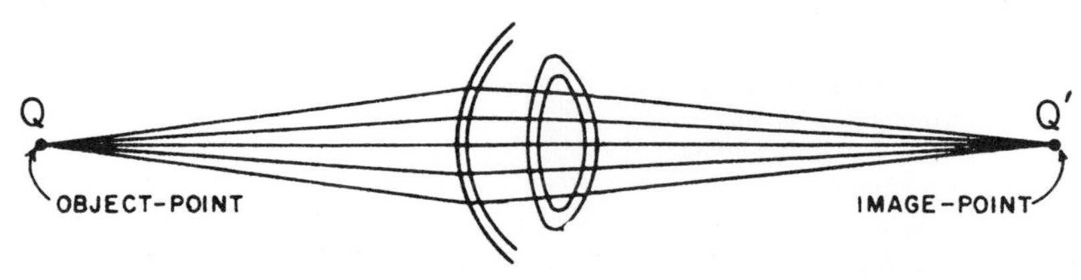

Fig. 12. Conjugate foci in object and image space (Fry, 1959).

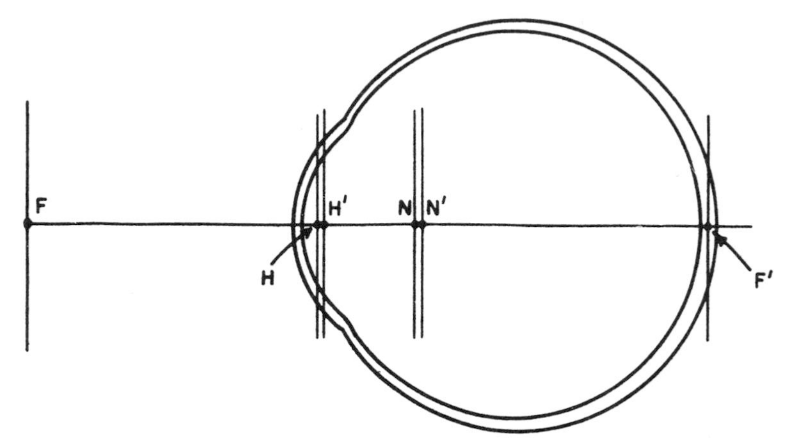

Fig. 13. Cardinal points of the Gullstrand schematic eye. The principal points H and H′ are located 1.348 mm and 1.602 mm from the front surface of the cornea, respectively. FH = N′F′ = 17.055 mm, and H′F′ = FN = 22.785 mm. The ratio H′F′/FH is equal to the index of the vitreous. The cardinal points lie on the optic axis. (Fry, 1959).

Once the principal points and the focal points are located, the primary focal length f and the secondary focal length f' can be determined. These are related to each other by

$$-n/f = n'/f' = F, \tag{2}$$

where F is the *refracting power* of the system, n is the index of the first medium, and n' is the index of the last medium. The distances from the nodal points to the focal points are

$$\overline{\mathrm{N'F'}} = -\overline{\mathrm{HF}} = -f \qquad \text{and} \qquad \overline{\mathrm{NF}} = \overline{\mathrm{H'F'}} = -f. \tag{3}$$

The use of the cardinal points and planes for ray tracing to locate the image Q′ formed by an object point Q off the axis is illustrated in Fig. 14. The ray from Q through the first nodal point N emerges from the system and is

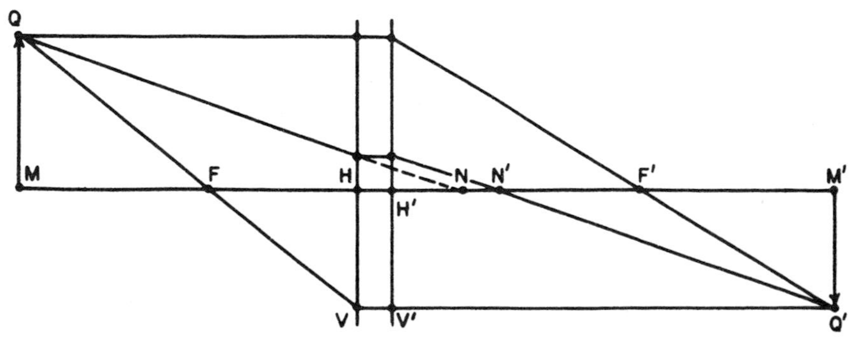

Fig. 14. Ray tracing. The distance from H to M is u, and that from H′ to M′ is u′. (Fry, 1959.)

directed through the second nodal point parallel to the line QN. The ray from Q parallel to the axis is redirected at the second principal plane through the secondary focal point F′. The ray from Q through F is redirected at the primary principal plane parallel to the axis. The points M and M′ in Fig. 14 are *conjugate foci.* The conjugate foci formula for location M′ when M is given and vice versa is

$$n'/u' = n/u + F. \tag{4}$$

The nodal points represent a pair of conjugate foci.

d. Pupil

The muscles in the iris can either constrict or dilate the pupil. The size of the pupil controls the amount of light entering the eye from each point in front of the eye and, therefore, affects the intensity of the stimulus applied to the retina. The pupil is the aperture stop of the system, but instead of dealing with the actual pupil, it is more important to deal with the *entrance* and *exit pupils.* The entrance pupil is conjugate to the actual pupil with respect to refraction at the cornea and lies about 3 mm behind the cornea. The exit pupil is conjugate to the entrance pupil with respect to the system as a whole. The entrance pupil is what a person sees when looking at another person's eye. Its position and diameter can be determined directly. When the actual pupil is centered on the optic axis, the entrance and the exit pupils are also centered on the axis.

e. Chief Ray

If we trace a bundle of rays from any given object point to its conjugate image point, there is one ray in the bundle of incident rays that is directed through the center of the entrance pupil. This is the *chief ray* of the bundle. The refracted portion of the chief ray passes through the image point and is directed through the center of the exit pupil. In Fig. 15 the object point M and

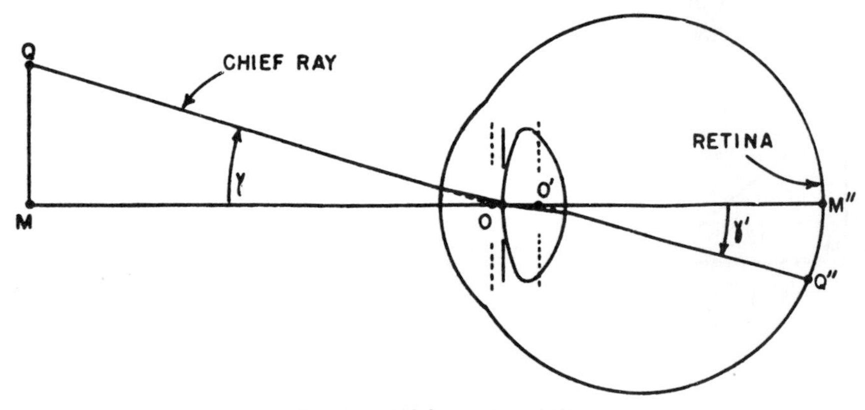

Fig. 15. Chief ray (Fry, 1959).

image point M′ lie on the optic axis, and the chief ray of the bundle coincides with the axis. The object point Q lies off the optic axis. QO represents the incident path of the chief ray, and O′Q′ represents the emerging path.

f. Out-of-Focus Blur Circle

In the schematic eye shown in Fig. 16, the bundle of rays that emerges into the vitreous come to a focus at Q′ and form a cone with the base at O′ and the apex at Q′. The image at Q′ is called the *optical image* and in terms of geometrical optics is a point image. Whenever an optical image falls in front or behind the retina, the image formed on the retina is said to be out of focus. When the image is out of focus, the rays that are intercepted by the retina are spread over an area that is disk-shaped, because it must have the same shape as the exit pupil. The disk-shaped area on the retina is called a *blur circle* and represents the *retinal image* of Q. The size of the blur circle depends on the size of the pupil and the extent to which the eye is out of focus. If the optical image Q′ falls at the retina, all the rays converge at the retina, and the blur circle reduces to a point.

g. Toroidal Astigmatism

In an actual eye the surfaces of the cornea may be *toroidal* instead of *spherical*, which makes the refracted bundle of rays *astigmatic*. A toric surface is generated by revolving an arc of a circle around an axis in the plane of the circle but not through the centre of the circle. When the axis passes through the center, the surface is spherical. The toric surface has two principal meridians, one parallel and the other perpendicular to the *axis of revolution*. The curvature varies from one meridian to the next, but it is maximal in one principal meridian and minimal in the other. Let us assume that the chief ray of a bundle is normal to a toric surface and also normal to the axis of

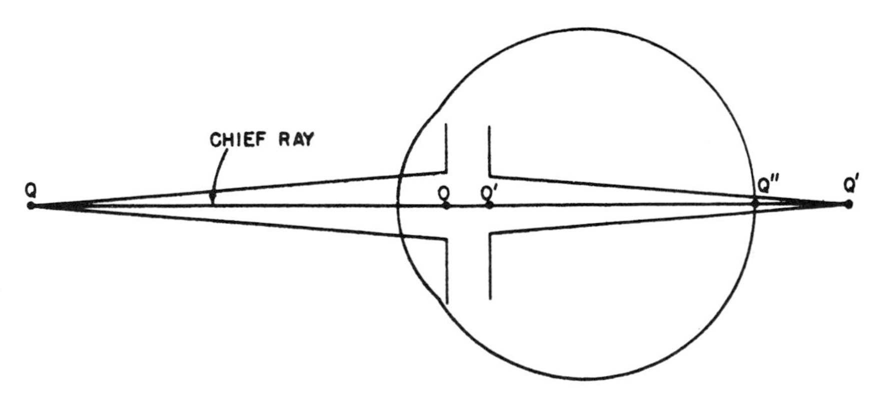

Fig. 16. Out-of-focus blur circle (Fry, 1959).

revolution. The incident rays verge at a common point, but the refracted bundle is astigmatic. As the bundle emerges from the exit pupil, it has the same shape as the exit pupil; but it then undergoes a change to ellipses of different sizes and shapes. As shown in Fig. 17 it becomes a horizontal focal line at Q′ and a vertical focal line at Q″. The distance from Q′ to Q″ is a measure of the amount of astigmatism.

h. Obliquity Astigmatism

If the chief ray of a bundle is obliquely incident at a surface, the emerging bundle is astigmatic. In this case the two focal lines are a short line that is oblique to the chief ray and a short arc that is concentric to the first focal line. The two types of astigmatism just described are the regular types. Deviations from these types are called *irregular astigmatism.*

i. The Primary Line of Sight and the Pupillary Axis

Lines of Sight Object points that lie at different places on a line directed through the center of the entrance pupil all produce blur circles on the retina that are concentric, and the perceived images are seen in the same direction. Therefore such lines are called *lines of sight.*

Primary Line of Sight If a normal person fixates a given point, the eye will be positioned so that the chief ray from that object point penetrates the *anchor point*, which falls within the fovea. This chief ray used for fixation is called the *foveal chief ray*. The incident path of this chief ray is the *primary line of sight.*

The Pupillary Axis and the Angle λ In an actual eye the refracting surfaces need not be centered on a common axis. We can still define an axis for the lens,

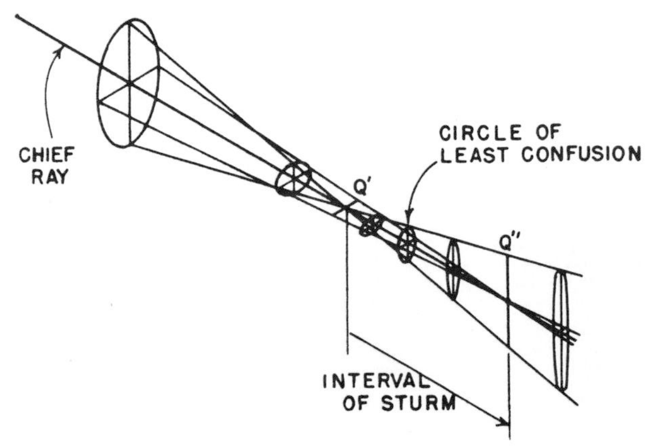

Fig. 17. Astigmatic blur ellipses (Fry, 1959).

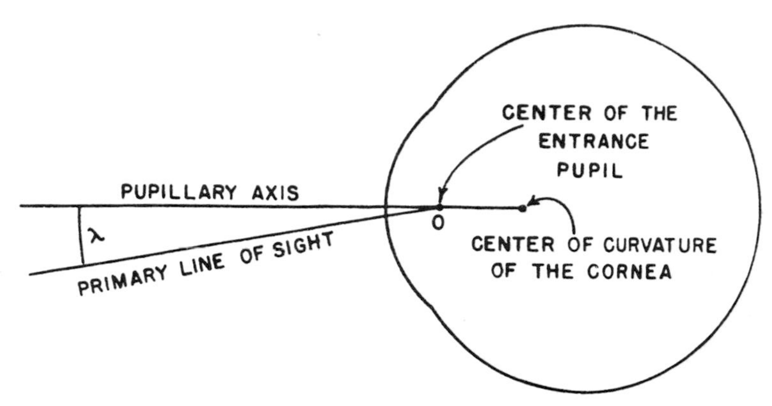

Fig. 18. The angle λ (Fry, 1959).

but this axis need not be normal to the front surface of the cornea. Instead of using the optic axis of the lens or the whole eye as a reference line for specifying directions, it is customary to use the pupillary axis. This is done because of the ease in establishing the position of the pupillary axis. It is the line that is normal to the front surface of the cornea and passes through the center of the entrance pupil. The primary line of sight also passes through the center of the entrance pupil and is displaced in a nasal direction from the pupillary axis through an angle λ as illustrated in Fig. 18*. The angle λ is about 5°.

j. Size of the Retinal Image

It is necessary to distinguish between the size of the blur circle produced by a single point (Fig. 16) and the linear distance between the centers of the images produced by two points. In Fig. 15, M″ and Q″ are the centers of the retinal images produced by the two points M and Q. The angle γ represents the angular size of the object and the angle γ' the size of the image. The ratio γ'/γ is the angular magnification. For the Gullstrand schematic eye the angular magnification is 0.87. The linear size of the retinal image is

$$\overline{M''Q''} = 0.87\gamma\overline{O'M''}, \tag{5}$$

where $\overline{O'M''}$ is the distance from the exit pupil to the retina. The angular magnification is related to the primary focal length f of the eye by

$$\gamma'/\gamma = -f/\overline{O'F'}, \tag{6}$$

where $\overline{O'F'}$ is the distance from the exit pupil O′ to the secondary focal point

* Sometime prior to 1941 I began using λ to designate the angle between the pupillary axis and the line of sight. This use of λ, described in Chapter VI of the mimeographed 1941 class notes for Optometry 615, The Ohio State University, has been adopted by Lancaster (1943) and Alpern (1969, p. 6).

F′. Another convenient relation that follows directly from the Smith–Helmholtz formula in geometrical optics is

$$(\text{size of the entrance pupil})/(\text{size of the exit pupil}) = n'\gamma'/\gamma. \qquad (7)$$

k. The Helmholtz Schematic Eye

Helmholtz (Southall, 1924, Vol. I, p. 354) created a schematic eye that is simpler than that of Gullstrand. For the cornea he substituted a single refracting surface that separates the aqueous humor from the air in front of the eye. He also assumed that the index of the lens is uniform, and so the lens can be represented by a two-surface lens. Laurance (1926, pp. 452–455) slightly modified the values given by Helmholtz for the constants so that they yield round numbers (−15 and +20 mm) for the primary and secondary focal lengths. The Helmholtz schematic eye is shown in Fig. 19, as are the constants as given by Laurance.

This schematic eye, useful for many calculations, is especially valuable in demonstrating why accommodation has little effect on the size of the retinal image. If we assume that accommodation is brought about by a change in the curvature of the front surface of the lens without a change in thickness, the angular magnification of the eye is not affected (Fig. 15).

The change in curvature of the front surface of the lens is often used as an index or measure of the changes in accommodation (Fincham, 1937; Allen, 1949). Nevertheless it must be acknowledged that in accommodation the lens does change in thickness, and the front surface advances toward the cornea. In a precise model of the eye these things must be taken into account.

l. Reduced Eye

Schematic eyes can be further simplified by removing the lens and compensating for this by increasing the curvature of the cornea. This is called a reduced eye (Fig. 20). The cornea becomes a single refracting surface that separates the air from the medium inside the eye, and the medium fills the entire space between the cornea and the retina. It is similar to an *aphakic* human eye, from which the lens has been removed. It can be made equivalent to the Helmholtz schematic eye by assuming that the index of the medium is 4/3, which is the same as that for the vitreous in the Helmholtz schematic eye, that the radius of the cornea is 5 mm, and that the pupil, which is also the exit pupil, lies 18.6 mm from the retina. The anterior focal length is −15 mm, and the secondary focal length is 20 mm.

The reduced eye has special uses. For example it is helpful in computing the effect of axial chromatic aberration on the blur of the retinal image. It is useful in dealing with *chromastereopsis* and stray light in the eye.

m. Determination of the Optical Constants

The optical constants used by Gullstrand and Helmholtz are based on relatively precise measurements and may be regarded as characteristic of a

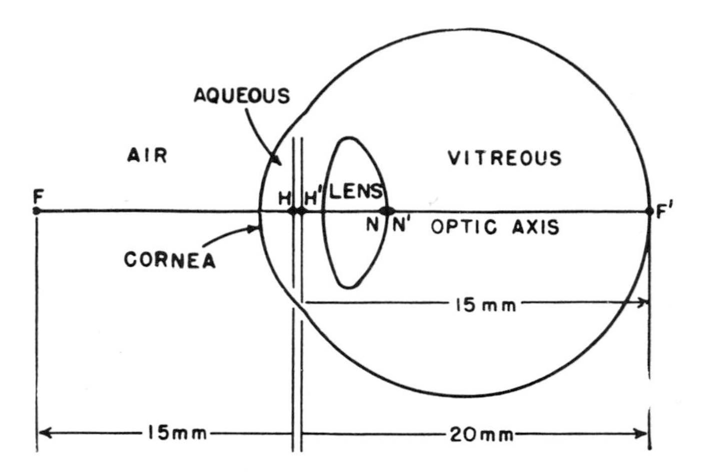

Distance from cornea to front of lens	3.6 mm
Thickness of lens	3.6 mm
Radius of curvature of cornea	8 mm
Radius of curvature of front surface of lens	10 mm
Radius of curvature of back surface of lens	6 mm
Index of aqueous (sodium light)	1.333
Index of lens (sodium light)	1.45
Index of vitreous (sodium light)	1.333
Distance from the cornea to H	1.95 mm
Distance from the cornea to H′	2.38 mm

Fig. 19. The Helmhotz schematic eye (Fry, 1959) with dimensions and indices as given by Laurance.

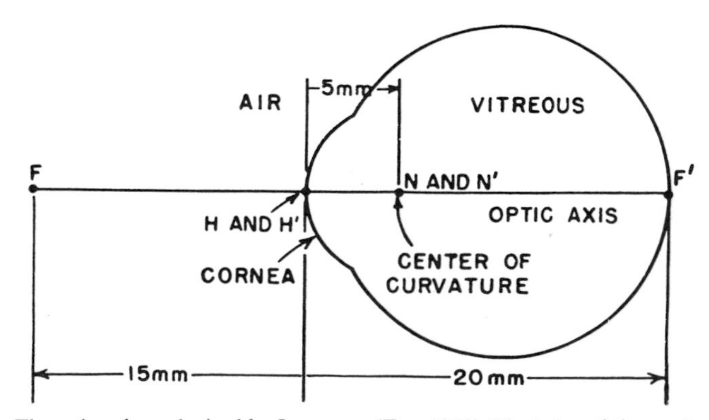

Fig. 20. The reduced eye devised by Laurance (Fry, 1959). The index of the medium is 1.333.

normal eye for the purpose of evaluating deviations from the normal. It must be noted, however, that assessment of the curvature of the back surface of the crystalline lens of an intact eye is subject to error, because the assessment has to be based on assumptions and approximations about the internal structure and the distribution of the isoindicial surfaces. The same thing applies to the thickness of the lens. Even the ultrasonic techniques for measuring the thickness of the lens depend upon uncertain assumptions about the speed of sound transmission through the lens.

The schematic eyes of Gullstrand and Helmholtz in Figs. 11 and 20 are supposed to represent static states of refraction. It must be assumed that the retina falls at the secondary focus, and this leads to an assessment of the distance from the cornea to the retina. The x-ray method of measuring the distance from the cornea to the retina provides a more precise assessment and may be used for determining the normal for this measurement. The average value found by Stenstrom (Duke-Elder, 1949) is 24.00 mm. An x-ray technique is also available for measuring the focal length of the eye.

Gullstrand (Southall, 1924, p. 392) attempted to deal with changes in accommodation by creating a schematic eye for the accommodated eye, involving a new set of constants. We can hope for a model of the eye with the specification of constants for all levels of accommodation, utilizing a multilayered lens that approximates the real lens.

A brief summary of the history of ocular measurements has been presented elsewhere (Fry, 1959).

2 Refraction and Accommodation

a. Spherical Errors of Refraction

When the eye is free from astigmatism and accommodation is relaxed, the point that is conjugate to the retina falls either at an infinite distance in front of the eye (*emmetropia*), at a finite distance in front of the eye (*myopia*), or at a finite distance behind the eye (*hyperopia*). These conditions are referred to as *refractive states.* To specify quantitatively the refractive state we make use of the concept of *refraction.* Refraction (expressed in diopters) is the reciprocal of the distance (in meters) from the point conjugate to the retina R to the spectacle point S, which lies on the primary line of sight 14 mm in front of the cornea (see Fig. 21). The refractive state when the accommodation is relaxed and, therefore, at its zero level is called static refraction, and R is called the far point (punctum remotum). Myopia and hyperopia are referred to as *ametropia.* They are also called spherical errors of refraction, because such errors can be corrected with a spherical plus or minus lens placed normal to the primary line of sight with its back surface at the spectacle point (see Fig. 22).

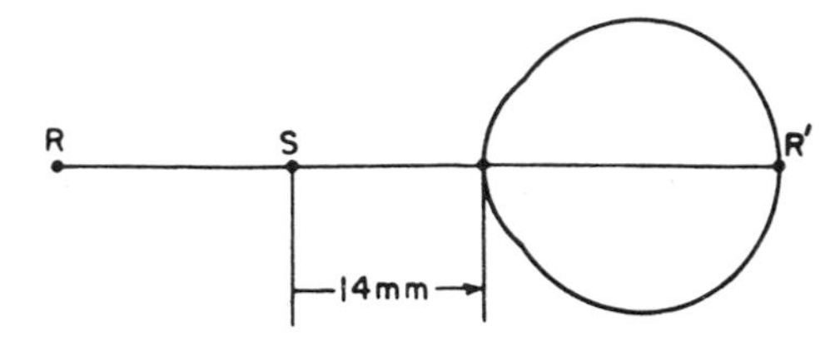

Fig. 21. Spectacle point (Fry, 1959).

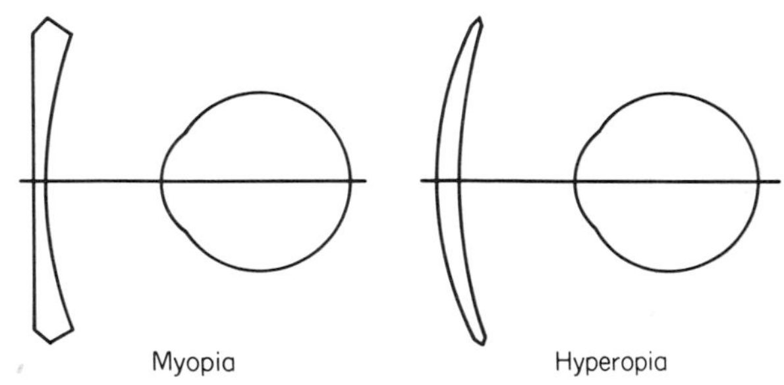

Fig. 22. Spectacle lenses for the correction of spherical ametropia (Fry, 1959).

The spectacle point falls close to the primary focal point, and for the purpose of computing the effect of the lens on the size of the retinal image one can assume that it actually falls there. If the primary focal point is used instead of the spectacle point to specify refraction, the refraction is called focal point refraction.

b. Accommodation

The eye can change its focus to bring objects that are at different distances into focus, one at a time. This change in refraction or refractive state is called *accommodation.* When the maximum amount of accommodation is in play, the refractive state is called *dynamic refraction,* and the point R which is conjugate to the retina is called the *punctum proximum.* The amount of accommodation that is brought into play is called the *amplitude of accommodation.*

When distance correction is worn and R is the point conjugate to the retina with the correcting lens, the reciprocal of the distance from R to S is the amount of accommodation in effect. For distant objects the amount of accommodation is zero. The near point of accommodation recedes with age as shown in Fig. 23. Between the ages of 40 and 50 years a person becomes presbyopic and can no longer accommodate with comfort on objects at the reading distance, and the loss of accommodation has to be supplemented by adding plus power to the lens worn before the eye, which leaves about half of

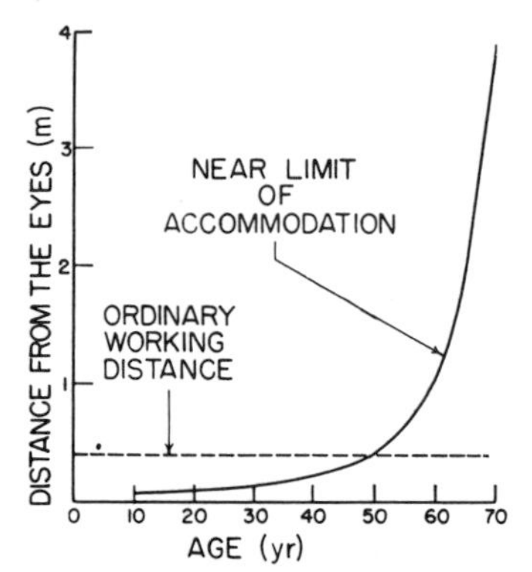

Fig. 23. Regression of the near point of accommodation with age, based on Donder's data (1864). It is assumed that the distance correction is worn. (Fry, 1959.)

the total amplitude in reserve. According to Hamasaki *et al.* (1956) a person becomes an *absolute presbyope* at the age of about 54, after which time the range of clear vision is determined entirely by the depth of focus of the eye.

The added plus power required by a presbyope, who wears a spectacle lens, is provided in the form of a bifocal segment that covers a large area in the lower portion of his field of view.

When the crystalline lens becomes opaque, it can be surgically removed. The eye is then said to be *aphakic*. The problem thus created can be corrected with a spectacle or contact lens, but it is also possible to replace the lens with a plastic insert, mounted inside the eye. A spectacle lens with a reading segment is still required.

c. Astigmatism

When the eye is astigmatic, the refraction varies from meridian to meridian, and the amount of astigmatism is the difference between the maximum and minimum refraction.

The meridians of an eye are planes that intersect at the primary line of sight. The 0–180° meridian lies in the plane of regard (see Fig. 24). The primary line

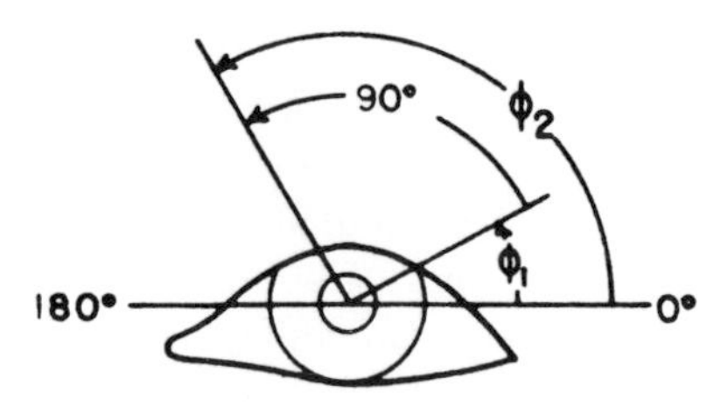

Fig. 24. Specification of meridians (Fry, 1959).

of sight is normal to the paper and penetrates the center of the pupil. The counterclockwise displacements from the 0–180° meridian of the principal meridians of astigmatism in a given case are represented by ϕ_1 and ϕ_2.

The astigmatism can be corrected with a minus cylinder with the axis of the cylinder in the meridian of least refraction.

An eye may have a combination of astigmatism and spherical ametropia. The correcting lens is equivalent to the combination of a sphere and a cylinder and is called a *spherocylindrical* lens.

d. Spectacle Lenses

For myopia or hyperopia the correcting lens has a spherical surface in front and behind; but in the case of a cylinder or spherocylinder, at least one surface has to be toroidal or cylindrical. The line that passes through the center of rotation and is normal to the base line and depressed downward 6° is normal to the front surface and penetrates it at the major reference point. If the prescription for the lens calls for zero prism power, the line normal to the front surface at the major reference point is also normal to the back surface, and thus it represents the optic axis of the lens. The lens must also have a specified thickness at this point. The index of refraction can vary from lens to lens but is usually 1.523 for glass spectacle lenses. The lenses can be tinted but are usually clear.

Various combinations of surfaces can provide the sphere and cylinder powers for correcting the refractive error, and the designer is free to select surfaces that will also provide prism power, or a specific amount of angular magnification, or freedom from unwanted sphere and cylinder in the peripheral parts of the lens, or freedom from reflections. In choosing the center thickness one must consider breakage, weight on the face, and transmittance. *Absorption lenses* can be used to control the overall transmittance or the selective transmittance for different wavelengths.

e. Contact Lenses

The outer portion of a scleral contact lens (Fig. 25) rides on the sclera, where its central portion covers the cornea with a layer of fluid which lies between the lens and the cornea. A corneal contact lens contacts the cornea.

Both surfaces of a hard corneal contact lens can be spherical, and because the liquid layer between the contact lens and the eye has about the same index as the cornea, the astigmatic front surface of the cornea as a refractive surface is eliminated, and hence the astigmatism is corrected. This is not possible with a soft contact lens which conforms to the cornea. The contact lens offers the possibility of correcting spherical aberration by the use of aspherical surfaces. A contact lens can be painted on the back side to produce an artificial pupil. Absorption contact lenses can also be used.

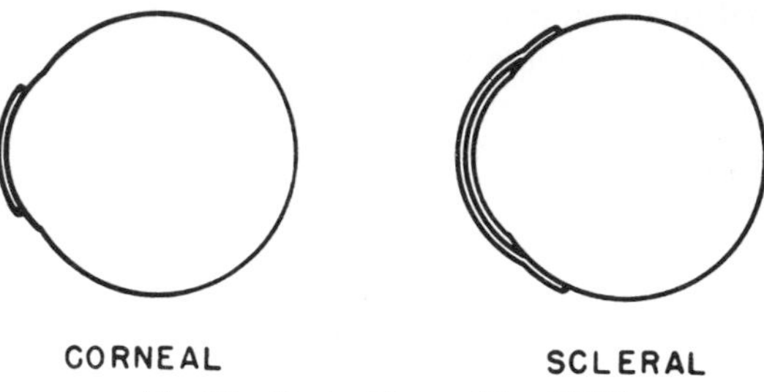

Fig. 25. Contact lenses (Fry, 1965b).

f. The Mechanism of Accommodation

The mechanism in the lens for adapting to changes in tension of the zonula (Fig. 1) and the changes with age make interesting topics but are of little concern to the matters at hand. They are discussed elsewhere (Alpern, 1969).

It suffices to point out here that the pupil size is largely controlled by the distribution of light on the retina. This response is ordinarily called the "*light reflex*." Pupil size is also affected by changes in accommodation. Changes in pupil size affect not only the amount of light falling on the retina but also the blur of the retinal image. Accommodation and the associated pupil constriction are also subject to voluntary control. In looking through a wire screen an observer can, at will, accommodate on the screen or on a distant object seen through it. In total darkness accommodation can be changed by concentrating on imaginary far and near points. One can also clear up an out-of-focus image of an object by watching the amount of blur without paying any attention to the perceived distance of the object. When one cannot tell from the nature of the blur or from other cues whether one is over accommodated or underaccommodated, one must make exploratory adjustments to verify the adjustments required to clear up the blur.

The accommodative mechanism is continuously making involuntary microadjustments of which a person is not aware; this is called *accommodative noise*. In general, a person is underaccommodated when looking at near objects and overaccommodated for distant objects. This is partly related to the amount of convergence in play, but it also occurs with one eye occluded. In darkness or in a field devoid of stimuli to accommodation, the eye is slightly myopic.

Information on the stimuli used in guiding the response to a need for clear vision can be found elsewhere (Alpern, 1969). This is also true for the neural mechanisms involved in the interrelations between accommodation, pupil constriction, and convergence. Furthermore, information on the use of drugs to control these mechanisms will have to be sought elsewhere.

D. FIELD OF VISION FOR NORMAL SEEING

1. Normal Seeing

The eye is normally in an environment in which it is looking at an array of *coherent point sources*. Some of these points sources may form a line as in the case of a heated wire filament; some form a uniform surface like a ribbon filament or a diffusely reflecting surface; and still others may be arrayed through the three-dimensional space in front of the eye as in the case of the sky, a smoke ring, or a sodium vapor source.

2. Elements of Solid Angle

The space in front of the eye can be divided into elements of *solid angle* $d\omega$ which can be specified in *steradians*. Each element of solid angle has its apex at the center of the entrance pupil and is displaced through an angle θ from the primary line of sight. Each element of solid angle contains an array of point sources that may be confined to a surface or distributed in three dimensions as in the case of the sky, uniformly illuminate the pupil and contribute to the image formed on the retina by that element of solid angle. The overlapping images produced by the different elements constitute the image formed by the total array of elements of solid angle. If the visible point sources within a given element of solid angle are confined to a surface that cuts across the element, this portion of the surface can be treated as a single point source. If the point sources are distributed throughout a part of the three-dimensional space within an element of solid angle, this space can be truncated into small elements of volume, and each of these elements can be treated as a point source.

It is customary to describe the field of view as an array of objects superimposed on a background, but from the point of view of optics each of these components has to be broken down into elements of solid angle. We can also describe the field of view as an array of points, lines, borders, gradients, and uniform areas.

a. LUMINANCE

For each element of solid angle in the field of view, one can specify the luminance L for that direction. Luminance is defined as the flux F per unit solid angle ω per unit projected area $A \cos\theta$ of the pupil,*

$$L = F/(\omega A \cos\theta). \quad (8)$$

* The symbol F is used for flux instead of ϕ, because ϕ is the symbol for the index of blur. The symbol for refracting power is also F, but this offers no chance for confusion.

Nit is the unit for specifying luminance:

$$\text{one nit} = \text{one lumen per steradian per square meter of projected pupillary area.} \quad (9)$$

The luminance in a given direction can be measured directly with a telephotometer.

3. Elementary Retinal Images

An element of solid angle in the visual field can be regarded as equivalent to a point source producing a certain amount of flux F that enters the pupil of the eye and forms an elementary retinal image. This distribution of flux is one in which $E(r)$, in lumens per square minute, is a function of r, the distance in minutes across the retina from the chief ray. It is known as a *point spread function.*

This distribution can be normalized by dividing the values of Er by F to obtain normalized values $G(r)$ which replace the actual values $E(r)$.

$$G(r) = \frac{E(r)}{F}. \quad (10)$$

F represents the total flux that enters the pupil and gets redistributed in the retinal image. The total flux (lumens) represents the rate of flow of energy through the pupil. The unit of energy is the *talbot*:

$$\text{one lumen} = \text{one talbot per second.} \quad (11)$$

Retinal illuminance at any part of the retinal image should be defined basically as flux (lumens) per unit area. It can also be expressed in *trolands.* When the distance from the second nodal point to the retina is 15 mm, which is its value in the Helmholtz schematic eye,

$$\text{one troland} = 4.44 \times 10^{-15} \text{ lumens per square micrometre.} \quad (12)$$

Retinal illuminance can also be defined in terms of flux per square minute of retinal area subtended at the second nodal point:

$$\text{one troland} = 8.46 \times 10^{-14} \text{ lumens per square minute.} \quad (13)$$

This makes it independent of having to specify the distance from the second nodal point to the retina.

The chief advantage in expressing retinal illuminance in terms of flux per square minute subtended at the second nodal point is that the angle subtended by two retinal points at the second nodal point may be assumed to be equal to the angle subtended by the corresponding lines of sight at the entrance pupil.

The value of defining things in this way becomes apparent when one has to compute the distribution of flux in the retinal image.

An array of elements of solid angles in the field of view produces an array of overlapping retinal images.

4. Distal and Proximal Stimuli

A distribution of luminance in the field of view is referred to as a distal stimulus, and the resulting distribution of illuminance on the retina as the proximal stimulus. People often use the expression *stimulus intensity* to refer either to the intensity of the distal stimulus (luminance) or to the intensity of the proximal stimulus (retinal illuminance), even though photometrists prefer to restrict the word *intensity* to designate the candlepower of a point source. One must be aware that the word *intensity* can be used in these different ways.

5. Specification of Stimulus Intensity

a. Large Solid Angles of Uniform Luminance

When the distal stimulus is a uniform distribution of luminance covering a solid angle one degree wide or larger, the retinal illuminance at the center of the image will be uniform in spite of the blurred borders; hence, one can relate the luminance L of the distal stimulus to the resulting retinal illuminance E_R as

$$E_R = LA, \tag{14}$$

where E_R is in trolands, L is the luminance in nits, and A is the area of the pupil in square millimetres.

The basic formula (Fry, 1955, p. 10) relating luminance L for a given solid angle $d\omega$ in the field of view and the retinal illuminance E_R in the corresponding retinal area dA is

$$E_R = \frac{d\omega}{dA} LAt \cos \theta, \tag{15}$$

where θ is the angular displacement of the center line of the solid angle from the line normal to the pupillary plane, t is the transmittance of the eye, dA varies with the size of the eye and optical constants and the angle of incidence at the retina, E_R and L are expressed in luxes and nits, and A and dA are in square meters. It is also assumed that the pupil is a hole in a thin flat diaphragm. A *lux* is a unit of illumination representing one lumen per square metre.

In deriving Eq. (14) from Eq. (12) and Eq. (15) it is necessary to assume that $\theta = 0$, $t = 1$, and $d\omega/dA = 4444$, where $d\omega$ is expressed in steradians.

b. Self-Luminous or Diffusely Reflecting Surface

In the case of a self-luminous surface (ribbon filament) or a diffusely reflecting surface (piece of paper) placed in front of the eye, the luminance in the direction of the eye can be specified in terms of candlepower (candelas) per unit of projected area (square metres). One nit of luminance represents one candela per square metre of projected area. Each element of area can be regarded as a point source.

c. Candlepower per Unit Length of Line Sources

In the case of a luminous line (heated wire filament) the stimulus intensity has to be specified in terms of candlepower per unit length D (candelas per metre). Each element of length can be considered equivalent to a point source.

d. Candlepower of a Point Source

In the case of a point source (stop) the intensity of the distal stimulus has to be specified in terms of *candlepower* (candelas).

E. RETINAL IMAGES

1. Point Sources

In this chapter we are concerned only with the prereceptor mechanisms that modify the image formed on the retina. The retina is treated as a hypothetical smooth surface. The impressions initiated in the photoreceptors are further modified by the fact that the receptors constitute a mosaic and by the interaction between the paths over which the impressions are transmitted to the brain. Thus we can differentiate between the quality of the *prereceptor retinal image* and the quality of the *perceived image* which involves also the neural mechanisms of the eye and brain.

a. Geometrical and Physical Images

Lines called *rays* are used in physics to depict the propagation of light. They are used as illustrated previously to trace rays in a bundle of rays from a point in front of the eye through the front part of the eye to form a point image on the retina. A section of the bundle that is normal to the chief ray at a point in front of or behind the retina yields a blur circle. Rays that are uniformly distributed across the entrance pupil are uniformly distributed in the blur circle; they are merely concentrated into a smaller area. The concentration of rays is roughly proportional to the flux density. If the beam is aberrated, the rays no longer focus at a point, but the distribution of flux can be assessed by looking at the concentration of rays at different parts of the image. Descriptions of the image, based on ray tracing, give useful approximations and for many problems provide adequate answers. An image described in this way is called a *geometrical image*.

An alternate approach to depicting the propagation of a beam is to use the rays to construct a wave front, which is a surface perpendicular to each of the rays. The wave front can be regarded as just another geometrical construction, but if the light emitted from the object point is coherent, the patches of light at different parts of the wave front are in phase, and the configuration of the wave front can be used to predict the flow of luminous energy along the beam. This is the pure physical optics approach to the propagation of a beam. The *physical image* has to be described as a distribution of flux over the surface of the retina.

The computation of the distribution of flux in an image can be very tedious; but because the computation is based on the wavelength of the light, the size and shape of the exit pupil, and the configuration of the wave front, we can make use of calculations that have already been made and formulas that have been derived to make quick assessment of the physical images formed by the eye.

b. The Image of a Point Source

Nowhere is the difference between the geometrical and the physical image more striking than in the case of a monochromatic point source. In the geometrical image the rays can converge to a point at the retina, and the image has zero radius and infinite illuminance; it is strictly a point image. In physical optics if the point source is coherent and the wave front emerging from the exit pupil is spherical and centered at the retina, the image formed is known as the *Fraunhofer image of a point source* (Fry, 1955, p. 69). The analytical expression for the distribution is

$$E \propto [(2/\bar{\gamma})J_1(\bar{\gamma})]^2, \tag{16}$$

where $J_1(\bar{\gamma})$ is the first order Bessel function of $\bar{\gamma}$ and where

$$\bar{\gamma} = (1/3438)(2\pi\bar{g}/\lambda)r. \tag{17}$$

In Eq. (17) $\bar{g}$ is the radius of the entrance pupil, and r is the distance from the center of the image. A cross section of the distribution is illustrated in Fig. 26; it is a distribution of flux over a considerable area. This distribution of flux is known as the *Fraunhofer point spread function*; it is radially symmetrical. It should be noted that the bright spot at the center is surrounded by a dark annulus. The area covered by the central bright spot, all the way out to the center of the first dark annulus, is called *Airy's disk*. When the pupil of the eye is 2 mm or smaller, Eq. (16) holds. When the pupil is larger, the spherical aberration of the eye has to be taken into consideration.

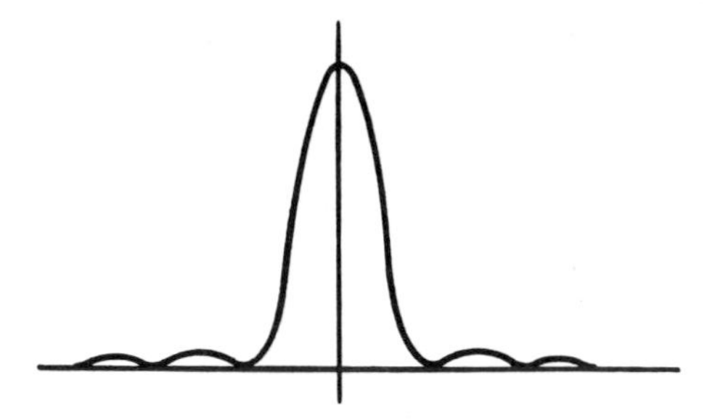

Fig. 26. Distribution of illuminance across the center of the physical image of a monochromatic point source in an eye free from spherical aberration and astigmatism and focused for the sharpest possible image (Fry, 1959).

c. Resolving Power

Rayleigh has arbitrarily stated that the eye can resolve two Fraunhofer images placed side by side when they overlap to the extent that the center of one image falls at the center of the first dark annulus of the other image (see Fig. 27). The resolution threshold (center-to-center separation) can be computed from the radius of the entrance pupil $\bar{g}$ with the equation

$$\text{resolution threshold in radians} = 0.61\ \lambda/\bar{g}. \tag{18}$$

This is known as Rayleigh's criterion (Jenkins and White, 1957, p. 300). The resolving power is the reciprocal of the resolution threshold.

The resolving power of the eye is usually measured, however, with a square-wave grating such as the five-bar pattern shown in Fig. 28a. It is called the Foucault pattern. In this case the resolving power is specified in terms of the number of black bars per millimetre at a certain distance or in terms of cycles

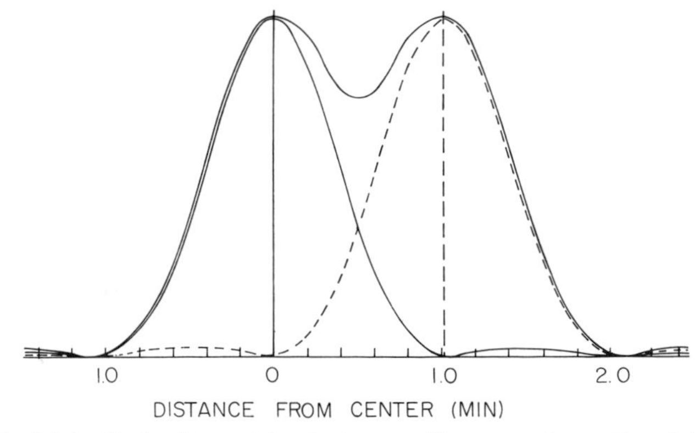

Fig. 27. Rayleigh criterion for resolving the images of two monochromatic points (Fry, 1970).

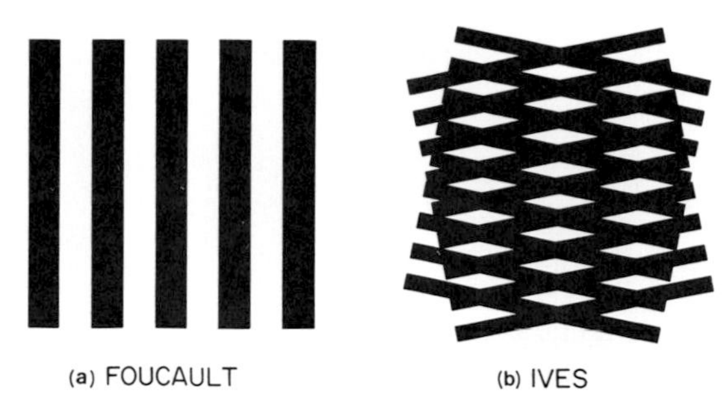

Fig. 28. Foucault and Ives crossed grating patterns for measuring visual acuity and resolving power.

per degree. This should not be confused with assessments expressed in terms of visual acuity, which is the reciprocal of the angular width of a single bar in minutes of arc. It is the number of half cycles per minute. The Ives pattern in Fig. 28b is also used for the measurement of visual acuity.

It should be noted that the resolving power of the eye, when measured in this way, includes the effect of the neural mechanisms of the eye and brain in degrading the quality of perceived images and, hence, is not a proper assessment of the quality of the optical image formed on the retina.

In many situations it is possible to use a Gaussian distribution (Fry, 1970, p. 72) as a substitute for the Fraunhofer image (Fig. 29). The analytical

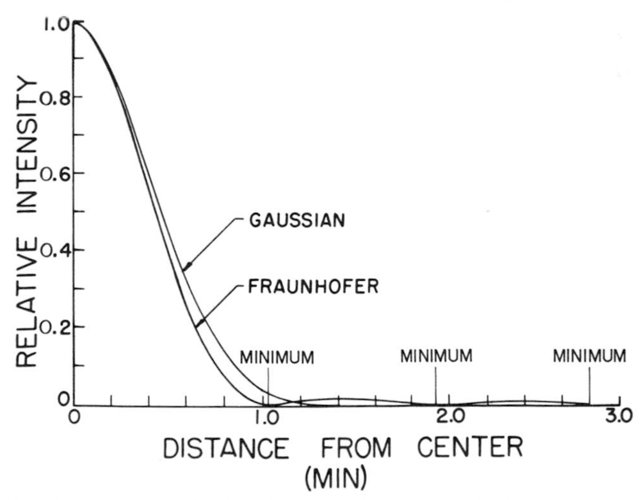

Fig. 29. Gaussian and Fraunhofer point spread functions which approximate each other; $\bar{g} = 1$ mm, $\lambda = 494$ nm, $\sigma = 0.4$ minutes. As explained in Section E4a of this chapter these two point spread functions have been determined to be equivalent, because they have the same index of blur (Fry, 1965a).

expression for the Gaussian curve is simpler to use than the more exact Fraunhofer equation,

$$E = F \frac{1}{2\pi\sigma^2} e^{-r^2/(2\sigma^2)}. \tag{19}$$

d. The Out-of-Focus Image

The simple out-of-focus image evaluated by geometrical optics is a *blur circle*, and the distribution of flux is called a *pill box distribution*. As the physical image of a point gradually goes out of focus, the bright spot at the center transforms to an annulus with a dark spot at its center. The dark spot, in turn, expands and develops into an annulus, and so on. The image becomes a series of concentric annuli, and the average retinal illuminance remains constant out to the last bright annulus; from this point the retinal illuminance tapers off to zero. The distribution across the center from edge to edge is illustrated in Fig. 30.

The physical out-of-focus image is called a *Fresnel image*. In the same figure is the pill box distribution for the geometrical image. The figure illustrates why the pill box distribution is a good approximation for the physical image.

The formula for the radius of the pill box distribution (Fry, 1970, p. 80) is

$$\bar{r} = (\bar{g}'/\overline{\mathrm{O'M'}})|v|. \tag{20}$$

Figure 31 shows the geometry involved in deriving this equation.

The simple out-of-focus image of a monchromatic coherent point source evaluated by physical optics is given by the following formula (Fry, 1955, p. 62):

$$E_{\mathrm{R}} = E_{0'}\left(\frac{n'}{\lambda c}\right)^2 (\mathrm{C}^2 + \mathrm{S}^2), \tag{21}$$

where E_{R} is the illuminance at the point P on the retina at a distance r from the point of penetration R′ of the chief ray. E_0, is the illuminance in the plane of the exit pupil. The wavelength is λ, and the index of the vitreous is n'. As shown in Fig. 32, c is the distance from the pole B of the wave front to the retina R′.

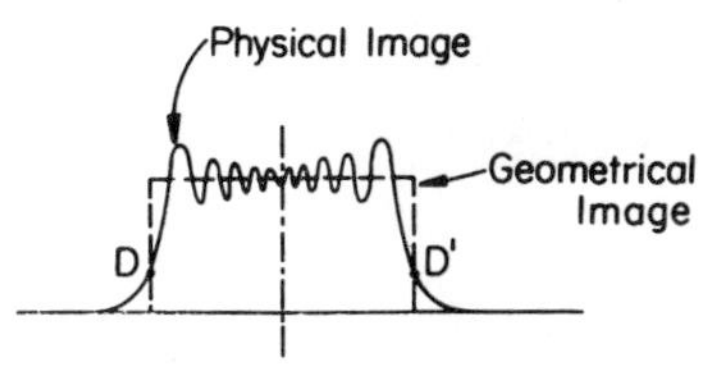

Fig. 30. Relation between the physical and geometric images of a monochromatic point that is out of focus (Fry, 1976).

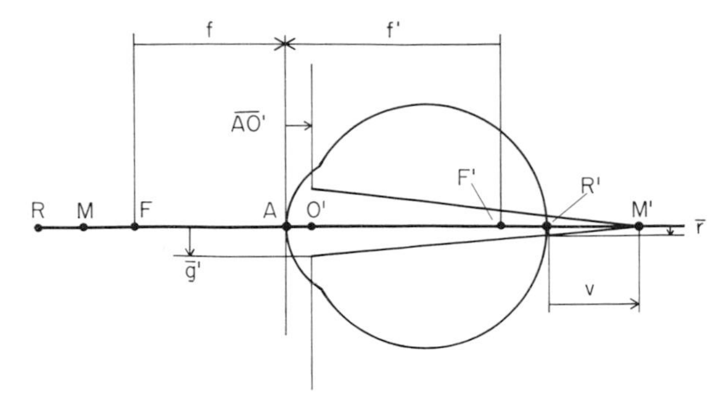

Fig. 31. Geometry involved in computing the size of the blur circle for an out-of focus eye (Fry, 1955).

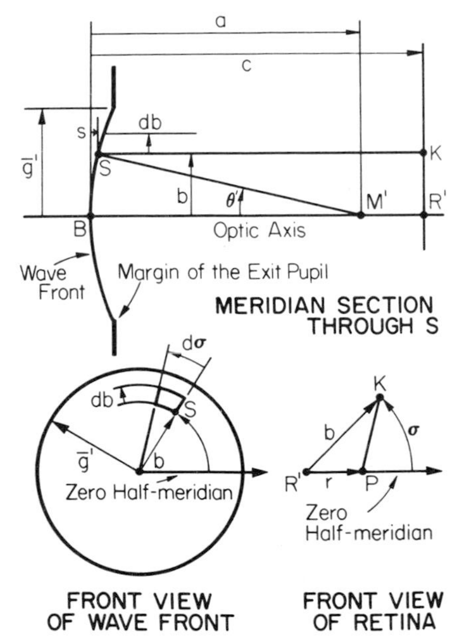

Fig. 32. Geometry involved in computing the physical image of a monochromatic point. The point lies on the optic axis, and the wave front is radially symmetrical with respect to the optic axis (Fry, 1955).

The wave front is assumed to be spherical and centered at M′. The exit pupil is round and normal to and centered on the chief ray.

The terms C and S are defined as

$$C = 2\pi \int_0^{\bar{g}'} [\cos(1/2)\omega] \cdot bJ_0(\gamma) db, \tag{22}$$

and

$$S = 2\pi \int_0^{\bar{g}'} [\sin(1/2)\omega] \cdot bJ_0(\gamma)\, db. \tag{23}$$

In these equations $J_0(\gamma)$ is a Bessel function defined as

$$J_0(\gamma) = (1/\pi) \int_0^{\pi} \cos(\gamma \cos \sigma)\, d\sigma. \tag{24}$$

The terms $1/2\omega$ and γ are defined as

$$(1/2)\omega = (\pi n' b^2/\lambda ac)(a - c) \tag{25}$$

and

$$\gamma = (2\pi n'/\lambda)(br/c). \tag{26}$$

The dimensions σ, $d\sigma$, b, db, c, r, and $\bar{g}'$ are indicated in Fig. 32.

Fry (1955) constructed an analog computer for solving Eqs. (15) and (16), but they can now be solved with a digital computer.

Lommel (Fry, 1955) has derived from Eqs. (22) and (23) the equations known as Lommel functions which can be used with a digital computer to assess the values of E_R. Hufford and Davis (1929) have modified Lommel's equations to speed up the calculations for large values of $\bar{\gamma}$ and $\bar{\omega}$. Epstein (1949) has derived equations for computing values of E_R near the border of an image when $\bar{\gamma}$ and $\bar{\omega}$ are nearly equal.

e. Spherical Aberration

In an emmetropic eye that is free from astigmatism, spherical aberration becomes a major factor in blurring the image formed at the fovea. The primary line of sight is displaced about 5° from the pupillary axis, which results in about 0.35 D of astigmatism. This astigmatism and the astigmatism produced by a toroidal cornea can be compensated by a cylindrical lens placed before the eye. The residual aberration is spherical aberration, which is the failure of rays entering the pupil at different distances from the center to converge at a common point; it may also be described as a failure of the wave front emerging from the exit pupil to have a spherical configuration. The central rays focus at the point on the chief ray called the *paraxial image*. An aberrated ray passing through a peripheral part of the pupil will cross the chief ray at a point in front of or behind the paraxial image, and the displacement is a measure of the spherical aberration. If the peripheral ray crosses in front of the paraxial focus, the aberration is positive; if it falls behind, it is negative. A plot of the height of the ray (distance of the ray from the chief ray in the plane of the pupil) versus the amount of aberration represents the spherical aberration for the meridian.

It is customary to average the values for the different meridians and to assume that the spherical aberration is radially symmetrical.

From the spherical aberration curve in Fig. 33 it is possible to reconstruct the wave front emerging from the exit pupil (Fry, 1955, pp. 95–98). One can then use Eqs. (21), (22), and (23) to compute the point spread function. In this case,

$$\frac{1}{2}\omega = \frac{2\pi n'}{\lambda}\left(\frac{s^2 + b^2}{2c} - s\right). \tag{27}$$

If the spherical aberration is not radially symmetrical, it is necessary to use Eqs. (28) and (29) instead of Eqs. (22) and (23) to compute C and S:

$$\mathrm{C} = \int_0^{2\pi}\int_0^{\bar{g}'} \cos[(1/2)\omega - \gamma\cos\sigma]b\,db\,d\sigma \tag{28}$$

and

$$\mathrm{S} = \int_0^{2\pi}\int_0^{\bar{g}'} \sin[(1/2)\omega - \gamma\cos\sigma]b\,db\,d\sigma. \tag{29}$$

Equations (22) and (23) are derived from Eqs. (28) and (29) by assuming the wave front to be radially symmetrical.

There are several ways to assess the spherical aberration of the eye. One is to determine the actual configuration of the cornea with a photokeratometer. Assume that the lens surfaces are spherical and centered on an axis normal to the cornea. Rays are traced that are uniformly distributed across the entrance pupil to determine where they penetrate the retina. This yields a *spot diagram.* The distribution of spots need not be radially symmetrical. At any part of the image the retinal illuminance is assumed to be proportional to the concentration of spots. The spot diagram can also be used to reconstruct the wave front emerging from the exit pupil and then compute the point spread function.

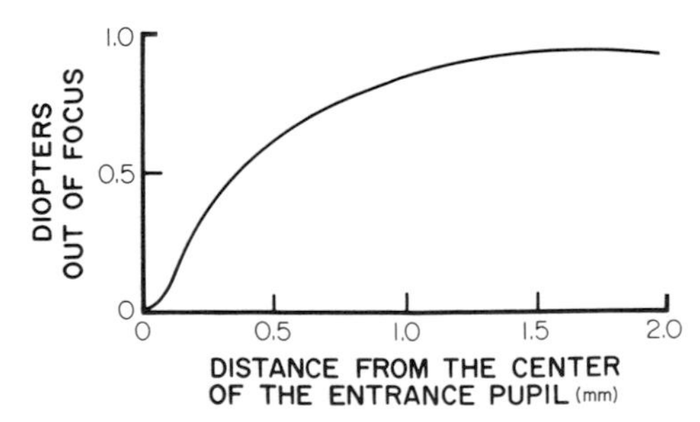

Fig. 33. Spherical aberration of the human eye. Ivanhof's refractive data for different zones of the pupil for relaxed accommodation. Each ordinate represents the number of diopters that the eye is overaccommodated for a distant point. (Fry, 1955.)

By locating the points conjugate to the retina, Koomen *et al.* (1949) acquired data for a series of annular pupils centered on the ray through the center of the real pupil. These measurements are based on the assumption that the spherical aberration is radially symmetrical. Their results for different amounts of accommodation are shown in Fig. 34. These data indicate that the spherical aberration reverses from positive to negative as the accommodation increases.

Ivanoff (1947) measured the spherical aberration by locating the conjugate focus of the retina for a narrow beam through the center of the pupil and a second beam through some peripheral part of the pupil (LeGrand, 1967, p. 29). His data for one meridian are shown in Fig. 33. Fry has assumed the spherical aberration to be radially symmetrical and used the data to compute point spread functions for various positions of the retina relative to the beam. Ivanoff's arrangement can also be used to generate a spot diagram.

Similar arrangements have been used by Ames and Proctor (1921) and by Fry (1949) and others for the measurement of spherical aberration.

Berny and Slansky (1970) have developed a procedure for assessing spherical aberration using variations in flux density in the plane of the pupil produced by light reflected from the retina. The amount of aberration can vary from point to point and meridian to meridian. From these data the point spread function can be computed.

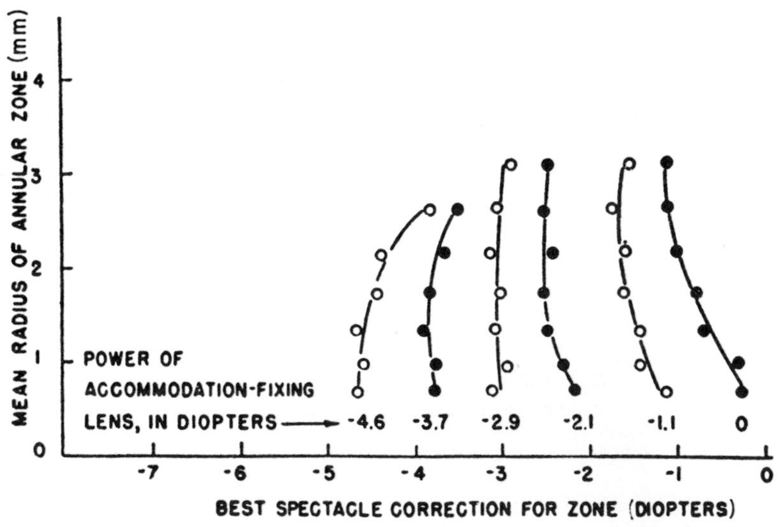

Fig. 34. Spherical aberration of an eye with various amounts of accommodation in play (Koomen *et al.*, 1949).

f. The Correction of Spherical Aberration with a Contact Lens

The spherical aberration changes with the level of accommodation, but for a given level it would be possible by using aspherical surfaces to provide a nearly complete correction for the spherical aberration of the eye. However, this is not widely practiced.

g. Artificial Pupils

Let us assume that an artificial pupil (pinhole) has been mounted in front of an eye and that it falls at the primary focal point. If the diameter is small, the beam entering the pupil is not affected by fluctuations in the size of the natural pupil. Furthermore, if it is small enough, it will reduce the effects of spherical aberration.

In the case of an artificial pupil placed at the primary focal point the exit pupil lies at infinity, and the angle subtended by the exit pupil at the image point M′ equals the angle subtended by the pupil at the first nodal point. For a round artificial pupil the radius $\bar{r}$ of the blur circle is related to the radius of the artificial pupil $\bar{g}$ as shown in Eq. (30) and Fig. 35.

$$\bar{r} = \bar{g}|v/f'|. \tag{30}$$

To assess the physical image when an artificial pupil is used, the approach is to compute the diffraction pattern formed by the pupil on the plane conjugate to the retina and then to use the principles of geometrical optics to transfer the image to the retina. In the special case of an object point located at an infinite distance, the Fraunhofer image lies in a plane at an infinite distance behind the pupil. If the eye behind the pupil is emmetropic and accommodation is relaxed, the Fraunhofer image is transferred to the retina.

Equation (9) serves as the formula for the Fraunhofer point spread function on the retina in the case of an artificial pupil and also for the natural pupil. The

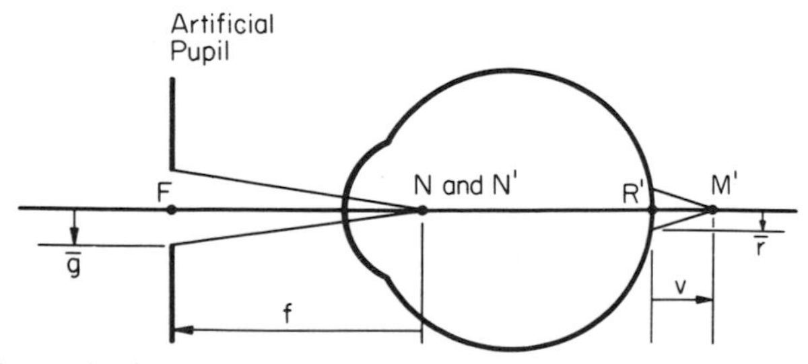

Fig. 35. Blur circle obtained with an artificial pupil. Geometry involved in computing the radius $\bar{r}$ of the blur circle (Fry, 1955).

term $\bar{g}$ is the radius of the artificial pupil, but for the natural pupil it is the radius of the entrance pupil. The distance of the pupil in front of the eye is not important. It may lie on the back surface of a contact lens.

Attention must be paid to other types of artificial pupils (Fig. 36) which serve special purposes. The most important is the *double slit* (Fig. 36a). If a line or slit, which may be regarded as an array of mono-chromatic coherent point sources of light, is viewed through a pair of slits, the image formed on the retina is a sine wave of 100% modulation as shown in Fig. 37 (Jenkins and White, 1957, pp. 311–320). The lines are referred to as a set of Young's fringes. Because it has 100% modulation, its quality is not degraded by defects in the image-forming mechanism; hence, it may be used to assess the degradation of the image as it is transferred from the retina to the brain. The spatial frequency of the fringes can be changed by regulating the separation of the slits. The center-to-center separation of the fringes, expressed in minutes of arc at the second nodal point, is given by

$$E_{\mathrm{R}} \propto \frac{\sin^2[(\pi/\lambda)b \sin\theta]}{[(\pi/\lambda)b \sin\theta]^2} \cos^2[(\pi/\lambda)d \sin\theta], \tag{31}$$

where b is the width of the slits, d is the center-to-center separation, and θ is the

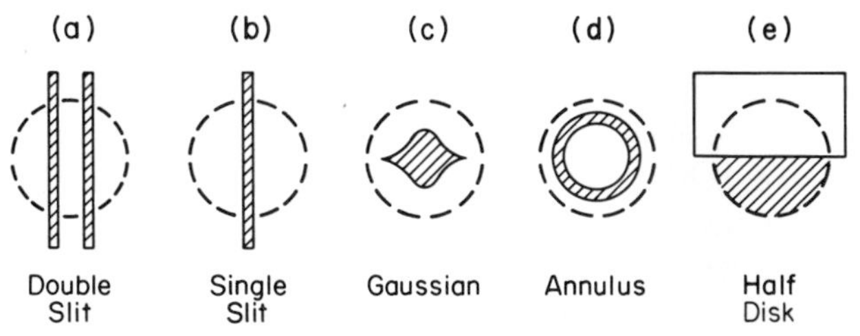

Fig. 36. Special types of artificial pupils. The dashed circle in each case represents the margin of the real pupil.

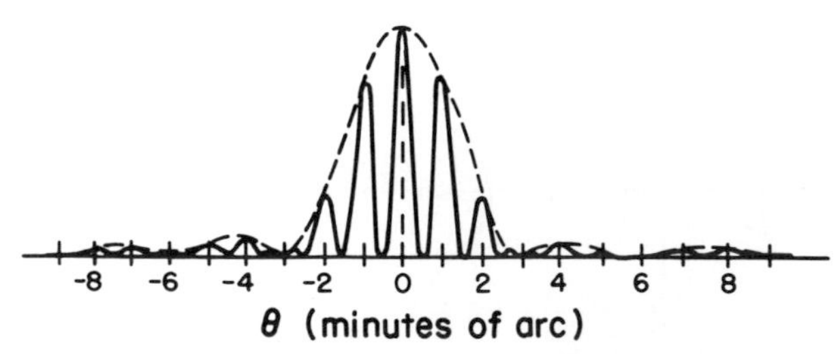

Fig. 37. Diffraction pattern produced by a double slit. The wavelength is 500 nm, the width of each of the two slits is 0.57 mm, and the center-to-center separation is 1.72 mm.

angular displacement from the center of the image. The shape of the envelope (dashed line in Fig. 37) is controlled by the width of the slits.

A single slit (Fig. 36b) called a *stenopaic slit* is used in testing for astigmatism. It is also useful for demonstrating diffraction effects, because slits of various widths are readily available. For a slit with a variable width one can observe the effect of varying the width directly. The formula (Jenkins and White, 1957, pp. 288–297) for the distribution of flux across the image of a line (Fig. 38) is

$$E_{\mathrm{R}} \propto \frac{\sin^2[(\pi/\lambda)b\sin\theta]}{[(\pi/\lambda)b\sin\theta]^2}, \tag{32}$$

where b is the width of the slit, and θ is angular displacement from the center of the image.

A *Gaussian aperture* (Fig. 36c) may be used to demonstrate freedom from spurious resolution (Fry, 1962).

Another important pupil is the *annular pupil* (Fig. 36d). It is useful in the analysis of the Stiles–Crawford effect and in the study of spherical aberration.

Still another kind of pupil to consider (Fig. 36e) is the case in which part of the natural pupil is covered with a diaphragm having a straight edge (Baraket, 1961, p. 77). An example of this can be found by lowering the upper eyelid to cover part of the natural pupil.

The process of changing the form and size of the point spread function by changing the size and shape of the pupil is known as *spatial filtering.*

2. Line Images

A line object such as a heated wire filament may be regarded as a linear array of points. Each element of length dh may be regarded as a point source. These points produce point spread images on the retina that, overlap and produce a *line image.* It can be described as a distribution of retinal illuminance E_{R} in which the value of E_{R} is a function of the distance t from the center line of the image. If the point images $G(r)$ are radially symmetrical, the *line spread*

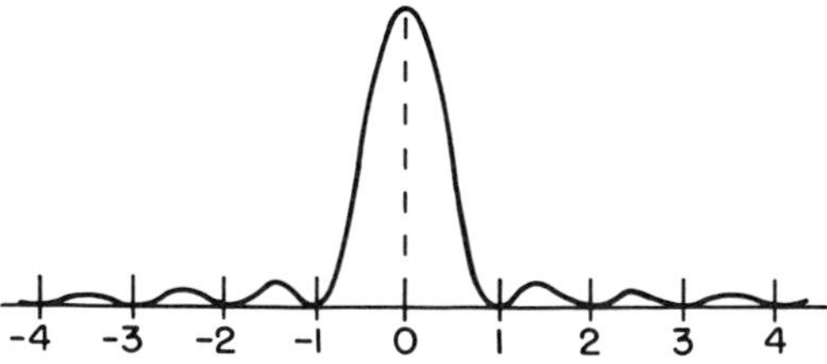

Fig. 38. Diffraction pattern produced by a single slit. The wavelength is 500 nm, and the width of the slit is 1.72 mm.

function can be computed from

$$E(t) = 2D \int_{|t|}^{\infty} [G(r)r/(r^2 - t^2)^{1/2}]dr, \tag{33}$$

where D is the flux (lumens) transmitted through the pupil from 1 minute of length of the line source measured at the center of the entrance pupil.

The values of $E(t)$ can be reduced to normalized values $H(t)$ by dividing by D.

Equation (33) has been used to derive line spread functions from various point spread functions.

a. Fraunhofer Line Spread Function for a Round Pupil

This is derived from the Fraunhofer point spread function, Eq. (16), for a round pupil (Fry, 1955, p. 77–78):

$$H(t) = \frac{8M}{\pi^2}\left[\frac{1}{1^2 \cdot 3} - \frac{2^2}{1^2 \cdot 3^2 \cdot 5}(Mt)^2 + \frac{2^4}{1^2 \cdot 3^2 \cdot 5^2 \cdot 7}(Mt)^4 - \cdots\right], \tag{34}$$

where $M = 2\pi \bar{g}' n'/\lambda c$, $\bar{g}'$ is the radius of the exit pupil, c is the distance from the exit pupil to the retina, λ is the wavelength, and n' is the index of refraction of the vitreous. This is known as *Struve's function*, for which values have been tabulated (Abramowitz and Stegun, 1965).

b. Fresnel Line Spread Functions with a Round Pupil

When the eye is out of focus, the physical image is complicated. Probably the simplest approach for assessing the line spread function is to use numerical integration in solving Eq. (32) for various values of t.

c. Physical Line Images from Asymmetrical Point Spread Functions

When the point spread function is not radially symmetrical, Eq. (33) can be used for computing the line spread function, but G has to be regarded as a function of both r and t. Even though the point spread function may be asymmetrical, it is possible for the line spread function to be bilaterally symmetrical for certain directions of the line with respect to the point spread function.

d. Geometrical Image of a Line Out of Focus

The line spread function (Fig. 39) is derived from the formula Eq. (13) for a blur circle (Fry, 1955, p. 49):

$$H(t) = (2/\pi\bar{r}^2)(\bar{r}^2 - t^2)^{1/2}. \tag{35}$$

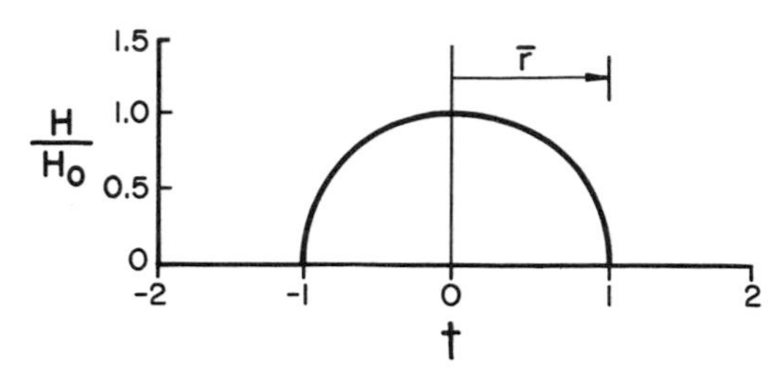

Fig. 39. Geometrical image of a line for an eye out of focus (Fry, 1955).

e. GAUSSIAN LINE SPREAD FUNCTION

$$H(t) = \frac{1}{\sigma\sqrt{2\pi}} e^{-t^2/(2\sigma^2)}. \tag{36}$$

It should be noted that the line spread function has the same form as the Gaussian point spread function from which it is derived [See Eq. (19)].

$$G(r) = \frac{1}{2\pi\sigma^2} e^{-r^2/(2\sigma^2)}. \tag{37}$$

The difference is that the intensity at the center of the line image is greater than at the center of the point image (Fry, 1955, pp. 14, 22).

f. LINE SPREAD FUNCTIONS AND BLUR ELLIPSES

For a blur ellipse the line spread function (Fry, 1955, pp. 50–55) is the same as for a blur circle except that, as shown in Fig. 40, $\bar{r}$ is the distance from the center to a line parallel to the line object and tangent to the blur ellipse,

$$\bar{r}^2 = \bar{a}^2 \sin^2(\phi - \sigma) + b^2 \cos^2(\phi - \sigma). \tag{38}$$

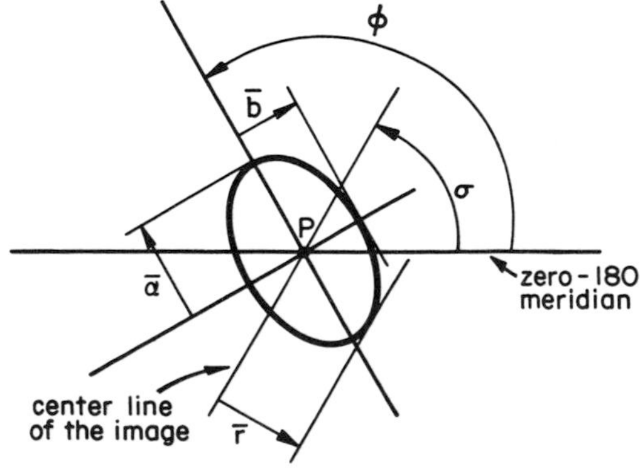

Fig. 40. Geometry involved in computing $\bar{r}$ for the image of a line seen by an astigmatic eye out of focus in both principal meridians.

It may be noted that the line spread function is bilaterally symmetrical in spite of the fact that the point spread function is radially asymmetrical.

g. Use of the Spot Diagram to Assess the Line Spread Function

A spot diagram such as shown in Fig. 41 is one kind of point spread function. As explained by Perrin (1960) and Smith (1966), this can be analyzed by dividing the array of points into parallel strips and counting the points in each strip. The number of points per strip is plotted against the distance t of the center of each strip from the chief ray, which gives relative values for the line spread function.

h. Objective Measurement of the Line Spread Function

Campbell and Gubisch (1966) used a coincidence optometer that forms an image of a narrow slit on the pigment epithelium, which is a diffusely reflecting surface. An image of this image was formed out in space by the image-forming mechanism of the eye. This was then scanned with a slit, and the light transmitted through the slit was measured with a sensing device. White light was used with artificial pupils of different sizes. In presenting the data (Fig. 42), allowance was made for the fact that the light twice traverses the optical system of the eye. The solid curves represent the line spread functions assessed in this way. The dotted curves represent theoretical Fraunhofer line spread functions for an eye corrected for chromatic aberration. For small pupils the empirical curves correspond to the theoretical curves except that the empirical

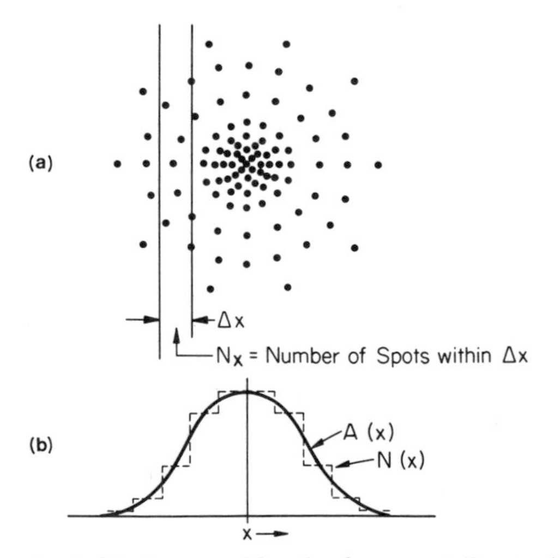

Fig. 41. The assessment of the line spread function from a spot diagram. (a) The spot diagram is summed in one direction by counting the number of spots (ray intersections) in each increment, Δx. (b) The number of spots is plotted against x to get the line spread function $A(x)$ (Smith, 1966).

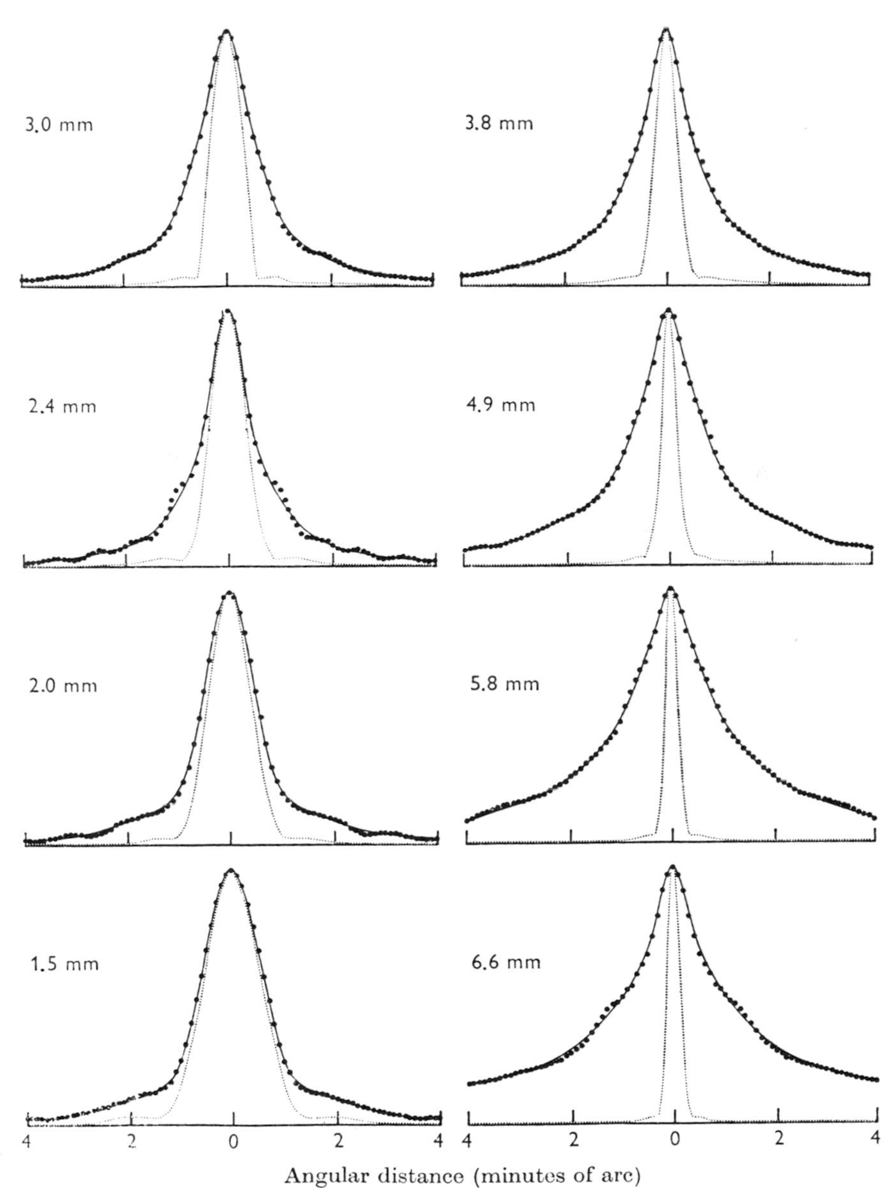

Fig. 42. Optical line spread functions of the human eye for various pupil diameters. The continuous curves are derived from measurements of light reflected from the retina. The dotted curves are theoretical curves for white light for an eye corrected for chromatic and spherical aberration and limited only by diffraction. (Campbell and Gubisch, 1966.)

curves flare out at the bottom. This is attributed to stray light in the eye. The failure to conform when the pupils are large is attributed to spherical aberration and irregular astigmatism.

3. Narrow Bars

a. Convolving a Pattern of Parallel Strips with a Line Spread Function

We have just completed our consideration of point and line spread functions and have shown how the line spread function is related to the point spread function. We now want to relate the line spread function to a new class of stimuli including bars and straight borders and gratings. Each of these patterns can be described as an array of narrow strips of uniform luminance. The effect of optical blur on such a stimulus can be assessed by convolving the array of strips with the line spread function representing the particular type of blur. Each of the strips is replaced by a line image, and the overlapping line images are summed to give the blurred image.

The luminance (nits) of the strips can be plotted as a function $L(s)$ of the angular displacement of the strips (minutes of arc) from some convenient starting point. The angular displacement is measured at the entrance pupil. This represents the unblurred distal stimulus.

The distribution of flux in the convolved image on the retina is given by

$$E(s) = A \int_{-\infty}^{\infty} H(t) L(s - t)\, dt, \tag{39}$$

where A is the area of the pupil in square millimeters, E is in trolands, and L is in nits. Here s represents the distance from the starting point (minutes of arc) measured at the second nodal point.

b. The Effect of Blur on the Image of a Narrow Bar

The effect of blur can be assessed by dividing the bar into a number of small strips and convolving it with the line spread function.

The formula for the intensity E at a distance w from the center of the image is

$$E = L\pi\bar{g}^2 \int_{-w-\bar{w}/2}^{-w+\bar{w}/2} H\, dt. \tag{40}$$

Figure 43 shows the images of bars of various widths $\bar{w}$ produced by convolving them with a Gaussian spread function having a sigma value of 5 μm (Fry, 1955, 27–29). For small bars the intensity at the center is proportional to the width of the bar (Ricco's law). But as the width increases,

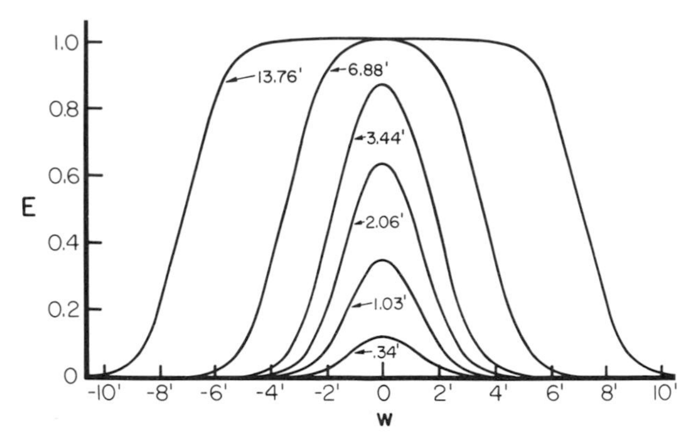

Fig. 43. Retinal images of bars of various widths. Each curve represents the illuminance E at a distance w from the center of the image of a bar of a given width $\bar{w}$. The curves are based on a Gaussian line spread function having a sigma value of 1.15 min. The values of E are relative to a maximal value of unity for wide bars. The values of w and $\bar{w}$ are expressed in terms of minutes of distance across the retina. (Fry, 1955.)

the intensity ceases to increase, the distribution becomes flat on top, and on each side the gradient is a symmetrical ogive that is independent of the width.

c. Subjective Measurement of the Line Spread Function

Fry (1946) used a subjective method to determine the line spread function. He measured the luminance difference threshold ΔL for bars of different widths. The white bars were 50 minutes high and placed on a white circular background 2.8° in diameter. The luminance of the background was 539 c/m^2. The display was viewed through an artificial pupil 2.33 mm in diameter.

For wide bars the threshold is constant and may be designated $\Delta L_{\min}$. To analyze the data, values of $\Delta L_{\min}/\Delta L$ are plotted as a function of $\bar{w}/2$ as shown in Fig. 44. The slope $d(\Delta L_{\min}/\Delta L)/d(\bar{w}/2)$ at any point on the curve represents the value of H for that value of $\bar{w}/2$. The values of H are normalized by dividing by H_0 at $\bar{w}/2 = 0$.

These values are then plotted as a function of t as shown in Fig. 45 (curve B) to give the line spread function. The values of $\bar{w}/2$ in Fig. 44 correspond to the values of t in Fig. 45.

The theory behind this assessment of the line spread function is as follows. The formula for the intensity at center of the image is

$$\Delta E = 2\Delta L\pi\bar{g}^2 \int_0^{\bar{w}/2} H\,dt; \tag{41}$$

see Eq. (40). For a wide bar L is at a minimum; hence,

$$\Delta E = \Delta L_{\min}\pi\bar{g}^2. \tag{42}$$

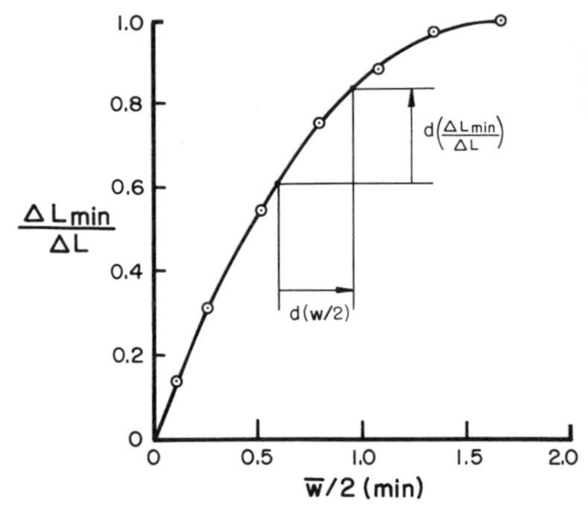

Fig. 44. The effect of varying the width $\bar{w}$ of a bar on the threshold luminance difference ΔL.

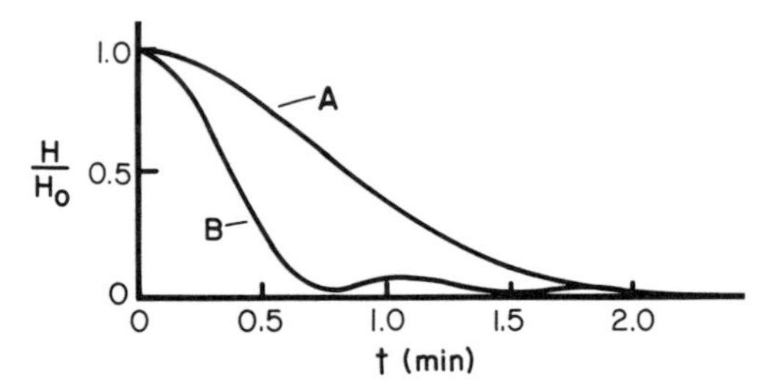

Fig. 45. Line spread function (curve A) for a human eye looking through a 2.33-mm diameter artificial pupil at a white point compared to the line spread function (curve B) for a schematic eye focused on a monochromatic point (555 nm) through a 2.33-mm diameter entrance pupil. The distance from the second nodal point to the retina was assumed to be 15 mm.

We can assume that at the threshold the intensity E at the center of the image is constant; hence,

$$\frac{\Delta L_{\text{min}}}{\Delta L} = 2\int_0^{\bar{w}/2} H\,dt. \tag{43}$$

Differentiating, we obtain

$$H = d\left(\frac{\Delta L_{\text{min}}}{\Delta L}\right)\Big/ dt = d\left(\frac{\Delta L_{\text{min}}}{\Delta L}\right)\Big/ d\left(\frac{\bar{w}}{2}\right). \tag{44}$$

The raw data in Fig. 44 have been fitted by a curve that represents the integral of the probability curve; hence, curve B in Fig. 45 is a Gaussian spread function with a value of 0.72 minutes for σ [see Eq. (36)].

Shown in the same figure is a theoretical Fraunhofer image of a line for monochromatic light (555 nm). Chromatic and spherical aberration can account for some of the difference. It is assumed in this approach that the threshold depends on the contrast between the intensity at the center of the image and the intensity of the background. One problem is that physiological irradiation may be involved as well as optical blur. The effect of physiological irradiation is kept to a minimum when the background has a high luminance.

4. Brightness Contrast Borders

A brightness contrast border is an abrupt transition at a border between a bright area and a dark area. It is known as a *step function*. Let us consider the case of a straight border. The effect of blur can be assessed by convolving the step function with a spread function for a line that is parallel to the border. Figure 46a shows a step function convolved with a Fraunhofer line spread function. This illustrates ordinary seeing. The border is perceived as sharp instead of being blurred.

If the eye is thrown out of focus, a pill-box distribution can be used instead of the Gaussian distribution to assess the effect of blur. Figure 46b shows a step function convolved with a pill-box distribution representing an out-of-focus blur circle. At the top and at the bottom of the gradient the slope abruptly changes to zero, and at these transition points light and dark Mach bands are

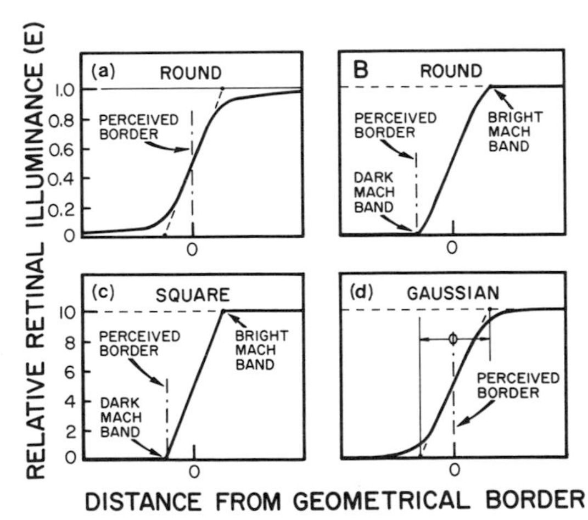

Fig. 46. Images of a brightness contrast border. (a) Physical image with the eye in focus and a round aperture. The horizontal scale is expanded to display the nature of the gradient. The other images are geometrical with the eye out of focus and with various apertures: (b) round, (c) square, and (d) Gaussian.

seen. This can be accentuated by using a square pupil (Fig. 46c), which generates a gradient with a uniform slope. On the other hand, if a Gaussian aperture (Fig. 46d) is used, one obtains an ogive gradient, which minimizes the Mach bands and accentuates the perceived sharpness of the contrast border at the center of the gradient.

If a pattern like that shown in Fig. 47 is viewed through a round pupil when the eye is out of focus, the white square is perceived to be larger than the black square. This occurs because the perceived edge of a blurred image is at the foot of the gradient. This phenomenon is known as *irradiation* (Southall, 1924, Vol. II, pp. 186–193). When the gradient is a symmetrical ogive, the perceived edge is at the middle of the ogive where it ought to be (Fry, 1931). Even when the eye is in focus, the white square may be perceived as larger than the black one. This is explained as an illusion based on the fact that the two squares are not perceived as lying at the same distance. In the case of narrow bars the gradient from the center to the edge is not a symmetrical ogive, hence, bright bars are seen as wider and more blurred than black bars of the same width (Wilcox, 1932).

a. Index of Blur

Fry and Cobb (1935) proposed the use of the ratio ϕ of the difference in luminance on the two sides of a border to the slope at the midpoint of the gradient as an *index of blur* (see Fig. 46); ϕ is also an indication of the width of the gradient zone. It is obvious that other features of the gradient have been ignored in this index of blur, but it can be used for assessing the effect of manipulating variables such as throwing the eye out of focus, changing the size of the pupil, or assessing spherical and axial chromatic aberration.

At the threshold of visibility the slope at the midpoint and the contrast between the flanking areas are the important factors; hence, the index of blur is an appropriate way of specifying the amount of blur. At suprathreshold levels

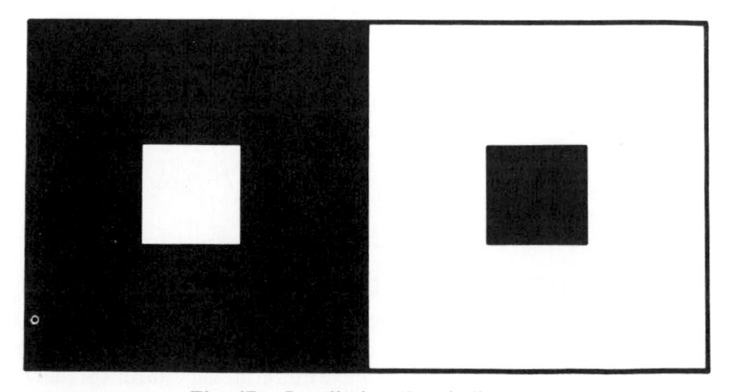

Fig. 47. Irradiation (Southall, 1924).

the shape of the gradient is also important, as is obvious from the Mach band effects. So the index of blur does not tell the whole story.

In photography, workers have to pay attention to the blur produced by the camera and by the film or print before the end product is ever viewed by the eye. They assess the role of the film and camera in terms of resolving power and the acutance (Perrin, 1960, p. 153) of a border, which is defined in terms of the densities in a transparency. They also consider the effect of blur on all aspects of the quality of the perceived image including perceived sharpness, clarity, and definition (Perrin, 1960, p. 152).

The index of blur is convenient, because it can be computed not only from the border spread function but also from the line spread function, the point spread function, or the modulation transfer function (MTF). (See Section E5b on MTF). It has a different meaning when applied to point and line spread functions and the modulation transfer function, but a numerical value for ϕ can be derived from each of these functions. In terms of the line spread function, it is the ratio of the area under the curve to the ordinate at the midpoint. In terms of the point spread function it is the ratio of the total flux in the image to the area of the cross section. If the distribution is not radially symmetrical, this varies from meridian to meridian. Ready-made formulas for computing the index of blur are available for various point spread functions.

Index of Blur for Fresnel Images Given below is an equation (Fry, 1976) for the index of blur ϕ for the image of a monochromatic coherent point source of light that is out of focus.

$$\phi = 3438\lambda c/(n'\bar{g}' V \cdot \overline{\mathrm{N'R'}}), \tag{45}$$

where $\overline{\mathrm{N'R'}}$ is the distance from the second nodal point to the retina and

$$V = 2\left[\frac{1}{0.5!\,1.5!} - \left(\frac{\bar{\omega}}{2}\right)^2 \frac{2!}{2.5!\,3.5!} + \left(\frac{\bar{\omega}}{2}\right)^4 \frac{4!}{4.5!\,5.5!} - \cdots\right], \tag{46}$$

and, as shown in the following equation, ω is a function of the size of the exit pupil and the extent to which the eye has been thrown out of focus.

$$\bar{\omega} = 2\pi\left(\frac{n'}{\lambda}\right)(\bar{g}')^2 \left|\frac{a - c}{ac}\right|. \tag{47}$$

The wavelength of the light is λ, and n' is the index of the vitreous. The dimensions a, c, and $\bar{g}'$ are indicated in Fig. 32.

For large values of ϕ the attempt to use Eq. (45) to compute V involves numbers having such a large array of digits that a computer cannot handle them. Alternate equations may be used for computing the values of V. For values of $\bar{\omega}$ smaller than eight,

$$1/V = 0.000497\bar{\omega}^{2.5} + 0.58903. \tag{48}$$

For values of $\bar{\omega}$ larger than eight,

$$1/V = C - 0.425 + 0.425\cos[360\bar{\omega}/(4\pi)], \tag{49}$$

where

$$C = (1/4)[(\bar{\omega} + 11.87)^2 + 32.7096]^{1/2} - 2.93 \tag{50}$$

The derivation of these equations is explained in the reference cited above.

Index of Blur for Fraunhofer Images The Fraunhofer image of a monochromatic coherent point source is a special case of Fresnel out-of-focus images, in which

$$a - c = 0; \tag{51}$$

hence, Eq. (46) reduces to

$$V = 1.6977. \tag{52}$$

For the case in which the retina falls at the secondary focus, the eye is emmetropic, and

$$(n'\bar{g}' \cdot \mathrm{N'R'}) = \bar{g}c, \tag{53}$$

where $\bar{g}$ is the radius of the entrance pupil; hence, Eq. (45) reduces to

$$\phi \text{ in radians} = 0.59\lambda/\bar{g}. \tag{54}$$

If we compare this equation with Eq. (11), we see that the value of ϕ is nearly equal to the radius of the first dark ring. Hence, according to the Rayleigh criterion, it may be said that two points can be resolved when they are separated by an amount equal to ϕ.

Index of Blur for Gaussian Images For Gaussian point spread functions used to approximate Fraunhofer images

$$\phi = \sigma\sqrt{2\pi}, \tag{55}$$

where σ is the standard deviation of the line spread function. The two point spread functions in Fig. 29 have been determined to be equivalent, because they have the same index of blur.

Index of Blur for Blur Circles and Ellipses For uniformly illuminated blur circles produced with round pupils

$$\phi = (1/2)\pi\bar{r}, \tag{56}$$

where $\bar{r}$ is the radius of a blur circle and one half of the distance between tangents in the case of an ellipse (see Fig. 40).

b. Method of Measuring the Index of Blur

The procedure involves the measurement of contrast thresholds for bars of various widths on a uniform background. To illustrate the procedure we have used the same set of data as used for assessing the line spread function [see Fig. (44)].

The data are plotted log ΔL against log width $\bar{w}$ as shown in Fig. 48. A line with a slope of minus one is fitted to the data for narrow bars. As the bars get wider, the threshold reaches a constant level, which is represented by a horizontal line. The abscissa value of the intersection of these two lines is the index of blur. In Fig. 48 the index of blur is 1.8 minutes.

At low luminance levels of the background the index decreases as the luminance level increases, which is taken to mean that the measurement includes both neural and optical spread. The value of the index found at high luminance levels comes closest to representing pure optical blur.

The following formula (Fry and Cobb, 1935) can be used for assessing the index of blur,

$$\phi = \Delta L_1 \bar{w}_1 / \Delta L_2, \tag{57}$$

where ΔL_1 is the brightness difference threshold for a bar that is so narrow ($\bar{w}_1$) that $\Delta L_1 \bar{w}_1$ is a constant and where ΔL_2 is the threshold for a bar that is so wide that the threshold is independent of the width.

c. The Effect of Pupil Size and Throwing the Eye Out of Focus on the Index of Blur

The index of blur can be used to demonstrate and explain the effect of using pinhole pupils on the perception of fine detail when the eye is thrown out of focus to various degrees (Fry, 1955, pp. 76–77). Equation (45) has been used to

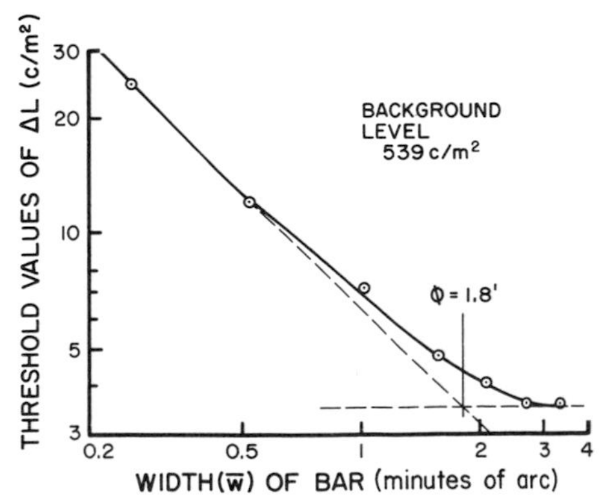

Fig. 48. The effect of varying the width $\bar{w}$ of a bar on the threshold luminance difference ΔL. The figure illustrates the procedure for analyzing the data to determine the index of blur ϕ.

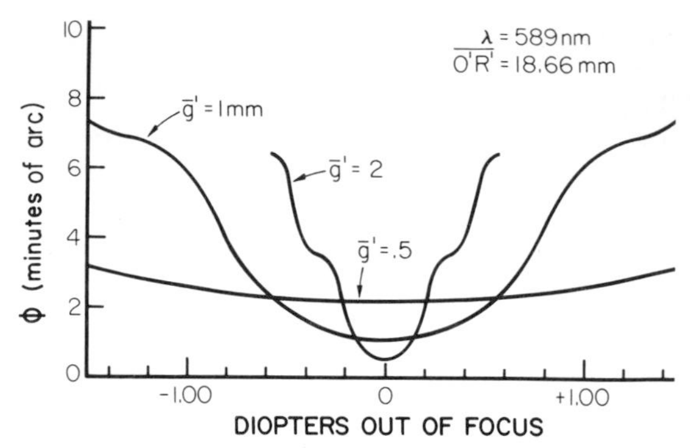

Fig. 49. The curves represent computed values of ϕ for different pupil sizes with the eye thrown out of focus to various degrees (Fry, 1955).

compute the values of ϕ shown in Fig. 49. The data apply to monochromatic light (589 nm) and an eye free from spherical aberration. The eye involved is the Helmholtz schematic eye thrown out of focus by placing thin lenses at the primary focal point. The retina falls at the secondary focal point, which lies 18.66 mm from the exit pupil. When the eye is in focus, reducing the size of the pupil impairs vision, but when the eye is considerably out of focus, the use of a pinhole improves it.

Figure 50 shows the effect of changing the size of the pupil when the eye is in focus. According to the graph, vision continues to improve as long as the size of the pupil increases. This does not occur when white light is used because of axial chromatic aberration. The value of $1/\phi$ increases as the pupil diameter increases up to about 3 mm and thereafter begins to decrease (see Fig. 65.) Cobb (1915) (Fry, 1970, p. 100) has also shown that with white light the

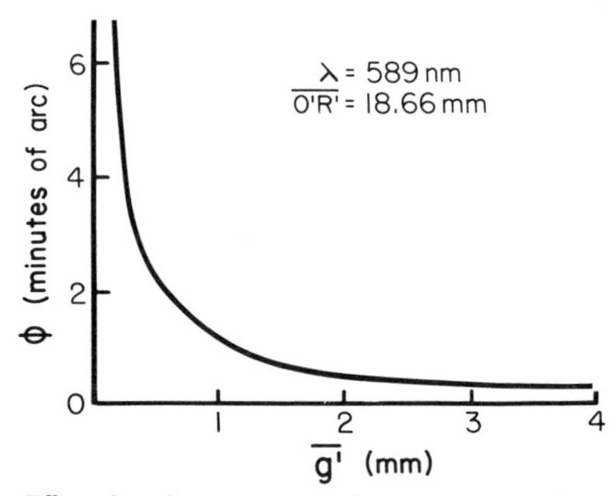

Fig. 50. Effect of pupil size on blur when the eye is in focus (Fry, 1955).

resolving power reaches a maximum at a pupil diameter of about 2 mm. Berger–Lheureux–Robardey (1965; Fry, 1970, p. 107) showed with monochromatic light (546.3 nm) that the resolving power in a real eye is not proportional to the size of the pupil and that this indicates that vision is impaired with a large pupil because of spherical aberration.

d. Other Indices

Gubisch (1967) has used *Strehl's ratio* as an index of the quality of an image formed by the eye. This is the ratio of the intensity at the central part of an image formed by the eye and the corresponding intensity in a model eye free from aberration and having the same focal length and aperture. Byram (1944, pp. 589–590) used a different index of image quality.

Still another index is that of Rayleigh (Section E1c of this chapter). A new version of this index is that proposed by Sparrow (Parent and Thompson, 1969, pp. 18–20).

5. Gratings

a. The Effect of Blur on the Image of a Sine-Wave Grating

The intensity distribution across a sine-wave grating is illustrated in Fig. 51. The center-to-center separation between the peaks is the *period*, and the reciprocal of the period is the *spatial frequency*. The minimum luminance L_{min} occurs at the troughs and the maximal luminance L_{max} at the peaks. The *modulation* is defined as

$$\text{modulation} = (L_{max} - L_{min})/(L_{max} + L_{min}). \tag{58}$$

This represents the ratio of the amplitude (ac component) to the average luminance (dc component).

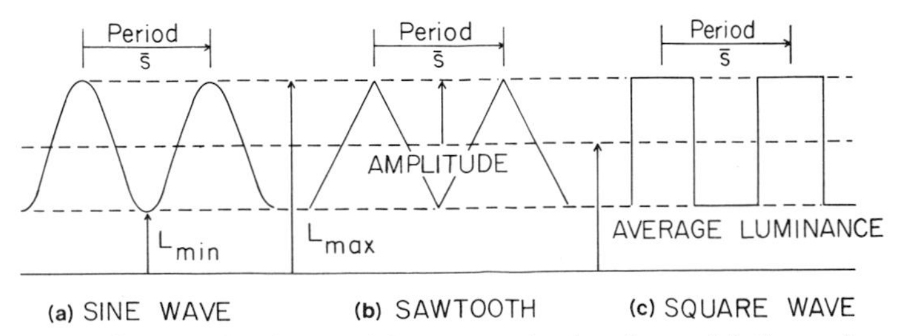

Fig. 51. Specification of period, resolving power, visual acuity, modulation, and contrast for various types of gratings (Fry, 1970).

It should be noted that *contrast* is defined as

$$\text{contrast} = (L_{\text{max}} - L_{\text{min}})/L_{\text{min}}; \tag{59}$$

hence, the value for contrast is more than twice as large as the value for modulation.

Sine-wave gratings that have a modulation as high as 0.73 can be generated by viewing a square-wave grating through a piece of ground glass. As will be explained later, this involves convolving a square wave with the point spread function of the ground glass. They can also be generated with a computer and a TV monitor. Sine waves of 100% modulation can be obtained with a Babinet compensator or a Newton wedge or a double slit. Various arrangements can be provided to regulate the amount of modulation.

If the line spread function is known for the image-forming mechanism of the eye, one can assess the effect of optical blur on the image of a sine-wave grating formed on the retina. This is a matter of convolving the sine-wave pattern with the line spread function. It can be demonstrated that convolving a sine-wave grating with any line spread function will result in a new sine wave of the same frequency but reduced in modulation. If the line spread function is bilaterally asymmetrical, there will be a shift in phase. But this is of no importance so far as the visibility is concerned; the change in modulation is the important effect.

At any given frequency one can measure the modulation threshold and use the threshold as a criterion for assessing the relative amounts of modulation reduction required by different kinds of blur to bring a sine wave to the threshold of visibility.

b. Modulation Transfer Function

When a sine wave is convolved with a line spread function, the factor by which the modulation is reduced is the modulation transfer factor. If the factors for different frequencies are plotted on a graph, the curve is called the *modulation transfer function* (MTF).

When the line spread function is bilaterally symmetrical, there is no change in phase, and the modulation transfer function T is the simple *cosine Fourier transform* of the line spread function,

$$T = \int_{-\infty}^{\infty} H(t)\cos(2\pi t/\bar{s})\,dt, \tag{60}$$

where t is the distance from the center of the line spread function in minutes and $\bar{s}$ is the center-to-center spacing of the lines of the grating. Hence, one can use Eq. (60) to compute the modulation transfer factors if the line spread function is known. One can also compute the line spread function from the MTF.

The formulas needed (Lamberts, 1958; Perrin, 1960) are more complicated, but one can compute the MTF from the line spread function even when the line image is bilaterally asymmetrical. At the same time the phase changes can be computed. One can also work backward from the MTF to the line spread function, but the phase changes have to be known (Lamberts, 1958; Perrin, 1960).

Fraunhofer and Gaussian MTFs Equation (60) has been used to derive the MTF from the Fraunhofer line spread function for a circular pupil (Epstein and Fry, 1973),

$$T = (2\pi)(\theta - \sin\theta\cos\theta), \tag{61}$$

where

$$\cos\theta = 3438\lambda/(2\bar{g}\bar{s}). \tag{62}$$

When we use the Gaussian line spread function [Eq. (36)] as a substitute for the Fraunhofer line spread function, the modulation transfer function is as follows,

$$T = \exp[-2(\pi\sigma/\bar{s})^2]. \tag{63}$$

The MTFs for the two point spread functions in Fig. 29 are shown in Fig. 52.

The Index of Blur and the MTF The index of blur ϕ can be defined in terms of the modulation transfer function,

$$1/\phi = 2\int_0^\infty T(1/\bar{s})\,d(1/\bar{s}). \tag{64}$$

This means that ϕ is the reciprocal of the area under the curve representing the modulation transfer function. For example, the two MTFs shown in Fig. 52 have the same ϕ value, because the areas under the two curves are equal.

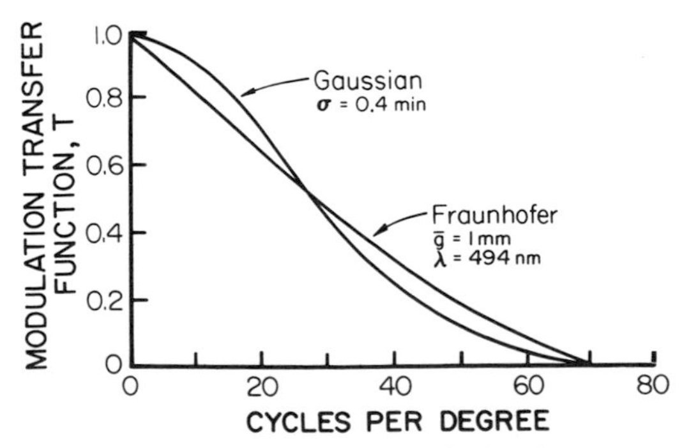

Fig. 52. Gaussian and Fraunhofer modulation transfer functions corresponding to the point spread functions in Fig. 29 (Fry, 1970).

The MTFs for the Geometrical Image of a Line Equation (60) has also been used to derive the MTF for the geometrical image of a line,

$$T = 2[J_1(2\pi\bar{r}/\bar{s})]/(2\pi\bar{r}/\bar{s}). \tag{65}$$

This MTF is illustrated in Fig. 53. The curve explains the phenomenon of *spurious resolution*, which occurs with an eye thrown out of focus. Siemens star in Fig. 54 (left) is a form of square-wave grating in which the frequency decreases as the distance from the center of the pattern increases (Perrin, 1960, p. 241). Figure 54 (right) is a photograph of the star made with an out-of-focus camera. The zones at which resolution is lost are the frequencies at which the modulation is zero. It may be noted that in going from one zone of resolution to the next, the phase changes by 180°. This is the significance of going from positive to negative in Fig. 53.

The pattern in Fig. 55b (Fry, 1962) can also be used to demonstrate spurious resolution. If this pattern is viewed through a round aperture with the eye out of focus, one sees the pattern in Fig. 55a. The phase changes are not easily observed. If Fig. 55b is observed through a small Gaussian aperture (see Fig. 36c) centered in the natural pupil, there is only one transition from fusion to resolution, as shown in Fig. 55c.

Other examples of spurious resolution can be found elsewhere (Fry, 1953).

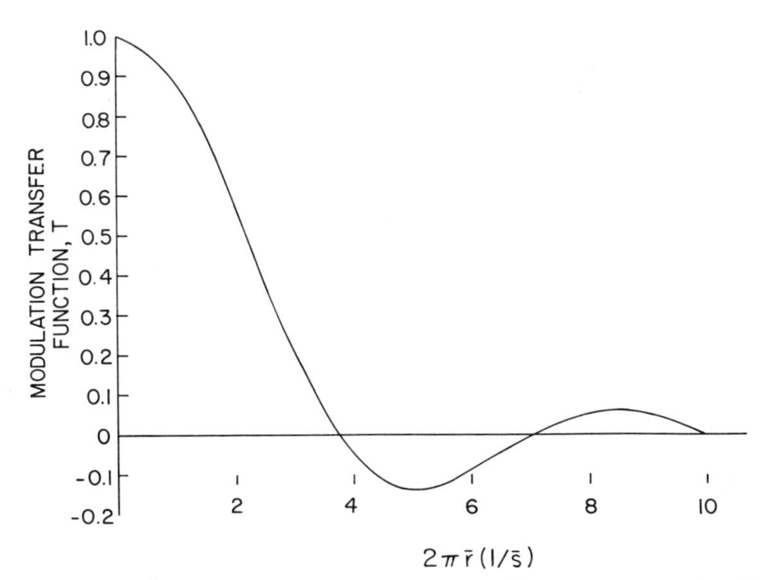

Fig. 53. Modulation transfer function for an out-of-focus blur circle; $\bar{r}$ is the radius of the blur circle (Fry, 1970).

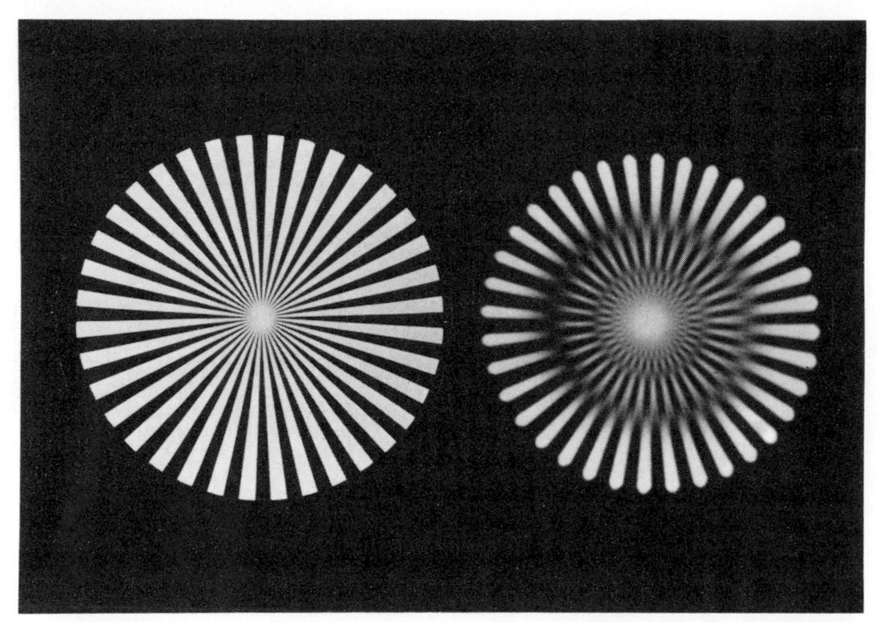

Fig. 54. Siemens star. Left, in focus. Right, out of focus with a round aperture. (Reprinted by permission of the Institut für Medizinische Optik der Univesität München.)

Self-Convolution of an Aperture A Fraunhofer line spread function produced by a round pupil, or a pupil of any other shape, can be derived by a process known as *self-convolution* of the aperture. The mathematical basis for the operation is explained elsewhere (Francon, 1963, pp. 52–64), but the procedure consists of drawing two outlines of the exit pupil as shown in Fig. 56, with the center of one displaced a certain distance m from the other. The area of the overlap is proportional to the modulation transfer factor, and the amount of separation m is proportional to the frequency. In the case of a circular pupil,

$$\text{area of overlap} = 2(\bar{g}')^2[\bar{\theta} - \sin\bar{\theta}\cos\bar{\theta}], \tag{66}$$

where

$$\cos\bar{\theta} = \tfrac{1}{2}m/\bar{g}, \tag{67}$$

when θ is expressed in radians and

$$m = 3438\lambda/\bar{s}. \tag{68}$$

Because the modulation transfer factor is always unity when the frequency is zero,

$$T = (\text{area of overlap})/\pi(\bar{g}')^2. \tag{69}$$

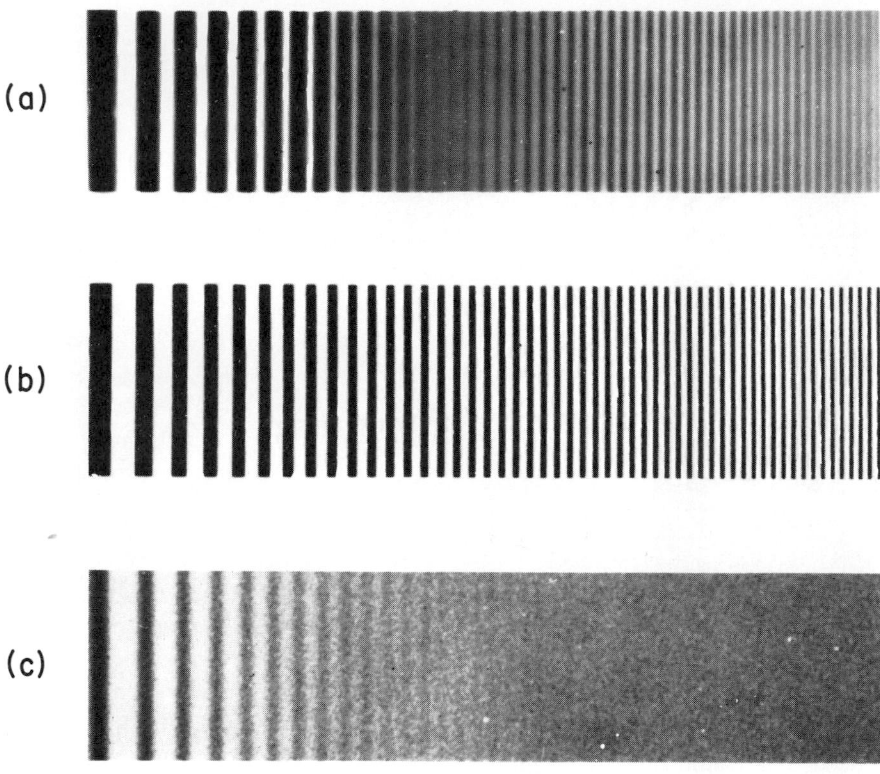

Fig. 55. The relation of blur to resolving power. The reproductions were made with the same camera, film, and development technique. (b) Camera in focus; circular aperture. (a) Camera out of focus; circular aperture giving spurious resolution. (c) Camera out of focus; Gaussian aperture shown in Fig. 36c, which eliminates the spurious resolution. (Fry, 1962.)

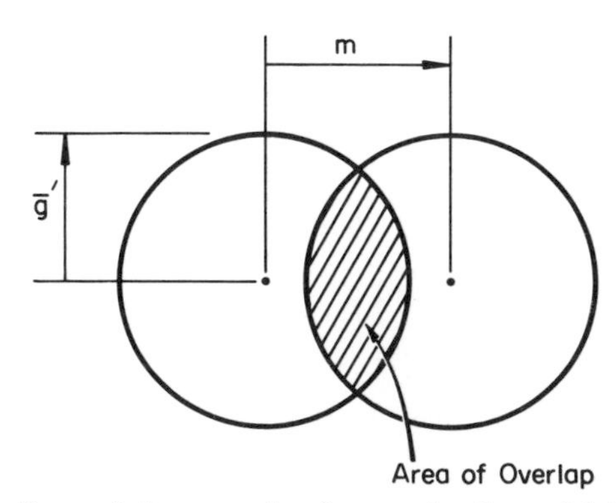

Fig. 56. Aperture self-convolution procedure for assessing the modulation transfer function.

Hence,

$$T = (2/\pi)(\bar{\theta} - \sin\bar{\theta}\cos\bar{\theta}). \tag{70}$$

One can now use Eq. (70) to derive the line spread function.

The same procedure may be used for a square aperture, or an aperture of any other shape.

Röhler's Method of Measuring the MTF Röhler (1962) has devised a method for measuring the MTF objectively. He forms an image of a point on the pigment epithelium. The light, diffusely reflected backward from the pigment epithelium through the lens and cornea, forms an image in front of the eye at the point conjugate to the retina. At this point it is scanned by a rotating square-wave grating that varies in frequency from one end of the grating to the other. The light transmitted by the grating generates a signal that is processed electronically to produce a recording of the modulation transfer function.

c. Square-Wave and Triangular-Wave Patterns

These are examples of periodic patterns that have been used for measuring resolving power. Square-wave gratings are easy to generate. The Ives acuity meter is a device that generates a moiré pattern of diamonds as shown in Fig. 28b. If the pattern is blurred in a vertical direction, it becomes transformed into a triangular pattern, illustrated in Fig. 51b.

The modulation of square waves and triangular waves can be specified the same as sine waves (see Fig. 51.) At spatial frequencies close to the limit of resolution, the optical transfer function of the eye converts square waves and triangular waves to sine waves; hence, it is easy to relate the modulation thresholds for square waves and triangular waves to those of sine waves. Periodic patterns such as square waves and triangular waves can be analyzed into a *Fourier series* of sine waves of different frequencies, which include a fundamental and various harmonics.

The mathematical procedures for analyzing a periodic pattern into its fundamental and harmonics are described by Stuart (1966). The Fourier equation for a square-wave grating with dark and bright bars of equal width is

$$\Delta L = \frac{4}{\pi}\cos\left(\frac{2\pi}{\bar{s}}x\right) - \frac{4}{3\pi}\cos\left(\frac{2\pi}{\bar{s}}3x\right) + \frac{4}{5\pi}\cos\left(\frac{2\pi}{\bar{s}}5x\right) - \cdots, \tag{71}$$

where ΔL is the displacement above or below the average value of L, and x is the lateral displacement from the midpoint of one of the bright bars. The resultant amplitude = 1.

The Fourier equation for a triangular grating is

$$\Delta L = \frac{8}{\pi^2}\cos\left(\frac{2\pi}{\bar{s}}x\right) + \frac{8}{9\pi^2}\cos\left(\frac{2\pi}{\bar{s}}3x\right) + \frac{8}{25\pi^2}\cos\left(\frac{2\pi}{\bar{s}}5x\right) + \cdots. \tag{72}$$

The first cosine term in each equation is the fundamental, and the succeeding terms represent the harmonics.

Convolving a square wave or a triangular wave with a Gaussian or Fraunhofer line spread function filters out the higher-frequency components and leaves, for the most part, only the fundamental.

This filtering process is illustrated by viewing a square-wave grating through a piece of ground glass. This is a convenient way of generating a sine-wave grating from a square-wave grating. Using this approach one can show that a square wave of unit amplitude is equivalent to a sine wave with an amplitude of $4/\pi$. A sine wave of $8/\pi^2$ amplitude is equivalent to a triangular wave of unit amplitude. The same kind of approach can be applied to other periodic patterns.

F. METHODS OF VIEWING

1. Ordinary Method of Viewing

Methods of viewing have been reviewed in two previous papers (Fry, 1965b, 1970).

Reference has been made to the fact that an ordinary field of view contains only coherent point sources, which produce Fraunhofer or Fresnel images on the retina. If such a point source is conjugate to the retina and if the eye is free from spherical aberration, Fraunhofer formulas can be used to compute the distribution of flux in the image. If the image is out of focus, the Fresnel formulas can be used. The ordinary method of viewing consists in placing an object with diffusely reflecting surfaces in front of the eye, with nothing between the eye and the object, and illuminating the object with a source of light (Fig. 57). Each point on the reflecting surfaces can be regarded as a coherent point source.

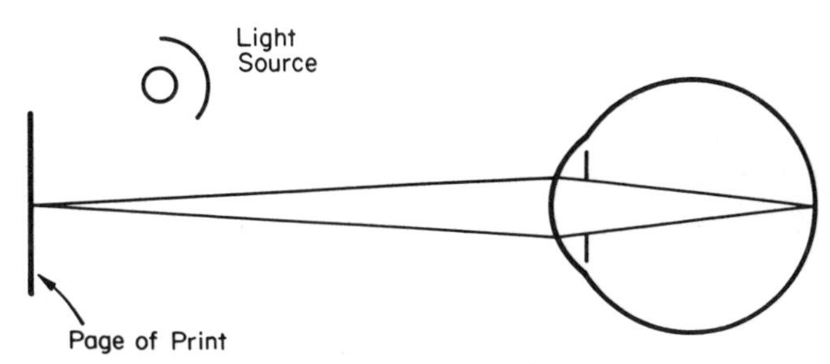

Fig. 57. Ordinary seeing.

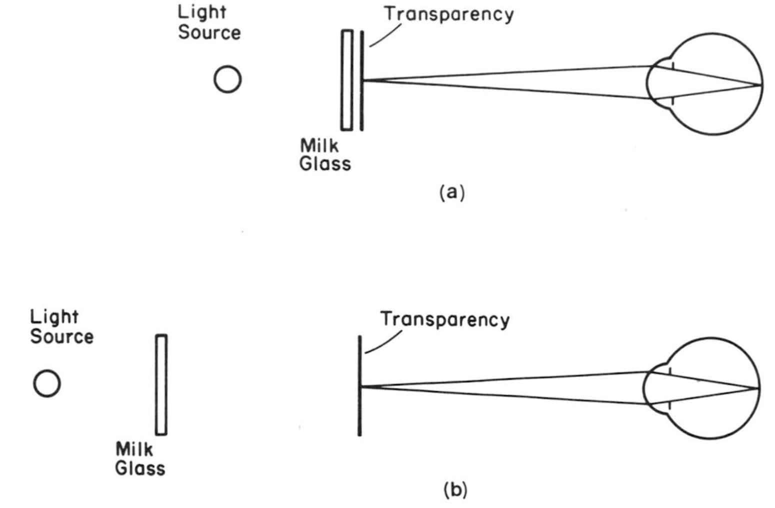

Fig. 58. Methods of viewing a transparency against a transilluminated piece of milk glass. (a) Transparency and milk glass in the same plane. (b) Milk glass at a finite distance behind the transparency.

Another procedure is to place a transparency in contact with a piece of opal flashed glass and illuminate it from behind with a source of light (Fig. 58a). This is equivalent to ordinary viewing, because each luminous point in the plane of the transparency can be regarded as a coherent point source. However, if the transparency is placed at some distance in front of the opal flashed glass (Fig. 58b), this is no longer ordinary viewing. If the eye looks directly at a heated filament, this also represents an array of coherent point sources and is an example of ordinary viewing.

2. Image Formation with an Array of Coherent Point Sources in the Plane of the Entrance Pupil

In Fig. 59 an image of a coil filament is formed in the plane of the pupil. Each point in the image of the filament is an independent coherent point

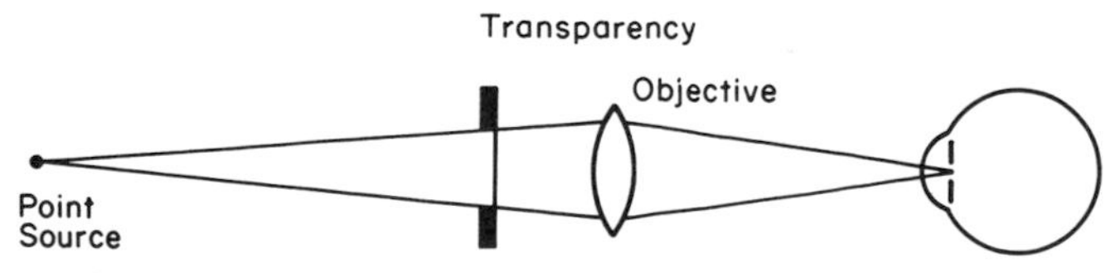

Fig. 59. Viewing a transparency with the source conjugate to the pupil.

source. The transparency is conjugate to the retina. The Fraunhofer and Fresnel formulas no longer apply.

Figure 60a is another example of the same thing. The ribbon filament, which is an array of independent point sources, is conjugate to the pupil. In Fig. 60b the ribbon filament and the condenser lens in Fig. 60a have been moved forward as a unit until the ribbon filament is conjugate to the transparency and to the retina, which converts the arrangement in Fig. 60a to one that is akin to ordinary viewing. In this way one can demonstrate the continuous range of changes between having the ribbon filament conjugate to the retina and having it conjugate to the pupil.

3. Directed Vision

If the aperture in Fig. 60a is made extremely small, the method of viewing is called *directed vision* by LeGrand (1936a, 1936c). The beam entering the pupil is not only smaller than the natural pupil so that fluctuations in the pupil do not affect the beam, but the beam is so narrow that the spherical aberration of the eye cannot affect the quality of the image. However, there are other complications.

If in this arrangement a transparent square-wave grating is used as the transparency and if a narrow slit parallel to the lines of the grating is substituted for the pinhole, a grating image of the slit is formed in the plane of

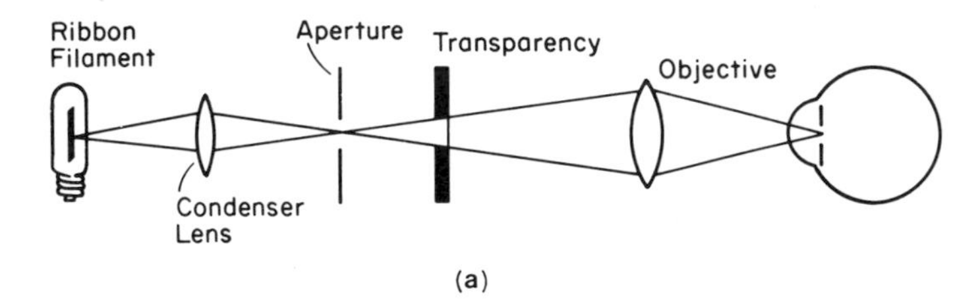

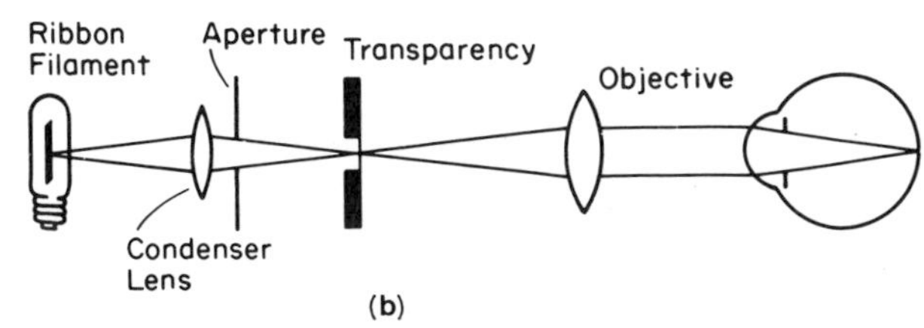

Fig. 60. Complex methods of viewing a transparency. (a) LeGrand's method of "directed vision" with the source conjugate to the pupil. (b) Source conjugate to the transparency.

the pupil, and part of this image falls beyond the margin of the pupil. Using this arrangement Westheimer (1960) placed a diaphragm in front of the eye to screen out certain lines of the grating image to produce various types of images on the retina. This is a form of spatial filtering.

4. Maxwellian Viewing

The term *Maxwellian viewing* means different things to different people; hence, to specify that a stimulus was seen in a Maxwellian view is not adequate. Perhaps the earliest reference to Maxwellian viewing was made by von Kries (Southall, 1924, Vol. II, p. 395). He referred to Maxwell's colorimeter (1860) by which an observer could see a uniform field of monochromatic light instead of a continuous spectrum. It is the difference between looking at the objective through the slit of a monochromator and looking at a spectrum through the eyepiece of a spectrometer.

The term Maxwellian viewing is now used to designate any arrangement in which a lens is used to produce a large uniform field of view. For example, it is used to describe an arrangement in which a lens is used to focus an image of a small source like a heated coil filament in the pupil of the eye (Fig. 59). The face of the lens is uniformly bright, and the beam entering the eye is smaller than the natural pupil; hence, the image formed on the retina is not affected by fluctuations of the pupil.

The fact that the beam entering the eye is smaller than the natural pupil was not a feature of Maxwell's arrangement. He looked through a slit that was longer than the width of his pupil, so what he saw was subject to fluctuations in the size of the pupil.

Let us consider the arrangement shown in Fig. 60a. The aperture can be changed in size so that the image in the plane of the pupil can be larger or smaller than the natural pupil. It is the use of the lens and not the size of the aperture that determines that this method of viewing should be called Maxwellian.

One can use an artificial pupil in front of the eye and centered on the primary line of sight to view a small object also centered on the primary line of sight. The beam of light from any point on the object is smaller than the natural pupil, but this does not qualify this method of viewing to be called Maxwellian. The writer does not know who first deliberately used a beam smaller than the natural pupil to avoid the effect of fluctuations of the pupil.

The term Maxwellian viewing does not imply anything about the coherence of the light that enters the eye and falls on the retina. The arrangements in Fig. 60 are both Maxwellian, but in one the ribbon filament is conjugate to the pupil and in the other is conjugate to the retina.

G. CHROMATIC ABERRATIONS

1. Axial Chromatic Aberration

This aberration results from the fact that the index of refraction of each medium varies with the wavelength of light. It does not occur with monochromatic light but is found whenever we have a point source that produces a mixture of wavelengths. One of the light components may focus on the retina, but the others will then focus in front of or behind the retina.

To illustrate the effect of this aberration on the blur of the retinal image we use the reduced eye (Fig. 61), which involves only one medium and a pupil, and an infinitely distant multichromatic point source centered on the optic axis.

We assume that the radius of curvature of the cornea is 5 mm and that the retina falls at the point where the 555-nm rays are focused, which lies 18.611 mm from the cornea. The pupil lies 1.34 mm from the cornea and has a radius of 1 mm.

a. Image of a Point Source

The image formed by a multichromatic source can be analyzed into separate images corresponding to the different parts of the spectrum (bands 10 nm wide). Only one of the components is focused on the retina. Each image is centered at the optic axis, and the intensity at any point P in that image is given by the equation for a Fresnel image, Eq. (21).

To assess the images produced by different wavelengths, it is necessary to know the indices of refraction of the vitreous for the various wavelengths. These have been deduced from the chromatic aberration data of Wald and

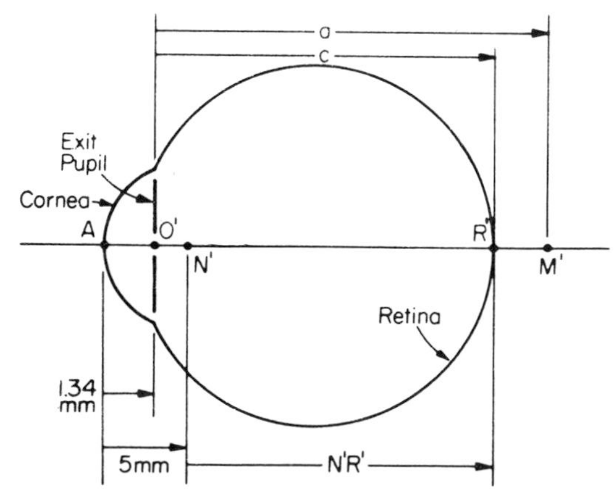

Fig. 61. Schematic eye used for computing longitudinal chromatic aberration. (Fry, 1976).

Griffin (1947). Bedford and Wyszecki (1957) have repeated the measurements with a large number of subjects and obtained results that are in good agreement. The data for the relation between n and λ can be fitted with Cauchy's equation (Fry, 1976),

$$n = 1.32546 + 0.002154\frac{10^{-6}}{\lambda^2} + 0.000176\frac{10^{-12}}{\lambda^4}. \tag{73}$$

This equation is represented by the curve in Fig. 62.

It is also necessary to locate for each wavelength the point M′ that is conjugate to the distant point source. Because the point source lies at an infinite distance ($\overline{AM} = \infty$), the formula for conjugate foci for a single refracting surface,

$$\frac{n'}{\overline{AM'}} = \frac{n}{\overline{AM}} + \frac{n' - n}{r}, \tag{74}$$

can be used to compute the distance $\overline{AM'}$ from the cornea to M′ for any wavelength. The point O′ in Fig. 61 has been used in this study, instead of B in

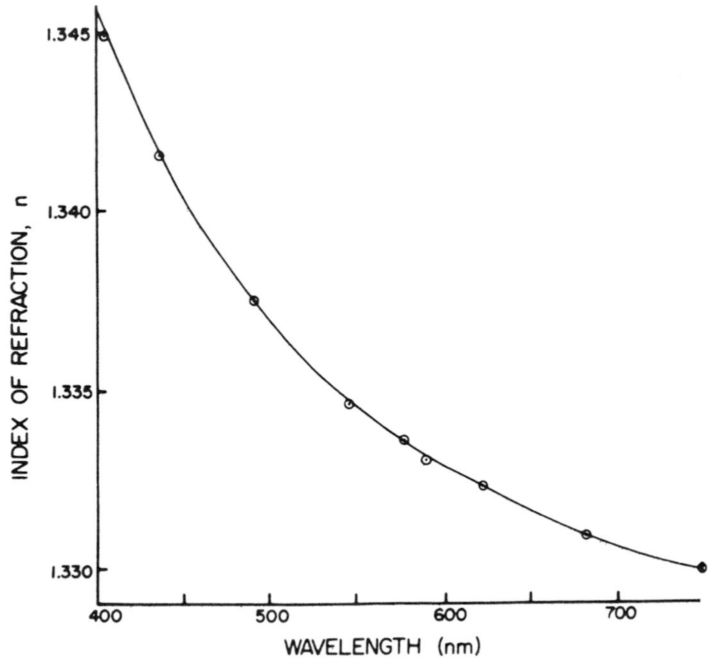

Fig. 62. Dispersion curve for the ocular medium (Fry, 1976).

Fig. 32, as the reference point for measuring the distances *c* and *a*. This approximation introduces little error as long as the pupil is small.

The total retinal illuminance $E_{p'}$ at any point P′ at a distance *r* from the chief ray is the simple sum of the contributions made by the different wavelengths.

In Fig. 63 one of the curves is the image produced by a white point source of light at an infinite distance when the eye is focused for 555 nm. It was assumed that the source has an equal energy spectrum; hence, the ordinates of the CIE photopic luminous efficiency curve were used as values for F_λ for the various wavelengths. Shown also in Fig. 63 is the Fraunhofer image for a monochromatic point source (555 nm). The total amount of light *F* in the multichromatic white source is given by

$$F = \int_0^\infty F_\lambda \, d\lambda \tag{75}$$

and is equal to the total amount of light in the monochromatic image. For white light the intensity is less concentrated at the center.

b. Index of Blur

We can use the index of blur to assess the wavelength for which the eye must be accommodated to achieve the sharpest image when a multichromatic point source is used, to assess the effect of pupil size on the *depth of focus*, and to assess the amount of degradation of image quality produced by chromatic

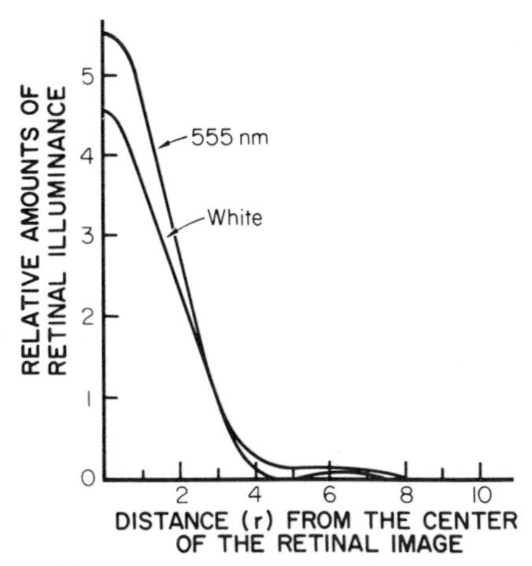

Fig. 63. The point spread function for monochromatic light (555 mλ) compared with that for white light. The two images contain the same amount of flux. The eye is free from spherical aberration, astigmatism, and chromatic dispersion. The distance from the center of the image is expressed in micrometres. (Fry, 1955.)

aberration when a multichromatic image is substituted for a monochromatic image.

With a monochromatic point source, one can compute the index of blur from Eq. (45) where V is a function of the size of the pupil and the extent to which the eye is out of focus. The index of blur for a multichromatic image is based upon the indices of blur for the separate monochromatic components as shown by

$$\frac{1}{\Phi} = \frac{1}{F} \int F_\lambda \left(\frac{1}{\phi}\right)_\lambda d\lambda, \tag{76}$$

where

$$F = \int F_\lambda \, d\lambda. \tag{77}$$

Figure 64 shows the effect of throwing the eye out of focus for a multichromatic source and a monochromatic source. It can be deduced from the curves that the depth of focus is wider for multichromatic light than for monochromatic light, but once the accommodation is adjusted to give the sharpest focus, the monochromatic image is slightly clearer than the multichromatic image. It can also be deduced that for an equal energy spectrum, the eye must be focused for 555 nm to achieve the sharpest image. The index of blur reaches its minimum at this wavelength. The Optical Society of America (1953, p. 297) recommends the spectral *centroid* of a source as an index of the wavelength for which to focus to achieve the sharpest vision. This applies to unimodal distributions and especially to narrow-band distributions.

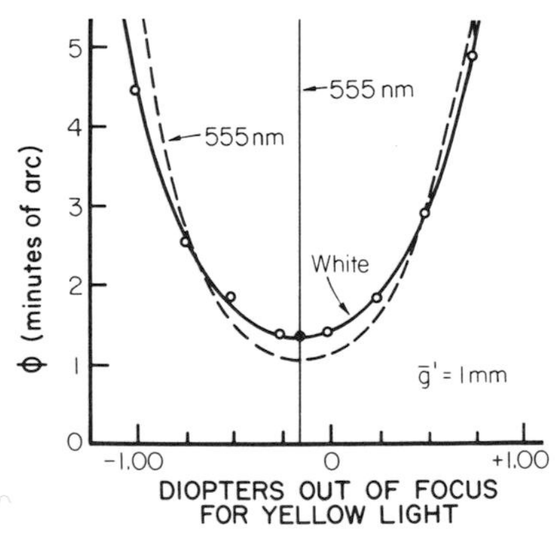

Fig. 64. The effect on blur (ϕ) from being out of focus. Curves for a monochromatic and a heterochromatic point source are presented for comparison (Fry, 1955).

The spectral centroid is a weighted average of the wavelengths in the spectral distribution.

$$\lambda = \int_0^\infty \lambda F \, d\lambda \Big/ \int_0^\infty F \, d\lambda. \tag{78}$$

For an equal energy source the spectral centroid is 560.2 nm.

A cobalt filter that transmits narrow bands at the two ends of the spectrum can be placed in front of a white point source, and the eye can focus on either the blue light or the red light and achieve a sharp red image with a blue fringe or vice versa. Such a filter can be used by a refractionist for assessing the refractive errors of an eye.

Similar comparisons have been made between high-pressure and low-pressure sodium sources (Fry, 1976) and between sources used for measuring visual acuity (Fry, 1978).

Figure 65 shows the effect of changing the diameter of the pupil on the index of blur for white light and monochromatic light. For monochromatic light the blur continues to decrease as the pupil size increases, but for white light the blur reduces to a minimum at about 3.0 mm and thereafter increases. The curves in Fig. 65 make no allowance for spherical aberration (see Fry, 1970, pp. 106–108).

It is possible to design a lens to compensate for the chromatic aberration of the eye. A lens of this type designed by Bedford and Wyszecki (1957) is shown in Fig. 66. This is very useful for correcting axial chromatic aberration for a

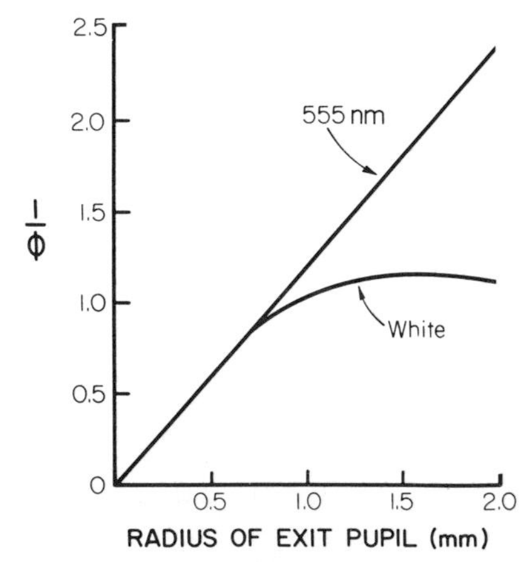

Fig. 65. The effect of pupil size on blur when the eye is in focus. The eye is focused for 555 nm. Curves for a monochromatic and for a heterochromatic point are included for comparison. $1/\phi$ is expressed in minutes.

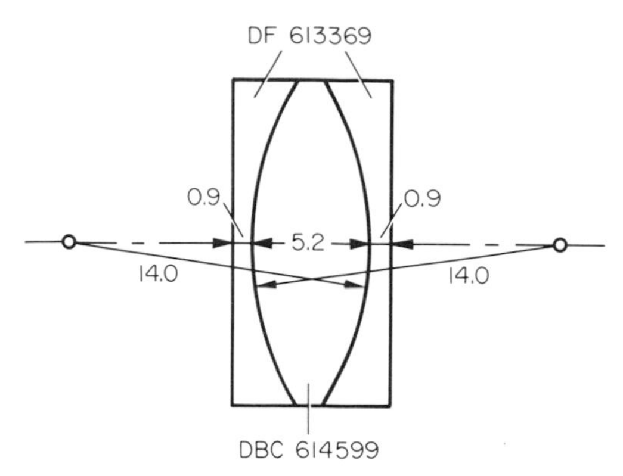

Fig. 66. Design data of a lens to correct axial chromatic aberration of the human eye (Bedford and Wyszecki, 1957).

point source and also for the dividing line of a bipartite field. Hence, it is widely used as part of the eyepiece of a colorimeter. It will not correct the color for lines that do not cross the field of view at the optic axes. It does not help for disk–annulus test objects where the border is a circle concentric with the optic axis. This is because the lens corrects only the *axial chromatic aberration* and not the *chromatic difference in magnification.*

2. Chromatic Dispersion

One form of chromatic dispersion occurs in connection with chromatic difference in magnification. In the reduced eye we can produce a change in the chromatic difference in magnification by displacing the pupil forward or backward from the nodal point. In Fig. 67a the light from each of the object points is a mixture of red and blue. The pupil lies in the same plane as the nodal point; hence, the blue and red rays from the points M and N are not dispersed at the cornea, and the magnification is the same for blue and red rays.

If the pupil is displaced forward from the nodal point, as in Fig. 67b, the blue and red rays are dispersed at the cornea. Hence, the blue images on the retina are more widely separated than the red images.

As long as we have two object points, we can talk about a difference in image size or a difference in magnification. But when we restrict attention to one point, we can say only that the red image is laterally displaced from the blue, that is, it is chromatic dispersion. One can also say that one image is displaced further from the optic axis than the other. This is more in line with the concept of difference in magnification.

In the reduced eye the chromatic difference in magnification is combined with a difference in focus, hence, the red and blue images cannot be seen clearly

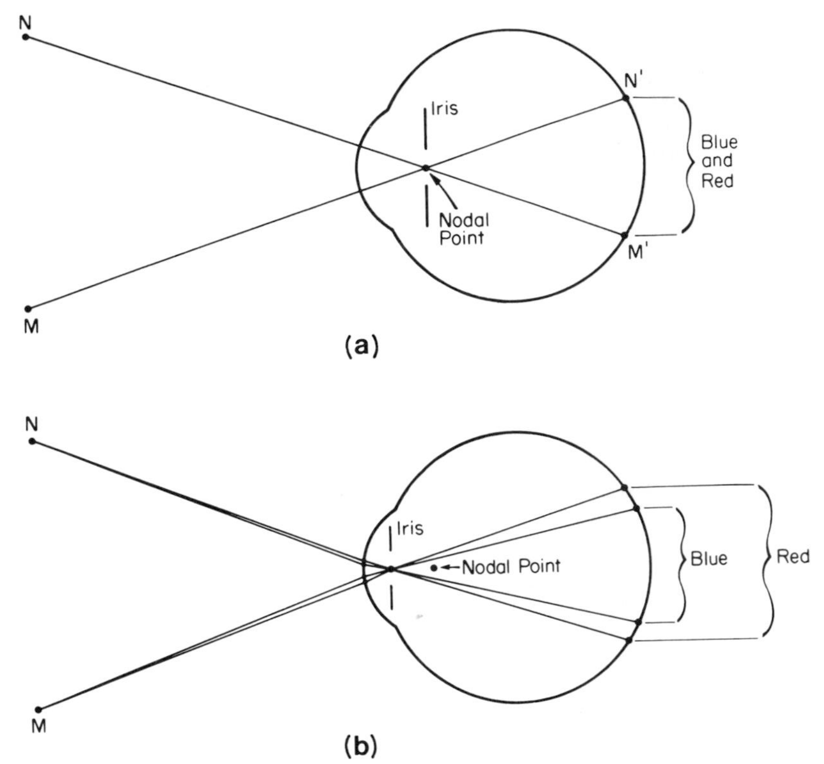

Fig. 67. Chromatic difference in magnification. When the iris and nodal point lie in the same plane, as in (a), the red and blue images have the same size. In (b), displacing the iris in front of the nodal point produces a difference in size.

at the same time. If the red image is clear, the blue will be blurred, or vice versa, or both may be blurred.

a. Decentered Pupils

Chromatic dispersion for a single point on the optic axis can be produced in the reduced eye by displacing the pupil in a lateral direction from the optic axis as illustrated in Fig. 68. The amount of dispersion is related to the amount of chromatic difference in focus. A noticeable amount of chromatic dispersion can be produced by covering one half of the pupil.

In normal eyes the center of the pupil is displaced in a temporal direction from the line connecting the point of fixation and the first nodal point. Because the dispersions are in opposite directions in the two eyes, they create the phenomenon of chromastereopsis when the two eyes are used. Red points in the field of view are seen as being more remote from the eyes than blue ones. If

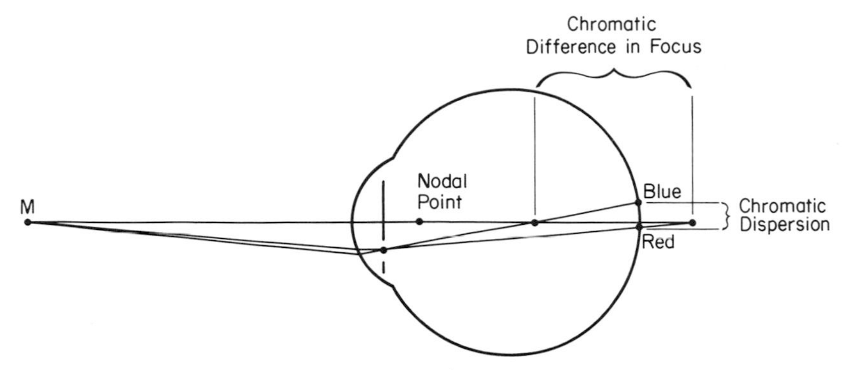

Fig. 68. Chromatic dispersion produced by displacing the pupil in a lateral direction.

a person looks at a pair of rectangles, one red and the other blue, placed one above the other, the right eye sees them as shown in Fig. 69b and the left eye as shown in Fig. 69a. It is easy, therefore, to see why the blue rectangle will be seen as more remote when both eyes are used.

If artificial pupils or slits are placed in front of the eyes, one can, by changing the separation, find the separation that will make the two rectangles be perceived as lying in the same plane. This gives a measure of the distance between the primary nodal points of the two eyes.

If a bipartite field consisting of one monochromatic stimulus on one side and a different one on the other is viewed through a vertical slit, one can make the borders separate or overlap by moving the head to one side or the other and can use perfect registration as the criterion for proper placement of the head. If white is presented on one side and a dichromatic stimulus on the other, one can use perfect registration of the two components of the dichromatic stimulus as the criterion for head placement.

b. Prisms and Decentered Lenses

Chromatic dispersion for a single point can also be produced by mounting a prism or decentered lens in front of the eye.

Fig. 69. Illustration of how chromastereopsis is produced by lateral displacement of the pupils.

Chromatic dispersion is encountered when an eye has to look through the periphery of a spectacle lens mounted in front of it. The lens is generally mounted with its optic axis coinciding with the line of sight, when the two lines of sight are both perpendicular to the base line and depressed about 6°. There is no chromatic dispersion when the eyes are pointing in this direction.

Bifocal segments are often made of glass of high index and also of high dispersion, thus the presence of color fringes at the borders of objects is a typical complaint.

Prisms or prism lenses are often worn to compensate muscle imbalance that makes it difficult for the two eyes to point in the same direction. In this case, points that lie straight ahead are subject to chromatic dispersion.

c. The Effect of Dispersion on the Image of a Point of White Light

So far the effects of dispersion have been described in terms of point sources that emit a mixture of red and blue light. When a point emits a continuous spectrum as in the case of white light, the image is an overlapping of images of different colors, spread out in the direction of the dispersion, as illustrated in Fig. 70 where the eye is focused for green light.

The computation of the amount of blur involves the same procedure as for axial chromatic aberration except that differently colored images are not only blurred by different amounts but are displaced laterally from each other in the direction of the dispersion.

The amount of dispersion can be specified quantitatively as the distance from the center of the 656-nm image to the center of the 486-nm image. The wavelengths 656, 486, and 589 nm correspond to the three Fraunhofer lines of the solar spectrum, in terms of which the ν-value of a medium is specified.

Chromatic dispersion will not affect the appearance of a line or border if the direction of the dispersion is parallel to that of the line or border but is maximal when the directions are perpendicular. The blur produced by dispersion is proportional to the cosine of the angle between the direction of dispersion and the direction of the line or border.

If we use a ring concentric to the optic axis instead of a pair of points to demonstrate chromatic difference in magnification, the ring will be uniformly

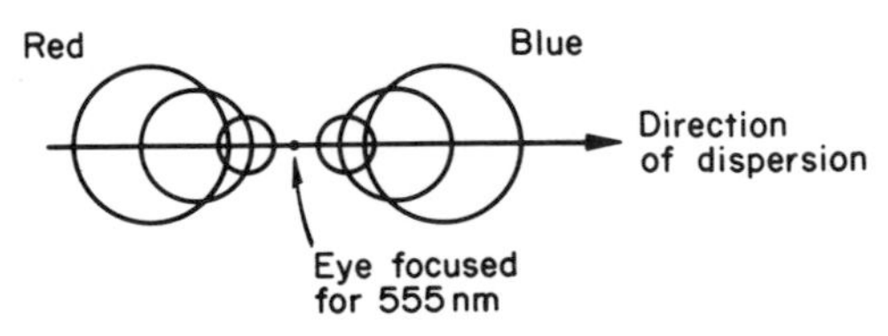

Fig. 70. Blurred image of a white point of light produced by chromatic dispersion.

blurred and will have colored fringes. By contrast, lines or borders radiating out from the optic axis will not be affected by chromatic differences in magnification.

H. PERIPHERAL VISION

Because of the sparseness of receptors and ganglion cells in the periphery of the retina, the ability to see peripheral detail is poor. To see fine detail, the observer must aim his primary line of sight at an object.

There are several reasons, however, why it is important to know about the quality of the image formed on the peripheral retina. In perimetry an effort is made to assess what the peripheral retina can see, and it is necessary to know about the blur of the retinal images of the targets used in perimetry for testing peripheral vision. These targets are usually mounted on the arc of a perimeter that has its center at the entrance pupil of the eye, or else they are projected on the inside wall of a sphere that is concentric with the entrance pupil.

The ophthalmic surgeon who uses a laser beam to photocoagulate the retina at certain spots to make it adhere to the choroid needs to know how to focus the beam on the retina.

The designer of a fundus camera needs to know about the curvature of the retina and the optics of the eye for peripheral vision to make the image of the retina fall properly on the film.

The major aberrations to which attention must be paid are *radial* (*obliquity*) *astigmatism* and *curvature of the field*. Chromatic dispersion resulting from the chromatic difference in magnification also contributes to the blur of a peripheral image.

The reader is referred to LeGrand's excellent review (1967, pp. 126–145) of the optics of peripheral vision. LeGrand (1967, pp. 127–128) has computed the astigmatic image surfaces using a schematic eye similar to that of Helmholtz. The ordinary ray-tracing formulas were used to trace the chief rays, and the classical formulas were used to locate the *tangential* and *sagittal* images. As can be seen in Fig. 71 the sagittal and tangential image surfaces fall on opposite sides of the retina, which is assumed to be a spherical surface centered on the optic axis. This is the best arrangement possible, because the astigmatism cannot be avoided.

It is possible to use a coincidence optometer or a skiascope to locate the point conjugate to the retina for any line of sight. Ferree, Rand, and Hardy (1931) made measurements of this type for a number of eyes.

The projected area of the entrance pupil varies with the angle of incidence of the chief ray. Spring and Stiles (1948) measured the apparent area of the pupil viewed from different angles η, and the mean data for thirteen subjects are shown in Table I which has been arranged by (LeGrand, 1967, p. 129). The

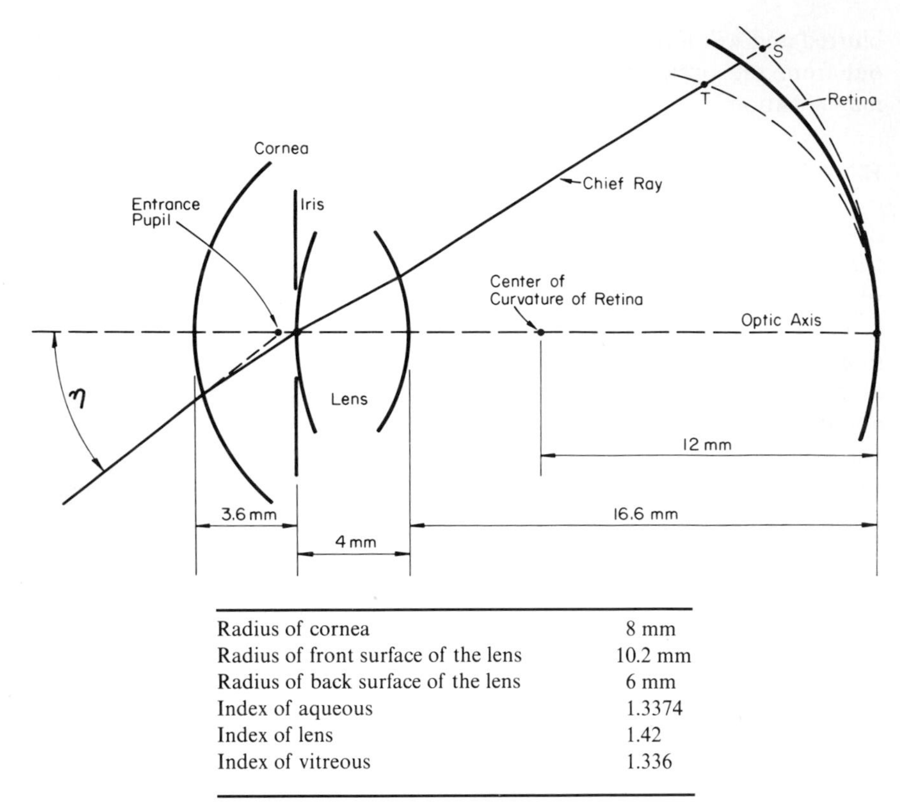

Radius of cornea	8 mm
Radius of front surface of the lens	10.2 mm
Radius of back surface of the lens	6 mm
Index of aqueous	1.3374
Index of lens	1.42
Index of vitreous	1.336

Fig. 71. LeGrand's calculation of the astigmatic image surfaces S and T of the eye relative to the retina and data for the schematic eye (LeGrand, 1967).

TABLE I. Area of the Entrance Pupil in Relative Values for Several Diameters d and Eccentricities η*.

	d(mm)			
η(deg)	8	6	4	2
0	1	1	1	1
25	0.94	0.94	0.94	0.94
50	0.76	0.76	0.76	0.76
75	0.44	0.44	0.44	0.44
85	0.25	0.27	0.29	0.30
95	0.094	0.114	0.132	0.15
100	0.030	0.042	0.053	0.064

* Le Grand (1967).

angle η is the displacement in a temporal direction of the peripheral chief ray from the primary line of sight.

The distance from the exit pupil to the retina is shorter for peripheral chief rays than for the central ray, and allowance for this needs to be made in assessing the ability of a peripheral beam to stimulate the retina. Peripheral rays are not perpendicular to the retina, but according to Laties and Enoch (1971) and Enoch and Laties (1971), the peripheral receptors are oriented toward the center of the exit pupil.

I. MAGNIFICATION AND DISTORTION

The optical magnification and distortion of the image formed on the retina are of minor importance compared to the directions in which perceived images produced by stimulating different points on the retina get projected out into space from the *center of projection.*

It will simplify matters if at the outset we limit consideration to a person with one eye. The perceived image produced by stimulating the anchor point will be perceived in the direction in which he perceives himself to be looking. This is in a fixed direction with respect to the head. If the observer has learned well how to use his eye he will be able to walk toward or point at the object that is producing the image.

Also the observer can learn to project in the proper direction relative to his anchor point image the images produced by stimulating peripheral parts of his retina. It is the magnification and distortion of these perceived arrays that becomes important and not the arrays formed on the retina.

If a person had a single eye, perception of relative directions would be limited to learning a fixed relation between points on the retina and perceived directions in visual space. A one-eyed person who wears glasses is faced with the fact that the eye rotates but the glasses are fixed with respect to the head. One can learn to turn the head rather than the eyes, or one can base the learned relations between physical space and perceived space on what one sees when looking straight ahead and then learn not to trust one's perception of space when looking through peripheral parts of the lens. A spectacle wearer with *gun barrel vision*, who has to be concerned only with the projection of foveal impressions, could learn to compensate the prismatic effects in the periphery of the lens. However, a normal one-eyed person cannot compensate the wide variety of distortion patterns for different directions of regard. Contact lenses do not present this problem, because they move with the eye.

The distortions that occur when one eye is used can be demonstrated by using an array of three points that lie on a line and are equally spaced. If they are not perceived as equally spaced, this indicates *nonuniform magnification.* If the three points do not appear aligned, the distortion is an *error in curvature*. If the three points form an angle with the middle point being the vertex, distortion can change the size of the angle; this is the *scissor effect*.

Distortions can also be assessed by using squares and circles (Fig. 72). Differential magnification in meridians at right angles will change a square into a diamond or a rectangle, and it will change a circle into an ellipse. Nonuniform magnification in the radial direction converts a square to a pin cushion or a barrel.

If a person has two eyes, the problem is much more complicated. The normal person with two eyes has two fields of view that overlap. The two eyes have to cope with the same objects at the same time. Various pairs of points on the two retinas are connected to the same part of the brain and are called *corresponding* points. Images of the same object formed on such a pair of points are seen in the same direction; in this case seeing one object with two eyes presents no problem. If we let the two eyes fixate a common point, it is possible to locate the points in the plane of regard that, stimulate corresponding points on the two retinas. The locus of these points is called the *horopter* (see Fig. 73). Points in front of or behind the horopter produce images on

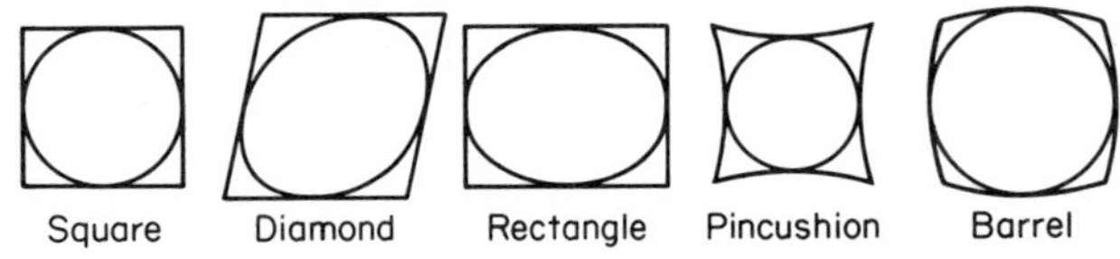

Fig. 72. Distorted squares with inscribed circles.

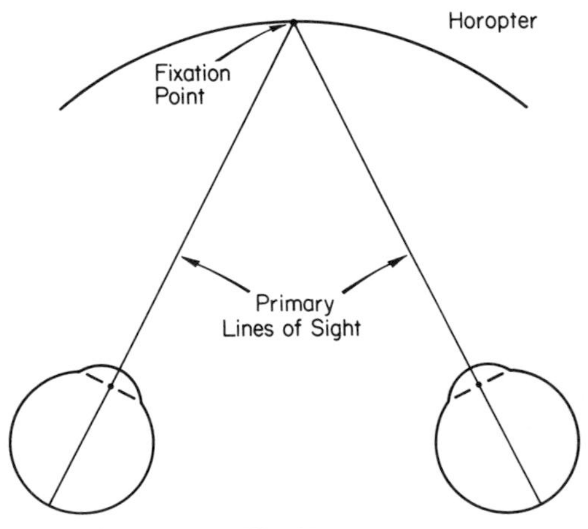

Fig. 73. Horopter.

disparate points, and the projected images are not seen in the same direction. If the disparity is small, a single image may be seen; but the image may be double if the disparity is large. If the object lies more remote than the horopter, the image on the left retina is seen to the left and the image on the right retina to the right. The disparity is described as *uncrossed.* If the object lies less remote than the horopter, the disparity is *crossed.* Disparity serves as a cue for perceived distance. This relationship has to be learned, and certain compromises have to be made between perceived distances and perceived directions.

If the images formed by an object on the two retinas differ in size and shape, errors in the perception of distance and the perceived curvature and orientation of surfaces can be introduced. These problems are collectively called *aniseikonia.* Partial relief from these distortions can be obtained by using aniseikonic lenses, which provide different magnifications for the two eyes. *Overall magnifiers* produce uniform magnification in all meridians, whereas *meridional magnifiers* vary from a minimum at one meridian to a maximum at the meridian at right angles. When lenses are worn to correct ametropia or aid accommodation, they may also correct aniseikonia, or make it worse, but the design can be manipulated to control the magnification and distortion as well as to provide refracting power. Prisms worn to compensate a phoria also distort the field of view differently for the two eyes and produce unique effects on space perception.

Differential prismatic displacements in the peripheries of the two lenses worn by a pair of eyes affects the phoria for peripheral directions of regard. This phenomenon is known as *anisophoria* and is quite separate from the problem of aniseikonia.

Aniseikonia and anisophoria produced by spectacle lenses can be avoided by using contact lenses, but new problems are produced by slippage and rotation on the cornea.

Telescopic spectacles are worn to achieve a large amount of magnification for seeing fine detail, and simple magnifiers are worn for the same purpose. It is difficult in this case to compensate for the effect of magnification on perceived space. The wearer must learn to be a head turner. An error in size can be reinterpreted as a change in distance.

J. TRANSMITTANCE AND STRAY LIGHT

Part of the light from a point source that is directed toward the entrance pupil of the eye is lost by reflection at the refracting surfaces. Another part is lost by absorption and scatter by the media. The remainder is transmitted to the retina to form the optical image. The importance of the loss in the amount of light transmitted is that the loss affects the efficiency of the eye as a receptor.

There are several sources of stray light in the eye, which we shall assess, but scatter by the media is the major source.

The light that is scattered falls as a patch of *veiling illuminance* on the retina and reduces the contrast of the retinal image. This is known as *disability glare*.

1. Disability Glare

Stiles (1929) measured the stray light in the eye by placing a glare source in the field of view at a certain glare angle θ from the primary line of sight (expressed in degrees). He adjusted the luminance of a test stimulus, A, centered on the primary line of sight to bring it to the threshold of visibility. He then turned off the glare source and superimposed on A a *patch of veiling luminance*, which was adjusted to bring A again to its threshold. The luminance L_V of the veiling patch (candles per square meter) required to achieve threshold is called the *equivalent veiling luminance*. Measurements were made at different glare angles from 1 to 10°, and the data were fitted by the formula

$$L_V = 4.16 E_0 / \theta^{1.5}, \tag{79}$$

where E_0 is the illuminance (lumens per square meter) in the plane of the entrance pupil.

Holladay (1926) used a similar procedure to obtain a set of data for glare angles from 5 to 25°, which conform to the equation

$$L_V = 9.2 E_0 / \theta^2. \tag{80}$$

These two equations are plotted as straight lines in Fig. 74. They cross at $\theta = 5°$.

Curve C, which conforms to the equation

$$L_V = \frac{9.2 E_0}{\theta(1.5 + \theta)}, \tag{81}$$

is tangent to both of the straight lines and at small angles also conforms to the requirement that L_V must be inversely proportional to θ. This requirement can be demonstrated by using a reduced eye to assess the distribution of stray light on the retina. Holladay (1926), Stiles (1929), and the author, used a reduced eye to calculate the distribution of stray light on the retina. It was assumed that the pupil falls at the cornea and that the medium is uniform from the cornea to the retina. It was arbitrarily assumed that the backward scatter was zero and that the forward scatter conforms to

$$\phi = R \cos^2 \beta \tag{82}$$

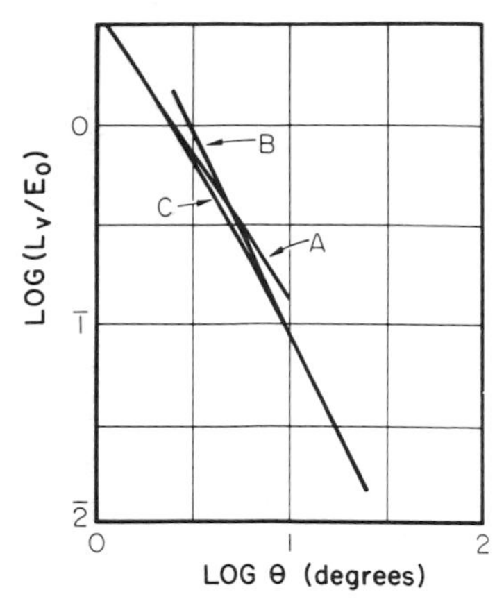

Fig. 74. Glare equations of Stiles (curve A) and Holladay (curve B). Curve C, which represents Eq. (81), is tangent to A and B and also conforms to the requirement that at small angles L_V is inversely proportional to θ (Fry, 1954).

and is not dependent on the wavelength of light. R is a constant, β is the *angular displacement* from the direction of the incident beam, and ϕ is the relative intensity of light scattered in various directions by an element of volume of the medium inside the eye. This leads to the following formula for the distribution of stray light on the retina:

$$L_V = 0.0075RE_0\left(\cot\frac{\theta}{n} - \cos\frac{\theta}{n}\right), \tag{83}$$

where E_0 is the illuminance in the plane of the pupil, n is the index of refraction, θ is the glare angle between the glare source and the primary line of sight, and R is a constant. The details of the derivation of Eq. (83) are presented in a separate paper (Fry, 1954). According to Eq. (83) L_V is inversely proportional to θ for small angles.

Almost any reasonable scatter function will result in the proposition that for small angles L_V must be inversely proportional to θ, and this proposition may be used in extrapolating empirical glare data to the zero glare angle.

The formula also demonstrates that if

$$R = 3.39, \tag{84}$$

L_V at a glare angle of 5° has the same value as for the empirical data of Stiles and Crawford. It follows that the fraction of light lost by scatter (13.3%) is less than the total attenuation, which according to Ludvigh and McCarthy (1938) is more than 30%. (See Fig. 93.)

a. Additivity of Disability Glare

If disability glare depends on stray light falling on the fovea, the effects produced by two or more glare sources must be additive. Holladay (1926) and Crawford (1936) tested the *additivity principle* and found that it held.

The natural pupil changes in size as more glare sources are added; but this similarly affects the light from the glare sources and the light from the test stimulus that enters the pupil. Change in pupil size also similarly affects the test stimulus and the veiling patch of light. The only complication is that at low levels of luminance the ratio of ΔL to L required for threshold is higher for the glare source than for the veiling patch, because of the difference in the size of the pupil. It would be better to use an artificial pupil (Fry, 1956).

Each element of solid angle in the field of view contributes to the stray light falling at the fovea. The integrated effect is given by

$$L_V = 0.107 \int_0^{2\pi} \int_0^{\pi/2} \frac{L \cos\theta \sin\theta \, d\theta \, d\phi}{\theta(1 + 38.2\,\theta)}, \tag{85}$$

where θ and ϕ are expressed in radians and represent angular displacement from the primary line of sight and meridianal displacement from the zero half-meridian, and L is the luminance in each of the various directions.

A glare lens has been designed that can be used in connection with the Pritchard telephotometer to assess L_V for any point of observation and any direction of regard in any field of view (Fry *et al.*, 1963).

For a bright disk that is on a dark background and centered on the primary line of sight the following formula can be used to compute L_V.

$$L_V = 0.0175L \log_e(1 + 38.2 \sin\theta), \tag{86}$$

where θ is the radius of the disk, and L is its luminance. If the disk is 5° in diameter, the value for L_V at the center is only 1.6% as large as L. If we increase the diameter to 120°, which includes a large part of the visual field, L_V is still only 5.9% as large as L. It may be concluded, therefore, that for a more or less uniform field, stray light at the fovea has little effect on visibility at the fovea. It is only in the case of high-intensity glare sources that stray light becomes a problem.

Note that Stiles' data do not extend to a glare angle of zero, and his approach cannot be used to extend the data to zero, because at small angles the effects produced by stray light cannot be differentiated from those produced by retinal interaction. It is necessary to use the theoretical approach to make the assessment for small angles.

Stiles was convinced that the effect of a glare source on the visibility of objects was too great to be explained in terms of stray light. He felt that the amount of stray light required to produce the measured effects was more than the amount of light known to be attenuated by the media. His judgement was based on attenuation measurements made on bovine eyes. According to Fry (1954) the total amount of stray light scattered by the media is only 13.3% of the total light entering the eye, but the data on human eyes indicate that the light lost by scatter and absorption is more than 30%.

One must recognize that a stimulus can inhibit responses to adjacent stimuli, but the distance across the retina over which such effects can occur is limited. Schouten and Ornstein (1939) claimed evidence that retinal interaction could involve the total distance from the ora serrata to the fovea. Fry and Alpern (1953) were not able to confirm this result. It should be noted that in the 1920s it was still widely held that the sensitivity of any part of the retina was dependent on the interaction effects from all parts of the retina.

Geldard (1931) claimed effects of retinal inhibition extending over a 5° range, but Schober and Fry (1968) showed this to be an artifact resulting from the misuse of an artificial pupil. However, it is well established that neural interaction can operate over a distance of one to two degrees. Hence, within this range it is difficult to differentiate between effects of stray light and retinal interaction.

b. Linearity Principle

It is a matter of principle that the amount of stray light at a given distance from the image of a glare source must be proportional to the intensity of the glare source. LeGrand (1937) found a discrepancy in testing this principle, which he took to be evidence of retinal interaction over large distances. However, Fry (1956) pointed out that this discrepancy can be eliminated by the use of an artificial pupil.

Vos (1962) refers to this problem as the problem of linearity. He used an artificial pupil 1 mm in diameter, but he placed the image of the glare source in the fovea and used an annular test stimulus surrounding the fovea and falling at some distance θ from the center of the fovea. He found for large glare angles that the relation between the intensity of the glare source and that of the test object was linear, but he brought to light that with small glare sources the ration L_V/E_0 is lower for small values of E_0. He attributed this to the Purkinje shift in dominance from rods to cones at the mesopic level of L_V and to the possibility that scatter by the lens and cornea is more dominant at large angles than scatter by the retina and the vitreous near the retina. Increasing the glare angle increases the angle of incidence for light scattered by the retina but has little effect on the incidence of light scattered by the cornea and lens.

c. Stray Light at Small Glare Angles

The problem of the distribution of stray light at small glare angles is illustrated in Fig. 75. The dashed curve in Fig. 75 represents the distribution of stray light for a point-sized light source, which conforms to Eq. (81).

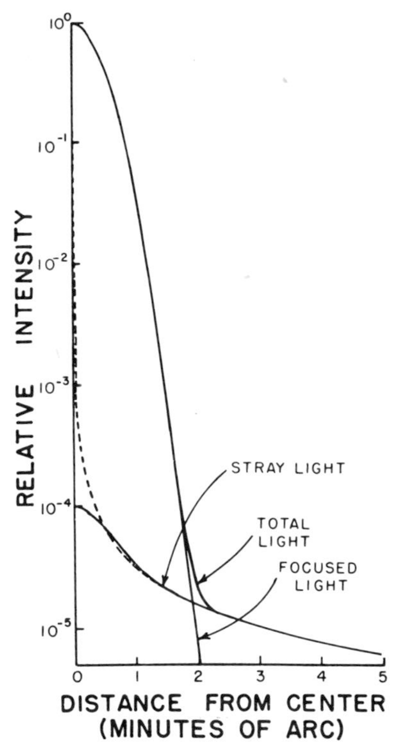

Fig. 75. Spread function for the retinal image of a point analyzed into its focused and stray-light components (Fry, 1965a).

Note that the distribution of stray light comes to a sharp focus at the zero glare angle. It is actually sharper than the image that is computed from diffraction theory for the focused or nonscattered light transmitted through the pupil. This paradox is the outcome of having assumed that the beam from a point-sized glare source comes to a focus at the retina. This artifact can be eliminated by using a beam that produces a Fraunhofer diffraction pattern instead of a point image (Fry, 1965a). The situation can be approximated by convolving the point spread function for stray light with the point spread function for a Fraunhofer image. Instead of doing exactly this, the author has substituted an equivalent Gaussian distribution for the Fraunhofer image. The result is shown in Fig. 75. The new curve (solid) for stray light is rounded off at the center and is the same as the old curve (dashed) for distances greater than 2 minutes of arc from the center. The new distribution can be used instead of the old for a more precise assessment of the distribution of stray light.

Walraven (1973) has used a different approach to assess glare at small angles.

2. Specular Reflection at the Surfaces

When a beam of light is obliquely incident at a refracting surface, the formula for reflectance is complicated and involves differences in polarization and transmittance for the different meridians. But in the case of normal incidence the formula is simple and involves no change in polarization,

$$r = \left(\frac{n' - n}{n' + n}\right)^2. \tag{87}$$

According to this formula the transmittance along the optic axis for the Helmholz schematic eye is

$$\begin{aligned} t &= (1 - r_{\text{cornea}})(1 - r_{\text{lens front surface}})(1 - r_{\text{lens back surface}}) \\ &= 0.913. \end{aligned} \tag{88}$$

No allowance is made for the specular reflectance at the interface between the vitreous and the retina, which is described as a sheen when the fundus is viewed with an ophthalmoscope.

The light transmitted at the refracting surfaces is called *useful light* and goes into the formation of the focused image. The light reflected back out of the eye is called *lost light*. If this light is from a point source and is reflected back out of the eye by the front and back surfaces of the cornea and the front and back surfaces of the lens, these beams form the first, second, third, and fourth *Purkinje images*, which may be used in the assessment of the configuration of the surfaces and the distances between them. If the light is reflected successively at an even number of surfaces, it is directed on toward the retina and is called *harmful light*. The images formed by inter-reflection are also called Purkinje images. If a point source is placed about 25 cm in front of the eye, one of these images falls on the retina and is called a *ghost image*. When the eye rotates, the ghost image moves across the retina in a direction opposite to that of the primary image; because of this it is used in the study of eye movements. This ghost image is formed by reflection at the back surface of the lens with a second reflection at the front surface of the cornea and is called the *sixth Purkinje image*.

If the light is reflected by the front surface of the lens and again by the front surface of the cornea, it forms the *fifth Purkinje image* near the back surface of the lens. Hence, it produces a large blur circle on the retina, which is part of the distribution of stray light on the retina. This generation of stray light by interreflection at the surfaces is called *flare*.

3. Reflection at the Iris

If a spot of light is focused on the iris, it forms a secondary source by diffuse reflection. The light reflected by the two surfaces of the cornea and transmitted backward through the pupil forms a nearly uniform circular patch of stray light. Ordinarily the illumination on the iris is the same as that covering the pupil, and the intensity of stray light from this source is very small.

4. Transmission through the Iris, Sclera, Choroid, and Closed Eyelid

a. Transmission through the Iris

The color of the iris may be blue or brown depending on the amount of pigment in its first two layers. The most absorption occurs in the pigment layer on the back side of the iris, which is an extension of the pigment epithelium of the retina. Boynton and Clarke (1964) tried to assess the amount of light transmitted through the iris and found it to be very small.

b. Transmission through the Sclera and Choroid

Light is diffusely transmitted by the sclera. The scatter produced by the sclera, compared with the clarity of the cornea, results from a higher content of water in the sclera. It is also true that the fibers in the stroma are larger and more randomly arranged than in the cornea.

The most absorption occurs at the vascular layers of the choroid and at the pigment epithelium, which covers the fundus. The reddish color of the blood is produced by hemoglobin. The pigment in the pigment epithelium and choroid is melanin. The pigment epithelium is a diffusely transmitting layer. The scatter functions for the sclera, choroid, and pigment epithelium are not known.

A bright point image or spot on the sclera will illuminate the interior of the eye; the general procedure is referred to as *transillumination* of the fundus. Such a spot will produce shadows of the retinal blood vessels on the layer of photoreceptors on the opposite side of the eye. Two such spots can be manipulated to produce parallax data for computing the distance of the vessels from the sensitive layers.

Fry and Alpern (1953) devised a method for assessing the stray light at the fovea produced by a spot of light on the sclera near the ora serrata.

c. Transmission through the Closed Eyelid

Light is diffusely transmitted through the closed eyelid. Part of this light is transmitted through the pupil, part through the iris, and part through the sclera, ciliary body, and choroid to reach the interior of the eye where it is scattered over the entire retina. Ihrig and Fry (1953), using the positive afterimage, have devised a method for assessing the capability of this stray light to stimulate the photoreceptors.

5. Transmission and Scatter by the Media

When the anterior segment of the lens is viewed with a slit lamp, one can see the light diffusely reflected from the vitreous and the layers of the cornea and lens. The various layers of the lens can be identified in the pattern of reflections, which provides much useful information about the configuration of internal surfaces of the lens and the role played by the changes in configuration during accommodation.

If the forward scatter is akin to the backward scatter, it would appear that the aqueous contributes very little to the stray light in the eye. On the other hand, it would appear that the cornea and the lens might be the major sources of stray light falling on the retina. But these bodies produce a wide distribution with a fairly flat peak at the retina, which does not conform to the distribution that is actually found. Scatter by the retina and the vitreous close to the retina proves to be more important. In the case of cataract, scatter by the lens increases significantly and does become a major source of scatter.

a. The Boettner–Wolter Data for Transmittance and Scatter of the Media

Boettner and Wolter (1962) have provided useful basic data on the transmittance and the scatter of the various media. They used the upper arrangement shown in Fig. 76 to measure the direct transmittance. This included the directly transmitted light as well as the scattered light within 1/2° of straight ahead. They used the lower arrangement shown in Fig. 76 for measurement of the total transmittance, which included the directly transmitted light and most of the light scattered in a forward direction. The difference represents the scattered light. In these experiments the cornea and the lens were each placed between a pair of parallel glass plates. The poles of the lens were flattened by pressure to give a finite area of contact with the two plates. Samples of aqueous and vitreous were also placed between parallel plates. The aqueous layer was 1.5 mm thick, and the vitreous layer was 17 mm thick. Measurements were made on nine human eyes excised at ages of 3 weeks and of 2, $4\frac{1}{2}$, 23, 42, 51, 53, 63, and 75 years.

b. Transmission and Scatter by the Cornea

Figure 77 shows data on the transmittance of the cornea. There is no age effect. The solid curve shows the total transmittance representative of six eyes.

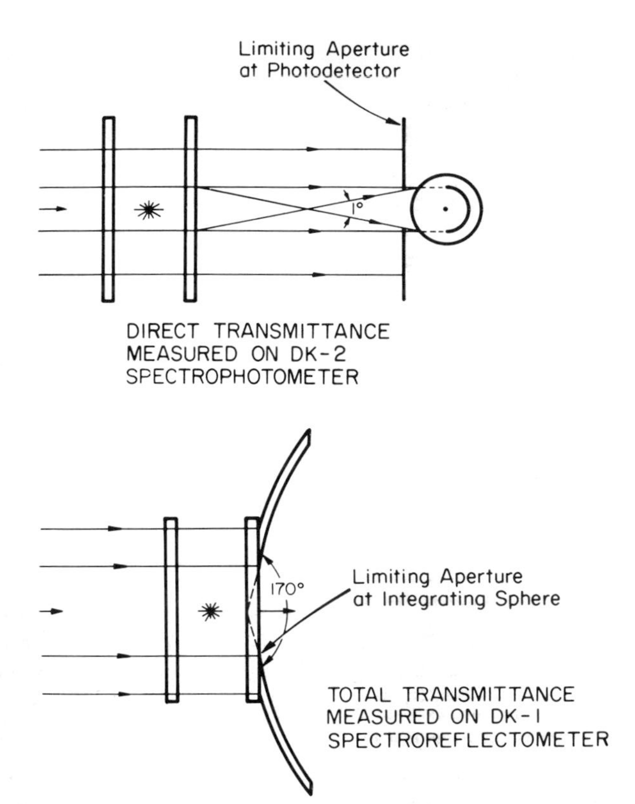

Fig. 76. Method of measuring direct and total transmittance (Boettner and Wolter, 1962).

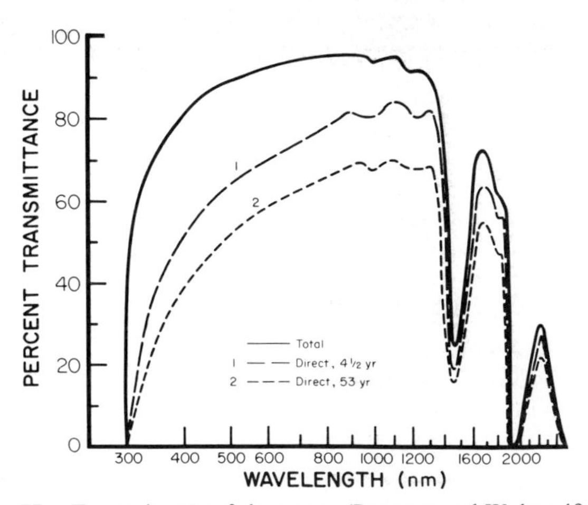

Fig. 77. Transmittance of the cornea (Boettner and Wolter, 1962).

Curve 2, showing direct transmittance for a 53-year-old eye, is close to the average for eight eyes. In the visible range of wavelengths a large part of the light attenuated is scattered in a forward direction.

A look at a slit-lamp picture (narrow slit) of the cornea indicates that nearly all of the layers yield back scatter, but the amounts differ, thus the various layers can be identified. The amount of scatter by the stroma can be increased by absorption of water. The fibers in the stroma are arranged in layers and in each layer are parallel and uniformly spaced. When water is absorbed the spacing becomes greater and less uniform. The pools or lakes of fluid that form increase the scatter. Scatter can also be increased by injecting air. Figure 78 shows the intensity of scattered light for various scattering angles produced by a beam transmitted through an edematous rabbit cornea devoid of epithelium and endothelium (Miller and Benedek, 1973).

An edematous epithelium generates a different kind of scatter. The epithelial cells become surrounded by interstitial fluid and produce diffraction effects. The edema may be caused by glaucoma, wearing contact lenses, or swimming in fresh water, or by endothelial dystrophy.

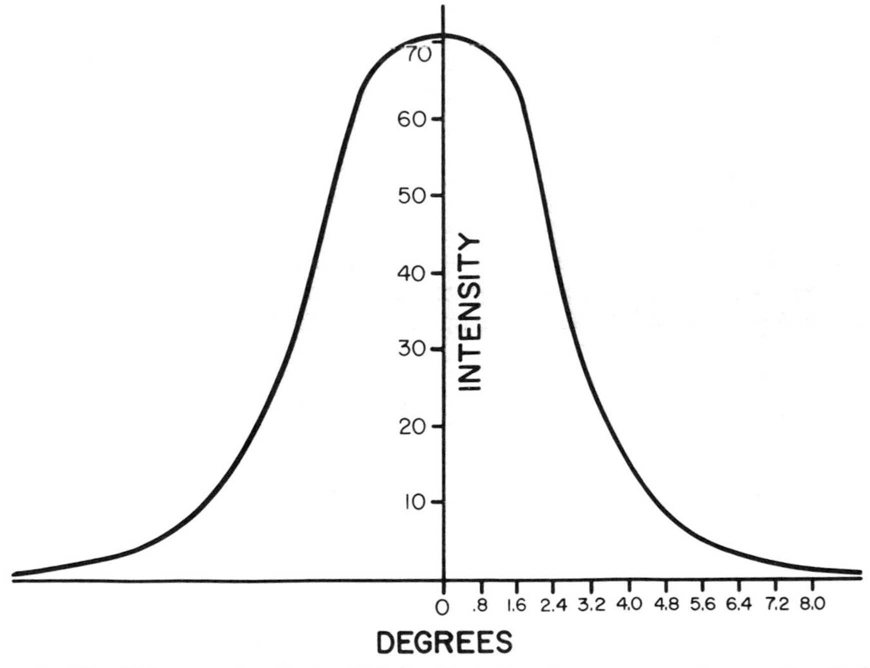

Fig. 78. Light-scattering plot, in which the effects of an edematous corneal stroma (devoid of epithelium and endothelium) on a distant point-1 source of light are measured at different positions from the central axis (Miller and Benedek, 1973). Courtesy of Charles C. Thomas, Publisher.

Finkelstein (1952) found that when a contact lens is worn for some time, and a small green source (546 mm) is fixated in the dark, the image of the small source has superimposed upon it a bright spot that is surrounded by two halos, one 3.18° in radius and the other 6.14°. Miller and Benedek (1973) assumed that in any direction the basilar layer of epithelial cells is equivalent to a diffraction grating with a center-to-center separation of 10 μm and that the corona and rings represent the zero-, first-, and second-order spectra. Miller and Benedek also assumed that the formula for a square-wave grating applies.

$$m\lambda = d \sin \theta, \tag{89}$$

where m is the order of the spectrum, θ is the radius of the halo, and d is the center-to-center distance between cells.

Miller and Benedek (1973) published a photograph of a white point taken through a rabbit cornea having an edematous epithelium. A densitometric trace across the image is reproduced in Fig. 79. It shows a white spot at the center surrounded by a single halo with a violet fringe inside and a red fringe

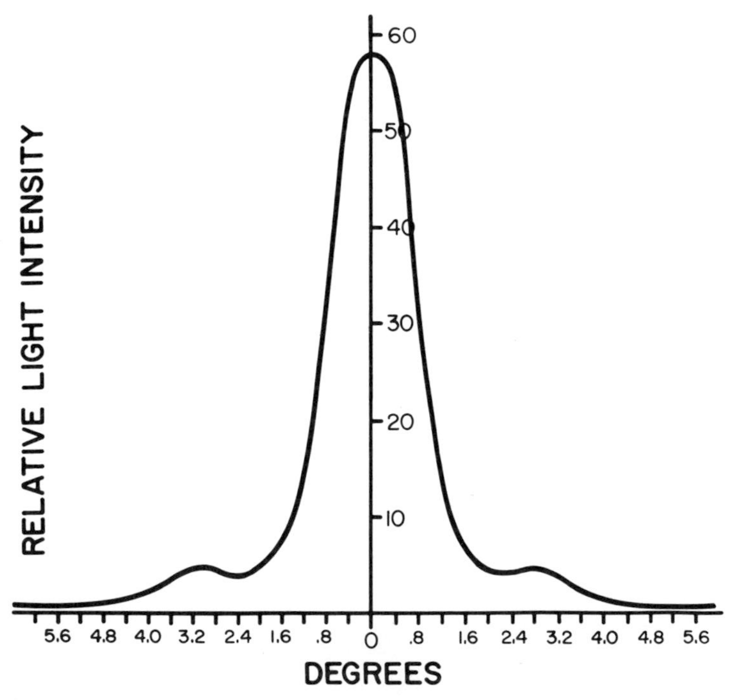

Fig. 79. A densitometric trace of a photograph of a white point through an excised edematous rabbit cornea, showing the angular position of the haloes which are represented as first-order maxima (Miller and Benedek, 1973). Courtesy of Charles C. Thomas, Publisher.

outside. They identified the central spot and halo as equivalent to those found by Finkelstein. Both the central spot and the halo disappeared when the epithelium was removed.

Because the outer ring took a longer time to develop than the inner, Finkelstein questioned whether the two rings had the same origin. Sheard (1919), who observed two rings around a white point under normal conditions of viewing, found sizes that are close to those found by Finkelstein. Sheard (1919) assumed that the two rings were generated by particles of different size having diameters of 18.4 and 8.3 μm. He used the following formula for relating particle size to ring size.

$$1.63 m\lambda = d \sin \theta, \tag{90}$$

where d is the diameter of the scattering particle, m is the order of the spectrum, and θ is the radius of the ring. Strictly speaking, the above formula applies to a monochromatic coherent point-source viewed through a single circular hole in a thin diaphragm. It applies also to a layer of multiple opaque spheres or transparent globules of index different from the medium in which they are imbedded, provided a source like a transilluminated piece of milk glass is imaged at the layer as shown in Fig. 80. The particles may be confined to a single thin layer or distributed throughout a thick layer. The formula applies also for ordinary seeing (Fig. 80) provided the particles are randomly arranged. The same formula applies to opaque particles or to air bubbles.

Sheard attributed his small ring (2.66° for 500 nm) to the external layer of epithelial cells and his large ring (5.66° for 500 nm) to the basilar layer or to the endothelium.

LeGrand (1937) used the arrangement shown in Fig. 81 for analyzing the scatter produced by the cornea. Green light (546 nm) is focused on the small aperture at R, which is conjugate to the cornea R′. If the head is moved sidewise, the beam passing through the cornea can be made to fall on the iris

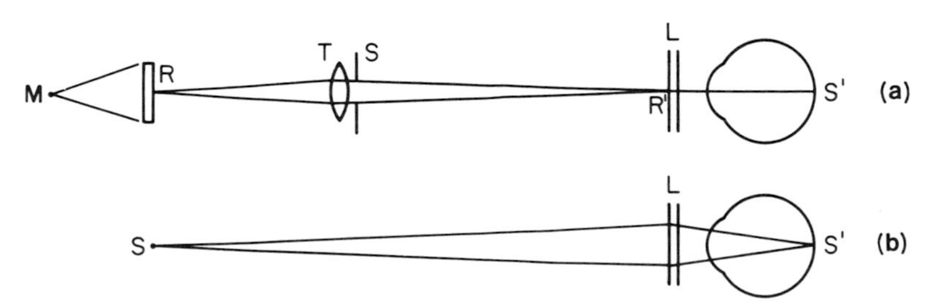

Fig. 80. Two ways of viewing a small source S through a layer L of scattering particles. In (a) the source is a small aperture S conjugate to the retina which permits the lens T to form an image R′ of a transilluminated plate of opal flashed glass R at the layer L of scattering particles. In (b) the small source S is conjugate to the retina.

Fig. 81. LeGrand's arrangement for differentiating scatter produced by the lens from that produced by the cornea. R is a small aperture illuminated by green light (546 nm) from a mercury source (LeGrand, 1937).

near the edge of the pupil so that it does not illuminate the lens. However, the illuminated part of the cornea will become a source of stray light, which will produce a shadow of the iris on the retina. The light inside the shadow is light diffusely scattered by the cornea.

When the beam is transmitted through the center of the pupil, the shadow of the iris is superimposed on the distribution of stray light produced by scatter at the lens and vitreous. Vos (1962) and Boynton and Clarke (1964) have used this arrangement to measure the amount of stray light produced by the cornea in a normal eye. Boynton and Clarke found that at an angle of 32° from the glare source about 25% of the stray light in the eye is produced at the cornea.

When the beam in Fig. 81 was allowed to fall on the iris near its edge, it was possible to see a part of a ring formed by diffraction at the cornea. LeGrand measured its radius to be 2.66° when monochromatic light (546 nm) was used instead of white. This is smaller than the ring produced when wearing a contact lens. It is not clear why the measurements are different, but in LeGrand's arrangement it is certain that the ring was produced by the cornea.

Simpson (1953) made an extensive study of a phenomenon that he found occurring in his own eyes for a period of time after waking up in the morning. He compared it to the corona that is seen by looking through a thin layer of lycopodium powder placed between two plates of glass (Fig. 80). This corona is due to scatter by a random array of opaque particles, to which Eq. (90) applies. For the first dark ring the formula may be restated as

$$1.22\lambda = d \sin \theta. \tag{91}$$

For white light he measured the diameter of the first dark ring and assumed it to be the same as for monochromatic light (571 nm). He then computed the diameter of the particles to be 29 μm, compared with 21 to 30 μm found by direct measurement. The whole display is called a *corona*, and the central spot is the *aureole*.

The early morning corona that he observed was produced by particles on the surface of his cornea; these could be washed away but could not be removed by rubbing the eyes. He could not identify the particles as blood or pus or moisture, but the diameter was computed to be about 8 μm. Descartes had described a similar corona, and Simpson proposed that the phenomenon be called *Descarte's corona.*

c. THE DIFFRACTION GRATING HALO PRODUCED BY THE LENS

The lens is made of layers of fibers that run a U-shaped course from a point near the axis on the front side of the lens toward the equator and back again to a point near the axis on the back of the lens (Fig. 3b). In an equatorial section (Fig. 3a) that cuts across all these fibers in a perpendicular direction, the cross sections are arranged in rows that radiate outward from the center, but the arrangement becomes quite irregular at the center. The radiating fibers and rows of cross sections produce optically a radial grating which, for a white point-source in front of the eye, generates on the retina a circular spectrum with an inner violet band and an outer red band. This corresponds to the first-order spectrum produced by a grating with parallel, uniformly spaced elements. Monochromatic light produces a narrow bright ring. The radius θ can be measured. The center-to-center separation of the lens fibers producing the ring is given by the formula

$$\text{center-to-center separation} = \lambda/\sin\theta, \tag{92}$$

where λ is the wavelength of the light. Simpson (1953) measured the radius of the ring for 570 nm to be about 3.3° and computed the center-to-center separation to be 9.7 μm. He failed to make allowance for the fact that the center of the lens lies 5 mm behind the cornea and that the apparent size of a fiber seen through the cornea is 1.23 times larger than it actually is. Hence, the computed width of a lens fiber should be 7.9 μm, which agrees with the assessment made from histological studies. Simpson used a small, round artificial pupil to study the effects obtained with beams passing through the lens near the edge of the pupil. Emsley and Fincham (1922) used a slit to study these effects. A knife edge may also be used to make the same test. The purpose of this test is to determine whether a ring is a diffraction grating spectrum or is produced by a random array of small scatterers as in the case of a lycopodium corona.

If a small beam is allowed to enter the eye near the edge of the pupil, as in LeGrand's arrangement (Fig. 81), only two small sectors of the circular spectrum are seen. These lie on opposite sides of the circle on the meridian perpendicular to the meridien in which the peripheral beam is located. LeGrand found the radius of the ring for 546 nm to be 3.23° Allowing for the fact that the lens lies behind the plane of the cornea, he computed the center-to-center spacing of the lens fibers to be 7.2 μm.

The circular spectrum is not seen if the beam entering the eye is narrow and confined to the center of the pupil. Simpson, who used artificial pupils, found that the diameter had to be as large as 3 mm for the circular spectrum to be seen. Sheard noted that the pupil had to be dilated before he could see his second ring, but he attributed the effect to the higher brightness of the spectrum. His second ring has the right radius (2.66° for 500 nm) to be due to the radial fibers of the lens, but instead Sheard attributed it to the basilar epithelium of the cornea.

d. The Ciliary Corona

Tscherning (1924, p. 188) called the bright central spot surrounding a bright point the *ciliary corona*. Simpson (1953) described this as an array of points and radiating lines, the details of which depend on the refractive state of the eye and the size of the point source. The point has to subtend an angle of less than 17 minutes of arc for the points and lines to be seen. The points and lines scintillate when the eye moves or changes focus.

Sheard (1919) attributed the fine structure to the irregular arrangement of fibers at the center of the lens. LeGrand (1937) used the arrangement in Fig. 81 to demonstrate that the central array of lines and points does not appear until the beam is allowed to pass through the pupil, thus proving that the origin lies in the lens.

e. Absorption and Scatter in the Lens

The Boettner and Wolter (1962) data on the total and direct transmittance for lenses of humans of various ages are shown in Fig. 82. It should be noted that the curves for direct transmittance T_D demonstrate an age effect; $(1 - T_D)$ includes the fraction of the light backscattered as well as the light absorbed. For the $4\frac{1}{2}$ year old subject the data for total transmittance T_T and direct transmittance T_D are both given. The difference $(T_T - T_D)$ is the fraction of the light scattered in a forward direction S. According to Hemenger (1982),

$$1 - S = e^{-\beta/\lambda^2}, \tag{93}$$

where β is a constant. Figure 83 shows the values of S plotted as a function of λ and fitted with Eq. (93).

Weale (1954), Lerman and Borkman (1976), and Wald (1949, pp. 121–123) have provided additional data on the direct transmittance of excised human lenses.

Another way of assessing the transmittance of the lens is to compare the luminosity curve of a normal observer with that of an aphakic observer. The ratio of the ordinates of the two curves at each wavelength should give the ratio of the transmittance of the lens to that of the aqueous replacing the lens. Wright (1951) and Wald (1945) made comparisons of this kind, and the results of Wald are shown in Fig. 84 for rod and cone vision, with the measurements confined to a point 8° above the fovea. Wright's data are shown in Fig. 85.

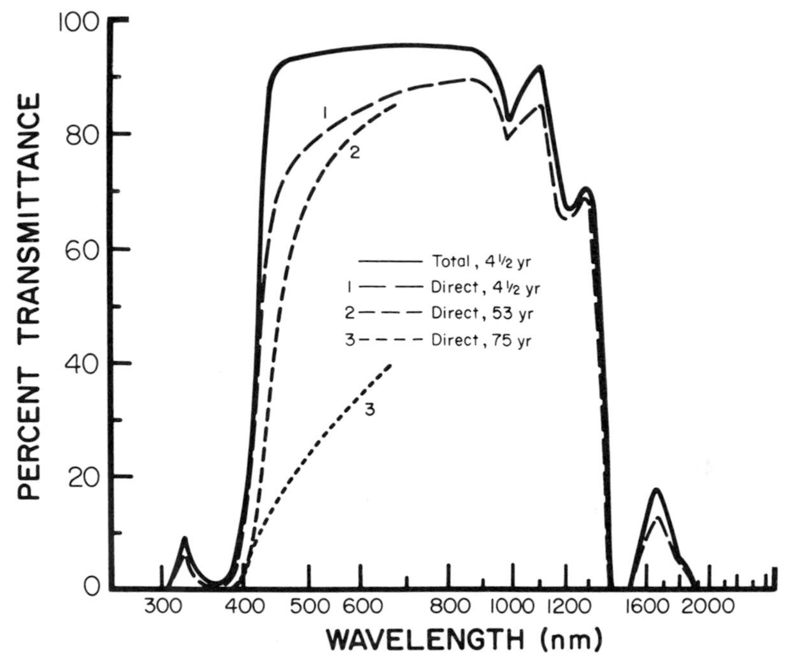

Fig. 82. Boettner–Wolter data for the transmittance of lenses. The solid curve represents the total transmittance for the lens of a 4-1/2 year old subject; the dashed curves represent the direct transmittance for the lenses of subjects of various ages (Boettner and Wolter, 1962).

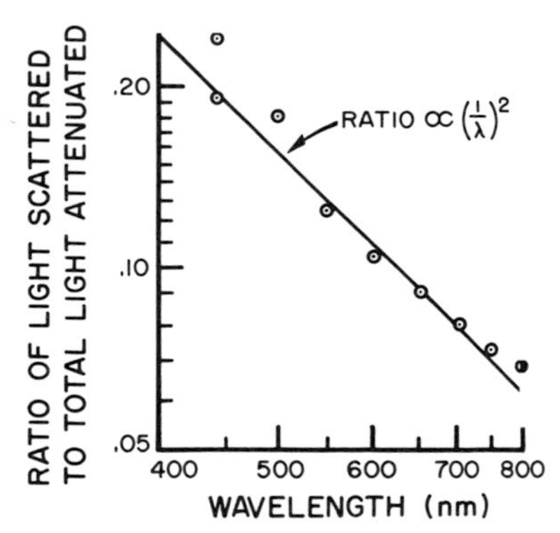

Fig. 83. Analysis of the Boettner and Wolter data for the lens of a 4-1/2 year old subject, which shows the dependence of the scattered light on wavelength.

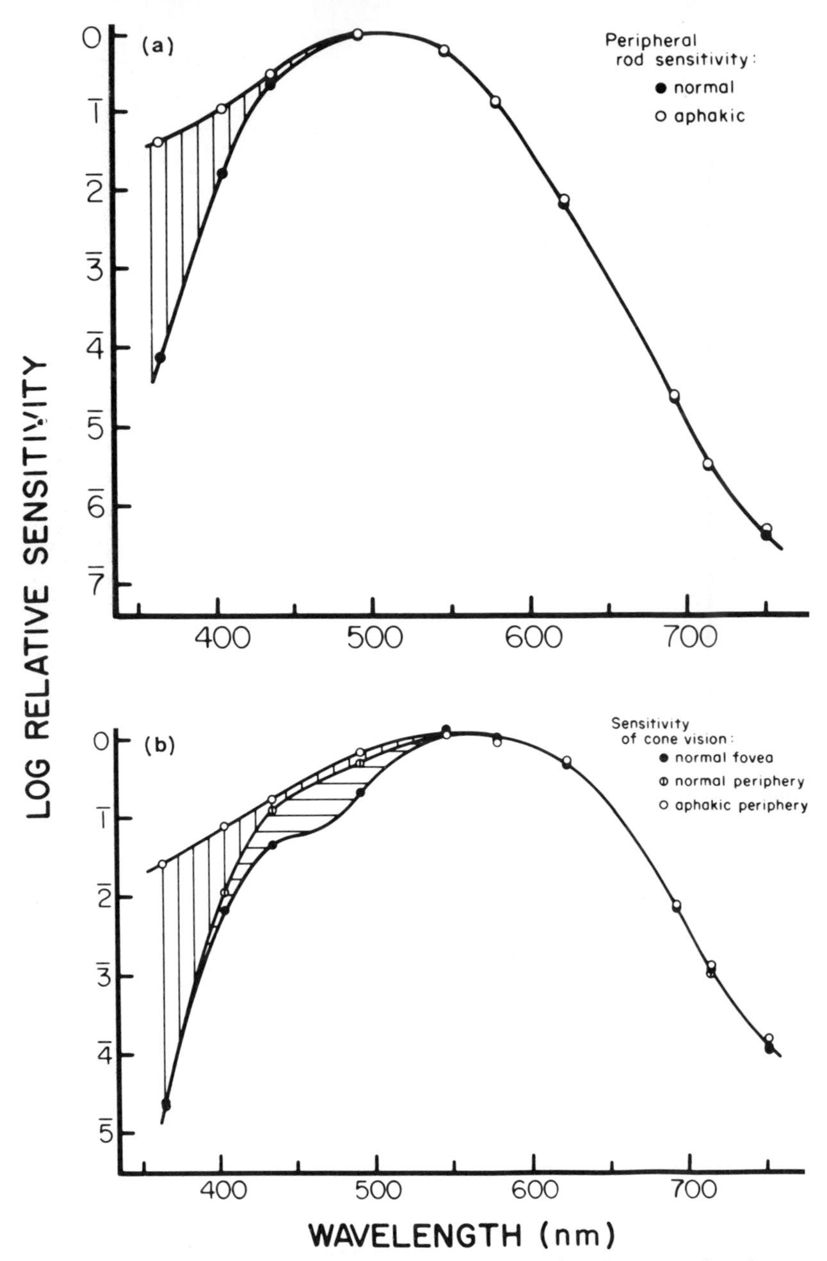

Fig. 84. Assessment of the transmittance of the human lens by comparing the spectral sensitivities of normal and aphakic subjects. The Graph (a) shows spectral sensitivities of rod vision 8° above the fovea. Both functions have been brought together above 546 nm, where they are parallel. Graph (b) compares spectral sensitivities in cone vision in the fovea and 8° above the fovea in normal and aphakic eyes. All the functions have been brought together above 578 mμ, where all of them are parallel. The horizontally hatched area represents the optical density (log 1/transmittance) of the macular pigment in the central fovea; the vertically hatched areas represents the optical density of the lens. (Wald, 1945.) Copyright 1945 by the AAAS.

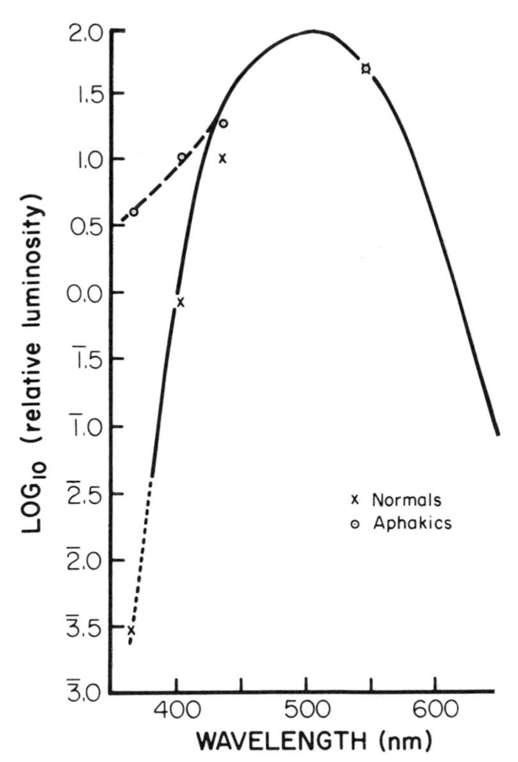

Fig. 85. Scotopic luminosity curve extended into the ultraviolet for normal and aphakic observers. The solid line represents the scotopic curve as measured by Crawford (1949); the dotted curve is the extension for normal observers and the dashed curve for aphakic observers (Wright, 1951).

It is possible to derive a certain amount of information about the yellowing of the lens with age by comparing the V_λ curves at different age levels. Ruddock (1965) has made such a study on 400 subjects and found that there is more absorptance at the blue end of the spectrum as age increases. It is difficult with these data to differentiate the yellowing due to scatter from that due to absorption.

Said and Weale (1959) measured the direct transmittance of the lens by comparing, for various wavelengths, the amounts of light reflected by the front and back surfaces of the lens. They showed that the transmittance is constant for ages up to 20, but with further aging there is a gradual increase in density. The data for subjects of different ages are shown in Fig. 86. Optical density is plotted as a function of λ.

Figure 87 shows the data replotted by Hemenger (1982) using log scales to show that at each age the scatter component is proportional to $1/\lambda^2$; but

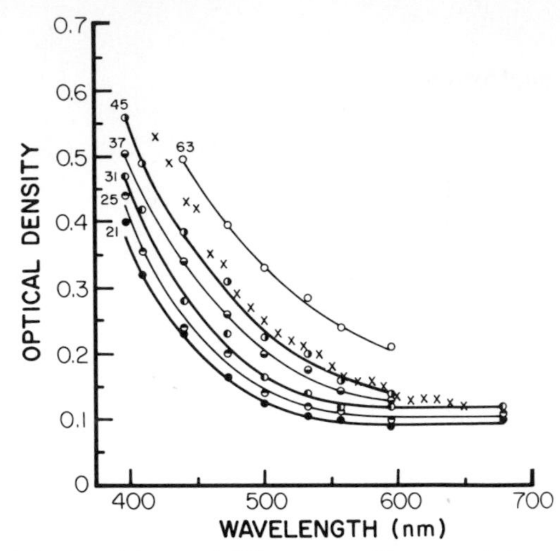

Fig. 86. Spectral density of lenses of different ages. The open circles are assessments made from light reflected at the front and back surfaces of the lenses of living eyes. The crosses represent average measurements made of direct transmission of lenses of enucleated eyes. (Said and Weale, 1959.)

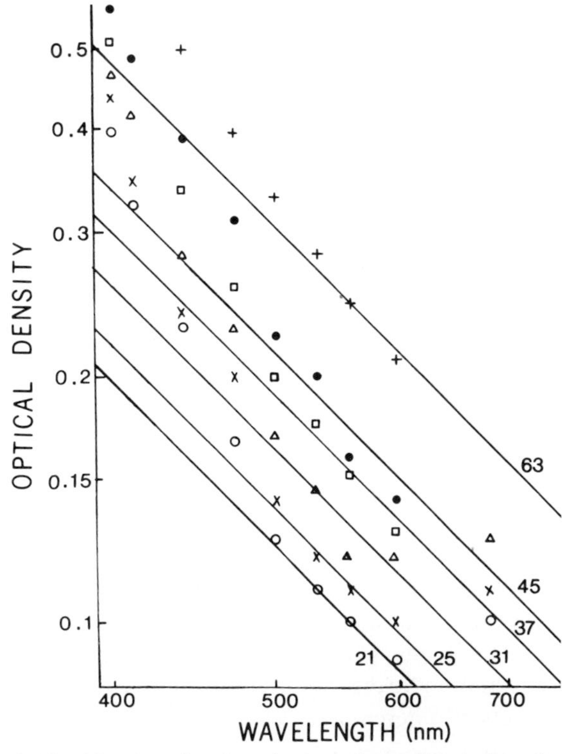

Fig. 87. Data in Fig. 86 replotted to show the dependence of optical density on wavelength in the range between 500 nm and 600 nm. The lines represent values of optical density that are inversely proportional to λ^2. (Hemenger, 1982.)

the amount of it depends on age. He also proposes that absorptance is independent of age and above 500 nm is constant for all wavelengths.

According to Hemenger (1982),

$$D = \beta/\lambda^2 + \gamma(\lambda), \tag{94}$$

where β is a constant dependent on age and $\gamma(\lambda)$ represents the component dependent on absorptance. The sloping lines (slope $= -2$) have been adjusted to fit the data from 500 to 600 nm. The data for 680 nm do not conform to the theory. The values of β at 600 nm have been plotted against age in Fig. 88.

The density due to absorption $\gamma(\lambda)$ is given by the displacement of the points above the lines with a slope of -2. Figure 89 shows the relation between $\gamma(\lambda)$ and λ as determined from the data in Fig. 87. It is scatter rather than absorptance that varies with age.

Bettleheim and Paunovic (1979) and Siew *et al.* (1981) used thin frontal slices of the lens (10 nm thick) to analyze the scatter produced by the lens. The slices were deliberately made thinner than the width of a lens fiber to minimize the role of the cell structure and to concentrate on fluctuations in density and optical anisotropy of the material inside the cells.

From the analysis of the data the researchers were able to determine that the size of the scattering units causing the fluctuations is about 300 nm. Hemenger (1982) has used this information in drawing his conclusion that the scatter produced by the lens must be proportional to $1/\lambda^2$.

They measured the scatter produced by a beam of 546-nm polarized light transmitted through the lens layer. The measurements were made in the planc perpendicular to the direction of polarization of the incident beam. They

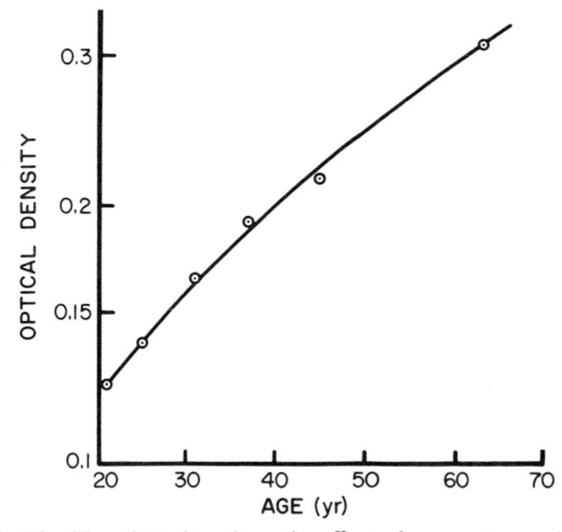

Fig. 88. Data in Fig. 87 replotted to show the effect of age on transmittance at 500 nm.

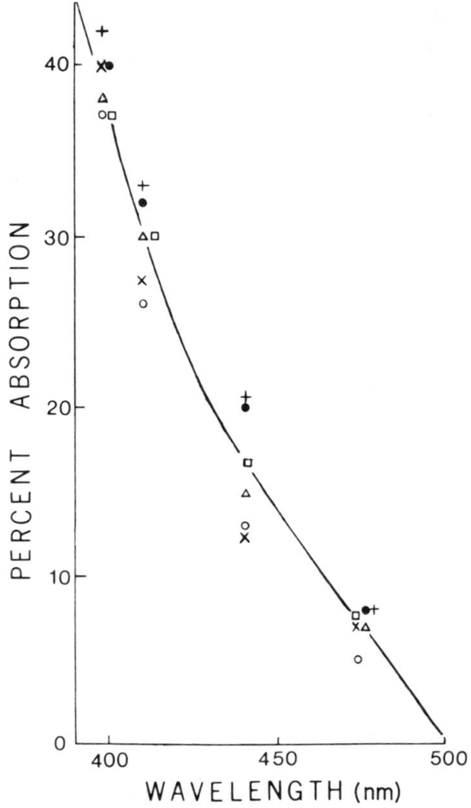

Fig. 89. Data in Fig. 87 in the range from 400 to 500 nm, replotted to show the effect of wavelength on the absorptance of the human lens. The data for the different ages have been brought together so that the diagonal lines in Fig. 87 coalesce. (Hemenger, 1982.)

assessed separately the light polarized parallel (‖) to the plane of measurement and the light polarized at right angles (⊥). It may be deduced from these data that the stray light from the lens falling on the retina is broadly distributed and that the sharp central peak in stray light found in the glare experiments must be attributed to the vitreous and the retina.

f. Scatter and Transmittance of the Vitreous and Retina

The vitreous contains scattering particles that are randomly oriented and larger than a wavelength of light; so the scatter is not wavelength dependent.

Figure 90 shows the data of Boettner and Wolter (1962) for the vitreous. There is no age effect. The curves are average data for several eyes. A large fraction of the light attenuated is scattered in a foreword direction. The data

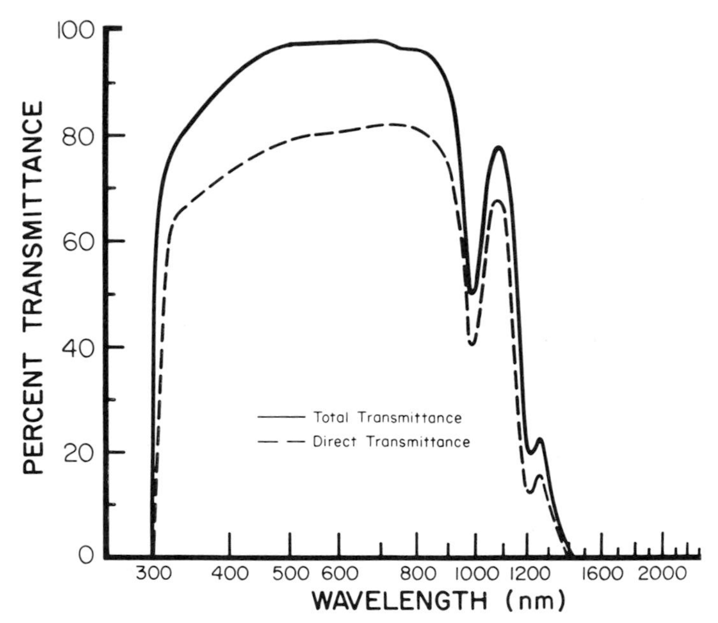

Fig. 90. Boettner and Wolter transmittance data for the vitreous humor (Boettner and Wolter, 1962).

do not include the light scattered by the retina, which Vos (1962) considered to be a major source of scatter. The retina must be dealt with separately as a transmitting and scattering medium. The thickness of the retina from the internal limiting membrane to the external limiting membrane is about 0.5 mm in the periphery but only about 20 μm at the center of the fovea. Except for the Henley layer of fibers and the cone cell bodies, there are no scattering elements in this region. The presence of the macular pigment in this region has a marked effect on the transmission.

Vos (1962) considered the neural elements in the retina to be important sources of scatter. Pokrowski (1926) paid more attention to the blood vessels on and near the surface of the retina.

DeMott and Boynton (1958) measured objectively the distribution of stray light falling at the back surface of the vitreous. They created a window in the back of the eye that exposed the vitreous and measured the stray light produced by beams directed at various parts of the retina. The data are shown in Fig. 91 and are compared with a curve based on Fry's formula (Eq. 86) for disability glare. The solid curve is based on the data in Table II of DeMott and Boynton (1958), who showed that 92.2% of the light entering the eye is transmitted either as focused light or stray light. The light at the zero scatter angle includes the light directly transmitted as well as the scattered light. The

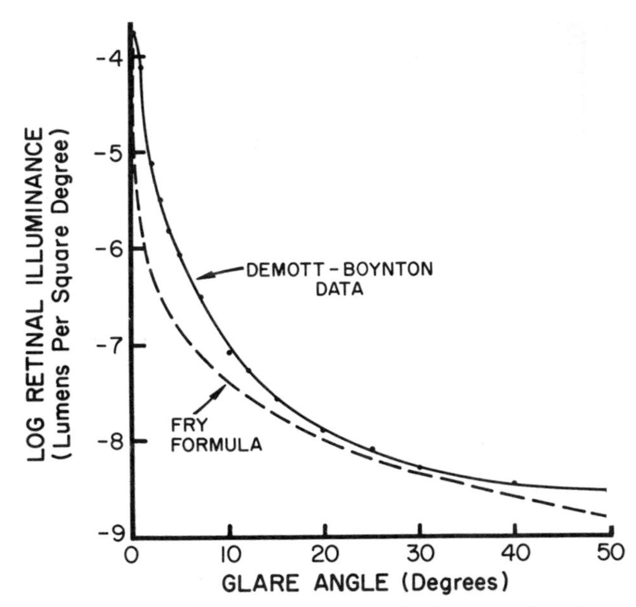

Fig. 91. Stray light on the retina for various angular displacements from the glare source. The solid curve is based on the DeMott and Boynton data (1958); the dashed curve is based on the Fry formula, Eq. (82).

source was assumed to be a disk 1.5° in diameter at one metre and having a candlepower of 100 candelas. The pupil was assumed to be 4 mm in diameter. One can use the procedure outlined by DeMott and Boynton to compute the lumens per square degree at a point 5° from the center of the image. This was found to be 9.25×10^{-7} lumens per square degree. The values at other glare angles are relative. The curve based on the Fry formula represents stray light for a point source (100 candelas) at one metre and for a pupil 4 mm in diameter. The values for veiling luminance have been converted to lumens per square degree. Instead of using the Fry formula to compute a curve for a point source, a distribution should have been computed for a 1.5° disk.

Because it is difficult to assess the difference between the response of the detector used by DeMott and Boynton and that of a cone for large angles of incidence, the subjective data should be more useful than the objective data in assessing disability glare. Note that the subjective data include light scattered by the retina and reflected by the pigment epithelium, choroid, and sclera in addition to the stray light arriving at the back of the vitreous. The major contribution made by DeMott and Boynton is demonstrating that the scatter by the media is sufficient to account for the stray light involved in disability glare and that the contributions made by the retina, pigment epithelium, sclera, and choroid can be ignored.

DeMott and Boynton also showed that the amount of stray light is not dependent on wavelength. It appears that the component scattered by the lens must depend on wavelength; but this component is broadly distributed, and near the center the stray light is dominated by the light scattered by the vitreous.

As Vos (1962) pointed out, the retina and the vitreous near the retina can be treated as a single component of the system. He used the *Boehm brushes* (1940) to demonstrate the role played by the retina in scattering, and the same arguments can be applied to the role played by the vitreous close to the retina. The Boehm brushes are seen when a point source is viewed through a polarizer: the veil acquires a wing structure with the maximum brightness perpendicular to the direction of polarization. The polarization of the scattered light requires a large angle of scatter. The angle is too small at the cornea and lens to produce polarization. The fact that the brushes are not wavelength dependent and that they occur in aphakic subjects is evidence that the brushes are not due to the lens. But if the scattering particles lie close to the layer of rods and cones, the stray light falling on the retina at some distance from the image of the glare source involves the large angle of scatter required for tthe Boehm effect.

Vos (1962) pointed out that when a narrow beam is transmitted through the pupil, the part of the light scattered by the cornea and lens is subject to the Stiles–Crawford effect when the image of the glare source falls in the periphery, and the glare is measured at the fovea. Scatter occurring in or near the retina is not affected by the point of entry through the pupil; howcver, it does affect the response of the foveal cones to stray light from the lens or cornea. On the other hand, if the glare source falls in the fovea, and the glare is measured in the periphery, the Stiles–Crawford effect is not pronounced, because the stray light is falling on the peripheral retina. This difference demonstrates that a part of the stray light originates close to the layer of photoreceptors.

g. Transmittance of the Combined Media

Boettner and Wolter (1962) used the data for the separate media to compute the direct transmittance for various combinations. Curve 1 in Fig. 92 represents transmittance measured at the front of the aqueous, 2 at the front of the lens, 3 at the front of the vitreous, and 4 at the front of the retina. Allowance was made for loss by reflection at the front of the cornea, but the other losses by specular reflection were ignored. The data were selected so that the curve shown in Fig. 92 represent young eyes.

Ludvigh and McCarthy (1938) measured the direct transmittance from the front of the cornea to the front of the retina; their data are shown in Fig. 93. These are the data usually used in assessing the effect of transmittance on

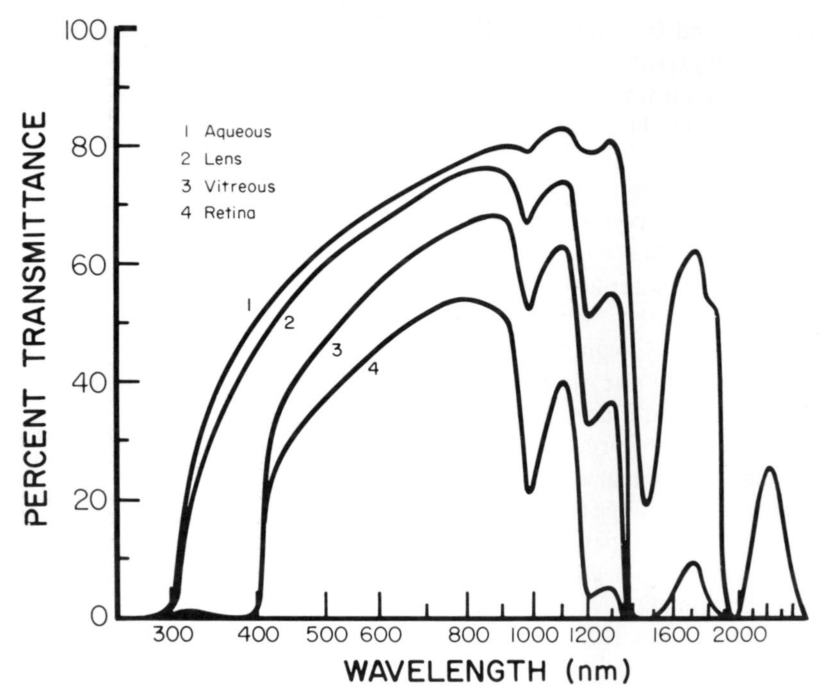

Fig. 92. Direct transmittance through the eye measured at various anterior surfaces, based on multiplying the transmittances for the different media (Boettner and Wolter, 1962).

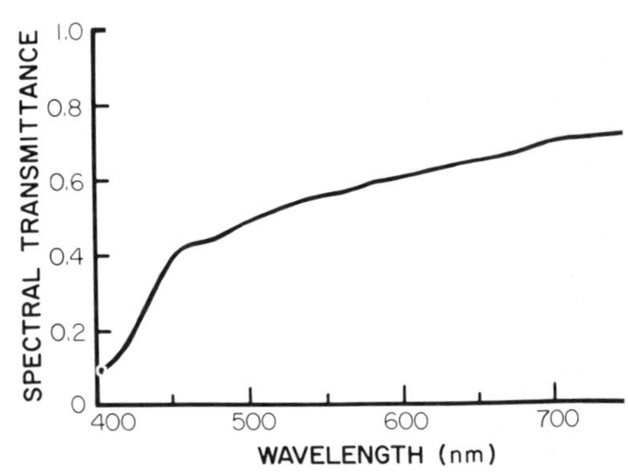

Fig. 93. Ludvigh and McCarthy (1938) values for the transmittance of the ocular media for various wavelengths (Fry, 1954).

stimuli applied to the photoreceptors. They exclude the effect of the retina and the macular pigment.

Geeraets and Berry (1968) have also measured the direct transmittance of the ocular media from the cornea to the back of the vitreous. Their data, which are shown in Fig. 94, make no allowance for the losses in reflectance at the refracting surfaces. They found much higher values for transmittance of the media than did either Ludvigh and McCarthy or Boettner and Wolters.

Alpern *et al.* (1965) have used eyes with a coloboma of the retina and choroid to assess the transmittance of the media in the living human eye.

Wyszecki and Stiles (1967, p. 219) have presented a table of factors that investigators can use to correct their data for transmittance of the ocular media. The variation from wavelength to wavelength is assumed to be dependent on the lens.

6. Transmittance of the Retina

Geeraets *et al.* (1960) measured the transmittance of the retina directly for 30 rabbit eyes and found that for the range from 350 to 1500 nm the transmittance did not vary more than 1% from unity. It was considered negligible for assessing the amount of light reaching the pigment epithelium and

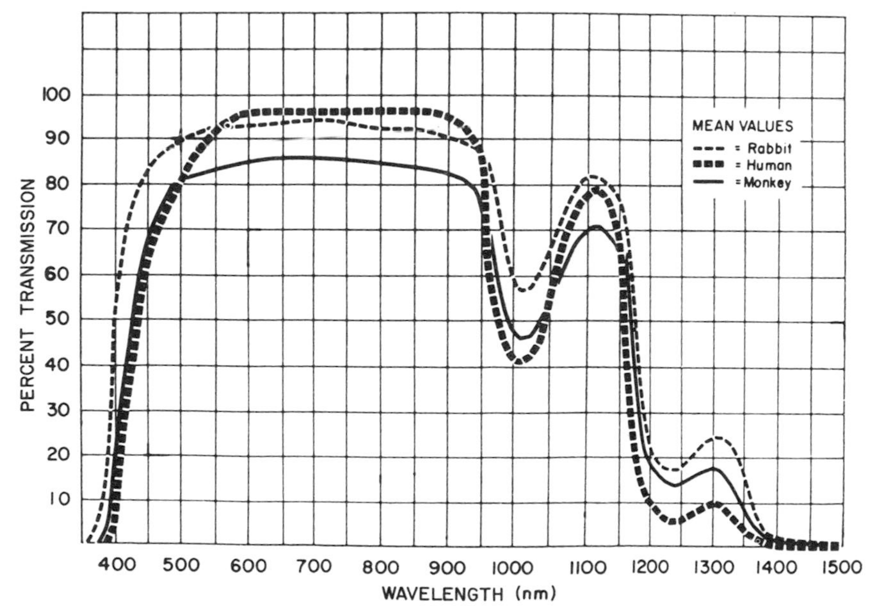

Fig. 94. Percent transmittance for light through the ocular media of human, monkey, and rabbit eyes (Geeraets and Berry, 1968). Published with permission from *The American Journal of Ophthalmology*, **66**, 15–20, 1968, copyright by the Ophthalmic Publishing Company.

producing retinal burns. Obviously the retinas measured by Geeraets *et al.* were not dark-adapted, and the measurements must have been made outside the macula. It follows from these findings that the retina cannot make a substantial contribution to the stray light falling on the photoreceptors.

Brown and Wald (1963, p. 6) measured the transmittance of a dark-adapted excised human retina for light (510 nm) at a point outside the fovea and once again after the photopigment was bleached. They found that the photopigment when dark-adapted absorbs about 20% more of the incident light.

a. MACULAR PIGMENT

The *macula lutea* is an oval-shaped area of the retina centered at the fovea and extending about 5° outward in the horizontal direction and 3° in the vertical. It contains a yellow pigment that can be extracted and has been identified by Wald (1949, pp. 117–121) as xanthophyll. The density is greatest in the area surrounding the foveal depression; it is lower at the center and, in the other direction, tapers toward the edge of the macula. The pigmentation is confined to the layers internal to the cell bodies of the photoreceptors.

The dashed curve in Fig. 95 represents the spectral density of a layer of given thickness of a solution of macular pigment in chloroform as measured by Wald. The solid curve is for leaf xanthophyll (lutein). The effect of macular pigment on vision has been investigated by Wald, Stiles, Schelling, and Ruddock, among others. Wald (1949, pp. 117–121) compared the absolute thresholds for a 1° disk at the center of the fovea and at a point 8° above the center.

The two sets of data are shown in Fig. 84b. The two curves have been made to coalesce at the red end of the spectrum, and the difference at the blue end is

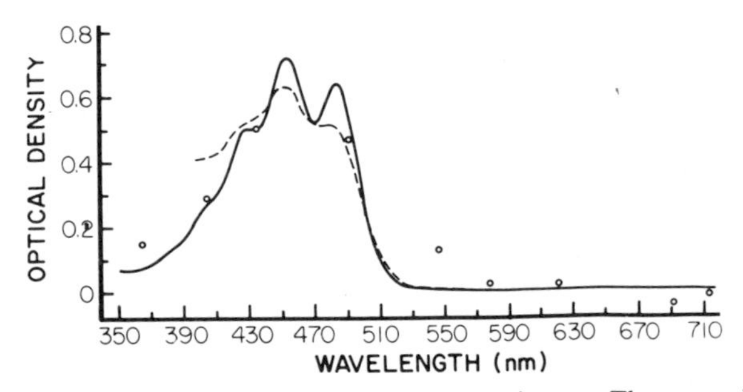

Fig. 95. Absorption spectrum of the human macular pigment. The open circles represent visual estimates of optical density, that is, differences in the log sensitivity of peripheral and foveal cones, comparable to the horizontally hatched area of Fig. 84. The broken line is the absorption spectrum of a preparation of xanthophyll extracted from human maculae, dissolved in chloroform. The solid line is the spectrum of crystalline lutein or leaf xanthophyll in chloroform. (Wald, 1949.)

assumed to represent the density of macular pigment. The open circles in Fig. 95 represent data obtained in this way. Stiles (1978, pp. 206–208) carried out a similar experiment but used a point 8° to the nasal side of the fixation point. His data are shown in Fig. 96 along with Wald's data for the density of a layer of a given thickness of a solution of xanthophyll in chloroform.

Wyszecki and Stiles (1967, p. 219) have used the data of Wald and Stiles to create a table of values for use in making allowance for macular pigment. Schelling (1950) and Ruddock (1963) have assessed the effect of macular pigment on color mixture data. According to Said and Weale (1959) there is no change in the macular pigment with age.

7. Reflection and Absorption at the Fundus

a. Light Reflected by the Pigment Epithelium

Reflected light from the pigment epithelium, a diffusely reflecting surface, makes retinoscopy and coincidence optometry possible. The important thing is that the pigment epithelium and the rod and cone layer are interdigitated to the extent that no distinction needs to be made between the layer of the rod and cone outer segments as an absorbing surface and the surface of the pigment epithelium as a reflecting surface.

The light reflected by the pigment epithelium can stimulate the photoreceptors on its way back through the retina. Light from a given point on the pigment epithelium is probably confined to one photoreceptor, or at most a cluster of seven or nineteen, and consequently has little effect on the perceived blurring of an image. This method of stimulating the photoreceptors is a form

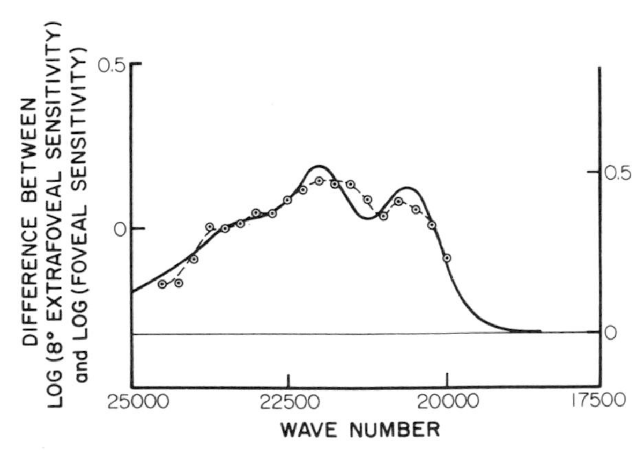

Fig. 96. Circles indicate difference between the log sensitivities in the fovea and at 8° extrafovea. Solid line is the optical density of a solution of xanthophyll in chloroform (Wald's measurements). Wave number is the reciprocal of the wavelength in centimeters. (Stiles, 1978.)

of halation. We need to know how much light is reflected and how much is absorbed by the photoreceptors.

b. Light Reflected by the Choroid and Sclera

The light transmitted by the pigment epithelium penetrates through the choroid all the way to the sclera. If the eye turns in and looks past the root of the nose at a point source, a diffuse image is formed on the surface of the sclera which can be seen by a second person.

The light scattered backward from the choroid and sclera is diffusely transmitted by the pigment epithelium. The proof of this is that the blood vessels in the choroid are clearly visible with an ophthalmoscope only when the pigment is absent, as in the case of an albino. Part of the light that is reflected by the choroid and sclera and passes back through the pigment epithelium is absorbed by the photoreceptors; this is a second form of halation. The light that passes back through the pigment epithelium adds itself to the light diffusely reflected from this layer.

c. Measurement of the Fundus Reflection

Brindley and Willmer (1952) measured the reflectance of the pigment epithelium by directing a beam through the pupil onto the pigment epithelium and measuring the light exiting from the pupil. Vos (1962) corrected the data for transmittance of the media but did not allow for absorption by the macular pigment or photopigment. The corrected data are shown in Fig. 97 Measurements of the fundus reflection have also been made by Campbell and Alpern (1962), Alpern and Campbell (1962), and Alpern *et al.* (1965).

The light reflected by the choroid and retina and diffusely transmitted through the pigment epithelium is important for several reasons, in particular in assessing its role in retinal densitometry (Rushton, 1965).

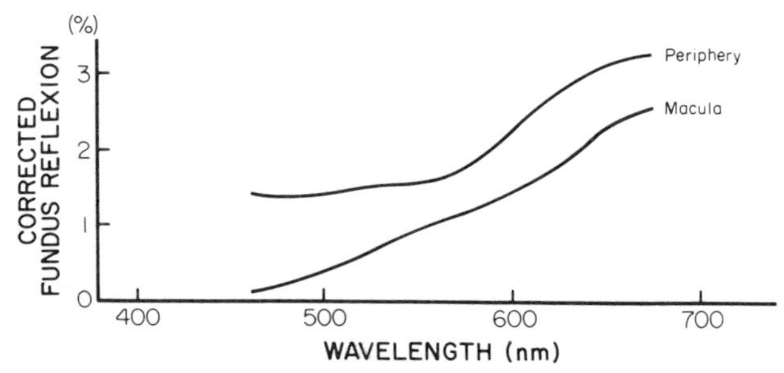

Fig. 97. Data of Brindley and Willmer (1952) for reflectance from the fundus, corrected by Vos for transmittance of the ocular media (Vos, 1962).

It is common now to assess the quality of the image formed by the eye on the pigment epithelium by measuring the distribution of flux reflected from the pigment epithelium (Campbell and Gubisch, 1966). In doing this, allowance has to be made for the light that gets transmitted through the pigment epithelium and is then reflected back by the choroid and sclera. It no doubt contributes to the color of the fundus reflex in retinoscopy but has no effect on determining the point conjugate to the retina (Fry, 1949).

Light emitted by the pigment epithelium can pass obliquely through the retina and vitreous and stimulate the photoreceptors on the opposite side of the fundus. This is a separate component of stray light. It is generally ignored in dealing with disability glare. However, if a spot on the pigment epithelium is illuminated, it can produce a shadow of the retinal blood vessels on the layer of rods and cones on the opposite side of the fundus (Tscherning, 1924, p. 184).

Toraldo di Francia and Ronchi (1952) have measured the amounts of light scattered in various directions from a spot illuminated with a beam normal to the retina.

Light reflected from a point on the pigment epithelium can emerge through the pupil, and part of it can be reflected back toward the retina. It comes to a focus [seventh Purkinje image (LeGrand, 1937)] about one millimeter behind the front surface of the lens and then spreads out over a circular area as it approaches the retina. The intensity is low enough to justify ignoring it.

d. Absorptance by the Choroid and Pigment Epithelium

The light reflected by the sclera is twice transmitted by the choroid and pigment epithelium. There are two pigments involved in absorbing light passing through the pigment epithelium and choroid: melanin and hemoglobin.

After removing a segment of the sclera from the back of an eye, Geeraets and Berry (1968) sent a beam of light through the eye from the cornea to the back of the choroid and measured the fraction of light transmitted. They then removed the choroid and pigment epithelium and the neural part of the retina and repeated the measurement. The difference gives the fraction of light absorbed by the pigment epithelium and the choroid and the neural part of the retina and also the light reflected at the pigment epithelium. They assumed there was no loss by transmittance through the retina.

To allow for the light reflected at the pigment epithelium they measured directly the reflection from the pigment epithelium backed up by the choroid and sclera. According to the data in Fig. 98, about 30% of the light transmitted by the pigment epithelium reaches the sclera, and 30% of that reflected back by the sclera returns to the pigment epithelium.

Figure 98 shows the absorptance by the choroid and the pigment epithelium when the data are corrected for loss by reflectance. This kind of information is needed in connection with the study of retinal burns.

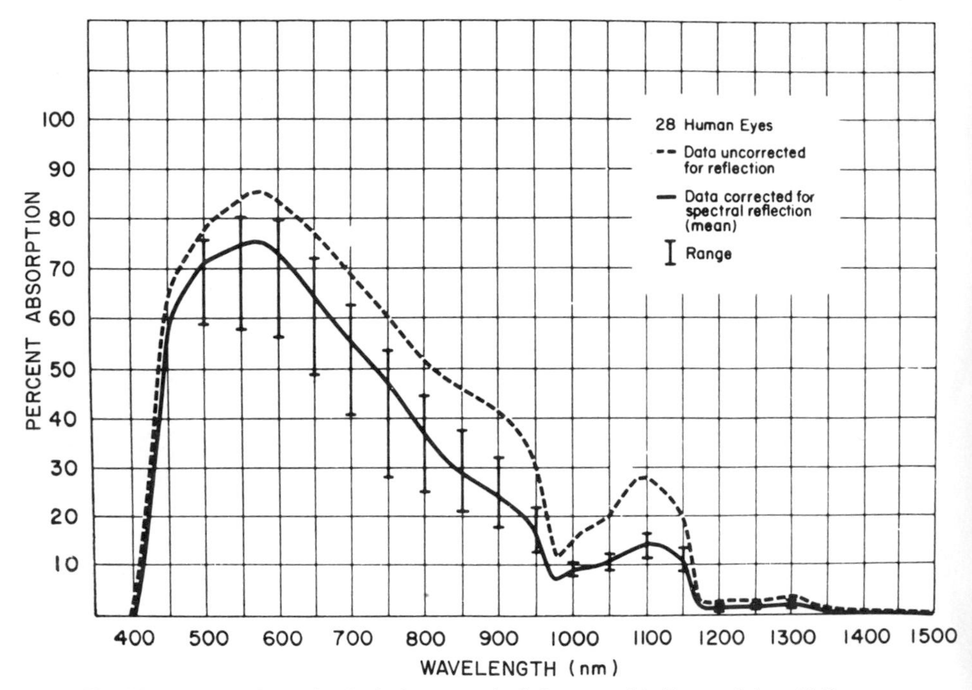

Fig. 98. Percent absorption in the human retinal pigment epithelium and choroid. Percentage of light incident on the cornea (Geeraets and Berry, 1968). Published with permission from *The American Journal of Ophthalmology*, **66**, 15–20, 1968, copyright by the Ophthalmic Publishing Company.

Geeraets *et al.* (1960) used excised rabbit eyes to show the effect of the degree of pigmentation on the transmittance of the pigment epithelium and choroid.

Curve OM in Fig. 99 shows the data for the transmittance of the ocular media from the cornea to the front of the retina. The remaining curves show the transmittance from the cornea to back side of the choroid. Curve IV is for albino eyes, and curves III, II, and I are for light, medium, and dark pigmented eyes, respectively. The difference between a pigmented eye and an albino eye shows the effect of melanin. The difference between curve OM and curve IV is very small because of the absence of melanin. Because the eyes in this study were excised, it was difficult to assess the role of hemoglobin.

Hunold and Malessa (1974) developed a technique for measuring the melanin pigmentation in the intact eye by means of reflectometry. The measurements are related to those made by Brindley and Willmer, which were previously described. The important thing is that they measured albinic, caucasion and negroid eyes, so the differences can be attributed to the amount of melanin. In the albino, the absorption can be attributed to hemoglobin.

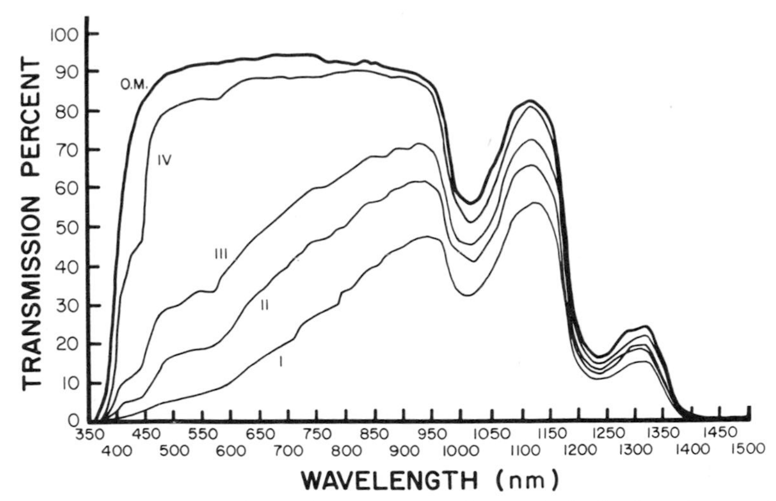

Fig. 99. Percentage transmittance in the ocular media and the fundus for rabbit eyes having varying degrees of pigmentation, for light incident on the cornea. Curve OM shows mean values through the ocular media. Curves I, II, and III give mean values of dark-pigmented fundi, medium-pigmented fundi, and light-pigmented fundi, respectively. Curve IV is for albino eyes, which have no fundus pigment. (Geeraets *et al.*, 1960.) Courtesy of *Archives of Ophthalmology*, **64**, 609, 1960, copyright American Medical Association.

K. FLUORESCENCE, BIOLUMINESCENCE, AND ELECTROLUMINESCENCE

1. Fluorescence in the Lens

Some of the light absorbed by the lens is reemitted as fluorescence. The intensity of fluorescence increases with age and with the development of a nuclear cataract. Lerman and Borkman (1976) identified one fluorogen that has an activation peak at 360 nm and an emission peak at 420–440 nm and another that has an activation peak at 435 nm and an emission peak at 500–520 nm.

Satoh *et al.* (1973) made an analysis of fluorescence for several different components of the lens substance, obtaining measurements for purple fluorescence (excitation at 290 and peak emission at 350 nm) and blue fluorescence (excitation at 360 and peak emission at 420–440 nm).

LeGrand (1938, 1948) measured the fluorescence produced by 365-nm ultraviolet light in the eyes of sheep, oxen, and rabbits. The average relative amounts of light emitted at various wavelengths are shown in Fig. 100. Included in the same figure are data procured by a visual method.

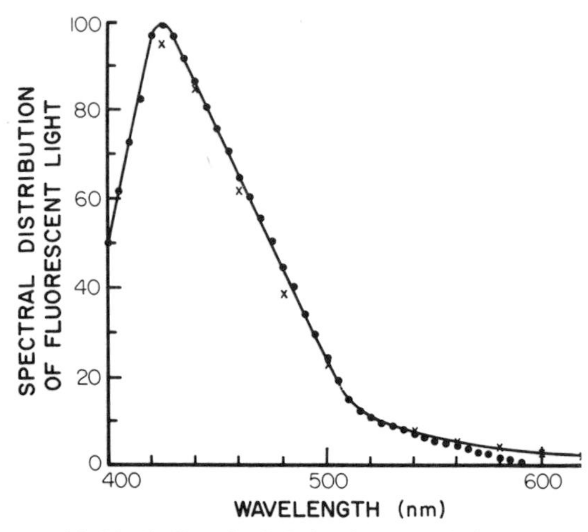

Fig. 100. Data provided by LeGrand (1948) for the spectral distribution of fluorescent light produced by irradiating lenses of oxen and rabbits with ultraviolet light (365 nm). The dots represent data obtained by a photographic method; the crosses are data obtained by a visual method, which is more precise for longer wavelengths. The curve is considered the best estimate of the mean results. (Wyszecki and Stiles, 1982.)

According to Boynton and Clarke (1964) the amount of fluorescence produced by ultraviolet light (400 nm) is negligible so far as disability glare is concerned.

The flux produced on the retina is broadly distributed. Fluorescence in a forward direction is included as part of the scattered light in the total transmission measurements.

2. Fluorescence of Rhodopsin

The fluorescence of rhodopsin can be demonstrated with an excised dark-adapted retina. The importance is that the rods respond not only to the ultraviolet light but also to the light produced by fluorescence. In the visible range the activating wavelength and the wavelengths produced by fluorescence can both contribute to scotopic brightness (LeGrand, 1957, pp. 93–94).

3. Bioluminescence and Electroluminescence

Judd (1927) deduced from his study of the Hering afterimage that it involves some form of bioluminescence.

The blue arcs of the retina (LeGrand, 1967, pp. 155–156) may be regarded as being produced by electrical stimulation of the photoreceptors by nerve impulses that are being transmitted across the retina over the ganglion cell axons. There is also the possibility, however, that these impulses generate a form of luminescence that in turn activates the photoreceptors.

L. POLARIZATION EFFECTS

1. Haidinger's Brushes

Haidinger's Brushes is a phenomenon that appears to depend on the presence in the upper retinal layers of dichroic blue-absorbing pigment molecules that have their long axes pointing toward the center of the fovea. These molecules absorb plane-polarized light maximally when the electric vector is parallel to the long axis. This creates the appearance of blue brushes on a yellow background when a white surface is viewed through a blue filter or through a slowly rotating array of alternate blue and neutral filters that represent sectors of a large disk. The rotating pattern is centered at the fovea, a fact that is used clinically to verify that the primary line of sight is pointing in a given direction. De Vries *et al.* (1953) and Naylor and Stanworth (1954) have shown that the density of the pigment producing the brushes has a spectral absorptance akin to that of the macular pigment.

2. Double Refraction of the Ocular Media

The ocular media have a double refraction equivalent to a retardation plate with the axis of the plate inclined downward and nasally (Boehm, 1940). This affects the polarization of light transmitted through the eye, and allowance for it must be made in assessing Haidinger's brushes (De Vries *et al.*, 1953; Naylor and Stanworth, 1954).

The phenomenon referred to as Boehm's brushes (1940) had been noted previously by LeGrand (1936b, 1967, p. 154) who pointed out that it might be related to the dichroic properties of the cornea. Vos (1962, pp. 29–39) attributed the brushes to scatter at large angles near the layer of rods and cones.

REFERENCES

Abramowitz, M., and Stegun, I. A. (1965). "Handbook of Mathematical Functions," pp. 495–502. Dover, New York.

Allen, M. J. (1949). An objective high speed photographic technique for simultaneously recording changes in accommodation and convergence. *Am. J. Optom. Arch. Am. Acad. Optom.* **26**, 279–289.

Alpern, M. (1969). *In* "The Eye" (H. Davson, ed.), Vol. III, pp. 217–254. Academic Press, New York.

Alpern, M., and Campbell, F. W. (1962). The spectral sensitivity of the consensual light reflex. *J. Physiol.* **164**, 478–507.

Alpern, M., Thompson, S., and Lee, M. S. (1965). Spectral transmittance of visible light by the human eye. *J. Opt. Soc. Am.* **55**, 723–727.

Ames, A., Jr., and Proctor, C. A. (1921). Aberrations of the eye. *J. Opt. Soc. Am.* **5**, 22–84.

Baraket, R. (1961). *In* "Progress in Optics" (E. Wolf, ed.), Vol. II, pp. 67–108. North-Holland, Amsterdam.

Bedford, R. E., and Wyszecki, G. (1957). Axial chromatic aberration of the human eye. *J. Opt. Soc. Am.* **47**, 564–565.

Berger-Lheureux-Robardey, S. (1965). Mesure de la fonction de transfert de modulation du système optique de l'oeil et des seuils de modulation rétiniens. *Rev. Opt.* **44**, 294–323.

Berny, F., and Slansky, S. (1970). *In* "Optical Instruments and Techniques" (J. H. Dickson, ed.), pp. 375–386. Oriel Press, Newcastle-upon-Tyne.

Bettelheim, F. A., and Paunovic, M. (1979). Light Scattering of Normal Lens I. *Biophys. J.* **26**, 85–99.

Boehm, G. (1940). Ueber ein neues entoptisches Phänomen im polarisierten Licht. *Acta. Ophthalmol.* **18**, 143–169.

Boettner, E. A., and Wolter, J. R. (1962). Transmission of the Ocular Media. *Invest. Ophthalmol.* **1**, 776–783.

Boynton, R. M., and Clarke, F. J. J. (1964). Sources of entopic scatter in the human eye. *J. Opt. Soc. Am.* **54**, 110–119.

Brindley, G. S., and Willmer, E. N. (1952). The reflection of light from the macular and peripheral fundus oculi in man. *J. Physiol.* **116**, 350–356.

Brown, P. K., and Wald, G. (1963). Visual pigments in human and monkey retinas. *Nature* **200**, 37–43.

Byram, G. M. (1944). Physical and photochemical basis of visual resolving power. *J. Opt. Soc. Am.* **34**, 571–591, 718–738.

Campbell, F. W., and Alpern, M. (1962). Pupillometer Spectral Sensitivity Curve and the Color of the Fundus. *J. Opt. Soc. Am.* **52**, 1084.

Campbell, F. W., and Gubisch, R. W. (1966). Optical quality of the human eye. *J. Physiol.* **186**, 558–578.

Cobb, P. W. (1915). The influence of pupillary diameter on visual acuity. *Am. J. Physiol.* **36**, 336–346.

Crawford, B. H. (1936). Integration of the glare effects from a number of sources. *Proc. Phys. Soc. London* **48**, 35–37.

Crawford, B. H. (1949). Scotopic visual function. *Proc. Phys. Soc.* **62** (series B), 321–334.

DeMott D. W., and Boynton, R. M. (1958). Retinal distribution of entoptic stray light. *J. Opt. Soc. Am.* **48**, 13–22.

DeVries, H. L., Spoor, A., and Jielof, R. (1953). Properties of the eye with respect to polarized light. *Physica.* **19**, 419–432.

Donders, F. C. (1864). "Accommodation and Refraction of the Eye," p. 204. New Sydenham Society, London.

Duke-Elder, W. S. (1944). "Text-book of Ophthalmology," Vol. I. Mosbey, St. Louis.

Duke-Elder, W. S. (1949). "Text-book of Ophthalmology," Vol. IV, pp. 4258–4259. Mosby, St. Louis.

Emsley, H. H., and Fincham, E. F. (1922). Diffraction halos in normal and glaucomatous eyes. *Trans. Opt. Soc. London* **23**, 225–336. (Reprinted in *Am. J. Physiol. Opt.* **4**, 247–272.)

Enoch, J. M., and Laties, A. M. (1971). An analysis of retinal receptor orientation, II. *Invest. Ophthalmol.* **10**, 959–970.

Epstein, L. I. (1949). Out of focus diffraction patterns, *J. Opt. Soc. Am.* **39**, 226–228.

Epstein, L. I., and Fry, G. A. (1973). Derivation of the MTF for an ideal optical system from Struve's equation for the line spread function. *Appl. Opt.* **12**, 132–133.

Ferree, C. E., Rand, G., and Hardy, C. (1931). Refraction for the peripheral field of view. *AMA Arch. Ophthalmol.* **5**, 717–731.

Fincham, E. F. (1937). "The Mechanism of Accommodation." *Br. J. Ophthalmol. Monograph* Supplement **VIII**, 1–80.

Finkelstein, I. S. (1952). The biophysics of corneal scatter and diffraction of light induced by contact lenses. *Am. J. Optom. Arch. Am. Acad. Optom.* **29**, 185–208, 231–259.

Françon, M. (1963). "Modern Applications of Physical Optics." Interscience Publishers, New York.

Fry, G. A. (1931). The relation of border contrast to the distinctness of vision. *Psych. Rev.* **38**, 542–549.

Fry, G. A. (1946) Blurredness of the retinal image. *Optom. Wkly.* Oct. 31, 1521–1523, 1537.

Fry, G. A. (1949). Factors contributing to the discrepancy between subjective and skiascopic determinations of the refraction of the eye. *O-Eye-O*, **15**, 8–12 (Autumn Quarter).

Fry, G. A. (1953). Targets and testing procedures for the measurement of visual acuity without glasses. *Am. J. Optom. Arch. Am. Acad. Optom.* **30**, 22–37.

Fry, G. A. (1954). A re-evaluation of the scattering theory of glare. *Ill. Eng.* **49**, 98–102.

Fry, G. A. (1955). "Blur of the Retinal Image." Ohio State University Press, Columbus.

Fry, G. A. (1956). Stray light and retinal interaction. *Am. J. Optom. Arch. Am. Acad. Optom.* **33**, 594–601.

Fry, G. A. (1959). *In* "Handbook of Physiology, Neurophysiology I" (J. Field, ed.), pp. 647–670. Amer. Physiol. Soc., Washington.

Fry, G. A. (1962). *In* "Visual Problems of the Armed Forces" (M. A. Whitcomb, ed.), pp. 79–91. Nat. Acad. of Sciences-Nat. Res. Council, Washington.

Fry, G. A. (1965a). Distribution of focused and stray light on the retina produced by a point source. *J. Opt. Soc. Am.* **55**, 333–334.

Fry, G. A. (1965b). The eye and vision. *In* "Applied Optics and Optical Engineering" (R. Kingslake, ed.), Vol. II, pp. 1–76. Academic Press, London.

Fry, G. A. (1969). "Geometrical Optics." Chilton, Philadelphia.

Fry, G. A. (1970). *In* "Progress in Optics." (E. Wolf, ed.), Vol. VIII, pp. 51–131. North-Holland, Amsterdam.

Fry, G. A. (1976). Blur of the retinal image of an object illuminated by low pressure and high pressure sodium lamps. *J. Ill. Eng. Soc.* **5**, 158–164.

Fry, G. A. (1978). *In* "Frontiers of Visual Science" (S. J. Cool and E. A. Smith III, eds.), pp. 253–263. Springer-Verlag, New York.

Fry, G. A., and Alpern, M. (1953). The effect on foveal vision produced by a spot of light on the sclera near the margin of the retina. *J. Opt. Soc. Am.* **43**, 187–188.

Fry, G. A., and Cobb, P. W. (1935). A new method for determining the blurredness of the retinal image. *Trans. Am. Acad. Ophthalmol. Otolaryngol.* pp. 423–428.

Fry, G. A., and Hill, W. W. (1962). The center of rotation of the eye. *Am. J. Optom. Arch. Am. Acad. Optom.* **39**, 581–595.

Fry, G. A., and Hill, W. W. (1963). The mechanics of elevating the eye. *Am. J. Optom. Arch. Am. Acad. Optom.* **40**, 707–716.

Fry, G. A., Treleaven, C. L., and Baxter, R. C. (1945). Specification of the direction of regard (Special Report NO. 1, Committee on Nomenclature and Standards, Am. Acad. Optom.). *Am. J. Optom. Arch. Am. Acad. Optom.* **22**, 351–360.

Fry, G. A., Pritchard, B. S., and Blackwell, H. R. (1963). Design and calibration of a disability glare lens. *Ill. Eng.* **58**, 120–123.

Geeraets, W. J., and Berry, E. R. (1968). Ocular spectral characteristics as related to hazards from lasers and other light sources. *Am. J. Ophthalmol.* **66**, 15–20.

Geeraets, W. J., Williams, R. C., Chan, G., Ham, W. T., Guerry, D., III, and Schmidt, F. H. (1960). The loss of light energy in retina and choroid. *Arch. Ophthalmol.* **64**, 606–615.

Geldard, F. A. (1931). Brightness contrast and Heyman's law. *J. Gen. Psych.* **5**, 191–206.

Gubisch, R. W. (1967). Optical performance of the human eye. *J. Opt. Soc. Am.* **57**, 407–415.

Hamasaki, D., Ong, J., and Marg, E. (1956). The amplitude of accommodation in presbyopia. *Am. J. Optom. Arch. Am. Acad. Optom.* **33**, 3–14.

Hemenger, R. P. (1982). "Optical density of the crystalline lens." Am. J. Optom. and Physiol. Opt., **59**, 34–42.

Holladay, L. L. (1926). Fundamentals of Glare and Visibility. *J. Opt. Soc. Am.* **12**, 271–319.

Hufford, M. E., and Davis, H. T. (1929). The diffraction of light by a circular opening and the Lommel wave theory. *Phys. Rev.* **33**, 589–597.

Hunold, W., and Malessa, P. (1974). Spectrophotometric determination of melanin pigmentation of the human fundus in vivo. *Ophthalmol. Res.* **6**, 355–362.

Ihrig, N., and Fry, G. A. (1953). "Effect of Flashes of Light through the Closed Eyelid," WADC Tech. Rep. 53–159, Part I, Part II. Wright Air Development Center.

Ivanoff, A. (1947). On the influence of accommodation on spherical aberration of the human eye. *J. Opt. Soc. Am.* **37**, 730–731.

Jenkins, F. A., and White, H. E. (1957). "Fundamentals of Optics." McGraw-Hill, New York.

Judd, D. B. (1927). A quantitative investigation of the Purkinje afterimage. *Am. J. Psychol.* **38**, 507–533.

Koomen, M. J., Tousey, R., and Scolnik, R. (1949). The spherical aberration of the eye. *J. Opt. Soc. Am.* **39**, 370–376.

Lamberts, R. L. (1958). Relationship between the sine-wave response and the distribution of energy in the optical image of a line. *J. Opt. Soc. Am.* **48**, 490–495.

Lancaster, W. B. (1943). Terminology in ocular motility and allied subjects. *Am. J. Ophthalmol.* **26**, 122–132.

Laties, A. M., and Enoch, J. M. (1971). An analysis of retinal receptor orientation. *Invest. Ophthalmol.* **10**, 69–77.

Laurance, L. (1926). "Visual Optics and Sight Testing." School of Optics, London.

Le Grand, Y. (1936a) Sur la vision en lumière dirigée. C. R. *Acad. Sci.* **202**, 592–594.

Le Grand, Y. (1936b). Sur deux propriétés de sources de lumière polarisee. *C. R. Acad. Sci.* **202**, 939–941.

Le Grand, Y. (1936c). *Réunions de l'Institute d'Optique.* **7th** Series, pp. 6–11.

Le Grand, Y. (1937). Recherches sur la diffusion de la lumière dans l'oeil humain. *Rev. Opt.* **16**, 201–214, 241–266.

Le Grand, Y. (1938). Sur la fluorescence du crystallin. *C. R. Acad. Sci.* **207**, 1128–1130.

Le Grand, Y. (1948). "Rescherches sur la fluorescence des milieux oculaires." Biofisica, Univ. de Brasil, Rio de Janieriro.

Le Grand, Y. (1957). "Light, Colour and Vision." Chapman and Hall, London.

Le Grand, Y. (1967). "Form and Space Vision." Indiana University Press, Bloomington.

Lerman, S., and Borkman, R. (1976). Spectroscopic evaluation and classification of the normal, aging and cataractous lens. *Ophthalmic Res.* **8**, 335–353.

Ludvigh, E., and McCarthy, E. F. (1938). Absorption of visible light by the refractive media of the human eye. *Arch. Ophthalmol.* **20**, 37–51.

Maxwell, J. C. (1860). On the theory of compound colours, and the relations of the colours of the spectrum. *Philos. Trans. R. Soc. London* **150**, 57–84.

Miller, D., and Benedek, G. (1973) "*Intraocular Light Scattering.*" Thomas, Springfield, *Ill.*

Naylor, E. J., and Stanworth, A. (1954). Retinal pigment and the Haidinger effect. *J. Physiol.* **124**, 543–552.

Optical Society of America Committee on Colorimetry. (1953). "The Science of Color," p. 297. Crowell, New York.

Østerberg, G. A. (1935). Topography of the layer of rods and cones in the human retina. *Acta. Ophthalmol.* Suppl. **VI**, pp. 1–103.

Parent, G. B., and Thompson, B. J. (1969). "Physical Optics Notebook." Soc. Photo-Opt. Instr. Eng., Redondo Beach.

Perrin, F. H. (1960). Methods of appraising photographic systems. *J.S.M.P.T.E.* **69**, 151–156, 239–249.

Pokrowski, G. L. (1926). Über die Lichtzerstreuung im Auge. *Z. Phys.* **35**, 776–782.

Riggs, L. A. (1965). *In* "Vision and Visual Perception" (C. H. Graham, ed.), pp. 341–345. J. Wiley, and Sons, Inc., New York.

Röhler, R. (1962). Die Abbildungseigens-eigenschaften der Augen medien. *Vision Res.* **2**, 391–429.

Ruddock, K. H. (1963). Evidence for macular pigmentation from colour matching data. *Vision Res.* **3**, 417–429.

Ruddock, K. H. (1965). The effect of age upon colour vision. *Vision Res.* **5**, 47–59.

Rushton, W. A. H. (1958). The cone pigments of the human fovea in colour blind and normal. *In* "Visual problems of colour" (Nal. Phys. Lab. Symposium 8). Vol. **I**, pp. 71–101. H.M.S.O., London.

Rushton, W. A. H. (1965). Stray light and the measurement of mixed pigments in the retina. *J. Physiol.* **176**, 46–55.

Said, F. S., and Weale, R. J. (1959). The variation with age of the spectral transmissivity of the living human crystalline lens. *Gerontolgia.* **3**, 213–231.

Satoh, K., Bando, M., and Nakajima, A. (1973). Flourescence in human lens. *Exp. Eye Res.* **16**, 167–172.

Schelling, H. von (1950). A method for calculating the effect of filters on color vision. *J. Opt. Soc. Am.* **40**, 419–423.

Schober, H. A. W., and Fry, G. A. (1968). The role of the pupil in disability glare measurements. *Vision Res.* **8**, 1107–1122.

Schouten, J. F., and Ornstein, L. S. (1939). Measurements on direct and indirect adaptation by means of a binocular method. *J. Opt. Soc. Am.* **29**, 168–182.

Sheard, C. (1919). Diffraction in the human eye and the phenomena of colored rings surrounding luminous sources. *Am. J. Ophthalmol.* **2**, 185–194.

Siew, E. L., Bettelheim, F. A., Chylack, L. T., Jr., and Tung, W. H. (1981). Studies on Human Cataracts II. *Invest. Ophthalmol. Vision Sci.* **20**, 334–347.

Simpson, G. C. (1953). Ocular halos and coronas. *B. J. Ophthalmol.* **37**, 450–486.

Smith, W. J. (1966). "Modern Optical Engineering," pp. 314–318. McGraw-Hill, N.Y.

Southall, J. P. C. (ed.) (1924). "Helmholtz's Treatise on Physiological Optics" (Translated from the 3rd German edition). Opt. Soc. of Am., Rochester.

Southall, J. P. C. (1933). *Mirrors, Prisms and Lenses* (3rd edition). MacMillian, New York.

Spring, K. H., and Stiles, W. S. (1948). Variation of pupil size with change in the angle at which the light strikes the retina. *Br. J. Ophthalmol.* **32**, 340–346.

Stiles, W. S. (1929). The scattering theory of the effect of glare upon the brightness difference threshold. *Proc. R. Soc. B.* **105**, 131–141.

Stiles, W. S. (1978). "Mechanisms of Colour Vision." Academic Press, London.

Stuart, R. D. (1966). "An Introduction to Fourier Analysis." Science Paperbacks, London.

Toraldo di Francia, G., and Ronchi, L. (1952). Directional scattering of light by the human retina. *J. Opt. Soc. Am.*, **42**, 782–783.

Tscherning, M. (1924). "Physiologic Optics." Keystone, Philadelphia.

Vilter, V. (1949). Biometric research on the synaptic organization of the human retina. *Societe Biologie* **143**, 830–832.

Vos, J. J. (1962). "On Mechanisms of Glare." Thesis, Institute for Perception, RVO-TNO, Soesterberg, Netherlands.

Wald, G. (1945). Human vision and the spectrum. *Science* **101**, 653–658.

Wald, G. (1949). The Photochemistry of Vision. *Doc. Ophthalmol.* **3**, 94–134.

Wald, G., and Griffin, D. R. (1947). Change in refractive power of the human eye in dim and bright light. *J. Opt. Soc. Am.* **37**, 321–336.

Walls, G. L. (1942). "The Vertebrate Eye." Cranbrook Institute of Science, Bloomfield Hills, Michigan.

Walraven, J. (1973). Spatial characteristics of chromatic induction; the segregation of lateral effects from stray light artifacts. *Vision Res.* **13**, 1739–1753.

Weale, R. A. (1954). Light absorption by the lens of the human eye. *Opt. Acta.* **5**, 107–110.

Westheimer, G. (1960). Modulation thresholds for sinusoidal light distributions on the retina. *J. Physiol.* **152**, 67–74.

Wilcox, W. W. (1932). The basis of the dependence of visual acuity on illumination. *Proc. Nat. Acad. Sci.* **18**, 47–56.

Wright, W. D. (1951). The visual sensitivity of normal and aphakic observers in the ultra-violet. *Anné Psychol.* **50**, 169–177.

Wyszecki, G., and Stiles, W. S. (1982). "Color Science." (2nd ed.) J. Wiley and Sons, New York.

3

The Eye as a Detector

G. A. FRY
College of Optometry
The Ohio State University
Columbus, Ohio

A. INTRODUCTION

Although we speak of the human eye as a receptor, we must realize that the sensitive layer of the retina is a mosaic of isolated receptors that respond individually to the light falling on them. Although the retina contains millions of these receptors, the eye has only one optical system which collects the quanta transmitted through the pupil and concentrates them on this or that part of the retina. We have to think in terms of the number of incident quanta per unit of area of the retina and the number being absorbed by each of the photoreceptors in that area. The absorbed quanta somehow generate potentials that are relayed to trains of transmitting elements in the retina which interact with each other. The disturbances initiated in the photoreceptors eventually generate nerve impulses in the axons of the ganglion cells at certain frequencies. These axons run across the surface of the retina to the

ISBN-0-12-304905-9

optic nerve head and become the fibers of the optic nerve. A cross section of the optic nerve represents a mosaic of elements that behave independently of each other in exactly the same way as the receptors.

We must explain how the pattern of impulses in a set of optic nerve fibers is related to the quanta being absorbed by the retinal receptors and how such a pattern of impulses can generate the perceived impressions of a border, a gradient, or a uniform patch of color.

At the outset we cover the problems of retinal optics, which deal with getting the light that forms the optical image funnelled into the photoreceptors where some of it is absorbed by the photopigment. We discuss the absolute and differential thresholds for rod and cone vision, the luminous efficiencies of the rods and cones and the different types of photopigments, the transmission of impressions through the retina, and the mechanisms of light and dark adaptation.

In describing the eye as a detector, we limited ourselves to mechanisms in one retina. The approach does not take into account that a normal human has two eyes and simultaneously processes information from both. The discussion in this chapter applies to a one-eyed human or a human with an opaque occluder over one eye.

We do not deal with the perception of such things as distance, direction, and motion. We do not cover the problems of color, because these problems are covered in another chapter, nor do we cover the response of the eye to intermittent and asynchronous stimulation.

Appreciation is expressed to the authors and publishers who have given permission to duplicate their figures. The source of each figure is indicated at the end of the figure legend.

B. OPTICS OF THE RETINA

The interior limiting membrane provides a smooth interface between the vitreous and the retina, which reflects light that can be seen as a sheen when the fundus is observed with an ophthalmoscope. The layers of the retina between the internal limiting membrane and the outer segments of the rods and cones do not represent a homogeneous medium. What the retina does to the light transmitted through it, aside from simple scatter, has not yet been assessed in detail. Eventually, part of the light reaches the outer limiting membrane where some of it enters the inner segments of the receptors and some of it enters the spaces in between. We have to be concerned about how the light that enters the receptors is transmitted through them and how quanta get absorbed by the photopigment.

1. Directional Sensitivity

One of the major aspects of retinal optics is that the efficiency of a photoreceptor in absorbing quanta is dependent on the direction from which a beam of light approaches the retina.

This effect can be demonstrated with the apparatus (Fig. 1) used by Stiles and Crawford (1933). One beam from A_1 is focused at the center of the pupil and enters as a circular beam 0.75 mm in diameter. The circular diaphragm at D_1 is imaged on the retina and subtends a visual angle of 1°. A second beam from A_2 enters the periphery of the pupil as a circular beam 0.5 mm in diameter. The diaphragm at D_2 is imaged on the retina so that it overlaps the image of D_1. The two beams can be alternated, and the retinal illuminances produced by the two beams can be adjusted to make a match by flicker photometry. Figure 2 shows the relative luminous efficiency of the peripheral beam for various displacements from the center of the pupil. The relative efficiency of the peripheral beam in stimulating the photoreceptors depends on the angle of incidence. When the point of entry is 4 mm from the center, the beam approaches the retina at a 10° angle from the central beam. It should be noted that the arrangement involves coherent sources focused in the plane of the pupil so that each beam as it approaches the retina has a wave front normal to its chief ray.

A review of the literature on the Stiles–Crawford (S–C) effect has been presented by Vos and Walraven (1962).

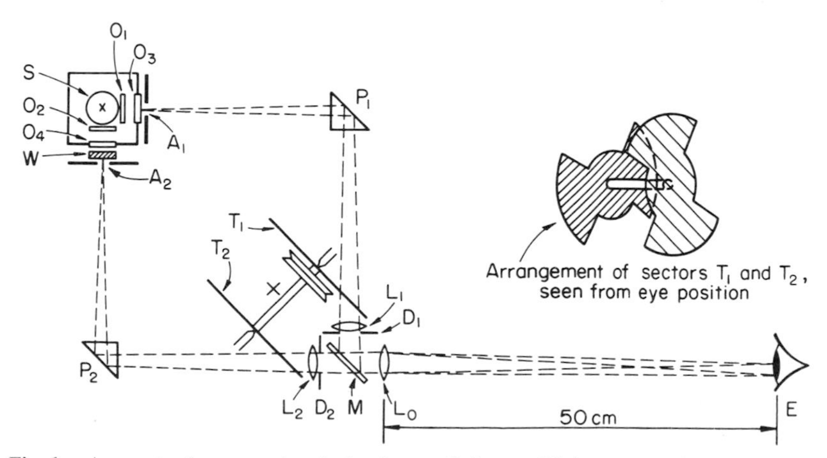

Fig. 1. Apparatus for measuring the luminous efficiency of light rays entering the eye pupil at different points (Stiles and Crawford, 1933).

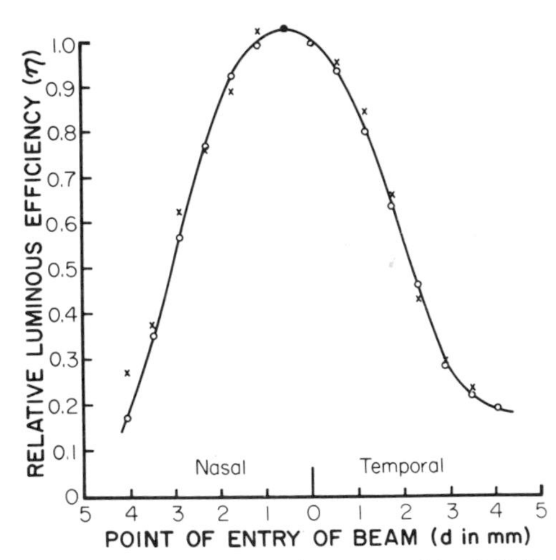

Fig. 2. The Stiles–Crawford effect. Horizontal traverse, subject B.H.C., ○ right eye, × left eye (Stiles and Crawford, 1933).

a. Type of Receptor

It is to be expected that the directional sensitivity is related to the structure of the photoreceptors as well as to the angle of incidence. In particular, attention can be called to three kinds of photoreceptors: cones in the fovea, cones in the periphery, and rods. These three types are illustrated in Fig. 5 in Chapter 2. The myoid portion of the rod is cylindrical, whereas that of a peripheral cone is cone-shaped. The cones in the fovea are more nearly like the rods. In terms of wave-guide theory, length of taper compensates for angle of taper.

The original experiments of Stiles and Crawford involved stimuli that were 1° in diameter centered on the foveal line of sight, and the response was a kind of average of the responses of all the receptors to which the stimulus was applied.

It is possible to isolate the three kinds of receptors. One may confine the stimulus to the rod-free region to test the foveal cones; low levels of stimulus intensity may be applied to the parafovea to isolate rods; or the rods may be bleached with blue light and high levels of intensity applied to the parafovea to isolate peripheral cones (Crawford, 1938; Flamant and Stiles, 1948).

b. Alignment with the Center of the Pupil

In demonstrating the directional sensitivity of the receptors, we use large areas of the retina and measure an average effect, but ultimately we have to assess the effect in terms of the individual receptors and the orientation of their axes relative to the center of the pupil. It should be noted that the most sensitive point of entry of a beam is generally at the center of the pupil, but there are some exceptions. O'Brien (1950) showed that by vibrating the point of entry it was possible to demonstrate for localized regions of the retina that the direction of maximum sensitivity varies from one region to the next. Enoch's study (1957) of amblyopes showed that these subjects had a center–marginal difference in sensitivity that was less than for a normal subject, and the point of maximum sensitivity was considerably displaced.

Bedell and Enoch (1979) showed that the receptors in the periphery of the retina 35° from the fovea were on the average aligned with the center of the exit pupil, although this makes their axes oblique to the surface of the retina. See also Laties and Enoch (1971) and Enoch and Laties (1971).

Enoch *et al.* (1979) occluded an eye for about a two-week period and showed that the directional effect was greatly reduced. Also, using a displaced aperture, they showed that the peak of the S–C function shifted toward the aperture. Thus there seems to be some gradual disarray in the dark, but the system favors alignment with the source of light in the pupillary aperture. When the eye is re-exposed to light admitted through the normal pupil, the normal directional sensitivity returns.

c. Response to Stray Light

Directional sensitivity is important from points of view other than that of light entering the eye through the pupil. Light scattered by the media can approach the retina with much larger angles of incidence. Light transmitted through the iris and sclera and light reflected by the retina can approach other parts of the retina with very large angles of incidence.

Part of the light transmitted through the sclera, choroid, and retina from the outside of the eye passes through the receptors parallel to their axes but in the opposite direction from normal.

d. Response to Monochromatic Stimuli

The directional sensitivity was originally measured with white light but has now been investigated with various monochromatic lights. It has been found for normal vision directional sensitivity is about the same for all parts of the spectrum, except that it is somewhat more pronounced at the center of the spectrum (Stiles, 1939).

e. The S–C Effect of the Second Kind

If monochromatic light is used to demonstrate the Stiles–Crawford effect, and the beams are compared directly, one beam will not be perceived as having the same hue as the other, but they can be matched by using different wavelengths (Vos and Walraven, 1962). The implication is that the effect is not the same for different cone types.

f. Experiments with Models

O'Brien (1951) postulated that the directional sensitivity was dependent on the loss of light by transmittance through the walls of the receptor and its failure to reach the pigment in the outer segments. The conical shape of the myoid portion of the receptor and the index inside and outside the receptor are important for this theory. He used a polystyrene foam cone to simulate a human cone.

Enoch and Fry (1958) have designed a similar cone, shown in Fig. 3. It is larger than a human cone by a factor of 70,000. The large end of the cone is 16.42 cm in diameter. At the point that it fits into the cone support the diameter is 5.47 cm. The apex angle is 7°18″. The polystyrene foam cone has an index of refraction of 1.0111 for a wavelength of 3.20 cm, whereas the index of an actual cone relative to the interstitial fluid is 1.032. Microwave radiation with a wavelength of 3.2, and also with a wavelength of 2.42, were used to illuminate the cone. According to the scale model of 70,000 to 1, 3.2 cm corresponds to a wavelength of 460 nm.

Provision was made to illuminate the cone with a point source at 8 m, which generates a spherical wavefront. With this arrangement it was shown that the cone does display directional sensitivity. Measurements were made with the foam cone in place and with the cone support only, and with the unit pointed toward the source and at various angles to the right and left.

In Fig. 4 the directional sensitivity of a polystyrene foam cone is compared with that of human cones.

The overall response for a wavelength of 2.41 cm was better than that for 3.20 cm, which raises the question of whether the configuration of a photoreceptor affects its luminous efficiency for the different wavelengths. It would appear that the response of the unit depends on the angle between the axis of the unit and the normal to the wave front.

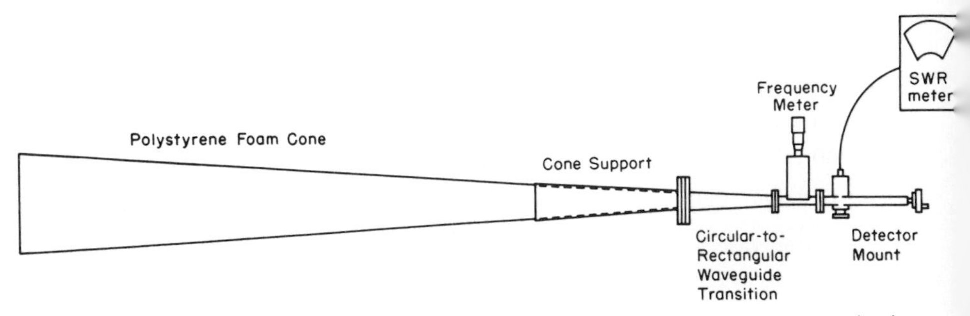

Fig. 3. Simulation of a human cone. Detector unit, polystyrene cone, cone support, circular-to-rectangular wave-guide transition, frequency meter, and tunable detector mount (Enoch and Fry, 1958).

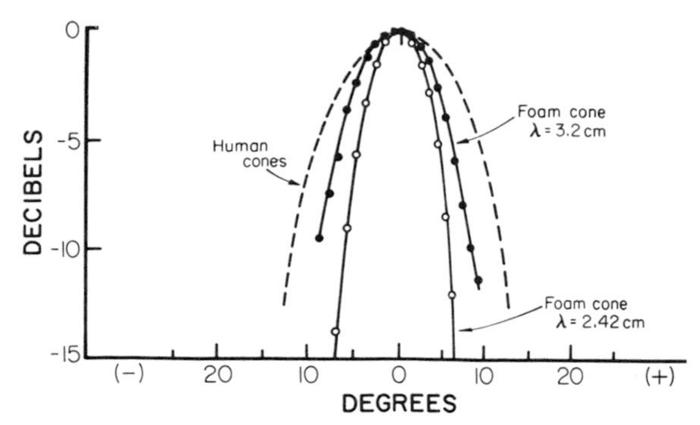

Fig. 4. A typical retinal directional sensitivity function (Stiles–Crawford effect). Data from model superimposed for comparison with maxima equated. (Enoch and Fry, 1958.)

g. Additivity of Separate Beams

These findings support the notion that the simultaneous responses to two or more point sources will be additive as long as the point sources are independent coherent sources.

This forms the foundation for the practice of making an allowance for reduced efficiency of light entering the periphery of the pupil in stimulating the retina. The theory is sound as long as we use the method of viewing illustrated in Fig. 5. The transilluminated plate of milk glass is focused in the plane of the pupil, and the pupil is uniformly illuminated. The image in the pupil represents an array of point sources. The aperture at M, which is centered on the primary line of sight, is conjugate to the retina. The light reaching a given part of the retina from a given part of the pupil can simply be added to the light coming from other parts of the pupil. But if we want to evaluate the effectiveness of the light in having quanta absorbed by the photopigment, we have to divide the pupil into zones and for each zone make allowance for the reduced efficiency based on the angle of incidence.

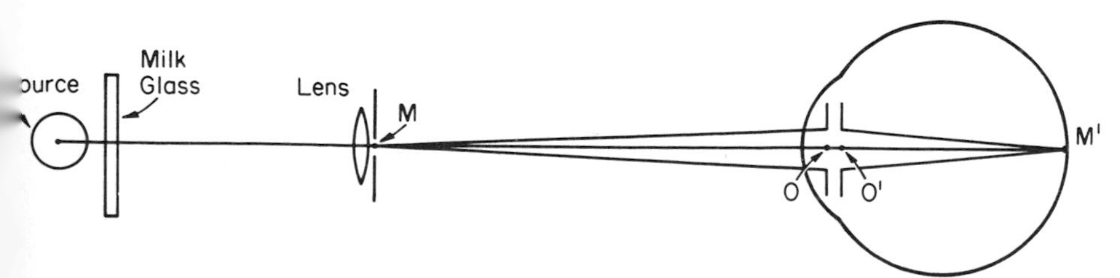

Fig. 5. Arrangement for producing an array of point sources in the plane of the exit pupil. The milk glass is conjugate to the plane of the entrance pupil.

This can be done by expressing the retinal illuminance in terms of *effective trolands*. The retinal illuminance E'_R expressed in this way is given by

$$E'_R = 2\pi L \int_0^{\bar{r}} \eta r \, dr, \tag{1}$$

where L is the luminance of the stimulus, $\bar{r}$ is the radius of the pupil, r is the distance (mm) from the center of the pupil, and η is the relative efficiency.

According to Stiles and Crawford (1933),

$$\eta = e^{-2.3ar^2}, \tag{2}$$

where a is of the order of 0.05.

According to Moon and Spencer (1944),

$$\eta = 1 - 0.085r^2 + 0.002r^4. \tag{3}$$

It may be noted that when η is unity for each zone of the pupil, Eq. (1) reduces to

$$E'_R = AL \tag{4}$$

where E'_R is expressed in trolands and A represents the area (mm^2) of the pupil.

h. The Stiles–Crawford Effect in Ordinary Seeing

The theory of reduced response because of oblique incidence does not apply to the case of an image of a point source focused on the retina as illustrated in Fig. 6. This is ordinary seeing. Although the energy converging on the retina approaches the retina obliquely from different parts of the pupil, it is collected to form a Fraunhofer image. At the retina the wave front of the beam is flat and parallel to the external limiting membrane. At any part of the image the energy flows through the outer segments parallel to their axes. There is no basis for a loss because of obliquity of incidence.

Drum (1975) has used annular pupils to study this problem and has found that contrary to the theory of reduced response there is loss in the situation

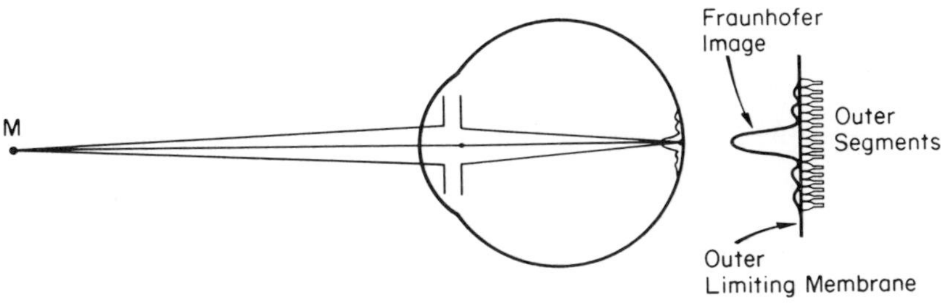

Fig. 6. Arrangement for producing the image of a coherent point source on the retina.

depicted in Fig. 6 which can be attributed to obliquity. Because of the uncertainty in reconciling theory and experimental findings, it is recommended that small artificial pupils be used to control the size of the beam entering the eye whenever an attempt is made to relate retinal response to stimulus intensity.

The method of viewing illustrated in Fig. 6 can be simulated with the microwave model shown in Fig. 3. Enoch and Fry (1958) constructed an elliptical mirror 1 m in diameter, placed a microwave source at the near focus, and formed a Fraunhofer image at the second focus 8 m from the mirror. The margin of the mirror corresponds to the margin of the exit pupil of the eye, and the Fraunhofer image corresponds to the image formed on the retina.

They demonstrated that the wave front at the Fraunhofer image is flat and normal to the axis of the beam and that no matter at what part of the image the mouth of the cone is placed, the maximum response occurs when the axis of the cone is normal to the wave front.

i. THEORIES OF THE STILES–CRAWFORD EFFECT

O'Brien (1945) formulated a theory of the Stiles–Crawford effect on the basis of the obliquity of the rays entering the cones and the leakage of energy when a ray is reflected at the walls of the ellipsoidal portion of the cones. This applies in the case of spherical wave fronts originating at points in the plane of the pupil but does not apply to flat wave fronts parallel to the retina.

Toraldo di Francia (1949) called attention to the fact that the diameter of the ellipsoid portion of the cone at the large end is somewhat less than the center-to-center separation of two cones, which is about 2.5 μm (about 5 wavelengths of green light), and hence, the receptors probably act as waveguides. Enoch (1960b, 1961) has explored this notion and has shown that some of the properties of receptors as waveguides show up in pictures of transilluminated excised retinas. Length of taper compensates angle of taper.

2. Blur Related to the Optics of the Retina

a. BLUR PRODUCED IN THE RETINA

Ohzu *et al.* (1972) excised the retinas of a squirrel and of a rat, placed each retina between two plates of glass, and focused an image of a grating on it. It was re-imaged on the plane of a scanning slit, which makes it possible to assess the modulation transfer factor for various frequencies. The data indicate that the transmission through the retina degrades the image, but it is difficult to estimate the effect on the distribution of flux applied to the outer segments.

It is not proper to try to assess the combined degradation of retinal optics and the image-forming mechanism by multiplying the two transfer functions.

It is probably best to sidestep the problem of disentangling the blur produced by the retinal structures from the blur produced by the image-forming mechanism of the eye. The effect of the retina can be included as part of the effect produced by the image-forming mechanism of the eye; then one can try to disentangle this combined effect from the blur produced by the neural irradiation, which occurs while the image is being transmitted through the retina and to the brain. The fact that the responses generated in the receptors are completely independent of each other makes it possible to regard these two parts of the system as independent blurring mechanisms that operate in tandem, and consequently the modulation transfer function of the whole system is the product of the functions for the two stages.

b. Blur Produced by Obliquity of Incidence at the Retina

Campbell (1958) has demonstrated that the sine-wave grating produced by a pair of slits in the periphery of the pupil is clearer when the slits are oriented in a radial direction rather than in the tangential direction. The path of the chief ray through the retina is oblique, but the significance of this finding is not clear.

c. Amblyopia and Retinal Optics

Enoch (1957, 1959, 1960a) has studied the relation between tilt of the photoreceptors and the quality of the image transferred to the brain of an amblyope.

d. The Effect of the Stiles–Crawford Effect on Blur

For an image of a point out of focus, the energy transmitted through the periphery of the pupil is concentrated in the periphery of the image. The same is true in the case of spherical aberration. If the rays transmitted through the periphery of the pupil are less effective in stimulating the retina, the amount of blur should be reduced. It has the same effect as reducing the size of the pupil or covering the pupil with a filter that is graded in density from the center out.

3. Self-Screening by the Photopigment

Electron microscopy reveals detail about the structure of an outer segment. It is a stack of platelets with the molecules of photopigment arranged in layers perpendicular to the axis of the receptor. The long axes of the individual molecules are parallel to the axis of the receptor. This means that a fair amount of self-screening occurs, which affects the relative luminous efficiency of the different wavelengths. Self-screening refers to the fact that when a substance has a high absorptance for a given wavelength, the opportunity for additional absorption becomes increasingly less as the beam penetrates further into the substance. The absorptance a is no longer proportional to the depth of penetration or the length of the path. Instead of a being proportional to the length of the path as in the following equation,

$$a = \alpha x, \tag{5}$$

$$a = 1 - e^{-\alpha x}, \tag{6}$$

where α is the attentuation coefficient, and x is the length of the path.

C. THRESHOLDS

1. Absolute Threshold for Rods

a. Minimum Energy Required for a Response

To assess how good the eye is at detecting radiant energy falling on the retina, we must know the least amount of energy required to produce a visible response.

Hecht *et al.* (1942) researched this problem by dark-adapting the retina and then exposing a spot, 10 min in diameter located 20° in a temporal direction from the fixation point, to flashes of green light (510 nm) 0.001 sec long. At each of several irradiances they determined the percentage of visible flashes and plotted the data. The data werc fitted with a Poisson probability curve as shown in Fig. 7, where the percentage of flashes seen is plotted against the logarithm of the average number of quanta per flash at the cornea.

The Poisson formula is

$$p_{na} = 1 - \sum_{i=0}^{n-1} \frac{a^i}{e^a i!}, \tag{7}$$

where a is the average number of quanta per flash, and p_{na} is the probability that at least n quanta will be absorbed. In Fig. 7 the value of n is 6, and 60% seeing occurs when the average number of quanta per flash measured at the cornea is 129. This represents the threshold of seeing. The average quanta per flash m at the cornea may be computed from the energy (ergs) per flash U as

$$m = U/h\nu, \tag{8}$$

where h is Plank's constant, and ν is the frequency of light waves per sec.

The results obtained over a period of months on seven subjects showed that the energy threshold ranged from 54 to 148 quanta at the cornea. Of the quanta incident at the cornea about 4% are lost by reflection at the cornea,

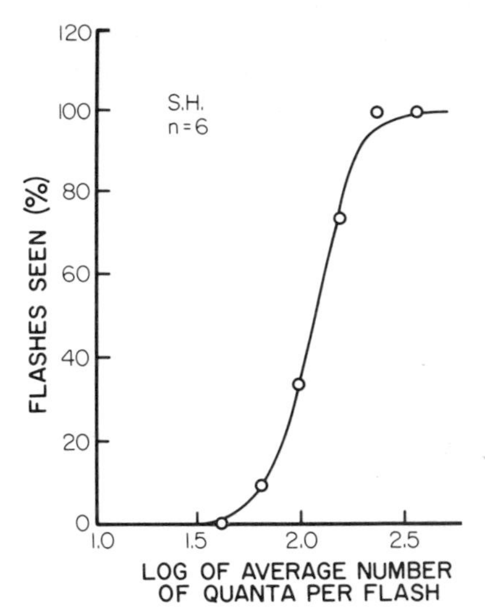

Fig. 7. Relation between the average number of quanta per flash and the frequency of seeing for subject S. H. (Bartlett, 1965).

about 50% are absorbed or scattered by the media, and it was estimated that only about 9.6% are absorbed by the rhodopsin. This means that about 5 to 14 quanta are absorbed per flash.

The 10-min diameter area contains about 500 rods, and the chance that any one rod absorbs more than one quantum is slim. Hence, seeing depends upon the responses of at least five rods to the absorption of one quantum each.

Bouman and van der Velden (1947) used smaller areas and deduced that two quanta can suffice for seeing. The two quanta must fall on photoreceptors separated by no more than 10 min of arc and no less than 0.10 sec apart in time. This is known as the two-quanta hypothesis.

b. The Effect of Area

Let us consider for the moment the case of flashes so short that the time interval between the two absorptions is short enough to get the full effect of temporal summation. If two absorptions occur at points 10 min apart, the summation effect halfway between must be larger than the neural response generated at either of the two receptors and large enough to drive a response through the retina.

If the two absorptions occur more than 10 min apart, the summated effect at the midpoint is subthreshold. If three absorptions occur simultaneously at the three corners of a triangle, a response will occur when any leg is shorter than 10 min or if the summation of the three at any point is greater than the threshold. If we wanted to formulate a general theory about thresholds for disk-shaped areas, we should assume a random distribution of absorptions and then assume that at some point within the area there must be a summated effect that exceeds the amount required for threshold. As long as the diameter is less than 10 min, it is a matter of the chance of two absorptions during the flash, and a 60% chance of two absorptions would be strictly proportional to the total energy. Hence, the threshold of radiance would be inversely proportional to the area. This reciprocity would fail for larger flashes. The size of the disk at which the threshold energy ceases to drop as the area increases is dependent on the neural spread of excitation and the summation of excitation.

If the area is increased until we include the whole field of view and if the stimulus intensity is gradually increased, the observer should see an array of small patches that increases in number as the radiance is raised. Eventually the frequency of absorptions per unit area becomes nearly uniform.

Figure 8 shows the data of Graham and Bartlett (1939) for the effect of area of a disk on the absolute threshold. Red and blue disks were used to make certain that cones were not involved. The luminances are given in photopic units for red light; the values for blue are lower by 3.3 log units for subject M and lower by 3.1 log units for subject G. The largest stimulus used has a diameter of 10°. The center of the disk was 15° from the center of the fovea.

Figure 9 shows the data of Shlaer (1937) for visual acuity measured with a square-wave grating with white and dark bars of equal width. The grating bars were vertical and covered an area 4° in diameter. At the lowest intensity level (0.0037 trolands), which approximates the absolute threshold for rods, bars as wide as 14 min will fuse to form a uniform field. Hence, if a large region of the retina were uniformly stimulated at a level of about 0.0037 troland, the perceived patch of light should be uniform even if the activity in the retina were splotchy. As the luminance of a coarse grating is decreased, the perceived

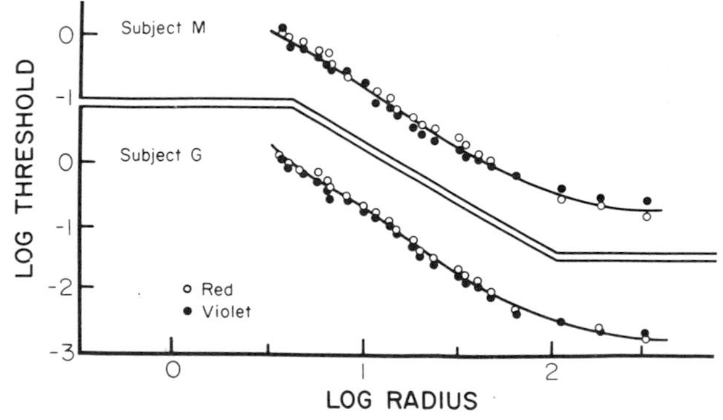

Fig. 8. Thresholds for red and for violet light in the periphery. The data for violet have been displaced upward as described in the text (Graham and Bartlett, 1939).

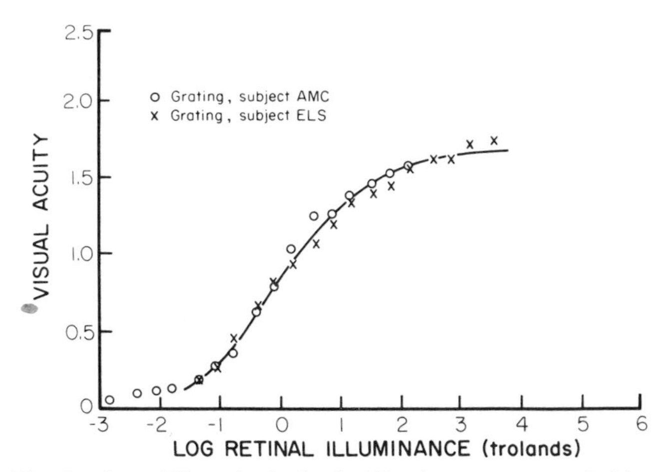

Fig. 9. Visual acuity at different levels of retinal illuminance as measured with resolution of a grating (Shlaer, 1937). Reproduced from *The Journal of General Physiology*, 1937, **21**, 165–188 by copyright permission of the Rockefeller University Press.

image becomes splotchy. Then these nonuniformities get washed out in the same way as for a uniform stimulus.

c. Temporal Summation

If we confine the flash to a small disk-shaped area 10 min in diameter and vary the duration, the energy required for threshold will vary inversely as the duration up to 0.1 sec, but beyond this the failure to reach the threshold will depend on the chance of not getting two absorptions less than 0.1 sec apart. As the duration continues to increase, the threshold flux will reach its minimum value when the chance of never getting two absorptions less than 0.1 sec apart becomes extremely high.

When the stimulus becomes continuous and the flux is near the threshold level, the occurence of two absorptions less than 0.1 sec apart happens only occasionally. Thus the visibility should be intermittent, and the 10-min spot should flicker.

If we now increase the luminance level, the rate of absorptions will increase, and the periods of visibility will fuse into each other. The question we must face is the rate of absorptions at which the fluctuations will fuse.

We can investigate this problem by using a random sequence of short flashes with a sizeable but nearly equal number of quanta in each flash. In order to create this kind of stimulus we have constructed the apparatus shown in Fig. 10. The random sequence was generated by drawing a spiral on a metal disk F and drilling along the three cycles of the spiral. The radius of the spiral gradually increased from 18.5 to 21.5 cm. The holes were 3.175 mm in diameter. Numbers were assigned to points 1° apart on the spiral, and holes

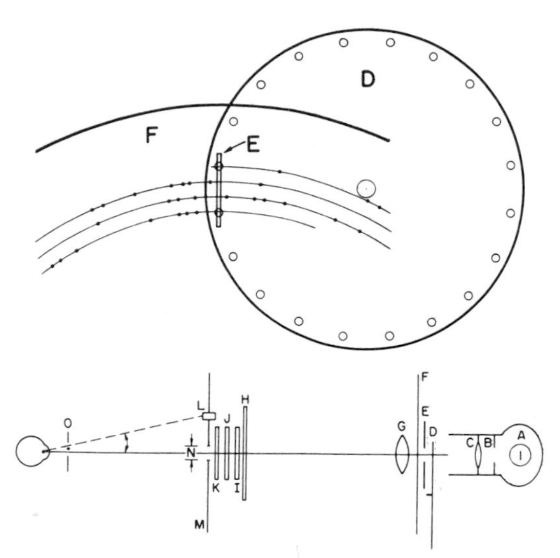

Fig. 10. Arrangement for generating a nonuniform sequence of flashes.

were drilled at those points corresponding to the sixes in a table of random digits. The total number of holes was 125.

This is slightly more frequent than one hole for each ten points, which is what the average would be if the sequence were long enough. The talbot intensity of the stimulus was computed on the assumption of an average number of 36 holes per turn of the disk. The disk was rotated in front of a wedge-shaped slit D designed so that as the holes move past the slit, a constant amount of energy is transmitted through the slit for each hole. A second rotating disk E with a series of larger holes determines which row of holes in the spiral is used during a given rotation of the big disk. Each set of 125 exposures is followed in tandem by a second set of exposures that is a repeat of the first. The pulses of light transmitted through the slit and rotating apertures illuminate a piece of milk glass K which covers a disk-shaped hole N in the screen M. This hole is 1-in. in diameter and subtends an angle of 3° at the eye. The eye views the disk through an artificial pupil 5 mm in diameter in the plane O. The eye fixates a small dim red fixation light at L which is 6.3° from the center of the disk. The source A is a ribbon filament driven by direct current and is conjugate to the slit at E. The aperture B is conjugate to the milk glass K. J is an interference filter with a peak transmittance at 555 nm. I is a fixed piece of Polaroid, and the Polaroid H is rotated to control the luminance. The front side of the screen at M can be illuminated briefly to keep the pupil of the eye centered on the artificial pupil in the screen at O.

When the eye was dark adapted, measurements of the absolute threshold were made at various speeds of the big disk. At slow speeds the spot flickers, but at higher speeds it becomes steady. The speed at which fusion occurs was also measured at various luminance levels above the threshold.

The results for the left eye of subject G.F. are shown in Fig. 11a. The same apparatus was used to study fusion with a steady rate of flashes. The spiral with random spacing and the small disk were replaced by a single disk having a circular row of 36 holes uniformly spaced. Each hole was 20 cm from the center of rotation. The results are shown in Fig. 11b. The transition from flicker to fusion is much smoother and occurs at a lower speed.

The spontaneous grouping of holes in the random array generates large peaks and troughs that survive an increase of speed that would wash out the small troughs and peaks that occur with uniform spacing. Eventually, temporal spreading washed out all the peaks and troughs.

In spite of the fact that the number of quanta per flash is quite large, the experiment shows that the retina will fuse random flashes near the threshold at an average rate of 29 flashes per sec. In the case of a 10-min disk, a 29 per sec average rate of absorptions would have the same kind of temporal arrangement as the set of random flashes, and this would represent the level at which the stimulus could be considered to be equivalent to a steady input. We now

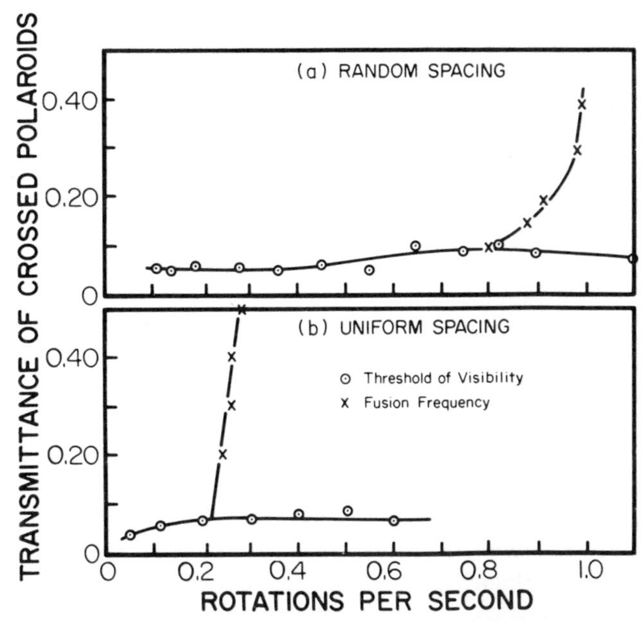

Fig. 11. Thresholds and fusion frequencies for random and uniform flashes. The average number of flashes per rotation is 36. When the Polaroid transmittance is 1.00, the talbot intensity is 0.0642 photopic trolands. When the speed is one rotation per second, each flash delivers on the average 0.87 quanta per square minute.

need to know what level of steady retinal illuminance would cause the average rate of absorptions to be 29 per sec in a 10-min spot and how this is related to the threshold of visibility.

According to Fig. 11b the Polaroid setting for threshold at high frequencies is about 0.07, and the talbot intensity of the stimulus is 0.0045 photopic trolands. It may be assumed that this is equivalent to steady stimulation. According to Graham and Bartlett the threshold for a 10-min disk is higher by a factor of about 143 than the threshold for a 3° disk; it may be assumed, therefore, that the threshold for a 10-min disk ($\lambda = 555$) is about 0.46 photopic trolands. At this level the 10-min disk would generate a flow of about 12500 quanta per sec into the eye. According to Hecht, Shlaer, and Pirenne (see Section C, 1, a) only 9.6% of the quanta incident at the cornea are absorbed by the rods; hence, the rate of absorptions in the 10-min area would be about 1200 per sec. The rate of absorptions at the threshold level required for fusion of the nonuniformities is only about 29 per sec; hence, the nonuniformities can be ignored for continuous exposure.

This explains why steady stimulation of a 10-min spot fails to produce flicker at the absolute threshold. Furthermore, if the mechanism for temporal fusion is able to wash out temporal nonuniformities in a given part of the retina, then in the case of a large stimulus or ganzfeld these nonuniformities will also be washed out in adjacent areas, which will lead to spatial uniformity. If there were any tendency to spatial nonuniformity, it would be washed out by the neural spread of excitation or the mechanism in the retina for equalizing the responses in adjacent regions of the retina. Hence, it is only in the case of very short flashes that we have to be concerned about the noise of the stimulus in trying to explain the threshold.

d. Retinal Gray

With the entire field of view dark the eye can see what is called retinal (or idioretinal) gray, which is due to spontaneous discharges in elements in the retina above the level of rods and cones. This is a noisy background but is not strong enough to affect the visibility of the signal generated by two absorptions in the rods.

If the eye is bright-adapted to a large disk of, say, about 1000 trolands and then placed again in darkness, an afterimage is seen in the form of a dark disk surrounded by retinal gray. The adaptation stimulus does not enhance the spontaneous discharges in the surrounding retina but depresses the activity in the area stimulated. This not only blocks the retinal gray signals from reaching the cortex but also blocks the responses to new stimuli applied to the photoreceptors. This effect involves neural adaptation, because the stimulus intensity is not strong enough to produce a positive afterimage or to bleach a significant amount of the photopigment.

If a very bright adapting stimulus is used, it not only produces neural adaptation but also bleaches a high percentage of the rhodopsin and produces a positive afterimage. The afterimage creates a veiling patch that acts like a background and converts the problem of the absolute threshold to one of a difference threshold. The depletion of the photopigment directly affects the threshold, because it takes more flux at the level of the cornea to produce the same rate of absorptions. Also, the free opsin generates a negative signal that raises the threshold.

The photopigment does not recover fully until after neural adaptation recovers and the positive afterimage disappears, and hence, during the later stages of recovery the negative signal generated by free opsin and the loss of sensitivity produced by the reduced amount of photopigment completely control the absolute threshold (Fry, 1965).

e. The Filling-In Phenomenon

An interesting case of the absolute threshold is that of a bright disk on a dark background that is centered on the primary line of sight and is large enough for its border to fall outside the rod-free area. In this case the visibility is dependent on the rods near the edge of the disk. The surprising thing is that the center of the disk is seen as uniform and bright as the peripheral parts of the disk. This involves a spread of excitation at the level of the bipolars or ganglion cells that fills in and covers up the fact that the cones near the fixation point are blind at this level of stimulation.

The demonstration can be made more striking by placing a dark disk at the center, which changes the stimulus to an annulus. The dark disk is not seen, and the annulus is seen as a uniform disk.

2. Differential Threshold for Rods

In order to study differential thresholds for rods it is necessary to avoid stimuli that are above the threshold level for cones. We could circumvent this problem by using subjects who are cone blind or by bleaching the cones with red light and testing the rods with green. Transmitting the green beam through the periphery of the pupil also makes it relatively more effective in stimulating rods compared with cones.

Fuortes *et al.* (1961) and Rushton (1961b) investigated the differential thresholds with different background intensities in the case of a photanope. Figure 12 shows the threshold data for a flashing beam (1/2 sec on and 1/2 sec off) 1° in diameter and 7° from the fovea. The background was 14° in diameter. The background intensity is specified in scotopic trolands. The absolute threshold is about 0.01 scotopic trolands. Over a considerable range, the ratio

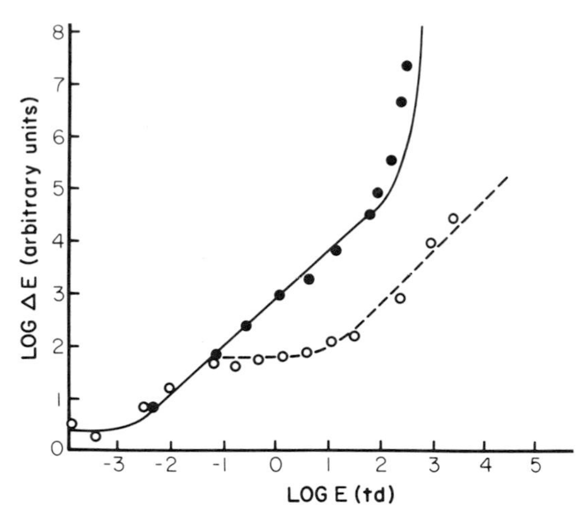

Fig. 12. Increment threshold 7° parafoveal. Ordinate, log increment flash intensity (arbitrary units); abscissa, log background field; ○, normal subject; ●, photanope; continuous curve, normal rod increment threshold curve from Aguilar and Stiles (1954). (Fuortes *et al.*, 1961.)

of the test flash intensity to the background intensity (Weber's fraction) is constant. Over the range in which Weber's fraction is constant, the amount of bleaching is small. The activity generated by the flash and the background both rise, and the flash and background intensity also rise; hence, we must explain why maintaining the ratio constant will keep the flashes at the threshold level.

Near the absolute threshold the visibility of a flash has to depend on a blob of activity at the center of the area stimulated, but at the higher levels the mechanisms responsible for the visibility of borders become more important. These mechanisms will be discussed later.

There is a background level (1000 scotopic trolands) at which the rods saturate. At this level only about one-half of the photopigment is bleached, and some special mechanism has to be found to explain the saturation. The saturation is not a peculiarity of the photanope, because the same phenomenon can be demonstrated with normal subjects (Aguilar and Stiles, 1954).

3. Absolute Threshold for Cones

a. Confining the Stimulus to the Rod-Free Area

The absolute threshold for cones can be measured with a small disk centered in the fovea. The center of a cluster of four small, barely visible points surrounding the test object can be used as the fixation target. This is a diamond-shaped constellation of four points located on a circle 2° in diameter. The eye needs to be dark adapted for 10 min. The method of adjustment can be used to measure the threshold.

b. Minimum Quanta Required for the Threshold

According to Hattwick's data (1954) for a dark-adapted fovea the absolute threshold for a 1° disk of white light is about 0.5 trolands. See Fig. 13. If we assume that the absolute threshold is 0.5 trolands for stimuli of other wavelength compositions, we can estimate that a monochromatic stimulus of 555 nm subtending a visual angle of 1° and viewed through a 2 mm artificial pupil produces 496,400 quanta per sec at the cornea at threshold. According to the data for retinal density [Rushton (1958); Brown and Wald (1963)] the cone photopigments in a dark-adapted eye will absorb about 20% of the 555-nm light falling on the fovea. Because about 40% of the 555-nm light falling on the cornea will be transmitted to the retina about 39,712 quanta per sec will be absorbed by the retina. According to Østerberg's data (1935) there are about 7860 cones in the disk-shaped area 1° in diameter at the center of the fovea. This means that the rate of absorption is about 50 per sec for each group of 10 cones, which is enough to ignore possible effects of scintillation and to consider the input to be steady.

c. Dependence on the Diameter of a Disk-Shaped Stimulus

Graham and Bartlett (1939) have generated a set of data showing the relation between the diameter of a disk-shaped test object and the absolute

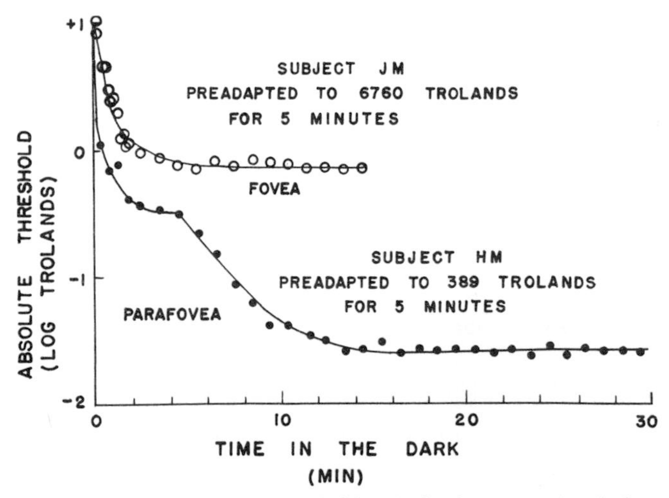

Fig. 13. Dark adaptation curves measured with a 1° stimulus centered at the fovea and 8° in a temporal direction from the center of the fovea (Hattwick, 1954).

threshold. They were aware of the fact that as a disk increases in size its border falls at increasing distances from the center of the fovea and eventually falls outside of the rod-free area where there is a mixture of rods and cones. To avoid having the data contaminated with rod responses they used red discs, which would depress the response of the rods compared with cones. It still needs to be pointed out, as shown in Fig. 7 in Chapter 2, that the concentration of cones decreases with the distance from the center of the fovea and that blue cones are missing or sparsely distributed at the center of the fovea. The data of Graham and Bartlett are shown in Fig. 14. The curve representing the data is one proposed by Nolan (1957),

$$\Delta L/\Delta L_{\min} = (1 + a/\mathrm{R})^2, \tag{9}$$

where $a = 10$ and $\Delta L_{\min} = 0.135$ f L. $\Delta L_{\min}$ is the value of ΔL for large values of R. R is the radius of the disk.

It should be noted that at small values of R the curve is asymptotic to a line with a slope of -2 and at large values is asymptotic to a horizontal line. The two asymptotes cross at the point at which $\mathrm{R} = a$.

Graham and Bartlett (1939) proposed the idea that the signals generated by the cones spread across the retina. At each ganglion cell the signals from nearby cones summate. The signal from each cone is a function of its distance from the ganglion cell. In the case of a disk the threshold depends on the summation at the center of the disk.

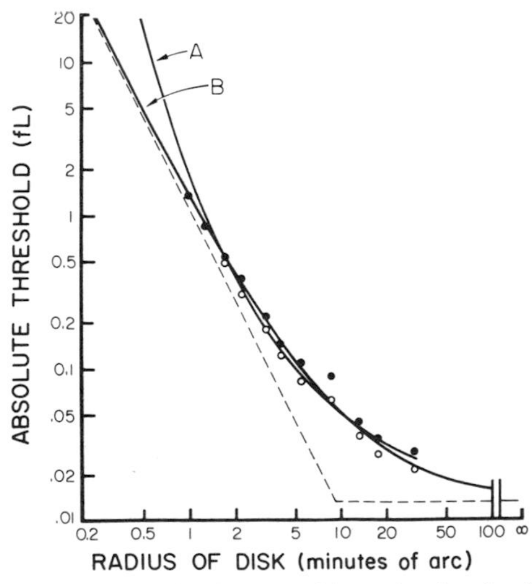

Fig. 14. Data of Graham and Bartlett for two subjects, showing the effect of varying the size of a centrally fixated disc upon its absolute luminance threshold. Curve A is the one used by Graham and Bartlett to fit the data. Curve B conforms to Eq. (9) with $a = 9$ and $\Delta L_{\min} = 0.0135$. (Fry, 1965b).

The general mathematical formulation of this theory is

$$\frac{1}{\Delta L} = k(2\pi) \int_0^R Gr\, dr, \tag{10}$$

where k is a constant, ΔL is the luminance of the disk at threshold, r is the distance from the center, R is the radius of the disk, and G is the point spread function representing both optical and neural spread.

Nolan's equation can be derived from Eq. (10) if

$$G = 1/(r + a)^3. \tag{11}$$

Figure 15a shows another set of data for Blackwell and Smith (1958). Natural pupils and white light were used. The data can also be fitted with Eq. (9). The constants $\Delta L_{\min}$ and a differ slightly; $a = 10$ and $\Delta L_{\min} =$ 0.0056 f L.

Neural and Optical Spread We need to consider the extent to which the spread function defined by Eq. (11) represents neural spread or optical spread.

The neural spread is superimposed on the optical spread. The optical image of a point in physical space is spread out over the retina, and at each receptor we begin with a different mechanism for spreading the image. The two systems are in tandem. It is equivalent to using a lens to form an image on a plate of ground glass and then using a second lens to form a second image. The blurred image of the first system is convolved with the spread function of the second system.

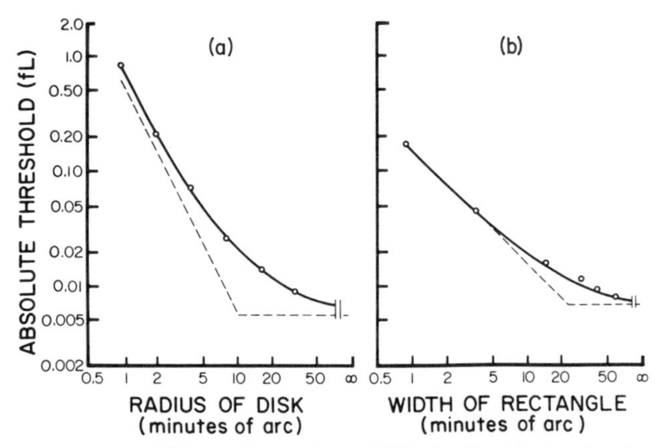

Fig. 15. Average data of Blackwell and Smith (1958) for 11 subjects, showing the effect of varying the size of a centrally fixated disk and the width of a centrally fixated rectangle on the absolute threshold. The length of the rectangle was kept constant at 64 min. The curve used in (a) conforms to Eq. (9), and the curve in (b) conforms to Eq. (12). For each of the two curves $a = 10$, but the value of k is 568 for (a) and 495 for (b). (Fry. 1965b.)

Neural inhibition is also involved, but this gets added in later at the level of the ganglion cells. It does not affect the convolution of the optical image with the neural spread function.

Although it is possible to convolve one point spread function directly with another (Fry, 1965b), it is not easy to deconvolve a composite point spread function into its neural and optical components. This is easy to do however in the case of line spread functions as will be demonstrated later.

d. Dependence of the Threshold of a Bar on Its Width

Let us explore how the absolute threshold for a narrow rectangle is related to that of a disk. Fortunately Blackwell and Smith (1958) have provided us with data for bars 64 min long and of various widths obtained under conditions similar to those used for disk-shaped targets. The results are plotted in Fig. 15b. The data have been fitted with the following formula, which involves the same point spread function $G(r)$ as for the absolute threshold of disk-shaped targets,

$$\frac{\Delta E_0}{\Delta E} = 4 \int_0^{w/2} \int_0^{32\,\text{min}} G(r)\, dh\, dt, \tag{12}$$

where t is the distance from the center of the bar in the direction of the width of the bar, and h is the distance from the center in the direction of the length. The symbol w represents the width of the bar, and 64 min represents the length of the bar which is fixed. The bar is long enough so that further increase in length does not affect the threshold. It has been assumed that the absolute threshold for a bar is dependent on the summation of excitation at the center of the bar. The absolute threshold for a wide bar is ΔE_0. This is somewhat lower than for a large disk. The length and curvature of the border makes the difference, but this will be discussed later. The important thing is that the constant a is the same for both sets of data.

Equation (12) may be rewritten as

$$\frac{\Delta E_0}{\Delta E} = 2 \int_0^{w/2} H(t)\, dt, \tag{13}$$

where

$$H(t) = 2 \int_0^{32\,\text{min}} G(r)\, dh = 2 \int_0^{32\,\text{min}} [a + (t^2 + h^2)^{\frac{1}{2}}]^{-3}\, dh. \tag{14}$$

$H(t)$ may be defined as the *line spread function* (LSF) and represents the distribution of excitation across the image of a line. It represents a combination of optical and neural spread. It is illustrated in Fig. 16 by the composite LSF curve.

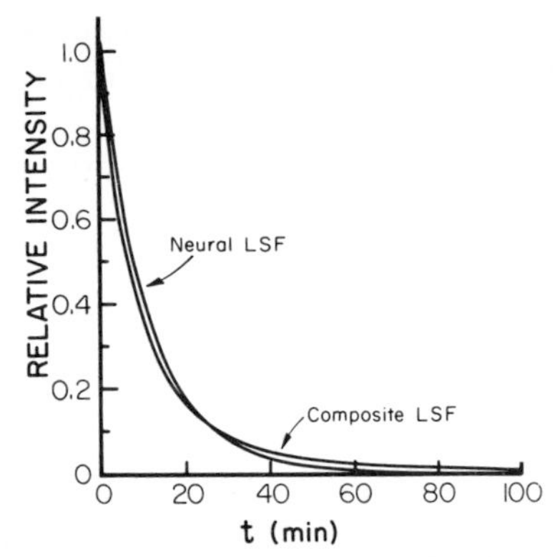

Fig. 16. Combination optical and neural line spread function compared with the isolated neural line spread function.

Analysis of the LSF into Its Neural and Optical Components We can use the following procedure to analyze the line spread function in Eq. (14) into its optical and neural components.

The first step is to make some assumption about the optical line spread function. In the analysis we assume that the optical point spread function approximates that computed for white light for an eye free from spherical aberration with an entrance pupil 2 mm in diameter. It is shown by the continuous curve in Fig. 17. This is only an approximation, because natural

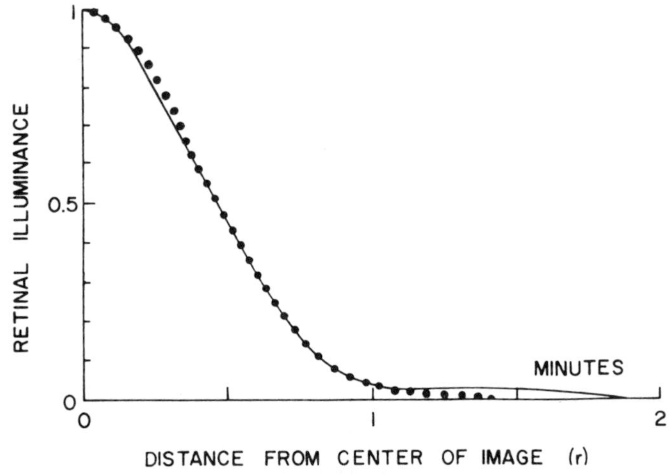

Fig. 17. Point spread function for optical blur of the retinal image. Point of white light (equal-energy spectrum) viewed through a 2 mm artificial pupil. Continuous curve is the spread function calculated from diffraction theory and chromatic aberration data. The dotted curve is the Gaussian approximation. (Fry, 1969b.)

pupils were used in the experiment by Blackwell and Smith. For convenience in calculation we have substituted the dotted curve in Fig. 17, which approximates the continuous curve and conforms to

$$G(r) = G_0 e^{-\frac{1}{2}(r/\sigma)^2}. \tag{15}$$

It represents the Gaussian point spread function. One of the properties of a Gaussian point spread function is that it yields a line spread function that has the same formula,

$$H(t) = H_0 e^{-\frac{1}{2}(t/\sigma)^2}, \tag{16}$$

with the same value of σ.

The next step is to compute the modulation transfer functions (MTF) corresponding to the composite and to the optical line spread function. The modulation transfer function is the one-dimensional Fourier transform of the corresponding line spread function. Because the line spread functions in question are bilaterally symmetrical, the cosine transfer suffices,

$$T(v) = 2 \int_0^\infty H(t) \cos(2\pi v t)\, dt, \tag{17}$$

where $T(v)$ is the transfer factor for a given frequency, and v is the frequency in cycles per min. In the case of the optical component of the blur,

$$T(v) = e^{-2(\pi\sigma v)^2}. \tag{18}$$

The MTF curves are shown in Fig. 18. In the case of the composite blur, numerical integration has been used to assess the MTF.

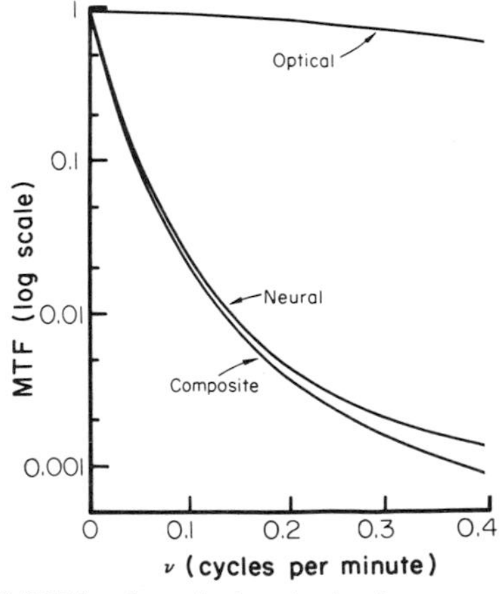

Fig. 18. Analysis of the MTF for the retinal excitation into neural and optical components.

Because the composite MTF is the product of the optical and neural MTFs, one can compute the neural MTF by dividing the composite MTF by the optical MTF. The values for the neural MTF found in this way are plotted as a third curve in Fig. 18. This curve, derived from the data of Blackwell and Smith, is also shown in Fig. 19. It turns out that this curve nearly coincides with the curve that conforms to

$$T(\nu) = \frac{1}{1 + (2\nu/k)^2}, \tag{19}$$

where the constant $k = 0.09$. This may be adopted as the formula for the MTF for neural excitation. Because the line spread function is the Fourier transform of the MTF,

$$H(t) = T(\nu)\cos 2\pi\nu t \, d\nu, \tag{20}$$

it follows that for neural spread

$$H(t) = H_0 e^{-kt}. \tag{21}$$

This curve is plotted in Fig. 16 to show the comparison between the composite line spread function and the neural line spread function. The values of H_0 for the two curves have been adjusted so the composite line spread function at $t = 0$ represents the optical line spread function [Eq. (16)] convolved with the neural line spread function [Eq. (21)]. The composite line spread function is more spread out and less peaked at the center than the neural line spread function. It is obvious, however, that at the absolute threshold the optical spread function is almost negligible compared with the neural spread function.

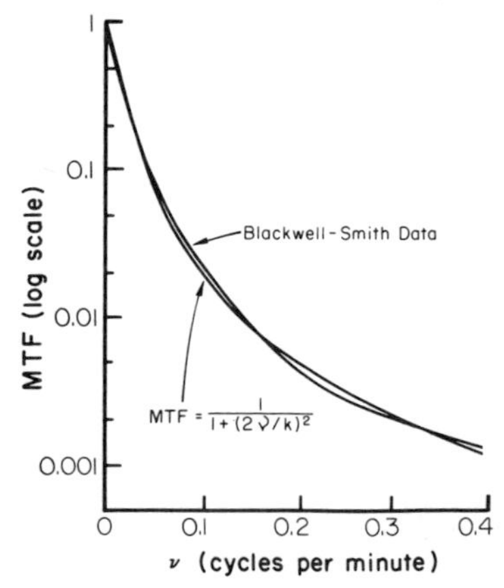

Fig. 19. The neural MTF for the retina derived from the Blackwell–Smith data and a simple curve conforming to Eq. (19) that approximates it.

e. Convolving a Bar with a Line Spread Function

Although it is possible to write an equation for convolving a bar with a line spread function, we can approach the problem by considering a black bar to be a gap between two large bright areas and derive the line spread function by convolving the separate borders and summating. In Fig. 20 the dashed lines represent the gradients for two blurred borders separated by the interval A. The solid line, which is the sum of the two curves, represents the complement of the spread function for a bar having a width A. This provides us with a much simpler mental picture of the blurred image of a bar.

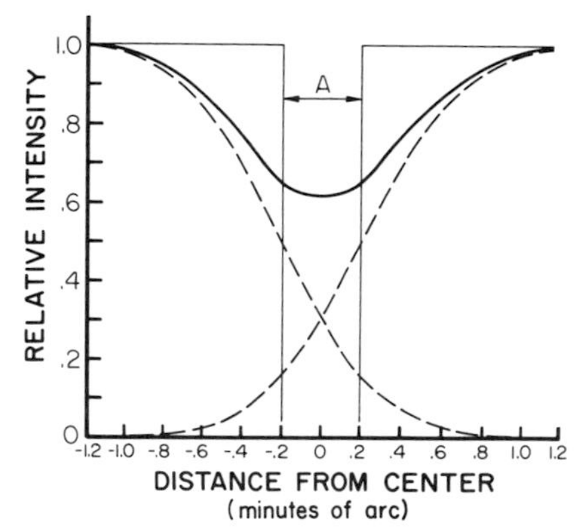

Fig. 20. Indirect method for deriving the spread function for a narrow bar.

4. Differential Thresholds for Cones

When a bright bar or disk is superimposed on a bright background the threshold of visibility is called the differential threshold or the contrast threshold. Contrast is defined as

$$\text{contrast} = \left| \frac{L_O - L}{L} \right|, \tag{22}$$

where L_O is the luminance of the object, and L is the luminance of the background. This equation also covers the case in which the background is darker than the object. One can think of the contrast as reversing when the background goes from brighter to darker.

a. Differential Threshold for Narrow Bars

Dependence on Background Luminance and Width of Bar Figure 21 shows threshold curves for bright bars on three backgrounds of different brightness.

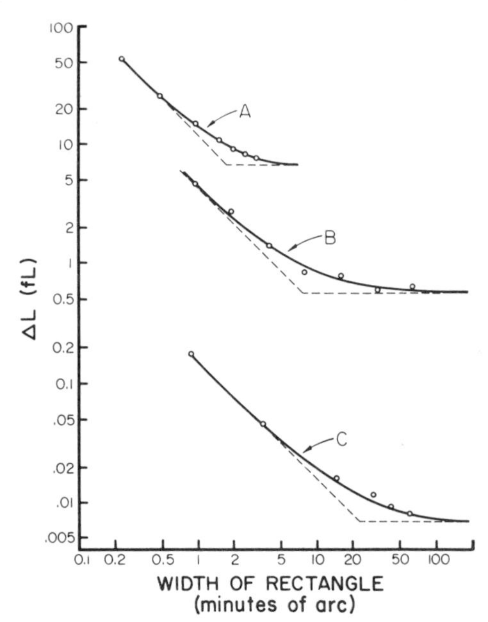

Fig. 21. The effect of varying the background luminance on the area intensity threshold curve for narrow rectangles. For curve A, $L = 157$ fL; for curve B, $L = 10$ fL; for curve C, $L = 0$. For curve A, the length of the rectangles is 50 min and for curves B and C the length is 64 min. Curve A represents data of Fry and Cobb (1935) for one subject; curve B average data of Blackwell and Kristofferson (1958) for four subjects; curve C average data of Blackwell and Smith (1958) for 11 subjects. Curve C is the same curve as that shown in Fig. 15 (Fry, 1965b).

The luminance of the background for curve A is 157 fL, that for curve B, 10 f L, and that for curve C is zero. Curve C represents the absolute threshold and curves A and B differential thresholds. Each is asymptotoic to a straight line at a slope of -1 and a second straight line that is horizontal. The asymptotes cross at the point whose abscissa value represents the index of blur. The index decreases as the background luminance increases.

Curve C is the same as the curve in Fig. 15b. It represents data obtained with natural pupils and binocular vision.

Curve B represents the data of Blackwell and Kristofferson (1958). The observations were made with natural pupils and binocular vision.

Curve A represents data of Fry and Cobb (1935). See Fig. 48 in Chapter 2. The measurements were made with one eye with a 2.33-mm artificial pupil.

Although the three curves were obtained with pupils of different size, and although B and C were obtained with two eyes and curve A with one, the important difference from curve to curve is the luminance level. It is to be expected that the optical blur would undergo little change; hence, the major factor in producing the change in the index of blur is physiological irradiation. What is needed is a careful analysis to determine if this change in the index of blur can be ascribed to a simple change in the value of k in Eq. (21). In what follows this is assumed to be the case.

Figure 21 also demonstrates that the differential threshold for wide bars depends upon the luminance level. But for the moment let us concentrate on how the threshold is affected by the width of the bar in the case of narrow bars.

The Role of Inhibition Nothing has been said so far about the processes of inhibition in the retina. In assessing its role we assume that each cone tends to inhibit the excitation initiated by that cone and by the adjacent cones. Hence, the inhibition I generated by cones in the region surrounding a given ganglion cell can be summated and subtracted from the excitation R at that point,

$$\text{net excitation} = R - I. \tag{23}$$

It is possible to assess the role of inhibition in the visibility of lines, borders, and gratings by assuming that the line spread function for inhibition is Gaussian,

$$I = I_0 e^{-0.5(t/\sigma)^2}. \tag{24}$$

Based on metacontrast data, we have estimated the value of σ to be about 8 min (Fry, 1948, 1970, 1973).

We will not be able to express the intensity of excitation and inhibition generated by a stimulus in terms of any known units until we know how to measure these intensities, but we can say that excitation and inhibition can be assessed in terms of the same units so that I can be subtracted from R. In the case of a uniform patch of luminance like the sky we can say that R and I are related to retinal illuminance E as

$$R = \mu E, \tag{25}$$

and

$$I = \psi E, \tag{26}$$

where μ and ψ are constants.

It has been derived from data related to the visibility of sine-wave gratings that the ratio $\psi/\mu = 0.8$ (Fry, 1969b, 1970). Hence, it has been assumed in what follows that $\psi = 0.8$ and $\mu = 1$.

We now show what happens in the case of bars of different width when L for the background is 100 c/m^2 and the value for k in Eq. (21) is assumed to be 0.8. The bar superimposed on the background has a ΔL value of 28.6 c/m^2. To simplify the calculation of retinal illuminance we assume the area of the entrance pupil to be 1 mm^2.

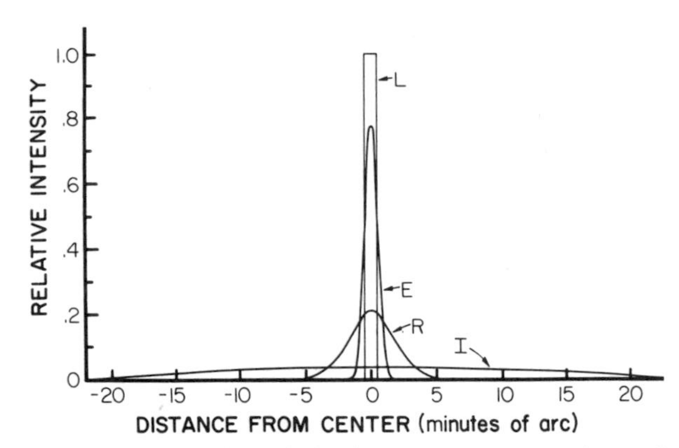

Fig. 22. Distributions representing *L* the luminance across a narrow bar used as a test object. *E* the retinal illuminance across the retina, and *R* and *I*, the distributions of excitation and inhibition at the level of the ganglion cells.

Figure 22 shows what happens to the distributions of *E*, *R*, and *I* when the bar, without reference to the background, is convolved with the line spread functions for optical blur [Eq. (16), $\sigma_E = 0.4$], neural blur [Eq. (21), $k = 0.8$], and inhibition [Eq. (24), $\sigma_I = 8$]. The bar in this case is one minute wide.

In Fig. 22, curve *L* represents the distribution of luminance across the bar in object space. Curve E is curve *L* convolved with the line spread function for optical blur and represents the optical image formed on the retina. Curve R is curve E convolved with the line spread function for neural blur. Curve I is curve E convolved with the line spread function for inhibition and multiplied by the factor ψ. Curve I has to be subtracted from curve R to obtain the net distribution of neural excitation produced by the bar. This in turn has to be added to the net excitation produced by the background.

Thresholds for Bars of Different Width Figure 23 shows composite curves for bars of different width. It is easy to see that for narrow bars the visibility depends on the amount of excitation at the center of the bar in comparison with the activity on the two sides of the bar. For narrow bars the amount of excitation at the center of the bar is proportional to the width of the bar, and this is the basis for Ricco's law. At this level of bar width, inhibition is too diffuse to interfere with the operation of Ricco's law.

As the bar increases in width, the distributions of inhibition and excitation at the two borders become more important. Above a certain width all that happens is the change in distance between the two borders. The contrast at the borders remains constant and determines the threshold.

At lower levels of luminance where the value of *k* may be as small as 0.09 [see Eq. (19)], the distribution of neural excitation is very diffuse, and there is

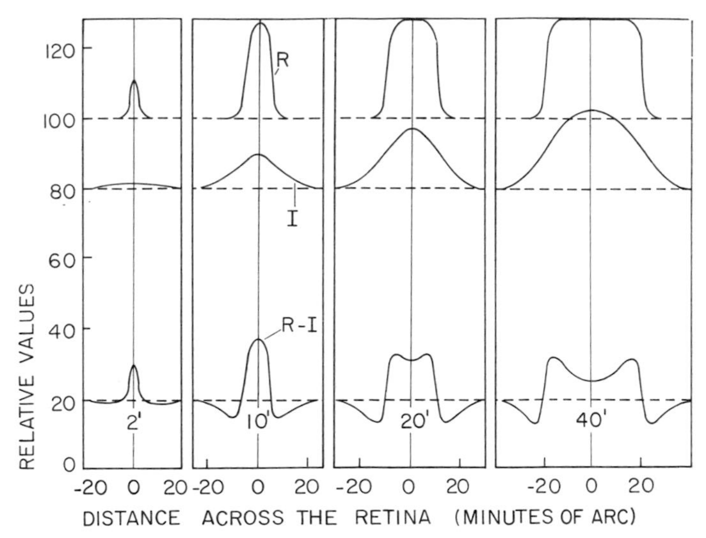

Fig. 23. Distributions of excitation R and inhibition I across images of bright bars of different widths on a bright background, $\Delta L L = 0.286$. The $R - I$ curves represent the net excitation transmitted through the retina (Fry, 1970).

little chance for border mechanisms to assert themselves; hence, the threshold with very wide bars would still be determined by the activity at the center.

Let us concentrate on bars and disks on bright backgrounds where, as shown in Fig. 23, the threshold comes to be dependent on border mechanisms.

The Effect of Varying the Length of a Bar Let us start with a bar like the one in Fig. 24a and put it on a background of 157 fL as in the case of curve A in Fig. 21. If the width is greater than 2 min the threshold is independent of the width and is determined by the border mechanisms at the two sides. If we start with a long bar, the reduction of its length begins to affect visibility at a length somewhat greater than 1°. This is too long for the length to have any effect on the excitation at the center, and we have to assume that the sheer length of the border affects the visibility (Fry, 1947).

b. The Effect of Varying the Size of a Disk

If we start with a small disk (Fig. 24b) and gradually increase its diameter the threshold continues to drop even beyond a diameter of 1°. We know from the data in Fig. 23a for narrow bars that increasing the diameter beyond about 5 min cannot affect the excitation at the center of the image; hence, the improved visibility of a disk has to depend on the increased length and radius of curvature of the border (Fry, 1947).

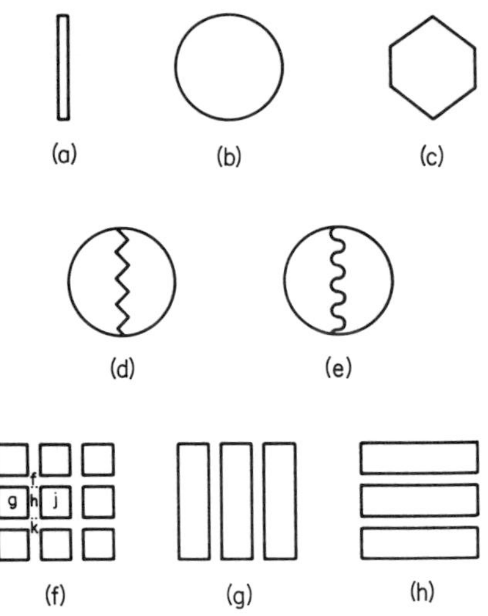

Fig. 24. Patterns used in the study of the visibility of borders.

c. SMOOTHING OUT OF SEGMENTED AND CRENULATED BORDERS

If we substitute a small regular polygon (Fig. 24c) for a small disk (Fig. 24b), we find that there are two thresholds, one at which the polygon is perceived as a disk and a higher threshold at which it is perceived as a polygon (Fry, 1947). In the case of a regular polygon the length of the segments and the angle between adjoining segments become the important variables as substitutes for the length of the circumference and the curvature of the border of the disk.

Bipartite patterns (Figs. 24d and e) can be used to investigate the effect of substituting a serrated or scalloped border for a straight border (Fry, 1947). If the number of cycles per unit length is large, one can demonstrate two brightness difference thresholds, one at which the serrations or scallops are seen, and one at which a blurred straight border is seen. A special mechanism has to be postulated to explain how the nonuniformities are ironed out.

d. THE FILLING-IN PHENOMENON

Another property of borders is illustrated by the positive afterimage induced by viewing nine squares such as shown in Fig. 24f (Fry and Robertson, 1935a). They are seen as uniform vertical or horizontal bars, as shown in Figs. 24g and h. These patterns alternate. This involves filling-in of the gaps, breakdown of existing borders, and creation of borders where none exist. For example (see Fig. 24f), when horizontal bars are seen, the gap h fills in, and the existing borders between g and h and between h and j break down; and nonexisting borders between h and f and between h and k come into being.

e. The Effect of Blur on the Visibility of a Border

A bipartite pattern like that shown in Fig. 25 can be used to study the effect of blur on the visibility of a border. Fry (1965a, p. 46) used a special arrangement to blur the distal stimulus. The eye was kept in focus and allowed to observe the target through a 2-mm pupil; hence, the blur produced in the optical system of the eye was negligible and was ignored. Provision was made for changing the contrast between the two halves of the field and the luminance gradient between the two halves which constitutes the test border. The results shown in Fig. 26 were obtained with a border with a Gaussian blur.

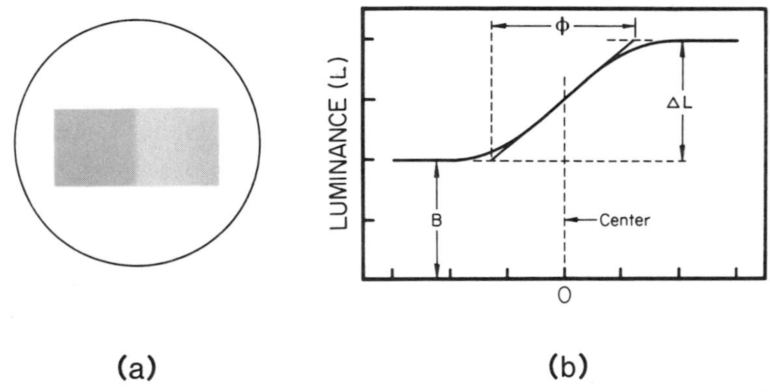

Fig. 25. Stimulus pattern used to study the effects of blur. The rectangle is 87 min high and 174 min wide (Fry, 1965a).

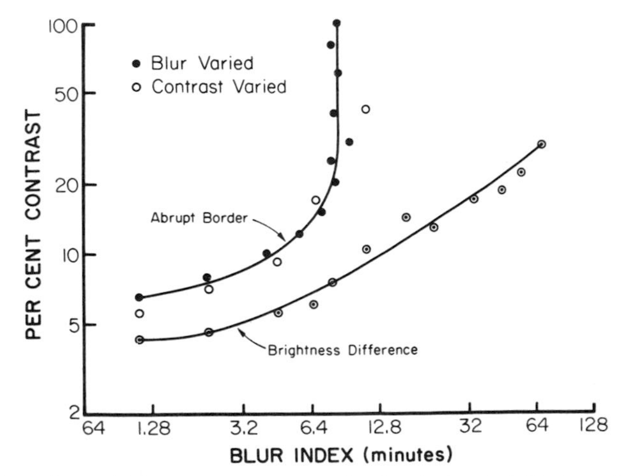

Fig. 26. The effect of blur and contrast on the threshold of visibility and the preceived abruptness of a border for subject JE (Fry, 1965a).

This is illustrated in Fig. 25. The equation for the gradient is

$$L = L_R - \Delta L \int_{-\infty}^{t} \frac{1}{\sigma\sqrt{2\pi}} e^{-0.5(t/\sigma)^2} \, dt, \tag{27}$$

where t represents the distance from the center and σ is a constant. The luminance L_R of the right half of the rectangle was 23 fL and that of the surround was 23 fL. The luminance of the left half was varied to change the contrast.

The amount of blur is specified by the index of blur ϕ, which is indicated in Fig. 25. In terms of the equation for the gradient,

$$\phi = \sigma\sqrt{2\pi}. \tag{28}$$

The visibility of the test border is affected slightly by the borders at the ends of the rectangle and also by the abuttment of the ends of the test border on the borders at the top and bottom of the rectangle. The effects of varying the gradient have to be superimposed on these effects.

The results in Fig. 26 show the effects of varying the index of blur and the luminance difference ΔL between the two halves. Two thresholds were noted, one at which the border became visible and divided the rectangle into two halves, one just noticeably brighter than the other, and a second threshold at which the border was no longer perceived as a blurred gradient but as a sharp border. As the index decreases, there are several degrees of sharpness between the border being first perceived as an abrupt border and being perceived as maximally sharp. This shows the tolerance of the eye for blurred borders. It also shows that a considerable amount of blurring can occur without affecting the differential threshold. In an earlier study (1931) Fry showed how the gradient is related to the perceived sharpness of a border.

Fry and Enoch (1959, pp. 15, 16) investigated the effect of using different types of blur as indicated by the gradients in Fig. 27. The stimulus pattern used is the same as that shown in Fig. 25. These three types of blur were produced by using three types of apertures in throwing the image of a border out of focus as illustrated in Fig. 27. They found that the threshold was not markedly affected by the type of blur and was almost completely dependent on the index of blur and contrast (Fig. 28). At suprathreshold levels, Mach band phenomena show up when there is an abrupt change in the slope at the top and bottom of the gradient.

f. Frequency Equalizing Mechanism

It is clear from Fig. 23 that, in the case of a wide bar, inhibition will enhance the brightness on the bright side of the border and depress it on the dark side, but this does not totally explain why an abrupt, sharp border is seen at the

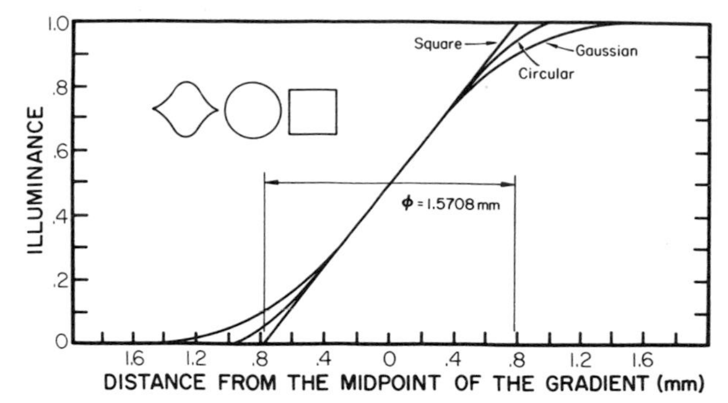

Fig. 27. The blur functions for the three kinds of apertures. The illuminance E on the screen as a function of the distance from the midpoint of the gradient (Fry and Enoch, 1959).

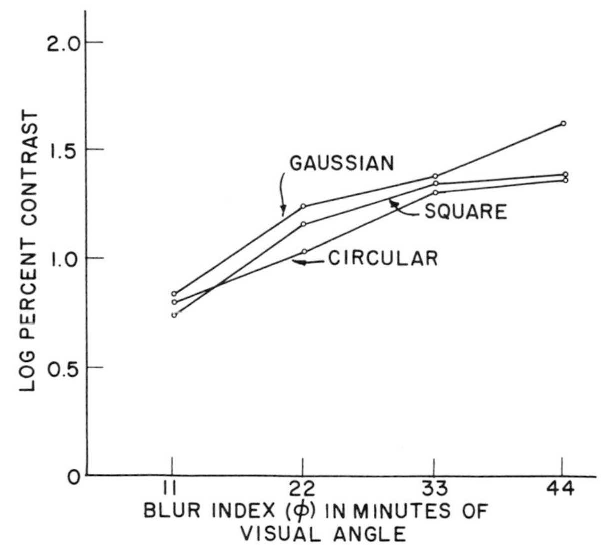

Fig. 28. Contrast threshold as a function of blur for three different types of blur resulting from the use of circular, square, and Gaussian apertures (Fry and Enoch, 1959).

midpoint of the gradient. It is still necessary to postulate a special mechanism that will cause an abrupt difference on the two sides of the border. In terms of optic nerve fibers this is a frequency-equalizing mechanism that will tolerate abrupt changes in frequency. This is best illustrated in terms of the Cornsweet border generated by rotating the disk in Fig. 29(a) at high speed. A perceived sharp border divides the disk into a central zone and a peripheral zone that are

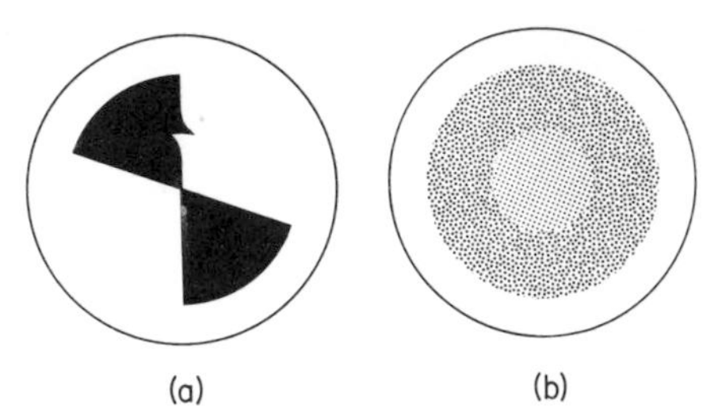

Fig. 29. Cornsweet's pattern (Ratliff, 1965) (a), which when rotated gives the illusion illustrated in (b) of a uniform disk surrounded by a uniform annulus of lower brightness (Fry, 1970).

perceptibly different in brightness (Ratliff, 1965, 1972; Fry, 1970, 1973a). This has to be due to the disturbance that is generated at the border and spreads out in both directions. The threshold has to be regarded as a struggle between the equalizing mechanism, which tends to wash out borders, and the mechanisms for generating and maintaining borders.

g. Intermittent Visibility

Another property of borders is that near the threshold they become intermittently visible. In the case of a small disk at the center of a cluster of four fixation dots, its transition from invisibility to visibility can be assessed by counting the frequency and lengths of visibility periods. These effects depend on independent processes in the separate retinas (Fry and Robertson, 1935b).

h. Subthreshold Seeing

A forced-choice method is used in Blackwell's measurements of visibility (1952) in which a flash occurs in one of four intervals and the subject has to specify or guess which. His percentage of correct answers may be better than those predicted from guessing, yet the subject cannot vouch for having seen a disk bounded by a border. This performance is still a mystery. It is only at a high level of correct calls that the subject can be sure of having seen a disk with borders.

i. The Effect of One Border on the Visibility of Another

Another important property of a border is that the presence of one border can interfere with the visibility of another (Fry and Bartley, 1935). This can be demonstrated with the stimulus pattern used by Dittmers (1920), shown in Fig. 30a. The disk a in the center is used to measure the contrast threshold between the disk and its background. The contrast of the border between zone b and zone c can be manipulated by raising or lowering the luminance of zone c. At one point zones c and b match, and the border disappears. Zone c

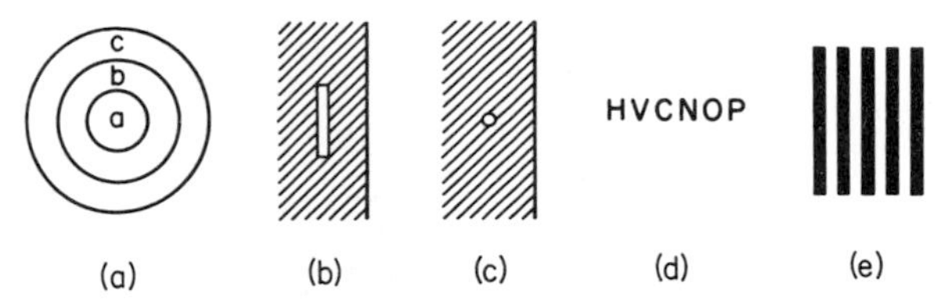

Fig. 30. Patterns for demonstrating the effect of one border on the visibility of another.

can be made either brighter or darker than b. The threshold increases as the contrast at the border between b and c increases (either positive or negative) or with the nearness of this border to the border of the central disk (Fry, 1953).

If the arrangement in Fig. 30b is used to test the effect of a border on the visibility of a narrow rectangle, the effect of stray light produced by the bright half of the field of view is included. In this case the amount of stray light at various points to the left of the border is easy to compute. If the arrangement in Fig. 30c is used, the effect is largely confined to stray light. The rectangle has a much lower threshold than the small disk and is more easily affected by border interaction.

If a row of letters is used (Fig. 30d) to test visual acuity, some subjects, if asked to name the second letter, will give the name of the third or fourth, which implies an inability to assess what is being looked at. This is a new problem, but to minimize this effect as well as the effect of interaction of borders (Flom *et al.*, 1963) the center-to-center spacing of the letters on a standard visual acuity chart (NRC, 1980) should be at least twice the width of the letters.

If we use a square-wave grating (Fig. 30e) as a contrast threshold target with a center-to-center separation that is large enough so that the gradients at the borders are independent of the amount of separation, the threshold remains at a constant level as the size of the pattern is increased. It is assumed that the visibility of none of the borders is affected by the adjacent borders. This follows from the Dittmer's effect (Fig. 30a) in which a border loses its effect on an adjacent border if it is itself at threshold level. For the same reason we can say, in the case of the threshold for a disk (Fig. 24b), that the border on one side is not affected by the border on the other, because both are at the threshold level. The length and curvature are the things of importance. In the case of the rectangle (Fig. 24a) the two vertical borders do not affect each other as long as both are at the threshold level.

The arrangement shown in Fig. 31 can be used to demonstrate (a) the effect of having two threshold borders converge at a common point (Brown, 1977), (b) the effect of having a threshold-contrast border terminate at a high-contrast border at right angles (Fry and Bartley, 1935), (c) the effect of having a threshold border and two suprathreshold borders converge at a common point, and (d) the effect of having two borders cross each other at right angles.

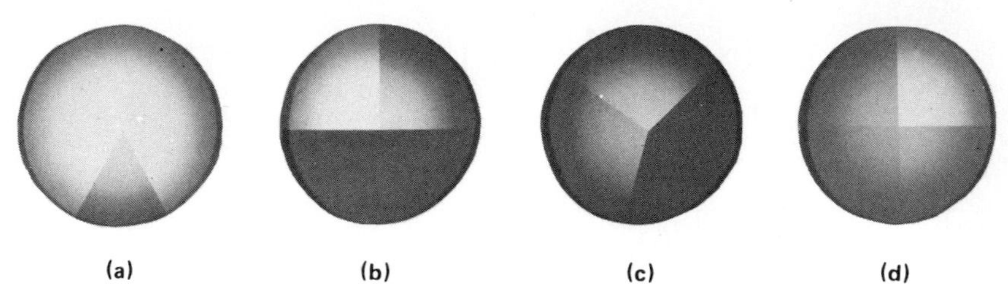

Fig. 31. Patterns for demonstrating the effect of one border on the visibility of another (Brown, 1977). The intensity gradients from the center out are Gaussian.

j. The Visibility of Gratings

The Visibility of a Sine-Wave Grating When the eye is exposed to a sine-wave grating, the distribution of luminance $L(s)$ across the pattern can be represented by a sine-wave formula,

$$L(s) = \bar{L} + \Delta L \cos(2\pi s/\bar{s}), \tag{29}$$

where as shown in Fig. 32, $\bar{L}$ is the average luminance (c/m^2), ΔL is the amplitude of modulation, $\bar{s}$ is the center-to-center distance between peaks, and s is the distance in min from the center of one of the peaks which is used as the starting point.

The modulation of a sine-wave grating is defined as the ratio of the amplitude ΔL to the average luminance $\bar{L}$,

$$\text{modulation} = \Delta L/\bar{L} = (L_{\text{max}} - L_{\text{min}})/(L_{\text{max}} + L_{\text{min}}).$$

The image formed on the retina is also a sine wave, and the distribution of flux across the image is

$$E(s) = \bar{E} + \Delta E \cos(2\pi s/\bar{s}), \tag{30}$$

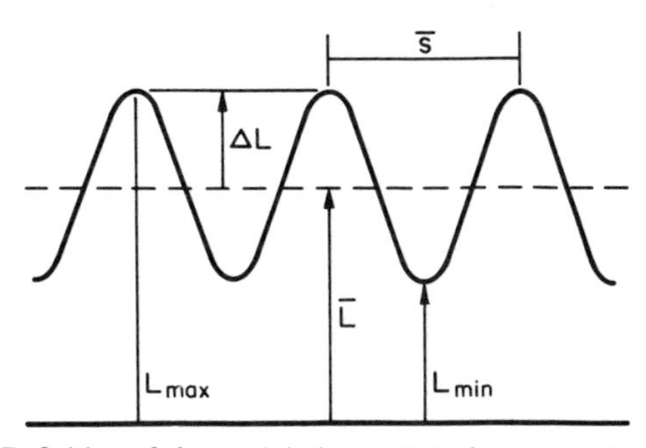

Fig. 32. Definition of the modulation and the frequency of a sine wave.

where $\bar{E} = \bar{A}\bar{L}$ expressed in trolands, $\Delta E = \bar{A}T_E\Delta L$, $\bar{A} =$ area of the pupil, and $T_E =$ modulation transfer factor for the image forming mechanism.

The factor T_E is the cosine Fourier transform of the line spread function of the image-forming mechanism when the LSF is bilaterally symmetrical. A plot of these factors as a function of $1/\bar{s}$ represents the modulation transfer function (MTF).

The fact that the retinal image is a sine wave and that T_E is the Fourier transform of the LSF can be demonstrated by convolving the external sine-wave stimulus with the line spread function of the image-forming mechanism (Fry, 1969b, 1970).

When the image of the sine wave is transmitted through the retina, it is modified by the spread of excitation and inhibition and the summation of these effects at the level of the ganglion cells. The prime effect is the change in the amount of modulation.

The distribution of excitation R across the retina at the level of the ganglion cells is

$$R = \mu[E + T_R\,\Delta E\cos(2\pi s/\bar{s})], \tag{31}$$

and the distribution of inhibition I is

$$I = \psi[E + T_I\,\Delta E\cos(2\pi s/\bar{s})], \tag{32}$$

where s is distance across the grating image beginning at the center of one of the bright bars, $\bar{s}$ is the center to center spacing, E is the average retinal illuminance, ΔE is the amplitude of the modulation, T_R and T_I are the transfer functions for excitation [Eq. (19)] and inhibition [Eq. (33)], and ψ and μ are constants.

$$T_I = e^{-2(\pi\sigma\bar{s})}. \tag{33}$$

The net result at the ganglion cell layer is $R - I$, and the modulation transfer function for the retina is

$$T_{(R-I)} = (\mu - \psi)^{-1}(\mu T_R - \psi T_I). \tag{34}$$

The modulation transfer function for the eye T including both optical blur and physiological irradiation is

$$T = T_E \cdot T_{(R-I)}. \tag{35}$$

If we assume that a constant value of modulation at the ganglion cell layer is required for the threshold of visibility, it follows that in terms of the external sine-wave stimulus,

$$\text{modulation threshold} = K/T, \tag{36}$$

where K is a constant. Figure 33 shows a comparison between the raw data and a plot of $0.01/T$.

It is to be noted that the two curves diverge at large values of $\bar{s}$. This can be explained on the ground that the threshold depends not solely on the amplitude of the modulation but also upon the slope of gradients, as presented in Fig. 26. In that study it was shown that the contrast threshold increases as the ϕ value of the blur increases. The basis for comparing the threshold for a single blurred border and the threshold for a sine-wave grating is illustrated in Fig. 34. The curve for a single border has the same amount of blur as the sine wave when dL/ds at the midpoint of the gradient is the same for both. The

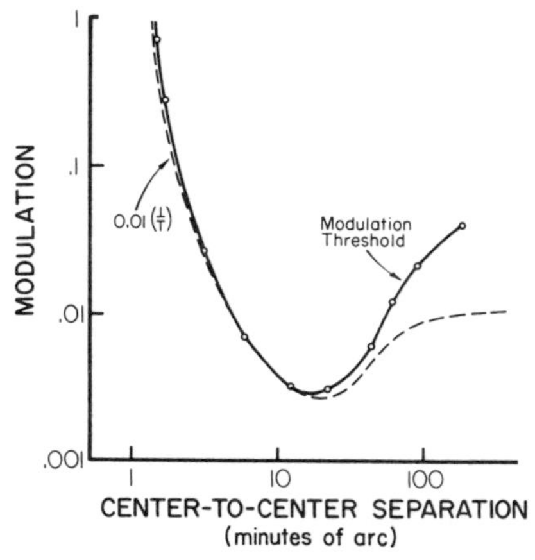

Fig. 33. The data for the modulation threshold curve were obtained with a sine-wave grating 87 min high and 174 min wide. The bars were vertical. The average luminance was 74.3 c/m^2, and the luminance of the background was the same. It was viewed through a 2-mm artificial pupil at a distance of 1 m. Equation (36) was used in computing values of T. The constants were as follows: $\mu/\psi = 0.8$, $\sigma_E = 0.4$, $k = 0.8$, and $\sigma_I = 8$ (Fry, 1969b).

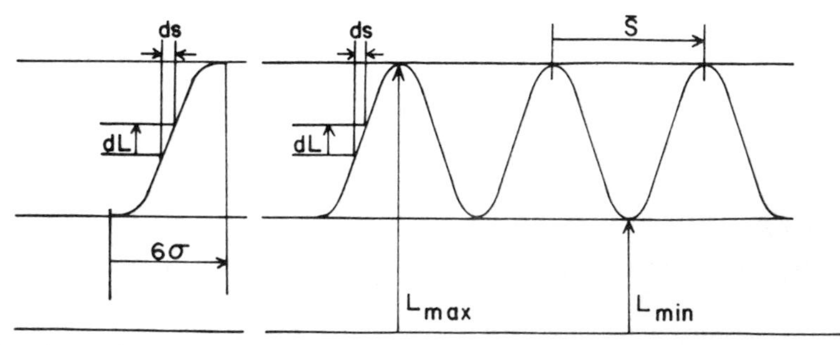

Fig. 34. Basis for comparing the modulation threshold for a single blurred border to the modulation threshold for a sine-wave grating (Fry, 1969b).

index of blur for the sine wave may be defined as

$$\phi = (L_{\text{max}} - L_{\text{min}})/(dL/ds) = \bar{s}/\pi. \tag{37}$$

Hence, ϕ increases as $\bar{s}$ increases, and the threshold rises, which accounts for the rise in the modulation threshold at low frequencies.

Figure 35 illustrates how excitation and inhibition interact at different frequencies for sine-wave gratings. The external stimulus illustrated in Fig. 35a is a sine-wave grating with its frequency increasing from one end to the other. Each half-cycle is 1.25 longer than the one to its left. The modulation of the waves is constant at 12%. The components of the response are shown in Fig. 35b. The R curve shows the stimulus convolved with the optical (Eq. 16) and neural (Eq. 21) line spread functions. The I curve shows the stimulus convolved with the optical (Eq. 16) and inhibition (Eq. 24) line spread functions. The $R - I$ curve represents the I curve subtracted from the R curve.

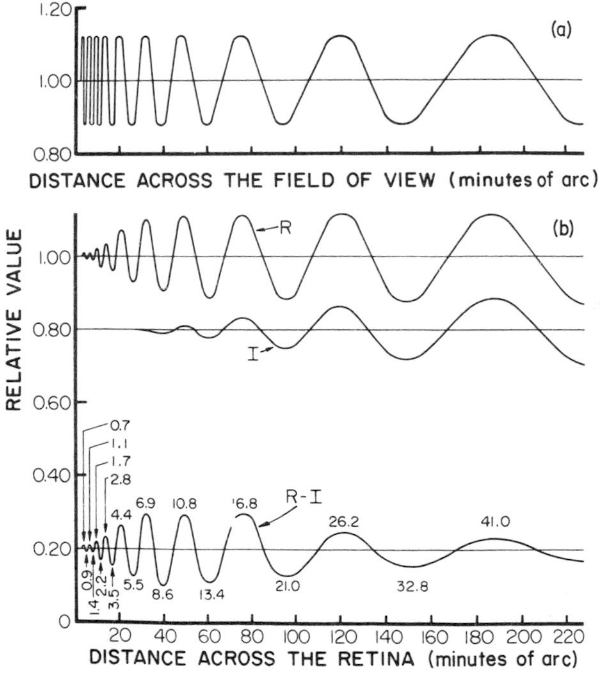

Fig. 35. Distributions (b) of excitation R and inhibition I across the retina produced by a sine-wave grating (a) of constant modulation (0.12) and variable frequency. Each half-cycle is 1.25 times wider than the preceding half-cycle. The numbers above and below the $R - I$ curve indicate the lengths of the half-cycles in minutes. The constants used in this analysis are as follows: $A = 1$ (Eq. 30), $\mu = 1$ (Eq. 31), $\psi = 0.8$ (Eq. 32), $\sigma_E = 0.4$ (Eq. 33), $k = 0.8$ (Eq. 19), and $\sigma_I = 8$ (Eq. 33) (Fry, 1970).

At high frequencies the amplitudes of excitation and inhibition both get subdued and eventually washed out, but the cut off is at a higher frequency for excitation, which produces a peak in the $R - I$ waves at a center-to-center spacing of about 20 min. At low frequencies the inhibitory waves tend to neutralize the excitatory waves. The amplitude drops, and the slopes at the borders between bars become more gradual. At medium frequencies at which the retina transmits the modulations of excitation but not the modulations of inhibition, inhibition enhances the modulation by reducing the dc component. This mechanism is also used in controlling the contrast of a TV display. The dc component is changed without affecting the ac component.

The Visibility of Square-Wave Gratings Figure 36 illustrates how excitation and inhibition interact at different frequencies for a square-wave grating. The external stimulus is shown in the upper graph. The spacing of the bars and average luminance and the modulation are the same as for the sine wave in Fig. 35. The same constants have also been used in computing the R, I, and

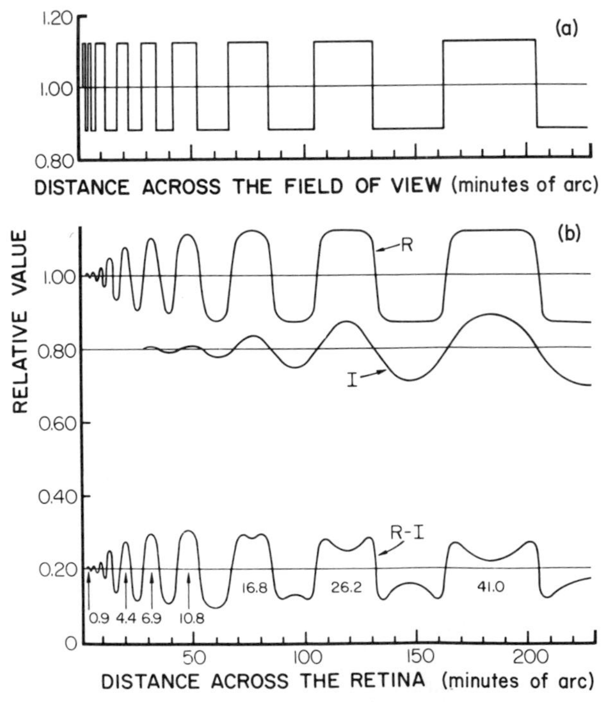

Fig. 36. Distributions (b) across the retina of excitation R and inhibition I produced by a square-wave grating (a) of constant modulation and variable frequency. Each half cycle is 1.25 times wider than the half cycle that precedes it. The numbers below the $R - I$ curve indicate the lengths of the half cycles (Fry, 1970).

$R - I$ components of the response in the lower graph. It may be noted that for high frequencies the square waves are reduced to sine waves, and except for the differences in amplitude, square waves may be treated the same as sine waves. At low frequencies borders at the edges of the bars become independent of the width of the bar in the sense that the contrast and the slope at the midpoint become constant, and this is somewhat lower than the contrast between peaks and troughs at the intermediate frequencies.

Figure 37 shows the modulation threshold curve for square-wave gratings of different frequency. In ordinary seeing the optical line spread function of the human eye is Gaussian and at high frequencies simply filters out the fundamental sine-wave component of the square-wave grating, and the response to a square wave is the same as to a sine wave with an amplitude $\frac{4}{3}$ times that of the square wave. At low frequencies (large values of $\bar{s}$) the modulation threshold should become independent of the width of the bars and depend upon the contrast at the borders separating the bright and dark bars. The threshold should be the same as for a single border. The dashed line represents the threshold level found for a single border. When the modulation is reduced to measure the threshold, the borders all reach the threshold at the same time, and no one of them can affect the threshold of its neighbors.

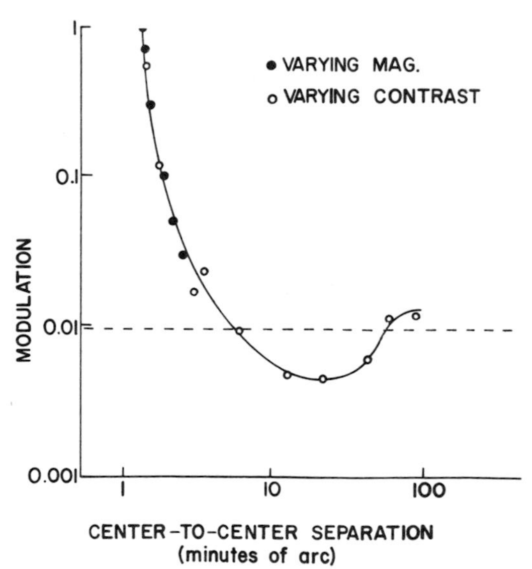

Fig. 37. Modulation threshold curve for a square-wave grating. Average data for two subjects, white light and a 2-mm entrance pupil. The dashed curve represents the modulation threshold for a single border. The conditions are the same as for the sine-wave gratings (Fig. 33) except that a border separating a dark bar on the left from a bright bar on the right always fell at the center of the pattern and a point on this line was fixated. (Fry, 1970).

The Effect of Average Luminance on the Visibility of a Grating It is not to be expected that the average luminance level will affect the LSF of the image-forming mechanism except indirectly through a change in the size of the pupil. The major effect is on the constant K. A second major effect is the change in the constant k. The amount of neural spread decreases as the luminance level increases. This has been demonstrated already in connection with the visibility of narrow bars (Fig. 21). It is possible also that the values of μ, ψ/μ, and σ_I change with the luminance level.

Studies of the effect of luminance level have been made by Van Nes (1968), Schober and Hiltz (1965), and Nachmias (1967).

It may be noted that Fig. 9 shows the effect of luminance level of the resolving power of the eye at 100% modulation. The upper limit of visual acuity for 100% modulation is about 1.7, which is equivalent to a center-to-center spacing of 1.2 min. This may be compared with the upper limit for resolving power at 100% modulation in Fig. 37.

Elimination of Blur Produced by the Image-Forming Mechanism of the Eye Campbell and Green (1965) and Le Grand (1935) used the double-slit technique to generate a sine-wave pattern on the retina with 100% modulation. This eliminates the blur produced by the image-forming mechanism of the eye. Campbell and Green used a demodulating device that can reduce the modulation to the threshold without changing the average retinal illuminance. The modulation factors required for threshold are represented by the solid curve in Fig. 38. The average retinal illuminance was kept constant at 500 trolands. The wavelength of the light was 632.8 nm.

The dashed curve in Fig. 38 is a plot of

$$\text{modulation threshold} = K[1/T_{(R-I)}], \tag{38}$$

where K is a constant (0.007) and $T_{(R-I)}$ is defined by Eq. (34). The constant k (0.8) is the same as in Fig. 33. The discrepancy at low spatial frequencies has to be explained as in the case of Fig. 33 by the fact that the threshold is dependent upon the slope of the gradients between the bars as well as the difference between peaks and troughs.

The discrepancy between the two curves in Fig. 38 at high frequencies has to be explained on the ground that the coarseness of the retinal mosaic imposes an upper limit on the resolving power. At 100% modulation the resolving power is shown in Fig. 38 to be about one cycle per minute. We cannot hope to isolate the neural component at high frequencies with this procedure.

Resolving Power at 100% Modulation In Fig. 33 the top edge of the graph represents the line of 100% modulation. If we move from right to left by changing the spacing, we find the point at which the lines of the grating can no longer be resolved. The resolving power (lines per degree) is slightly better for

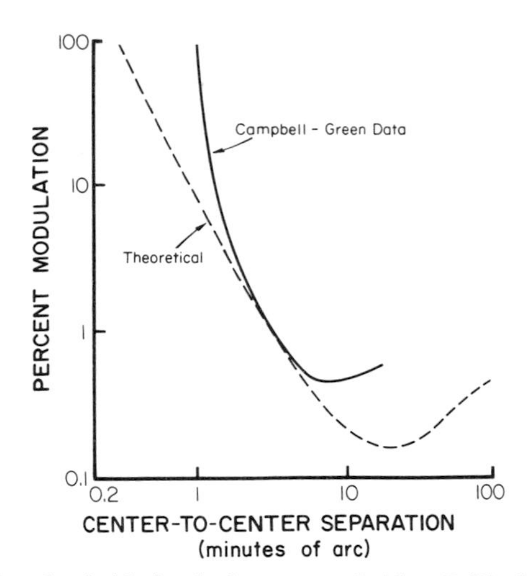

Fig. 38. Modulation thresholds for the human eye (subject D.G.) when the image-forming mechanism is by passed by using Young's fringes. The theoretical curve, which represents a plot of Eq. (38), shows that the resolving power at high levels of modulation is limited by the coarseness of the retinal mosaic.

square waves (Fig. 37) than sine waves. If we go on further to the left in Fig. 38, we find the resolving power that is to be expected when we form a 100% modulation sine-wave grating on the retina with the double-slit technique. Efforts to measure the ultimate resolving power by this method have been carried out by various investigators. Thc data for several investigators are found in Table I. Byram (1944) reported that although the transition from resolution to fusion was smooth in ordinary seeing, it was not smooth with a double-slit pattern; as the lines approached the limit of resolution, they broke up into segments and appeared to shimmer or flutter. Campbell and Green (1965) pointed out that just before fusion the central patch scintillates and appears brighter and more desaturated than the surround. These effects may

TABLE I. Resolution Thresholds Measured with a Double Slit Diffraction Pattern (Fry, 1970)

Investigator	Resolution threshold (min.)	Retinal illuminance (trolands)	Wavelength (nm)
Le Grand	1.25	13,600	546
Le Grand	1.45	2.72	546
Campbell	0.98	500	633
Green	0.95	500	633
Westheimer	1.25	2200	555
Byram	0.50	not specified	not specified

be taken as evidence that the structure of the retinal mosaic is involved. Helmholtz (1924) reported seeing similar effects when a fine grating is viewed with the natural pupil. Westheimer (1960) reported that his subjects found the transition to be smooth.

The fact that the resolving power found by the use of external square-wave gratings is slightly poorer than that found with double-slit patterns may be taken as evidence that the resolving power of the image-forming mechanism determines the resolving power of the eye in ordinary seeing. The fact that the point spread function for neural spread has a sharp peak explains why the transmitting mechanism can impose no limit on the resolving power. It is remarkable that the center-to-center spacing of the receptors at the center of the fovea is about 1/2 min measured at the second nodal point. It is one of the wonders of nature that blur of the optical image formed on the retina imposes its limit at the same level at which the coarseness of the retinal mosaic begins to be of importance.

k. Retinal Astigmatism

There is the question whether the perceived sharpness of a line can depend on whether it is aligned with parallel rows of receptors on the retina, which is a property of the retinal mosaic.

l. Vernier Acuity

Vernier acuity involving alignment of two lines placed end to end with a precision of about 2 to 10 sec does make it necessary to take into account the coarseness of the retinal mosaic and the nature of the arrangement, that is, whether hexagonal, rectangular, or random. In general the two lines to be aligned may be perceived as having a finite width, but the adjustment can be made by making the center of the one perceived line to be aligned with the center of the other.

m. Relative Perceived Directions

The problem is illustrated by a row of three points. Is the middle point perceived as lying on a line connecting the other two? Is it perceived as lying midway between the other two? The precision in making this kind of judgement does not appear to be limited by anything at the retinal level. The points may be perceived as blobs of finite size, but the perceived direction of the blob may be assessed in terms of the direction of the center of the blob.

D. VISIBILITY

The concept of visibility is important, because in visual science we have to differentiate between the ability of the eye or the observer to see standard types of test objects and the ability of a given visual display to be seen by a standard observer. In the first case we refer to the *sensitivity of the observer* and in the second to the *visibility of a display*. These are matters of practical importance, because we may be interested in how sensitivity varies from one observer to the next or how we can adapt optical aids to help an observer see what needs to be seen. On the other hand, we may try to modify a display or change the illumination in an environment to make a display visible to a standard observer; here we are concerned about the visibility of some critical detail or feature of the display.

1. Definition of Visibility

The word visibility can be used to describe the visibility of an object bounded by a contrast border or the visibility of the border itself. In this context a point is a small area surrounded by a border, and a line is a narrow bar flanked by two borders. The object or the border may be defined as the detail to be seen. For an object on a background the contrast C is defined (IES Lighting Handbook, 1981, pp. 3–19) as

$$C = \left| \frac{L_D - L}{L} \right|, \tag{39}$$

where L_D is the luminance of the object, and L is the luminance of the background. For a bipartite field neither half can be identified as the background, but we can arbitrarily choose the dark half to be the background. We can do the same thing for a square-wave grating, but in this case we can also use the average luminance for the denominator.

The contrast can be reduced to the threshold of visibility by reducing the difference between L_D and L while L is kept constant until the threshold is reached. At this level the contrast is called the threshold contrast $\bar{C}$. The visibility is the ratio of C to $\bar{C}$,

$$V = C/\bar{C}. \tag{40}$$

2. Measurement of Visibility

The visibility of a given display can be measured directly with a visibility meter that superimposes a white patch of veiling luminance L_V and at the same time reduces the luminance of both the disk and the background by a factor F such that

$$L_V + FL = L. \tag{41}$$

At the threshold

$$\bar{C} = \left|\frac{L_D F - LF}{LF + L_V}\right| = F\left|\frac{L_D - L}{L}\right| = FC. \tag{42}$$

Hence,

$$V = 1/F. \tag{43}$$

The visibility meter can also be used to measure the visibility of any complex display involving a combination of crossing borders or parallel borders. The borders may be sharp or blurred; they may involve luminance contrast or color contrast In the case of a complex display the background luminance is defined as the average luminance of the area on which the veiling patch is superimposed. For a disk on a darker background the measurement of the background luminance is based on a portion of the background that does not include the disk.

It must be noted that measurement of visibility does not have to include the measurement and specification of contrast.

In some instances the visibility of a contrast border involves a difference in chromaticity on the two sides of the border. In this case the contrast-reducing device changes the chromaticity on each side toward white, but nothing is assumed about how the difference in chromaticity contributes to the visibility.

The CIE report No. 19/2 (CIE, 1979) approved by the 3.4.2 Committee on Visual Performance and also approved by the CIE Council has endorsed the above concept of visibility and, furthermore, has adopted the concept that the level of visibility for a given display can be regarded as equivalent to that of a bright 4-min disk on a darker background when both have the same background luminance and the same suprathreshold level of visibility.

Based on the data of Blackwell and Blackwell (1971), the same group has adopted a standard curve (Fig. 39) for the contrast thresholds $\bar{C}$ of a 4-min disk at different levels of background luminance L for observers from 20 to 30 years of age,

$$\bar{C} = 0.05936[(1.639/L)^{0.4} + 1]^{2.5}. \tag{44}$$

The threshold curve defined by Eq. (44) has the property that at low values of background luminance it approaches a slope of -1, and at high values it becomes independent of the background.

In the CIE system suprathreshold levels of visibility are defined in terms of log units of contrast above the threshold level. Other approaches may also be used. For example, the perceived contrast of a given display can be compared with the perceived contrast of a standard test object of a specifiable contrast (Yonemura and Tibbott, 1981).

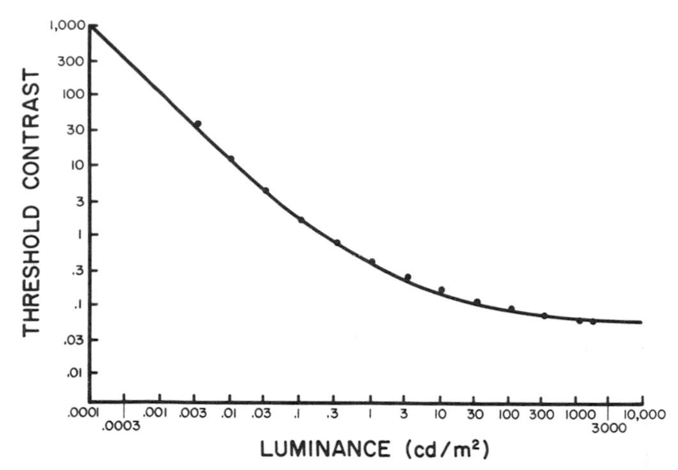

Fig. 39. Effect of luminance of the background on the contrast threshold for a bright 4-min disk on a darker background. Data of Blackwell and Blackwell, 1971, for 68 observers aged 20–30 yr. On-axis location, 0.2-sec exposures, adjustment to visibility threshold. (CIE, 1979.)

REFERENCES

Aguilar, M., and Stiles, W. S. (1954). Saturation of rod mechanism of the retina at high levels of stimulation. *Opt. Acta* **1**, 59–65.

Bartlett, N. R. (1965). *In* "Vision and Visual Perception" (C. H. Graham, ed.), pp. 154–184. John Wiley and Sons, New York.

Bedell, H. E., and Enoch, J. M. (1979). A study of the Stiles-Crawford (S-C) function at 35° in the temporal field and the stability of the foveal S-C peak over time. *J. Opt. Soc. Am.* **69**, 435–442.

Blackwell, H. R. (1952). Brightness discrimination data for the specification of quantity and quality of illumination. *Ill. Eng.* **47**, 602–609.

Blackwell, H. R., and Kristofferson, A. B. (1958). "The Effects of Size and Shape on Visual Detection for Continuous Foveal Targets at Moderate Background Luminance." Report of the University of Mich. Eng. Res. Inst. (Project 2455) to the Bureau of Ships, Department of the Navy (under Contract No. Nobs–72038).

Blackwell, H. R., and Smith, S. W. (1958). "The Effect of Target Size and Shape on Visual Detection." Project Michigan Report 2144–46T, Ann Arbor.

Blackwell, O. M., and Blackwell, H. R. (1971). Visual performance data for 156 normal observers of various ages. *J. Ill. Eng. Soc.* **1**, 3–13.

Bouman, M. A., and van der Velden, H. A. (1947). The two-quanta explanation of the threshold values and visual acuity on the visual angle and the time of observation. *J. Opt. Soc. Am.* **37**, 908–919.

Brown, W. W. (1977). "The Effect of a Border in the Visual Field on the Visibility of a Nearby Border." Doctoral Dissertation, Ohio State University, Columbus, Ohio.

Byram, G. M. (1944). Physical and photochemical basis of visual resolving power. *J. Opt. Soc. Am.* **34**, 718–724.

Campbell, F. W. (1958). A retinal acuity direction effect. *J. Physiol.* **144**, 25P–26P.

Campbell, F. W., and Green, D. G. (1965). Optical and retinal factors affecting visual resolution. *J. Physiol.* **181**, 576–593.

CIE (1979). "An Analytic Model for Describing the Influence of Lighting Parameters upon Visual Performance." Publication CIE No. 19/2 (TC.-3.1), 1980.

Crawford, B. H. (1938). The luminous efficiency of light entering the eye pupil at different points and its relation to brightness threshold measurements. *Proc. R. Soc.* **124B**, 81–96.

Dittmers, F. (1920). Ueber die Abhängigkeit der Unterschiedsschwelle für Helligkeiten von der Antagonistischen Induktion. *Z. Psychol. Physiol. Sinnesorg.* **51**, 214–232.

Drum, B. (1975). Additivity of the Stiles-Crawford effect for a Fraunhofer image. *Vision Res.* **15**, 291–298.

Enoch, J. M. (1957). Amblyopia and the Stiles-Crawford effect. *Am. J. Optom. Arch. Am. Acad. Optom.* **34**, 298–308.

Enoch, J. M. (1959). Receptor Amblyopia. *Am. J. Ophthalmol.* **48**, 262–274.

Enoch, J. M. (1960a). Optical interaction effects in models of parts of the visual receptors. *Arch. Ophthalmol.* **63**, 548–558.

Enoch, J. M. (1960b). Waveguide modes: Are they present and what is their possible role in the visual mechanism? *J. Opt. Soc. Am.* **50**, 1025–1026.

Enoch, J. M. (1961). Nature of transmission of energy in the retinal receptors. *J. Opt. Soc. Am.* **51**, 1122–1126.

Enoch, J. M., and Fry, G. A. (1958). Characteristics of a model retinal receptor studied at microwave frequencies. *J. Opt. Soc. Am.* **48**, 899–911.

Enoch, J. M., and Laties, A. M. (1971). An analysis of retinal receptor orientation, II. *Invest. Ophthalmol.* **10**, 959–970.

Enoch, J. M., Birch, D. G., and Birch, E. E. (1979). Monocular light exclusion for a period of days reduces directional sensitivity of human retina. *Science* **206**, 705–707.

Flamant, F., and Stiles, W. S. (1948). The directional and spectral sensitivities of the retinal rods to adapting fields of different wavelengths. *J. Physiol.* **107**, 187–202.

Flom, M. C., Weymouth, F. W., and Kahneman, D. (1963). Visual resolution and contour interaction. *J. Opt. Soc. Am.* **53**, 1026–1032.

Fry, G. A. (1931). The relation of border contrast to the distinctness of vision. *Psych. Rev.* **38**, 542–549.

Fry, G. A. (1947). The relation of the configuration of a brightness contrast border to its visibility. *J. Opt. Soc. Am.* **37**, 166–175.

Fry, G. A. (1948). Mechanisms subserving simultaneous brightness contrast. *Am. J. Optom. Arch. Am. Acad. Optom.* **25**, 162–175.

Fry, G. A. (1953). Effect of brightness distribution in the entire field of view upon the performance of a task. *Eng. Exp. Stat. News.* **25**, 22–28.

Fry, G. A. (1965a). The eye and vision. *In* "Applied Optics and Optical Engineering" Academic Press, London. (R. Kingslake, ed.), Vol. II, pp. 1–76.

Fry, G. A. (1965b). Physiological irradiation across the retina. *J. Opt. Soc. Am.* **55**, 108–111.

Fry, G. A. (1969a). Mechanisms subserving bright and dark adaptation. *Am. J. Optom. Arch. Am. Acad. Optom.* **46**, 319–338.

Fry, G. A. (1969b). Visibility of Sine-wave Gratings. *J. Opt. Soc. Am.* **59**, 610–617.

Fry, G. A. (1970). The optical performance of the human eye. *In* "Progress in Optics" (E. Wolf, ed.), Vol. VIII, pp. 51–131. North Holland, Amsterdam.

Fry, G. A. (1973). Mechanisms subserving surface and border brightness contrast. *Am. J. Optom. Arch. Am. Acad. Optom.* **50**, 17–33.

Fry, G. A., and Bartley, S. H. (1935). The effect of one border in the visual field upon the threshold of another. *Am. J. Physiol.* **112**, 414–421.

Fry, G. A., and Cobb, P. W. (1935). A new method of determining the blurredness of the retinal image. *Trans. Am. Acad. Ophth. Otolaryn.* pp. 423–428.

Fry, G. A., and Enoch, J. M. (1959). "The Relation of Blurr and Grain to the Visibility of Contrast Borders and Gratings" (Report to Rome Air Development Center under Contract No. AF30(602)1580). The Ohio State University Research Foundation, Columbus, Ohio.

Fry, G. A., and Robertson, V. M. (1935a). Alleged effects of figure-ground upon hue and brilliance. *Am. J. Psych.* **47**, 424–435.

Fry, G. A., and Robertson, V. M. (1935b). The physiological basis for the periodic merging of an area into background. *Am. J. Psych.* **47**, 644–655.

Fuortes, M. G. F., Gunkel, R. D., and Rushton, W. A. H. (1961). Increment thresholds in a subject deficient in cone vision. *J. Physiol.* **156**, 179–192.

Graham, C. H., and Bartlett, N. R. (1939). The relation of size of stimulus and intensity in the human eye: II. Intensity thresholds for red and violet light. *J. Exp. Psychol.* **24**, 574–587.

Harris, C. (1980). "Visual Coding and Adaptability," pp. 115–136. Erlbaum, Hillsdale, N.J.

Hattwick, R. G. (1954). Dark adaptation to intermediate levels and to complete darkness. *J. Opt. Soc. Am.* **44**, 223–228.

Hecht, S., Shlaer, S., and Pirenne, M. H. (1942). Energy, quanta and vision. *J. Gen. Physiol.* **25**, 819–840.

Helmholtz, H. von. (1924). "Helmholtz's Treatise on Physiological Optics" (J. P. C. Southall, ed.), Vol. II, pp. 34–35 (Am. ed., Op. Soc. Am.).

Ill. Eng. Soc. of North America (1981). "IES Lighting Handbook" (J. Kaufman, ed.), Reference Volume, pp. 3–14. Ill. Eng. Soc. N. A.

Laties, A. M., and Enoch, J. M. (1971). An analysis of retinal receptor orientation. *Invest. Ophthalmol.* **10**, 69–77.

Le Grand, Y. (1935). Sur le mesure de l'acuité visuelle au moyen de franges d'interférence. *Comptes Rendu Acad. Sci.* **200**, 490–491.

Moon, P., and Spencer, D. E. (1944). On the Stiles-Crawford Effect. *J. Opt. Soc. Am.* **34**, 319–329.

Nachmias, J. (1967). Effect of exposure duration on visual contrast sensitivity with square wave gratings. *J. Opt. Soc. Am.* **57**, 421–427.

Nolan, G. F. (1957). On the functional relation between luminous energy, target size and duration for foveal stimuli. *J. Opt. Soc. Am.* **47**, 394–397.

NRC Committee on Vision (1980). Recommended Standard Procedures for the Clinical Measurement and Specification of Visual Acuity (Report of Working Group 39). *Adv. Ophthalmol.* **41**, 103–141.

O'Brien, B. (1945). A theory of the Stiles-Crawford Effect. *J. Opt. Soc. Am.* **36**, 506–509.

O'Brien, B. (1950). Local variations of the Stiles and Crawford effect. *J. Opt. Soc. Am.* **40**, 796.

O'Brien, B. (1951). Vision and resolution in the central retina. *J. Opt. Soc. Am.* **41**, 882–894.

Ohzu, H., Enoch, J. M., and O'Hair, J. (1972). Optical modulation by isolated retina and retinal receptors. *Vision Res.* **12**, 231–244.

Østerberg, G. A. (1955). Topography of the layer of rods and cones in the human retina. *Acta. Ophthalmol. Suppl.* **VI**. pp. 1–103.

Ratliff, F. (1965). "Mach Bands." Holden-Day, San Francisco.

Ratliff, F. (1972). Contour and Contrast. *Sci. Am.* **226**, 90–101.

Rushton, W. A. H. (1961b). Rhodopsin measurement and dark adaptation in a subject deficient in cone vision. *J. Physiol.* **156**, 193–205.

Schober, H. A. W., and Hiltz, R. (1965). Contrast sensitivity of the human eye for square-wave gratings. *J. Opt. Soc. Am.* **55**, 1086–1091.

Shlaer, S. (1937). The relation between visual acuity and illumination. *J. Gen. Physiol.* **21**, 165–188.

Stiles, W. S. (1939). The directional sensitivity of the retina and the spectral sensitivities of the rods and cones. *Proc. R. Soc.* **B127**, 64–105.

Stiles, W. S., and Crawford, B. H. (1933). Luminous efficiency of rays entering the eye pupil at different points. *Proc. R. Soc.* **B112**, 428–450.

Toraldo di Francia, G. (1949). Retina cones as dielectric antennas. *J. Opt. Soc. Am.* **39**, 324.

Van Nes, F. L. (1968). "Experimental studies in Spatiotemporal Contrast Transfer by the Human Eye." Doctoral Thesis, University of Utrecht.

Vos, J. J., and Walraven, P. L. (1962). The Stiles-Crawford Effect. A survey. *Atti Fond. Giorgio Ronchi* **17**, 302–318.

Westheimer, G. (1960). Modulation thresholds for sinusoidal light distributions on the retina. *J. Physiol.* **152**, 67–74.

Yonemura, G. T., and Tibbott, R. L. (1981). Equal apparent conspicuity contours with five-bar grating stimuli. *J. Ill. Eng. Soc.* **10**, 155–163.

OPTICAL RADIATION MEASUREMENTS, VOL. 5

4

Visual Pigments and Sensitivity

ROBERT M. BOYNTON
Department of Psychology,
University of California, San Diego
La Jolla, California

A. INTRODUCTION

1. General Remarks

Most of the topics of this chapter have been treated elsewhere in a more quantitative fashion. For research purposes such detail is important. For this survey, however, the reader might miss the forest—not so much by overlooking its trees—but rather because of a preoccupation with intricate

ISBN-0-12-304905-9

description of their limbs. The trees are here, and the interested reader should have little trouble following their branches by way of references that are provided; many citations to secondary sources include a page number where the start of a relevant discussion occurs and additional primary references may be located.

2. Photopic versus Scotopic Spectral Sensitivity

The psychological historian E. G. Boring (1942) characterized the work of Johannes Purkinje, a 19th-century Austrian physiologist, as representing "... the phenomenology of vision in its best and most effective form." Purkinje (1823) was first to report that on the transition from daylight to twilight the relative brightnesses of different-colored objects fail to maintain a constant relation. A modern example is found in some of the license plates issued from California's Department of Motor Vehicles. By daylight these are seen as having bright yellow alphanumeric characters upon a relatively dark blue background. At sunset the figures dim faster than their background; under starlight they disappear altogether, and the plate assumes a uniform gray appearance.

The explanation for this kind of intensity-dependent change in the relative brightnesses of colors, appropriately known as the *Purkinje shift*, is found in the differing action-spectra of the two principal classes of human photoreceptors, the rods and cones. A comparison between them can be made psychophysically by using threshold (see Chapters 6 and 7) as a criterion. By prior adaptation to a bright field, from which rods recover more slowly than cones, exclusive excitation of fully recovered cones can be achieved for a few minutes during the period following the termination of the bleaching field. (In the vision laboratory this is known as working on the *cone plateau* of the dark adaptation curve.) The resulting threshold versus wavelength function for cone vision as obtained by Wald (1945) is shown in Fig. 1.

Except for the longest wavelengths, rod sensitivity is easily measured using weak stimulation of the peripheral retina, where rod vision dominates. The procedure succeeds not because individual rods are more sensitive than individual cones but because the summation pools of rods are far greater than those of cones. Consequently, photons absorbed by peripheral cones arouse a negligible response of the visual system under conditions in which enough rods are excited to elicit a threshold sensation (see Brindley, 1970, p. 177).

The term *photopic* characterizes visual function under conditions of high-intensity illumination in which cone photoreceptors dominate and the rods contribute relatively little, either because their responses are saturated (Aguilar and Stiles, 1954) or because they are inhibited by signals from cones

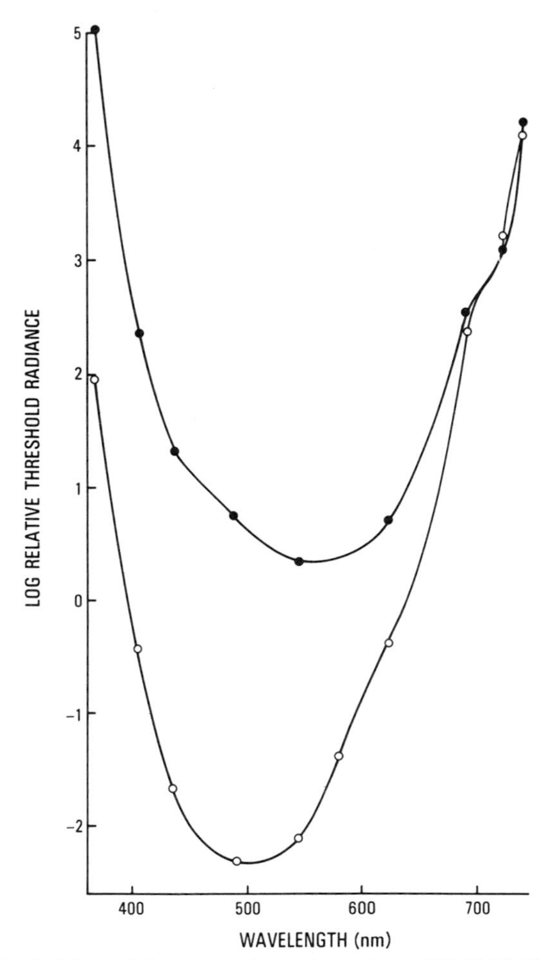

Fig. 1. Log threshold sensitivity curves plotted from data of Wald (1945). Stimuli were 1° in visual angle, presented for 40 msec, 8° above the fovea. The plot shows log relative threshold (flash radiance required for threshold visibility) versus wavelength. The open circles are for the dark-adapted eye (rod vision); the closed circles were obtained on the cone plateau following a high-intensity bleach (cone vision). Wald's data for foveal stimulation lie close to the lower curve.

(Makous and Boothe, 1974). *Scotopic* refers to dim conditions in which rods alone mediate visual response. Sometimes these terms are intended to characterize the state of adaptation of the eye. If so, it is important to realize that illumination levels normally regarded as photopic (for example, interior illumination) nevertheless allow a significant rod contribution; conversely, cones are readily stimulated in the dark-adapted eye by bright flashes that by themselves are too small or transient to significantly alter the adaptive state.

Vision also occurs under intermediate *mesopic* conditions in which rods and cones contribute to visual sensations more or less equally.

Threshold functions of wavelength are also plotted in Fig. 1 for conditions favoring the selective excitation of rods. A logarithmic ordinate is used, because an enormous range of threshold radiances must be represented and also because the variance of threshold measurements tends toward homogeneity on this scale. (In other words, experimental error approximates a constant percentage of mean threshold value.) The use of threshold as a response criterion permits a valid comparison, on an absolute basis, of rod and cone sensitivity functions for the particular size and duration of the test flash used.

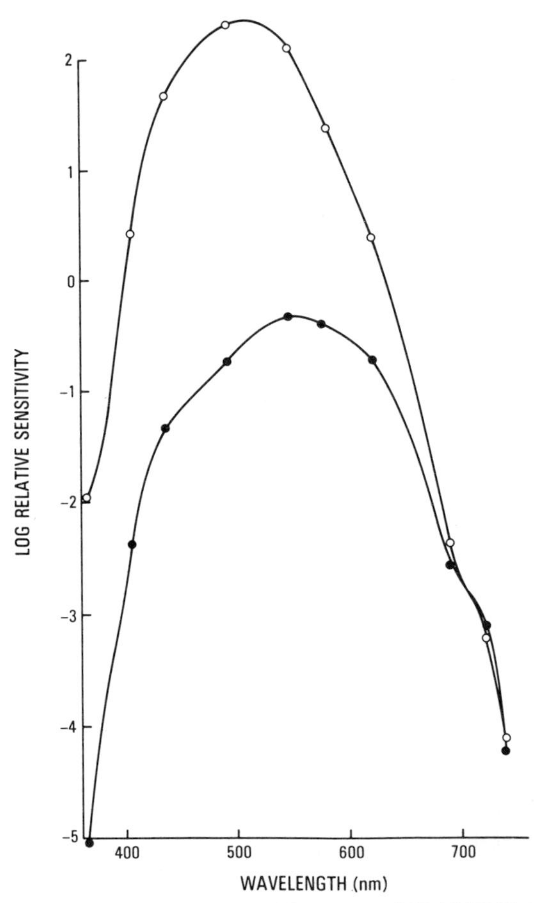

Fig. 2. Log relative sensitivity curves plotted from data of Wald (1945). The plot shows log relative sensitivity versus wavelength (reciprocal flash radiance required for threshold visibility). These are the same data shown in Fig. 1, inverted to convert from log threshold to log sensitivity.

A vertical slice through the region between the rod and cone functions defines a *photochromatic interval* within which light is seen using rods but color perception (which depends on cones) is not evident. This interval is greatest at about 480 nm, where rod-mediated thresholds are roughly 1000 times lower than those of cones. At the longest wavelengths, at which rod and cone sensitivities are nearly equal, there is no photochromatic interval. In the neutrally adapted eye it is not possible to excite enough rods to produce a visual sensation without also exciting cones; such stimuli appear reddish.

It is common to plot data of this kind on a sensitivity basis, in which sensitivity is defined as the reciprocal of threshold. When a logarithmic scale is used on the ordinate, the sensitivity plot has the same shape as the threshold plot but is inverted, as shown in Fig. 2. Figure 3 results from normalizing rod and cone sensitivities independently (setting their peak values equal to 1.0) and plotting the resulting curves on a linear ordinate. This common way of comparing rod and cone sensitivity functions suffers from two deficiencies. First the independent normalization of the two functions easily leads to a false impression that cones are much more sensitive than rods at long wavelengths. Second the curves seem to reach zero values when they approach about 1% of their peaks. The tails of such curves, as is evident from the logarithmic plot of Fig. 2, actually consist of monotonically decreasing values which can be measured accurately by psychophysical methods and which are important for color vision.

Spectral sensitivity may also be measured using suprathreshold criteria. For example, with two fields of different colors placed side by side, an observer may be instructed to adjust the intensity of one field so that it will match the other one for brightness. Or a flickering stimulus, whose two components are alternated, may be used. These methods, which are described in detail in Section C, do not permit an absolute comparison between rod and cone sensitivities. For rod vision the choice of method does not affect the shape of

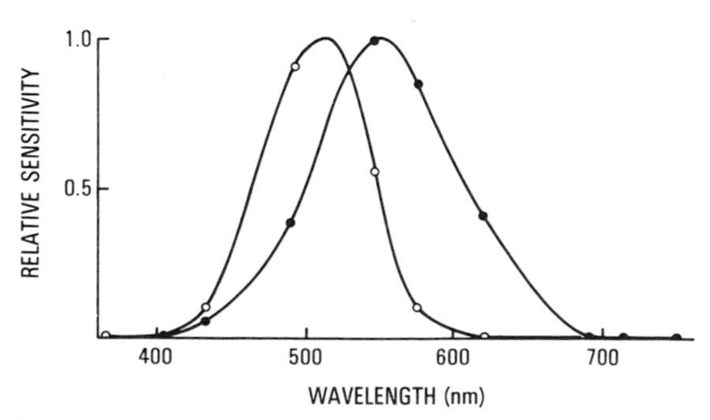

Fig. 3. Data of Fig. 2, normalized at peak sensitivity and plotted on a linear sensitivity basis.

the resulting sensitivity function in any important way, because only one type of visual photopigment is found in rods (see Section B). As wavelength is varied, the measurement of sensitivity depends on adjusting the rate of photon incidence on rods to compensate for a varying probability of photon absorption. As a result, if the intensities of two stimuli that produce the same threshold excitation of rods are increased in the same ratio, both fields will be brighter than before, but they will continue to match.

Although spectral sensitivity for rod vision is, therefore, independent of the response criterion, this is not true for cone-mediated spectral sensitivity, because three different types of cones contribute to the overall function, and their relative contributions vary with the response criterion.

3. Factors on Which Visual Sensitivity Depends

a. Photochemical Factors

All vision depends on the absorption of photons of light by molecules of visual photopigments. The absorption of each photon bleaches a molecule, leading to an electrochemical element of visual response (Baylor *et al.*, 1979). Each incident photon vibrates at a characteristic frequency inversely proportional to its wavelength. Although every molecule of photopigment has the potential for being excited by a photon of any frequency, this probability is low at the molecular level even for the optimal frequency. A high level of visual sensitivity nevertheless is achieved, because tens of millions of photopigment molecules are stacked in specialized layers, called disks, that make up the receptor outer segment (Sjöstrand, 1953; Ripps and Weale, 1976, p. 16; Rodieck, 1973, p. 70). The excitation of a single pigment molecule anywhere in any disk of the outer segment arouses an element of excitation of the receptor as a whole; thus excited, the receptor is capable of transmitting a signal to the next neural elements in the visual chain, the bipolar and horizontal cells of the retina. The probability that a photon will be absorbed by a photopigment molecule depends on the frequency of the incident light. For each type of visual pigment there is some particular frequency that is most effective on it; the corresponding wavelength, in nanometers, is called λ_{max} (lambda max). Quantum efficiency is high in the visual system; about two thirds of the absorbed photons give rise to an element of photoreceptor response. Quantum efficiency is independent of wavelength across the visible spectrum.

Many visual pigments have been extracted and studied chemically (Lythgoe, 1972). Dartnall (1953) found that a wide variety of these, obtained from many animal species, could be characterized by absorption spectra that (if plotted as a function of light frequency) have the same shape but differ only in λ_{max}. Dartnall developed a nomogram, shown in Fig. 4, that interrelates

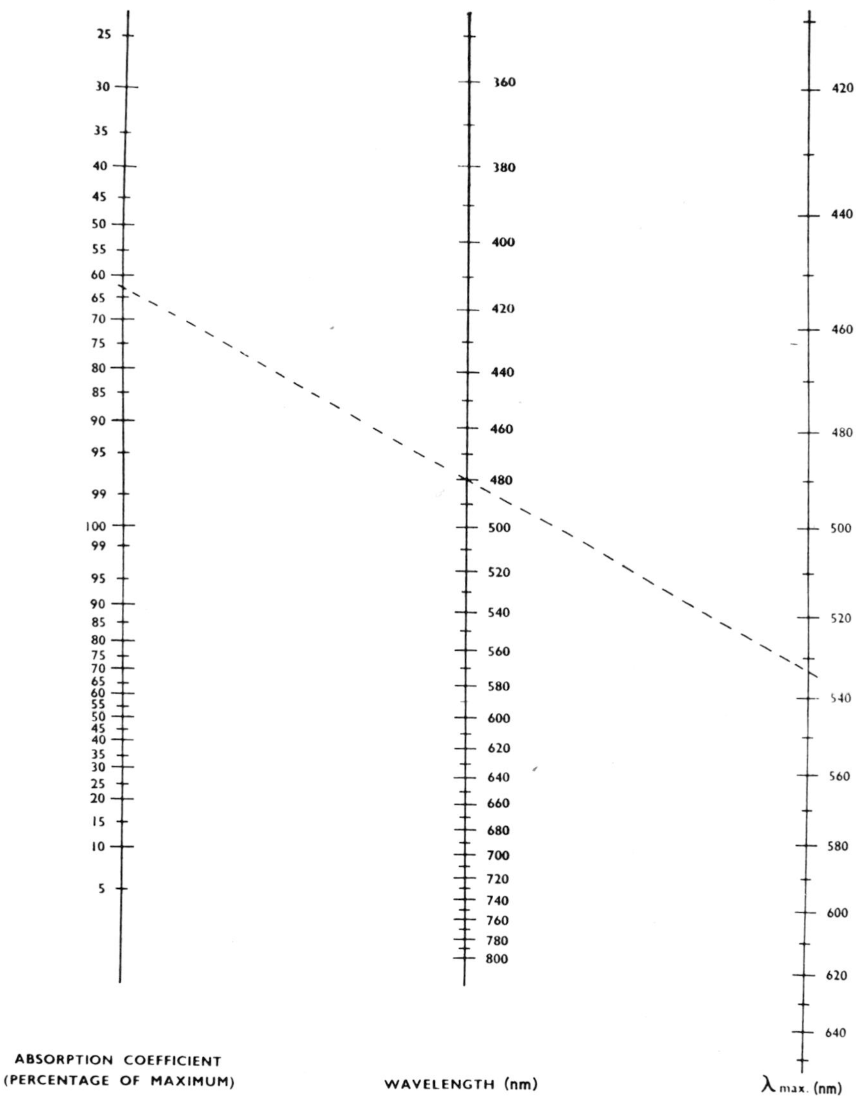

Fig. 4. The Dartnall nomogram. By laying a straight edge across the chart, so that it intersects the three vertical scales, the relations between percent absorption, wavelength and λ_{max} are indicated. For the example shown, a pigment with $\lambda_{max} = 535$ nm will have an absorption coefficient 63% of maximum at a test wavelength of 480 nm. From Dartnall (1953).

λ_{max}, relative percent absorption, and the wavelength of a test light. Given any two of these quantities, the third can be obtained by laying a straight edge across the chart. Photopigments having action spectra characterized by this diagram are sometimes called *nomogram pigments* and may be said to follow *Dartnall's rule.*

Little theoretical significance can be attached to Dartnall's generalization, because it is not yet well understood just what kind of interaction between the frequency of photon vibration and the molecular structure of visual photopigments accounts for Dartnall's observation (LeGrand, 1972). Theory aside, Dartnall's generalization, if strictly true, would make possible the specification of the entire action spectrum of a photopigment from measurements confined to the accurate determination of λ_{max}.

Once a photon is absorbed and bleaches a photopigment molecule, the resulting rod signal is probably the same no matter what the frequency of the absorbed photon or where in the outer segment it is absorbed (Rodieck, 1973, p. 262; Baylor *et al.*, 1979). On this basis one can understand why any two scotopic stimuli can be made to look the same no matter how different their spectral composition. The sole requirement for equivalence is that the same rate of photon absorption be produced for each stimulus; this is always possible by adjusting the relative radiances. These ideas are subsumed under the *principle of univariance* elucidated by Naka and Rushton (1966).

b. Physiological Factors

Scotopic vision depends upon light absorption in rods, all of which are fundamentally alike. Because, by the principle of univariance, any two stimuli of different wavelength that are adjusted to be equally effective will produce the same rate of photon absorption in all of the rods that are illuminated by them, no change in rod-mediated spectral sensitivity would be expected; and none is found as a function of variables such as field size, stimulus duration, or retinal location.

Absolute sensitivity is another matter. Rods do not function independently of one another. At an initial stage of pooling of rod signals, rods mutually augment their responses by a factor of at least five, compared with their response in isolation (Fain, 1975). In each human eye there are more than one hundred million rods; but there are only about one million ganglion cells, whose axons constitute the fibers of the optic nerve that carries impulses from eye to brain. The neural convergence implied by these figures occurs in overlapping summation pools.

At threshold the absorption of only one photon in each of a few rods, perhaps half a dozen, is sufficient to arouse a visual sensation (Hecht *et al.*, 1942). In the periphery of the dark-adapted eye such a threshold stimulus can be spread without loss of efficiency over an area containing several thousands of rods, only a tiny fraction of which will absorb a photon.

Because visual stimuli are ordinarily continuous in time, radiance (or its photometric counterpart, luminance) is usually the relevant measure of light intensity. Increasing the area of a stimulus of fixed radiance, as can be accomplished by enlarging a variable diaphragm placed before a uniform field, increases the number of photons per unit time supplied to the summation pool underlying the retinal image. When this is done, the stimulus becomes progressively more effective until a limiting area is reaching, beyond which area has no effect (Brindley, 1970, p. 163). Similarly, stimulus effectiveness can be increased by lengthening the duration of a flash of fixed radiance up to a limit (Brindley, 1970, p. 169). The reciprocal relation between radiance and area is known as *Ricco's law*; that between radiance and duration is *Bloch's law*. The most efficient visual stimulus, measured in terms of numbers of photons (related to the energy in a flash), must be sufficiently small and brief to fall within the spatial and temporal intervals in which photon integration is complete. For rod vision these limits are approximately 1° and 100 msec.

Because of the presence of three classes of cones, spectral sensitivity for photopic vision depends on interactions among the outputs of the disparate cone types, each of which has its own action spectrum. This interaction is reflected in the wide variety of spectral sensitivities found among the single cells of the visual system that receive their inputs from cones, often complicated by rod inputs as well. Psychophysical procedures also lead to a variety of spectral sensitivity curves. This evidence suggests, with support from electrophysiology (DeValois and DeValois, 1975), that the visual pathways are organized into separate chromatic and achromatic "channels." Depending on which of these predominates under given conditions, the wide variety of spectral sensitivity functions that can be obtained in a psychophysical experiment (see Section C) seldom reflect the action spectra of cone photopigments.

c. Methodological Factors

Thresholds cannot be regarded as occurring abruptly as a function of stimulus radiance; instead the threshold is a statistically based concept. The human observer is not a machine; his concept of what should be judged as visible can vary and is subject to various experimental manipulations (see Chapter 6). A variety of psychophysical methods can be employed, each of which will give a different absolute result. The same kinds of concerns exist for comparing suprathreshold lights, for example by brightness matching.

The influence of methodological variables, like physiological ones, is not great when relative scotopic spectral sensitivity is being evaluated. Because of the univariant action of the test stimulus on a homogeneous collection of rods, the same relative spectral sensitivity will be obtained, one that relates closely to the action spectrum of rhodopsin, so long as a given methodology is consistently employed. Again this is not true for photopic vision. The relative contributions of the three cone types to visual signals are intensity dependent; methodological variations also can alter the average radiance levels corresponding to threshold or suprathreshold determinations. Moreover, the criterion chosen can have a direct impact. For example, very different spectral sensitivities will result depending on whether the subject is instructed to judge the presence of a minimally perceptible stimulus or one for which hue just barely disappears. For a stimulus delivered to the peripheral retina containing a mixed rod and cone population, the use of the first criterion leads to a scotopic spectral sensitivity curve, whereas the second yields a photopic function not much different from that for cones in Fig. 2.

B. SCOTOPIC SPECTRAL SENSITIVITY

1. Rhodopsin

Many of our present concepts about rod photopigments have resulted from efforts in two laboratories during the 1950s and 1960s, those of George Wald at Harvard of and H. J. A. Dartnall at the Institute of Ophthalmology in London. In 1962 Dartnall contributed four chapters to the first edition of Davson's *The Eye* that are strongly recommended; ten years later he edited an 800-page volume in the *Handbook of Sensory Physiology* series (Dartnall, 1972) on the photochemistry of vision, where even more detail is available. Excellent reviews of the work from Wald's laboratory can be found in Wald *et al.*, (1963) and Wald (1968), the latter based on Wald's Nobel address. In 1977 an issue of *Vision Research* commemorated the 100th anniversary of the discovery of rhodopsin, the photopigment found in rods, by publishing translations by Ruth Hubbard of classic German papers by Boll (1877/1977) and Kühne (1879). Also recommended are three chapters on visual photopigments in Rodieck's (1973) textbook *The Vertebrate Retina.*

A molecule of rhodopsin consists of the protein *opsin*, which has a molecular weight estimated at 40,000, and a much smaller prosthetic (added) structure known as the *chromophore.* An estimated 400 million rhodopsin molecules are located in each rod receptor. These molecules are not free to move along the length of a receptor, because they are located in and constitute part of the structure of the receptor outer segments.

To study rhodopsin *in vitro* the eyes of frogs have often been used; these contain the same kind of rhodopsin as that found in humans. The retina is removed from the eye, and an attempt is made to separate the outer segments of the receptors from the remainder of the material. Many eyes are needed to obtain enough outer segments for analysis. Rhodopsin is not soluble in water, but by using an aqueous solution of a detergent, such as digitonin, one can examine rhodopsin in solution. (Most common solvents cannot be used, because they would bleach or even destroy rhodopsin molecules.)

Vision depends, of course, upon the absorption spectrum of rhodopsin *in situ*, not in the test tube. There are various complications in making comparisons between the two. The absorption spectrum depends on the concentration of the solution and, for a given concentration, on its thickness; as either or both of these increases, the relative absorption spectrum broadens. For this reason chemists use an *extinction spectrum* that is equivalent to the relative absorption spectrum of an infinitely thin or dilute solution. The extinction spectrum can easily be calculated even when solutions of medium concentration are used experimentally. The extinction spectrum can be related in turn to the probability that a single photon of light will be absorbed by a molecule of rhodopsin. Relative spectral absorption for an infinitely dilute solution is known as relative *absorbance*.

When path lengths and concentrations are finite, the change in the relative absorption spectrum results from a phenomenon known as *self-screening*. Although a long path does not alter the probability that a photon will be absorbed once it reaches a given molecule of photopigment, the probability that the photon will reach such a "buried" molecule in the first place is significantly reduced because of the increased probability that it will have already been absorbed by a molecule located somewhere nearer the source of light. If the path length is long enough, virtually all photons will be absorbed eventually, even for those wavelengths for which absorption probability is very much lower for λ_{max}. In general, the longer the path length the broader the absorption spectrum (see Fig. 5). Self-screening can be a significant effect in cone-mediated vision, but it is not important for rod vision.

Many technical difficulties cloud the accurate measurement of the extinction spectrum of a photopigment. Chief among these is the need to obtain as pure a solution as possible. One way to deal with remaining impurities is to obtain a difference spectrum. This procedure depends on the fact that visual photopigments are labile, that is, when irradiated with light, they bleach. Because impurities do not bleach, absorption due to their presence remains constant. However, because the photoproducts caused by bleaching also absorb light (though they themselves do not bleach), the raw difference spectrum is seldom equivalent to an absorption spectrum. Another problem is that most eyes contain more than one pigment; for example, there are three cone pigments in addition to rhodopsin. A method of dealing with this situation uses partial bleaching with spectral lights that have selective action upon the separate pigments.

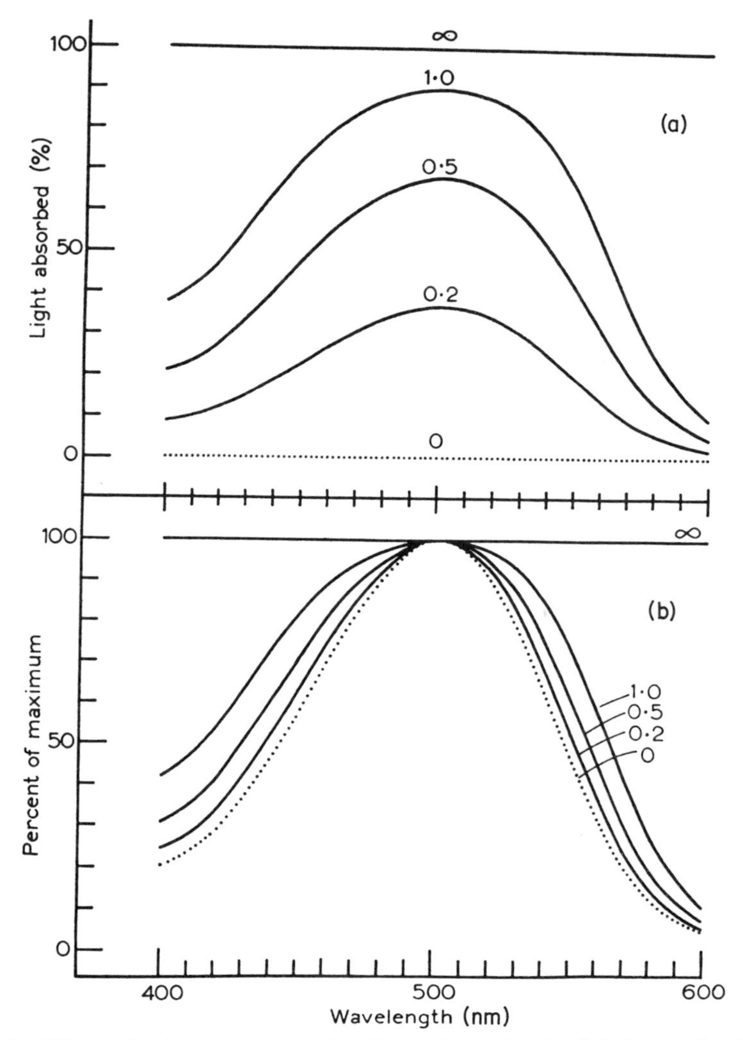

Fig. 5. Effects of self-screening. In (a), relative absorption is plotted as a function of wavelength for a pigment with maximum absorption at 500 nm. If the density of the solution (or receptor) through which the light passes approaches zero, no light is absorbed at any wavelength (dotted line). If the density of the solution is infinite, all light is absorbed at all wavelengths (top line). The shapes of the curves are more easily compared if each is normalized to percent of maximum as in (b). This shows the broadening of the relative absorption curves as density is increased in steps from nearly zero to 1.0. Further increases in density would eventually result in the flat curve at the top, which is for infinite density. The curve labelled 0 is also called the *relative absorbance* of the pigment, which can be regarded as the relative probability that a photon will be absorbed as a function of wavelength. (Dartnall, 1962.)

2. Bleaching

The chemical structure of the extensively studied rhodopsin chromophore is the aldehyde of vitamin A, known as *retinaldehyde* (or *retinal*). The chromophore is the active site of photon absorption. In the unbleached rhodopsin molecule, retinal assumes a *cis* form (see Fig. 6) that permits it to lock onto the opsin base. When a photon of light is absorbed, retinal straightens to another of its steric configurations, the *all-trans* form, in a process called *photoisomerization*, the initial stage of bleaching. This process unlocks the retinal within 10 psec and allows it to float free. Isomerization somehow causes an increase in the conductance of the receptor outer-segment membrane, which decreases the extracellular current flow around the receptor (which is maximal in the dark). Associated with this signal, perhaps caused by it, a packet of *neurotransmitter* is released at the proximal end of the receptor, informing the horizontal and bipolar cells of the retina, with which the rod makes synaptic contact, that a photon has been captured in their neighborhood.

A bleached molecule of opsin, from which the retinal has been detached, is no longer reactive to light; neither is the disassociated chromophore. Bleaching, therefore, causes a reduction of the density of photosensitive rhodopsin within the rod receptor. Although this partially desensitizes the receptor by reducing the probability of photon absorption, this depletion effect of bleaching is negligible for human scotopic vision, beacuse rod signals become saturated at field intensities that bleach only about 3% of their rhodopsin molecules (Aguilar and Stiles, 1954). Once isomerization occurs, a variety of other reactions quickly follow. Regeneration of rhodopsin is important, because vision would cease without it. These processes have been

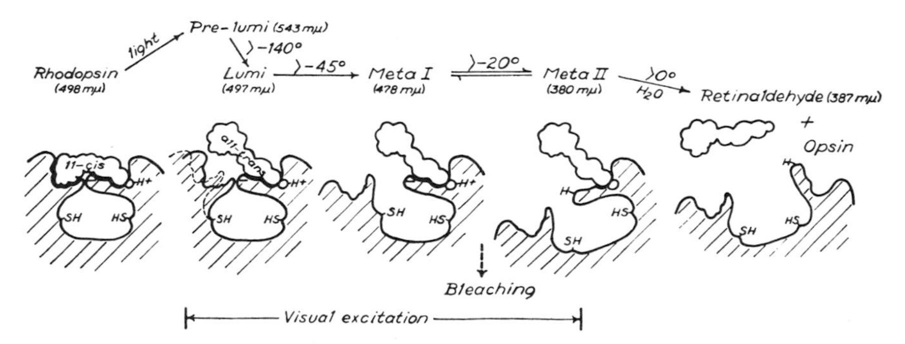

Fig. 6. Photoisomerization and bleaching. In this schematic diagram from Wald (1968, p. 234), the photosensitive molecule has its chromophore in the 11-*cis* form attached to the protein base (far left) until excited by light, after which the chromophore straightens to the *all-trans* form and then is dissociated from the protein base.

fully studied, and although they are a crucial part of the total photochemical cycle upon which vision depends, they are not a direct part of the visual signalling process and will not be further discussed here.

3. The Rhodopsin Absorption Spectrum and Scotopic Sensitivity

The relative absorption spectrum of rhodopsin in dilute solution shows the fraction of incoming light absorbed (a dimensionless quantity) plotted as a function of wavelength, normalized to 1.0 at its peak. The units in which the light is measured do not matter. A *scotopic spectral sensitivity curve*, on the other hand, shows the reciprocal of the quantity of light required to cause a criterion visual effect, also plotted as a function of wavelength. Here the shape of the sensitivity curve will depend on whether the amount of light is measured in energy or in quantum units. (The shape of the curve depends on this choice, because the energy contained in a photon of light is inveresely proportional to the wavelength of the photon.) Consequently, if two scotopic spectral sensitivity curves, each representing the same data but plotted in these two different ways, are normalized at the wavelength of peak sensitivity near 505 nm, then the photon-based sensitivity curve will be higher for shorter wavelengths and the energy-based curve will be higher for the longer wavelengths. Vision depends on photons, not energy, because the absorption of a photon elicits an identical response in the receptor regardless of the energy that it contains. Therefore, it is photon-based sensitivity that must be used for a valid comparison with the chemical data.

4. Prereceptoral Absorption

In a psychophysical experiment one can only measure the amount of light reaching the cornea of the eye, only some of which reaches the receptors. If the fraction transmitted from cornea to retina were a constant function of wavelength, then relative sensitivity measured at the cornea would not differ from that which would be obtained if the measurements could be made directly at the retina. However, it is well established that this fraction is wavelength dependent Wald, 1945; Said and Weale, 1959; Wyszecki and Stiles, 1967).

There are two major sources of such wavelength dependency. These are the spectral absorptions of the *macular pigment* and of the *lens*. Macular pigment lies within the retina above the receptors and is confined to the central region of vision, probably extending no further than 10° from the foveal center. Although, for rod sensitivity measurements, the effects of the macular pigment could be avoided by using peripheral stimuli, this strategy would not work for photopic measurements requiring the use of the rod-free foveal area. Brown and Wald (1963) have shown that the macular pigment is the xanthophyll *lutein*; it has also been measured psychophysically by at least two methods. The density spectrum of the macular pigment is shown in Fig. 7. Although the shape of this curve would apply reasonably well to any observer, there are large differences in the density of the macular pigment from one individual to another.

The second major source of prereceptoral light absorption is the lens of the eye. Lens absorption curves for eyes of various ages are shown in Fig. 8 in which it is seen that all lenses absorb much more light in the short visible wavelengths than in the long ones, that absorption at all wavelengths increases with age, and that the increase with age is selective as a function of wavelength. Data of this kind have been obtained by measuring lenses extracted from human eyes and also by comparing spectral sensitivity measurements of aphakic individuals (whose lenses have been surgically removed) with those of normal observers.

Some animals have additional filters and reflectors in their eyes that are not found in humans. Examples are the *oil droplets* of the inner segments of many birds (Rodieck, 1973, p. 48) and the directionally reflecting *tepetum* found in various animals (including cats and dogs, but not humans). The tapetum allows reflected photons a second chance for absorption by receptors, one that is wavelengths dependent. Some of the light that is not absorbed passes out through the pupil causing the "eye shine" seen when the observer is nearly in line with the source of light that is irradiating the animal's eye.

There is also a good deal of scattered light in all eyes. Consequently some of the light that otherwise would reach the primary retinal image is visually

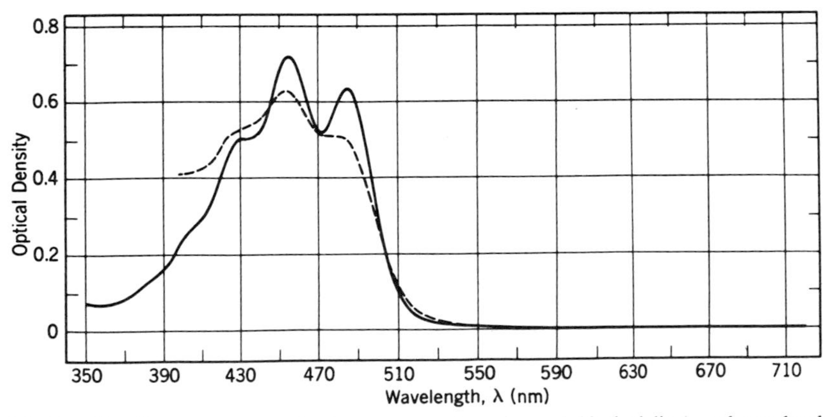

Fig. 7. Density spectrum of the human macular pigment (dashed line) and xanthophyll (continuous line). (Wyszecki and Stiles, 1967, p. 217, after Wald.)

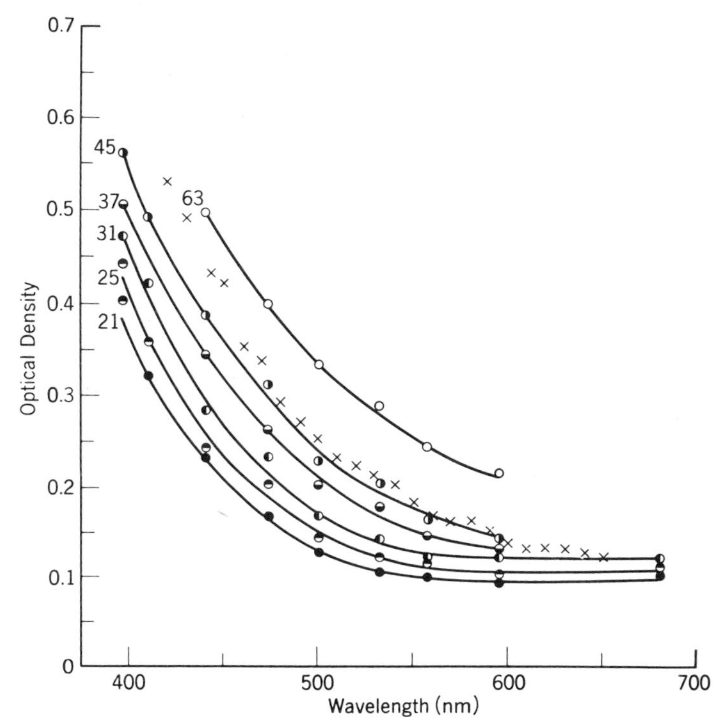

Fig. 8. Optical density of human eye lenses as a function of wavelength, with the age of the lens as a parameter. The crosses refer to mean densities for two eyes (ages 48 and 53) measured by a different method after surgical removal. (Wyszecki and Stiles, 1967, p. 217, after Said and Weale, 1959.)

ineffective (or may cause deleterious glare), because it is lost to irrelevant regions of the eye. Such scatter is, however, not strongly wavelength dependent (DeMott and Davis, 1960). The macular pigment and lens account for almost all of the selective absorption of light by the optical media of humans.

5. Red Shift

When artifacts introduced by impurities and the photoproducts of bleaching are taken into account as well as possible and the effects of screening pigments on the psychophysical measurements are also dealt with, there remains a small but nevertheless significant difference between the two functions being compared. The psychophysically determined sensitivity curve peaks slightly toward longer wavelengths than the *in vitro* data would predict. This *red shift* is probably related to the highly structured arrangement of the rhodopsin molecules in the receptor outer segments, which differs markedly from their random arrangement in the test tube (Brindley, 1970, p. 162).

C. PHOTOPIC SPECTRAL SENSITIVITY

The spectral sensitivity of the human eye at photopic levels of illumination depends upon three different classes of cone receptors, often further complicated by the activity of rods. Depending on specific stimulus parameters, including those that determine the state of adaptation of the eye, the relative excitation of the three kinds of cones will vary, as will the state of neural organization of the pathways that carry signals from cones to higher visual centers. For these reasons, as already noted, there is, for cone vision, no unitary photopic spectral sensitivity curve that transcends differences in conditions or method. Figure 9 shows, as an example, that variations in the state of chromatic adaptation of the eye can produce enormous changes in the shapes of spectral sensitivity curves that are based on foveal threshold measurements.

1. Theory

Photopic spectral sensitivity will be easier to understand if placed in an appropriate theoretical context. The concepts now to be presented to help elucidate the relation between cone pigments and photopic spectral sensitivity should be regarded as a useful model only. For more detail the reader is referred to *Human Color Vision* (Boynton, 1979, p. 211) and to Hurvich (1978) where further discussion and appropriate references will be found.

The three kinds of cones may be called ρ, γ, and β. The cones of different types are differentially sensitive to wavelength, because they contain different photopigments. When test stimulus radiance is within $\pm 60\%$ of a given adaptation level, signals from cones are nearly linear monotonic functions of retinal irradiance weighted according to the action spectrum of the cone. Signals from ρ and γ cones to double duty (Fig. 10). On the one hand they are summed and delivered to a luminance channel. A signal in the luminance channel, therefore, can arise from any ratio of excitations of ρ and γ cone signals that sum to produce it; the luminance channel receives no input from β cones and is color blind. On the other hand, an opponent-color signal is extracted from ρ and γ cones; it depends upon the difference between their signals. The sign of the opponent-color signal is regarded as negative for wavelengths shorter than the one producing chromatic balance (about 570 nm) and positive for wavelengths longer than this. The red—green opponent-color signal carries crude color information. For monochromatic

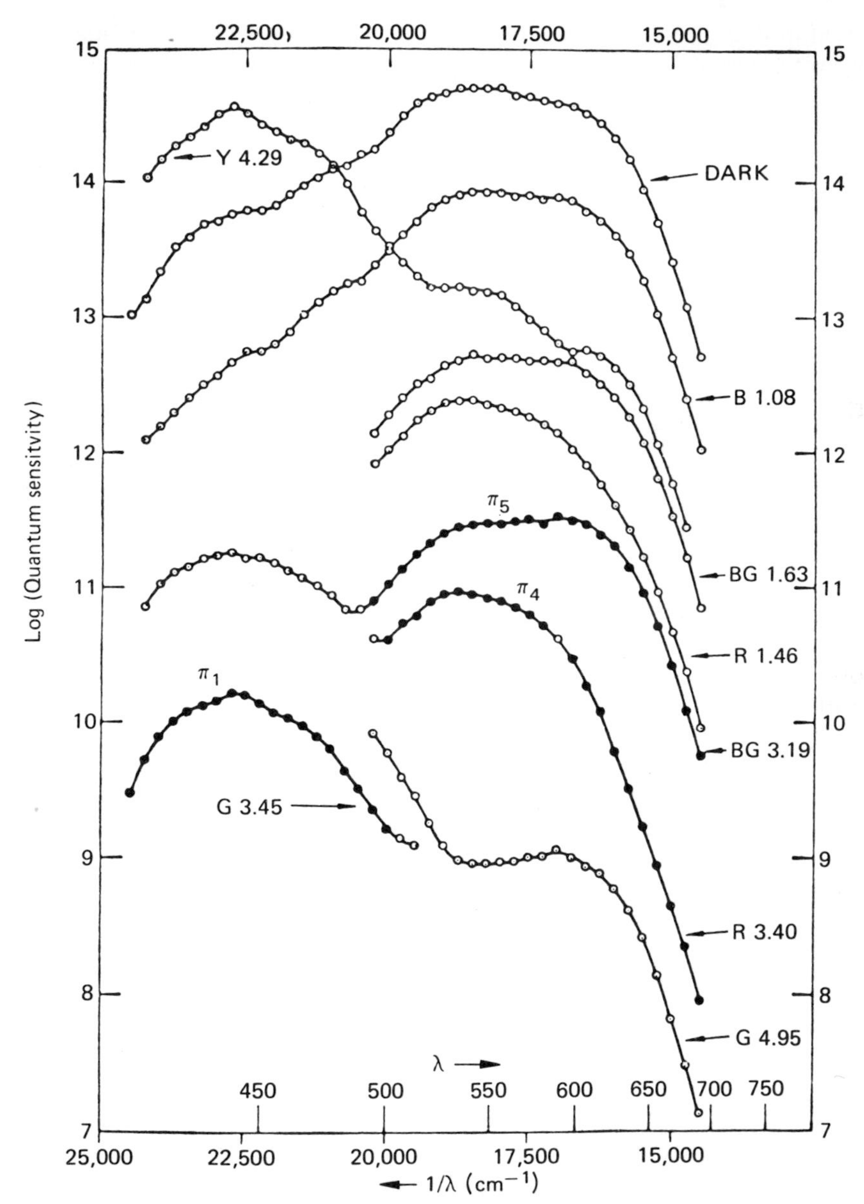

Fig. 9. Spectral sensitivity curves of human subjects adapted to blue, blue-green, green, yellow, and red (B, BG, G, Y, and R) backgrounds of the retinal illuminances indicated (in trolands). Characteristic shapes of π-mechanisms are labeled π_1, π_4, and π_5. Vertical displacements are arbitrary. (Boynton, 1979, p. 189, after Stiles, 1964.)

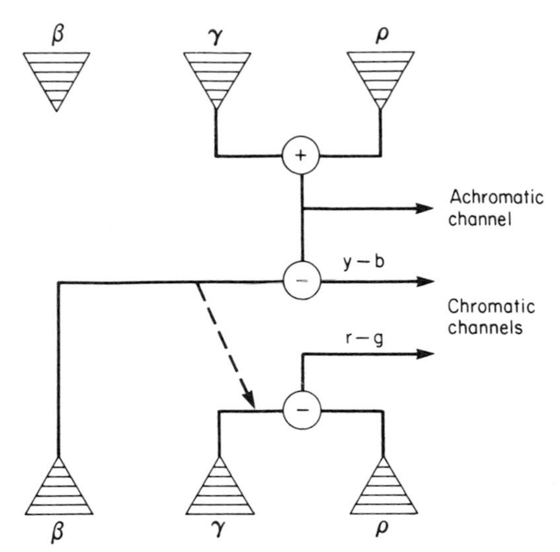

Fig. 10. Opponent-color model of human color vision. The β, γ, and ρ cones at the top represent the same three cones shown at the bottom (Boynton, 1979).

light this signal is restricted to an indication of which part of the spectrum has provided the stimulus. In the case of a mixture of wavelengths, the opponent-color signal can also show whether it is the excitation of ρ or γ cones that predominates.

A second opponent-color signal (yellow–blue) arises from a difference between the output of β cones and the $\rho + \gamma$ luminance signal. The neutral wavelength for this channel is near 500 nm.

Because no two spectral stimuli elicit exactly the same values of red–green and yellow–blue difference signals, they usually can be discriminated from one another. It is also true that no two wavelengths excite the ρ, γ, and β cones in exactly the same ratios. Together with the luminance channels, the two kinds of opponent pathways allow a reworking of the basic trichromatic code that begins with signals from three classes of photopigments housed in three kinds of cones. The appearance of the spectrum is easy to understand its terms of opponent-color theory, provided that one assumes that signals from β cones also feed to some extent into the red–green opponent channel to produce the reddish-blue (violet) seen at very short wavelengths.

See Fig. 10 for a schematic of the "wiring diagram" for the model just discussed. Keep in mind, however, that there are about 7 million cones in the retina, unevenly distributed; for simplicity the spatial role of the cones, which interacts in interesting ways with their role for color vision, has been ignored here.

2. Heterochromatic Photometry

Photometry, a subject of practical importance that is covered in detail in Chapter 11, is also a subject with some interesting theoretical twists. Except for metameric matches (Section D and also Chapter 12), no two stimuli at photopic intensity levels will look alike, no matter what their relative radiances. How then can it be said that there is, visually, a greater quantity of one light than of another? Heterochromatic photometry attempts to answer this question by putting the two lights side by side and then asking the observer to adjust their relative radiance until they look equally bright. This is not easy to do; the results are quite variable. Nevertheless, with enough measurements stable average results can be obtained. A more serious problem is posed by additivity failures (discussed in Chapters 5 and 11), which make it impossible to write a single linear equation that can be used to calculate values that predict relative brightnesses from spectroradiometric measurements.

a. Flicker

Flicker photometry was originally developed to meet the need for a method that would yield additive (linear) results (Rood, 1893; LeGrand, 1968, pp. 64, 109; Boynton, 1979, p. 299). Lights of different chromaticity are exchanged for one another at a moderate flicker rate of 10 to 15 Hz. Except at low frequencies the chromatic visual channels have a poor response to flicker and react to the flickering stimulus as if it were steady. The color of the light becomes that of an additive mixture of the two components, as if by simultaneous optical superposition. The luminance channels, on the other hand, exhibit an excellent response to flicker and can follow frequencies at least as high as 50 Hz. Thus, they react vigorously to flicker at the test rate of 10 to 15 Hz, provided that the inputs to the luminance channels from the two components of the flickering field are different. Therefore, if luminance is adjusted to produce the same $\rho + \gamma$ cone output from each component, the luminance channels will see this exchange as a steady light; because the chromatic channels cannot follow at this rate, there will be no perceived flicker.

Photometric evaluation by this method yields results that to a first approximation are not only additive but also transitive. Consequently, a spectral sensitivity curve can be obtained using an arbitrarily chosen reference stimulus. The resulting function relates to the sum of underlying sensitivities of ρ and γ cones. The β cones make no contribution (Eisner and MacLeod, 1980); therefore, two lights that are equated by flicker photometry may produce very different excitations of β cones, so that one of the lights may look very much bluer (and brighter) than the other. This might seem to invalidate the method, but it does not, because the excess β cone excitation caused by one of the stimuli does not seem to contribute importantly to the most critical aspects of visual performance. For example, visual acuity improves with luminance, yet there is little effect of wavelength on visual acuity for stimuli that have been equated by flicker photometry (Brown *et al.*, 1957; Riggs, 1965, p. 33; Meyers *et al.*, 1973).

b. Minimally Distinct Border

The method of *minimally distinct border*, which yields results very similar to those of heterochromatic flicker photometry, depends upon a critical juxtaposition of two fields being compared so that they share a common edge (Boynton and Kaiser, 1968; Boynton, 1978, p. 193). As an observer adjusts the intensity of one part of the field relative to the other, it is easy to find the setting where the border appears minimally distinct. The contour that remains depends only upon signals in the red-green opponent channels, whose strength depends upon the difference between the ratios of ρ and γ cone excitation for the two parts of the field. Just as for flicker photometry, β cones have been shown not to contribute to the distinctness of foveal borders. Both methods may be regarded as determining the condition of equal $\rho + \gamma$ signals for the lights being compared.

3. Increment Thresholds

Threshold probes are very useful for assessing the state of a system, because they provide the minimal perturbation of the system being assessed. In Fig. 9 we have already seen an cxample of the use of the threshold method; there is a clear suppression of sensitivity to long wavelengths by a red adapting stimulus and to short wavelengths by a blue one. This is an example of selective chromatic adaptation.

It might be thought that curves like this would be comprised of a linear sum of cone sensitivity curves, appropriately weighted, and that the effect of adaption would be to alter the weights attached to the curves. Instead, it seems that the threshold is jointly determined by activity of the luminance and opponent-color channels (Harwerth and Sperling, 1971; Krauskopf and Mollon, 1971; King-Smith and Carden, 1976). The rules of combination are complex and are not yet precisely known. Certain procedures can be used to favor selectively the contribution of one class of channel over another. Related to the poor high-frequency response of the chromatic channels is a longer integration time. Whereas the intensity–time reciprocity (Bloch's law) begins to fail for the luminance channels at around 20 msec, light is integrated by chromatic channels for at least 100 msec. The same sort of difference exists for spatial integration. Therefore, the use of a relatively long and large test probe favors its detection by the chromatic channels.

White light is ineffective for exciting chromatic channels, because it causes a balanced condition in both kinds of opponent-color channels (this is why the light appears achromatic). Instead, white light is selectively effective upon achromatic channels. Adaptation to white light and the use of a relatively large and long test probe favors the chromatic channels and helps to produce three-humped curves, as shown in Fig. 11. The use of small and brief test probes without adaptation tends to tap the luminance channels and produces a curve more similar to the obtained by flicker photometry.

The full story is more complicated than this, because adapting stimuli have direct effects upon receptors as well as channels. The main point to be made here is that, in attempting to deduce facts about receptors or photopigment sensitivities using psychophysical methods, one is dealing with the entire visual system. The sensitivities that are measured may or may not be dominated by peripheral events closely related to cone action spectra. (Sensitivity curves based on increment thresholds will be considered further in Section E.)

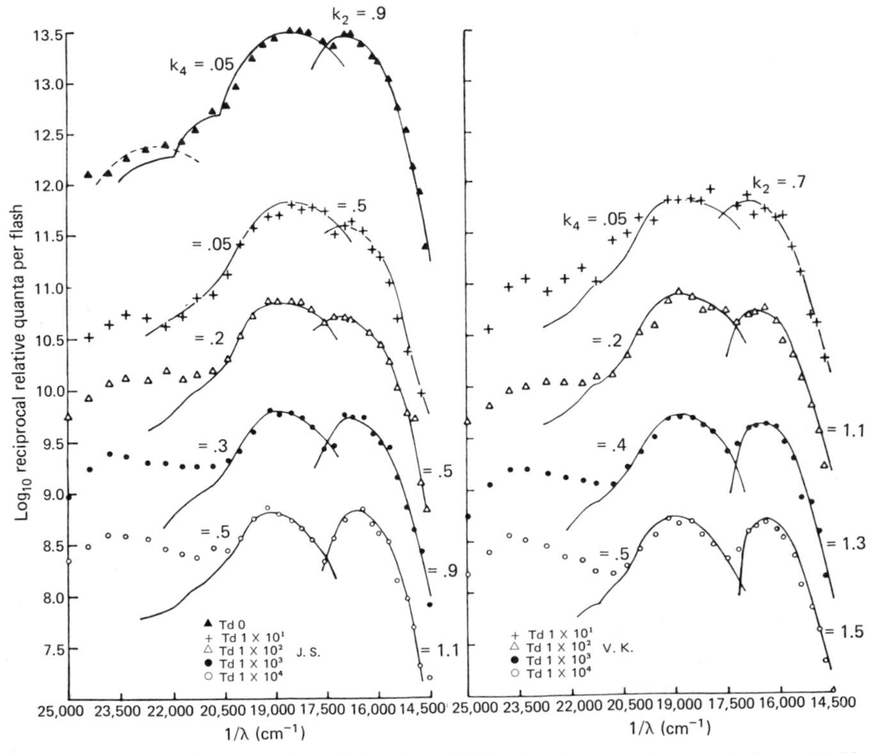

Fig. 11. Data of Harwerth and Sperling (1971) showing complex spectral sensitivities obtained against white backgrounds at various levels of retinal illuminance (Boynton, 1979).

D. CONE PHOTOPIGMENTS

1. Relation to Color-Mixture Data

a. METAMERISM

Metamerism (see Chapter 12) refers to a condition in which two physically dissimilar stimuli have the same color appearance. Metameric stimulus pairs are believed to exist because the flux reaching the retina from each member of the pair nevertheless excites each class of cone in the same way (Brindley, 1970, p. 205). As a result none of the cones can distinguish between the lights. As a further consequence messages arising from cone signals and sent to the visual brain will also be identical for both stimuli, and the same sensations will be produced from each member of the stimulus pair, so a color match results. Clarke (1960) proposed the expression *physiological identity* to capture the idea that two stimuli look alike because they produce the same rates of photon absorption in the photoreceptors.

b. WAVEGUIDE EFFECTS

The dimensions of photoreceptors, in particular those of foveal cones, are on the same order as the wavelength of light. This gives the receptors antenna like properties that produce uneven distributions of light within the receptors that are characteristic of all optical waveguides (Enoch, 1961). These effects vary with wavelength and undoubtedly alter the probability of photon absorption. However, it appears very unlikely that waveguide effects are involved in the color-coding process; more likely they merely add noise, which tends to average out over the very large numbers of cones that are used in color vision (Boynton, 1979, p. 249). Evidence supporting this interpretation includes (1) the observation that colors change little when monochromatic light is transmitted through the sclera, reaching the receptors from behind (Brindley and Rushton, 1959), (2) the survival of the action spectra of ρ and γ cone photopigments in selective bleaching of suspensions of cones where normal structure has been totally destroyed by the use of ultrasound (Murray, 1968), and (3) good agreement between rhodopsin chemistry and scotopic psychophysics (given that rods also have dimensions consistent with waveguide effects).

c. COLOR EQUATIONS AND TRANSFORMATIONS

An equation can be written to characterize a color match (Maxwell, 1855; Wright, 1958, p. 69; Wyszecki and Stiles, 1967, p. 228),

$$c(\mathrm{C}) = r(\mathrm{R}) + g(\mathrm{G}) + b(\mathrm{B}).$$

The symbols in parentheses represent qualitatively different lights. For example, C might be a white light, and R, G, and B are monochromatic primaries used to match the white in a colorimeter. The lowercase symbols represent the amounts of the primaries required for a match: c units of C match r units of R added to g units of G and b units of B.

The equation as written makes a mathematical and not a physiological statement. To characterize the results of a color-matching experiment in this way implies an isomorphism between the operations of the formal mathematical domain and the empirical colorimetric one. For example, *equals* translates into *matches with*, and *plus* translates into *additively combined with*, as for example by the superposition of projected lights.

In linear algebra, rules such as transitivity and the distributive law are obeyed. A set of tests of such rules in the empirical domain of color mixture, known collectively as *Grassman's laws* (Grassman, 1854; Wyszecki and Stiles, 1967, p. 233) has been made and found to be true within reasonable limits for small, centrally viewed fields. By far the simplest way to explain an isomorphism between the domain of algebra and that of color-matching experiments is to suppose that the matches are based upon a physiological identity at the first stage—equal photon catches in three classes of photopigments. Obedience to Grassman's laws would, if attributable to later stages of processing, imply a linearity of the visual system that is unlikely and is, in fact, refuted by electrophysiological experiment (Arden, 1976, p. 265).

In the basic small-field color-mixture experiment, spectral stimuli are matched using additive combinations of three spectral primaries. Color-mixture curves resulting from such experiments can be used to predict metameric matches in general. All spectral test wavelengths (except the primaries themselves) require that a negative value be assigned to the amount of one of the three primaries required for a match. These negative values correspond experimentally to adding a small amount of one of the primaries to the wavelength under test.

Experimental color-matching functions can be transformed mathematically into an all-positive set where the primaries, because they cannot physically be realized, are said to be imaginary. These all-positive functions make the same predictions about color matches as do the experimental ones. Among an infinitely large population of such all-positive functions, there is a unique set for which each curve is proportional to the action spectrum of one of the three kinds of cones. This special set predicts color matches, because such matches are determined by cone action spectra. Each of the imaginary primaries that correspond to this set can be regarded as a hypothetical stimulus that would be uniquely absorbed by one and only one, of the three kinds of cones. (The primaries are unrealizable because cone absorption spectra overlap, meaning that all real primaries are absorbed by at least two of the three kinds of cone photopigments.)

Rushton (1963a, 1965) proposed the names *erythrolabe*, *chlorolabe*, and *cyanolabe* (meaning red-, green-, and blue-catching) to describe these cone pigments. If uniquely contained in the ρ, γ, and β cones, the action spectrum of a cone would be the same as that of its constituent pigment. The weight of existing evidence strongly suggests that there exists only one photopigment for each type of cone.

If measured at the cornea, the action spectra of the three classes of cone photopigments, however they may be distributed in the cones, must predict color matches. No set of action spectra failing to meet this criterion can be the correct set.

2. Physical Measures

For reasons not yet understood, cone pigments have so far proved highly resistant to extraction and *in vitro* chemical analysis. More than 100 years after the first successful extraction of rod pigments, only a couple of instances of successful cone pigment extraction have been reported, neither with primate material. Two physical methods that have been used successfully to study cone pigments will now be described.

a. Retinal Densitometry

This method (Brindley and Willmer, 1952; Weale, 1953a; Rushton, 1972) has the important advantage that it can be employed using living human subjects. It is based on the fact that a dark-adapted retina, in which none of the photopigments has been bleached, absorbs more light than a light-adapted retina, in which some or all of the pigment has been bleached. Light passing through the retina is reflected from the fundal layers behind, and some of this light, after passing in reverse through the retina, will emerge through the pupil of the eye and can be measured with a physical detector. The basic idea of the method is that there should be less light emerging from the dark-adapted than from the light-adapted eye. By choosing a criterion percentage of bleaching as indicated by a certain amount of test light emerging from the eye, one can determine an action spectrum. This is accomplished by varying the radiance of a bleaching light at various wavelengths until a criterion amount of emerging test light, restricted to the fovea and corresponding to some fixed amount of bleaching, is returned (Mitchell and Rushton, 1971). The weak test light must be of sufficient intensity to make the measurement, yet not so intense as to cause appreciable bleaching. (Also, the bleaching light must briefly be occluded during the measurement; otherwise it would saturate the response of the photomultiplier tube that is used to measure it.)

The method has not proved successful for studying cyanolabe, probably because there are too few β cones to allow detection of any change in the reflected light due to bleaching. The action spectrum determined in the normal eye is, therefore, determined jointly by ρ and γ cones. To measure these independently, color-deficient observers have been employed. When small fields are used, one class of deficients, the dichromats, behave almost as if they were lacking one or another of the long-wavelength sensitive pigments. The absorption spectrum of chloralabe can be assessed by studying protanopes, a class of dichromat lacking the ρ-cone pigment, erythrolabe. Deuteranopes, who lack cholorlabe, can be used to gauge the spectral properties of erythrolabe. Provided that the test stimulus is confined to the rod-free foveal area, the method confirms the presence of liabile (bleachable) pigments with peak sensitivities near 540 and 565 nm. Moreover, lights of differing wavelengths, when adjusted to bleach equal amounts of the remaining measurable photopigment, appear equally bright to the dichromat (Rushton, 1963b, 1965).

b. MICROSPECTROPHOTOMETRY

This method, first used successfully with goldfish cones (Marks *et al.*, 1964; Liebman, 1972), consists of the direct measurement of light absorption in the outer segments of photoreceptors. A section of retina is sandwiched between microscope slides and observed under moderately high magnification. A receptor is identified (usually lying on its side), and a very tiny test beam is passed through its outer segment. Another light beam is delivered through a clear area of the slide to act as a reference. Using a sensitive photomultiplier tube one determines the amounts of lights of various test wavelengths that pass through the outer segment. From such data an absorption spectrum can be calculated. The best and most recent work verifies that in macaque monkeys there are three classes of these cones with peak sensitivities near 440, 540, and 565 nm, although the 440-nm (β) cones are very hard to find (Bowmaker *et al.*, 1978, and 1980).

Limited data are available from one human retina (Bowmaker and Dartnall, 1980), based on 19 ρ cones, 11 γ cones, 3 β cones, with peak sensitivities at about 563, 534, and 420 nm. There is a considerable spread (with a standard deviation of about 4 or 5 nm) of λ_{max} within each class, which appear to make up trimodal rather than normal distribution. Agreement of the average ρ and γ sensitivities with psychophysical estimates is reasonably good.

Taken together, these two physical methods have provided direct confirmation of the idea that the first stage of chromatic processions is based on cones of three classes with action spectra that overlap but peak in different parts of the visible spectrum. Because both methods are technically difficult, they yield data of somewhat limited precision, particularly in the tails of the absorption spectra. If the Dartnall template could be used safely to characterize the shape of cone action spectra, then an accurate measurement of the wavelength of peak sensitivity would establish the remainder of the function, including the difficult-to-measure tails. Unfortunately, however, the Dartnall nomogram cannot be so used, for it has been observed that the absorption spectra for most pigments peaking at long wavelengths, including erythrolabe and chlorolabe, are narrower than that of rhodopsin (Smith and Pokorny, 1972; Bowmaker *et al.*, 1978).

3. Psychophysical Estimates of Cone Action Spectra

a. Use of Data from Dichromats

If the eye contained only one class of cone, the psychophysical determination of the action spectrum of such cones would be a relatively easy matter, comparable to that for rods. The overlapping sensitivity functions of three classes of cones ensure that at least two of these will contribute to a psychophysically obtained spectral sensitivity curve under ordinary circumstances.

A partial solution to the problem is to take psychophysical advantage, as Rushton did with his densitometry, of the presence in the population of substantial numbers of color-deficient dichromats. When tested on a conventional anomaloscope (Rayleigh, 1881; LeGrand, 1968, p. 336), which requires a match between a mixture of red and green spectral lights and a spectral yellow, these subjects behave as if one or another of the two long-wavelength pigments is missing. They accept any ratio of red to green as a match to yellow, provided that they can adjust the radiance of the yellow to make the fields equally bright. Protanopes appear to be lacking erythrolabe; deuteranopes lack chlorolabe. At least to a first approximation, both classes of dichromats accept metameric color matches that are made by normal observers. Such behavior implies that the remaining pigments are in each case normal. (The β cones are not involved in the anomaloscope matches because of their negligible sensitivity to long wavelengths.)

Given that these observers have only one functional long-wave-sensitive pigment, then the measurement of the action spectrum of that pigment is possible using thresholds or brightness matches, if care is taken to keep the stimulus fields in the fovea where there are no rods. Below about 520 nm, where β cones begin to make a significant contribution, the short-wavelength tail of the ρ or γ cones sensitivity curve can be uncovered only if the β cones can somehow be kept out of the way. One procedure for doing this is to use heterochromatic flicker photometry in which β cones make a negligible contribution. The minimally distinct border method gives similar results. Because there are two classes of dichromats, each having one of the two normal long-wave-sensitive photopigments, their action spectra can be measured in either way.

Another method utilizing dichromats has a longer history and is somewhat more complicated to explain. Because dichromacy is a reduced form of normal color vision (Wyszecki and Stiles, 1967, p. 404; LeGrand, 1968, p. 346), it is legitimate to represent the color matches of dichromats on the chromaticity diagram of the normal observer. Doing so results in lines of confusion on the normal diagram, where all stimuli that plot along any such a line are metameric matches for the dichromat. An implication is that all such stimuli would match were it not for the differential excitation of β cones, which differs for each line. These lines are approximately straight, and each set converges toward a different point (see Fig. 12). From a knowledge of the location of these confusion points and the system of primaries on which a given chromaticity diagram is based, it is possible to calculate a unique action spectrum that is representative of the missing pigment. König and Dieterici (1893) understood this in the 19th century and were the first to make such calculations, which yielded sensitivity curves that differ rather little from more modern estimates. These are known as *König fundamentals.*

With the same strategy, estimates of the β cone action spectrum have been made from the data of tritanopes, a third (but extremely rare) class of dichromat. Another procedure (for normals) is to use long-wavelength light to adapt ρ and γ cones selectively; doing so reveals the sensitivity of β cones at

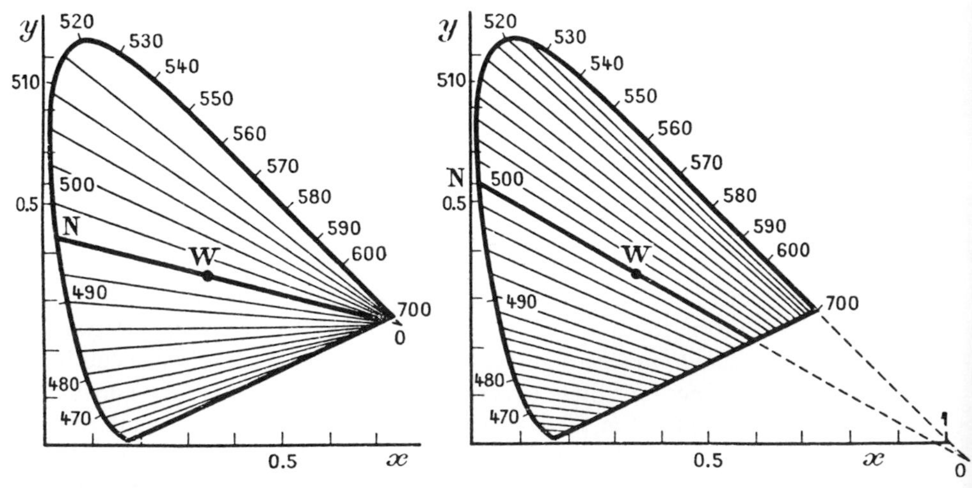

Fig. 12. Lines of confusion, represented on the CIE chromaticity diagram, for protanopes (left) and deuteranopes (right) (LeGrand, 1968).

short wavelengths where contributions from the long-wavelength cones become negligible.

A set of modern data based on psychophysical procedures of the sort just discussed is that of Smith and Pokorny (1975). These are shown in Fig. 13, where the curves drawn are based on empirical equations published by Boynton and Wisowaty (1980). These curves, which are in reasonable agreement with the results of retinal densitometry and microspectrophotometry, are probably as accurate as any of the estimates of the action spectra of human cone pigments currently available.

One subtle feature not evident in the physical measures is the separation of the ρ and γ curves at short wavelengths. These curves are maximally separated at 465 nm; a monotonic decrease in this separation occurs for wavelengths both longer and shorter than this. As a consequence, for each wavelength shorter than 465 nm, there is a longer one for which the same separation exists between the ρ and γ functions. On a log sensitivity plot, equal separations imply equal ratios of γ and ρ cone excitation which, in turn, implies that if the radiances of such stimulus pairs are adjusted for equal luminance, then both members of the pair should elicit equal excitations of ρ cones as well as of γ

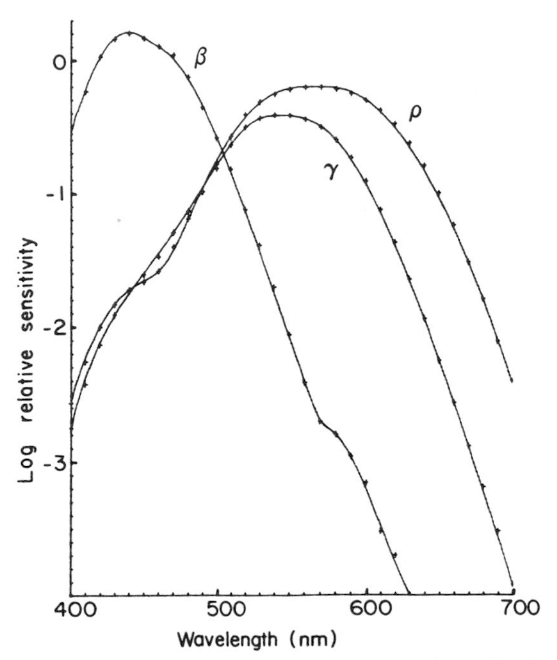

Fig. 13. The action spectra of human ρ, γ, and β cones as estimated by Smith and Pokorny (1975). The curves are based on equations derived by Boynton and Wisowaty (1980).

cones. Because such pairs would differ only in their capacity to excite β cones, they would match exactly for a tritanope; so they are known as *tritanopic metamers*. This same relationship reveals itself in a different way on a chromaticity diagram, where straight lines drawn from the tritanopic confusion point intersect the curved short-wavelength locus on the spectrum at two points. The wavelengths corresponding to these points are tritanopic metamers, as are all nonspectral colors along the line connecting them.

The methods discussed in this section do not allow a determination of the relative heights of the three cone functions, which ideally should be related at any wavelength to the probabilities of photon absorption by an ρ, γ, or β cone. There is, however, some reason to believe that the ρ and γ curves should be positioned so that they cross at 491 nm, because if this is done, the linear sum of their sensitivities corresponds to the photopic luminous efficiency function as obtained by flicker photometry (Vos and Walraven, 1971). Although the β curve should be positioned very low to indicate that these cones make a negligible contribution in this respect, it should be positioned higher if the contribution of β cones to color balance is represented.

b. π-MECHANISMS

Selective chromatic adaptation can be used to help isolate cone sensitivities psychophysically in normal observers. However, the ρ and γ sensitivity curves lie so close together that it becomes difficult and probably impossible, by measuring test-stimulus sensitivity curves, to isolate one of these from the other in normal observers (although Wald, 1968, claims to have done so; however, see Norren and Bouman, 1976, and Boynton, 1979, p. 193 for contrary opinions). Starting in the 1930s Stiles, whose researches on this topic have recently been reprinted in a single volume (Stiles, 1978), hit upon the idea of using a criterion *adaptive* effect to gauge the spectral sensitivity of cone mechanisms. Consider, for example, a long-wavelength test flash (667 nm) presented foveally, which excites ρ cones almost exclusively. An observer adjusts it to threshold. It is then raised by an arbitrary factor above threshold, and a determination is made of the radiance of steady adapting fields, presented at about 40 spectral wavelengths, that is required in each case to reduce the visibility of the test stimulus back to threshold. Stiles operationally defined a red mechanism, called π_5 (*pi* for *process*), on the basis of its spectral sensitivity obtained in this way (and certain other properties). By changing the test wavelength to middle or short wavelengths, he also was able to deduce other mechanisms. Of these π_5, π_1 (a blue mechanism), and π_4 (a green one) have been most widely investigated. There is much more to Stiles's methodology than this, including consideration of the various assumptions pertaining to the procedure that have been tested and sometimes have failed. These details are available in Stiles (1978) and in summaries by Brindley (1970), Boynton (1979, p. 187), Enoch (1972), and Rodieck (1973, p. 726).

Interest in π-mechanisms has peaked in recent years. Of particular concern is whether they might, in fact, reflect the action spectra of the three kinds of cones. On the basis of curve shapes alone, this is difficult to decide because of individual differences and experimental errors. It is, however, clear that π_4 and π_5 do not correctly predict tritanopic metamers for short wavelengths; also, they are a little broader than the Smith–Pokorny estimates. Although the three principal π-mechanisms may account adequately for color matching (Cavonius and Estévez, 1978; Pugh and Siegel, 1978), this fact is not sufficient for them necessarily to represent cone action spectra.

Under some circumstances π-mechanisms exhibit nonlinear properties that rule out the hypothesis that they represent simple cone sensitivities. For example, if the amounts of two adapting fields that are required to produce the criterion effect are independently scaled to unity, any amounts of mixtures of radiances of these two fields summing to unity should produce equivalent effects. Very often they do not (Boynton *et al.*, 1966; Wandell and Pugh, 1980).

E. RETINAL LOCATION AND VISUAL SENSITIVITY

When a light falls on the peripheral retina, relations between visual sensitivity and visual photopigments becomes more complex than for foveal vision. A psychophysical spectral sensitivity curve for eccentric vision is not only more difficult to obtain than for the fovea, but it is also harder to interpret, because it usually results from the excitation of rods as well as of three kinds of cones; to complicate matters further, the relative distributions of these four classes of receptors vary as a function of retinal eccentricity. There is also the possibility, which must be tested, that the absorption spectra of the cone pigments might vary with retinal location. Despite these problems considerable progress has been made within the last decade toward clarifying these relationships.

1. Receptor Distributions

Because human rods and cones are morphologically distinct, it has been possible to count them and make plots of receptor density as a function of retinal eccentricity (Øesterberg, 1935). Human cones, though of three types, all look the same under the microscope; for this reason, present concepts about their relative distributions are based mainly on psychophysical evidence.

Rods are absent in the foveal center; they appear gradually starting at about 20 min eccentricity and then increase rapidly in density to an eccentricity of about 10° to 20°, decreasing gradually beyond that (Fig. 14). Their morphology differs little with retinal location. Foveal cones are rodlike in

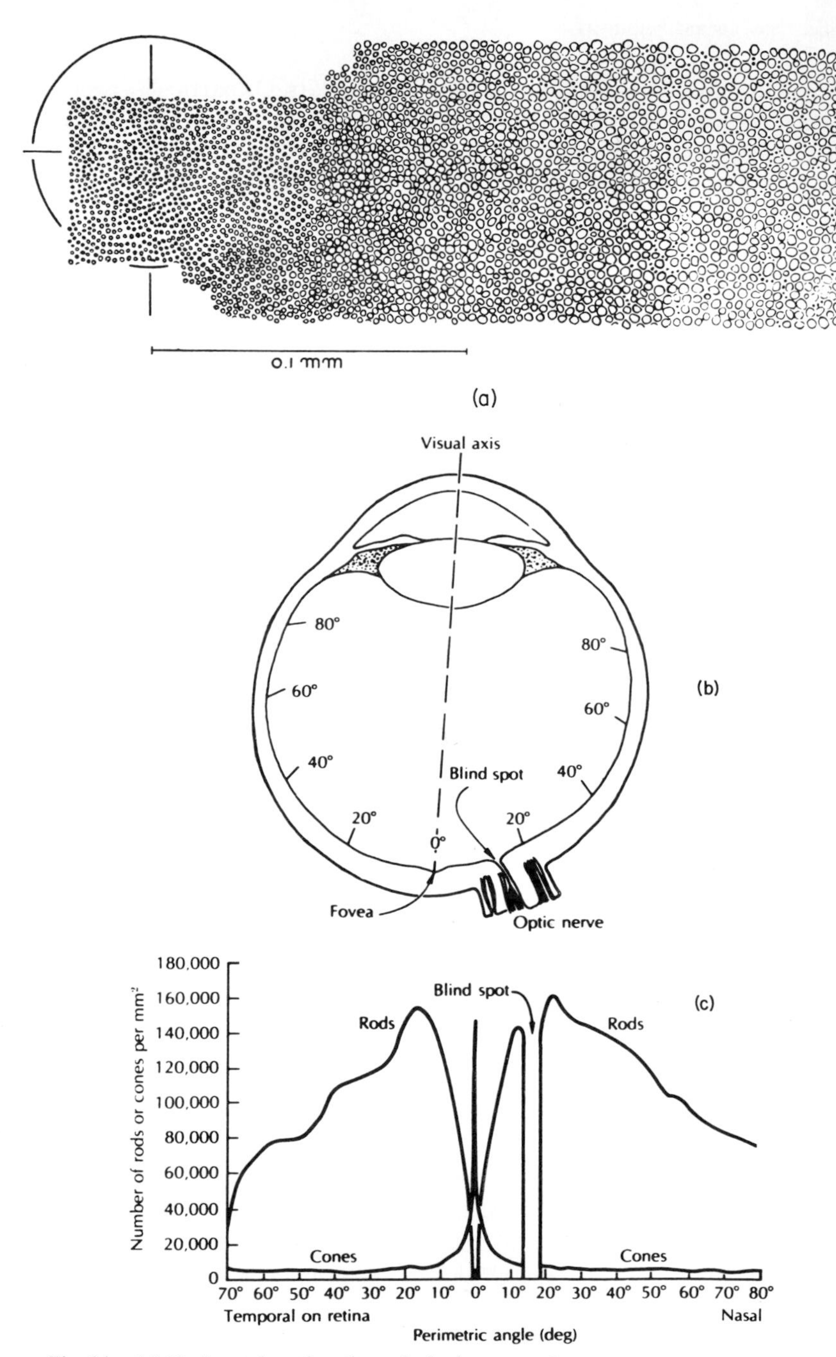

Fig. 14. (a) Horizontal section through the human retina containing the fovea (centered in cross hairs). Cones appear as dots in the central region and then as open circles whose diameters enlarge toward the parafovea; rods are seen as small black dots. (b) and (c) Rod and cone densities as a function of retinal location. (Boynton, 1979, pp. 88, 89, after Pirenne, 1948, and Cornsweet, 1970.)

appearance and are very densely packed. Their density decreases very rapidly and becomes more or less constant beyond a few degrees. Because their cross-sectional area increases markedly over the range where cone density decreases, cones occupy more retinal area in the periphery relative to rods than their much lower density might suggest (Pirenne, 1948, p. 28).

Little is known about the relative numbers of ρ and γ cones or whether this varies with eccentricity. There is evidence to indicate that their ratio varies considerably among observers (Rushton and Baker, 1964); ρ cones probably outnumber γ cones, but this is very uncertain. The β cones are absent in the very center of the fovea; their numbers increase with eccentricity to about 5° and then decrease again, resulting in an annular region of maximal β-cone density. Possibly as few as 1 in 50 of the cones are of the β variety; this, combined with their absence in the central fovea (Wald, 1967), helps to account for their failure to contribute to spectral sensitivity when the flicker or minimum-border methods are used. The fact that the influence of β cones is great in color matching and increment threshold experiments suggests that their signals undergo special amplification for purposes of color vision, part of which may result from pooling over very large areas. The ρ and γ cones seem to differ little from one another except for the pigments they contain and the connections they make; β cones differ in a very large number of ways from the other two kinds, and the β cone system is rodlike in a number of respects (Willmer, 1946; Brindley, 1970, p. 257; Mollon, 1980).

Direct evidence of the distribution of ρ, γ, and β cones in the primate retina was reported by Marc and Sperling (1977) based on a selective reaction of chemically treated cones in mounted retinae to stimulation with lights from various spectral regions. For the peripheral baboon retina, they found a rhomboidal pattern of β cones spaced about 10 min apart. Using a psychophysical technique, Williams *et al.* (1978) have confirmed that this pattern exists also in humans. The recent microspectrophotometry discussed in Section D2 is consistent with the idea that β cones are few in number and not found in the central fovea.

2. Psychophysics in the Peripheral Visual Field

We do our most critical seeing at or near the point of fixation, corresponding to a foveal retinal image. At the fovea the visual system exhibits its finest spatial acuity and optimal sensitivity to chromatic differences. Such excellent performance relates to an extraordinarily high density of foveal cones and a correspondingly large number of optic nerve fibers that carry foveal information to the visual cortex, whose foveal projection area is very large compared with the area of the brain receiving input from a corresponding amount of peripheral retina (see Chapter 5).

Peripheral vision is nevertheless vital for providing the "big picture". Foveal vision can occur only for the tiny region of the visual field at which the eyes are aimed at a given moment. It is peripheral vision that provides the information needed to help determine the pattern of eye movements that allow foveal vision to sample large external areas sequentially. Peripheral cones are importantly involved, and color plays a key role, as shown by experiments in visual search (Christ, 1975).

Basic experiments in the visual periphery have probably been avoided to some degree, because they are intrinsically difficult to execute. To study the perception of a stimulus in a known location of the peripheral visual field, an observer must be trained to look at a fixation point which he otherwise must ignore, while paying attention to what is indistinctly seen "out of the corner of the eye." In normal vision, where visual attention and fixation are almost the same thing, such deliberate concentration on events in the peripheral visual field is virtually unknown. In general, discriminations that are easy and comfortable to make in foveal vision become indistinct and disagreeable in the periphery, especially when small stimuli are used. To approach the degree of experimental reliability obtainable in foveal experiments, many more peripheral observations must be made, which greatly increases the experimental time required to obtain data of comparable accuracy.

3. Cone Sensitivities in the Periphery

a. Color Matches

Foveal color matches do not seem to fail in eccentric vision (LeGrand, 1968, p. 214). At first glance this observation might appear to indicate that peripheral color matches are determined by equal rates of photon absorption in the same three kinds of cones as those that populate the fovea. However, because peripheral color matching with small, contiguous fields is a very insensitive procedure, such a conclusion would be wrong. As already noted, peripheral vision is physiologically integrative and subjectively indistinct. Consequently, small, contiguous fields, even if grossly mismatched foveally, tend in peripheral vision to appear homogeneous over their entire extent. The method of contiguous small-field matching is too insensitive to prove anything.

If small fields are separated to minimize the integrative effect, more accurate color matches are possible in restricted peripheral regions. Experiments using such fields (Clarke, 1960) reveal that the additive character of foveal color matches is lost. For example, mixtures of colors plot along curved lines in a chromaticity diagram, indicating that the peripheral color matches so represented must depend on something more complicated than equal photon catches by three kinds of cones. As it turns out, the major source of difficulty relates to rod excitation by peripheral stimuli. When the differential activity of the rods associated with the two parts of the field upsets photopic color matches, this complication is known as *rod intrusion.*

By taking advantage of the integrative nature of peripheral vision, one can actually increase experimental precision. The trick is simply to use large fields and view them centrally. In experiments of this sort, observers are free to fixate anywhere within the large stimulus area. Despite the physiological complexity of the procedure, which causes a large retinal image to dance across a very inhomogeneous region of the retina, it is an easy and natural task from the observer's viewpoint, and important insights have resulted from its use.

With such large fields, nonadditivity reveals itself most obviously by the observation that metameric color matches achieved at moderately high luminances often fail when a neutral filter is placed before the eye (for small fields such a match is, of course, retained). Additivity failure implies that matches do not depend upon identical receptor excitations. The matches occur, instead, because identical signals exist somewhere beyond the receptor level, in pathways between the receptors and higher centers of the brain. Reducing field intensity increases rod activity; this will have the same effect on the two parts of the field only if they also match initially for rods. Signals from rods and cones share common pathways to the brain, accounting for the fact that differential rod activity, although it upsets large-field color matches, introduces no new sensations.

Given that there are three classes of cones and one of rods, it should be possible to produce stimuli that are metameric for all four receptor types. By extension from foveal color matching, in which three kinds of photopigments imply the need for three primaries in the colorimeter, one might expect that four primaries should be used to achieve *tetrachromatic matches* in the periphery. However, because rods introduce no new sensations, three primaries prove sufficient for unique color matches with large fields. Adding a fourth primary introduces an unnecessary degree of freedom and leads to color matches that are not unique; that is, many different proportions of the same four primaries can be used to match a particular reference stimulus.

The problem can be addressed by remembering that, at scotopic intensities, matches are one-dimensional and are determined by rods alone. Therefore, a photopic match that does not excite rods differentially should be one that remains a match at scotopic levels. Trezona (1970, 1976) developed a procedure for testing such matches experimentally and was able to achieve unique tetrachromatic stimulus pairs. To red, yellow, and blue primaries, which by themselves would be sufficient to allow a trichromatic large-field match, she added a fourth primary near 500 nm, chosen to excite rods selectively relative to cones. The experimental procedure is an iterative one. Starting with an arbitrary amount of the 500-nm primary, fixed by the experimenter, the observer makes a trichromatic color match at a photopic intensity by manipulating the amounts of the red, yellow, and blue primaries. Taking great care not to alter the relative spectral distributions of the fields, intensities are then reduced to a scotopic level. Except by chance, such fields will not match, because they will excite rods differentially. To achieve such a match the subject adjusts the 500-nm primary without touching the other three controls. When field intensities are subsequently increased to the original level, the previous photopic match fails because of the change just made in the fourth primary, which, although maximally effective upon rods, nevertheless excites cones as well. Without adjusting the 500-nm primary, the observer then alters the other three to regain a photopic match. Because of the change in the 500-nm primary made scotopically, the new photopic match differs from the original one, and when the intensity level is reduced again, it proves to be a little closer to a scotopic match. Trezona found that by continuing this procedure over several trials, settings could eventually be achieved to determine matches that were true at all intensity levels (see Fig. 15). These final settings were the unique ones required to cause equal photon absorptions in the four types of underlying photopigments.

In a CIE-sponsored program of large-field color matching, Stiles (1958) was troubled by rod intrusion despite his ability to work at high field levels using a Maxwellian-view tricolorimeter. Replacing the usual green primary (as Trezona did also) with a yellow one helped by exciting rods less. Most observers were still troubled by *Maxwell's spot*, a difference in color appearance in the central part of the field often attributed to macular pigment, which covers only the central 4° or so of the retina (see color illustration in Arden 1976, p. 492). Rather than block out the central region and rely on annular field matches, the experimenters instructed their observers to ignore the central region. Based upon these experiments, despite failure of field tests (Stiles and Wyszecki, 1962), the CIE adopted in 1964 a set of color-matching functions that differed from the set developed in 1931 for small fields and recommended that colorimetric specifications and calculations based on the new functions should be carried out for fields larger than those subtending 4° of visual angle.

As Trezona points out, the procedure adopted by the CIE for large-field color specification is fundamentally unsound. Although the specification of large-field color matches is possible using only three variables, such specifications apply only to the specific set of experimental primaries chosen. Calculations of the sort required to transform data to other sets of primaries, whether real or imaginary, are illegitimate (see Chapter 3, Volume 2, of the present series). The all-positive large-field color mixture curves adopted by the

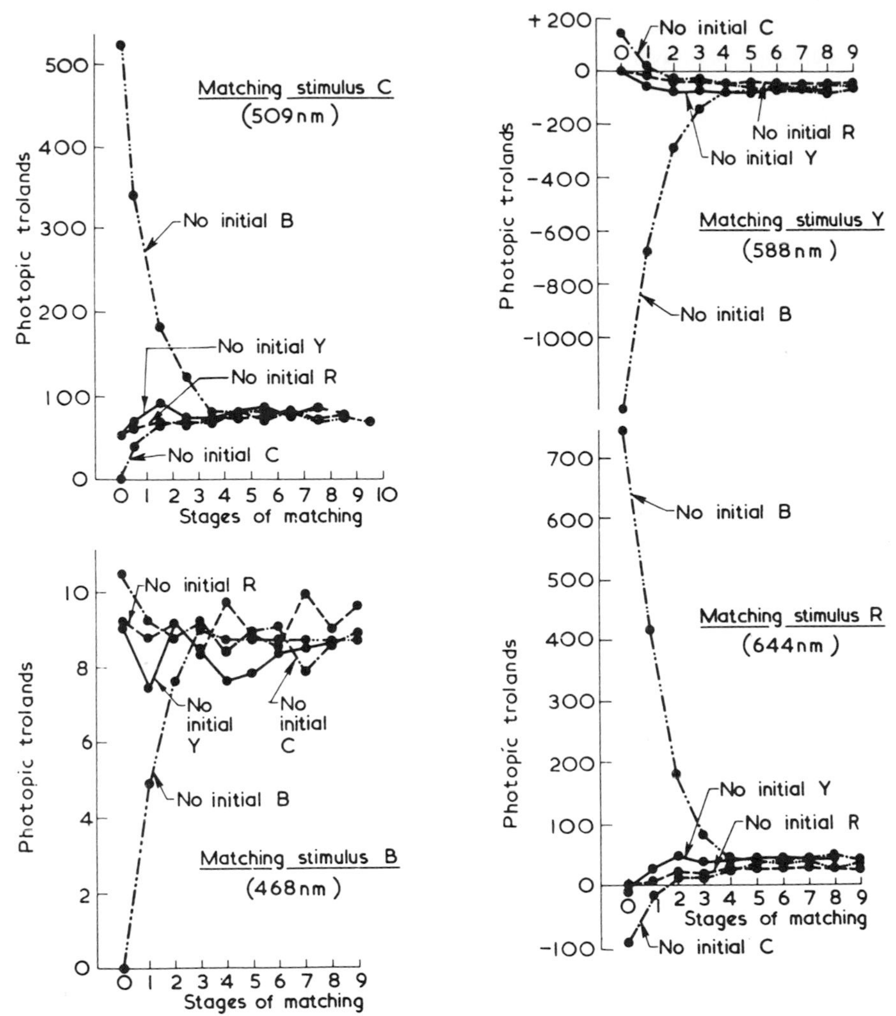

Fig. 15. Data from tetrachromatic color-matching experiments of Trezona (1974). The final matches are represented by the asymptotic values of curves at the far right of each panel, showing the amounts of the four primaries (R = 644 nm, C = 509 nm, B = 468 nm, and Y = 588 nm), one in each panel, required to match a test color of 485 nm at 55 photopic trolands. The iterative matching procedure was started on separate occasions with one of the four primaries extinguished. The final match does not depend upon the starting condition.

CIE required such transformations from instrumental primaries. Only if color matches are specified in a tetrachromatic system, to ensure that the computed match is metameric for rods as well as for cones, can such calculations legitimately be made.

Trezona also reports that the troublesome Maxwell's spot disappears for a tetrachromatic match. If the spot were due entirely to macular pigment, as is widely assumed, this result would not be predicted. However, if it results instead from the selective influence of rods, such influence, which would be great in the periphery and minimal in central vision, would predict a differential color appearance.

Rod intrusion is not the only possible cause of difference between color matches in fovea and periphery. Pokorny and Smith (1976) have shown, under conditions in which rod function was eliminated by working on the cone plateau and long-wavelength test fields were used (the classic Rayleigh match), that small differences nevertheless are observed. Their calculations suggest that the responsible agent is a differential amount of self-screening by the cone pigments, which is consistent with cone morphology, because cones are thin and elongated only in the fovea and are short, squat, and conelike in the periphery. Their interpretation is consistent with the idea that there is no difference in the cone pigments as a function of retinal eccentricity.

b. Spectral Sensitivity Measurements

Another approach to the psychophysical evaluation of peripheral cone action spectra is to measure spectral sensitivity curves of various kinds by use of peripheral vision. (An example of such measurements has already been given in Fig. 1). At scotopic levels, in which vision is controlled by rods, the meaning of such a function is clear. At photopic levels, in which four kinds of receptors may contribute in complicated ways to the measured result, spectral sensitivity curves in the periphery have proved difficult to interpret.

Weale (1953b), in an early study of this kind, used heterochromatic brightness matching and concluded that β cones dominate in the periphery. This conclusion has been criticized on the grounds that his white adapting field, used to minimize the influence of rods, was actually yellow, which may have selectively depressed the sensitivities of ρ and γ cones. Wooten and Wald (1973) attempted to look at one cone type at a time by means of high-intensity selective adaptation, an extension of Wald's earlier foveal experiments. They concluded, exactly contrary to Weale, that the participation of β cones drops off selectively in the periphery. They offered no real proof that the spectral sensitivities that they isolated were those of single classes of receptors; in fact, they admitted that this could not be so in certain cases, including a long-wavelength tail on their β function.

Using heterochromatic flicker photometry under conditions that should have minimized rod participation, Abramov and Gordon (1977) obtained functions peaking near 500 nm; such peaks are often taken as *prima facie* evidence of rod involvement. Although Abramov and Gordon did not favor this interpretation, they could not absolutely rule it out. Recently Stabell and

Stabell (1981), using a brightness matching technique, have also found a peak near 500 nm that they ascribe to rods, even at high luminance levels. Using a threshold method, on the other hand, these authors found a long-wavelength peak suggestive of constant ρ and γ cone contributions at all retinal eccentricities and a short-wavelength shoulder or peak indicating, in agreement with Wooten and Wald, that β cone activity is optimal in the near periphery and falls off in the far periphery.

Safe conclusions, based both on color mixture and on spectral sensitivity measurements, are that (1) peripheral cones probably contain the same three photopigments found in the fovea, (2) their action spectra may differ a little because of differential self-screening, and (3) unless unusual precautions are taken, rod activity will enter into peripheral photopic vision.

REFERENCES

Abramov, I., and Gordon, J. (1977). Color vision in the peripheral retina I. Spectral sensitivity. *J. Opt. Soc. Am.* **67**, 195–202.

Aguilar, M., and Stiles, W. S. (1954). Saturation of the rod mechanism at high levels of stimulation. *Opt. Acta* **1**, 59–65.

Arden, G. (1976). The retina-neurophysiology. *In* "The eye," (H. Davson, ed.), Vol. 2A (2nd ed.). Academic Press, New York.

Baylor, D. A., Lamb, T. D., and Yau, K. W. (1979). Responses of retinal rods to single photons. *J. Physiol.* **288**, 613–634.

Boll, F. (1877/1977). On the anatomy and physiology of the frog retina (Ruth Hubbard, transl., 1977). *Vision Res.* **17**, 1249–1265.

Boring, E. G. (1942). "Sensation and perception in the history of experimental psychology." Appleton-Century-Crofts, New York.

Bowmaker, J. K., and Dartnall, H. J. A. (1980). Visual pigments of rods and cones in a human retina. *J. Physiol.* **298**, 501–511.

Bowmaker, J. K., Dartnall, H. J. A., Lythgoe, J. N., and Mollon, J. D. (1978). The visual pigments of rods and cones in the rhesus monkey, *Macaca mulatta. J. Physiol.* **274**, 329–348.

Bowmaker, J. K., Dartnall, H. J. A., and Mollon, J. D. (1980). Microspectrophotometric demonstration of four classes of photoreceptor in an old world primate maccaca fascisularis. *J. Physiol.* **298**, 131–143.

Boynton, R. M. (1978). Ten Years of research with the minimally distinct border. *In* "Visual Psychophysics and Physiology" (J. C. Armington, J. Krauskopf, and B. Wooten, eds.). Academic Press, New York.

Boynton, R. M. (1979). "Human color vision." Holt, New York.

Boynton, R. M., and P. K. Kaiser. (1968). Vision: The additivity law made to work for heterochromatic photometry with bipartite fields. *Science* **161**, 366–368.

Boynton, R. M., and Wisowaty, J. J. (1980). Equations for chromatic discrimination models. *J. Opt. Soc. Am.* **70**, 1471–1476.

Boynton, R. M., Das, S. R., and Gardiner, J. (1966). Interactions among visual mechanisms revealed by mixing conditioning fields. *J. Opt. Soc. Am.* **56**, 1775–1780.

Brindley, G. S. (1970). "Physiology of the retina and visual pathway" (2nd ed.). Williams & Wilkins, Baltimore.

Brindley, G. S., and Rushton, W. A. H. (1959). The colour of monochromatic light when passed into the human retina from behind. *J. Physiol.* **147**, 204–208.

Brindley, G. S., and Willmer, E. N. (1952). The reflexion of light from the macular and peripheral fundus oculi in man. *J. Physiol.* **116**, 350–356.

Brown, J. L., Kuhns, M. P., and Adler, H. E. (1957). Relation of threshold criterion to the functional receptors of the eye. *J. Opt. Soc. Am.* **47**, 198–204.

Brown, P. K., and Wald, G. (1963). Visual pigments in human and monkey retinas. *Nature* **200**, 37–43.

Cavonius, C. R., and Estèvez, O. (1978). π-mechanisms and cone fundamentals. *In* "Visual psychophysics and physiology" (J. C. Armington, J. Krauskopf and B. Wooten, eds.). Academic Press, New York.

Christ, R. W. (1975). Review and analysis of color coding research for visual displays. *Human Factors* **17**(6), 542–570.

Clarke, F. J. J. (1960). Extra-foveal colour metrics. *Opt. Acta* **7**, 355–384.

Cornsweet, T. N. (1970). "Visual Perception." Academic Press, New York.

Dartnall, H. J. A. (1953). The interpretation of spectral sensitivity curves. *Br. Med. Bull.* **9**, 24–30.

Dartnall, H. J. A. (1962). The photobiology of visual processing. *In* "The Eye" (H. Davson, ed.). Academic Press, New York.

Dartnall, H. J. A. (ed.) (1972). Photochemistry of vision. "Handbook of Sensory Physiology" v. VII/1. v. VII/1. Springer-Verlag, New York.

DeMott, D. W., and Davis, T. P. (1960). Entoptic scatter as a function of wavelength. *J. Opt. Soc. Am.* **50**, 495–601.

DeValois, R. L., and DeValois, K. K. (1975). Neural coding of color. *In* "Handbook of Perception" (E. C. Carterette and M. P. Friedman, eds.), Vol. 5. Academic Press, New York.

Eisner, A., and MacLeod, D. I. A. (1980) Blue-sensitive cones do not contribute to luminance. *J. Opt. Soc. Am.* **70**, 121–123.

Enoch, J. M. (1961). Wave-guide modes in retinal receptors. *Science* **133**, 1353–1354.

Enoch, J. M. (1972). The two-color threshold technique of Stiles and derived component color mechanisms. *In* "Handbook of Sensory Physiology" (D. Jameson and L. M. Hurvich, eds.), VII/4. Springer-Verlag, New York.

Fain, G. L. (1975). Quantum sensitivity of rods in the toad retina. *Science* **187**, 838–841.

Grassman, H. (1854). On the thoery of compound colours. *Philos. Mag.* **7**(4), 254–264.

Harwerth, R. S., and Sperling, H. G. (1971). Prolonged color blindness induced by intense spectral lights in rhesus monkeys. *Science* **174**, 520–523.

Hecht, S., Shlaer, S., and Pirenne, M. H. (1942). Energy, quanta, and vision. *J. Gen. Physiol.* **25**, 819–840.

Hurvich, L. M. (1978). Two decades of opponent processes. *In* "AIC Color 77" (F. W. Billmeyer and G. Wyszecki, eds.). Adam Hilger, Bristol.

King-Smith, P. E., and Carden, D. (1976). Luminance and opponent-color contributions to visual detection and adaptation and to temporal and spatial integration. *J. Opt. Soc. Am.* **66**, 709–717.

König, A., and Dieterici, C. (1893). Die Grunempfindungen in normalen and anomalen Farbensystemen und ihre Intensitatsverteilung im Spektrum. *Z. Psychol. Physiol. Sinnesorgane*, **4**, 241–347.

Krauskopf, J., and Mollon, J. D. (1971). The independence of the temporal integration properties of individual chromatic mechanisms in the human eye. *J. Physiol.* **219**, 611–623.

Kühne, W. (1879). Chemical processes in the retina (G. Wald, R. Hubbard, and H. Hoffman, transl., 1977). *Vision Res.* **17**, 1269–1316.

LeGrand, Y. (1968). "Light, Colour, and Vision" (2nd ed.). Translated by R. W. G. Hunt, J. W. T. Walsh, and F. R. W. Hunt. Halsted Press, Somerset, N.J.

LeGrand, Y. (1972). About the photopigments of colour vision. *Mod. Probl. Ophthalmol.* **11**, 186–192.

Liebman, P. (1972). Microspectrophotometry of photoreceptors. *In* "Handbook of Sensory Physiology" (H. J. A. Dartnall, ed.), Vol. VI/1. Photochemistry of Vision. Springer-Verlag, Berlin/Heidelberg/New York.

Lythgoe, J. N. (1972). List of vertebrate visual pigments. *In* "Handbook of Sensory Physiology" (H. J. A. Dartnall, ed.), Vol. VII/2. Photochemistry of vision. Springer-Verlag, New York.

Makous, W., and Boothe, R. (1974). Cones block signals from rods. *Vision Res.* **14**, 285–294.

Marc, R. E., and Sperling, H. G. (1977). Chromatic organization of primate cones. *Science* **196**, 454–456.

Marks, W. B., Dobelle, W. H., and MacNichol, E. G. (1964). Visual pigments of single primate cones. *Science* **143**, 1181–1183.

Maxwell, J. C. (1855). Experiments on colour, as perceived by the eye, with remarks on colour-blindness. *Trans. R. Soc. Edinburgh* **21**, 275–298.

Meyers, K. J., Ingling, C. R., Jr., and Drum, B. A. (1973). Brightness additivity for a grating target. *Vision Res.* **13**, 1165–1173.

Mitchell, D. E., and Rushton, W. A. H. (1971). Visual pigments in dichromats. *Vision Res.* **11**, 1033–1043.

Mollon, J. D. (1980). Post-receptoral processes in colour vision. *Nature* **283**, 623–624.

Murray, G. C. (1968). "Visual pigment multiplicity in cones of the primate fovea." Ph.D. dissertation, John Hopkins University, Baltimore.

Naka, K.-I., and Rushton, W. A. H. (1966). An attempt to analyse colour reception by electrophysiology. *J. Physiol.* **185**, 556–586.

Norren, D. V., and Bouman, M. A. (1976). Is it possible to isolate fundamental cone mechanisms with Wald's method of chromatic adaptation? *Mod. Probl. Ophthalmol.* **17**, 27–32.

Øesterberg, G. (1935). Topography of the layer of rods and cones in the human retina. *Acta Ophthalmol. Suppl.* **6**.

Pirenne, M. H. (1948). "Vision and the eye." Chapman & Hall, London.

Pokorny, J., and Smith, V. C. (1976). Effect of field size on red-green color mixture equations. *J. Opt. Soc. Am.* **66**, 705–708.

Pugh, E. N. Jr., and Siegel, C. (1978). Evaluation of the candidacy of the pi-mechanisms of Stiles for color-matching fundamentals. *Vision Res.* **18**, 317–330.

Purkinje, J. (1823). "Beobachtungen und Versuche zur Physiolgie der Sinne" (1st ed.). J. G. Calve, Prague, [cited in Brindley (1970)].

Rayleigh, Lord (J. W. Strutt) (1881). Experiments on colour. *Nature London* **25**, 64–66.

Riggs, L. A. (1965). Visual acuity. *In* "Vision and Visual Perception" (C. H., Graham, ed.). Wiley, New York.

Ripps, H., and Weale, R. A. (1976). The visual photoreceptors. *In* "The Eye" (H.Davson, ed.), Vol. 2A (Visual functions in man). Academic Press, New York.

Rodieck, R. W. (1973). "The vertebrate retina." San Francisco, Freeman.

Rood, O. N. (1893). On a photometric method which is independent of color. *Am. J. Sci.* **46**, 173–176.

Rushton, W. A. H. (1963a). A cone pigment in the protanope. *J. Physiol.* **168**, 345–359.

Rushton, W. A. H. (1963b). The density of chlorolabe in the foveal cone in the protanope. *J. Physiol.* **168**, 360–373.

Rushton, W. A. H. (1965). A foveal pigment in the deuteranope. *J. Physiol.* **176**, 24–37.

Rushton, W. A. H. (1972). Visual pigments in man. *In* "Handbook of Sensory Physiology" (H. J. A. Dartnall, ed.), Vol. VII/1. Photochemistry of vision. Springer-Verlag, New York.

Rushton, W. A. H., and Baker, H. D. (1976). Red/Green sensitivity in normal vision. *Vision Res.* **4**, 75–85.

Said, F. S., and Weale, R. A. (1959). The variation with age of the spectral transmissitivity of the living human crystalline lens. *Gerontologia* **3**, 213–231.

Smith, V. C., and Pokorny, J. (1972). Spectral sensitivity of color-blind observers and the cone photopigments. *Vision Res.* **12**, 2059–2071.

Smith, V. C., and Pokorny, J. (1975). Spectral sensitivity of the foveal cone photopigments between 400 and 500 nm. *Vision Res.* **15**, 161–171.

Sjöstrand, F. S. (1953). The ultrastructure of the outer segments of rods and cones of the eye as revealed by the electron microscope. *J. Cell. Comp. Physiol.* **42**, 15–44.

Stabell, B., and Stabell, V. (1981). Absolute spectral sensitivity at different eccentricities. *J. Opt. Soc. Am.* **71**, 836–840.

Stiles, W. S. (1958). The average colour-matching functions for a large matching field. *In* "Visual Problems of Colour," Vol. 1, NPL Symposium No. 8. Her Majesty's Stationery Office, London.

Stiles, W. S. (1964). Appendix: Foveal threshold sensitivity of fields of different colors. *Science*, **145**, 1016–1017.

Stiles, W. S. (1978). "Mechanisms of colour vision." Constable, Edinburgh.

Stiles, W. S., and Burch, J. M. (1959). N.P.L. Colour-matching investigation: Final report, 1958. *Opt. Acta* **6**, 1–26.

Stiles, W. S., and Wyszecki, G. (1962). Field trials of color-mixture functions. *J. Opt. Soc. Am.* **52**, 58–75.

Trezona, P. W. (1970). Rod participation in the "blue" mechanism and its effect on colour matching. *Vision Res.* **10**, 317–332.

Trezona, P. W. (1974). Additivity in the tetrachromatic colour matching system. *Vision Res.* **14**, 1291–1303.

Trezona, P. W. (1976). Aspects of peripheral color vision. *Mod. Probl. Ophthalmol.* **17**, 52–70.

Vos, J. J., and Walraven, P. L. (1971). On the derivation of the foveal receptor primaries. *Vision Res.* **11**, 799–818.

Wald, G. (1945). Human vision and the spectrum. *Science* **101**, 653–686.

Wald, G. (1967). Blue-blindness in the normal fovea. *J. Opt. Soc. Am.* **57**, 1289–1301.

Wald, G. (1968). Molecular basis of visual excitation. *Science* **162**, 230–239.

Wald, G., Brown, P. K., and Gibbons, I. R. (1963). The problems of visual excitation. *J. Opt. Soc. Am.* **53**, 20–35.

Wandell, B. A., and Pugh, E. N., Jr. (1980). Detection of long-duration, long-wavelength incremental flashes by a chromatically-coded pathway. *Vision Res.* **20**, 613–624.

Weale, R. A. (1953a). Photochemical reactions in the living cat's retina. *J. Physiol.* **122**, 322–331.

Weale, R. A. (1953b). Spectral sensitivity and wave-length discrimination of the peripheral retina. *J. Physiol. London* **119**, 170–190.

Williams, D. R., MacLeod, D. I. A., and Hayhoe, M. M. (1978). Distribution of blue-sensitive cones in the fovea. *Investigative Ophthalmology Visual Sci.* Supplement **17**, 177.

Willmer, E. N. (1946). "Retinal Structure and Color Vision." Cambridge University Press, Cambridge.

Wooten, B. R., and Wald, G. (1973). Color-vision mechanisms in the peripheral retinas of normal and dichromatic observers. *J. Gen. Physiol.* **61**, 125–145.

Wright, W. D. (1958). "The Measurement of Colour." Macmillan, New York.

Wyszecki, G., and Stiles, W. S. (1967). "Color Science." Wiley, New York.

5

Mechanisms of Vision

C. J. BARTLESON

Research Laboratories
Eastman Kodak Company
Rochester, New York

A. INTRODUCTION

The purpose of this chapter is to present a brief, simplified summary of the major features of the visual mechanism. In doing so, many simplifications will be made. A number of aspects of the visual mechanism will be treated superficially, if at all, because the subject is too complex to be described in detail in the space available here. We hope that the presentation will be helpful to those who are not familiar with the mechanism of vision by providing some insight into its workings.

The eye is an extension of the brain. Its structure and physiology are specialized to provide efficient vision, but its mechanism cannot be considered

ISBN-0-12-304905-9

in isolation from that of the brain. Although the eye plays an important and initial role in mediating vision, it is only one of the stages involved in "seeing." A discussion of the mechanisms of vision must include the specialized function of the entire neural network that is responsible for vision. This chapter will attempt to do that by describing briefly the major aspects of visual function that are known or generally surmised.

The human brain contains about as many neurons (nerve cells) as there are stars in our galaxy, of the order of 10^{11} to 10^{13}. It has been estimated that about 38% of all man's sensory input is mediated through his optic nerves (Bruesch and Arey, 1942). The extent to which we rely on our visual perceptions is probably greater than that for any of our other senses. For these reasons, and because the well-ordered extremities of the optic tract are somewhat remote from the brain proper (and, therefore, more readily accessible to investigation), more research has been devoted to the study of vision than to any other single, specialized area of neurophysiology. Still, the mechanism of vision is only partially understood. Advances in techniques of electrophysiological measurement, which permit much more detailed study of receptor behavior and retinal circuitry than has ever before been possible, have provided many answers to long-standing questions about the visual mechanism but, at the same time, have uncovered more new questions than answers. Necessarily, then, an introductory discussion such as this must deal with general relationships, leaving detailed treatments to the burgeoning current literature of brain and vision research. Inevitably, some of the material presented here will be superseded in the coming years by more exact knowledge. However, the purpose of this chapter is to present a brief summary of what is presently known about the mechanism of vision with particular emphasis on those aspects of the subject that will help the reader to understand the uses and limitations of visual measurements.

The following sections will provide general information about some basic principles of neurophysiology in order to help the reader understand the specific neurophysiological functions of retinal processing and of signal integration and elaboration throughout higher centers of the brain. Additional sections will relate this information to color vision, brightness perception, and spatial vision.

B. THE BASIC NEUROPHYSIOLOGY OF VISION

1. General Concepts

Nerve cells called *neurons* make up the elemental transmission networks of our nervous systems. Neurons are genetically and biochemically similar to other cells of the body, but they have properties that make them specialized and distinct from other living cells. They are capable of generating electrical signals in response to stimulation. Neurons communicate with one another by transmitting and receiving signals. They may excite or inhibit. Some, called *receptors*, may be stimulated by external energy; for example, photoreceptor cells may be excited by light impinging upon them. Others may be excited (or inhibited) by chemical or electrical stimulation by other neurons.

Although neurons differ greatly in anatomical shape, depending on their location and function, they all communicate with the same stereotyped electrical signals. There are basically two forms of these signals, one for transmission over short distances and the other for transmission over relatively long distances. When neurons communicate with cells that are close, they transmit *graded potentials*; the voltage amplitude depends directly on the extent to which the neuron is stimulated. When transmission distances are relatively long, *frequency-modulated pulses* are used to prevent "line losses," that is, to maintain a satisfactory signal-to-noise ratio in the communication channel. Magnitude of stimulation is often represented in such pulses as variations in the frequency of voltage spikes (Adrian, 1946).

Since all neurons produce stereotyped coded electrical signals, the nature of the information they convey must be determined in some way other than the form of electrical signal. That is, a firing neuron may signal the extent to which it is excited but not the quality or content of whatever causes the excitation. Although neurons for vision, audition, olfaction, tactile sense, and so forth all respond with stereotyped signals, somehow sensory content must also be identified. The key to content lies in the specific circuitry by which neurons are interconnected. The origin and destination of a neural pathway determine the content of its signals. Visual information is transmitted from the retina of the eye to the occipital cortex at the rear of the brain through certain prescribed pathways; audition information travels through different but equally prescribed circuits from ear to a more lateral aspect of the brain; and so on. Neural circuitry is, then, as important as neuron response in the mechanism of vision.

Although neurons differ in anatomical structure and circuit function, they are similar in certain basic respects. Each neuron consists of four general parts: (1) the *cell body*, (2) *dendrites*, (3) *axons*, and (4) *synapses*. Figure 1 shows a schematic representation of a neuron with these four parts labeled.

The cell body has a nucleus containing deoxyribonucleic acid (DNA), the genetic material responsible for location and operation of the cell. Immediately surrounding the nucleus is a layer of cytoplasm containing the mitochondria, which provide the cell's nourishment and energy requirements. An outer sheath of membrane connects the cell to other cell processes. The control center of the neuron is located in the cell body.

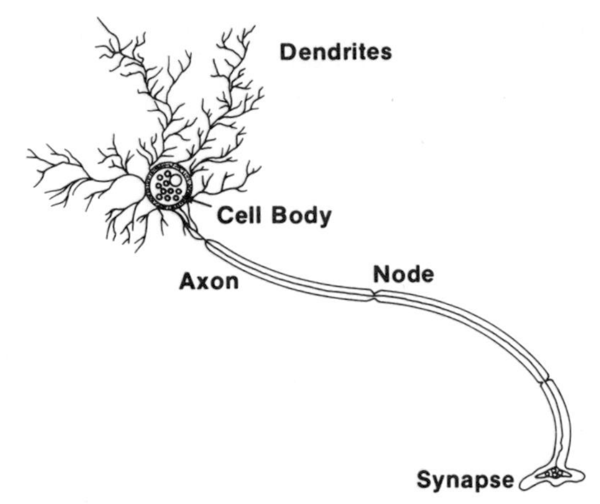

Fig. 1. Highly schematic representation of a neuron showing receptive dendrites, the cell body, a myelinated axon interrupted by nodes of Ranvier, and terminating in a synaptic bulb.

An arboreal network of processes radiate from the cell body. These are called *dendrites*, and they are largely, but not exclusively, responsible for receiving signals from other neurons. Their membranes are permeated by chemical *transmitter* substances from synaptic junctions (*vide infra*) with other cells. These substances cause a chemical change in the *receiver* material of the postsynaptic matter, which initiates an electrical potential that is passed along the dendrite to the cell body.

Usually a cell will have one long *axon* that is specialized for propagation of signals to other cells. The liquid material inside the axon, called *axoplasm*, is rich in potassium ions; other cations and anions are involved, but the concentration of potassium is 10–20 times as high as that in the saline blood solution surrounding the axon. Conversely, the exterior environment of the axon is about 10–20 times as rich in sodium ions as the axoplasm. The axon membrane is selectively permeable to these ions. At rest the membrane is much more permeable to potassium than to sodium. Potassium, therefore, tends to leak out from the axon. This sets up a concentration gradient across the membrane. Because interior anions cannot move out of the membrane with the positively charged potassium ions, the axoplasm becomes negative with respect to the axon exterior. This, in turn, tends to push positively charged sodium ions into the axon, but the cell membrane is less permeable to them than to outward-flowing potassium ions; so an equilibrium condition is reached where the electrical potential exactly balances the concentration gradient. This equilibrium potential for the rest state is about -70 mV. When this equilibrium potential is upset (reduced or depolarized from -70 mV or increased or hyperpolarized to higher negative values) an enzymatic reaction involving certain kinds of proteins takes place that acts as a selective channel for preferentially passing sodium or potassium through the membrane. Adenosine triphosphate is a compound that tends to "pump" ions into or out of the axon. When the potential depolarizes, a sodium channel opens, and these ions are transported into the axon. This action changes both concentration gradient and accompanying membrane potential until, at some point, a new equilibrium condition is attained, at which time the sodium channel closes and a potassium channel opens; potassium ions then flow into the axon until the rest-state equilibrium is reached again. This sequence of events is referred to as a *sodium pump* (or more properly, a sodium–potassium–adenosine triphosphatase pump). The sodium pump's action changes the electrical potential between axon interior and exterior.

Figure 2 illustrates the principal ways in which axon potentials change. The curves labeled B and C represent changes over time at a single point along the axon when the stimulating signal is small. Curve B is a depolarizing change, and C is a hyperpolarizing one. In both cases the small current results in a passive fluctuation in membrane potential that decays with time. These are called *graded potentials*; the duration of the potential change varies with duration of stimulating current but rises and decays more slowly than the initiating current because of the electrical capacitance of the membrane. Such localized, graded potentials are used for transmission over short distances, usually less than approximately 1 or 2 mm. They are found at sensory endings (such as receptors) where they are called *generator potentials* and at junctions between cells where they are called *synaptic potentials*. When the stimulating signal is large enough to depolarize the membrane to a certain critical level, called the *threshold*, the sodium pump passes over to a positive feedback mode. The potential then forms a rapid, transient pulse, reversing its electrical polarity, after which the membrane returns to its original condition (in some nerve cells there is a transient hyperpolarization, or negative undershoot). These impulses last 1 to 2 msec. Their amplitudes bear no relation to the magnitude of stimulation, unlike those of the graded potentials. Instead, the frequencies at which the impulses occur are related to signal magnitude. Thus the impulse, called an *action potential*, is a pulse-coded representation of signal information. These, then, are the two kinds of transmission signals found in neurons: graded potentials and action potentials. The graded potentials vary with signal current approximately according to Ohm's law; they therefore sum together. The action potential is of fixed amplitude, independent of signal current, and is followed by a brief period of enforced "silence" during which another impulse cannot be generated; this all-or-nothing behavior precludes the possibility of simple summation.

Long-distance transmission by action potentials is further facilitated by certain anatomical and physiological characteristics of axons. Many axon

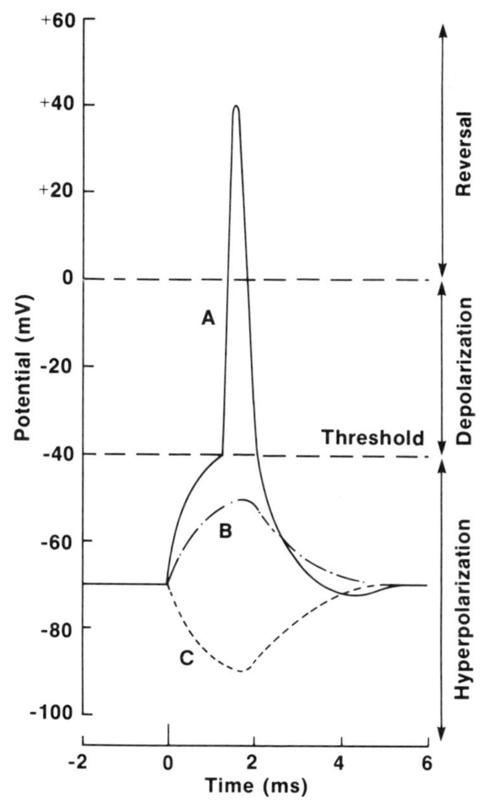

Fig. 2. Three of the ways in which a neuronal potential may change as a result of stimulation. Curve A shows a graded decrease in membrane potential from the rest state (−70 mV) to the threshold (−40 mV here), at which point a pulse or action potential is triggered, peaking at a positive value (about +40 mV) and then quickly returning to the resting potential. Curve B shows the time course of a change in membrane potential in response to lower levels of stimulation. The depolarization is not enough to drive the graded potential to threshold; consequently, it gradually returns to the resting potential. Curve C shows a graded increase in negative potential (hyperpolarization) and gradual return to the resting potential. Hyperpolarized action potentials (not shown) may also be generated.

membranes are surrounded by satellite cells: *neuroglial cells* in the brain and *Schwann cells* in the periphery. They form a material called myelin which wraps around the axon membrane to form a high-resistance, low-capacity insulation. At intervals of about 1 mm along the axon, the myelin fails to enclose the axon membrane, exposing it directly to its external environment. These uninsulated patches are called *nodes of Ranvier*, or simply nodes. Because ions cannot easily flow in and out of the insulating myelin, ionic currents of the action potential take the lower resistance path through the nodes. Figure 3 illustrates the preferential location of sites for action potentials in myelinated axons. The impulses propagate along the axon, jumping quickly from one node to another. The result is twofold: it speeds up conduction along an an axon, and it increases efficiency by reducing the total number of sodium ions that must be pumped to propagate a signal along the axon.

At the end of the axon there is an enlargement forming a terminal bulb (or button, or bouton) called a *synapse*. The synapses transmit information from one cell to another. There are excitatory and inhibitory chemical synaptic connections, excitatory and inhibitory electrical transmissions, and presynaptic inhibition at certain excitatory nerve terminals. They all perform the same kinds of functions, excitation or inhibition, in different ways depending upon their locations in the circuit. Figure 4 is a general schematic illustration of a synaptic bulb. The synapse contains many tiny globules, called *synaptic vesicles*, each carrying many thousands of molecules of chemical transmitter substances (whose effects are either excitatory or inhibitory). When the signal potential arrives at the synapse, some of the vesicles release their molecules. The transmitter molecules migrate through the presynaptic membrane, across a narrow cleft (the *synaptic gap*), and into the postsynaptic material of the adjacent dendrite of another cell. The effect of the transmitter is produced by its interaction with specific chemoreceptor molecules in the postsynaptic membrane material of the receiving cell. Neither the transmitter nor receiver

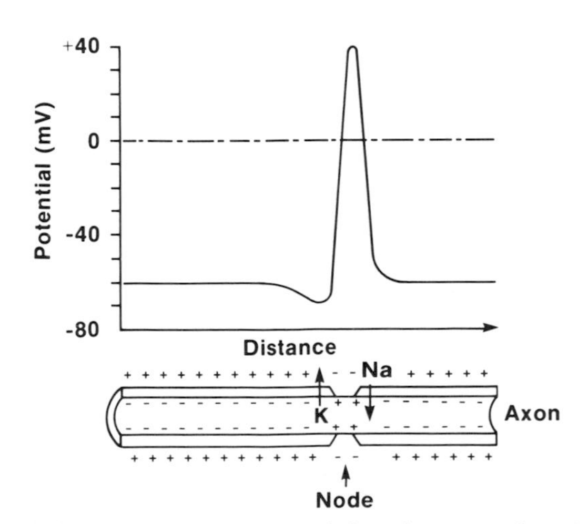

Fig. 3. Schematic illustration of signal transmission along a myelinated axon. The lowest resistance path is from one node to another. Action potentials tend to jump from one node to the next along the length of the axon.

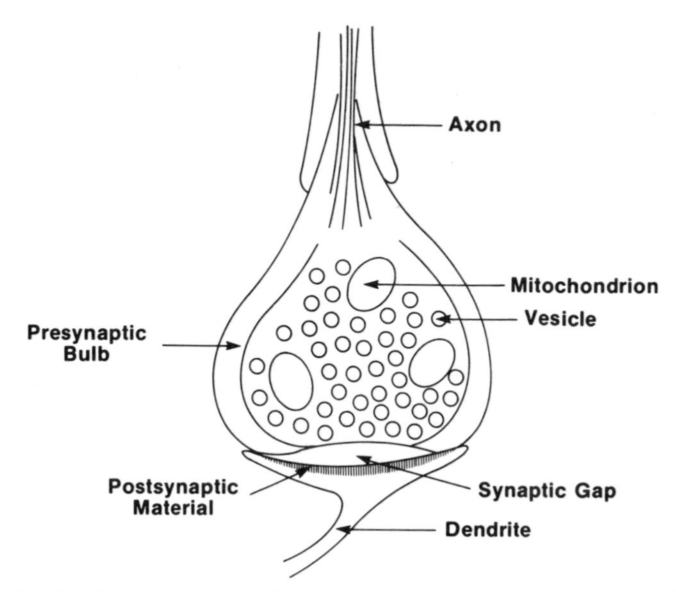

Fig. 4. Highly schematic representation of a synapse at an axon terminal. Potential changes relayed through the axon cause some synaptic vesicles to release transmitter molecules, which migrate across the synaptic gap to the adjoining dendrite process where they interact with receptor molecules to cause a change, which the dendrite signals to its cell body.

chemical constitution is known with certainty for synaptic transmission in the visual mechanism.

Although the receiving cell is usually coupled to the transmitting cell by its dendrites, synapses may attach to other cell bodies and to axons as well. Excitation or inhibition may also be electrical when cell bodies or axons are in close proximity (of the order of 1 to 2 nm separation). In their connections, neurons both converge and diverge. Each cell receives from many others and transmits to many others. Their patterns of interconnection and the details of their operations are much more complex than suggested by the general schema presented here. However, the discussion of basic principles in the foregoing paragraphs should acquaint the reader with the general plan and function of neurons so that the following discussion of specific neural function throughout the visual pathways can be appreciated as a special case of the overall nervous system. For those nonspecialists who wish to pursue some of the details of the nervous system in greater depth, there are a number of publications that may be helpful (for example, Llinás, 1975; Kuffler and Nicholls, 1976; Rose, 1976; Iversen, 1979; Kandel, 1979; Nauta and Feirtag, 1979; Stevens, 1979; and Schwartz, 1980).

2. Retinal Neurophysiology

The retina contains only five or six kinds of neural cells. Cells processing related information tend to be grouped together for simplified (short-distance) communication. Both graded potentials and action potentials are found in the retina; the graded potentials are observed at the distal periphery where cells tend to be closest together. The pattern of interconnection appears to be quite specific. Although both convergence and divergence exist, convergence tends to dominate. In each eye, more than 170 million receptor cells (the most peripheral nerve cells in the retina) ultimately converge on about 1 million ganglion-cell axons (the most proximal retinal cells). Considerable integration and signal coding takes place in the retina. Figure 5 has been prepared after

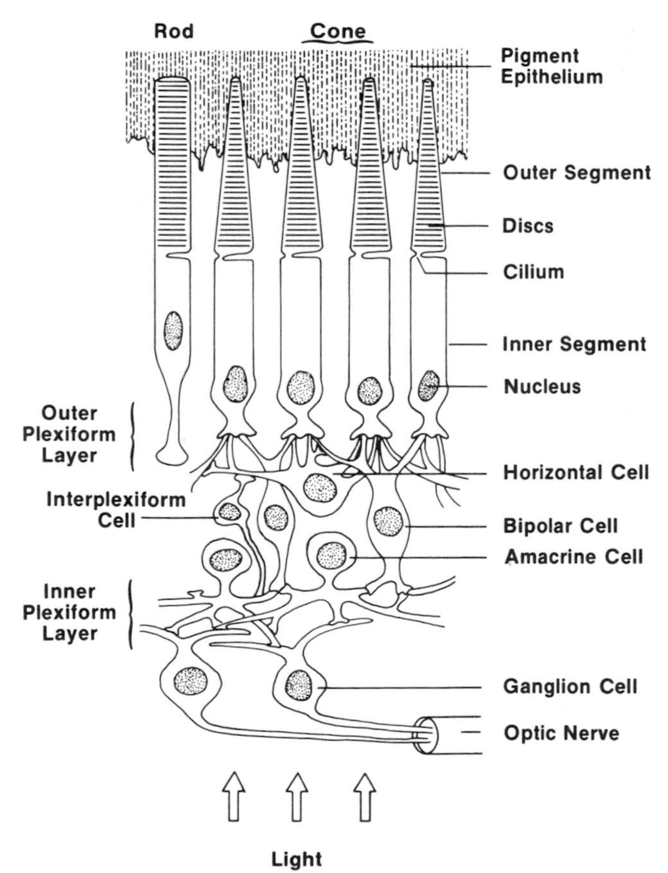

Fig. 5. Schematic "wiring diagram" (not intended to be anatomically and morphologically correct in detail) illustrating the relationships of various neural receptors and interneurons in the retina. Modeled after proposals by Dowling and Boycott (1966), Michael (1969), Werblin (1973), and Boynton (1979). See text for explanation.

proposals of several workers (Dowling and Boycott, 1966; Michael, 1969; Werblin, 1973; and Boynton, 1979) to provide a highly schematic indication of the principal neural components of the human retina. Much of the picture is surmised from studies of infra-human vertebrates and, particularly, primate retinae, which seem to be most similar to those of man. The specific structure of *Homo sapiens* retinae remains to be determined with certainty, however.

The photoreceptors of the retina are of two kinds, rods and cones, so named because of their shapes (although cones may also be rodlike in shape at the fovea of the retina). They are buried in a layer of light-absorbing material called the *pigment epithelium*, which serves as a kind of antihalation material and also provides nourishment to the receptor cells. Instead of arboreal dendritic structures, the photoreceptors have specialized, elongated receiver structures. The *outer segment* contains many laminar discs, in and around which is found one of four kinds of photolabile pigments. The rods contain *rhodopsin*. There may be as many as 10^4 molecules of rhodopsin in each of about 10^3 discs within each of 10^8 rods (Sjöstrand, 1953; Wald, *et al.*, 1963). Each molecule can absorb a single photon of light. The rods are primarily useful for vision at very low luminances, from about 10^{-3} cd $\cdot$ m^{-2} down to around 10^{-6} cd $\cdot$ m^{-2}. The cones contain one of four pigments that have been called *erythrolabe*, *chlorolabe*, and *cyanolabe* (Rushton, 1972). Less is known about the makeup of these pigments than about rhodopsin, but it is assumed that there are certain similarities as well as differences (see Chapter 4). The cones are distributed much more densely in and around the fovea and are responsible for both high-acuity and color vision. When photons are absorbed by photopigments, they induce a stereochemical rearrangement of the pigment molecule, proceeding under many conditions to complete bleaching. This somehow gives rise to change in the cell's electrical potential. This generator potential is passed along through the *cilium* to the *inner segment*, which contains the cell's nucleus. The inner segment terminates in synaptic junctions that are called *spherules* in rods and *pedicles* in cones. Cone pedicles tend to be located closer to the cell nucleus and are generally larger and more complex than rod spherules.

Unlike most other neurons, reception of stimulus information (that is, absorption of photons) in photoreceptors leads to *hyper*polarized graded potentials. Apparently, rods and cones continually release transmitters in the dark and cease to do so when they absorb photons; it is as if the stimulus were the *absence* of light. Although photon absorption is the primary stimulus, photoreceptors may also be stimulated by neighboring receptors through *horizontal cell* connections and at *gap junctions* where axons are very close.

The *outer plexiform layer* of the retina contains the synaptic junctions between photoreceptors and *horizontal cells*. Cone pedicles may have up to 25 or 30 indentations (*invaginations*) on their bases, to which the dendrites of horizontal cells and some *bipolar cells* couple. The horizontal cells interconnect photoreceptors. In some cases there is also direct contact between rod spherules and cone pedicles, but the principal interconnections among photoreceptors are through these horizontal cells. Some may connect rods to cones through very long axons, but most appear to provide connections among cones (Kolb, 1970). The area of interconnection of a single horizontal cell may be as much as 1 mm^2 (corresponding to more than 3° of visual subtense). A typical horizontal cell may contact about seven cones, and a single cone may connect to three or four horizontal cells. Their potentials are of the graded variety and are sustained hyperpolarization or depolarization in form.

Each of the invaginations of the cone pedicle accommodates three dendritic connections. The outer ones generally are taken by the interconnecting horizontal cells, but the center seems to be reserved for one of two types of bipolar cell: the *midget bipolar cells* (not shown in Fig. 5). These cells appear to connect exclusively with the *midget ganglion* cell processes in the *inner plexiform layer*. A second kind of bipolar cell, the *flat bipolar cell*, synapses with cone pedicles in the flat regions between invaginations. These flat bipolar cells synapse with processes from amacrine and *diffuse ganglion cells*. Although bipolar cells connect only to rods or to cones, they may merge at amacrine cells, thus providing still another potential point of interaction between rods and cones. The bipolar output is, again, a graded potential that seems mainly to be a kind of difference signal, representing some function of the contrast between a small area and its surround. Because no recordings have yet been made from primate horizontal, bipolar, or amacrine cells, the foregoing has been established only by infra-primate studies.

A recently discovered cell, the *interplexiform cell*, provides connections between junctions in the two plexifom layers (Dowling and Ehringer, 1976). Its precise function is not understood, but it is well located to perform a feedback control function.

Amacrine cells have no direct contact with photoreceptors. They synapse with bipolar cells, ganglion cells, and other amacrine cells. Apparently they help to define the fields of ganglion cells and feed back onto bipolar cells. In addition, certain *efferent fibers* (neural fibers providing a flow of information toward the periphery, or in an *e*fferent direction rather than *a*fferent, proximal, direction) return from the brain and terminate on the amacrine cells, perhaps providing central feedback from the isthmo-optic nucleus of the brain to modulate cell properties, but this function is controversial and has not been found in primates. The amacrine output, for the first time in the chain from receptor to brain, consists of pulsed action potentials.

Similarly the output of the ganglion cells also consists of action potentials. They are fed by bipolar cells (midget bipolar cells connecting to midget ganglia) and by amacrine cells. A striking feature of the output of many ganglion cells is that they do not appear to signal much information about levels of illumination. Primarily they signal information related to contrast in light levels over space or time. The ganglion cells seem to have two different functions, one to signal spatial light contrast and the second to signal temporal changes in light. Certain amacrine cells may provide the necessary temporal sensitivity. Thus at the level of the ganglion cells, whose axons are bundled together to form the optic nerve, processed information is available about contrast and about change.

The net balance between divergence and convergence favors convergence by a factor of about 100:1, although this ratio varies greatly with retinal eccentricity. The roughly 100 million receptor elements converge by integration and processing to the approximately 1 million axons of the ganglion cells that make up the optic nerve leading out of the eyeball and on to the next higher center of vision. Early in this retinal chain, perhaps at the horizontal-cell level but certainly at the level of the bipolar cells, chromatic and contrast information has been generated. It is refined and recoded in the amacrine and ganglion cell stages, where information signaling temporal change is also added. In short, a reasonably efficient amalgamation of available information has been made and is represented in the pattern of optic nerve fiber impulses to provide the input for effective higher stages of refinement and processing aimed at a congruent perceptual relation to the variations in light arrayed over space and time throughout the physical world under examination.

Schematic representations of the signals from each of these kinds of neural cells are shown in Fig. 6, which illustrates idealized response signals from a small area of the foveal retina under two conditions of stimulation, with a dark surround and with a surround illuminated to a somewhat lower level than the focal spot. It should be obvious from the foregoing discussion of interconnections that we cannot ignore spatial interactions when discussing cell output. Although an isolated spot of light may give rise to a distinct hyperpolarized, graded potential in a receptor, the same spot of light leads to a much diminished potential when it is surrounded by an illuminated area (compare the uppermost left and right graphs of Fig. 6). After two stages of transformation (at the bipolar level), the signal for the spot with an illuminated surround may even begin to depolarize. These graded potentials give rise to pulsed action potentials in the amacrine cells that may show little change in frequency of spikes from the rest state with induction (on the right), whereas a definite increase in spikes may occur with onset of the signal without induction (on the left). Finally there are three general forms of action potential in ganglion cells. Some cells, called *off cells*, tend to be silent during stimulation, but signal pulses when the light is turned off; again, the rate of signalling is different with and without induction. Other units, called *on cells*, signal when

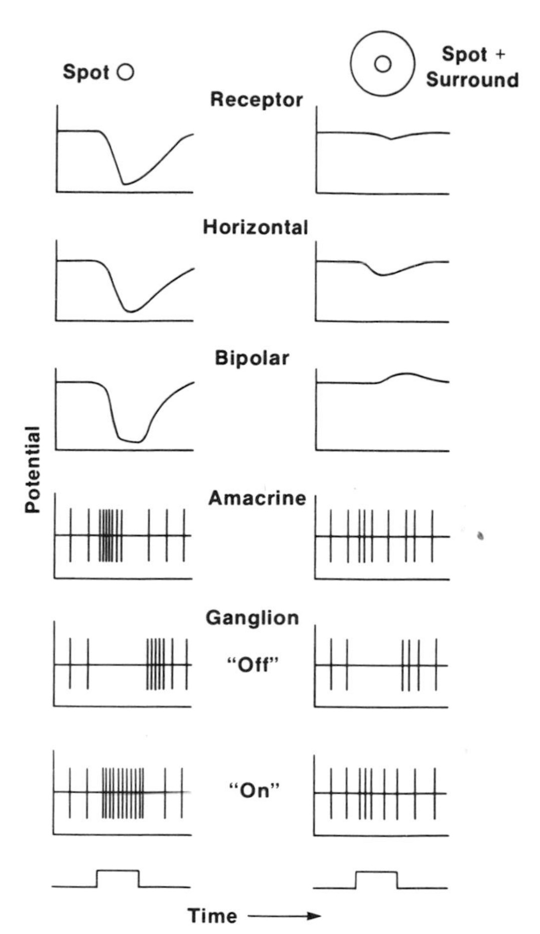

Fig. 6. Idealized representation of responses of retinal neurons to stimulation. Response to an isolated, small spot of light is shown at the left, and responses to the same spot surrounded by an illuminated annulus are shown on the right. Receptor, horizontal, and bipolar cells generate graded potentials. Amacrine and ganglion cells respond with action potentials.

the light is turned on. Finally, some ganglia signal both onset and termination of stimulation and are called *on–off cells.*

Some of these cells provide a transient response to onset or extinction of light stimuli, and some exhibit sustained responses. The transient type have been called *phasic* (Gouras, 1968), or simply *Y,* cells (Enroth–Cugell and Robson, 1966; Hochstein and Shapley, 1976a,b). The sustained response variety have been labeled *tonic*, or *X*, cells. Distinctions between these two cell types are much more subtle than the simple division into sustained and transient response characteristics, but these terms will serve here as a useful, if approximate, shorthand notation for their differences. Gouras (1968) has suggested that X cells reflect a rather direct signal from the photoreceptors by connections through the midget bipolar cells that connect cone pedicles and midget ganglia, whereas Y cells may derive their stimulation from diffuse bipolar cells that interconnect many photoreceptors. Possibly the most distinctive features of these neuron classes are that X cells tend to exhibit linear spatial summation and Y cells do not. The X cells tend to signal in a spatially and spectrally opponent (*vide infra*) manner, and Y cells are nonopponent. These differences (in ganglion and higher-order cells) may help to explain some of the differences between chromatic and achromatic pathways of the visual mechanism, which will be discussed in a later section of this chapter.

It should be clear by now that three aspects of cell signalling are important in attempts to evaluate neural transmission throughout the visual pathways: response magnitude, antagonistic spatial and spectral feedback, and temporal response characteristics. Just as it is impossible to ignore spatial relationships, so also it is unrealistic to ignore temporal response characteristics of cells. The sensitivity of the retina (that is, the capability of the retinal network of neurons to respond to stimulation) is a result of three distinct kinds of cell properties associated with response magnitude, antagonistic feedback, and temporal response. Interestingly, the sensitivities of these three aspects of retinal response seem to be located at distinct sites within the retina. Figure 7 (after Werblin, 1973) provides a schematic guide to the organization of retinal sensitivity adjustment.

Retinal response to a test flash is determined by the existing cell sensitivity to overall background luminance, surround luminance, and temporal change in luminance in the schematic diagram of Fig. 7. The photochemical processes in the receptor cell are affected by the background luminance. The surround seems to affect horizontal cells, causing them to regulate sensitivity, probably by some form of feedforward to the bipolar cells and feedback to receptors. In the inner plexiform layer some amacrine cells respond only weakly to steady light but are strongly excited by changes in light; they often provide on–off responses signalling both onset and termination of light (Werblin and Copenhagen, 1974). Thus some ganglion cells signal level and some signal change, some signal chromatic contrast and some achromatic contrast, some are opponent and some nonopponent.

The complexities of these interactive combinations make it difficult to speak of single receptor cell output by the time our discussion has progressed through the retina to the layer of ganglion cells. It will be helpful to consider an organizational construct called *receptive fields* at this point. A receptive field of a neuron is simply the area on the retina that influences the activity of the

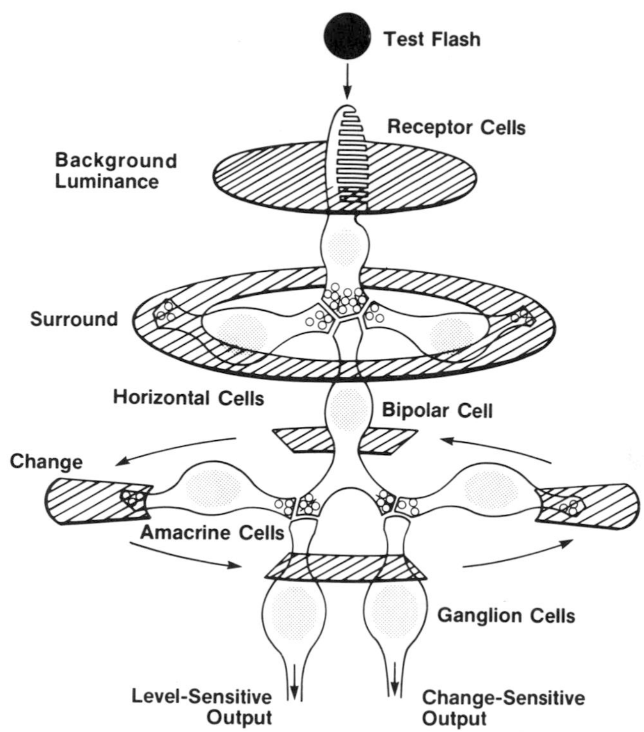

Fig. 7. Sensitivity to a test flash of light is probably determined principally by three properties of scene luminance, each activating specific neural mechanisms. Average background luminance affects receptor sensitivity. Surround luminance affects horizontal cell sensitivity. Changes in luminance affect amacrine cell sensitivity. (After F. S. Werblin, "The Control of Sensitivity in the Retina", Copyright © 1973, *Scientific American*, Inc. All rights reserved.) Redrawn with permission.

neuron in question. Several photoreceptors converge on more proximal cells which, in turn, converge on a single ganglion cell. Thus the retinal area represented by a single ganglion cell is extensive enough to include many cells. When any of those cells are stimulated, the ganglion cell's activities are affected. When an area of the retina is stimulated, and the ganglion cell activity is not influenced, that retinal area is outside the ganglion cell's receptive field. This useful construct was proposed by Hartline (1938) and is now commonly used to describe response characteristics of neural cells. It provides us with useful insights into cell organization and response characteristics.

The concept of *opponent* and *nonopponent* responses, mentioned above, is easily defined in terms of receptive fields, for example. Figure 8 illustrates a difference between spectrally opponent and nonopponent receptive fields. In the nonopponent field on the right, a spot of light incident in the center of the

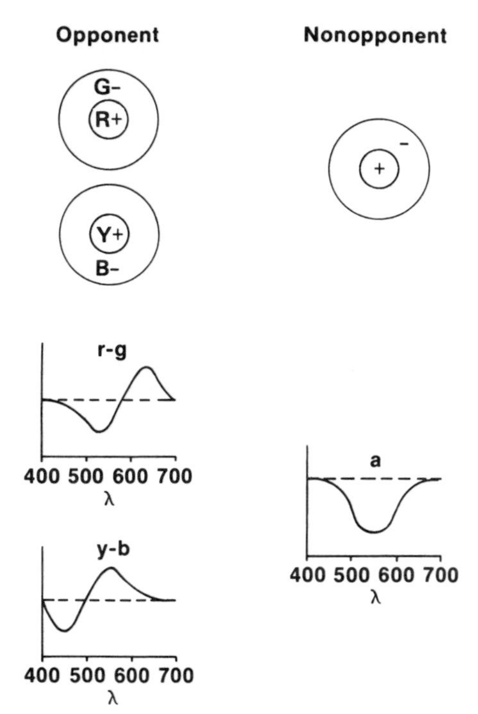

Fig. 8. Receptive fields for spectrally opponent mechanisms (upper left) and spectrally nonopponent mechanisms (upper right) together with their response characteristics as a function of wavelength (lower part of the figure). See text for discussion.

receptive field evokes a hyperpolarized response. When the wavelength of monochromatic light is varied, the magnitude of the response also varies. A graph of that variation, shown in the lower right portion of the figure, describes the spectral response of the center of the receptive field. It is hyperpolarized for all wavelengths, the strength of signal varying with wavelength; some spectrally nonopponent cells are of the kind that depolarize. The surround also varies nonspecifically with wavelength but with opposite sign. A large field is relatively ineffective, because the center and surround signals tend to cancel. Such fields are thought to signal achromatic information, and the field is labeled *a* for achromatic (some workers use the symbol *Y* for luminance). The two receptive fields on the left are opponent ones, distinguished in the figure by small, central subfields marked with polarity sign opposite to that of the larger, annular subfield. When a spot of monochromatic light is incident on the center, only certain wavelengths of light elicit a response that varies with wavelength; other wavelengths of light elicit no response. When the spot of monochromatic light is incident on the surround, again only certain wavelengths will evoke a response, but the wavelengths that give rise to a response are different from those that cause a response in the center. Such fields are said to have center-surround opponent chromatic responses. Large fields are ineffective unless their wavelength of stimulation is near the point of crossover where the polarity of response signals reverses sign. The lower left-hand portion of the figure shows two types of opponent chromatic responses related to the two examples of opponent polarity in the upper part of the figure. In one the center signals depolarization for long wavelengths, and the surround signals hyperpolarization for middle and short wavelengths. These kinds of receptive fields are often labeled $r - g$ (for red minus green), because they resemble psychophysical functions that represent spectral sensitivity relationships (for instance, Hurvich and Jameson, 1955). The lower graph illustrates a receptive field with a center that also hyperpolarizes for short-wavelength stimulation but depolarizes when stimulated by middle- and long- wavelength monochromatic light. Such cells are labeled $y - b$ (for yellow minus blue). Spectrally opponent cells' receptive fields may have a number of polarization combinations, but they all tend to fall into one of these two categories: r, g and y, b opponent chromatic organization. Usually, the cell will also have *on-center* and *off-surround* properties (or vice versa) as well. Spectrally nonopponent fields appear to provide achromatic (simple light-level) information.

Most opponent cells are also spatially opponent. The receptive field of such a cell is illustrated in Fig. 9. Its center responds with a signal having a polarity opposite to that of its surround. When a small spot of light moves across the diameter of the field, the cell responds negatively (in the example) until the spot reaches the center; then the cell fires positively. After the spot passes through the center, back into the surround, the cell again has a negative output.

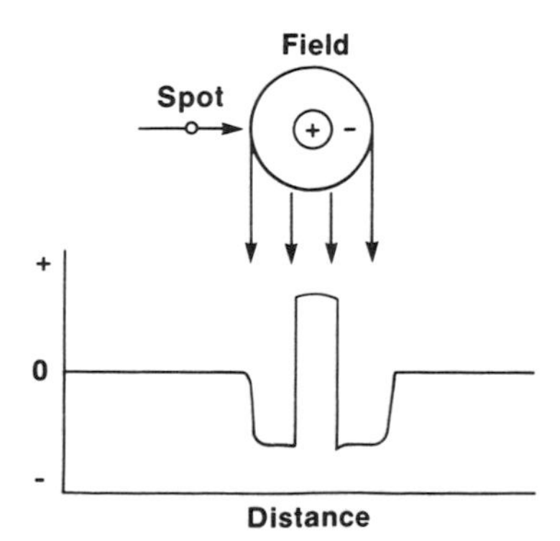

Fig. 9. Receptive field for spatially opponent mechanism (upper part of figure) and its response as a function of the position of a small spot of light traversing the field.

Properties of on-center, off-surround (or its complement) are common in such cells.

Thus the organization of signals in the ganglion cell axons is coded to reflect chromatic information, level, contrast, and both spatial and temporal relationships. These axons are contained in the sheath of the optic nerve, which passes through and out the rear of the eyeball to higher brain centers. Figure 10 shows a schematic representation of the human brain and the route of some of the visual pathways through it.

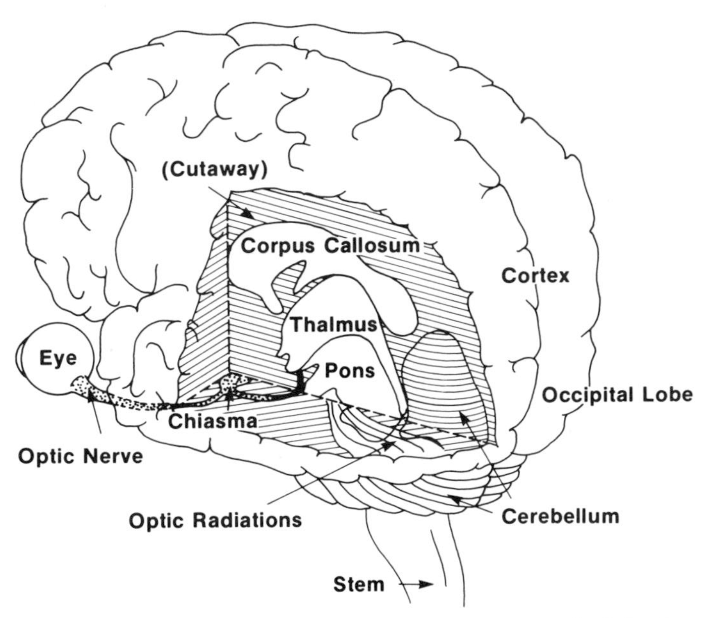

Fig. 10. Schematic representation of a human brain showing the path of signal transmission from the eye, through the optic nerve, to the geniculate bodies of the thalmus, and through the optic radiations to the striate cortex in the occipital lobes.

3. The Lateral Geniculate Nucleus

The optic nerve fibers from the two eyes converge at the *optic chiasma* where fibers carrying information from the left and right halves of both eyes cross over and join to proceed to higher centers. This decussation results in the left visual field being projected onto the right striate cortex of the brain and vice versa. Some fibers go to the *pons*, a folded sheet of nerve cells that is part of the hind brain below the *cerebellum* and is useful in coordinating muscle fibers of the limbs with visual images. Some fibers go to the *corpus callosum*, a transverse commissure that connects the cerebral hemispheres, where feedback to control eye movements and pupil size is provided and, especially, where integration and coordination between cerebral hemispheres is thought to take place. First, however, the fibers of the optic nerve enter the left and right *lateral geniculate nuclei* of the *thalmus*. Each lateral geniculate nucleus is a distinctly layered structure; in the monkey there are six layers of cells, each layer supplied by signals from one or the other eye. There appears to be a highly ordered sequence of topographical connections, even to the extent that adjacent-layer cells are in register with respect to their receptive fields. A striking feature of the lateral geniculate nucleus is that most of the projected area of the layers corresponds to the retinal fovea centralis, with relatively little area projecting to the periphery of the retina. There is, in other words, a kind of "magnification" of the fovea centralis area effected in the lateral geniculate nucleus. Each geniculate cell receives input from several optic nerve fibers and also synapses with several geniculate neurons (see Hubel and Wiesel, 1961; Szentágothai, 1973). Responses from geniculate neurons are much the same as those from retinal ganglion cells. The receptive-field organization of the ganglion cells in the retina is largely mirrored in the lateral geniculate nucleus, with the addition of the magnification effected for receptive fields corresponding to the fovea centralis.

4. The Visual Cortex

Visual information from the lateral geniculate nucleus passes through the *optic radiations* to the *cortex* of the brain. Incoming radiations form characteristic stripes that give this visual area of the brain its name: the *striate cortex* (sometimes identified hierarchically as Area 17). The immediately adjacent cortical areas in the occipital lobe (Areas 18 and 19) are also concerned with vision. All three areas contain many different neural cells, of which many can be classified structurally as *stellate* and *pyramidal cells*. Stellate cells have axons that terminate locally, but pyramidal cells have long axons that penetrate to deep inner layers and leave the cortex (Ramón y Cajal, 1955; Hubel and Wiesel, 1965; Zeki, 1974). Each cortical neuron has a corresponding receptive field (projected onto the retina), and as with retinal ganglia and geniculate neurons, responses occur on a background of continuous neural activity. Cortical neurons respond only slightly or not at all to diffuse illumination of the eye but are triggered by spatial and temporal changes in illumination. Most or all are sensitive to the orientation of stimulation. Figure 11 illustrates orientation sensitivity of a simple cell. When a bar of light passes across the receptive field, the extent of the response of the orientation-sensitive cell depends on the position of the bar with respect to the

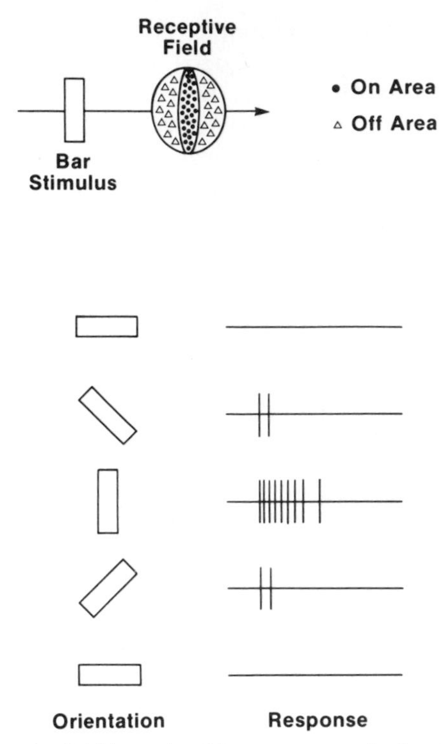

Fig. 11. Simple receptive field (upper) and its response to a horizontally moving bar of light (lower). Magnitude of response depends on orientation of the bar. This receptive field responds most to vertially oriented bars; that is, the frequency of pulses is greatest when bar is vertical and least when bar is horizontal.

on and off areas of the receptive field. In the illustration of Fig. 11 the field is most sensitive to vertical orientation. It does not respond to horizontal bars, responds only slightly to diagonal orientations, and responds maximally to vertical orientation of the bar. In some such cells there is also a preferred direction of movement for the bar. That is, the cell will fire when the vertically oriented bar travels from left to right, for example, but not when it travels from right to left. It has been suggested that processing for such selectivity may occur early in the chain of neurons, even within the retina itself (Barlow and Levick, 1965; Michael, 1969). Consider, for example, the wiring diagram of Fig. 12. In that diagram the bipolar cell is connected to two groups of receptors, to one group directly and to another through an inhibitory horizontal cell. A stimulus moving in the preferred direction (to the right here) excites a bipolar cell which, in turn, excites a ganglion cell. The ganglion cell fires. When the direction of motion is reversed (to the left), the stimulus first

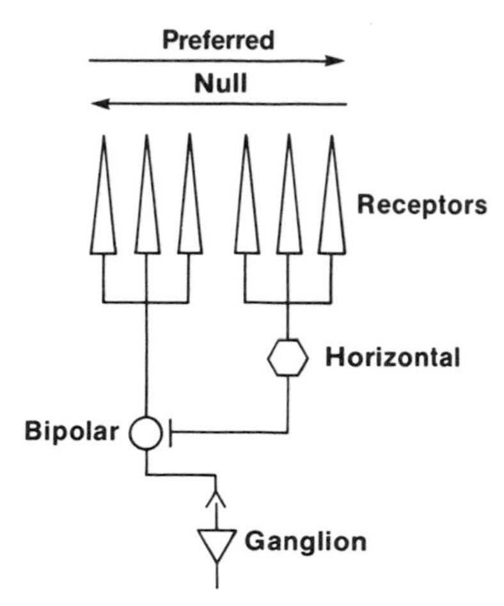

Fig. 12. One proposal for a retinal mechanism for directional sensitivity. When a stimulus moves in the preferred direction, it excites receptors connected directly to a bipolar cell which, in turn, excites a ganglion cell to fire. When the stimulus moves in the null direction, the receptors excite a horizontal cell, which inhibits the bipolar cell, and there is no firing of the ganglion cell.

excites a horizontal cell which inhibits the bipolar cell, and, consequently, the ganglion cell does not fire.

Whether or not such highly specialized retinal circuitry exists as an initial stage of orientation sensitivity (and the question is one of considerable debate), the fact is that most cortical cells seem to be sensitive to orientation, and some are also sensitive to direction (Hubel and Wiesel, 1959, 1962, 1968). Hubel and Wiesel have classified such cells according to their receptive-field characteristics into three general categories: simple cells, complex cells, and hypercomplex cells. Simple cells tend to respond to edges or lines and are sensitive to both orientation and position. Complex cells tend to respond to edges or bars, and although they are orientation-sensitive, they tend not to be sensitive to position. Hypercomplex cells tend to respond to ends of lines or bars or to corners of stimulus configurations; they are generally sensitive to both position and orientation. Table I has been prepared after the manner of Kuffler and Nicholls (1976) and may help to provide a general perspective on the various types of receptive fields associated with different neural stages in the visual mechanism.

Throughout the visual mechanism, functionally related cells seem to form interconnected aggregates at successive stages. The retina is the most distal functional unit. It consists of receptors, bipolar cells, and ganglion cells which, together with their interconnecting horizontal and amacrine cells, provide the first stage of integration of visual information. The second stage of neural

TABLE I. Summary of Receptive Field Characteristics[a]

Cell type	Field shape	Optimal stimulus	Response to diffuse light	Orientation sensitivity	Position sensitivity	Direction sensitivity
Receptor	⊕	Light	Good	No	Yes	No
Ganglion		Spot or bar	Moderate	No	Yes	No
Geniculate		Spot or bar	Poor	No	Yes	No
Simple		Edge or line	None	Yes	Yes	Some yes; some no
Complex	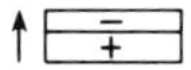	Edge or bar	None	Yes	No	Some yes; some no
Hypercomplex	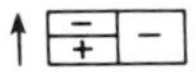	Line end; bar end, corner	None	Yes	Yes	Some yes; some no

[a] Based on Table 1, Chapter 2, of Kuffler and Nicholls (1976)

elaboration is in the lateral geniculate nucleus. A registered, laminar organization of cells begins to transform the abstraction of retinal information. Important retinal areas, such as the fovea centralis, are given more weight than less important peripheral areas; a kind of magnification takes place in the lateral geniculate nucleus as a result of its structural organization. The striate cortex (Area 17) may be considered the third stage of elaboration and integration. A given striate area receives input principally from a group of geniculate axons with receptive fields in the same part of the retina. Cortical neurons are stacked on top of one another, forming columns in which the neurons perform similar functions (Hubel and Wiesel, 1962, 1963, 1968, 1974). That is, simple, complex, and hypercomplex cells with similar receptive-field orientations form columns that are aligned roughly perpendicular to the cortical surface (Mountcastle, 1957). Each column receives information from overlapping receptive fields. Separate columns exist for each axis of orientation; some deal with color, some with movement, some with ocular dominance, and so on. The different-function columns are grouped to elaborate hierarchically ordered fields. In this way each retinal area is analyzed over and over, in column after column, to evaluate many different attributes of the stimulus. Interaction among cortical cells is accomplished by electrical potentials set up between neuroglial cells, axons, and the intracellular fluid (*cerebrospinal fluid*) of the brain (Kuffler and Nicholls, 1976). In this way individual cells "talk" to one another. Interactions among the columns in the primary visual area (Area 17) provide integrated inputs to higher-order cells in the adjacent Areas 18 and 19 of the cortex. A huge number of neural cells analyze many aspects of the stimulus configuration in space and time.

The details of the neural connections and function throughout the visual mechanism are more complicated than the generalizations set forth here. Our understanding of them is not as complete as might be inferred from such simple statements. Nonetheless, these generalizations should help to convey to the reader some sense of system and hierarchy in the processing of visual information. In addition, it should be obvious by now that the visual mechanism includes *all* of the neural pathways from retina to cortex; it is not limited to the eye. We cannot simply consider responses to isolated points of light. We must consider the effect of surrounding stimulation (spatial interactions) and the influence of changing stimulation (temporal effects) when we discuss the sensitivity and response of a particular point on the retina to a particular stimulus configuration. Understanding and applying existing knowledge of the ways in which the visual mechanism operates is greatly facilitated if we think in terms of receptive-field constructs rather than point-by-point receptor responses. The following sections will use these constructs to summarize some of the salient features of visual function.

C. VISUAL RESPONSE RANGE

1. Compression, Adaptation, and Specialization

The visual mechanism can respond to light over an intensity range of about one trillion to one ($10^{12}:1$). We can see luminances from approximately 10^{-6} cd · m^{-2} up to roughly 10^{6} cd · m^{-2}. However, no single cell or group of cells has a dynamic operating range as large as 10^{12}. The key to our ability to function under such widely differing conditions resides in three characteristics of the visual mechanism: compression, adaptation, and specialization. These characteristics are controlled primarily by the combined result of the cells' transfer functions and feedback control through intercellular connections. Both factors tend to provide maximum contrast at any given light level.

The process begins with the absorption of photons (light quanta) by the photopigments located in the receptor outer segments. The *law of photochemical equivalence* states that when one photon is absorbed by a molecule (atom, ion, etc.) of absorbing substance, one light-activated molecule (atom, ion, etc.) is produced. This response to light absorption is called the *primary photochemical reaction.* What the light-activated molecule does thereafter depends on its nature and environment. The molecule may transfer its energy to another molecule, it may re-emit the light, or it may undergo a variety of *secondary reactions* such as isomerization, polymerization, oxidation, and photolysis. The secondary reaction of visual pigments is *isomerization* (a change in stereochemical arrangement). In most cases the secondary isomerization reaction will not be induced by every photon absorbed. Therefore, the efficiency of the process, or the *quantum yield*, is of primary concern; it is defined as

$$\text{quantum yield} = \frac{\text{number of molecules isomerized}}{\text{number of photons absorbed}}.$$

Quantum yield is one measure of sensitivity. Because the energy of a photon varies with its frequency of vibration, most photochemical reactions are wavelength dependent (to use the inverse measure of frequency that is commonly employed in visual science). Therefore, the quantum yield or sensitivity of the photopigment differs from one wavelength to another. The amount of unbleached pigment available to absorb photons of a given frequency is maximum at threshold under conditions of dark adaptation. However, at higher background levels of illumination, some of the pigment molecules will have been bleached and cannot take part in the reaction (until after they have been regenerated). As more and more pigment molecules are bleached, fewer and fewer are available to be isomerized in the secondary photochemical reaction. The result is a transduction process (converting photon energy to isomerized pigment molecule) that tends to saturate. Because the process also has a threshold, the complete pigment transfer function is nonlinear. When the proportion of pigment bleached is plotted against the logarithm of stimulus intensity, as in Fig. 13, an approximately S-shaped curve results.

Naka and Rushton (1966) have proposed that a general hyperbolic tangent function can be made to fit the relationship between proportion of bleached pigment and log intensity of stimulation. When the range of stimulus intensities corresponding to 0 and 1.0 bleaching proportions is normalized, the relationship may be expressed as

$$p = \frac{1}{2} + \frac{1}{2}\left[\frac{e^{\alpha} - e^{-\alpha}}{e^{\alpha} + e^{-\alpha}}\right], \quad (1)$$

where p stands for proportion of pigment bleached after prolonged exposure to the conditioning intensity, α is a linear normalization function of logarithm of intensity, and e is the natural base ($\cong 2.71828\ldots$). In arithmetic units Eq. (1) reduces to something close to

$$p = \beta\left[\frac{L}{L + k}\right], \quad (2)$$

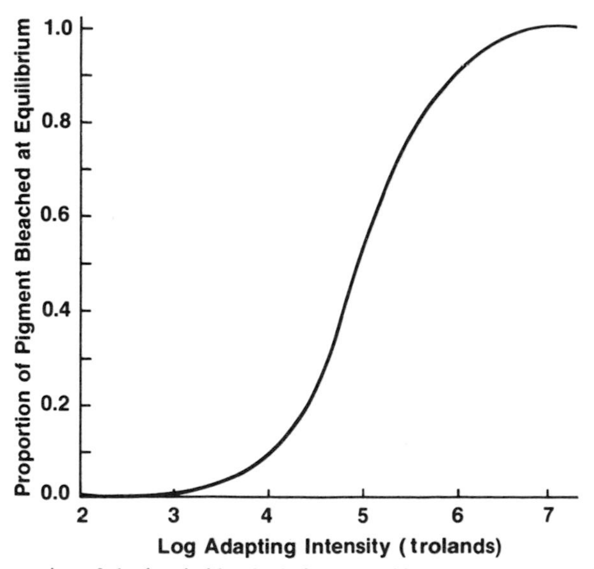

Fig. 13. Proportion of rhodopsin bleached after several hours exposure to various intensities of illumination. Curve follows the form of a hyperbolic tangent function of the logarithm of intensity. (After Naka and Rushton, 1966, *Journal of Physiology*, London.)

where L is light intensity, β is a scale coefficient, and k is the stimulus intensity that corresponds to half saturation, that is, when $p = 0.5$.

The transfer function of the receptor, although it is largely affected by the pigment transfer function, also includes the effects of electrical and chemical mechanisms by which the cell's generator potential is produced. Both the efficiency and the linearity of the process by which the number of isomerized molecules is reflected as a change in graded potential at the rod spherules and cone pedicles affect the photoreceptor transfer function. Boynton and Whitten (1970; Boynton, 1979) have found the receptor response for monkeys to be nonlinear. They have proposed an equation for steady-state viewing conditions (extended conditioning by spatially extensive, unchanging stimulation) that also incorporates a factor to mirror the effect of pupil dilation (which varies with level of stimulation). Their equation is

$$R = \beta\left[\frac{(LAp)^{0.7}}{(LAp)^{0.7} + k^{0.7}}\right], \tag{3}$$

where R represents response, A stands for pupil area, and the other symbols are the same as for Eq. (2). Equation (3) is similar to Eq. (2) although it takes more factors into account. The shape of the curve corresponding to Eq. (3) is very similar to that of the hyperbolic tangent function of Eq. (1).

Equation (3) represents receptor response compression with increasing intensity of adaptation. In addition to compression, adaptation intensity also significantly affects the *time* of receptor response. Baylor and Hodgkin (1974) studied the effect of increasing adaptation intensity on the time course of receptor responses in turtle cone receptors. Figure 14 illustrates their findings. As intensity increases, both rise time (the time required to reach a criterion response) and time to peak response, decrease. In other words the receptor responds more rapidly to more intense light. These results are consistent with a receptor-adjustment process, proposed from time to time, that is different from the photoisomerization of pigments and that would influence the membrane potential of the receptor (for example, Ives, 1922; Kelly, 1969; Baylor and Hodgkin, 1974). In essence the hypothesis involves triggered release of some kind of material as a result of photon absorption. The material is thought to diffuse to the receptor's outer segment membrane where it blocks the flow of sodium ions across the membrane. The process would have to be reversible as well to maintain response to steady illumination. If true, such a process would provide one means of receptor self-adjustment or adaptation. The effect of such adaptation would be to translate the curve of Fig. 13 laterally along the abscissa. In other words there would be a tendency to shift the transfer function horizontally with adapting intensity.

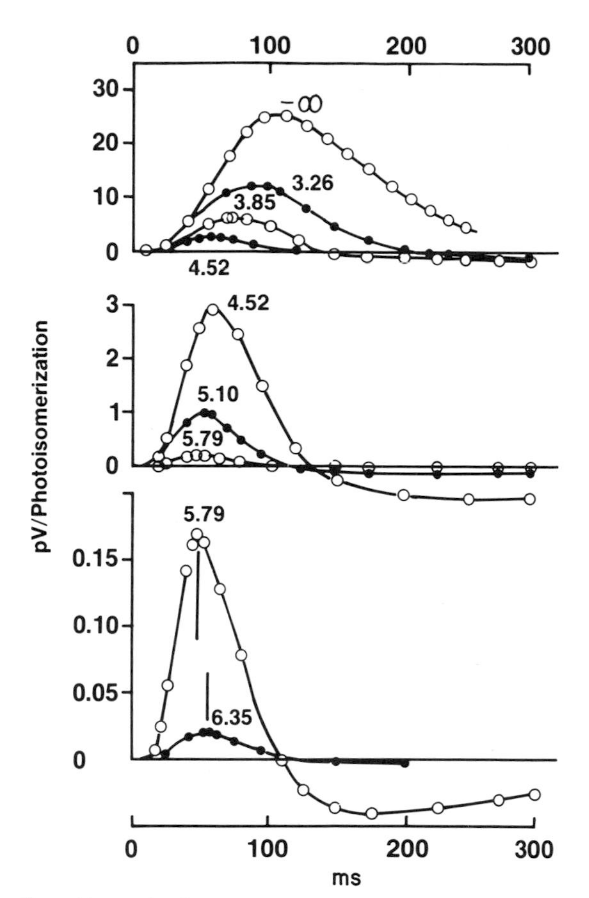

Fig. 14. The effect of increasing background intensity on the response of turtle cones to a test light of constant intensity. Curves are labeled with the logarithms of the background intensity (expressed as the number of photoisomerizations per cone per second). Responses are expressed as microvolts per photoisomerization. Separate expanded response scales are used only for clarity. (After Baylor and Hodgkin, 1974, *Journal of Physiology*.)

Such shifts need not depend wholly (nor even at all) on receptor adaptation. The visual mechanism, as we have seen, consists of both vertical and horizontal signalling processes. That is, in addition to signals flowing from the distal periphery toward the cortex (an afferent flow), there are also signals laterally connecting various cells, so that the output of a particular cell is influenced not only by the afferent signal level but also by signal levels of neighboring cells. Apparently, horizontal and amacrine cell lateral connections influence the "bias" of bipolar and ganglion cells. In effect, this process shifts the bipolar and ganglion cell transfer functions with respect to adapting intensity. The result can be imagined as a lateral series of transfer functions such as those shown in Fig. 15.

In Fig. 15 high sensitivity is represented by curves positioned toward the left and low sensitivity by curves farther to the right. The upper graph is intended to represent photoreceptor sensitivity functions. Rod functions are located well to the left of those for cones, indicating higher sensitivity. It is likely that rods are not inherently more sensitive than cones but owe their apparent higher sensitivity to signal summation over broader field areas than is the case with cones. This suggests a conferred enhancement in sensitivity from interconnections in the postreceptoral neural pathways. However, Fig. 15 follows the common convention of showing a rod receptor process of higher sensitivity than that for the cones. As the level of adapting intensity increases, the rod processes begin to saturate, and cone processes come into play. At some intermediate, low levels of adaptation both rods and cones operate; these levels correspond to *mesopic vision*, "in between" exclusively rod action (*scotopic vision*) and exclusively cone action (*photopic vision*). As adaptation increases still more, cone functions shift to the right, or lower-sensitivity,

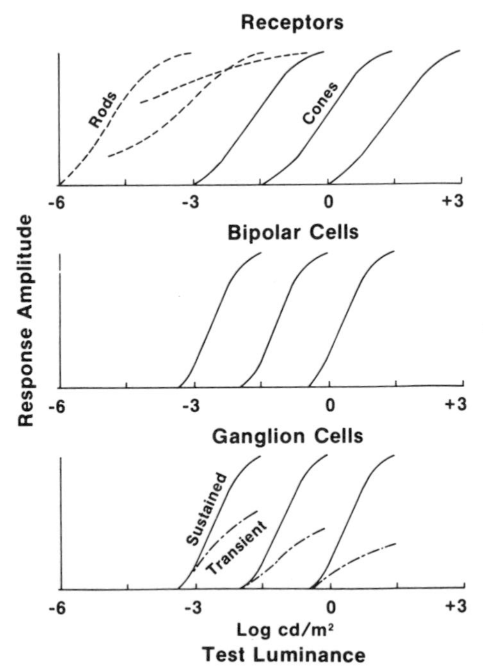

Fig. 15. Idealized transfer functions for three neural levels: receptors, top graph; bipolar cells, middle graph; and ganglion cells, lower graph. See text for discussion.

portion of the graph. The luminance range over which the receptors function under any one adaptation condition is about 2 log units (that is, about 100:1). The transfer function shifts to the right as adaptation intensity increases, covering the same range of 2 log units at higher absolute levels.

Transfer functions for bipolar cells are illustrated in the middle graph of Fig. 15. These functions also shift along the abscissa as adapting intensity changes. However, they cover a shorter range of test luminances, about 1 log unit instead of roughly 2 log units. Thus the bipolar cell signals contrast more effectively than the receptors do; the transfer functions of bipolar cells are steeper. This contrast enhancement is probably a result of antagonistic interactions from interconnecting horizontal cells that may oppose the receptor signals, not by attenuation but by linear subtraction (Werblin, 1974; Shantz and Naka, 1976). At the same time the steeper slope of the bipolar transfer function requires that sensitivity be adjusted more critically to maintain optimum contrast response. That is, the transfer function must be well centered for each adaptation level. Possibly, light adaptation reduces the effectiveness of all stimuli (central and surround) by nearly the same factor (Normann and Werblin, 1974); this would help to account for the approximate constancy of appearances of objects as illuminance changes. Adaptation level adjustments also influence the time course of response; rise time varies with adaptation intensity such that initial response may be linear with light level even though sustained responses are nonlinearly related to intensity (Baylor *et al.*, 1974a,b). An important consequence of these spatial interactions is that stimulus areas providing equal irradiances may evoke different electrophysiological and visual responses, depending on their surrounding irradiances.

Idealized transfer functions for ganglion cells are illustrated in the lower part of Fig. 15. Recall that some ganglion cells simply pass along sustained signals from the bipolar cells, and others respond to changes in signals. There are, in effect, two kinds of transfer functions at the ganglion level, one for sustained signals and one for changes. Both transfer functions shift laterally with adaptation. In addition, the transient or change-detecting functions have reduced slopes as adaptation intensity increases. When certain amacrine cells are activated by transient stimuli, they tend to reduce the effectiveness of the bipolar-to-ganglion signals. Thus there is an enhancement of spatiotemporal antagonism with increasing adaptation level. Central changes in stimulation activate the ganglion cells, but surrounding changes activate amacrine cells which inhibit the ganglion cell activity. The process places maximum emphasis on local change and tends to diminish the effect of steady or broad-field changes. The result is an enhanced ability to detect movement of small objects (or, in the obverse, to see fine stationary detail with small eye movements).

As with the bipolar transfer functions the ganglion transfer functions cover a stimulus range of only about 1 log unit. Here, too, the functions must be well centered to provide maximum contrast; that is, adaptational adjustments must be more finely tuned than would be necessary if the intensity range of the transfer function were greater. There is some evidence from lower vertebrates that the postreceptoral neural levels of bipolar and ganglion cells are much more influenced by adaptation than earlier retinal stages (Green *et al.*, 1975). Thus the useful range of action potentials from a given cell, which is roughly from 0 to 100 pulses a second, is translated by the process of adaptation so as to provide maximum information about differences in the visual field. Information about spatial differences and about temporal changes is enhanced by the visual mechanism's ability to compress, adapt, and specialize.

2. Physical Limits

Normal visual responses are elicited by light energy impinging on the retina. To elicit a response, that light must have a certain minimum energy and must not exceed a certain maximum energy; exceeding the maximum level may lead to temporary embarrassment of the tissues or even permanent pathological damage. The limits of the ranges of normal light stimulation are not sharply defined, and they depend on a number of factors relating to the conditions of viewing and the state of the organism. These will be summarized briefly in the following sections.

a. Energy Levels

A classic study of the absolute threshold of vision (Hecht *et al.*, 1942) involved determining responses of dark-adapted (30-min) observers to 0.001-sec flashes of 510-nm spectral light directed to a circular 10-min diameter area of the retina located 20° (horizontally) off-axis from the fovea centralis. These conditions are specified in some detail, because the energy required to elicit a visual response varies with each of the factors alluded to in the specification. Under these conditions, chosen to yield maximum sensitivity, the amount of energy corresponding to a 60% probability of seeing the flash ranged from 2.1 to 5.7×10^{-10} erg. From Table II the energy of one photon at a wavelength in air of 510 nm is approximately 3.89×10^{-12} erg. Dividing the threshold energy values by the energy of a single photon, we see that between 54 and 146 photons are required at the cornea to elicit a threshold response to such brief exposure with a stimulus of this size and location.

There are energy losses, by absorption and scattering, between the cornea and the retinal receptors, and therefore the number of photons available for absorption by the photopigments is considerably smaller than that of photons incident on the cornea. It is estimated that not more than 9–10% of the light incident on the cornea is absorbed by the retinal cells (Ludvigh and McCarthy, 1938; Wald, 1938; Crawford, 1949). This would mean that between 5 and 14

TABLE II. Relations between Frequency or Wavelength of Light in Air and Important Physical and Psychophysical Corresponding Values

Frequency[a] (10^{14} Hz)	Wave number[b] (cm^{-1})	Wavelength (nm)	Energy/photon[c] (10^{-12} erg)	Millions of photons per sec per mm^2 of retina to produce 1 troland	Number of photons per sec per receptor to produce 1 troland[d]
7.494813	25,000.00	400	4.9658	24,660	165,000
7.312012	24,390.24	410	4.8446	9,247	62,000
7.137917	23,809.52	420	4.7293	2,791	18,600
6.971919	23,255.81	430	4.6193	986	6,580
6.813466	22,727.27	440	4.5143	510	3,400
6.662056	22,222.22	450	4.4140	315	2,100
6.517228	21,739.13	460	4.3181	204	1,360
6.378564	21,276.60	470	4.2262	137	918
6.245677	20,833.33	480	4.1381	92.1	613
6.118214	20,408.16	490	4.0537	62.8	419
5.995850	20,000.00	500	3.9726	41.2	275
5.878284	19,607.84	510	3.8947	27.0	180
5.765240	19,230.77	520	3.8198	19.5	130
5.656462	18,867.92	530	3.7477	16.4	109
5.551713	18,518.52	540	3.6783	15.1	101
5.450773	18,181.82	550	3.6115	14.7	98
5.353438	17,857.14	560	3.5470	15.0	100
5.249518	17,543.86	570	3.4847	16.0	106
5.168836	17,241.38	580	3.4247	17.8	118
5.081223	16,949.15	590	3.3666	20.8	139
4.996542	16,666.67	600	3.3105	25.4	169
4.914631	16,393.44	610	3.2562	32.3	216
4.835363	16,129.03	620	3.2037	43.3	289
4.758611	15,873.02	630	3.1529	63.4	423
4.684258	15,625.00	640	3.1036	94.4	650
4.612192	15,384.62	650	3.0559	162	1,080
4.542311	15,151.52	660	3.0096	288	1,920
4.474515	14,925.37	670	2.9646	558	3,720
4.408713	14,705.88	680	2.9210	1,064	7,100
4.344819	14,492.75	690	2.8787	2,242	15,000
4.282750	14,285.71	700	2.8376	4,623	31,000

[a] $\nu = c\lambda^{-1}$; where $c = 2.997925 \times 10^{8}$ m $\cdot$ sec^{-1} and $\lambda = 10^{-9}$ m

[b] $cm^{-1} = 10^{7}$ nm^{-1}

[c] $E = h\nu$, where $h = 6.6256 \times 10^{-27}$ erg sec

[d] Receptor distribution density of 1.5×10^{5} mm^{-2} assumed.

photons are available at the site of the retinal receptors for the conditions of the Hecht *et al.* (1942) experiment. Of the roughly 500 rods in the region of the exposed retina, those authors reasoned from probability calculations that a single photon is absorbed by each of from 5 to 14 rods. Wald and Brown (1953) concluded that the molar extinction of rhodopsin, the photopigment of the rods, leads to a primary photochemical reaction with a quantum efficiency of 1.0; therefore, a single molecule in each of 5 to 14 rods provides the minimal arousal of a visual response.

When the duration of flashes is increased, the rate of irradiation rather than the number of irradiating photons becomes critical to determining a threshold response. Graham and Margaria (1935) reported that threshold data for 1-msec exposures should be corrected by a factor of 10^{-2} to predict threshold energy levels for longer exposures. The power of 2.1×10^{-10} erg to 5.7×10^{-10} erg for a period of 1 msec is 3.9×10^{-7} erg $\cdot$ sec^{-1} on average. When multiplied by 10^{-2}, the threshold rate for longer exposures becomes 3.9×10^{-9} erg $\cdot$ sec^{-1} or an average of 10^3 photons $\cdot$ sec^{-1} for the stimuli of the Hecht *et al.* experiment. The corresponding rate for a foveal image of a point source of 555-nm monochromatic light is about 2×10^4 photons $\cdot$ sec^{-1} (Davson, 1962). These values vary over a range of about 40,000:1, depending on wavelength of irradiation between 400 and 700 nm. A representative threshold figure for heterochromatic light is 0.75 to 1.0×10^{-6} cd $\cdot$ m^{-2} scotopic luminance. If the pupil area is assumed to be 50 mm^2, this luminance is equivalent to a retinal illuminance of about 4.4×10^{-5} scotopic trolands.

Both the time of exposure and the area exposed influence the energy required for the threshold of vision. Bloch (1885) showed that the threshold is constant if exposure time is less than a critical period. Various workers (for example, Herrick, 1956; Johnson and Bartlett, 1956; Barlow, 1958; Clark and Blackwell, 1959; Thomas, 1965) have measured the critical period to be between 80 and 100 msec for foveal stimulation. According to Bloch's law, then, there is complete temporal summation of energy for stimulation of 100 msec and less. For longer stimulation, temporal summation is incomplete. There have been a number of empirical expressions suggested to describe this incomplete summation (for instance, Blondel and Rey, 1911; Van der Velden, 1944). All involve nonlinear functions of summation with time. They may be roughly summarized as indicating that the threshold varies in proportion to the time of exposure raised to a power less than unity, generally in the range of $\frac{1}{2}$ to $\frac{2}{3}$ and differing between fovea and periphery.

Spatial summation is complete for stimuli of small angular subtenses according to Ricco's law (1877). The threshold does not vary when the angle subtended by the test flash does not exceed about 7 to 10 min of arc in the fovea and up to about 1° in the periphery (Baumgardt, 1949). Above these subtenses, summation is not complete (Piper, 1903). Piper's law states that the threshold is proportional to the square root of the stimulus area, but that relationship does not hold for extra-foveal stimuli or for areas greater than those corresponding to diameters of about 8° in angular subtense. In general the simple temporal and spatial summation rules do not seem to hold well for large and long-lasting stimuli (Barlow, 1958).

Accordingly, the energy at the retina required for threshold with a 10-min subtense and best observational conditions may be as low as about 10 photons for a 1-msec flash or as many as 10^3 photons $\cdot$ sec^{-1} for extended exposure; the value varies greatly with time and area of exposure, the state of the organism (Denton and Pirenne, 1954), and wavelength of the stimulating light. Measuring corneal energy at peak-sensitivity wavelengths, 555 nm for cones and 507 nm for rods, we find that approximately 600 photons are required for cones and 80 for rods; other measurements indicate a ratio of rod-to-cone sensitivity of about 10:1 (Baumgardt and Smith, 1967).

When these thresholds are expressed as luminances (rather than energy or power), we find that heterochromatic light (often loosely called white light) of visual angles as large as 45° may elicit a visual response with as little as about 10^{-6} cd $\cdot$ m^{-2} scotopic luminance, although this figure may vary among observers from as low as 0.4×10^{-6} cd $\cdot$ m^{-2} up to 2×10^{-6} cd $\cdot$ m^{-2} scotopic luminance (Davson, 1962). Above the absolute threshold, rod vision predominates (scotopic vision) up to a luminance of about 10^{-3} cd $\cdot$ m^{-2} scotopic luminance. Above about 1 to 3 cd $\cdot$ m^{-2}, cone vision predominates (photopic vision). The intermediate range (mesopic vision) involves both rod and cone vision to various extents, depending on the level of illumination.

Color vision, involving hue responses of all kinds, takes place in the range of photopic vision, above about 1 to 3 cd $\cdot$ m^{-2} or 10–20 td (trolands) in retinal illuminance. Trichromatic color matches made with foveal (cone) vision are essentially invariant when luminance is varied between this lower limit and an upper limit of several thousand cd $\cdot$ m^{-2} or 8,000–10,000 td of retinal illuminance (see Fig. 16). Such color matches break down at lower and higher levels.

Color vision obtains to much higher levels (even though color matches break down). However, in the region of 10^6 cd $\cdot$ m^{-2}, or about 3×10^6 td, retinal tissues may be damaged. The exact level at which damage may occur depends on a number of factors including exposure time, retinal image area, and level of fundus pigmentation; for this reason it is well to avoid exposures to light above about 30,000 cd $\cdot$ m^{-2}, or 100,000 td, a conservative but safe upper limit.

b. Spectral Range

The range of spectral wavelengths over which light may stimulate vision by absorption in the photoreceptors is great. Visual images have been recorded

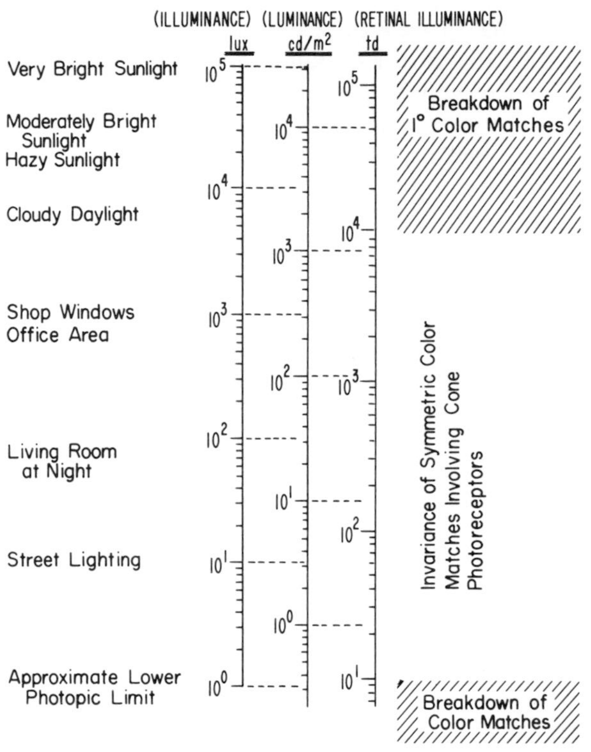

Fig. 16. Comparison of illuminance (lux), luminance (cd/m^2), and retinal illuminance (td) values. Luminance values are computed by assuming a spectrally nonselective, diffuse reflector of about 95% reflectance. Retinal illuminances are computed by assuming an average consensual pupil response. (Bartleson, 1981b, *Color Research and Application*); reproduced with permission from John Wiley and Sons, Inc.

for spectral stimuli of wavelengths as low as 300 nm and as high as 1000 nm (Burnham *et al.*, 1963). However, the sensitivities of the photoreceptors are so low for the extreme wavelengths outside the range of 380 to 780 nm that the CIE does not regard such energy as being within the "visible spectrum". The amount of energy required to see such spectral stimuli is very large, and there is a high likelihood that tissue damage may ensue. Fortunately, common light sources (except for the sun) do not generally provide such high radiant power at these extreme wavelengths. In addition, the ocular media through which the light must pass to reach the retinal receptors absorb most light of wavelengths shorter than 360–380 nm and longer than 1400–1600 nm.

c. Acuity

Visual acuity is the capability of the visual mechanism to discriminate among fine details, that is to detect, resolve, or recognize small stimuli or differences among stimuli. The most common clinical measure of acuity is one made by requiring the observer to read lines of high-contrast print displayed on a *Snellen chart* (Snellen, 1862; Cowan, 1928). This measure is based on the assumption that 1 min of arc is the standard for minimum visual angle; it also requires *recognition* of the stimulus, a more demanding criterion than either resolution or detection. The typical Snellen chart contains an array of letters consisting of stroke widths one-fifth as large in visual angle as the heights of the letters and arranged in rows of decreasing visual angle. One row, usually near the middle of the chart, contains letters whose stroke widths subtend 1 min of arc at 20 ft (or 6 m in many countries). An observer is said to have normal visual acuity if he can recognize all the letters whose stroke widths subtend 1 min of arc when he views the chart from a distance of 20 ft; his acuity is specified as 20/20, meaning that he recognizes the 20-ft standard letters at a distance of 20 ft. If the observer can recognize letters of smaller visual subtense at 20 ft, then his acuity is greater than normal (20/15, 20/10, and so forth, meaning that he sees the 15- or 10-ft standard at a distance of 20 ft). If, on the other hand, he can see only letters of visual subtenses larger than 1 min of arc when he views the chart at 20 ft (20/30, 20/40, and so forth), then his acuity is said to be less than normal. Other charts are similarly designed to measure acuity with accommodation to shorter viewing distances.

Visual acuity measured in this way is then simply

$$\text{acuity} = \frac{\text{standard distance}}{\text{test distance}}.$$

This relation corresponds to the ratio of object size to image size. Figure 17 illustrates that correspondence. The angle, a, in minutes of arc, subtended by a test object is equal to 3450 times the object width (w in millimetres) divided by the viewing distance (D in millimetres). The reciprocal of this minimum angle is usually taken as the specification of visual resolution in scientific and other nonclinical applications,

$$v = a^{-1} = (D)(3450w)^{-1}. \tag{4}$$

If a focal distance n of 17 mm is assumed from nodal point to retina of the eye, an object 1 min of arc in width will correspond to a retinal image width of 4.9 μm. Because cone photoreceptors in the foveola have diameters of 1.0–1.5 μm, the light from a 1 min of arc stimulus must impinge upon 4 or 5 cones for an eye with no aberrations or diffraction. However, all optical systems are limited by diffraction. When a luminous object is presented against a dark background, diffraction of the eye causes the pattern of stimuli on the retina to be the same for all objects subtending less than about 10 sec of arc. The retinal image of a point source of light is about 1.5 min of arc with a 3-mm pupil. This limitation is not the same for a dark object presented against a light

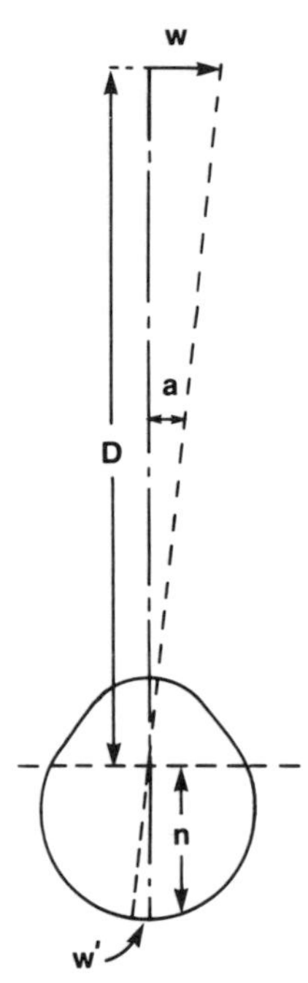

Fig. 17. Diagram to illustrate the specification of visual acuity. The width w of the test object is so small that it can just barely be discriminated at a distance D from the eye. The width (or other appropriate critical dimension) of the object so discriminated then subtends an angle a at the eye. This angle, provided it is small, is given approximately by the relation $a = w/D$ in radians or $a = 3450\ w/D$ in minutes of arc. Visual acuity v is defined as the reciprocal of this angle in minutes. Thus $v = 1/a = 0.00029\ D/w$. The geometrical image on the retina has a width $w' = nw/D$, where n and D are the distances of the retina and test object, respectively, from the nodel point of the eye.

background, and for this reason, most tests of visual acuity involve such stimuli. Under favorable conditions a dark line that subtends about 1 sec of arc can be detected (Hecht and Mintz, 1939). Under these conditions a retinal illuminance difference of as little as 1% can cause a sufficient differential along a row of cones with respect to their immediate neighbors that will lead to detection of the line. With disc-shaped or square targets there is no integration of signals along an extended path; consequently, larger angular subtenses are required for threshold detection. A dark disc whose diameter subtends about 30 sec of arc is a reasonable representation of threshold acuity for such targets (Pickering, 1915; Walls, 1943).

Detection of a target is generally a less stringent criterion than resolution of the elements of the target; still greater elaboration is involved in the conscious response that recognizes the target. One would expect, therefore, that acuity thresholds for resolution would tend to be higher than those for detection, and generally they are found to be so. Whereas dark lines on a light background may be detected for line widths as low as 1 sec of arc, two fine dark lines must be separated by about 30 sec of arc to be resolved individually (Wilcox and Purdy, 1933). When a grating of such lines is presented to an observer, an alternate light-dark angular subtense of about 1 min of arc is required for resolution. This threshold value is approximately the same as that required for the different task of recognition of letters in a Snellen chart. A second kind of recognition target is called the *Landolt ring* or *C* (Landolt, 1889; Cowan, 1928). It is a ring with a gap that may occur at any one of four locations separated by 90°; the task of the observer is to indicate where the gap is located. This recognition task yields thresholds of the order of 25 sec of arc under the best conditions. The task probably involves less elaboration than is required for recognition of alphabetical characters.

Table III is a summary of approximate thresholds of normal visual acuity expressed as minimum visual angles (in seconds of arc) required to satisfy particular criteria for different kinds of tasks. These values are intended to be roughly representative of what might be found under ideal conditions. The values vary according to the effects of a number of factors. Location of the image on the retina will affect threshold acuity, because the size and distribution of receptors vary over the retina, and the areas of spatial summation in receptive fields also vary across the retina. The size of the pupil is important, because it affects optical aberrations and diffraction as well as retinal illuminance. The latter is important to thresholds of acuity, because it determines both level and physical contrast of the retinal image. Physical contrast of the target will also influence the acuity threshold, as will the exposure or time of presentation of the target. In addition, the spectral distribution of irradiance from the target has an effect on the acuity threshold, because chromatic aberration of the eye's optical system varies with wavelength. All of these factors combine to emphasize that the data of Table III have no absolute meaning; but they do provide use with some relative information of practical interest.

Table III indicates that detection is finer with dark targets presented on a light ground than with light or luminous targets against a dark field. It also shows that detection of lines is easier than detection of small, symmetrical targets such as circles or squares. Dark lines can be detected when their widths are only a fraction of the diameters of individual foveal cones. Integration and elaboration of signals across arrays of cones are presumed to enhance our ability to detect such fine targets. A dark-line width of 1 sec of arc may be detected at moderate to high retinal illuminances. A disruption of linearity or offset of as little as 2 sec of arc can be seen for measurements of vernier acuity (Berry, 1948). When equal-width light and dark bars are arrayed to form a square-wave grating, about 25 sec of arc separation is required to detect the

TABLE III. Approximate Thresholds of Visual Acuity under Ideal Viewing Conditions for Several Tasks and High-Contrast Targets (Thresholds in Seconds of Arc Visual Subtense).

Tasks	Targets							
	Light lines (width)	Light circles (diameter)	Dark lines (width)	Dark circles (diameter)	Square-wave gratings (separation)	Vernier adjustment (displacement)	Landolt rings (gap diameter)	Snellen chart (stroke width)
Detection	10	90	1	30	25	2	—	—
Resolution	60	90	—	—	60	—	—	—
Recognition	—	—	—	—	—	—	25	60

Note: Data are approximate, and values are subject to considerable variation with viewing and target condition.

fact that the target consists of a grating, and a separation of approximately 60 sec of arc is needed for a resolution task such as counting the dark bars (discussions of resolution with sine-wave gratings will be set forth in the section of this chapter that deals with spatial vision). In general Table III shows that the more stringent criteria of resolution and recognition yield higher thresholds than those found for detection. In any case there is no single, simple answer to the question of the acuity limit for the visual mechanism.

The physical limits for normal functioning of the visual mechanism have been shown in the preceding sections to depend on optical properties of the eye, receptor sensitivity, anatomical organization, and neural processing characteristics. The threshold level of energy or power that stimulates vision is a function of the number of receptors irradiated, their sensitivities and those of the receptive fields in which they participate, and the projected sizes of those receptive fields. The spectral limits of vision are determined principally by sensitivity factors; accordingly, the limits of spectral stimulation are functions of the same factors that govern overall sensitivity of the visual mechanism. Finally, the limits of acuity depend strongly on spatial integration and elaboration throughout neural receptive fields as well as on sensitivities to differences among elements of stimulus arrays; in acuity measurements involving recognition, higher-order cerebral elaboration is also a factor that may contribute to determination of the threshold limit. In short the physical limits within which the mechanism of vision functions normally are rather elastic and depend significantly upon all the characteristics of the components of the visual system. There are no hard and fast specifications for these limits, but the foregoing should provide some appreciation for the general domain of normal operation.

D. COLOR VISION

1. Opponent Response Considerations

Ewald Hering presented a series of reports to the Imperial Academy of Sciences in Vienna from 1872–1874 (Hering, 1872, 1874a,b,c,d, 1875) that were later published as a monograph (Hering, 1878). These reports offered an opponent-processing, neurophysiological view of color response that was taken to run counter to the accepted trichromatic theory of color expounded by Thomas Young (1807a,b) and elaborated by Hermann Ludwig von Helmholtz (1860). Hering's views were largely disregarded by color scientists for about 80 years until quantitative psychophysical measurements of coded *chromatic opponent-response* sensitivities were measured by Leo M. Hurvich and Dorothea Jameson (1955). Red, green, yellow, and blue sensitivities (*chromatic*

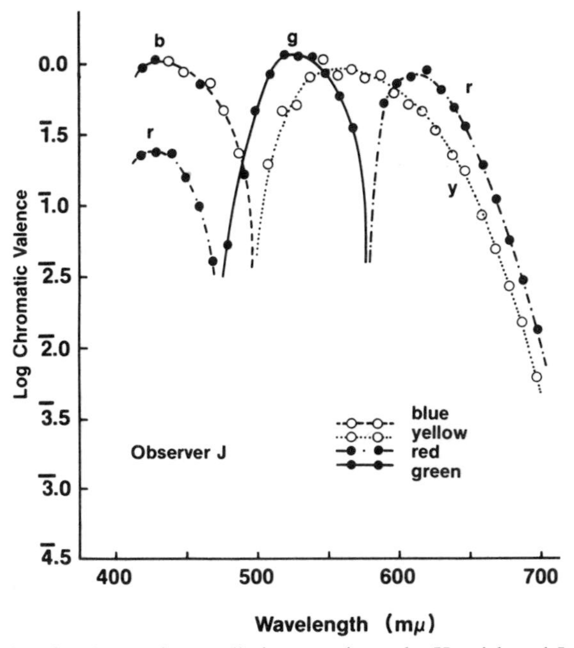

Fig. 18. Results of a chromatic cancellation experiment by Hurvich and Jameson (1955) in which the perceived amounts of red, green, yellow, and blue hues in spectral colors were assessed in terms of the amounts of complementary colors required to cancel each hue component. (Hurvich and Jameson, 1955, *Journal of the Optical Society of America*).

valences) measured for one of their observers are illustrated in Fig. 18. These measurements were made by a method called *chromatic cancellation.* According to Hering's concept there are four *unitary hues*—red, green, yellow, and blue—none of which contain any perceptible trace of any other hue. In addition, it was recognized that we cannot perceive both red and green or both yellow and blue hues in the same image point at the same time; in other words, red and green are opponent responses, and so are yellow and blue. When lights of the wavelengths (or combination of wavelengths in the case of red) that elicit unitary hues are determined for a particular adaptation, set of viewing conditions, and observer, they may be additively combined with light of other wavelengths (that elicit *binary hues* under the specified conditions) to cancel one of the unitary hue components of the binary hue. For example, a spectral stimulus that elicits an orange perception may be combined with light of the wavelength that elicits unitary green in an amount sufficient to cancel the red component of hue appearance, leaving the test stimulus appearing as unitary yellow. The minimum amount of the light corresponding to the green unitary hue that is required to cancel the red may be taken as a measure of the chromatic strength (or valence) of red for that wavelength of the spectral test stimulus (and under those conditions). The chromatic valence of yellow may be determined by adding light of the wavelength that elicits unitary blue in an amount just enough to cancel the yellowness of the test stimulus. When these amounts of light for unitary green and unitary blue are added to the light that appears binary orange, its appearance becomes neutral or hueless. In this way, using appropriate pairs of unitary hue-evoking stimuli, one can determine the chromatic valences for test stimuli of each wavelength throughout the spectrum. The experiment has been replicated in other laboratories with results very close to those originally published by Hurvich and Jameson (Romeskie, 1976, 1978; Werner and Wooten, 1979).

Hering's theory and the cited psychophysical results involve chromatic coding into opponent pairs: red opposed to green (*r–g*) and yellow opposed to blue (*y–b*). A stimulus cannot appear both red and green or both yellow and blue at the same time in the same place. Binary hues (the alternative to unitary hues) may appear red–yellow, red–blue, green–yellow, or green–blue in any proportion.* Accordingly, the data of Hurvich and Jameson are usually plotted to form a graph of relative chromatic sensitivity, where the *r–g* function of wavelength changes polarity or sign at the wavelength that evokes a unitary yellow response, and the *y–b* function changes sign at the wavelength that evokes a unitary green response (see Fig. 19). Under certain conditions the red chromatic sensitivity is present at short wavelengths, shown as the dashed extension of the *r–g* curve in Fig. 19; this will be discussed later in this section. Considerable experimentation has gone into the elaboration of this opponent-response theory, and it has recently been summarized by Hurvich (1978, 1981).

Direct electrophysiological support for the opponent-response model came shortly after the publication of Hurvich and Jameson's results (Svaetichin, 1956). The first intracellular, polarity-reversing, graded potentials to be recorded in a visual system were made by inserting microelectrodes into fish retinae, exposing the retinae to light modified by interference filters of rapidly and systematically varying spectral centroids, and recording cell output. Svaetichin found three classes of cellular output: one resembling a nonopponent achromatic response, which was either depolarized or hyperpolarized at all wavelengths and two resembling the *r–g* and *y–b* opponent chromatic processes that changed polarization with wavelength but differed in the wavelengths at which these changes occurred.

Shortly thereafter DeValois and his co-workers began a series of experiments in which electrophysiological records were made from cells in the lateral geniculate nucleus of macaque monkeys whose eyes were irradiated with

* Color perceptions also differ in brightness and lightness, but the hues seen are either unitary or binary, even for color sensations that we call brown, olive, pink, and so forth (Bartleson, 1976).

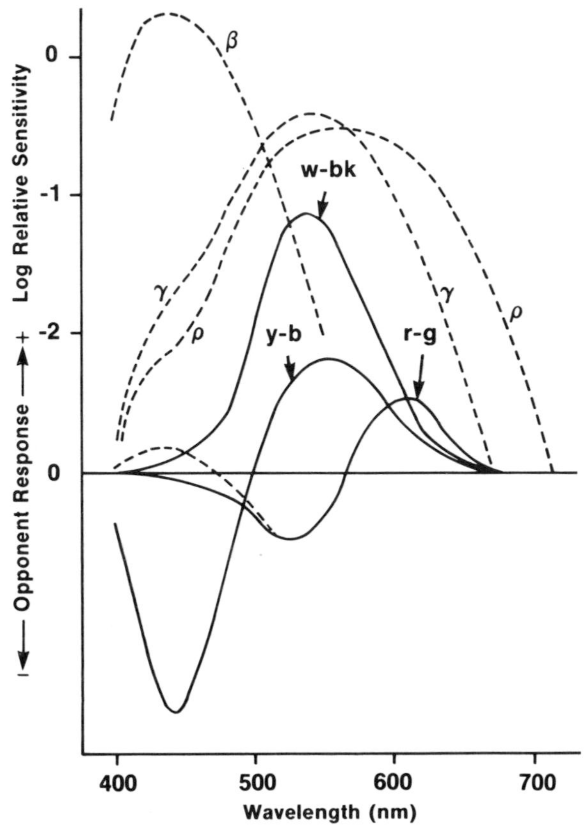

Fig. 19. Assumed cone sensitivity functions (ρ, γ, β) adjusted so that ρ and γ cross at 570 nm, and β is equal to the sum of ρ and γ at 500 nm. Opponent-response functions (w–bk, r–g, y–b) computed as $w{-}bk = \rho + \gamma$; $y{-}b = \beta - (\rho + \gamma)$; $r{-}g = \rho - \gamma$ (solid curve) or $r{-}g = \rho - \gamma + 0.05\beta$ for dashed extension at short wavelengths. (Adapted from Boynton, 1979.)

near-monochromatic, extensive, diffuse light of different spectral centroids (DeValois *et al.*, 1958; DeValois, 1965a,b, DeValois and DeValois, 1975). They have found nonopponent cells that either depolarize or hyperpolarize and four varieties of spectrally opponent cells that may be characterized as $r - g$, $g - r$, $y - b$, and $b - y$. Many related experiments have been reported in which ample physiological bases and additional psychophysical evidence have been accumulated to suppport the opponent-colors processing scheme; these results are summarized in various textbooks on color vision (for instance, Boynton, 1979; Hurvich, 1981; Wyszecki and Stiles, 1982).

The Hurvich and Jameson general model may be expressed in the following way:

$$\begin{aligned} w\text{–}bk &= f_1(a_1\rho + a_2\gamma + a_3\beta) - i_{w-bk} \\ r\text{–}g &= f_2(a_4\rho + a_5\gamma + a_6\beta) - i_{r-g} \\ y\text{–}b &= f_3(a_7\rho + a_8\gamma + a_9\beta) - i_{y-b} \end{aligned} \tag{5}$$

where ρ, γ, β are initial-stage trichromatic sensitivities, and $w\text{–}bk$, $r\text{–}g$, and $y\text{–}b$ are opponent neural sensitivities. The functions f_j admit the possibility of nonlinearities, and the coefficients a_j may be positive or negative as appropriate; Hurvich and Jameson set a_5 and a_9 negative. Finally, the additive terms i_j are response functions related to nonfocal stimuli providing induction to the response evoked by the focal stimulus. This induction term may be used in two ways, to characterize adaptation effects and to represent induction or contrast effects. In the latter sense, expanded equations have been derived (Hurvich and Jameson, 1961, 1974; Jameson and Hurvich, 1959, 1961a,b, 1964, 1972, 1975) to show that

> ... redness activity generated by focal stimulation of one retinal region produces opposite greenness activity in surrounding or functionally related tissue, and vice versa; and similarly for the yellow–blue and black–white neural processes. The strength of the induced response is related to the strength of the directly elicited activities in each of the mutually interacting areas and it is also related to the spatial contiguity of the areas in question [Hurvich, 1978, page 40].

There are data that suggest that although the $r\text{–}g$ opponent-process term of Eq. (5) is linearly related to the photon absorptions of the ρ, γ, and β processes, the $y\text{–}b$ term may not be (Larimer *et al.*, 1974, 1975; Cicerone, *et al.*, 1975; Ikeda and Ayama, 1980; Werner and Wooten, 1979). Apparently, there are large differences among observers in the degree of nonlinearity of the $y\text{–}b$ system, and these differences make it difficult to determine just how the $y\text{–}b$ differs from the linear $r\text{–}g$ opponent process. Larimer *et al.*, (1975) proposed the following form for the $y\text{–}b$ process using ρ, γ, and β primaries published by Vos and Walraven (1971):

$$y\text{–}b = k_1\gamma \pm k_2|\rho - \gamma|^n - k_3\beta. \tag{6}$$

The value of k_2 is positive when $\beta - \gamma > 0$ and negative when $\beta - \gamma < 0$; the exponent n is of the order of 1/2. Werner and Wooten (1979) have proposed a somewhat different form for the same primaries:

$$y\text{–}b = |c_1\rho + c_2\gamma|^n - c_3\beta, \tag{7}$$

where the exponent n is of the order of 3. Applying regression analysis to the determination of n for their own data and that of Hurvich and Jameson

(1955) and Romeskie (1976), Werner and Wooten found that n varied between 0.76 and 4.39 among individual observers. Those authors concluded that "the yellow-blue opponent channel is indeed best described as nonlinear, although we acknowledge marked individual differences, with some observers approaching nearly perfect linearity" (Werner and Wooten, 1979).

2. A Model for Color Vision

Despite uncertainties about the precise details of human visual opponent-response mechanisms, a number of models have been proposed for such mechanisms (for example, Hurvich and Jameson, 1955; Walraven, 1962, 1976; Abramov, 1968; Jameson and Hurvich, 1968; Guth and Lodge, 1973; DeValois and DeValois, 1975; Ingling and Tsou, 1977; Boynton, 1979; Guth *et al.*, 1980). There is reasonable agreement among vision scientists that the general form of such models is appropriate, but there is significant disagreement about the details of models. Each of the models derives chromatic opponent-response functions (in various ways) from a set of trichromatic fundamental primaries; generally r–g, y–b, and w–bk are assumed to be linear transforms of some set of fundamentals to a first-order approximation. There have been numerous sets of fundamental primaries proposed for over 100 years, and there are differences among the various models at this stage of selecting fundamental primaries.

Although fundamental primaries may be derived in a number of ways, they may be classed in two quite broad categories: those that are linear transforms of a set of color-mixture functions and those that are not. The advantage associated with fundamentals that are a linear transform of color-mixture data is that they describe a closed system in which the so-called laws of color mixture are not violated. These rules are discussed in detail in Chapter 3 of Volume 2 of this treatise and will be reviewed briefly in Chapter 12. Basically the rules are those of mathematical linearity involving additivity, transitivity, and validity of substitution for metamers. Although these rules apply only under certain conditions and over restricted ranges of stimulation, they form the cornerstone of the practice of colorimetry; therefore, it is desirable that a model of color vision not violate them. Accordingly, only those fundamental primaries that are a linear transform of some reasonably valid set of color-mixture functions will be discussed.

A commonly used set of color-mixture functions is that recommended by the Commission Internationale de l'Éclairage (CIE) in 1931, derived mainly from the chromaticity matching experiments of Guild (1925–1926) and Wright (1928–1929). Because these experiments determined only chromaticity relations, it was necessary in deriving color-mixture functions to invoke the

CIE 1924 luminous efficiency function $V(\lambda)$, then called the visibility curve, which was intended to represent the relative efficiency of various spectral stimuli of equal radiant power to elicit brightness responses in an observer having normal color vision. The mathematics by which $V(\lambda)$ is invoked to derived color-mixture coefficients at each wavelength from the data of chromaticity matches is such that all three of the CIE color-mixture functions $\bar{x}(\lambda)$, $\bar{y}(\lambda)$, and $\bar{z}(\lambda)$ depend on values of $V(\lambda)$. Unfortunately, measurements made since the CIE adopted $V(\lambda)$ in 1924 show a significant discrepancy between $V(\lambda)$ and results of flicker photometry at wavelengths shorter than about 460 nm (see Chapter 11). Judd (1951) has proposed a correction to the $V(\lambda)$ function for data in that wavelength range and has computed the modified color-mixture functions that result from invoking this corrected luminous efficiency function. The modified color-mixture functions are symbolized as $\bar{x}'(\lambda)$, $\bar{y}'(\lambda)$, and $\bar{z}'(\lambda)$. They are nonlinearly related to the CIE 1931 color-mixture functions.

Stiles and Burch (1955, 1959) determined all three color-mixture functions directly for both 2° and 10° diameter central fields. Their 10° data were adopted by the CIE as part of the basis for the CIE 1964 Supplementary Colorimetric Observer. Their 2° data are used only infrequently, but one application has been to define a set of linear transform, fundamental primaries (Estévez, 1979).

Table IV lists the color-mixture coefficients at 10-nm intervals for all three sets of color-mixture functions. Both the CIE 1931 and the Judd 1951 functions are normalized in Table IV so that $\bar{x}$ or $\bar{x}'$, $\bar{y}$ or $\bar{y}'$, and $\bar{z}$ or $\bar{z}'$ sum to about 10.75. The $\bar{r}$, $\bar{g}$, $\bar{b}$ data are the amounts of the three mixture-primaries used by Stiles and Burch (1955) to match unit amounts of monochromatic light after modification for minor experimental calibration error (Stiles and Burch, 1959).

A partial list of the linear transform equations that have been proposed for these color-mixture functions is set forth in Table V. Included are relationships to CIE 1931 $\bar{x}$, $\bar{y}$, $\bar{z}$ for fundamentals tendered by Koenig (Koenig and Dieterici, 1892), by Fick (Wyszecki and Stiles, 1967), by Pitt (1935), and by Judd (1944). Linear transforms based on the Judd-corrected $\bar{x}'$, $\bar{y}'$, $\bar{z}'$ functions or on Vos' (1978) refinement of them are shown for proposals of Vos (1978; which is not materially different from that of Vos and Walraven, 1971) and of Smith and Pokorny (Boynton, 1979). A similar proposal (not shown) by Guth and co-workers (Guth, 1972; Guth and Lodge, 1973; Guth *et al.*, 1980) is related to that labeled "Smith and Pokorny" in Table V as follow: $\bar{R} \cong 0.637\rho$, $G \cong 0.392\gamma$, and $B \cong 0.00258\beta$, where R, G, and B are the symbols used for the Guth primaries. Finally, the transformation from Stiles and Burch's corrected 2° color-mixture functions to a set of fundamentals offered by Estévez (1979) appears at the bottom of the table.

TABLE IV. 2° Foveal Color-Mixture Functions Expressed at Corneal Level

Wavelength (nm)	1931 CIE			Judd 1951			Stiles–Burch, 1955/59		
λ	$\bar{x}$	$\bar{y}$	$\bar{z}$	$\bar{x}'$	$\bar{y}'$	$\bar{z}'$	$\bar{r}$	$\bar{g}$	$\bar{b}$
400	0.0143	0.0004	0.0679	0.0611	0.0045	0.2799	0.0097	−0.0022	0.0619
410	0.0435	0.0012	0.2074	0.1267	0.0093	0.5835	0.0312	−0.0078	0.2284
420	0.1344	0.0040	0.6456	0.2285	0.0175	1.0622	0.0522	−0.0173	0.5246
430	0.2839	0.0116	1.3856	0.3081	0.0273	1.4526	0.0449	−0.0210	0.7958
440	0.3483	0.0230	1.7471	0.3312	0.0379	1.6064	0.0157	−0.0089	0.9656
450	0.3362	0.0380	1.7721	0.2888	0.0468	1.4717	−0.0286	0.0186	0.9161
460	0.2908	0.0600	1.6692	0.2323	0.0600	1.2880	−0.0960	0.0710	0.7823
470	0.1954	0.0910	1.2876	0.1745	0.0910	1.1133	−0.1746	0.1500	0.6116
480	0.0956	0.1390	0.8130	0.0920	0.1390	0.7552	−0.2376	0.2370	0.3635
490	0.0320	0.2080	0.4652	0.0318	0.2080	0.4461	−0.2787	0.3334	0.1963
500	0.0049	0.3230	0.2720	0.0048	0.3230	0.2644	−0.2958	0.4909	0.1072
510	0.0093	0.5030	0.1582	0.0093	0.5030	0.1541	−0.2685	0.7000	0.0495
520	0.0633	0.7100	0.0782	0.0636	0.7100	0.0763	−0.1480	0.9107	0.0124
530	0.1655	0.8620	0.0422	0.1668	0.8620	0.0412	0.1048	1.0352	−0.0048
540	0.2904	0.9540	0.0203	0.2936	0.9540	0.0200	0.4915	1.0523	0.0131
550	0.4334	0.9950	0.0087	0.4364	0.9950	0.0088	0.7900	1.0365	−0.0156
560	0.5945	0.9950	0.0039	0.5970	0.9950	0.0039	1.2300	0.9369	−0.0154
570	0.7621	0.9520	0.0021	0.7642	0.9520	0.0020	1.7492	0.8294	−0.0134
580	0.9163	0.8700	0.0017	0.9159	0.8700	0.0016	2.2731	0.6499	−0.0105
590	1.0263	0.7570	0.0011	1.0225	0.7570	0.0011	2.6728	0.4749	−0.0076
600	1.0622	0.6310	0.0008	1.0544	0.6310	0.0007	2.8750	0.3013	−0.0045
610	1.0026	0.5030	0.0003	0.9922	0.5030	0.0003	2.7680	0.1651	−0.0026
620	0.8544	0.3810	0.0002	0.8432	0.3810	0.0002	2.3790	0.0751	−0.0016
630	0.6424	0.2650	0.0000	0.6327	0.2650	0.0001	1.8184	0.0263	−0.0007
640	0.4479	0.1750	0.0000	0.4404	0.1750	0.0000	1.2552	0.0046	−0.0002
650	0.2835	0.1070	0.0000	0.2787	0.1070	0.0000	0.7864	−0.0020	0.0001
660	0.1649	0.0610	0.0000	0.1619	0.0610	0.0000	0.4436	−0.0027	0.0002
670	0.0874	0.0320	0.0000	0.8580	0.0320	0.0000	0.2350	−0.0019	0.0002
680	0.0468	0.0170	0.0000	0.0459	0.0170	0.0000	0.1210	−0.0011	0.0001
690	0.0227	0.0082	0.0000	0.0222	0.0082	0.0000	0.0603	−0.006	0.0000
700	0.0114	0.0041	0.0000	0.0113	0.0041	0.0000	0.0281	−0.003	0.0000

These (or other) fundamental primaries are used in opponent-response equations of the form suggested by Hurvich and Jameson [see Eq. (5)] to derive a pair of chromatic and one achromatic expressions.* The details of the methods by which these chromatic and achromatic opponent-response functions are derived from the fundamental primaries represent a second area of difference among the various color-vision models that have been suggested by different workers. Both the coefficients of Eq. (5) and the implied neural interconnections vary from one model to another. Although there is substantial agreement about the general form of the model, these unresolved differences over details make it necessary, in an exposition such as this, either to present all variations or to select what might be a representative model for discussion. The latter course has been chosen, although an attempt will be made to mention some of the more important factors that distinguish the model selected here for tutorial purposes from some of the other models most often discussed among vision scientists.

Figure 20 is a schematic diagram of the retinal circuitry implied by the model that will be discussed here. It is primarily drawn from proposals of Ingling and Tsou (1977) and Ingling (1977). In the model a total color response consists of the vector sum of two chromatic and one achromatic processes, which will be symbolized here as $C1$, $C2$, and A, with R standing for total response;

$$R = (C1^2 + C2^2 + A^2)^{\frac{1}{2}}, \tag{8}$$

where $C1$ corresponds to r–g, $C2$ to y–b, and A to w–bk.

The model involves two classes of receptors, rods and cones. The rod-mediated process is symbolized as α. The cone-mediated processes are shown in the diagram as ρ, γ, and β. Each has a subscript c or s, standing for center (focal) or surround, respectively. The surround process is taken to induce some change in the output of the focal process. In addition a broken line connecting all the interaction junctions among cone processes is intended to represent the interplay of modified outputs of the focal processes of different and similar

* The terms chromatic and achromatic are somewhat unfortunate choices for descriptors of the opponent-response functions, because those words imply perceptual qualities when, in fact, the processes are physiological, not perceptual. It cannot be denied that the three processes relate in some way to different perceptual qualities, but there is no indication that the relationships are simple or linear. Similarly the terms red–green (r–g) and yellow–blue (y–b) should not be taken to imply a one-to-one correspondence with perceived qualities except under special conditions. Most unfortunate is the use of R, G, and B (standing for red, green, and blue) in connection with photoreceptors. In no sense are such neural cells red, green, or blue; neither in appearance, in sensitivity, nor in signaling characteristics are they properly characterized as R, G, and B. The first first two sets of terms (chromatic/achromatic and r–g/y–b) are used here, because they are ubiquitous throughout the literarure on color vision. However, the symbols R, G, and B are avoided except where reference is made to their use by other authors.

TABLE V. Trichromatic Fundamental Primaries (ρ, γ, β) based on Linear Transformations of Color-Mixture Functions

I. Based on $\bar{x}$, $\bar{y}$, $\bar{z}$		
	Koenig	$\rho = 0.0713\bar{x} + 0.9625\bar{y} - 0.0147\bar{z}$
		$\gamma = -0.3952\bar{x} + 1.1668\bar{y} + 0.0815\bar{z}$
		$\beta = 0.000x\bar{x} + 0.0000\bar{y} + 0.5610\bar{z}$
	Fick	$\rho = 0.5960\bar{x} + 0.5151\bar{y} - 0.1229\bar{z}$
		$\gamma = -0.3952\bar{x} + 1.1668\bar{y} + 0.0815\bar{z}$
		$\beta = 0.0000\bar{x} + 0.0000\bar{y} + 0.5610\bar{z}$
	Pitt	$\rho = 0.3550\bar{x} + 4.7250\bar{y} - 0.0800\bar{z}$
		$\gamma = -2.3000\bar{x} + 6.7950\bar{y} + 0.5050\bar{z}$
		$\beta = 0.0000\bar{x} + 0.0000\bar{y} + 5.0000\bar{z}$
	Judd	$\rho = 0.00\bar{x} + 1.00\bar{y} + 0.00\bar{z}$
		$\gamma = -0.46\bar{x} + 1.36\bar{y} + 0.10\bar{z}$
		$\beta = 0.00\bar{x} + 0.00\bar{y} + 1.00\bar{z}$
II. Based on $\bar{x}'$, $\bar{y}'$, $\bar{z}'$		
	Vos	$\rho = 0.1551646\bar{x}' + 0.5430763\bar{y}' - 0.0370161\bar{z}'$
		$\gamma = -0.1551646\bar{x}' + 0.4569237\bar{y}' + 0.0296946\bar{z}'$
		$\beta = 0.0000000\bar{x}' + 0.0000000\bar{y}' + 0.0073215\bar{z}'$
	Smith and Pokorny	$\rho = 0.15514\bar{x}' + 0.54312\bar{y}' - 0.03286\bar{z}'$
		$\gamma = -0.15514\bar{x}' + 0.45684\bar{y}' - 0.03286\bar{z}'$
		$\beta = 0.00000\bar{x}' + 0.00000\bar{y}' + 0.001608\bar{z}'$
III. Based on $\bar{r}$, $\bar{g}$, $\bar{b}$		
	Estévez	$\rho = 0.3845\bar{r} + 1.0062\bar{g} + 0.0512\bar{b}$
		$\gamma = 0.1248\bar{r} + 1.3478\bar{g} + 0.1039\bar{b}$
		$\beta = 0.0000\bar{r} + 0.0305\bar{g} + 1.5551\bar{b}$

kinds of cone processes. These outputs are simply symbolized as e_o, without regard to whether they may be linear or nonlinear or, in later stages, without regard to the form of combination of afferent signals.

The $C1$ opponent-response signal is formed by taking a difference between ρ and γ. The model proposed by Ingling and Tsou suggests that for unrelated stimuli (that is, with a dark surround) this difference is of the form

$$C1 = 1.2\rho - 1.6\gamma. \tag{9}$$

However, under related or illuminated surround conditions, some contribution from the β component is added to $C1$ by a conditional process that operates only when $\gamma > \beta$ to take the difference between the two $(\gamma - \beta)$. This is

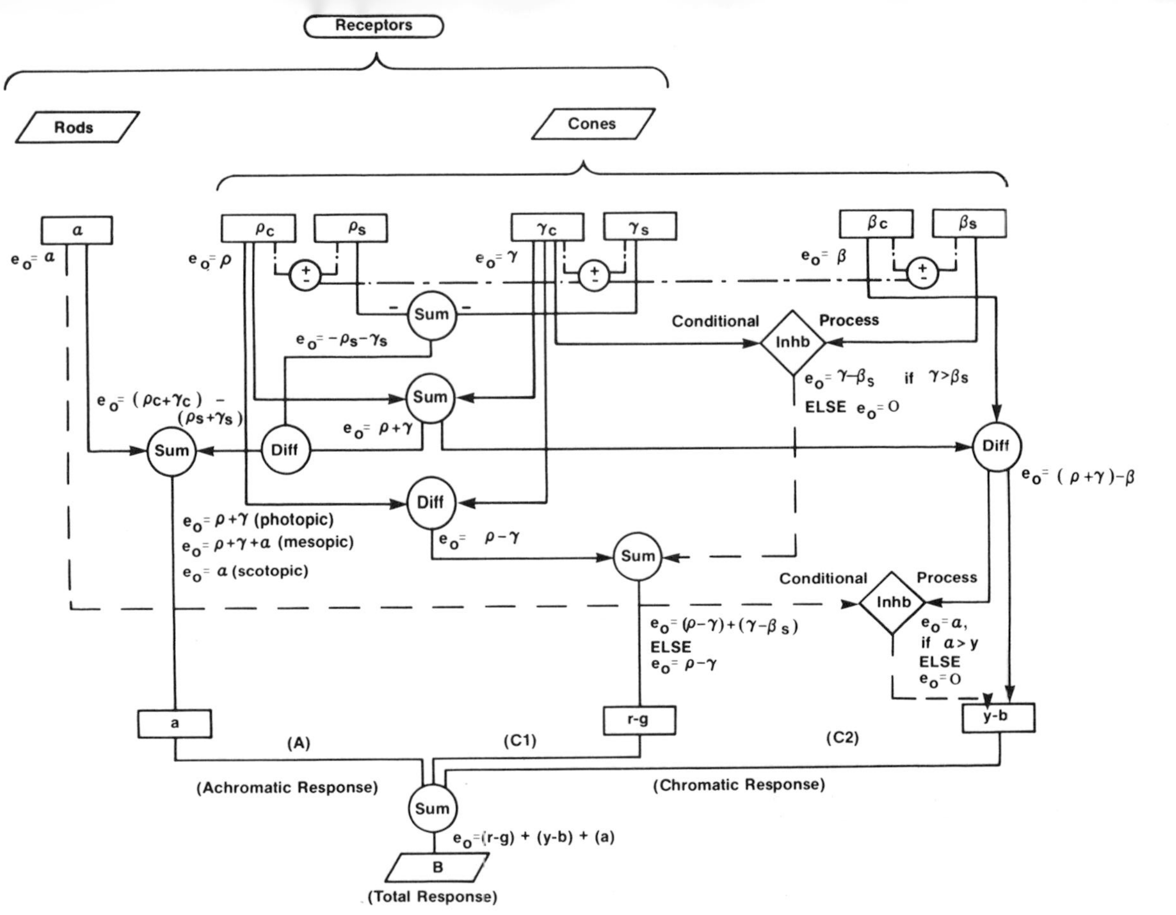

Fig. 20. Highly schematic diagram of a basic model of color and brightness vision circuitry. See text for explanation. (Based on Ingling and Tsou, 1977, *Vision Research.*)

equivalent to suggestions of others (for instance, Jameson and Hurvich, 1968; Boynton, 1979; Werner and Wooten, 1979) that the β cone process influences the $C1$ signal at suprathreshold conditions. Accordingly, Eq. (9) assumes the following form, using coefficients proposed by Ingling and Tsou,

$$C1 = 1.2\rho - 1.6\gamma + 0.4\beta, \tag{10a}$$

based on the Smith and Pokorny fundamentals (see Table V). Werner and Wooten (1979) using fundamentals derived from an iodopsin pigment nomogram (see Chapter 4) found the following coefficients:

$$C1 = 1.89\rho - 2.80\gamma + 0.45\beta. \tag{10b}$$

For comparison with Eq. (9) their coefficients of ρ and γ were 1.78 and 2.63, respectively (Werner and Wooten, 1979).

The difference in chromatic valence functions implied by Eqs. (9) and (10) is that illustrated in Fig. 19 for the solid r–g curve and its dashed extension, which reverses polarity at short wavelengths. In substance this conditional, interactive process means that short-wavelength stimuli, shorter than the wavelength of the stimulus that elicits a unitary blue response, take on a reddish-blue binary hue under normal viewing conditions. In fact, this is just what is observed when such stimuli are viewed. Psychophysical results indicate that the $C1$ opponent process is qualitatively different for related and unrelated colors. No coefficient or single additive constant can be used in Eq. (9) to match the short-wavelength reversal of the $C1$ process at all suprathreshold conditions. The suggestion that the β cones contribute to $C1$ seems to solve a number of theoretical problems (see Ingling, 1977).

In the model of Fig. 20 the $C2$ process is formed initially by combination of output from all three cone processes. In particular the β process is subtracted from the combined $\rho + \gamma$ processes. The relative amounts of these components may vary with level of illumination; this would mean, again, that the opponent-response function $C2$ takes on different shapes at threshold and suprathreshold conditions. Ingling and Tsou (1977) represent these differences by a change of coefficients when going from threshold to suprathreshold conditions.

Threshold:

$$C2 = 0.20(0.24\rho + 0.18\gamma) - 0.06(0.70\beta) - 0.075\gamma \tag{11}$$

Suprathreshold:

$$C2 = 1.00(0.24\rho + 0.18\gamma) - 1.00(0.70\beta) - 0.075\gamma \tag{12}$$

Equations 11 and 12 also include a subtractive term for γ. This additional linking is intended to account for certain experimental results that indicate differences among observers in their selections of the wavelengths corresponding to unitary green in the presence of a desaturating (white-appearing) stimulus, particularly at low levels of illumination (Wright, 1946; Rubin, 1961; Jacobs and Wascher, 1967; Richards, 1967). In the absence of the desaturating stimulus and with carefully controlled dark (physiologically neutral) adaptation, the distribution of spectral loci of stimuli evoking unitary green is unimodal (Hurvich *et al.*, 1968). The additional linkage has been left out of the diagram of Fig. 20, because its effect is small and it is somewhat controversial.

After the initial combining of $(\rho + \gamma) - \beta$, the $C2$ mechanism is subjected to a conditional processing stage representing rod–cone interactions that may occur under certain conditions. At low levels, where $\alpha > (\rho + \gamma)$, the initial $C2$ signal may be inhibited or combined with rod signals (α). At low photopic and mesopic levels, rod responses tend to intrude on the *y–b* process so that, effectively, the scotopic rod sensitivity function becomes confounded with the short-wavelength cone sensitivity (Trezona, 1970; Richards and Luria, 1964; Ingling and Drum, 1977; Ingling *et al.*, 1977). The low-level conditional linking of Fig. 20 is intended to represent such interactions.

The achromatic (A) process also assumes different forms depending on level of illumination. At photopic levels, where only or primarily cones are operating, the A process is formed by combining the output of the ρ and γ cone processes. Ingling and Tsou represent this combination as

$$A = 0.6\rho + 0.4\gamma. \tag{13}$$

Together, the terms of Eq. (13) form a spectral distribution that is essentially the same as Judd's 1951 corrected $V(\lambda)$ function when the Smith and Pokorny (Boynton, 1979) peak-normalized fundamental primaries are used.

At scotopic, or rod vision, levels of illumination, only the rod output, or α process, contributes to the A signal. This is represented in Fig. 20 as a summing stage connecting $(\rho + \gamma)$ and α. Because at high illuminances the α output is saturated and therefore inoperative, only cones contribute to the photopic achromatic process. The individual cone processes are summed and have a short time constant, probably as a result of a direct or short-circuit path. At low illuminances, in the scotopic range, the rods are operative, but the signal is below the cone thresholds; accordingly, only the rods contribute to A. In between, in the mesopic range, rods are not yet saturated, and the cone threshold has been exceeded by some amount; accordingly, both rods and cones contribute to the mesopic achromatic process. The relative amounts of their contributions depend on where in the dynamic range their operating functions are being tapped by the particular level of illumination involved. This combination method accounts for the gradual change from scotopic sensitivity to cone sensitivity as illuminance increases throughout the mesopic range. In addition, the relatively shorter time constant of these rather simpler circuit connections probably accounts for the difference in temporal response between the achromatic and chromatic processes. That is, fast-reacting Y cells (see Section B.2) probably form the achromatic signal, and the slower X cells may make up the chromatic signals (although this supposition is a controversial one).

Finally, the total response is represented in Fig. 20 by the combination of all three processes, A, $C1$, and $C2$. In terms of the simplified model illustrated in Fig. 20, there are four basic attributes of color (although only three are independent) and about 56 simple ratio combinations of them when complementary and relative expressions are considered. Some of these correspond quite obviously to commonly discussed attributes of visual perception; others are obscure and probably not very useful relations. The basic attributes that are common to all color perceptions may be expressed as follows:

$$\text{Total Response} = \text{Brightness} = (C1^2 + C2^2 + A^2)^{1/2}$$
$$\text{Achromaticness} = (A^2)^{1/2}$$
$$\text{Chromaticness} = (C1^2 + C2^2)^{1/2}$$
$$\text{Hue: } H_r = (C1^2)^{1/2} \text{ when } C1 \text{ is positive}$$
$$H_g = (C1^2)^{1/2} \text{ when } C1 \text{ is negative}$$
$$H_y = (C2^2)^{1/2} \text{ when } C2 \text{ is positive}$$
$$H_b = (C2^2)^{1/2} \text{ when } C2 \text{ is negative}$$

Some of the relations derived from these basic attributes are listed in Table VI. The four basic attributes range from zero to some maximum value corresponding to the response-saturation limit of the visual process. (The symbol for infinity in Table VI is merely a mathematical limit; responses saturate well below that theoretical limit). These attributes apply to all colors under all viewing conditions throughout the photopic range. That is, we may see brightness, chromaticness, and hue as a result of any photopic stimulation that elicits a chromatic response; if there is no chromaticness elicited (and, hence, no hue as well), we see only a brightness consisting of achromaticness.

Some of the relations among the four basic attributes may also be seen for any stimulus under any condition of viewing. However, because these relations are relative (involving ratios in the examples of Table VI), their magnitudes range only from zero to unity. For example, *saturation* is the relative chromaticness in the total response; its maximum corresponds to a case in which there is no achromaticness in the response, meaning that the response is maximally chromatic and thus is represented as a unity proportion. On the other hand, when the response consists of only an achromatic component, saturation is zero. Chromaticness K increases with level of illumination, but saturation s is essentially independent of level, because it is a proportion of chromaticness in the total response K/B.

TABLE VI. Some Attributes of Color Implied by a Model of Color Vision

Attribute name	Symbol	Relation	Conditions	Constraints
I. Basic attributes of all colors				
[a] Brightness	B	$B = (C1^2 + C2^2 + A^2)^{\frac{1}{2}}$	all	$0 \leq B \leq \infty$
Achromaticness	A	$A = (A^2)^{\frac{1}{2}}$	all	$0 \leq A \leq \infty$
[a] Chromaticness	K	$K = (C1^2 + C2^2)^{\frac{1}{2}}$	all	$0 \leq K \leq \infty$
[a] Hue	H	$H_r = (C1^2)^{\frac{1}{2}}$	$C1 > 0$	
		$H_g = (C1^2)^{\frac{1}{2}}$	$C1 < 0$	
		$H_y = (C2^2)^{\frac{1}{2}}$	$C2 > 0$	
		$H_b = (C2^2)^{\frac{1}{2}}$	$C2 < 0$	
II. Relative attributes of all colors				
[a] Hue proportion	h	$h = H/K$	all	$0 \leq h \leq 1$
[a] Saturation	s	$s = K/B$	all	$0 \leq s \leq 1$
Paleness	p	$p = 1 - K/B$	all	$0 \leq p \leq 1$
Achromatic strength	a	$a = A/B$	all	$0 \leq a \leq 1$
III. Relative attributes of related colors (where B_w = brightness of a reference color)				
[a] Lightness	ℓ	$\ell = B/B_w$	$B \leq B_w$	$0 \leq \ell \leq 1$
Darkness	d	$d = 1 - B/B_w$	$B \leq B_w$	$0 \leq d \leq 1$
Gray content	g	$g = A/B_w$	$B \leq B_w$	$0 \leq g \leq 1$
White content	w	$w = 1 - A/B_w$	$B \leq B_w$	$0 \leq w \leq 1$
[a] Chroma	k	$k = K/B_w$	$B \leq B_w$	$0 \leq k \leq 1$

[a] Indicates reasonably common terms.

Certain other attributes that we may abstract from a color response can be seen only when the sample in question is part of an array of stimuli, that is, when we deal with related colors. These derived relative attributes are also proportions of the four basic attributes (B, A, K, and H), but in this case the ratios are taken among different stimuli in the same field of view. *Lightness* is a good example of such an attribute. It is the ratio of the brightness of a focal object (B) to that of a white-appearing reference object (B_w). Here again, not only does lightness vary between 0 and 1, it tends to remain constant with changes in level of illumination over the photopic range of illuminances.

The point is simply that the model of Fig. 20 provides response relationships that (conceptually) correspond well to what we see when we look at objects. In fact, the utility of the model is much greater than the qualitative correspondences set forth in Table VI. In a number of instances, prototypes of the model have provided quantitative predictions of visual phenomena that agree closely with independent experimental results (for instance, Guth, 1972; Vos and Walraven, 1972; Ingling and Tsou, 1977; Guth *et al.*, 1980). Figure 21 shows receptor stage and opponent sensitivities of the model. The latter agree closely with those measured by Hurvich and Jameson (1955), Romeskie (1976,

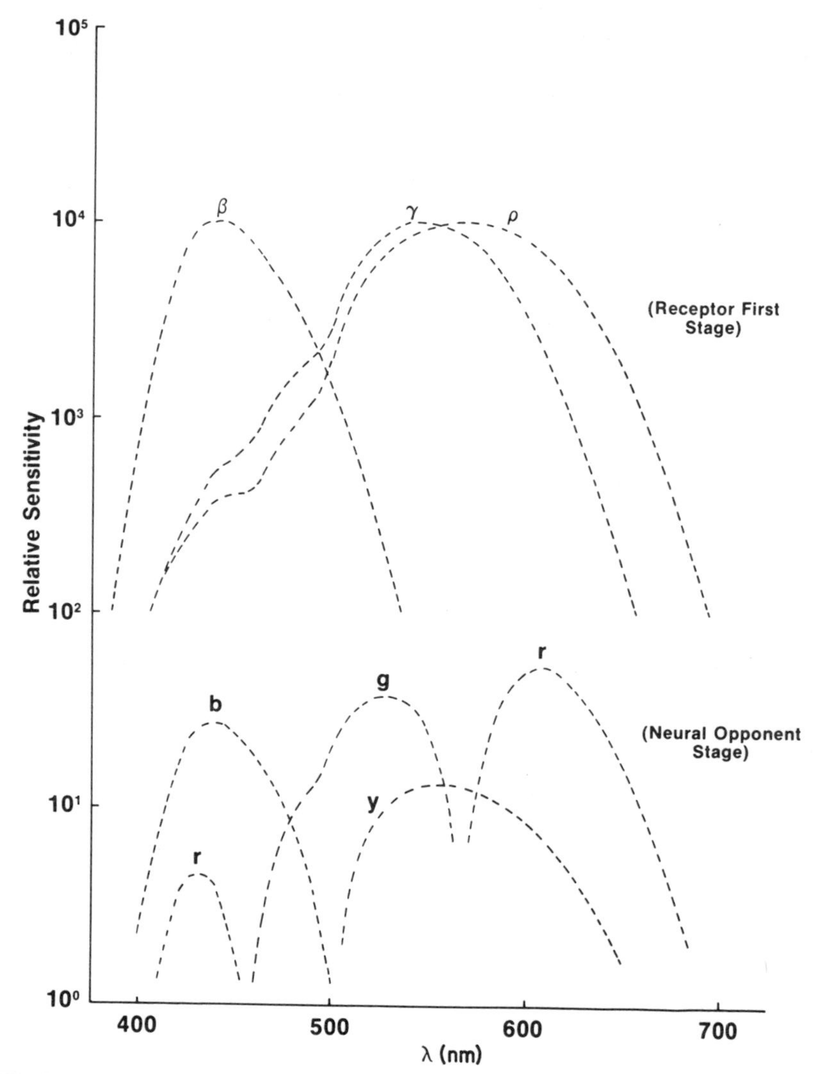

Fig. 21. Spectral sensitivities for suprathreshold photopic, daylight adaptation. Receptor-stage (ρ, γ, β) sensitivities (normalized at peak) at top; opponent (r, g, y, b) neural sensitivities at bottom (normalized with respect to r and arbitrarily positioned on the ordinate).

1978), and Werner and Wooten (1979). Figure 22 shows two different forms of opponent (r, g, y, b) sensitivites found at threshold and suprathreshold chromatic conditions (see, for example, Ingling, 1977) predicted by the model of Fig. 20. These two forms lead to different chromatic valence functions, as in

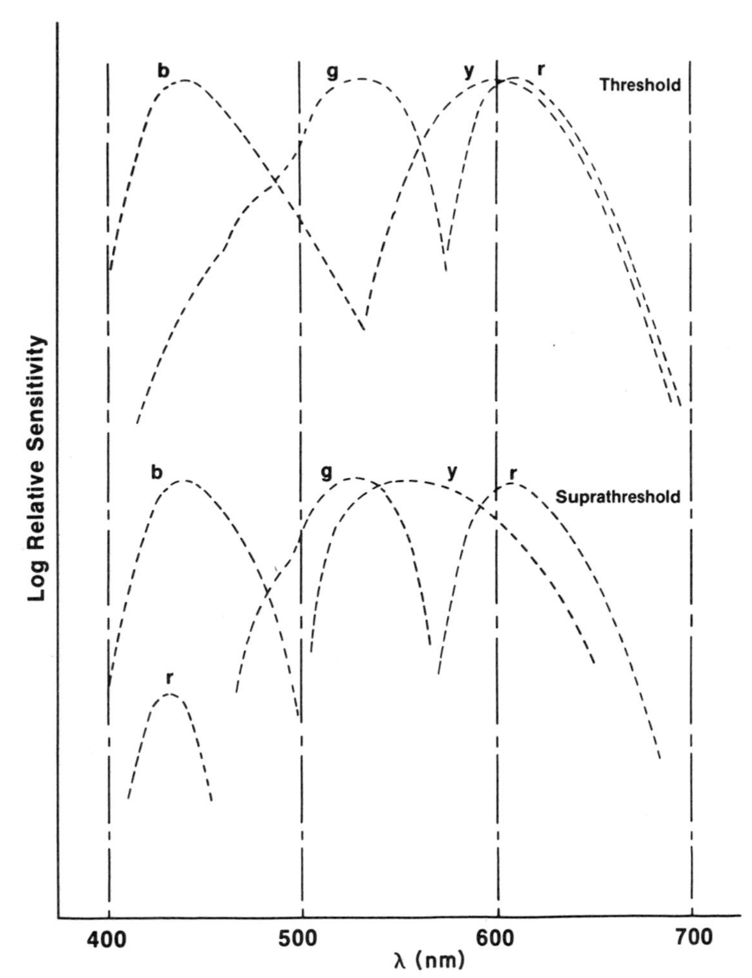

Fig. 22. Spectral sensitivities of opponent processes at threshold and suprathreshold levels. Opponent-response sensitivities (r, g, y, b) for threshold or dark-surround conditions at top; same for suprathreshold or "white"-surround conditions at bottom, arbitrarily displaced along the ordinate.

the left of Fig. 23. When Eq. (8) is used to compute total responses, they also lead to two distinctly different total sensitivity functions, as shown on the right of Fig. 23. The upper total sensitivity function (a) is of the kind found when brightnesses of maximum-purity samples of different dominant wavelengths are matched against a dark surround (for instance, Hsia and Graham, 1957). The lower (b) corresponds to total sensitivity functions determined from increment thresholds found with "white" adaptation and illuminated

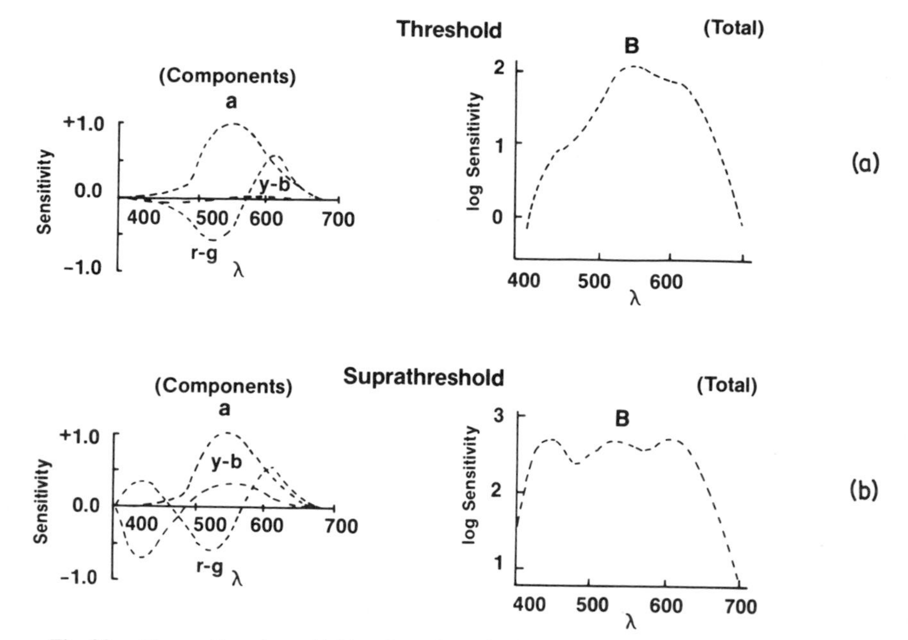

Fig. 23. Chromatic valence (left) and total sensitivities (right) for threshold (dark-surround) and suprathreshold ("white"-surround) conditions as predicted by the model of Fig. 20.

surrounds (for instance, Sperling and Harwerth, 1971). Figure 24 shows a suprathreshold, total sensitivity function (top) and an achromatic sensitivity function (middle). Except for wavelengths below about 460 nm, the latter agrees with the CIE 1924 $V(\lambda)$ function, which is also indicated in the middle of the figure. The achromatic function agrees with those found consistently by flicker photometry. It is represented in Table VI as A. Its chromatic counterpart, K in Table VI, leads to the function in Fig. 24 labeled "chromatic sensitivity". The ratio at each wavelength of total sensitivity B to achromatic sensitivity A yields the bottom curve of Fig. 24. This curve agrees closely with measurements of purity thresholds (as summarized in Hurvich and Jameson, 1955), of the Helmholtz–Kohlrausch effect (for instance, Kaiser and Comerford, 1975; Bartleson, 1981a), and what Evans (1974) has called zero gray-content. Guth *et al.* (1980) have shown that by varying the coefficients of $C1$ and $C2$ in the chromatic response K, it is possible to match the many different wavelength discrimination functions that have been reported for different viewing and adaptation conditions and for observers exhibiting different forms of anomalous color vision.

In short the model form illustrated in Fig. 20 can be used to predict the results of a wide variety of experimental results from studies of color vision.

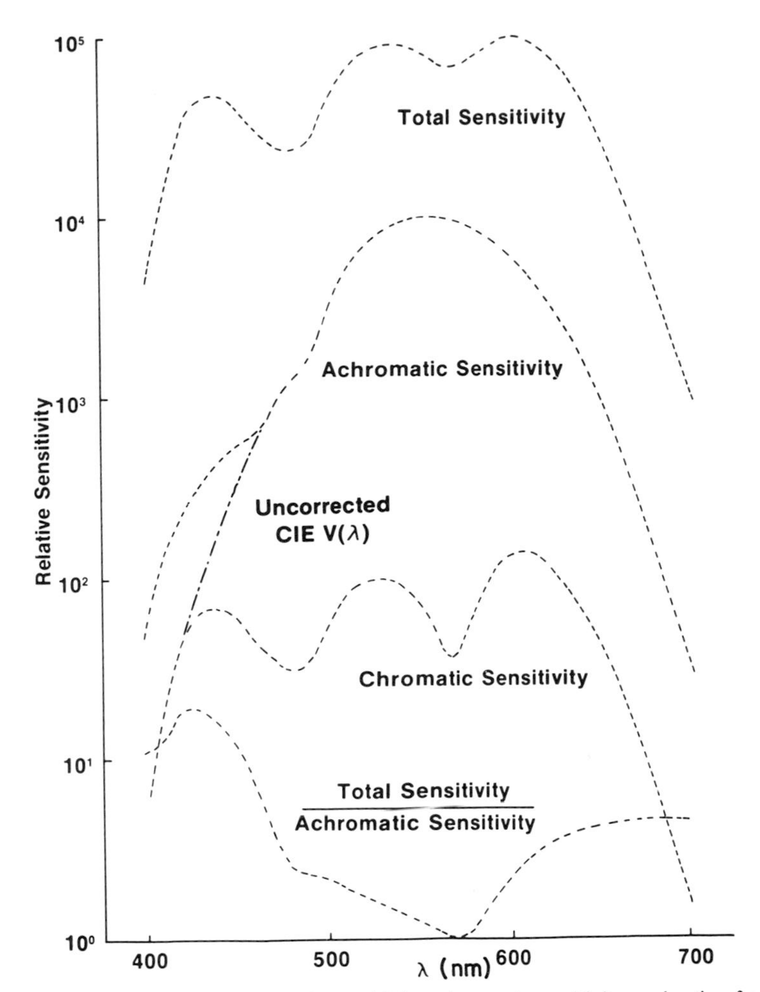

Fig. 24. Total sensitivity, achromatic sensitivity, chromatic sensitivity, and ratio of total to achromatic sensitivity calculated from model of Fig. 20. CIE 1924 $V(\lambda)$ function is also shown. Suprathreshold, photopic, daylight adaptation.

Only a few examples have been cited here. Others, such as chromatic adaptation, will be treated in Chapter 12. Our purpose has not been to show the many details of these comparisons between experimental results and model predictions but merely to emphasize that the model has both qualitative and quantitative utility. Still, it is a gross oversimplification of the underlying complexities of the mechanism of color vision. In time the model will be refined and extended to provide more general and more accurate predictions. It seems doubtful, however, that the quadruplex of receptors and the basic opponent-response form of neural signal-processing will be superseded in the process of developing a more accurate model. That basic scheme seems well established and will likely survive with only refinements and elaborations.

E. BRIGHTNESS

The model of Fig. 20 depicts the photopic achromatic process as a combination of ρ and γ; the β system does not contribute to the achromatic process in that model. This scheme is consistent with generally prevailing opinion in the vision-science community (for example, Eisner and McLeod, 1980), although there have been proposals that the β system does contribute to the achromatic process (for example, Vos and Walraven, 1971). It is sometimes assumed that the achromatic process derives from Y cells, whereas the chromatic process results from activity of the slower X cells. This difference in temporal response (slower-acting chromatic X cells and faster-acting achromatic Y cells) makes it possible to isolate activities of the chromatic and achromatic processes of photopic vision. One of the ways in which this can be accomplished is by the use of flicker photometry in which two lights presented to the same area of the retina fluctuate over time. The task of an observer in a flicker-photometry experiment is illustrated with Fig. 25.

When two lights A and B of different luminances and chromaticities are alternated very slowly, as in the top graph of Fig. 25, the observer sees both the intensity and the color of the display fluctuate.* When the alternation rate is increased, as in the middle graph, there is a point (or region of fluctuation rates) at which color differences blend, but intensity differences still elicit the sensation of flicker. The slower response characteristics of the chromatic process prevent the visual mechanism from tracking fluctuations in color as represented by the unchanging horizontal line in the figure, although the faster-acting achromatic process still follows variations in intensity. At some higher alternation rate that depends on average luminance, area, position on the retina, and other factors, the achromatic process can no longer track changes in intensity, and the display appears uniform in time. At this condition, represented in the lower graph of Fig. 25, there is complete fusion of the two lights; they appear as an intensity-time average of A and B. In the flicker region where the chromatic process yields fusion but the achromatic process does not, there will be an adjustment of flicker rate that will minimize or just eliminate all flicker. This criterion condition is taken to mean that the achromatic components of the lights A and B are matched. The point at which

* Phase shifts resulting from different induction and response times have been normalized in Fig. 25 to simplify comparison of the relative amplitudes of flicker for chromatic and achromatic processes.

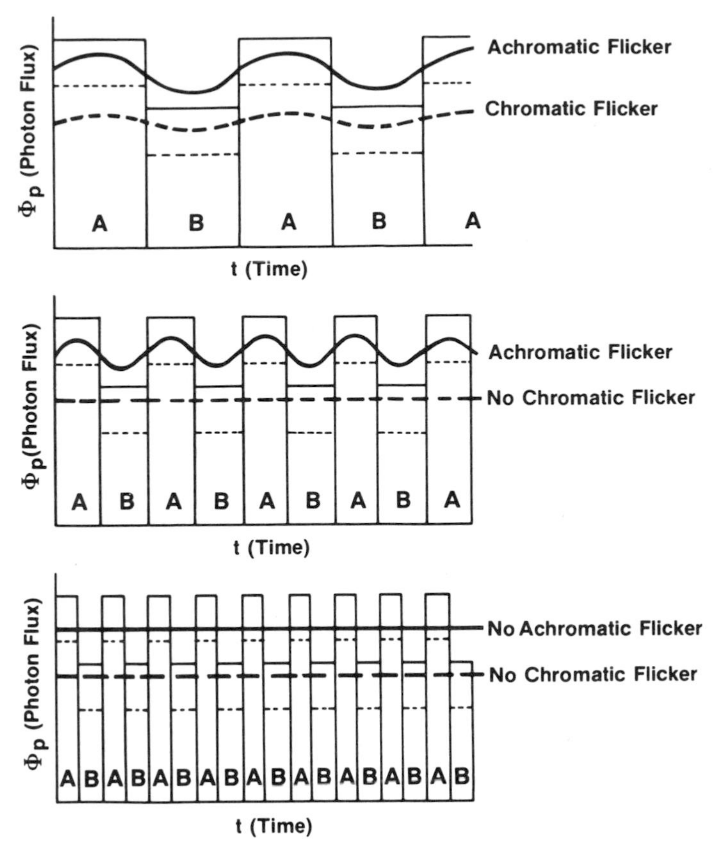

Fig. 25. Schematic illustration of the effect of different time constants of the chromatic and achromatic processes on the sensation of flicker when two lights A and B of different intensities and chromaticities are alternated. Dashed bars represent the amount of chromatic content, and solid bars correspond to achromatic content. Solid and dashed alternating curves refer to achromatic and chromatic sensations, respectively. Phase differences are normalized.

chromatic flicker just disappears may be taken to indicate that the chromatic components of the lights are matched. However, that point is very difficult to determine with precision, because the remaining achromatic flicker intrudes on the confuses the perception. On the other hand, the point at which all flicker just disappears or reaches a minimum is relatively easy to determine with high precision. For this reason flicker photometry is frequently used to determine the relative efficiency of different lights to evoke an achromatic response. Other methods that also isolate the achromatic process, such as adjustment of the two halves of a bipartite field to produce a minimum border contrast between them, are discussed in Chapter 11.

The model of Fig. 20 shows the scotopic achromatic process consisting of only α system activity. Thus at absolute threshold only the achromatic process is active and is the result of stimulation of rod or α systems. Figure 26 illustrates the relative amounts of energy required at different wavelengths of light to stimulate scotopic and photopic achromatic processes. The lower curve, labeled "rods," corresponds approximately to the absolute threshold. As the intensity of light is increased above the absolute threshold, lights of all wavelengths shorter than about 620 nm appear chromatically the same (near-neutral). This range of intensities is referred to as the *photochromatic interval.* The upper limit of the photochromatic interval is called the *photochromatic threshold*; there is no photochromatic interval for lights of wavelengths longer than about 620 nm, where the photochromatic threshold is essentially equal to the absolute threshold. Above the photochromatic threshold, lights of different wavelengths generally appear to have different hues. The photochromatic threshold corresponds to the lower limit of the chromatic process. The levels at which the scotopic process saturates and only the photopic process operates are represented by the curve labeled "cones" in Fig. 26. Intensities in the region of the photochromatic interval, involving only the α

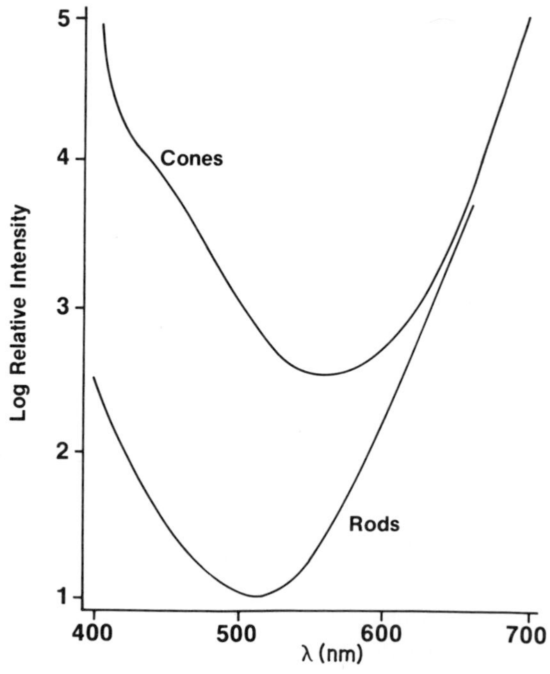

Fig. 26. Relative energy required to produce a threshold response for foveal stimuli. The distance between the curves for rods and cones represents the photochromatic interval.

process of Fig. 20, have also been used to equate achromatic responses by flicker photometry in order to derive a scotopic luminous efficiency function of wavelength.

Luminance, a measure involving the achromatic process, has been defined by the CIE for both photopic and scotopic vision* as

$$L_v(\text{photopic}) = (683\ \text{lm} \cdot \text{w}^{-1}) \int_{360\ \text{nm}}^{830\ \text{nm}} L_{e\lambda} \cdot V(\lambda) \cdot d\lambda \tag{14}$$

and

$$L'_v(\text{scotopic}) = (1700\ \text{lm} \cdot \text{w}^{-1}) \int_{360\ \text{nm}}^{830\ \text{nm}} L_{e\lambda} \cdot V'(\lambda) \cdot d\lambda \tag{15}$$

where

L_v = photopic luminance in $\text{cd} \cdot \text{m}^{-2}$
L'_v = scotopic luminance in $\text{cd} \cdot \text{m}^{-2}$
L_e = spectral radiance in $\text{w} \cdot \text{m}^{-2} \cdot \text{sr}^{-1} \cdot \text{nm}^{-1}$
$V(\lambda)$ = spectral luminous efficiency function for photopic vision,
$V'(\lambda)$ = spectral luminous efficiency function for scotopic vision,
$683\ \text{lm} \cdot \text{w}^{-1}$ = maximum spectral luminous efficiency for photopic vision, occurring at a wavelength of 555 nm with neutral adaptation,
$1700\ \text{lm} \cdot \text{w}^{-1}$ = maximum spectral luminous efficiency for scotopic vision, occurring at a wavelength of 507 nm with neutral adaptation.

Table VII lists values of $V(\lambda)$ and $V'(\lambda)$ at 10-nm intervals from 380 to 700 nm; these values are currently recommended by the CIE. The data for $V(\lambda)$ were adopted by the CIE in 1924. They represent a combination of results from several experiments (see Chapter 11 for details). Data vary among observers. A representative range of variation among observers with normal color vision is illustrated in Fig. 27. The curves of that figure were constructed from data of Gibson and Tyndall (1923). All the data were collected from experiments in which a test field was shown against a dark background to observers who were neutrally adapted. When adaptation varies from neutral, the luminous efficiency functions also vary. They vary with both chromaticity and luminance of the adapting field. Figure 28 illustrates those variations found by Boynton *et al.* (1959) for three adapting levels in each of four chromatic adaptation conditions. The curves of Fig. 28 differ systematically as adapting luminance increases from the absolute threshold. In addition, the curves for low ($8\ \text{cd} \cdot \text{m}^{-2}$) to moderate ($80\ \text{cd} \cdot \text{m}^{-2}$) adapting luminances differ among the four adapting chromaticities. There are, then, a number of factors that may influence experimental results to cause differences among

* No convention exists for defining luminance under mesopic levels, those levels involving both the α and $\rho + \gamma$ processes.

TABLE VII. Spectral Luminous Efficiency for the CIE Standard Observers

λ (nm)	Photopic vision $V(\lambda)$	Scotopic vision $V'(\lambda)$
380	0.000039	0.000589
390	0.000120	0.00221
400	0.000396	0.00929
410	0.00121	0.0348
420	0.00400	0.966
430	0.0116	0.1998
440	0.0230	0.328
450	0.0380	0.455
460	0.0600	0.567
470	0.910	0.676
480	0.129	0.793
490	0.208	0.904
500	0.323	0.982
510	0.503	0.997
520	0.710	0.935
530	0.862	0.811
540	0.954	0.650
550	0.995	0.481
560	0.995	0.329
570	0.952	0.208
580	0.870	0.121
590	0.757	0.0655
600	0.631	0.0332
610	0.503	0.0159
620	0.381	0.00737
630	0.265	0.00334
640	0.175	0.00150
650	0.107	0.000677
660	0.0610	0.000313
670	0.0320	0.000148
680	0.0170	0.0000715
690	0.00821	0.0000353
700	0.00410	0.0000178

data intended to represent relative spectral luminous efficiency at photopic levels. These factors were apparently not fully appreciated when the CIE adopted the $V(\lambda)$ function in 1924. At that time the $V(\lambda)$ function was called the visibility function, because it was thought that it described the sensitivity of observers with normal color vision to the brightnesses of lights of different dominant wavelengths. We now know that this is not true. Not only do the

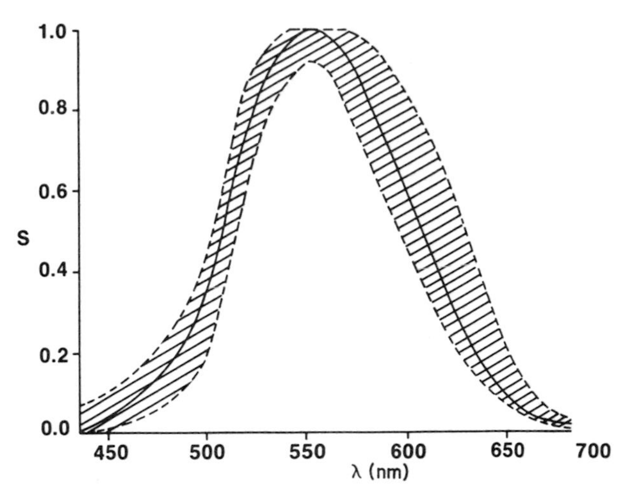

Fig. 27. Graph of data from Gibson and Tyndall (1923), *Bulletin of the National Bureau of Standards*, showing average luminosity function for observers with normal color vision (solid curve) and approximate limits of experimental variation (shaded area).

conditions of observation influence the resultant functions, but the experimental method by which they are determined significantly influences their shapes. Figure 29 shows the CIE $V(\lambda)$ function and three other curves that represent sensitivity over wavelength. Curve B differs from the $V(\lambda)$ function (curve A) only below about 460 nm; it is the Judd 1951 correction to $V(\lambda)$. The differences between curves A and B are important in vision science experiments and also in a number of practical applications in which brightnesses of white objects are compared, for example, in matching panels of white painted appliances such as refrigerators. Curve C is derived from a different experimental method. Instead of flicker photometry, the data for curve C were obtained by direct, heterochromatic brightness matching of spectral stimuli. The curve is significantly broader than the $V(\lambda)$ function. It indicates that the amounts of energy of different wavelengths required to maintain constant brightness are more nearly the same, especially over the central portion of the function, than is the case for $V(\lambda)$. Curve D is the result of an experimental determination of the threshold increment that must be superimposed on a larger white-appearing background to change the perceived brightness of the test area. That curve differs greatly from the $V(\lambda)$ function. From its shape one sees that the chromatic process plays a much greater role under these conditions than in either direct heterochromatic brightness matching (curve C) or flicker photometry (curve B). Hurvich (1960) cites experimental results as far back as 1904 that suggest that the relative contribution of the chromatic systems increases under white adaptation conditions.

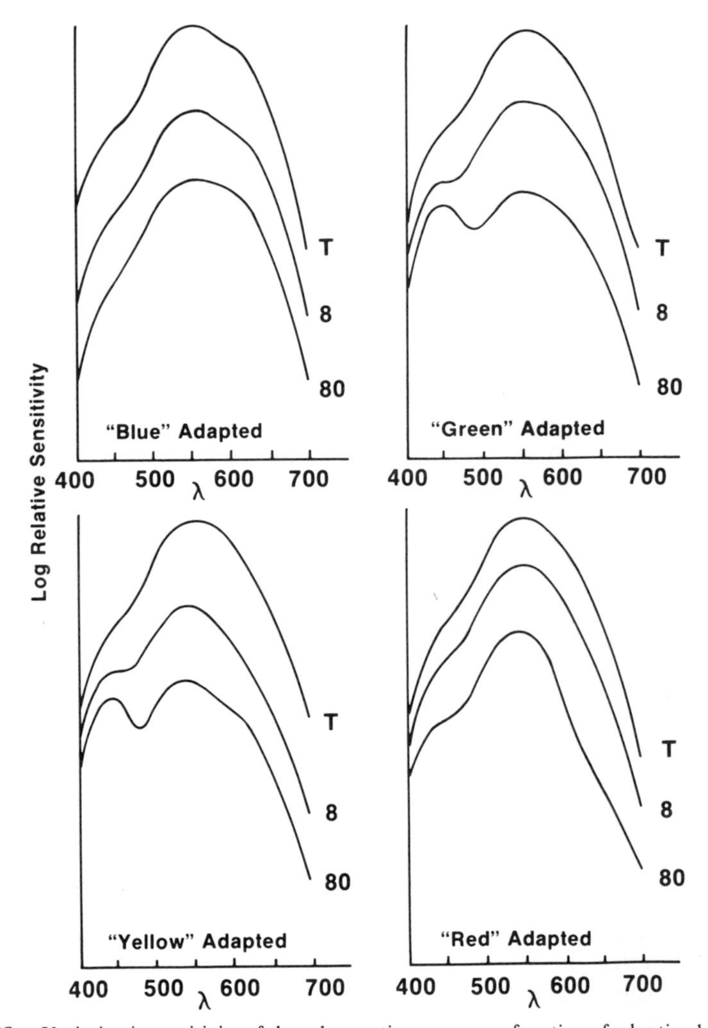

Fig. 28. Variation in sensitivity of the achromatic process as a function of adapting luminance and chromaticity. Four adapting chromaticities are identified by color names associated with their appearances under the experimental conditions. Adapting luminances are for threshold T, 8, and 80 cd · m^{-2}. Graphs represent some of the data reported by Boynton, Kandel, and Onley, 1959, *Journal of the Optical Society of America.*

In one way or another these curves all describe sensitivity to brightness for dark or white adaptation. Their shapes are quite different. They also differ in another important respect; curves A and B represent methods that tap only the achromatic process, and curves C and D are from methods that involve both achromatic and chromatic processes. The achromatic process seems to be

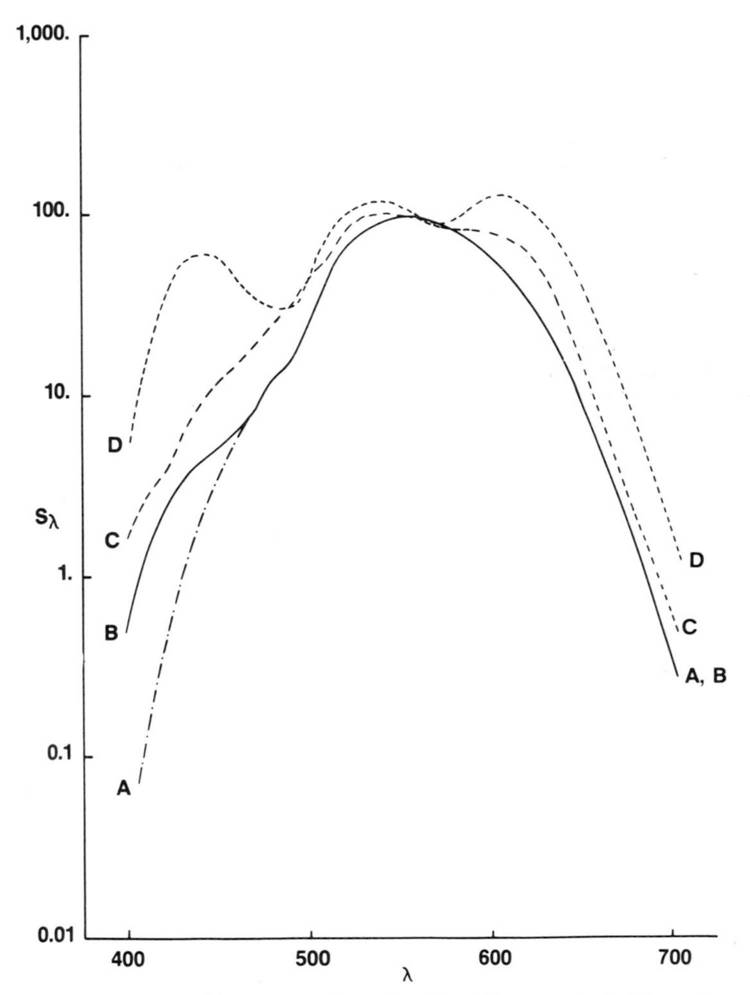

Fig. 29. Achromatic sensitivity curves determined by different methods. Curve A represents the CIE (1924) V(λ) function based primarily on flicker photometry. Curve B reflects Judd's (1951) correction to V(λ). Curve C is for direct heterochromatic brightness matching. Curve D is for increment threshold data against a white-appearing field.

additive; that is, it has the property that the luminance of the combined mixture of two lights is equal to the sum of their individual luminances, a property called *additivity*. The chromatic process does not have this property. Additivity may be illustrated by the examples in Fig. 30. Suppose that two lights A and B of different chromaticities, are matched by adjusting the intensity of B until its brightness matches that of A (Fig. 30a). Also C is made to match A (Fig. 30b). We now have three lights that elicit the same brightness

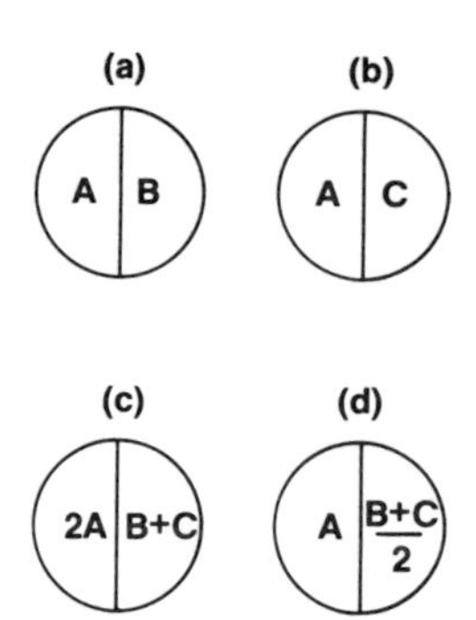

Fig. 30. Method for testing additivity using three lights A, B, and C in colorimetric mixtures. Adjust B to match A in (a) and C to match A in (b). The system is linear and additive if 2A then matches B + C, in (c). The system is additive if A matches (B + C)/2 in (d). If the conditions of (c), and (d) do not yield a match, the system is not additive.

sensation in an observer under the specified viewing conditions. If the system being measured is linear, we should be able to add the lights B and C to produce a mixture that will be equally as bright as A added to itself (Fig. 30c). If B + C is equal in brightness to 2A, then the system is both linear and additive. We should also be able to match the brightness of (B + C)/2 to the original A (Fig. 30d). This operation describes an additive system when it results in a successful brightness match.

When the lights A, B, and C have the same chromaticity, the results from the paradigms of Fig. 30 will be additive. In all three cases the lights will elicit responses in both achromatic and chromatic mechanisms, but the chromatic activities will be the same in parts (a), (b), and (c) of the figure. However, when the chromaticities of the lights are not the same, chromatic activity will be different for each of the paradigms; A will not, therefore, match the brightness of (B + C)/2. In short, additivity failure is found whenever there is an imbalance of either chromatic activity or achromatic activity. Additivity failures have been measured in a number of experiments. Two series of such determinations are summarized in Fig. 31. The experiments illustrated were reported by Guth and co-workers (Guth *et al.*, 1969; Guth, 1970; Guth and Lodge, 1973) and by Kaiser and Wyszecki (1978). The common experimental paradigm is illustrated in Fig. 31. The brightnesses of a white-appearing light W and a near-monochromatic light λ were matched as in Step 1. If additivity held, then the brightness of W would also match $0.5\text{W} + 0.5\lambda$. Therefore, a test of additivity would be to place W in the upper field and 0.5W in the lower field, and then add a proportion a of λ to provide a brightness match as in Step 2. This was done in both sets of experiments. The results, shown as ranges of a in the lower part of the figure, reveal that a differed from the value of 0.5 required for additivity throughout most of the range of spectral stimuli used for λ. The

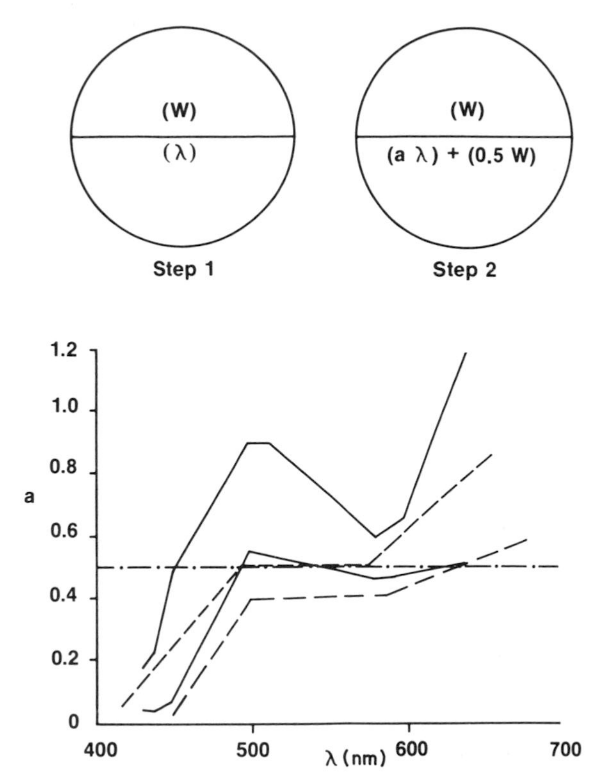

Fig. 31. Experimental determinations of additivity failure. In Step 1 a white-appearing light W is matched to a near-monochromatic light λ. In Step 2, half the intensity of the "white" light (0.5W) is mixed with an amount of the monochromatic light ($a\lambda$) to make the mixture match the "white" light W. If the system were additive, a would equal 0.5. Values of a determined by Guth and co-workers (Guth, Donley, and Marrocco, 1969; Guth, 1970; Guth and Lodge, 1973) are shown as solid lines, and values of a determined by Kaiser and Wyszecki (1978) are shown as broken lines.

solid lines illustrate the range of results found in the Guth *et al.* experiments, and the broken lines define that range for the Kaiser and Wyszecki experiment.

Two types of additivity failure may be noted in Fig. 31; at some wavelengths a is greater than 0.5, but at others a is less than the 0.5 value required for additivity. The first type (in which $a > 0.5$) has been called *subadditivity* by Guth and *additivity failure of the cancellation type* by Kaiser and Wyszecki. The second type in which observers require a smaller proportion of λ ($a < 0.5$), has been called *superadditivity* by Guth and *additivity failure of the enhancement type* by Kaiser and Wyszecki. The enhancement type of additivity failure tends to occur with stimuli of relatively short wavelength, below about 500 nm. The cancellation type of additivity failure seems more common for stimuli of relatively long wavelength, above about 600 nm. At least for the Kaiser and Wyszecki data, stimuli with intermediate wavelengths tend more often to provide additivity. Kaiser and Wyszecki were able to show that these results could also be deduced by calculation from other experimental results where direct brightness matching procedures have been used (for instance, Sanders and Wyszecki, 1964; Kaiser and Smith, 1972).

The underlying mechanistic reasons for these additivity failures seem to be that although the achromatic process of the visual mechanism may be linear and additive, the chromatic process is not. Therefore, whenever the chromatic process is differentially activated, we should not expect additivity to obtain. Luminance, as defined in Eq. (14), is really a measure of relative achromatic activity. Consequently, equating stimuli for luminance normalizes only achromatic activity of the visual mechanism. If chromatic activity differs between two stimuli, then the total or brightness response will be unequal [see Eq. (8)]. Figure 32 has been prepared along the lines of an illustration used by Kaiser and Comerford (1975) to demonstrate differences between equal luminance and equal brightness for spectral stimuli. Total activity (that is, brightness) is plotted against the dominant wavelengths of narrow spectral bands of light in both graphs of Fig. 32. Achromatic activity is shown as the shaded areas, and chromatic activity is designated by unshaded areas. The upper graph (a) shows that achromatic activity has been adjusted to the same value for all spectral bands; the shaded bars are all of the same height. This situation corresponds to one of equal luminances for all the stimuli. However, under those conditions the model of Eq. (8) and Fig. 20 provides unequal amounts of chromatic activity. This is shown in Fig. 32a as the different heights of the unshaded bars. The outline of the total (shaded plus unshaded) bar heights over wavelength resembles the shape of threshold purity functions and curves of the *Helmholtz–Kohlrausch effect*, an effect in which less luminance is generally required at all wavelengths to match the brightnesses of chromatic stimuli to those of achromatic stimuli. In fact, we can examine the model's predictions of the different luminances required to elicit equal brightness at all wavelengths by equating total activity over wavelength rather than equating achromatic activity. This has been done in the lower graph (b) of the figure. There, the total heights of bars are all the same; however, the heights of the shaded bars vary with wavelength as the inverse of the way that the unshaded bars vary in the upper graph (a). It is clear from Fig. 32 that the two criteria represent different conditions of the state of the visual mechanism. Equating brightnesses (total responses) is not the same task as equating activity in the achromatic process for stimuli that differ in dominant wavelength. Consequently, when the luminances of two chromatic stimuli match, we should not necessarily expect them to be equally bright. Conversely,

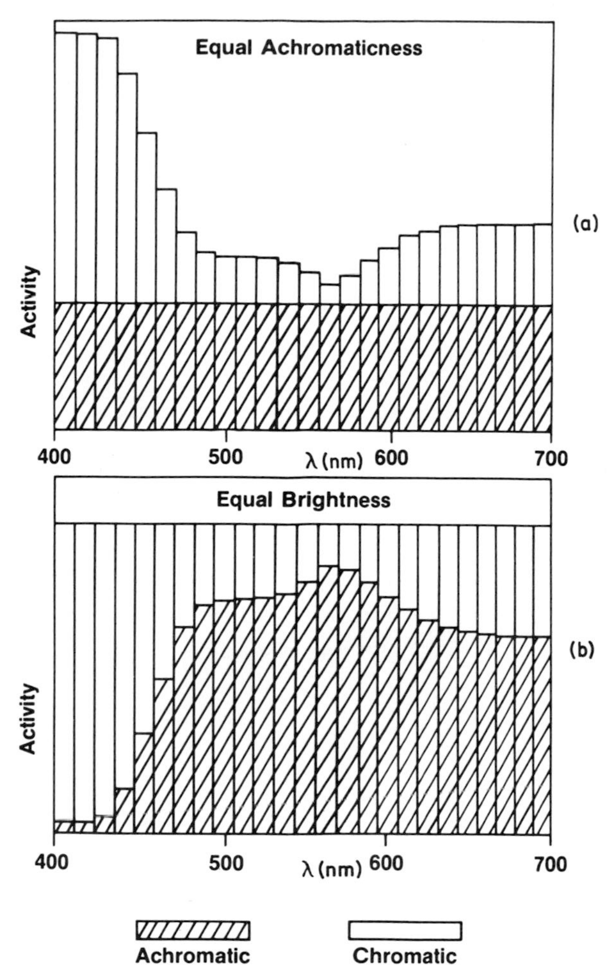

Fig. 32. Schematic examples of the difference between equality of luminance (a) and equality of brightness (b) for different narrow bands of spectral light. See text for explanation. (Based on Kaiser and Comerford, 1975, *Vision Research.*)

when two chromatic stimuli are equally bright, we should not necessarily expect their luminances to be identical.

The extent to which the luminance of an achromatic stimulus must be changed in order to effect a brightness match with a chromatic stimulus of the same original luminance (the Helmholtz–Kohlrausch coefficient k) is a function of the chromaticness elicited by the stimulus. Figure 33 illustrates the ways in which log k varies with CIE 1931 excitation purity p_e for related object-color stimuli of 15 different dominant wavelengths under conditions of

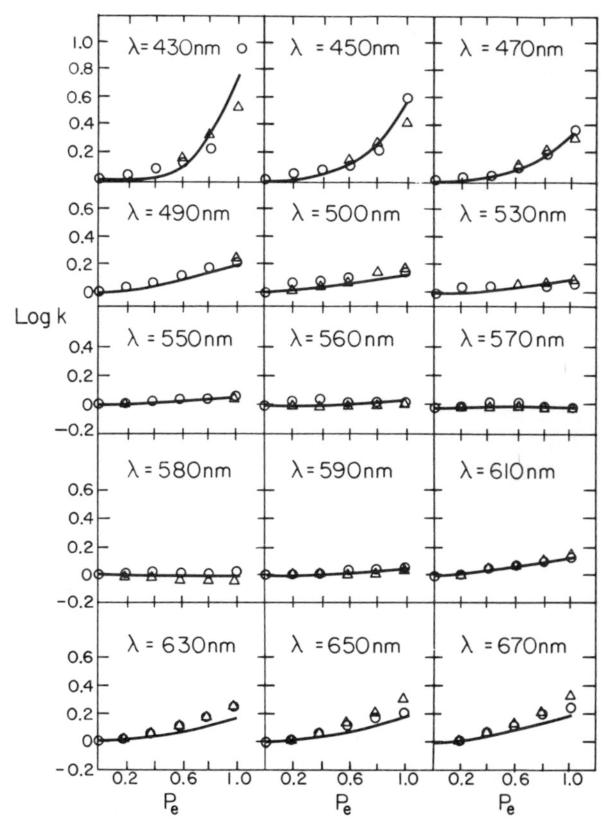

Fig. 33. Graphs indicate the common logarithms of the amount k by which the luminance of a zero purity stimulus must be multiplied in order that its brightness will match that of a nonzero purity stimulus, as indicated on the excitation purity p_e abscissa. Circles are data from Kaiser and Smith (1972); triangles are data from Bartleson (1979); curves are constructed from an empirical model proposed by Bartleson. Reprinted with permission, from Bartleson, 1981a, *Farbe*.

"white" adaptation. Open circles represent data from an experiment of Kaiser and Smith (1972), and open triangles are for data from an experiment reported by Bartleson (1979). The solid curves are predictions from an empirical model of brightness prediction from luminance values (Bartleson, 1981a). The shapes of these kinds of curves vary with chromatic adaptation. The meaning of the data in Fig. 33 may be more obvious from the projection of Fig. 34, which shows a surface of constant perceived lightness (relative brightness from Table VI) in a space where luminance Y is plotted against chromaticity (x, y), as reported by Sanders and Wyszecki (1957). The figure shows that the luminances required for equal lightnesses generally decrease as purity increases. When the surface curvature is extrapolated to the spectrum locus,

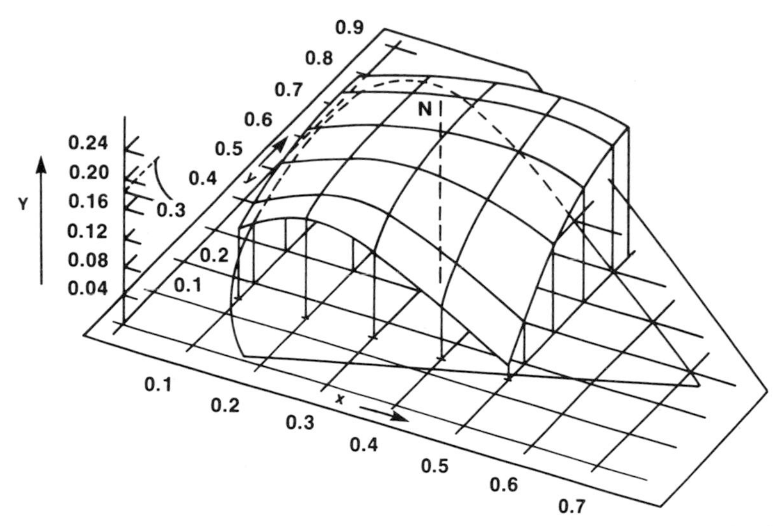

Fig. 34. Surface of constant perceived lightness in CIE 1931 x, y, Y space. The luminance required for constant lightness generally decreases as purity increases. (After Sanders and Wyszecki, 1957, *Journal of the Optical Society of America.*)

the required luminances follow the form of the shape delineated by the shaded bars in Fig. 32b.

There have been many empirical and a few theoretical formulae proposed to attempt to predict brightness from luminances. One recent formula represents an empirical model that appears to account reasonably well for the factors discussed in this section and also provides predictions that agree well with published experimental data where perceived brightnesses and lightnesses have been determined (Bartleson, 1981a). The formula is

$$Q^{**} = a(k \cdot L^{\frac{1}{3}}) - b, \tag{16}$$

where Q^{**} stands for predicted brightness, L is luminance in $\text{cd} \cdot \text{m}^{-2}$, k is a Helmholtz–Kohlrausch coefficient, and a and b are constants whose values depend on viewing conditions. For unrelated colors or dark adaptation when an isolated stimulus is viewed, $a = 10$ and $b = 0$. The situation is more complicated when related colors are viewed. Then both the luminance of a reference white object and a term corresponding to induction from surrounding receptive fields must be determined in order to calculate the values of a and b. The empirical induction term I is

$$I = \frac{(\mu_f \cdot L_f) + (\mu_s \cdot L_s)}{(\mu_f \cdot L_f)}, \tag{17}$$

where

μ_f = the fractional area of the focal stimulus relative to a congruent circular area with a diameter of 40°,
μ_s = the fractional area of the surround (that is, all nonfocal elements) relative to a congruent circular area with a diameter of 40°,
L_f = the luminance of a white-appearing focal reference,
L_s = the luminance of a uniform surround or the arithmetic mean luminance of a nonuniform surround.

Given the value of I, two other coefficients, q_a and q_b, are determined as

$$q_a = 35.2(I)^{-0.130}, \tag{18a}$$

$$q_b = 0.389(I)^{0.362} \tag{18b}$$

The constants a and b of Eq. (16) are then determined as a power function of L_f(the luminance of a white-appearing reference object):

$$a = q_a(L_f)^{-0.164}, \tag{19a}$$

$$b = q_b(L_f)^{0.215} \tag{19b}$$

This is all that need be done for achromatic stimuli. However, for chromatic stimuli, the value of k ($\neq 1.0$) must also be determined for use in Eq. (16). That value derives from the relation of chromatic to achromatic activity in the visual mechanism. First, relative fundamental tristimulus values are calculated from the CIE tristimulus values of the sample using the Koenig fundamental primaries of Table V:

$$\begin{aligned} R &= 0.713(X/Y) + 0.963 - 0.0147(Z/Y), \\ G &= -0.395(X/Y) + 1.17 + 0.0815(Z/Y), \\ B &= 0.561(Z/Y). \end{aligned} \tag{20}$$

The opponent-response values ($C1, C2, A$) are then computed from the relative fundamental tristimulus values of Eq. (20) as follows:

$$\begin{aligned} C1 &= 1.66R - 2.23G + 0.37B, \\ C2 &= 0.34R + 0.06G - 0.71B. \\ A &= 0.85R + 0.15G. \end{aligned} \tag{21}$$

These opponent-response values are used to calculate a chromatic purity term P that will be used to determine the parameters of the equation for calculating k from excitation purity:

$$P = \frac{(\mathrm{C1}^2 + \mathrm{C2}^2)^{1/2}}{\mathrm{A}}. \tag{22}$$

The term P is the value of K/A from Table VI, the ratio of chromatic to achromatic activity. It is used to determine values m and n:

$$m = 0.2983\,[\log(10\text{P})] - 0.1834, \tag{23a}$$

$$n = 1.3268\,[\log(10\text{P})] - 0.0304. \tag{23b}$$

Finally, the value of k for related colors is calculated:

$$k = \text{antilog}(m \cdot p_e^n). \tag{24}$$

Because the tristimulus values R, G, and B depend on the state of chromatic adaptation, k will vary with adaptation as well. A model for chromatic adaptation will be discussed in Chapter 12.

Lightness is simply the brightness of a test object relative to that of a reference white object (B/B_w in Table VI) and is calculated as

$$L^{**} = (Q^{**})(Q_w{}^{**})^{-1}. \tag{25}$$

The calculations for chromatic colors are more complex than those for achromatic colors. This is because both the achromatic and chromatic processes contribute to their brightnesses. In addition, the calculations for related colors are more complicated than those for unrelated colors, and this results from the interactions of contiguous receptive fields, previously referred to as *induction*.

Even this level of complexity represents a simplification, however. In an earlier section of this chapter, it was stressed that at the level of the ganglion cells and higher stages in the visual mechanism, signals correspond primarily to contrast, that is, to differences in stimulation rather than principally to the level of stimulation. This suggests that the simplified induction term [I in Eq. (17)] may represent one of the most important factors involved in any attempt to predict brightnesses or lightnesses of related colors. Hurvich and Jameson (1961, 1974; Jameson and Hurvich, 1959, 1961a,b, 1964, 1972, 1975) have pointed out repeatedly that the cells and tissues of the visual mechanism form an overlapping mosaic of functionally interrelated elements whose activities interact spatially in a *mutually* antagonistic fashion. That is, "... the strength of the induced response is related to the strength of the directly elicited activities in each of the mutually interacting areas and it is also related to the spatial contiguity of the areas in question" (Hurvich, 1978, p. 40). Those workers have developed a model that embodies assumptions about the sizes of receptive fields, the relative proportions of excitatory centers and inhibitory surrounds of the receptive fields, and the relative scaling factors of the excitatory and inhibitory inputs to the brightness response (Hurvich and Jameson, 1974; Jameson and Hurvich, 1975; Hurvich, 1981). Mutual interactions are treated by iterative mathematical methods. Their model predicts well both contrast, in which responses tend to be different, and assimilation, in which they tend to be similar.

The complexities of brightness and lightness responses are sizeable. The simple concept of luminance relates only to the achromatic component of brightness and, therefore, is most useful for samples that elicit no chromatic activity or where the chromatic activities evoked by stimuli are identical. When this is true, samples of equal luminance will be equally bright. However, even in these restricted cases, luminance relations do not accord with brightness relations; an increase in luminance may actually evoke a decrease in brightness, so in general there is not even a monotonic relationship between luminance and brightness. To obtain an estimate of brightness that is more congruent with observation, models that are more complex, structured after the model of the visual mechanism set forth in Fig. 20, are required. The example of such a model cited here treats both achromatic and chromatic components of response and takes some cognizance of induction for related colors. Even so, this model provides only estimates that might be described as first-order approximations of contrast and assimilation. To obtain better approximations of contrast and assimilation, a model that treats mutually antagonistic interactions among receptive fields of unequal size and different center-surround organization must be used. It seems likely that even more complex models will be developed in the coming years. They will probably evolve from an increased understanding of the physiological characteristics of the visual mechanism, combined with interpretation of the results of psychophysical experimentation.

F. SPATIAL VISION

Earlier in this chapter, it was stated that the perceived brightness of an achromatic element in the field of view depends not only on the intensity or the luminance of that point but also on the relationship of its luminance to those of other points in the field. In other words the spatial distribution of luminances within the field of view will affect the contrast perceived. Accordingly, many workers have investigated the relationships between brightnesses and the distributions of intensities that elicit them. Some have measured inhibition of brightness responses as a function of spatial and luminance field-characteristics such as shape, location, contiguity, luminance, and chromaticity of areas surrounding a focal element. An infinite variety of such combinations could be studied. Partly for this reason some other workers have adopted a technique that requires fewer combinations to specify spatial frequency response characteristics of systems. It is called *linear systems analysis* and has been used for many years to assess the temporal properties of communication systems. In recent years linear systems analysis has also been used to study certain spatial properties of imaging systems, including some aspects of the visual mechanism, following the initial work of deLange (1952).

The underlying rationale for the application of linear systems analysis to the domain of spatial information rests with the fact that, for a linear transfer system, any distribution of intensities over space can be analyzed as some specific combination of sinusoidal variations of intensity over distance. By combining sine waves of different frequencies and amplitudes, it is possible to produce any distribution or shape of intensity variation as a function of distance. In other words any wave form can be generated, or *synthesized*, by summing appropriate sine waves, and any distribution can be *analyzed*, or completely described, by specifying a particular set of sine waves and their amplitudes or intensities that can be added together to reproduce the distribution. The former is called *Fourier synthesis*, and the latter is known as *Fourier analysis*; both are consequences of a theorem proposed by Jean Baptiste Joseph Fourier (1822).

The Fourier transform $F(v)$ of a function $f(x)$ may be defined as

$$F(v) = \int_{-\infty}^{+\infty} f(x)[\exp(-2\pi i v x)]\,dx,$$

where the kernel, $[\exp(-2\pi i v x)]$, of the function represents a normal distribution, such as might be found to describe the point spread function of an optical system. The Fourier theorem transforms the distance function to the domain of frequencies, rather than the original space or time domains. Simply by reversing the sign of the kernel, one defines an inverse transform that converts information in the frequency domain to space or time:

$$f(x) = \int_{-\infty}^{-\infty} F(v)[\exp(+2\pi i v x)]\,dv.$$

Mathematical operations in the frequency domain are often more straightforward than in spatial or temporal domains. This is where the power of the Fourier transform method lies. Certain relationships involving the function are rendered relatively simple in the transform. For example, the following are a few illustrations of properties or relations in both domains:

Property	Function	Transform
Similarity	$f(ax)$	$(a^{-1})F(v \cdot a^{-1})$
Linearity	$af(x) + bg(x)$	$aF(v) + bG(v)$
Convolution	$f(x) \odot g(x)$	$F(v) \cdot G(v)$
Autocorrelation	$f(x) \star f(x)$	$\lvert F(v)\rvert^2$

The following paragraphs will provide a restatement and examples of the mathematical statements set forth above.

To illustrate how Fourier analysis can be used to specify the spatial frequency response of a simple imaging system such as an optical lens, imagine that a series of sample cards is made, on each of which the reflectance varies sinusoidally in one direction. Figure 35 illustrates changes in luminance along one such card. The spatial frequency ν may be expressed in any convenient measure; in the illustration ν is set forth as the number of cycles contained in each millimetre of distance (mm^{-1}). The spatial frequency at the image plane of the lens will be different from that on the card according to the magnification or reduction provided by the focal length of the lens. Another measure of frequency that incorporates this magnification factor (that is, a measure that avoids the need to multiply ν by the optical magnification) is the the number of cycles in each unit of angular subtense, for instance, deg^{-1}. Conversion between the two forms for expressing spatial frequency is simply

$$\mathrm{deg}^{-1} = \mathrm{mm}^{-1}\left[\frac{360}{2\pi d}\right], \tag{26}$$

where d is the distance in millimetres from the nodal point of the lens to the plane of the object, called here "viewing distance." In effect, the expression of

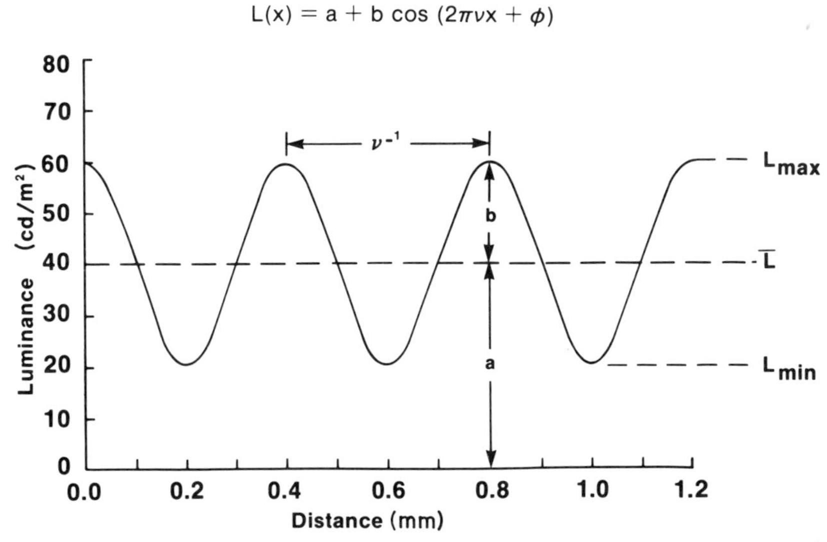

Fig. 35. Graph of luminance versus distance according to the function shown at the top of the graph. Here ν stands for spatial frequency and ν^{-1} for wavelength or period of one cycle; ϕ is a measure of phase. Note that the function is identical to a sine function except for ϕ and that it is symmetrical about the mean luminance. See text for discussion.

spatial frequency in mm^{-1} implies a concern with what is at the object, and expression in deg^{-1} concerns what is at the image.

The amplitude of variation in reflectance along the appropriate dimension of each of the cards is the same on all cards, but each card exhibits a different frequency of variation. Under any one condition of illumination, the luminances L at various points along the card will vary from some maximum (L_{max}) to some minimum (L_{min}). Photometric measurements of the luminance at each point can be made. We may symbolize each of these luminances as L_o, where the subscript identifies it as an object luminance. The image of each of the cards, as focused by the lens, may also be measured photometrically, point by point over distance. The luminance so measured may be symbolized as L_i, where the subscript identifies it as an image luminance. Values of L_i will also vary sinusoidally with distance from a minimum to a maximum. In general, however, values of L_o and L_i for corresponding points on the target will not be identical because of absorptions, flare, and imperfections of the lens. As a result the difference between $(L_i)_{max}$ and $(L_i)_{min}$ will tend to be less than that between $(L_o)_{max}$ and $(L_o)_{min}$. That is, the amplitude of the luminance variation in the image will be smaller than the amplitude variation in the original object. As the frequency increases, producing finer and finer patterns, this luminance amplitude will decrease more and more because of the effect of lens aberrations. A measure of the change in amplitude ($A = L_{max} - L_{min}$) between the original card and its image could be

$$A_t(v) = \left[\frac{(L_i)_{max} - (L_i)_{min}}{(L_o)_{max} - (L_o)_{min}}\right](v), \tag{27}$$

where $A_t(v)$ stands for *amplitude transfer function*, and v stands for spatial frequency in some appropriate unit. $A_t(v)$ is simply the ratio of maximum minus minimum luminance of the image to that of the original target. Note that because the lens transfer is linear with intensity for incoherent light, no change in wave form of the target takes place between original and image; only the luminance amplitude may vary as a result of flare, absorption, and optical imperfections of the lens.

If a neutral-density filter that absorbed 50% of the light were to be placed between the lens and the target, then $A_t(v)$ would decrease by half at each frequency. In other words the amplitude response function is sensitive to level of illumination and to light attentuation from any cause. Another form of transfer measure that is not sensitive to light attenuation is one called *modulation transfer*; it is sensitive only to (or reflects changes in) the shape of the response function over spatial frequency. Modulation is a specification of the amplitude of luminance variation relative to the mean luminance. It may be exemplified by relations among the curve positions labeled *a* and *b* in Fig. 35. The mean luminance $\bar{L}$ is indicated as the distance a from the zero luminance line in the figure. Half the amplitude is indicated by the distance b. Modulation is simply defined as b/a. Figure 35 shows that $b = (L_{\max} - L_{\min})/2$ and $a = (L_{\max} + L_{\min})/2$. The factors of $\frac{1}{2}$ cancel so that b/a may be expressed as

$$M = \left[\frac{(L_{\max} - L_{\min})}{(L_{\max} + L_{\min})}\right]. \tag{28}$$

If we symbolize the modulation of the target as M_o and that of the image as M_i, we may express the modulation transfer as a function of spatial frequency as the ratio of the two:

$$M_t(v) = M_i(v)M_o(v)^{-1}. \tag{29}$$

Note that the difference between $A_t(v)$ and $M_t(v)$ is that the latter is insensitive to level; that is, $M_t(v)$ is a normalized measure and reflects only changes in the shape of the curve that describes output as a function of frequency. It is called the *modulation transfer function* (of spatial frequency) and is usually abbreviated MTF. A graph of M_t versus v indicates the relative frequency response of the system. In our illustration it may be obtained by plotting the ratio of M_t determined for each of the cards in the series at the position corresponding to the spatial frequency represented on the card. The resulting graphical representation of M_t versus v is known as an MTF curve.

An underlying assumption in the method of linear systems analysis, as emphasized by its very name, is that the transfer system under analysis is one with a linear transfer characteristic. Additivity of sine-wave powers obtains only when the system is linear.* Fourier's theorem does not apply to nonlinear systems. When the transfer characteristic is not linear, sine waves will not be transferred without distorting their shapes. Figure 36 illustrates differences between the undistorted output of a linear system and the distorted output of a nonlinear system. This kind of difference between the two systems is important, because it bears on the validity of the method of Fourier analysis. The method is mathematically very powerful for a linear system. It provides means for addressing directly the questions of acuity and contrast sensitivity of imaging systems. The image or response to a line is known as the *line spread function*, because there is some spreading of the image of a line in any imperfect imaging system. Actually all imaging systems are imperfect; they differ mainly in their degrees and kinds of imperfection. The line spread function may be described by a series of sine waves according to Fourier's theory. The modulation transfer function $M_t(v)$ of a linear system such as a lens is related to

* Intensities (powers) of light are additive for spatially and temporally incoherent light, such as we encounter in everyday situations with unaided viewing, but amplitudes are additive with coherent light. This discussion will be limited to the more common case of incoherent light.

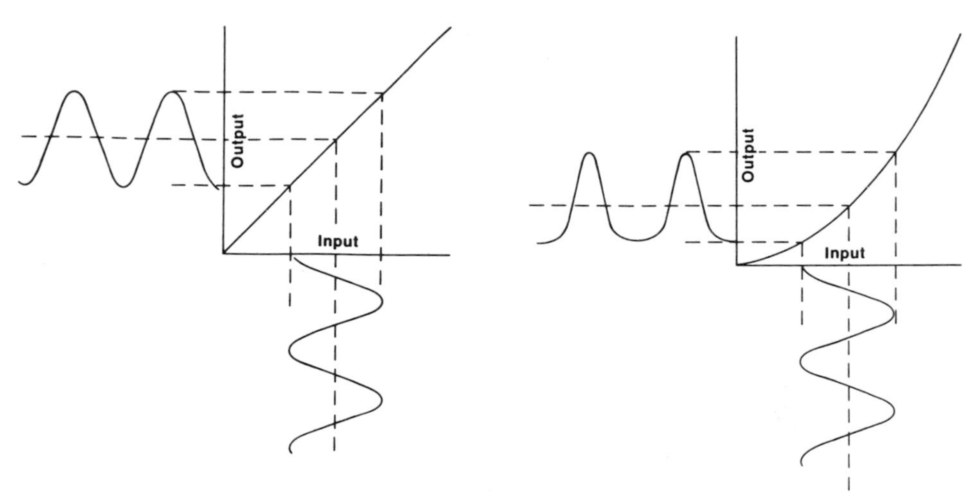

Fig. 36. Illustration of the transfer of sinusoidal variation in luminance over space through a linear system (left) and a nonlinear system (right). The linear system leaves the input sine wave undistorted, but the nonlinear system introduces harmonic distortion.

the line spread function $\ell(x)$ through a Fourier transform:

$$M_t(v) = \int_{-\infty}^{+\infty} \ell(x) \cdot e^{-2\pi vix} \cdot dx, \tag{30}$$

and the inverse relation is

$$\ell(x) = \int_{-\infty}^{+\infty} M_t(v) \cdot e^{+2\pi vix} \cdot dv, \tag{31}$$

where

x = distance;
v = spatial frequency;
e = the Naperian base, defined as $\lim_{n\to\infty}[1 + 1/n]^n$, the first six figures of which are 2.71828;
π = the transcendental number representing the ratio of the circumference of a circle to its diameter, the first six figures of which are 3.14159;
$i = \sqrt{-1}$, the imaginary unit of mathematics.

For images of points the MTF is related to the *point spread function* $h(r)$ by a Hankel transform (Jones, 1958):

$$M_t(v) = 2\pi \int_0^{\infty} h(r) \cdot J_0 \cdot (2\pi r) \cdot dr, \tag{32}$$

where r is the radius, J_0 is the zero-order Bessel function of the first kind (see, for example, Abramowitz and Stegun, 1965), and other symbols are as in Eq. (31).

The line spread function $\ell(x)$ is, in turn, related to the point spread function by an Abel transform (Jones, 1958):

$$\ell(x) = 2\int_{x}^{\infty} h(r) \cdot (r^2 - x^2)^{-1/2} \cdot r \cdot dr, \tag{33}$$

and the inverses of Eqs. (32) and (33) define the opposite relationships.

Use of these relationships assumes that each of the Fourier components adds together with others in a simple algebraic fashion over the transform. This will generally be true of a simple system such as a lens. The system is said to be linear, homogeneous, and isotropic; these are necessary conditions for the validity of a Fourier analysis. A linear system is one that does not distort the waveform of any input. A homogeneous system is one in which the transfer characteristics are the same in all locations. A system is isotropic if its transfer characteristics are the same in all directions. As we have seen, the visual mechanism does not satisfy these criteria. Responses are not linearly related to input intensities. The lens of the eye is not homogeneous, the density distribution of rod and cone photoreceptors is far from homogeneous over the retina, and the topology of neurons in the layers of the lateral geniculate nucleus is inhomogeneous in a manner that favors processing of high-acuity information. Even if the anisotropy produced by the natural astigmatism of the lens of the eye is optically corrected, neural processing elements apparently are anisotropic (Alpern and David, 1959; Campbell *et al.*, 1966). The visual mechanism fails to meet the necessary criteria for Fourier analysis on all counts, therefore.

However, Fourier analysis is used to measure certain characteristics of the contrast sensitivity of the visual mechanism. Its use requires various constraints and assumptions. The results are generally recognized as approximations. Cornsweet (1970) has adopted the phrase "describing function of the human visual system" to distinguish such approximations from MTF measurements of linear transfer systems that are homogeneous and isotropic. The intent is commendable, but the phrase seems too general; instead, in this chapter these approximations will be termed "contrast sensitivity functions," a phrase often used by vision scientists to make the same distinction. As with all psychophysical functions, contrast sensitivity functions apply strictly to the viewing and experimental conditions used in their determination. The choice of a descriptor that differs from the abbreviation MTF is intended to imply that contrast sensitivity functions differ from MTFs in a number of ways; in particular, they do not involve additivity and, therefore, do not lend themselves to convolution intended to describe a complete system from the integration of its parts. In short, a contrast sensitivity function of spatial frequency is just one more kind of psychophysical measurement and does not have the special properties required for valid Fourier analysis.

1. Threshold Contrast Sensitivity

That being so, there is no reason why any one experimental method should be preferred over another. However, most contrast sensitivity functions have been measured for *threshold contrast*, primarily in attempts to bestow what is thought to be more generality on the data and to permit their use in a close approximation to Fourier's theorem. The question of linearity is dealt with by measuring the threshold of contrast sensitivity. The underlying rationale is that any nonlinear system approximates a linear one as the dynamic range is reduced to very small excursions. Homogeneity is assumed for small, central visual areas such as the fovea centralis or the fovea. Although the anisotropy of the visual mechanism is such as to reduce sensitivity to modulation between vertical and 45° oblique targets by about 15% at 10 cycles $\cdot$ deg^{-1} and 50% at 30 cycles $\cdot$ deg^{-1} (Campbell *et al.*, 1966), it is usually ignored by measuring a single contrast sensitivity threshold function with a vertically oriented grating. The procedure involves presenting the observer with a sine-wave (or other periodic) grating of a particular spatial frequency. The amplitude of the intensity variation can be adjusted until the observer can just detect the presence of a grating. When this procedure is repeated for a sufficient number of gratings with different spatial frequencies, the modulation required for threshold contrast detection can be plotted against spatial frequency, as in the illustration of Fig. 37(a); Fig. 37(b) shows the same data where sensitivity is

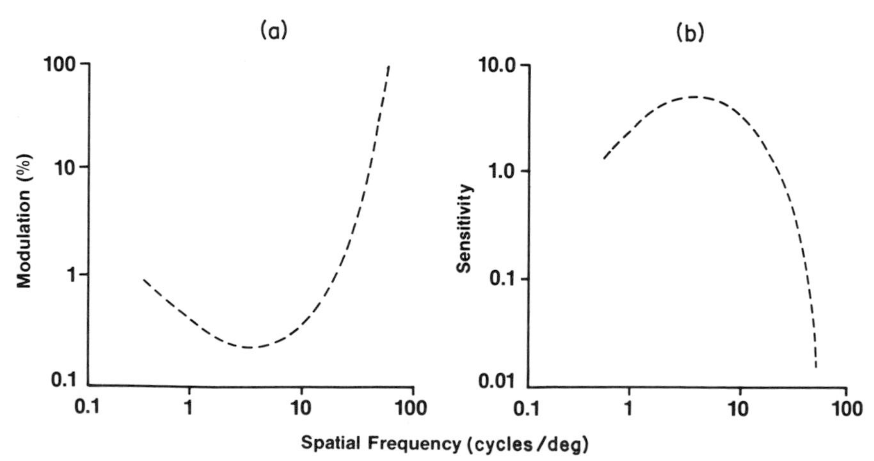

Fig. 37. Representative results for determination of threshold contrast sensitivity for sine-wave gratings with light of 525 nm at 90 trolands: (a) shows threshold modulation, defined in Eq. (27) of the text, and (b) shows sensitivity calculated as the reciprocal of threshold modulation.

calculated as the reciprocal of the threshold modulation. For measurements of threshold contrast sensitivity with sine waves, the functions determined by various workers tend to reach a peak at 3–6 cycles $\cdot$ deg^{-1}. Sensitivity is less at both higher and lower spatial frequencies. The reduction in sensitivity at higher frequencies is taken to reflect characteristics of the eye's lens. The reduced sensitivities at lower frequencies usually have been thought to be a consequence of the spatial frequency characteristics of the neural system, although it has been proposed recently (Campbell *et al.*, 1981) that low-frequency structure may be perceived by detection of contrast gradients rather than by harmonic analysis as is thought to be appropriate to high-frequency structure. If these assumptions are used to perform a Fourier transform of the kind of data illustrated in Fig. 37, both a positive, optical, line spread function and a negative, neural, spread function can be calculated from the relation of Eq. 31. The result is similar to that illustrated in Fig. 38. The spread function of

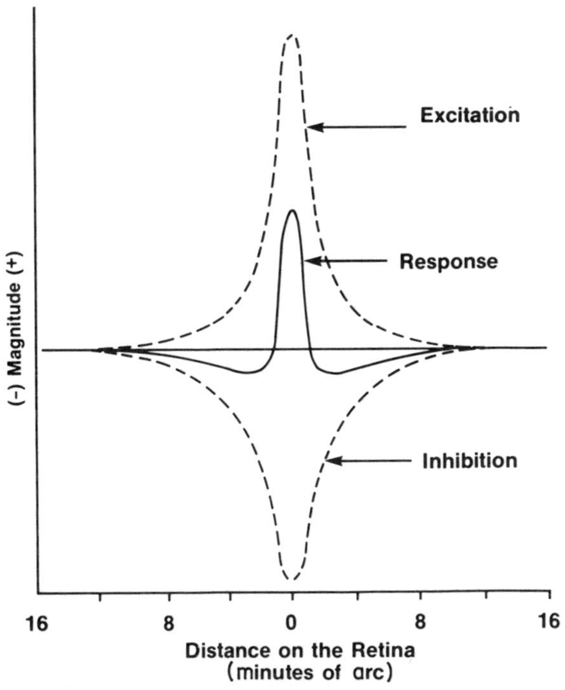

Fig. 38. Effect of a narrow line of light on the visual mechanism as implied by the threshold contrast sensitivity function of Fig. 37. Incident flux in the retinal image is spread by lens diffraction, aberrations, and so forth. Inhibition varies with spatial contiguity. The resultant visual response tends to enhance contrast by narrowing the positive response lobe and incorporating negative responses at the line's outer edges.

the optical excitation is shown as the all-positive curve and that of the neural inhibition is the all-negative curve. The resulting integration of spread functions yields a response curve that has a positive central element and graded negative "skirts," implying an enhancement of the contrast perceived for a narrow line of the dimensions represented by the figure. That prediction accords well with the enhancement observed in viewing such stimulus arrays, and largely for that reason, threshold contrast sensitivity functions are often accepted as reasonably, albeit approximate and qualitative, methods for analyzing the contrast sensitivity of the visual mechanism.

There is, however, no single function that adequately represents threshold contrast sensitivity as measured with sine-wave gratings. The functions differ according to retinal eccentricity and the diameter of the pupil (for instance, Campbell and Gubisch, 1966). Threshold functions also differ with the level of retinal illuminance (for instance, Van Nes and Bouman, 1967) and with the dominant wavelength of the stimulus (for instance, Kelly, 1974). In the latter instance, stimuli that excite mainly the achromatic or γ processes seem to yield highest sensitivity to relatively high spatial frequencies, with ρ processes showing somewhat less sensitivity and β processes considerably less sensitivity to relatively high spatial frequencies. Orientation of the target significantly affects results with sine-wave gratings; this is illustrated in Fig. 39, which is patterned after results published by Watanabe *et al.* (1968).

When square waves are used rather than sine waves, the contrast sensitivity function has a quite different shape at low spatial frequencies. This is illustrated in Fig. 40(a), based on the work of Campbell *et al.* (1978). The reduction in sensitivity to low frequencies found with sine-wave gratings is not in evidence when square-wave periodic targets are used. This is also true even

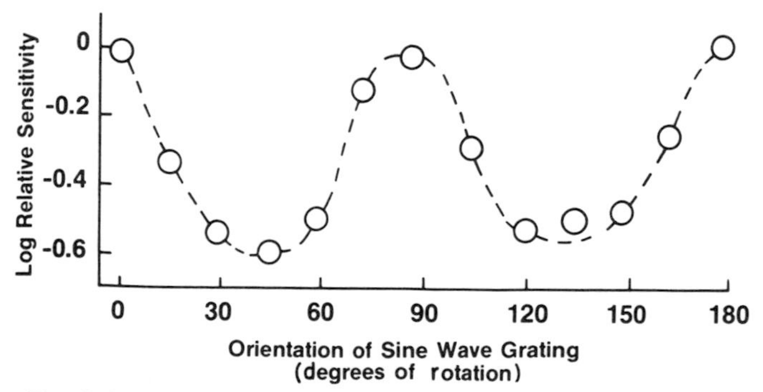

Fig. 39. Threshold contrast sensitivity to an 11.4 cycles · deg^{-1} sine-wave grating rotated (from vertical = 0°) through 180°; average luminance of about 34 cd · m^{-2}; pattern subtending about 6° visual subtense. [After Watanabe *et al.*, (1968), *Vision Research*, redrawn with permission.]

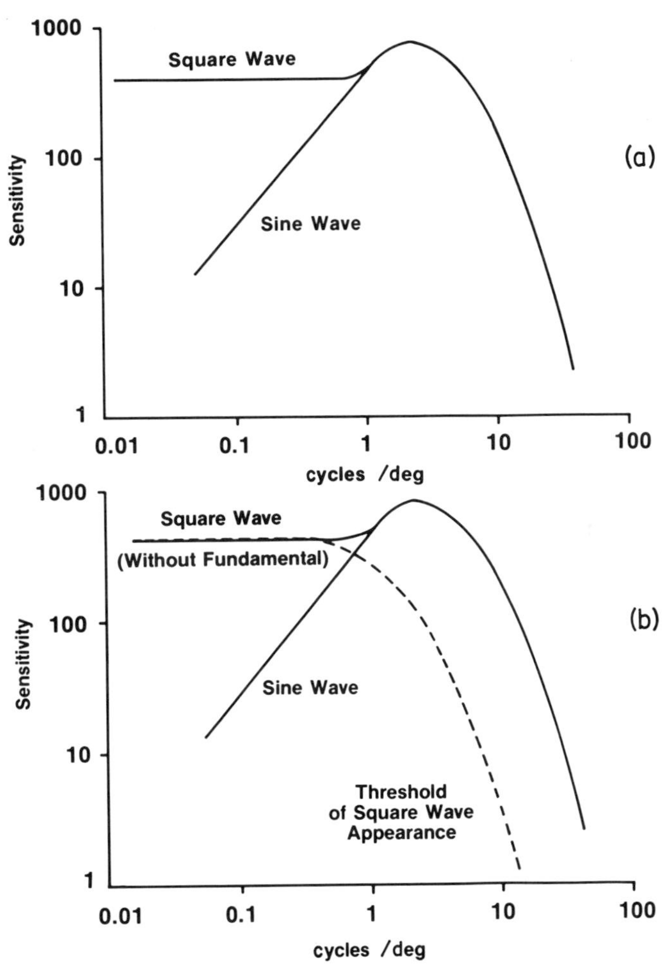

Fig. 40. Experimental results for threshold contrast sensitivity with sine- and square-wave gratings (a). Results for sine and missing fundamental square waves (b). Dashed extension of missing fundamental square-wave function is for threshold of appearance difference between square wave with and without fundamentals. In both (a) and (b) the threshold contrast sensitivity function for sine waves has been multiplied by $4/\pi$ to facilitate comparison of results from sine- and square-wave gratings. [After Campbell and Robson, (1968), *Journal of Physiology, London.*]

when the fundamental is missing from the square wave, as in Fig. 40(b). The square-wave grating appears to be a square wave when its contrast reaches threshold provided its frequency is less than 1 cycle $\cdot$ deg^{-1}. Above that frequency, however, the square and sine waves are indistinguishable at threshold. Campbell and Robson (1968) showed that contrast must be increased to the point where the third harmonic could be detected in order that the square wave would be seen as such. The dashed extension of the square-wave function shown in Fig. 40(b) represents the threshold for detection of the third harmonic.

These data illustrate two important facts. First, the threshold contrast sensitivity varies with wave form of the periodic stimulus. Second, the tasks of detection and recognition differ in that they involve different perceptual judgments, as was noted in the earlier section on visual acuity. We see, then, that contrast sensitivity functions determined at threshold vary considerably depending on stimulus characteristics. Even if this were not so, there is still a question about the validity of threshold data to explain or analyze spatial response characteristics of the visual mechanism for suprathreshold conditions. Because much, if not most, practical interest centers on suprathreshold conditions, it is important to determine whether contrast sensitivity functions measured under these conditions differ significantly from those found for threshold conditions.

2. Suprathreshold Contrast Sensitivity

Fewer experiments have been conducted for suprathreshold contrast than for threshold, but they confirm that sensitivity functions are not the same at threshold and above it. Although the threshold of detection may be determined for threshold experiments, a different method must be used for gathering data at suprathreshold. Often a method of contrast matching is used. One such method, developed by Davidson (1968), involves comparing a grating of each spatial frequency to one of a fixed reference frequency. The observer alternately views the test and reference gratings and adjusts the physical contrast of the reference until its perceived contrast matches that of the test grating (or vice versa). Lowry and DePalma (1961) used a luminance-matching method, in which the luminance of a narrow adjacent reference strip was changed to match that perceived brightness at each point along a periodic grating. Others have used direct scaling techniques to estimate perceived contrast (for instance, Cannon, 1979; Ginsburg *et al.*, 1980). Indirect measures such as reaction time (Harwerth and Levi, 1978) and electrophysiological measurements of visually evoked responses (Jones and Keck, 1978) have also been used.

An impressive number of studies agree well in the finding that contrast sensitivities for various frequencies are lower and more nearly equal as the physical contrast of sine-wave targets increases (for instance, Watanabe *et al.*, 1968; Blakemore *et al.*, 1973; Georgeson and Sullivan, 1975; Bryngdahl, 1966; Cannon, 1979; Ginsburg, 1977). Cannon (1979) has suggested that perceived contrast or visual response R is proportional to the difference between the suprathreshold physical contrast or modulation C and the contrast at threshold C_{th}:

$$R(v) = k[C(v) - C_{th}(v)]. \tag{34}$$

Figure 41 has been prepared by using Eq. (34) according to the relationship $C = (R/k) + C_{th}$ where $k = 0.14$ and $C_{th}(v)$ is a threshold contrast sensitivity function from Watanabe *et al.* (1968). In addition to the threshold condition, four levels of contrast response are shown in the figure. Each of these conditions yields a curve of suprathreshold contrast C required at each frequency v to elicit the constant criterion response. As in Fig. 37 the inverses of the functions shown define contrast sensitivity. Figure 41 shows clearly that sensitivity decreases with increasing physical contrast and that sensitivity tends toward constant logarithmic values as physical contrast increases, except at high spatial frequencies where, most likely, the aberrations of the eye's lens represent the limiting factor in determining sensitivity. Figure 42, taken from Cannon (1979), illustrates a surface of contrast sensation over variation of physical contrast and spatial frequency.

The implication of these findings, which are consistent with the experimental results cited above, is that threshold contrast sensitivity functions are probably not appropriate for suprathreshold conditions. Kulikowski (1976) reported that gratings appeared to have the same contrast when the difference

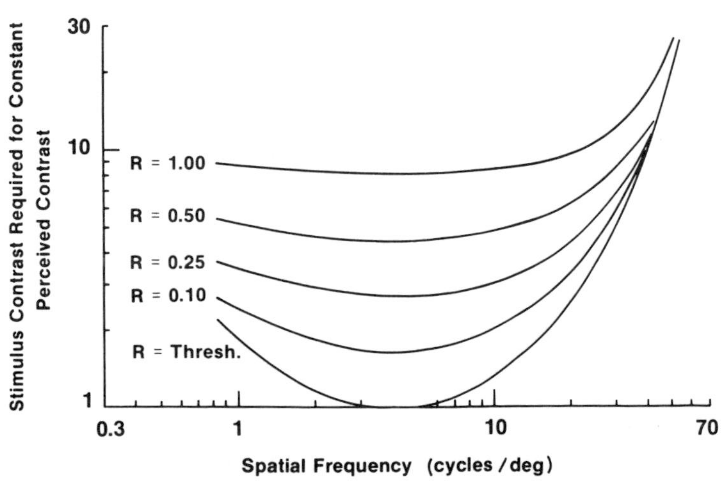

Fig. 41. Curves of stimulus contrast required for constant perceived contrast calculated according to the relation $C = (R/k) + C_{th}$, where $k = 0.14$, and the threshold contrast function C_{th} is one of those reported by Watanabe *et al.* (1968).

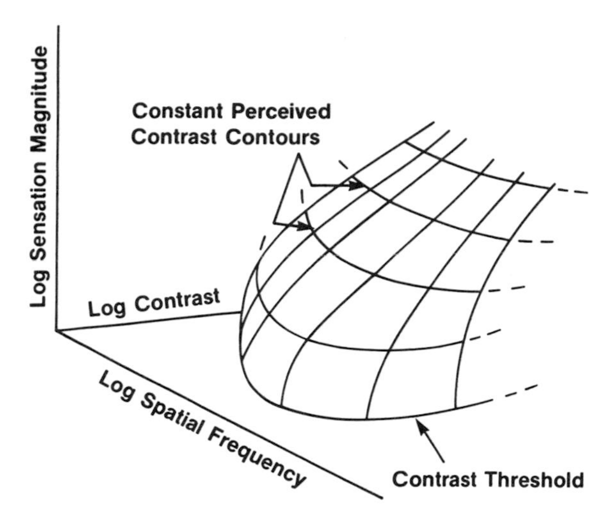

Fig. 42. Idealized contrast sensation surface derived from Eq. (34) of the text. Note that coordinates are logarithmic on all axes. [After Cannon, (1979), *Vision Research*, redrawn with permission.]

between their physical contrast and that corresponding to threshold was equal. The data of Figs. 41 and 42 are close to having that property. The experimental method used by Kulikowski in which all observers, experienced and inexperienced alike, yielded the same results* was a matching method. Accordingly, questions regarding the validity of scaling techniques do not enter into the matter of inferences that may justifiably be drawn from the experiment. One implication of these results is that although the various suprathreshold curves of Fig. 41 show diminishing frequency selectivity as perceived contrast increases, they should all be congruent when plotted in linear coordinates. In this respect, both Cannon's proposal, Eq. (34), and Kulikowski's are the same. Inspection of Eq. (34) indicates that $C - C_{th}$ is equal to the constant R/k for any single value of R. When the data of Fig. 41, which are based on Eq. (34), are replotted as the increment above threshold $(C - C_{th})$ required for constant response R, the horizontal lines of Fig. 43 result. Thus contours of perceived isocontrast may be simply represented by a constant stimulus increment above threshold, at least for spatial frequencies in the middle region of the sensitivity band.

* Kulikowski also used a psychophysical fractionation method, but results differed between experienced and inexperienced observers for that method, probably because of difficulties encountered by the inexperienced observers; results for experienced observers were the same by both matching and fractionation methods.

Recall that certain cortical neurons respond only to intensity contrast of properly oriented stimulus arrays. Casual consideration of the data in Fig. 43 might seem to suggest that these contrast-detecting receptive fields operate in a linear manner. However, the same result could be had by subtracting outputs of nonlinear processes. Such difference signals could represent ratios of stimulus intensities (approximately or exactly, depending upon the form of nonlinearity involved). Ernst Mach pointed out in 1868 (Mach, 1868; Ratliff, 1965) that it is important that we be able to detect a given level of contrast at all levels of illumination for effective functioning of the organism as a whole. Because of the nonlinear operating characteristics of neural transmission from ganglion cells through cortical cells noted in an earlier section of this chapter, a given *stimulus difference* assumes different response magnitudes as illuminance varies. However, a given *stimulus ratio* tends to be maintained at more nearly the same difference in logarithmic or power function response outputs (the difference between the two forms of nonlinearity is rarely, if ever, apparent from experimental data). It seems more likely, then, that contrast detectors respond to intensity ratios and that differences among nonlinear response outputs are responsible for the constancy illustrated in Fig. 43.

Perceived contrast appears to increase as a power function of physical contrast. Unfortunately, however, there are conflicting results regarding the nature of these power functions for different spatial frequencies. Figure 44 schematically summarizes the different kinds of results found by various experimenters. One form of result suggests that the exponent of the contrast power function increases as frequency increases (for instance, Franzen and

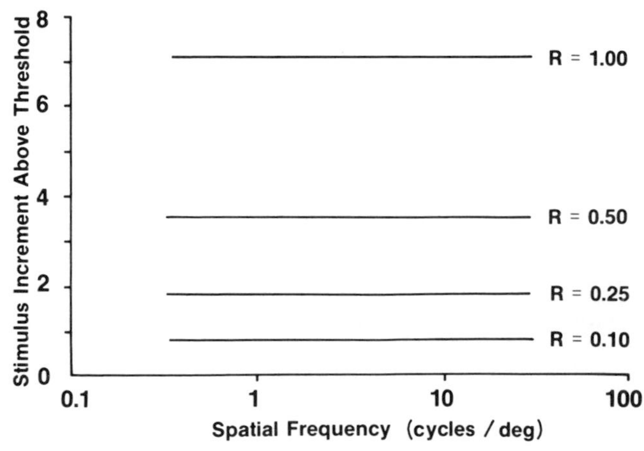

Fig. 43. Curves of contrast increment above threshold ($C' = C - C_{th}$) required for constant perceived contrast; data as in Fig. 41.

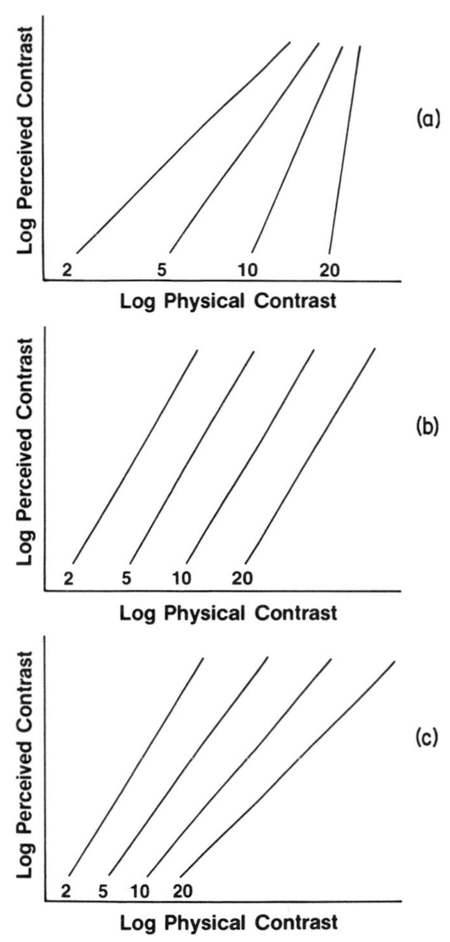

Fig. 44. Schematic representation of three kinds of results from assessments of suprathreshold contrast of sine-wave gratings of 2, 5, 10, and 20 cycles · deg^{-1}. See text for explanation.

Berkley, 1975). This is shown in Fig. 44(a). Because the coordinates are both logarithmic, the increasing slopes of the lines for higher spatial frequencies indicate that the exponent of the power function relating perceived contrast to physical contrast increases as frequency increases. A second kind of result is shown in Fig. 44(b). This graph represents results from which there is no indication that the exponent changes with frequency (for instance, Cannon, 1979). Finally, Fig. 44(c) schematically illustrates results suggesting that the power decreases with frequency (for instance, Hamerly *et al.*, 1977). The disagreement among these kinds of results for the assessment of suprathreshold contrast indicates clearly that the question of how perceived contrast increases at different spatial frequencies as physical contrast rises remains an unsolved problem.

3. Spatial Frequency Channels

Additional insight into the visual mechanism for detecting or seeing contrast has been obtained by experiments with wave forms that differ from those of single-frequency sine waves. Ginsburg *et al.* (1980) have used magnitude estimation to scale the perceived contrast of sine and square waves. They found that the magnitude estimates were linear functions of stimulus contrast. In addition, the psychophysical functions for square waves were steeper than those for sine waves by a factor of $4/\pi$, which is the ratio of stimulus power for the two kinds of wave forms. If these findings are generally true, the implication is that there is a reasonably high degree of linearity in suprathreshold processing of visual contrast. It would be helpful to be able to model the contrast mechanism linearly at suprathreshold as well as at threshold but there is, as yet, far from general agreement that the suprathreshold process is linear.

Experiments using complex arrays of sine waves that vary in frequency, amplitude, and phase have yielded valuable additional information about possible mechanisms for contrast processing. Contrast-detection experiments of several kinds have suggested the existence of numerous spatial channels, each tuned to a narrow range of spatial frequencies (for instance, Campbell and Robson, 1968; Blakemore and Campbell, 1969; Sachs *et al.*, 1971). These channels seem to be both narrow and adaptive. Adaptation to a high-spatial-frequency grating reduces sensitivity to contrast of gratings over a range of nearby frequencies (for instance, Blakemore *et al.*, 1973). Both contrast matches (Quick *et al.*, 1976) and contrast-scaling data (Hamerly *et al.*, 1977) have shown perceived contrast to be insensitive to interactions of combined stimulus contrast of patterns widely separated in frequency; that is, the contrast perceived in one sine-wave grating was not influenced by a second grating of much different frequency and phase. These results suggest that the visual mechanism acts not as a single spatial filter but rather as a combination of numerous spatial filters.

Sachs *et al.* (1971) addressed the question of whether two components of a complex grating could be detected independently. The bulk of their data relate to combinations of spatial frequencies with a reference frequency of 14 cycles$\cdot$deg^{-1}. They found that components could not be detected independently when the ratio of frequencies was between about $\frac{4}{5}$ and $\frac{5}{4}$ but that independent detection occurred for frequencies that differed by more than these ratios. This finding held true as well for other reference frequencies ranging from 1.9 to 22.4 cycles · deg^{-1}. They concluded (Sachs *et al.*, 1971):

> (1) The human visual system contains several sensory channels, each selectively sensitive to a different, moderately narrow range of spatial frequencies; (2) the outputs of the various channels are stochastically independent; (3) the output of each channel is passed through a separate threshold device; (4) a grating is detected when the critical level of at least one of the threshold devices is exceeded, or when a "yes" (*response*) is generated by an independent guessing mechanism. [p. 1183]

Their model can be represented functionally as

$$R(v) = k\pi\{k_c \cdot r_c^2 \cdot \exp[-(\pi \cdot r_s \cdot v)^2] - k_s \cdot r_s^2 \cdot \exp[-(\pi \cdot r_s \cdot v)^2]\}, \quad (35)$$

where R is the contrast response, r_c is the response of receptive field center, r_s is the response of receptive field surround, and k is the relative gain of receptive field center k_c and surround k_s.

Equation (35) describes a family of curves whose characteristics depend on values of r_s, r_c, k_s, and k_c. The narrowest member of that family, centered on 14 cycles · deg^{-1}, has been plotted as curve B in Fig. 45. It may be compared with the experimentally determined channel sensitivity for 14 cycles · deg^{-1}, which is plotted as the solid curve A in Fig. 45. Obviously, curve B is far too broad to be a good representation of the experimental results. However, recall that cortical columns of neurons analyze signals over and over, effectively providing a cascading of operations. The cascaded response of n filters is the same as that of a single filter raised to the nth power, yielding a much narrower-band effective filter. If the interactions represented in Eq. (35) are cascaded ten times, that is, if the signal is successively passed through ten

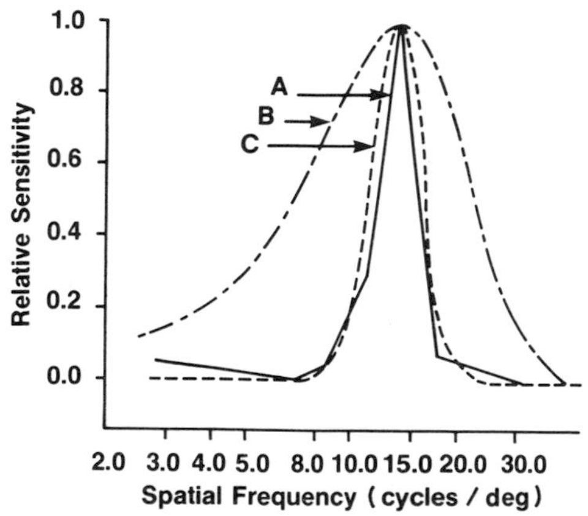

Fig. 45. Curve A, experimentally determined sensitivity of a spatial frequency channel centered at 14 cycles · deg^{-1}, curve B, narrowest member of the family of curves corresponding to Eq. (35) of the text; curve C, tenfold cascaded result from curve B. See text for details. (Sachs *et al.*, 1971, *Journal of the Optical Society of America.*)

such filters, the narrow-band curve labeled C in Fig. 45 results. Thus far, electrophysiological measurements have not uncovered such cascading processes, although such a mechanism remains a possibility for explaining the kinds of results obtained by Sachs *et al.* At least two psychophysical experiments have indicated multiple tuning of spatial frequency channels for depth and motion perception as well (Chase and Smith, 1981; Anstis and Harris, 1974). Other results indicate multiple tuning for frequency and depth (for example, Blakemore and Hague, 1973) and for frequency and motion (for example, Tolhurst, 1973). It is not unreasonable to speculate that multiple-frequency tuning may also occur. In any case it seems reasonably clear that the visual mechanism is sensitive to relatively narrow bands of spatial frequencies and that the output signals of these spatially selective channels are combined in some manner to provide independent response attributes for spatial frequencies, much as is the case in processing temporal frequency signals in audition. It is possible that such response functions are formed by linear combinations of modulation, nonlinear combinations, amplitude (peak-to-trough) detectors, or "line" detectors of some kind (Kulikowski and Kingsmith, 1973; Kingsmith and Kulikowski, 1975; Arend and Lange, 1980). The theoretical problem involved in the interpretation of such data is to discover a single model that can satisfactorily explain narrow-spatial-frequency tuning at all contrast levels, threshold and suprathreshold.

4. Spatio–temporal Interactions

Kelly (1966, 1974) has pointed out a striking analogy between spatial and temporal frequency characteristics of the visual mechanism. In both cases there is a high-frequency cut off and a marked decrease in sensitivity to low frequencies. The latter is generally attributed to lateral or temporal inhibitions associated with neural processing networks. It is also of considerable theoretical significance that these inhibitions can be quenched in whole or in part by interactions between spatial and temporal frequency properties of stimulus targets. Temporal flicker sensitivity is greatly enhanced (that is, inhibition is quenched) by presenting an observer with a sinusoidally varying luminance target of about 3 $\text{cycles} \cdot \text{deg}^{-1}$ rather than a spatially uniform target. Similarly, contrast sensitivity is enhanced when the grating is presented by varying its luminances sinusoidally over time at a rate of about 8 Hz ($\text{cycles} \cdot \text{sec}^{-1}$).

These kinds of results are illustrated in Fig. 46. The graph in (a) shows maximum and minimum achromatic flicker-sensitivity functions at spatial frequencies of zero (open circles) and 3 $\text{cycles} \cdot \text{deg}^{-1}$ (filled circles); an artificial pupil of 2.3-mm diameter and a mean retinal illuminance of 10^3 td was used in

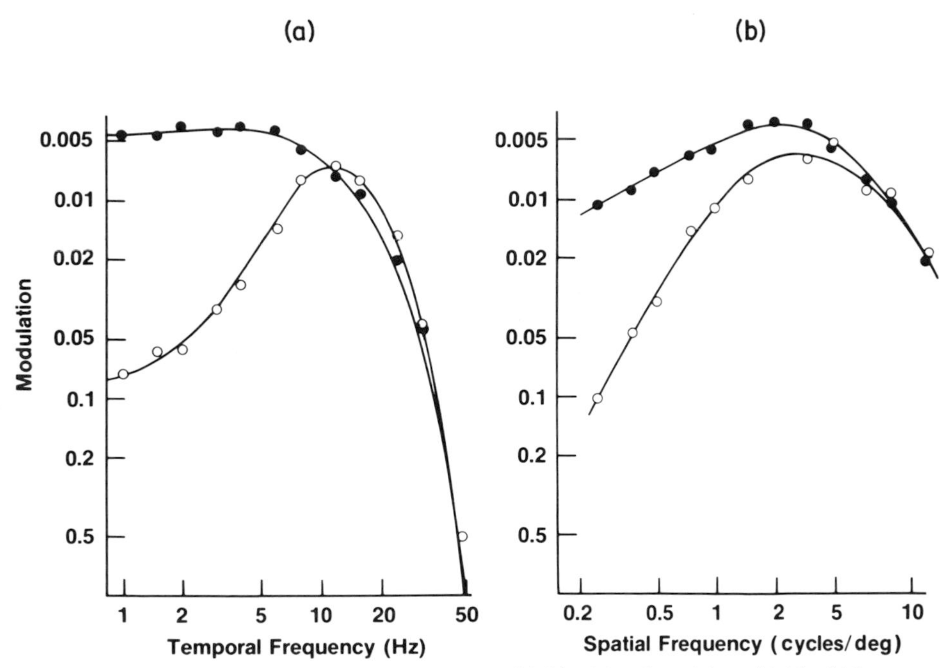

Fig. 46. Maximum and minimum temporal sensitivities (a) and spatial sensitivities (b) for an achromatic stimulus array when using a 2.3-mm diameter artificial pupil and 10^3 td retinal illuminance. See text for details. (Kelly, 1974, *Journal of the Optical Society of America.*)

the experiment illustrated. The uniform target permits considerable inhibition below about 8 Hz, as is evident from the steeply decreasing sensitivity function at those frequencies. On the other hand there appears to be little or no inhibition with the 3 cycles$\cdot$deg^{-1} target at low temporal frequencies, as is evident from the fact that the sensitivity function is nearly horizontal below about 8 Hz. Similarly the graph in (b) illustrates a reduction in inhibition at low spatial frequencies when the target is temporally modulated at 8 Hz (closed circles) compared with the more inhibited sensitivities for low frequencies when the target is invariant over time (open circles). Note that with spatial frequencies [Fig. 46(b)] inhibition is not completely quenched as it is with temporal frequencies but is merely considerably reduced.

Kelly (1974) suggests that physiological spatio-temporal resolution may be limited at high frequencies by response times of direct centripetal pathways and at low frequencies by lateral pathways that integrate over relatively large areas. Because of their narrow bandwidths these inhibiting pathways cannot respond fast enough to the 8-Hz or 3-cycles$\cdot$deg^{-1} modulation, and their inhibitory actions are, therefore, reduced or eliminated. The curves in Fig. 46 may be taken to be members, from different positions, of a surface of frequency

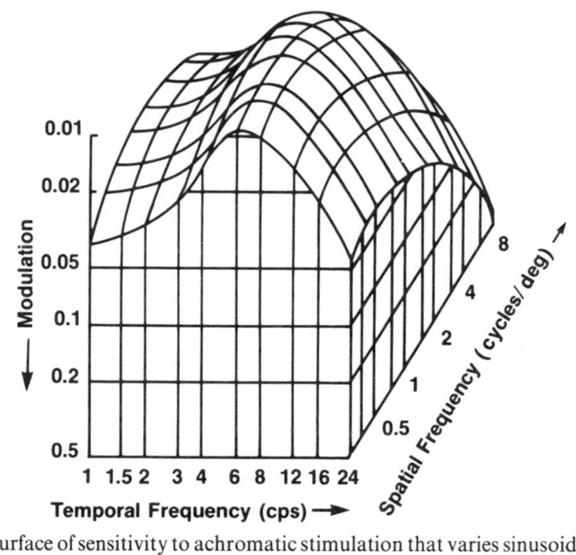

Fig. 47. A surface of sensitivity to achromatic stimulation that varies sinusoidally in space and time (Kelly, 1966, *Journal of the Optical Society of America.*)

sensitivities, both spatial and temporal. Figure 47 illustrates such a surface. Modulation required for threshold perception is arrayed logarithmically against temporal frequency and spatial frequency, also arrayed logarithmically, in the perspective graph of Fig. 47. The shape of this surface varies with adaptation, illuminance, and chromaticity.

Much the same phenomena can be seen with chromatic stimuli. Figure 48 illustrates patternless flicker thresholds for "red", "green", and "blue" stimuli (a) and contrast thresholds for flickerless patterned stimuli of the same chromaticities (b). The chromaticities of the stimuli were produced by superimposing Kodak Wratten gelatin filters No. 45A, 58, and 23A over a P24-phosphor cathode ray tube. These graphs show the threshold functions for Kelly's observers and conditions without interaction between spatial and temporal domains. Figure 49 illustrates what happens to the chromatic thresholds with interaction. In (a), flicker sensitivity is shown for red, green, and blue phases when the spatial frequency is set at optimum: 3 cycles $\cdot$ deg^{-1} for red and green and 1 cycle $\cdot$ deg^{-1} for blue. In (b), the temporal frequencies used were 8 Hz for red and green and 1.5 Hz for blue. At these frequencies (spatial and temporal) the flicker and contrast sensitivity curves should be the same as in Fig. 48, and the filled points, transferred from Fig. 48, indicate that the data are consistent in this respect.

It is interesting that all three flicker-sensitivity curves in Fig. 49(a) resemble low-pass filters of about the same shape. They differ primarily in their absolute

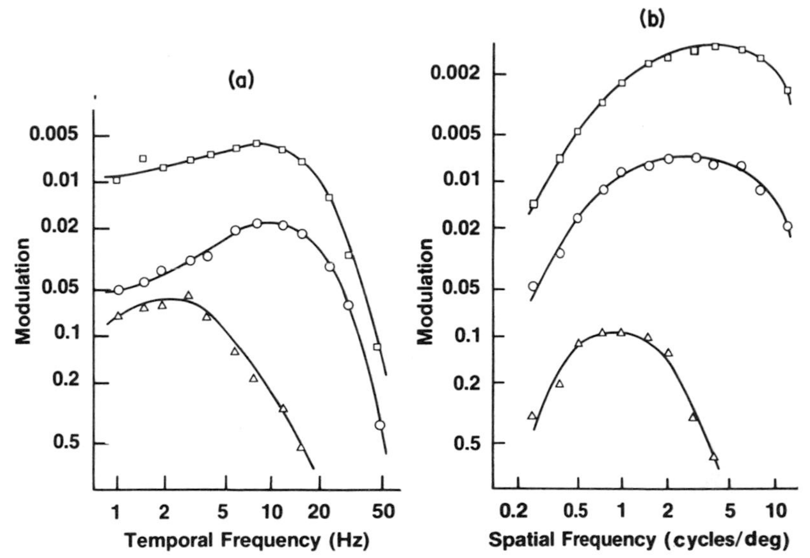

Fig. 48. (a) Patternless flicker sensitivity for "green" (squares), "red" (circles), and "blue" (triangles) targets. (b) Flickerless contrast sensitivity for "green" (squares), "red" (circles), and "blue" (triangles) targets. (Kelly, 1974, *Journal of the Optical Society of America.*)

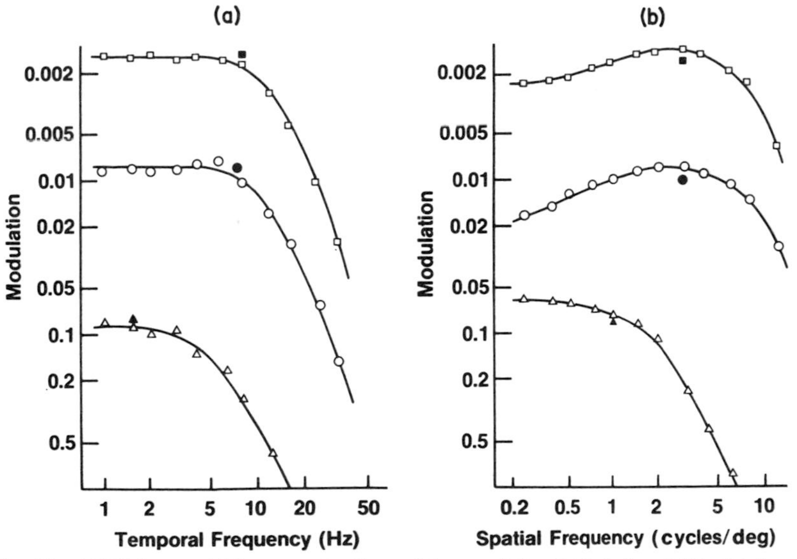

Fig. 49. (a) Flicker sensitivity curves for "green" (squares), "red" (circles), and "blue" (triangles) targets using optimum grating patterns of 3 cycles $\cdot$ deg^{-1} for "red" and "green" and 1 cycle $\cdot$ deg^{-1} for "blue". (b) Contrast sensitivities for "green" (squares), "red" (circles), and "blue" (triangles) targets using optimum flicker of 8 Hz for "red" and "green" and 1.5 Hz for "blue". Filled points represent data transferred from Fig. 48 for the spatial and temporal frequencies common to both illustrations. (Kelly, 1974, *Journal of the Optical Society of America.*)

sensitivities, green being most sensitive and blue least by about a factor of 60. If the horizontal portions at low temporal frequencies do indicate that the effects of inhibitory pathways have been eliminated, then the shapes of these flicker functions may represent the dynamic characteristics of the underlying excitatory mechanisms. Except for the blue case, the shapes of the spatial frequency contrast functions still show some effects of inhibition, and tests with other frequencies did not eliminate them. Apparently, even though 8 Hz is the optimum flicker rate for the red and green cases, there is still some inhibition, suggesting that at least a few lateral pathways can respond at this rate. The blue curve does not show inhibition, and presumably the lateral pathways have been sufficiently inactivated by the 1.5-Hz presentation. In the blue it would appear that contrast sensitivity may be governed by the spatial distribution of β receptor processes, which are fewer than ρ and γ processes and, at least in the fovea, are more sparsely distributed.

As indicated above, the low-pass filters *may* represent the dynamic characteristics of the underlying excitatory mechanisms. The similarity among their shapes has also been noted. Figure 50 illustrates the degree of similarity. The experimental data from Figs. 46 and 49 (that is, Kelly's 1974

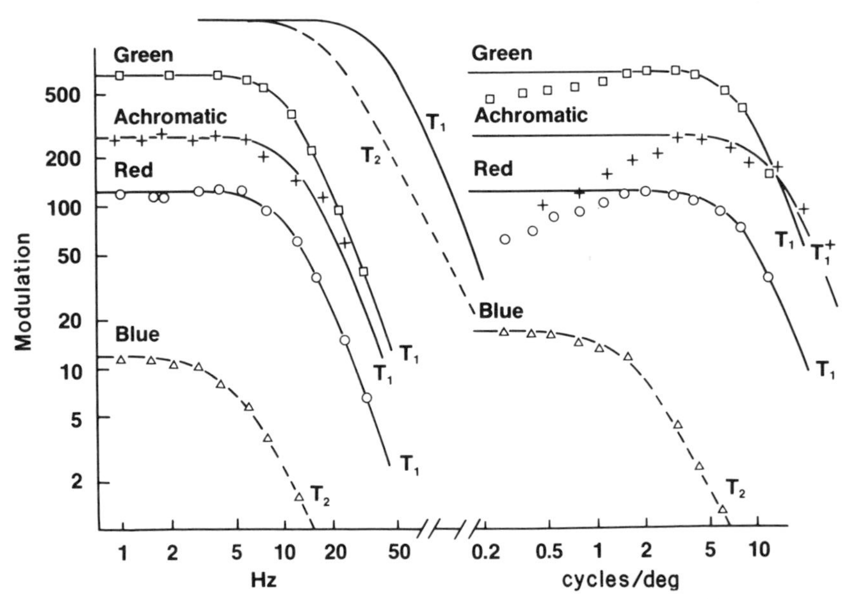

Fig. 50. The same data as are shown in Figs. 46 and 49 as sensitivities but all fitted with one of two template curves, T_1 and T_2, which are shown between the two sets of data. Neutral, "red," and "green" temporal and spatial sensitivity data are fitted with template curve T_1. Data for both temporal and spatial "blue" conditions are fitted with template curve T_2. Template curves represent two low-pass filters.

data for achromatic, red, green, and blue stimulation) have been replotted in Fig. 50. It is most likely that the achromatic sensitivity curve should be positioned at or above the level of the green sensitivity curve. However, because Fig. 50 is intended to illustrate only relative shapes of the sensitivity functions, the green and achromatic sensitivity curves are shown in the relation originally published by Kelly (1974). Two template curves appear to fit all the data, one for blue and one for achromatic, red, and green. The two template curves (labeled T_1 and T_2) are shown in the middle of Fig. 50. They have been drawn through the data points by allowing translation along ordinate and abscissa. Where there is essentially no inhibition, the template curve fits the data well. Where there is residual inhibition, with the achromatic, red, and green spatial frequency contrast thresholds, the data depart from the template curve. At relatively high frequencies, however, the fit is good even in these cases. Possibly, then, these templates suggest the shape of the low-pass filter that describes sensitivities of the excitatory processes of spatial and temporal vision.

All of the foregoing provides a reasonably cohesive picture of the spatio–temporal characteristics of the visual mechanism. Low-frequency inhibition occurs for both chromatic and achromatic stimulation. These effects can be manipulated by appropriate combinations of spatial and temporal patterns. Under certain conditions inhibition can be quenched; and when this happens, the underlying mechanisms appear similar to low-pass filters, so similar, in fact, that one template fits all but the blue data. Throughout most of the spatio–temporal domain, achromatic and green sensitivities are greatest, then red, and blue least of all. The bandwidths of the green and red spatial mechanisms do not differ much from the achromatic spatial bandwidth or from the corresponding temporal bandwidths. The blue mechanism is much narrower in both spatial and temporal dimensions, however. In this, as in most psychophysical results, blue is very much the "odd man out"; it is slow in response and gross in acuity.

G. CONCLUSIONS

The purpose of this chapter has been to provide a brief and simplified summary of the major features of the visual mechanism. To cover so broad a subject in a relatively short space, it has been necessary to indulge in many simplifications. We hope that what is presented will help those who are not familiar with the complexities of the visual mechanism to gain some useful insight into its workings. Some aspects of vision have been left out; a particularly important omission is a discussion of what is known about visual adaptation. However, this is a subject that will be treated in Chapter 12 of this volume. There is, therefore, some justification in treating the matter only by reference in this chapter.

The reader will appreciate by now that the eye is an extension of the brain and cannot be treated separately from the neural mechanisms of the brain as a whole. Through these neural mechanisms, reactions to light stimuli are processed in a variety of wondrous ways. Although the details of that processing elude precise description, a clear enough picture of the mechanism of vision emerges to reveal something of the care that must be exercised in making and evaluating measurements of light by visual means. Visual assessments, both casual and the most scientifically disciplined, provide us with valuable information. Their utility is highest when their meanings are clear, when the person making and evaluating visual measurements knows something of how they are formed. That, we hope, is where whatever value this chapter may have resides.

The reader who has followed the exposition carefully will not mistake a measure of CIE luminance for a measure of brightness. He will not expect to predict exactly the sharpness he perceives when viewing a photograph or a television image from convolutions of modulation transfer functions. He will not be puzzled when a textile sample appears differently colored when imbedded in different surroundings. In short, the reader should, by now, have learned to assess the conditions under which visual measurements are made and to try to interpret their influence on the visual mechanism in order to avoid drawing misleading conclusions about the meaning of those measurements.

The science of colorimetry has, for many years, provided a technique for making useful measurements: the determination of equality of appearance between two fields or samples displayed in the same or nearly the same visual field. That measurement taps only the most distal portion of the visual mechanism. When the photon catches are the same in the photoreceptor outer segments, we can see that all subsequent, more proximal stages must also yield matching responses. In effect, colorimetry cancels out the complexity of the visual mechanism, and in doing so it fails to provide additional useful information. One can obtain that additional information, however, by carefully designing measurement methods that allow us to assess precisely and quantitatively the relations between inequalities of distal stimulation and differences in visual responses. The science of linking stimulus and response is called *psychophysics.* It is a science that can deal with matters of matching, differences, and magnitudes or appearance resulting from visual responses. Psychophysics takes cognizance of the various factors that may affect operation of the visual mechanism and, properly practiced, incorporates methods for controlling those factors that are not of interest while permitting those that are of interest to vary systematically. It involves the derivation of scales along which responses are arrayed in isomorphism with attributes of physical stimulation. In short, psychophysics offers a systematic way of making visual measurements of all kinds.

For that reason Part II of this book will deal in some detail with psychophysical methodology and techniques for scaling visual perceptions. The four chapters in Part I have provided information about the eye, its sensitivity characteristics, and the mechanism of vision. That information will aid in understanding many of the reasons for the development of the psychophysical techniques that are the subject of the following chapters and will provide the bases for interpreting results of psychophysical and scaling experiments.

REFERENCES

Abramov, I. (1968). Further analysis of the responses of LGN cells. *J. Opt. Soc. Am.* **58**, 574–579.

Abramowitz, M., and Stegun, I. (1965). "Handbook of Mathematical Functions." Dover Press, New York.

Adrian, E. D. (1946). "The Physical Background of Perception." Clarendon Press, Oxford.

Alpern, M., and David, H. (1959). The additivity of contrast in the human eye. *J. Gen. Physiol.* **43**, 109–126.

Anstis, S. M., and Harris, J. P. (1974). Movement after effects contingent on binocular disparity. *Perception* **3**, 153–168.

Arend, L. E., Jr., and Lange, R. V. (1980). Narrow-band spatial mechanisms in apparent contrast matching. *Vision Res.* **20**, 143–147.

Barlow, H. B. (1958). Temporal and spatial summation in human vision at different background intensities. *J. Physiol. London* **141**, 337–350.

Barlow, H. B., and Levick, W. R. (1965). The mechanism of directionally selective units in rabbits' retina. *J. Physiol.* **178**, 477–504.

Bartleson, C. J. (1976) Brown. *Color Res. Appl.* **1**, 181–191.

Bartleson, C. J. (1979). Changes in color appearance with variations in chromatic adaptation. *Color Res. Appl.* **4**, 167–186.

Bartleson, C. J. (1981a). Measures of brightness and lightness. *Farbe* **28**, 132–148. (*NB*: No. 3/6 issued in 1981 but dated 1980.)

Bartleson, C. J. (1981b). On chromatic adaptation and persistence. *Color Res. Appl.* **6**, 153–160.

Baumgardt, E. (1949). Les théories photochimiques classiques et quantiques de la vision et l'inhibition nerveuse en vision liminaire. *Rev. Opt.* **28**, 453–478, 661–690.

Baumgardt, E., and Smith, S. W. (1967). Comparaison de la sensibilité des cônes et des bâtonnets de l'oeil humain. *Compte Rendu Acad. Sci. Paris* **264**, 3041–3044.

Baylor, D. A., and Hodgkin, A. L. (1974). Change in time scale and sensitivity in turtle photoreceptors. *J. Physiol.* **242**, 729–758.

Baylor, D. A., Hodgkin, A. L., and Lamb, T. (1974a). Reconstruction of the electrical responses of turtle cones to flashes and steps of light. *J. Physiol.* **242**, 759–791.

Baylor, D. A., Hodgkin, A. L., and Lamb, T. (1974b). The electrical response of turtle cones to flashes and steps of light. *J. Physiol.* **242**, 685–727.

Berry, R. N. (1948). Quantitative relations among vernier, real depth, and stereoscopic depth acuities. *J. Exp. Psychol.* **38**, 708–721.

Blakemore, C., and Campbell, F. W. (1969). On the existence of neurones in the human visual system selectively sensitive to the orientation and size of retinal images. *J. Physiol.* **203**, 237–260.
Blakemore, C., and Hague, B. (1973). Evidence for disparity detecting neurones in the human visual system. *J. Physiol.* **225**, 437–455.
Blakemore, C., Muncey, J. P. J., and Ridley, R. M. (1973). Stimulus specificity in the human visual system. *Vision Res.* **13**, 1915–1931.
Bloch, A. M. (1885). Expériences sur la vision. *Compte Rendu Soc. Biol. Paris* **37**, 493–495.
Blondel, A. M., and Rey, J. (1911). Sur la perception des lumières brèves à la limite de leur portée. *J. Phys.* **1**, 530–550.
Boynton, R. M. (1979). "*Human Color Vision.*" Holt, Rinehart and Winston, New York.
Boynton, R. M., and Whitten, D. N. (1970). Visual adaptation in monkey cones: recordings of late receptor potentials. *Science* **170**, 1423–1426.
Boynton, R. M., Kandel, G., and Onley, J. W. (1959). Rapid chromatic adaptation of normal and dichromatic observers. *J. Opt. Soc. Am.* **49**, 654–666.
Bruesch, S. R., and Arey, L. B. (1942). The number of myelinated and unmyelinated fibers in the optic nerve of vertebrates. *J. Comp. Neurol.* **77**, 631–635.
Bryngdahl, O. (1966). Perceived contrast variation with eccentricity and spatial sine-wave stimuli: Size determination of receptive field centres. *Vision Res.* **6**, 553–565.
Burnham, R. W., Hanes, R. M., and Bartleson, C. J. (1963). "Color: A Guide to Basic Facts and Concepts." J. Wiley and Sons, New York.
Campbell, F. W., and Gubisch, R. W. (1966). Optical quality of the human eye. *J. Physiol. London* **186**, 558–578.
Campbell, F. W. and Robson, J. G. (1968). Application of Fourier analysis to the visibility of gratings. *J. Physiol. London* **197**, 551–566.
Campbell, F. W., Howell, E. R., and Johnstone, J. R. (1978). A comparison of threshold and suprathreshold appearance of gratings with components in the low and high spatial frequency range. *J. Physiol. London* **284**, 193–201.
Campbell, F. W., Johnstone, J. R., and Ross, J. (1981). An explanation for the visibility of low-frequency gratings. *Vision Res.* **21**, 723–730.
Campbell, F. W., Kulikowski, J. J., and Levinson, J. (1966). The effect of orientation on the visual resolution of gratings. *J. Physiol. London* **187**, 427–436.
Cannon, M. W., Jr. (1979). Contrast sensation: a linear function of stimulus contrast. *Vision Res.* **19**, 1045–1052.
Chase, W., and Smith, R. (1981). Spatial frequency channels tuned for depth and motion. *Vision Res.* **21**, 621–625.
Cicerone, C. M., Krantz, D. H., and Larimer, J. (1975). Opponent-process additivity. III. Effect of moderate chromatic adaptation. *Vision Res.* **15**, 1125–1135.
Clark, W. C., and Blackwell, H. R. (1959). "Relations between visibility thresholds for single and double pulses." Univ. Michigan Report No. 2, pp. 144–343-T.
Cornsweet, T. N. (1970). "Visual Perception." Academic Press, New York.
Cowan, A. (1928). Test cards for determination of visual acuity. *Br. J. Psychol.* **29**, 252–266.
Crawford, B. H. (1949). The scotopic visibility function. *Phys. Soc. Proc.* **62**, 321–334.
Davidson, M. L. (1968). Perturbation approach to spatial brightness interaction in human vision. *J. Opt. Soc. Am.* **58**, 1300–1309.
Davson, H. (1962). "The Eye," Vol. 2. Academic Press, New York.
deLange, H. (1952). Experiments on flicker and some calculations on an electrical analogue of the foveal systems. *Physica* **18**, 935–950.
Denton, E. J., and Pirenne, M. H. (1954). The absolute sensitivity and functional stability of the human eye. *J. Physiol. London* **123**, 417–442.

DeValois, R. L. (1965a). Behavioral and electrophysiological studies of primate vision. *In* "Contributions to Sensory Physiology" (W. D. Neff, ed.), Vol. 1, Academic Press, New York.

DeValois, R. L. (1965b). Analysis and coding of color vision in the primate visual system. *Cold Spring Harbor Symp. Quant. Biol.* **30**, 567–579.

DeValois, R. L., and DeValois, K. K. (1975). Neural coding of color. *In* "Handbook of Perception" (E. C. Carterette and M. P. Friedman, eds.), Vol. 5. Academic Press, New York.

DeValois, R. L., Smith, C. J., Kitai, S. T., and Karoly, S. J. (1958). Responses of single cells in different layers of the primate lateral geniculate nucleus to monochromatic light. *Science* **127**, 238–239.

Dowling, J. E., and Boycott, B. B. (1966). Organization of the primate retina: Electron microscopy. *Proc. R. Soc. London B* **166**, 80–111.

Dowling, J. E., and Ehringer, B. (1976). The interplexiform cell: A new type of retinal neuron. *Invest. Ophthalmol.* **15**, 916–926.

Eisner, A., and McLeod, D. I. A. (1980). Blue sensitive cones do not contribute to luminance. *J. Opt. Soc. Am.* **70**, 121–123.

Enroth-Cugell, C., and Robson, J. G. (1966). The contrast sensitivity of retinal ganglion cells of the cat. *J. Physiol.* **187**, 517–552.

Estévez, O. (1979). "On the Fundamental Data-Base of Normal and Dichromatic Color Vision." Ph.D. dissertation, University of Amsterdam, The Netherlands.

Evans, R. M. (1974). "The Perception of Color." J. Wiley and Sons, New York.

Fourier, J. B. J. (1822). "Théorie Analytique de la Chaleur." F. Didot, Paris.

Franzen, O., and Berkley, M. (1975). Apparent contrast as a function of modulation depth and spatial frequency: A comparison between perceptual and electrophysiological measures. *Vision Res.* **15**, 655–660.

Georgeson, M. A., and Sullivan, G. D. (1975). Contrast constancy: Deblurring in human vision by spatial frequency channels. *J. Physiol. London* **252**, 627–656.

Gibson, K. S., and Tyndall, E. P. T. (1923). Visibility of radiant energy. *Bull. Natl. Bur. Stand.* **19**, 131–191.

Ginsburg, A. P. (1977). "Visual Information Processing Based on Spatial Filters Constrained by Biological Data." Ph.D. Dissertation, University of Cambridge, England.

Ginsburg, A. P., Cannon, M. W., and Nelson, M. A. (1980). Suprathreshold processing of complex visual stimuli: Evidence for linearity in contrast perception. *Science* **208**, 619–621.

Gouras, P. (1968). Identification of cone mechanisms in monkey ganglion cells. *J. Physiol.* **199**, 533–547.

Graham, C. H., and Margaria, R. (1935). Area and the intensity-time relation in the peripheral retina. *Am. J. Physiol.* **113**, 299–305.

Green, D. G., Dowling, J. E., Siegel, I. M., and Ripps, H. (1975). Retinal mechanisms of visual adaptation in the skate. *J. Gen. Physiol.* **65**, 483–502.

Guild, J. (1925–1926). The colorimetric properties of the spectrum. *Philos. Trans. Roy. Soc. London* **230A**, 149–187.

Guth, S. L. (1970). Photometric and colorimetric additivity at various intensities. *In* M. Richter (ed.), *AIC Proceedings Color 69*. Musterschmidt-Verlag, Gottingen.

Guth, S. L. (1972). A new color model. *In* "Color Metrics: Proceedings of the Helmholtz Memorial Symposium on Color Metrics 1971" (J. J. Vos, L. F. C. Friele, and P. L. Walraven, eds.). AIC/Holland, Soesterberg, The Netherlands.

Guth, S. L., and Lodge, H. R. (1973). Heterochromatic additivity, foveal spectral sensitivity, and a new color model. *J. Opt. Soc. Am.* **63**, 450–462.

Guth, S. L., Donley, N. V., and Marrocco, R. T. (1969). On luminance additivity and related topics. *Vision Res.* **9**, 537–575.

Guth, S. L., Massof, R. W., and Benzschawel, T. (1980). A vector model for normal and dichromatic color vision. *J. Opt. Soc. Am.* **70**, 197–212.

Hamerly, J. R., Quick, R. F., Jr., and Reichert, R. A. (1977). A study of grating contrast judgment. *Vision Res.* **17**, 201–207.

Hartline, H. K. (1938). The response of single optic nerve fibers of the vertebrate eye to illumination of the retina. *Am. J. Physiol.* **121**, 400–415.

Harwerth, R. S., and Levi, D. M. (1978). Reaction time as a measure of suprathreshold grating detection. *Vision Res.* **18**, 1579–1586.

Hecht, S., and Mintz, E. U. (1939). The visibility of single lines at various illuminations and the retinal basis of visual resolution. *J. Gen. Physiol.* **22**, 593–612.

Hecht, S., Shlaer, S., and Pirenne, M. H. (1942). Energy quanta, and vision. *J. Gen. Physiol.* **25**, 819–840.

Helmholtz, H. von (1860). Handbuch der physiologischen Optic. Band 2, Voss, Hamburg. (2nd ed. in 1896, 3rd. ed. in 1911; 2nd ed. translated by J. P. C. Southall as Handbook of Physiological Optics." Optical Society of America, Rochester, N.Y. 1924).

Hering, E. (1872). Zur Lehre vom Lichtsinne, I. Über successiven Lichtinduction. *S.-B. Akad. Wiss. Wien, Math.-Nat. Kl. Part III* **66**, 5–24.

Hering, E. (1874a). Zur Lehre vom Lichtsinne. II. Über simultanen Lichtcontrast. *S.-B. Akad. Wiss. Wien, Math.-Nat. Kl. Part III* **68**, 186–201.

Hering, E. (1874b). Zur Lehre vom Lichtsinne. III. Über simultanen Lichtinduction und über successiven Contrast. *S.-B. Akad. Wiss. Wien, Math.-Nat. Kl. Part III* **68**, 229–244.

Hering, E. (1874c). Zur Lehre vom Lichtsinne. IV. Über die sogenannte Intensität der Lichtempfindung und über die Empfindung des Schwarzen. *S.-B. Akad. Wiss. Wien, Math.-Nat. Kl. Part III* **69**, 85–104.

Hering, E. (1874d). Zur Lehre vom. Lichtsinne. V. Grundzüge einer Theorie des Lichtsinnes *S.-B. Akad. Wiss. Wien, Math.-Nat. Kl. Part III* **69**, 179–217.

Hering, E. (1875). Zur Lehre vom Lichtsinne. VI. Grundzüge einer Theorie des Farbensinnes *S.-B. Akad. Wiss. Wien, Math.-Nat. Kl. Part III* **70**, 169–204.

Hering, E. (1878). "Zur Lehre vom Lichtsinne." Carl Gerold's Sohn, Wien.

Herick, R. M. (1956). Foveal luminance discrimination as a function of the duration of the decrement or increment in luminance. *J. Comp. Physiol. Psychol.* **49**, 437–443.

Hochstein, S., and Shapley, R. M. (1976a). Quantitative analysis of retinal ganglion cell classifications. *J. Physiol.* **262**, 237–264.

Hochstein, S., and Shapley, R. M. (1976b). Linear and nonlinear spatial subunits in Y cat retinal ganglion cells. *J. Physiol.* **262**, 265–284.

Hsia, Y., and Graham, C. H. (1957). Spectral luminosity curves of protanopic, deuteranopic, and normal subjects. *Proc. Natl. Acad. Sci. U.S.A.* **43**, 1011–1019.

Hubel, D. H., and Wiesel, T. N. (1959). Receptive fields of single neurones in the cat's striate cortex. *J. Physiol.* **148**, 574–591.

Hubel, D. H., and Wiesel, T. N. (1961). Integrative action in the cat's lateral geniculate body. *J. Physiol.* **155**, 385–398.

Hubel, D. H., and Wiesel, T. N. (1962). Receptive fields, binocular interaction and functional architecture in the cat's visual cortex. *J. Physiol.* **160**, 106–154.

Hubel, D. H., and Wiesel, T. N. (1963). Shape and arrangement of columns in cat's striate cortex. *J. Physiol.* **165**, 559–568.

Hubel, D. H., and Wiesel, T. N. (1965). Receptive fields and functional architecture in two non-striate visual areas (18 and 19) of the cat. *J. Neurophysiol.* **28**, 229–289.

Hubel, D. H., and Wiesel, T. N. (1968). Receptive fields and functional architecture of the monkey striate cortex. *J. Physiol.* **195**, 215–243.

Hubel, D. H., and Wiesel, T. N. (1974). Sequence regularity and geometry of orientation columns in the monkey striate cortex. *J. Comp. Neurol.* **158**, 267–294.

Hurvich, L. M. (1960). The opponent-pairs scheme. *In* "Mechanisms of Colour Discrimination" (Y. Galifret, ed.), Pergamon Press, London.

Hurvich, L. M. (1978). Two decades of opponent processes. *In* "AIC Color 77" (F. W. Billmeyer, Jr. and G. Wyszecki, eds.). Adam Hilger, Bristol.

Hurvich, L. M. (1981). "Color Vision." Sinauer Assoc., Sunderland, Mass.

Hurvich, L. M., and Jameson, D. (1955). Some quantitative aspects of an opponent-colors theory. II. Brightness, saturation, and hue in normal and dichromatic vision. *J. Opt. Soc. Am.* **45**, 602–616.

Hurvich, L. M., and Jameson, D. (1961). Opponent chromatic induction and wavelength discrimination. *In* "The Visual System: Neurophysiology and Psychophysics" (R. Jung and H. Kornhuber, eds.). Springer-Verlag, Berlin.

Hurvich, L. M., and Jameson, D. (1974). Opponent processes as a model of neural organization. *Am. Psychol.* **29**, 88–102.

Hurvich, L. M., Jameson, D., and Cohen, J. D. (1968). The experimental determination of unique green in the spectrum. *Percept. Psychophys.* **4**, 65–68.

Ikeda, M., and Ayama, M. (1980). Additivity of opponent chromatic valence. *Vision Res.* **20**, 995–999.

Ingling, C., Jr. (1977). The spectral sensitivity of the opponent-color channels. *Vision Res.* **17**, 1083–1089.

Ingling, C. R., Jr., and Drum, B. A. (1977). Why the blue arcs of retina are blue. *Vision Res.* **17**, 498–500.

Ingling, C. R., Jr., and Tsou, B. H-P. (1977). Orthogonal combinations of the three visual channels. *Vision Res.* **17**, 1075–1082.

Ingling, C. R., Jr. Burns, S. A., and Drum, B. S. (1977). Desaturating blue increases only chromatic brightness. *Vision Res.* **17**, 501–503.

Iversen, L. L. (1979). The chemistry of the brain. *Sci. Am.* **241**, 134–149.

Ives, H. E. (1922). A theory of intermittent vision. *J. Opt. Soc. Am. Rev. Sci. Instrum.* **6**, 343–361.

Jacobs, G. H., and Wascher, T. C. (1967). Bezold-Brücke hue shift: Further measurements. *J. Opt. Soc. Am.* **57**, 1155–1156.

Jameson, D., and Hurvich, L. M. (1959). Perceived color and its dependence on focal, surrounding, and preceding stimulus variables. *J. Opt. Soc. Am.* **49**, 890–898.

Jameson, D., and Hurvich, L. M. (1961a). Opponent chromatic induction: Experimental evaluation and theoretical account. *J. Opt. Soc. Am.* **51**, 46–53.

Jameson, D., and Hurvich, L. M. (1961b). Complexities of perceived brightness. *Science* **133**, 174–179.

Jameson, D., and Hurvich, L. M. (1964). Theory of brightness and color contrast in human vision. *Vision Res.* **4**, 135–154.

Jameson, D., and Hurvich, L. M. (1968). Opponent-response functions related to measured cone photopigments. *J. Opt. Soc. Am.* **58**, 429–430.

Jameson, D., and Hurvich, L. M. (1972). Color adaptation: Sensitivity, contrast, after-images. *In* "Handbook of Sensory Physiology: Visual Psychophysics" VII/4. (D. Jameson and L. M. Hurvich, eds.). Springer-Verlag, Berlin.

Jameson, D., and Hurvich, L. M. (1975). From contrast to assimilation in art and in the eye. *Leonardo* **8**, 125–131.

Johnson, E. P., and Bartlett, N. R. (1956). Effect of stimulus duration on electrical response of the human retina. *J. Opt. Soc. Am.* **46**, 167–170.

Jones, R. C. (1958). On the point and line spread functions of photographic images. *J. Opt. Soc. Am.* **48**, 934–937.

Jones, R. and Keck, M. J. (1978). Visual evoked response as a function of grating spatial frequency. *Invest. Ophthalmol. Visual Sci.* **17**, 652–659.

Judd, D. B. (1944). Standard response functions for protanopic and deuteranopic vision. *J. Res. Natl. Bur. Stand.* **33**, 407–437.

Judd, D. B. (1951). International Commission on Illumination, Technical Committee No. 7, colorimetry and artificial daylight. Report of Secretariat, United States Committe. *In* "CIE Compte rendu douzième session. Stockholm, CIE, Paris.

Kaiser, P. K., and Comerford, J. P. (1975). Flicker photometry of equally bright lights. *Vision Res.* **15**, 1399–1402.

Kaiser, P. K., and Smith, P. (1972). The luminance of equally bright colors. *Compte rendu CIE 1971, Barcelona*, pp. 143–144.

Kaiser, P. K., and Wyszecki, G. (1978). Additivity failures in heterochromatic brightness matching. *Color Res. Appl.* **3**, 177–182.

Kandel, E. R. (1979). Small systems of neurons. *Sci. Am.* **241**, 67–76.

Kelly, D. H. (1966). Frequency doubling in visual responses. *J. Opt. Soc. Am.* **56**, 1628–1633.

Kelly, D. H. (1969). Diffusion model of linear flicker responses. *J. Opt. Soc. Am.* **59**, 1665–1670.

Kelly, D. H. (1974). Spatio-temporal frequency characteristics of color-vision mechanisms. *J. Opt. Soc. Am.* **64**, 983–990.

Kingsmith, P. E., and Kulikowski, J. J. (1975). The detection of gratings by independent activation of line detectors. *J. Physiol.* **247**, 237–271.

Koenig, A., and Dieterici, C. (1982). Die Grundempfindungen in normalen und anomalen Farbensystemen und ihre Intensitäsverteilung im Spektrum. *Z. Psychol. Physiol. Sinnesorg* **4**, 241–347.

Kolb, H. (1970). Organization of the outer plexiform layer of the primate retina: Electron microscopy of Golgi-impregnated cells. *Philos. Trans. Roy. Soc. London B* **258** 261–283.

Kuffler, S. W., and Nicholls, J. S. (1976). "From Neuron to Brain." Sinauer Assoc., Sunderland, Mass.

Kulikowski, J. J. (1976). Effective contrast constancy and linearity of contrast sensation. *Vision Res.* **16**, 1419–1431.

Kulikowski, J. J., and Kingsmith, P. E. (1973). Spatial arrangement of line, edge and grating detectors revealed by subthreshold summation. *Vision Res.* **13**, 1455–1478.

Landolt, E. (1889). Tableau d'optotypes pour la determination de l'acuité visuelle. *Soc. Francais Ophthalmol.* p. 157.

Larimer, J., Krantz, D. H., and Cicerone, C. M. (1974). Opponent-process additivity. I. Red/green equilibria. *Vision Res.* **14**, 1127–1140.

Larimer, J., Krantz, D. H., and Cicerone, C. M. (1975). Opponent-process additivity. II. Yellow/blue equilibria and non-linear models. *Vision Res.* **15**, 723–731.

Llinás, R. R. (1975). The cortex of the cerebellum. *Sci. Am.* **37**, 56–71.

Lowry, E. M., and DePalma, J. J. (1961). Sine-wave response of the visual system. I. The Mach phenomenon. *J. Opt. Soc. Am.* **51**, 740–746.

Ludvigh, E., and McCarthy, E. F. (1938). Absorption of visible light by the refractive media of the human eye. *Arch. Ophthalmol.* **20**, 37–51.

Mach, E. (1868). Über die physiologische Wirkung räumlich vertheilter Lichtreize (Vierte Abhandlung). *Sitzungsber. Math. Naturwiss. Kl. Kaiserlichen Akad. Wiss. Wien* **57/2**, 11–19.

Michael, C. R. (1969). Retinal processing of visual images. *Sci. Am.* **220**, 104–114.

Mountcastle, V. B. (1957). Modality and topographic properties of single neurons of cat's somatic sensory cortex. *J. Neurophysiol.* **20**, 408–434.

Naka, K. I., and Rushton, W. A. H. (1966). S-potentials from colour units in the retina of fish (Cyprinidae). *J. Physiol. London* **185**, 536–555.

Nauta, W. J. H., and Feirtag, M. (1979). The organization of the brain. *Sci. Am.* **241**, 88–111.

Normann, R. A., and Werblin, F. S. (1974). Control of retinal sensitivity. I. Light and dark adaptation of vertebrate rods and cones. *J. Gen. Physiol.* **63**, 37–61.

Pickering, W. A. (1915). Report on Mars. *Pop. Astron.* **23**, 569–588.

Piper, H. (1903). Über die Abhängigkeit des Reizwertes leuchtender Objekte von ihrer Flächenbezw. Winkelgrosse. *Z. Psych. Physiol. Sinnesorg.* **32**, 98–112.

Pitt, F. H. G. (1935). "Characteristics of Dichromatic Vision" (Medical Research Council, Report of the Committee on Physiology of Vision No. XIV). His Majesty's Stationery Office, London.

Quick, Jr., R. F., Hamerly, J. R., and Reichert, T. A. (1976). The absence of a measurable "critical band" at low suprathreshold contrasts. *Vision Res.* **16**, 351–355.

Ramón y Cajal, S. (1955). "Histologie du système nerveux," Vol. II. C. S. I. C., Madrid.

Ratliff, F. (1965). "Mach Bands: Quantitative Studies on Neural Networks in the Retina." Holden-Day, San Francisco.

Ricco, A. (1877). Relazione fra il minimo angolo visuale e l'intensità luminosa. *Ann. Ottal.* **6**, 373–479.

Richards, W. (1967). Differences among color normals: Classes I and II. *J. Opt. Soc. Am.* **57**, 1047–1055.

Richards, W., and Luria, S. M. (1964) Color-mixture functions at low luminance levels. *Vision Res.* **4**, 281–313.

Romeskie, M. I. (1976). "Chromatic Opponent-Response Functions of Anomalous Trichromats." Ph.D. dissertation, Brown University, Providence, R. I. (University Microfilms, Ann Arbor, Michigan.)

Romeskie, M. I. (1978). Chromatic opponent-response functions of anomalous trichromats. *Vision Res.* **18**, 1521–1532.

Rose, S. (1976). "The Conscious Brain," rev. ed. Penguin, London.

Rubin, M. L. (1961). Spectral hue loci of normal and anomalous trichromats. *Am. J. Ophthalmol.* **52**, 166–172.

Rushton, W. A. H. (1972). Visual pigments in man. *In* "Handbook of Sensory Physiology" (H. J. A. Dartnall, ed.), Vol. VII/1. Springer-Verlag, New York.

Sachs, M. B., Nachmias, J., and Robson, J. G. (1971). Spatial-frequency channels in human vision. *J. Opt. Soc. Am.* **61**, 1176–1186.

Sanders, C. L., and Wyszecki, G. (1957). Correlate for lightness in terms of C. I. E. tristimulus values, Part I. *J. Opt. Soc. Am.* **47**, 398–404.

Sanders, C. L., and Wyszecki, G. (1964). Correlate for brightness in terms of CIE colour-matching data. *Compte rendu CIE Wien, 1963* 221–230.

Schwartz, J. H. (1980). The transport substances in nerve cells. *Sci. Am.* **242**, 152–171.

Shantz, M., and Naka, K. I. (1976). The bipolar cell. *Vision Res.* **16**, 1517–1518.

Sjöstrand, F. S. (1953). The ultrastructure of the outer segments of rods and cones of the eye as revealed by the electron microscope. *J. Cell. Comp. Physiol.* **42**, 15–44.

Snellen, H. (1862). "Probebuchstaben zur Bestimmung der Sehschärfe." P. W. van de Wiejer, Utrecht.

Sperling, H. G., and Harwerth, R. S. (1971). Red-green cone interactions in the increment-threshold spectral sensitivity of primates. *Science* **172**, 180–184.

Stevens, C. F. (1979). The neuron. *Sci. Am.* **241**, 55–65.

Stiles, W. S., and Burch, J. M. (1955). Interim report to the Commission Internationale de l'Eclairage, Zurich, 1955, on the National Physical Laboratory's investigation of colour-matching. *Opt. Acta* **2**, 168–174.

Stiles, W. S., and Burch, J. M. (1959). N. P. L. colour-matching investigation: Final report, 1958. *Opt. Acta* **6**, 1–26.

Svaetichin, G. (1956). Spectral response curves of single cones. *Acta Physiol. Scand.* **39** (suppl. 134), 17–46.

Szentágothai, J. (1973). Neuronal and synaptic architecture of the lateral geniculate nucleus. *In* "Handbook of Sensory Physiology: Central Visual Information" (H. H. Kornhuber, ed.), Vol. VI. Springer-Verlag, Berlin.

Thomas, J. G. (1965). Threshold measurements of Mach bands. *J. Opt. Soc. Am.* **55**, 521–524.

Tolhurst, D. J. (1973). Separate channels for the analysis of the shape and the movement of a moving visual stimulus. *J. Physiol.* **231**, 385–403.

Trezona, P. W. (1970). Rod participation in the "blue" mechanism and its effect on colour matching. *Vision Res.* **10**, 317–332.

Van der Velden, H. (1944). Over het antaal lichtquanta dat nodig es for en licht prickel bij dat menselyk oog. *Physica* **11**, 179–189.

Van Nes, F. L., and Bouman, M. A. (1967). Spatial modulation transfer in the human eye. *J. Opt. Soc. Am.* **57**, 401–406.

Vos, J. J. (1978). Colorimetric and photometric properties of a 2° fundamental observed. *Color Res. Appl.* **3**, 125–128.

Vos, J. J., and Walraven, P. L. (1971). On the derivation of the foveal receptor primaries. *Vision Res.* **11**, 799–818.

Vos, J. J., and Walraven, P. L. (1972). A zone-fluctuation line element describing colour discrimination. *In* "Color Metrics, Proceedings of the Helmholtz Memorial Symposium on Color Metrics, 1971" (J. J. Vos, L. F. C. Friele, and P. L. Walraven, eds.). AIC/Holland, Soesterberg, The Netherlands.

Walls, G. L. (1943). Factors in human visual resolution. *J. Opt. Soc. Am.* **33**, 487–505.

Wald, G. (1938). Area and visual threshold. *J. Gen. Physiol.* **21**, 269–287.

Wald, G., and Brown, P. K. (1953). The molar extinction of rhodopsin. *J. Gen. Physiol.* **37**, 189–200.

Wald, G., Brown, P. K., and Gibbons, I. R. (1963). The problem of visual excitation. *J. Opt. Soc. Am.* **53**, 20–35.

Walraven, P. L. (1962). "On the Mechanisms of Colour Vision." Ph.D. dissertation, Institute for Perception RVO-TNO, Soesterberg, The Netherlands.

Walraven, P. L. (1976). Basic mechanisms of defective colour vision. *In* "Modern Problems in Ophthalmology" (E. B. Streiff, ed.), Vol. 17. Karger, Basel.

Watanabe, A., Mori, T., Nagata, S., and Hiwatashi, K. (1968). Spatial sine-wave responses of the human visual system. *Vision Res.* **8**, 1245–1263.

Werblin, F. S. (1973). The control of sensitivity in the retina. *Sci. Am.* **228**, 70–79.

Werblin, F. S. (1974). Control of retinal sensitivity-II. Lateral interactions at the outer plexiform layer. *J. Gen. Physiol.* **63**, 62–87.

Werblin, F. S., and Copenhagen, D. R. (1974). Control of retinal sensitivity-III. Lateral interactions at the inner plexiform layer. *J. Gen. Physiol.* **63**, 88–110.

Werner, J. S., and Wooten, B. R. (1979). Opponent chromatic mechanisms: Relation to photopigments and hue naming. *J. Opt. Soc. Am.* **69**, 422–434.

Wilcox, W. W., and Purdy, D. M. (1933). Visual acuity and its physiological basis. *Br. J. Psychol.* **23**, 233–261.

Wright, W. D. (1928/1929). A re-determination of the mixture curves of the spectrum. *Trans. Opt. Soc. London* **30**, 141–164.

Wright, W. D. (1946). "Researches on Normal and Defective Colour Vision." Kimpton, London.

Wyszecki, G., and Stiles, W. S. (1967). "Color Science," J. Wiley & Sons, New York.

Wyszecki, G., and Stiles, W. S. (1982). "Color Science," 2nd ed. Wiley-Interscience, New York.

Young, R. W. (1970). Visual cells. *Sci. Am.* **223**, 81–91.

Young, T. (1807a). On the theory of light and colours. *In* T. Young, "Lectures in Natural Philosophy" Vol. 2. London, privately printed for Joseph Johnson, St. Paul's Churchyard, by William Savage. (Reprinted in *Philos. Trans. Roy. Soc. London* **92**, 20–71, 1820.)

Young, T. (1807b). An account of some cases of the production of colours. *In* T. Young, "Lectures in Natural Philosophy" Vol. 2. London, privately printed for Joseph Johnson, St. Paul's Churchyard, by William Savage. (Reprinted in *Philos. Trans. Roy. Soc. London* **92**, 387–397, 1820.)

Zeki, S. M. (1974). Functional organization of a visual area in the posterior bank of the superior temporal sulcus of the rhesus monkey. *J. Physiol.* **236**, 549–573.

PART

Psychophysical Methods

6

Psychophysics

ROBERT M. BOYNTON
Department of Psychology
University of California, San Diego
La Jolla, California

A. THE PROBLEMS OF SUBJECTIVE MEASUREMENT

1. Introduction

While lying in bed on an October morning in 1850, Gustav Theodor Fechner, a German physicist and philosopher, was suddenly inspired by the thought that it might be possible to develop "an exact theory of the relation of

ISBN-0-12-304905-9

body and mind" (Boring, 1955). Ten years later, in his *Elements of Psychophysics*, Fechner (1966) clearly demonstrated that a kind of subjective measurement is possible. His book provides a detailed record of his ideas and concerns about psychophysical methodology. It also includes descriptions of his own extensive experiments and anticipates many of the problems to be discussed in this chapter, more than a century later.

Today visual psychophysics, like experimental psychology, is both a collection of methods and a subject matter. That visual psychophysics has a substantive, nonmethodological content is illustrated by the fact that one of the twelve sections of the Association for Research in Vision and Ophthalmology, the largest organization of visual scientists in the world, is called Visual Psychophysics and Physiological Optics. Each year, papers are read in section meetings on topics such as motion and perception, chromatic mechanisms, visual adaptation, and binocular vision.

From this substantive viewpoint visual psychophysics concerns the study of lawful stimulus–response relationships and theoretical concepts about explanatory mechanisms. For the purpose of developing such knowledge the methods of psychophysics are seen as one means (among many) toward understanding vision, but not as ends in themselves. To the psychometrician or mathematical psychologist, by contrast, methodology for its own sake can become a primary and sometimes obsessive interest, leading to a concern with procedural and calculational details that may not matter very much for those who actually apply psychophysical methods to the investigation of visual problems.

As a result, those with substantive visual interests may become bored and impatient with the efforts of the methodologists. This is nothing new. For example, Guilford (1954, p. 9) quotes William James as saying in 1890 that "Fechner's book was the starting point of a new department of literature, which it would perhaps be impossible to match for the qualities of thoroughness and subtlety, but of which, in the humble opinion of the present writer, the proper psychological outcome is just *nothing*." To the extent that James's negative viewpoint is directed at a brand of psychophysics concerned with method at the expense of any substance at all, his statement is correct by definition. But this is seldom the case. The author's interest lies in visual mechanisms first, but methodology is a not-too-distant second. Actually, these cannot be completely divorced.

Psychophysical methods, when properly employed, have proven their capacity to produce a valuable corpus of accurate information about the senses. These wholly noninvasive procedures have a special value, because they can be freely applied to human subjects. Moreover, these methods have had an important impact by pointing the way toward improved stimulus control and experimental design in electrophysiology, in which there once was a tendency to concentrate almost entirely on recording methods (Boynton, 1961).

Fechner invented a method that we shall call the "null instrument approach to classical psychophysics." His inspired idea was to define sensitivity in physical units, specifying it for a particular condition as the reciprocal of an amount of a physical quantity required to elicit a criterion subjective response. The method requires that a continuous and monotonic relation exist, as it typically does, between a measurable physical quantity and a correlated subjective intensity. Although no attempt need be made to estimate (let alone measure) the amount of a sensation serving as the response criterion, some means must be found to allow the observer to find and maintain a stable criterion of subjective intensity over the course of an experiment.

The development of various null-instrument procedures constitutes Fechner's greatest contribution to psychophysics. However, Fechner is much better known for the result of his inspiration on that October morning, namely, his controversial attempt to measure sensation itself. By 1850 it had already been established, especially in experiments with lifted weights conducted by E. H. Weber, that thresholds of sensory difference could be measured in physical units at various points along a continuum of physical intensity. Fechner added the critical assumption that such physical threshold steps should correspond to equal sensory steps. If so, a scale of increasing sensation can be created by summing these steps as a function of stimulus intensity. The idea behind this is illustrated for discrete steps in Fig. 1.

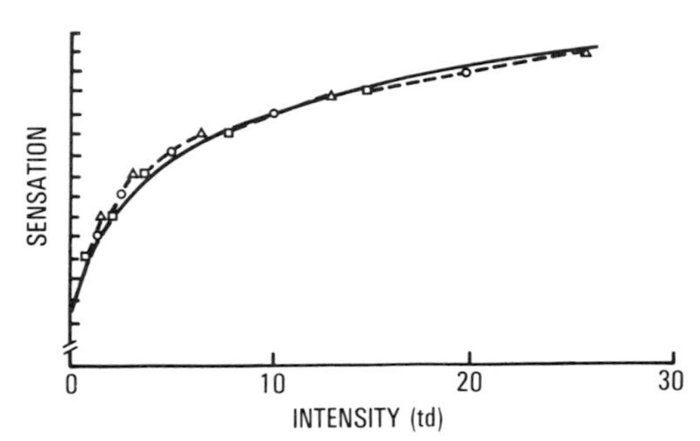

Fig. 1. An example of a Fechnerian-type construction of a sensation versus intensity function using data of Boynton and Kambe (1980). The experiment consisted of comparing two fields of equal intensity and then determining the value of intensities required to produce discriminable steps in both upward and downward directions. The ordinate consists of a series of equal intervals of arbitrary size. Construction starts with the point plotted above 20 td (circle). The triangle to the right is plotted on ordinate one interval higher, the square to the left, one interval lower. Starting at 10 td, the upward triangle is at an intensity nearly coincident with the downward square from 20 td; one assumes the abscissa difference to be insignificant, so they are plotted at the same ordinate value. The construction continues in similar fashion downward and to the left. The fitted curve is $y = 8 \log x$.

Weber's law (expressed in fractional form as $\Delta I/I = K$) was adopted by Fechner and was assumed to apply in the limit ($dI/I = K$), from which it follows by mathematical integration that sensation increases linearly as a function of the logarithm of stimulus intensity. This is Fechner's law; that is, equal stimulus ratios are predicted to elicit equal sensory differences. An avalanche of controversy followed, much of it concerned with the unproven yet crucial assumption that equal thresholds of difference, wherever they occur along a physical continuum, necessarily imply equal steps of sensation. Also under fire was the related question of whether it is possible in principle to develop a method capable of testing such an assumption (Johnson, 1929, 1945; Cobb, 1932).

2. Viewpoints Concerning Subjective Measurement

The conditions of measurement that are met in the physical sciences, sometimes called the "exact" sciences precisely because they are not particularly controversial, were fully understood in Fechner's day. Formal statements of what constitutes proper measurement are widely available (for example, see Cohen and Nagel, 1934, p. 297). The measurement of light intensity provides a relevant example. The amount of light in a monochromatic flash is proportional to the number of photons that the flash contains, and these can be physically counted. Light flashes, therefore, can be ordered according to how many photons are represented. The ordering is transitive, and unequivocal statements can also be made about differences and ratios anywhere along the scale of light intensity.

Some physical scales of measurement, though yielding transitive values, are much cruder than this. For example, hardness scales have been developed according to the ability of materials to scratch one another. These determinations may constitute a form of measurement, but in a much more limited sense than for light intensity. The scratch procedure does not permit a quantitative statement concerning the relative hardnesses of two classes of materials such as chalk and glass. To measure properly the hardness of a material requires that a unit of hardness be specified; moreover, some operation must be found to permit a determination of how many hardness units are to be counted before the hardness that is characteristic of a material is reached.

It might be thought that a solution to the hardness problem would be to take a diamond as the hardest substance, fracture it to a specified point, and then draw it across the smooth surfaces of other substances at a controlled rate and pressure. The softer a surface, the deeper would be the groove scratched into it by the diamond point. To provide a measure of hardness, materials could then be ordered according to depths of grooves. But if the groove in glass were one tenth that in chalk, could one make the claim that glass is 10 times as hard as chalk? The claim might *seem* to be legitimate because what is measured is distance, and there is no controversy about its standing as a fundamental and unequivocally measurable concept. But relating distance to hardness is another matter. If there is a concept of hardness having a more fundamental significance than that implied by the arbitrary operational definition provided by the scratch test, how does one know that this property might not be nonlinearly related to scratch depth (logarithmically, for example) so that glass is "really" only twice as hard as chalk?

Trying to scale a subjective quality like brightness presents even more difficulties than trying to scale hardness. If we wish, we can *say* that a light judged as "100" by a human observer is "ten times as bright" as one judged to be a "10," but how can we hope to verify that the numbers given by the observer are linearly related to perceived brightness? We could know this only if we had some way to lay brightness end to end, some way to add them up. (This much, at least, could be done for the depths of the scratches related to hardness.) But no one has yet contrived a noncontroversial way to do this for something as subjective as brightness.

Two additional reservations concerning the possibility of subjective measurement have been advanced by Newman (1974): (1) Unlike physical concepts, which are universal, subjective facts may be idiosyncratic; (2) different procedures for assessing the same presumed psychological magnitude give different measurements, and there are no rules for choosing among them. The first of these objections concerns both individual differences and the inherently private nature of sensory experience. There is but one physical world: given the right tools and training, anyone can measure a particular aspect of it and get a correct answer. By contrast, each person experiences a very private subjective world. If one person were to insist that a light appears ten times as bright as another light although they nearly match for another person, on what basis could either be doubted? Each of us can evaluate only the strength of our own sensations. By contrast, anyone can scratch chalk with a diamond and objectively measure depth.

The second objection applies equally to the scratch and the sensation. Taking the hardness test first, suppose that some other operation for measuring hardness were developed, for example, the amount of pressure from an extended flat diamond plate required to significantly alter the shape of a cubic centimeter of a substance under test. Suppose that 10,000 times more pressure were required to deform glass as to crush chalk. By this criterion, glass would then be 10,000 times as hard as chalk. Yet by the scratch test they differed only by a factor of 10. There seems to be no right or wrong here, merely an arbitrary choice of operational definitions, and there is no fundamental procedure by which to specify how much harder a chunk of glass is than a piece of chalk. Similarly, there may be no fundamental sense in which a light can be specified "twice as bright" as another.

This problem does not arise for the measurement of a physical property such as length. If a metre is defined (as it once was) as the length of a standard metal bar in Paris, the metric distance from home plate to the right-field wall in San Diego Stadium depends on how many such bars must be placed end to end before reaching the wall. It does not matter in a practical sense that the Parisian standard is not available, because there are many other rulers whose calibration refers ultimately to it; any of these will serve the purpose. Furthermore, transits, range finders, or laser beams may be employed, and within experimental error all will give the same answer. By contrast, it is difficult to imagine an amount of brightness locked up somewhere in a vault, let alone that there could exist, as for length, a variety of alternative operations by which to measure it, all of which give the same answer.

Proponents of a negative veiw concerning subjective measurement have included philosopher N. R. Campbell and colorimetrist J. Guild, who seem to have been the most influential members of a committee appointed in 1925 by the British Association for the Advancement of Science "to consider and report upon the possibility of quantitative estimates of sensory events." The committee, which included an approximately equal representation of psychologists (including visual scientist K. J. W. Craik) and physical scientists, debated off and on for seven years and ended as a hung jury. The intensity of their deliberations is suggested by the following statement from the report of the committee (Ferguson, 1932), authored by Guild, who was for many years a staff physicist at the National Physical Laboratory:

> To insist on calling these other processes [of subjective scaling] measurement adds nothing to their actual significance but merely debases the coinage of verbal intercourse. Measurement is not a term with some mysterious inherent meaning, part of which may have been overlooked by physicists and may be in the course of discovery by psychologists. It is merely a word conventionally employed to denote certain ideas. To use it to denote other ideas does not broaden its meaning but destroys it: we cease to know what is understood by the term and our pockets have been picked of useful coin. [p. 345]

The main point on which the physicists refused to yield concerned the operation of addition. This matter is discussed at some length in the Introduction and third chapter of Volume 2 of this treatise (Bartleson, 1980a,b). No one had yet discovered a method whereby a large number of small sensations could be added to produce a sensation equal to a larger one. (Fifty years later, there has been no progress in this regard.)

Viewpoints about subjective measurement are somewhat akin to attitudes about political matters. For example, the political conservative may be convinced that governmental programs designed to redress societal inequities do not work and, therefore, should be abandoned. Similarly, a measurement conservative like Guild may feel that because subjective measurement is not possible, efforts along these lines should be abandoned (or, at least, called by some other name). The political liberal, even if somewhat skeptical of the power of government to remedy social imbalances, nevertheless may feel that the issue is so important that it is wrong not to try. The measurement liberal similarly feels that hope must not be abandoned. Therefore, *some* attempt at subjective measurement is regarded as better than none; and if in making the effort the rigorous criteria that apply to physical measurement cannot be met, then they should be relaxed.

Where measurement is concerned, a continuum of viewpoints exists between the extremes. If Guild represents a conservative endpoint, the liberal anchor has been well defined by S. S. Stevens, an experimental psychologist. In 1951 Stevens edited the *Handbook of Experimental Psychology* to which he contributed an initial chapter on measurement that proved to be very influential.

According to Stevens (1946) measurement is the act of assigning numbers to objects according to rules. If accepted, this broad definition opens the door for the specification of various kinds of measurement scales. Stevens identified four of these, which he called nominal, ordinal, interval, and ratio. These conceptions are illustrated in Fig. 2, which also uses some ideas of Metfessel (1947).

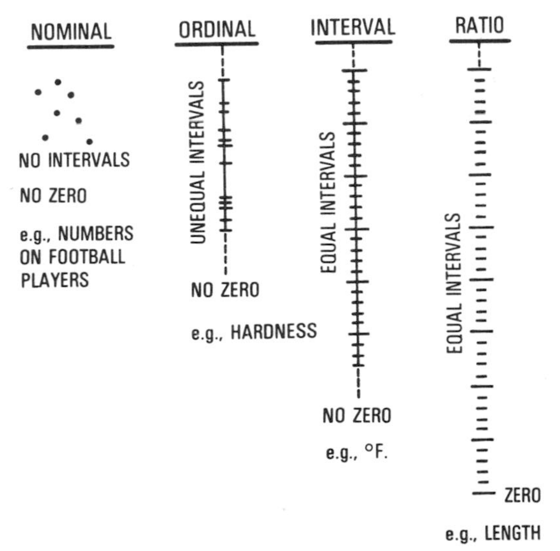

Fig. 2. Various measurement scales. A *nominal scale* merely uses numbers instead of names to distinguish among members of a group. An *ordinal scale* is derived from the operating of rank-ordering and gives no information about the meaning of distances along it. An *interval scale* is derived from the use of operations capable of specifying the equality of intervals and differences, so that equal distances anywhere along the scale have the same significance. A *ratio scale* is an interval scale with a zero point; equal ratios as well as equal intervals have the same meaning everywhere along this scale.

Ratio scales, which are the only satisfactory kind for a conservative like Guild, exist in the domain of physical measurement where length, mass, and energy are well-known exemplars. The numbers along any such scale of measurement apply to certain measurable properties of objects, not to the objects themselves. Thus it makes sense to say that one cylinder is two meters longer than another although this says nothing about their circumferences or weights. Even without a unit of measurement for weight, it would be meaningful to say that John weighs 20% more than Henry, whatever their relative heights or hair color. These properties of length or weight are measured along ratio scales, ordinarily determined indirectly with instruments whose calibrations must be checked from time to time and ultimately referred to basic standards.

Because the scale of numbers is itself a ratio scale, a major advantage of using a ratio scale of measurement is that a very useful isomorphism potentially exists between values determined empirically and the well developed mathematical systems of arithmetic and algebra. If experimental operations can also be found that correspond to the fundamental operations of algebra (addition, subtraction, equality, less than, greater than), then this possible isomorphism can be tested experimentally. If it holds, then experimentally determined quantities can legitimately be manipulated mathematically. This procedure permits quantitative predictions that are likely to prove valid for untested conditions. Probably the best visual example of the usefulness of this procedure lies in the domain of colorimetric measurement, which is based upon null-instrument psychophysics. See Chapters 7, 11, and 12 of this volume and Chapter 3 in Volume 2 of this treatise (Bartleson, 1980b). Given a very limited set of measurements, namely, the matching of spectral wavelengths with additive amounts of three primaries, the results can, if desired, be translated mathematically to an entirely new system of primaries (as was done by the CIE). As a result, any color, including those never tested experimentally, may be meaningfully specified by the relative amounts of the three primaries.

There is, however, a serious limitation to the elegant system just mentioned: it provides no information at all about the appearance of colors. (Many persons, including some physicists, seem to overlook this as they publish chromaticity diagrams in full color.) To paraphrase Newton, there are no colors, properly speaking, in a chromaticity diagram. There is only a *specification* of colors such that all possible colors that match each other will plot at a particular point. Their *appearance*, on the other hand, depends on many unspecified parameters such as luminance, area, retinal region stimulated, duration of presentation, state of adaptation of the eye, and the influence of surrounding colors. Although procedures are also needed to measure how colors appear, these cannot be null-instrument methods. Efforts of this other sort constitute the craft of psychological scaling.

It might as well be admitted that no unequivocal additive operation exists, or is ever likely to exist, for subjective measurement. Therefore, if measurement is said to depend on an additive operation, then subjective measurement is impossible. However, there are useful *physical* scales that also lack such an operational basis. The hardness scale has already been cited as one example. To take another: How does one add two pressures together? To point out that pressure is a derived measure whose two components themselves are measurable along ratio scales does not diminish the fact that pressure measurement would be useful anyway; its utility does not seem to depend very much on the presence or absence of fundamental properties. One could still fill tires to the recommended pressure, and it would not matter if the scale were no better than ordinal, provided that it were used in a consistent way by tire manufacturers and those who make pressure gauges. Similarly, though it may be unfortunate that the system of algebra cannot be freely applied to the products of subjective measurement, such measurements may nevertheless prove to be useful.

B. CLASSICAL (NULL-INSTRUMENT) PSYCHOPHYSICS

1. Introduction

The following (with a few unimportant words omitted) is from Fechner (1966, p. 39): "The magnitude of stimuli can be approached by exact measurement, and the equality of sensation may well be found by taking the necessary steps. We therefore take stimulus sensitivity as inversely proportional to the magnitude of stimuli that cause equally intense sensations."

To describe the psychophysical methods that constitute the "necessary steps" to which Fechner alludes, it will be helpful to have in mind two experimental situations.

a. Situation 1

Assume that an observer in a visual threshold experiment is presented with a fixation target consisting of a dim ring of light whose diameter subtends 2° of visual angle. He is told to fixate the center of the ring where the test stimulus will be presented in the form of a circular disc of light subtending 1° in visual angle, either steadily or as a flash that occurs once every 5 sec. The radiance of the test flash is varied by means of a circular neutral wedge (variable-density filter) whose density relates to its angular position.

Depending on the psychophysical method chosen, the wedge may be varied either continuously or in discrete steps by the observer or by the experimenter. The wedge is connected to a shaft encoder that signals its rotary position and delivers this information to a computer in which calibration data are stored. The output of the computer is delivered to a digital display that is visible only to the experimenter, and that reads directly in trolands (td) of retinal illumination.

The experimental question is: What is the amount of light (number of trolands) required for a just barely visible sensation? The experiment can be conducted in the dark (except for the fixation ring) or against a large uniform background (in which case the intensity of the fixation ring must be increased to keep it visible).

b. SITUATION 2

Another way to measure differential sensitivity, one having a long history extending back to Weber's experiments with lifted weights, is to introduce a second stimulus and require that it be compared in some way with the first one. An example is brightness-matching using a divided field. One half of the field is set at a fixed luminance (the reference or standard field), and the other is manipulated in luminance (the comparison or variable field.) When two stimuli are compared, all of the psychophysical methods applicable to Situation 1 can be used. In addition, some other methods that are not applicable to the single-stimulus paradigm can be employed.

2. The Basic Fechnerian Methods: Overview

The psychophysical methods to be discussed in this section are the traditional ones originally defined by Fechner. Three main methods will be sketched briefly and characterized in terms of various general aspects of experimental procedure, with special emphasis on the advantages and disadvantages of each.

The quantitative details of how thresholds are calculated are available elsewhere (see, for example, Chapter 7; Woodworth, 1938; Corso, 1967; and Kling and Riggs, 1971). Our discussion here will be oriented toward the operations that are required to obtain a valid answer to the experimental question that has been posed.

a. THE METHOD OF LIMITS

Using the *ascending method of limits*, the experimenter begins the stimulus series well below threshold and gradually increases intensity (for example, by moving the neutral wedge) until the observer reports that the stimulus is seen. With the *descending method of limits*, the series starts with a clearly visible stimulus whose intensity is gradually reduced until the observer signals that it is no longer visible.

b. The Method of Adjustment

The observer has control of stimulus intensity, for example, by means of a knob connected to a shaft that turns the neutral density wedge. By rotating the knob, the observer makes the stimulus appear and disappear, until a setting is found where the stimulus is just barely discernable. The procedure is repeated a number of times to determine an average value.

c. The Method of Single Stimuli

A set of five or six stimulus intensities is chosen in advance by the experimenter. These are presented as discrete flashes. Prior to each presentation, the observer is given a warning signal. The weakest stimulus of the series is selected so that the observer will seldom see it, the strongest so that he usually will. Stimuli are typically presented in random order of intensity until each one has been seen many times. Threshold is calculated as the intensity that yields a visible flash on a certain fraction of the trials. (If two stimuli are presented to be compared, this procedure is called the *method of constant stimuli.*)

The method of single stimuli generates a frequency-of-seeing curve; that is, given enough trials there will be a smooth monotonic increase in the fraction of stimuli reported as seen, as a function of stimulus intensity. The maximum slope of the frequency-of-seeing curve is reciprocally related to differential sensitivity. In other words, the smaller the stimulus range over which the function ascends a criterion amount, the more sensitive the observer is to a change in stimulus intensity. A measure of differential sensitivity can also be obtained using the method of adjustment by calculating the standard deviation of the observer's settings.

When presented against a steady background, the incremental threshold of a superimposed flash is itself a measure of differential sensitivity. In this case the slope of the frequency-of-seeing curve represents a second-order differential sensitivity.

3. Discussion of the Null-Instrument Methods

a. Method of Limits

The method of limits is sometimes used to get a rough estimate of threshold to determine the range within which to choose the stimuli to use with the constant-stimulus procedure. With sufficient repetitions, however, results obtained with the method of limits can stand on their own. Decisions must be made concerning how to define a threshold and where to start and end a particular series. With the descending method, the observer will initially report *yes* to each flash; but eventually, as the stimulus weakens, a *no* response will be given. It is not safe to stop at this point, because a stimulus that might have been seen could have been missed because of a blink or wandering fixation. Moreover, owing to various fluctuations in the sensitivity of the visual system, the observer's attention, or the stimulus, reversals will occur; that is, after having said *no* the observer will report *yes* to a stimulus one or more steps lower in intensity. Although there is no standard method, a typical procedure is to continue the series until two or three consecutive *no* responses have been given.

It is highly advisable that the size of the descending steps be decided in advance and that these be employed without deviation. It is a serious mistake to allow informality to creep into such procedures. For example, in an effort to gain sensitivity one might be tempted to decrease the size of the steps when it was felt that the intensity of the stimulus was in the neighborhood of the expected final threshold. This procedure can seriously bias the result toward the experimenter's expectation.

The steps normally used differ by equal ratios (that is, they are equally spaced on a logarithmic scale of intensity). These are chosen because, as noted earlier, the variance of visual sensitivity measurements tends to be constant on this basis over a wide range of stimulus intensities. A step size of about 0.1 log unit (about 25%) is often a good choice.

The experimenter must be careful not to bias the observer by his choice of the level at which each series is started. This must be varied unpredictably; otherwise a devious observer could give highly reliable but utterly meaningless data, for example, by saying no to the fourth flash in each descending series. Consequently some runs must start very far above the expected reversal point, a procedure that consumes a great deal of experimental time without providing useful information. We have sometimes used the procedure of allowing the observer to decide how many steps the intensity should be reduced for each next trial, with the proviso that the run be restarted if the subsequent flash following a decrease of more than one step is reported as not seen. Here, because the amount of stimulus change from one trial to the next is determined by the observer and not by the experimenter, no bias is introduced and runs can start far above the reversal level without wasting so much time.

When the ascending method of limits is used, the first flash should never be seen, and the starting points must be varied from trial to trial for reasons similar to those already discussed for the descending case. Unfortunately, there can be no short-cut procedure for approaching the threshold rapidly on a prolonged ascending run. A series can be terminated after the observer has given two or three yes responses; step size should be the same as for the descending series.

So long as a given criterion is adopted and maintained, it matters little exactly how the threshold is computed. In an effort to obtain a "true" threshold, both ascending and descending series often are used and the results averaged. Errors of anticipation or habituation have been described to account for the fact that thresholds based upon ascending and descending series often disagree, but these terms seem to explain very little.

The method of limits embodies an inherent inefficiency, because large numbers of stimuli must be presented that are virtually always seen (or unseen), whereas the greatest amount of information is obtained for flashes that are seen about half the time. From the observer's perspective the ascending method is somewhat frustrating because of the large numbers of flashes that are not seen. On the other hand, with the descending method there may be a serious concern about habituation to a long series of visible stimuli.

A variation on the method of limits, which has had a long history without acquiring a special name, could be called the *method of inequality*. It does not apply to the prototypical threshold situation but can be used in Situation 2 when the two stimuli to be compared can be set physically equal at the start of each trial. An example would be two fields of the same spectral content and luminance. At the start of a trial one of the fields begins to change, for example, becoming physically more intense, and the subject signals when the change is just perceived. Changes of chromaticity at constant luminance can also be gauged this way. Although this is probably the most direct way to determine a step of perceived difference, the observer may have difficulty maintaining a stable criterion. A procedure that helps to accomplish this has been described by Boynton and Kambe (1980).

b. METHOD OF SINGLE STIMULI

If humans were machines, the method of single stimuli would be chosen for the most accurate measurements. It yields results in the form of a probability-of-seeing function that constitutes an indirect measure of response magnitude at the bottom end of the sensation scale.

In using the method there is an initial problem concerning the choice of stimulus intensities to be used. Some preliminary work is always required before making this choice, perhaps by using the method of limits or adjustment. A stimulus range of 0.4 $\log_{10}$ units (comprised of five intensities separated by 0.1 log unit) often proves to be satisfactory. If, say, 50 trials at each intensity are used, the 250 trials would be presented randomly without replacement. The result is then plotted as a frequency-of-seeing curve or, more accurately, as five data points, each of which represents the percentage of stimuli reported as seen at each of the five stimulus intensity levels used.

If threshold is defined as that intensity for which a flash is reported as seen exactly half the time, only by chance would any of the experimental conditions produce exactly that result; even if this were to happen, a more accurate estimate of the 50% value could be achieved by using the entire data set because of the larger measurement error associated with any single point. To do this requires that some sort of function be fitted to the five data points. How this is done usually matters little for practical purposes, but an objective method is often preferred. An ogive (the integral of the normal distribution curve) usually fits such data rather well and is consistent with the theoretical view that sensitivity, from trial to trial, is normally distributed. A standard procedure called *probit analysis* (Finney, 1947) exists for doing the curve fitting.

There are serious problems with the method that make further discussion of the details of curve fitting academic. The major problem is that the estimate of threshold is significantly biased by the particular stimulus range selected by the experimenter. This bias, called a *range effect*, tends to shift the calculated 50% point toward the center of the tested range, as if the observer had unconsciously attempted to shift the criterion so as to equalize the number of yes and no responses given over the full set of trials. Forced choice procedures (see Section 4c) reduce the range effect somewhat but cannot eliminate it.

Another problem with the method of single stimuli is that it consumes a great deal of time and is taxing for the observer, who must judge large numbers of stimuli that are just on the borderline between visibility and invisibility; these are difficult judgments to make. And because of the range effect, the method of single stimuli seems more accurate than it is. Even if the data points rise neatly with intensity and are very well fitted by an ogive, thereby giving the impression of a very precise threshold determination, the range effect means that the threshold estimate is not valid. Reliability must not be confused with validity. Repeating the experiment using the same stimuli could give almost exactly the same result; yet by dropping the weakest stimulus of the set and replacing it by another at the upper end of the range, one could shift the threshold significantly in the direction of that change. On the other hand, the steepness of the frequency-of-seeing function, which provides a measure of differential sensitivity, is less affected by the range of stimuli chosen. (Steepness can be evaluated by the slope of the tangent to the best-fitting ogive at the 50% point.)

The randomization of stimuli is intended to minimize sequential effects, because it has been shown that observers are incapable of ignoring their previous responses. Nevertheless, data obtained by a block method in which all stimuli at one intensity are presented, followed by all of those at another intensity, can sometimes yield data superior to the randomized procedure (Blackwell, 1953). A version of this method is to let the order of levels be influenced by performance on the first block. For example, if 60% *yes* responses are recorded at the initial intensity level, a lower level is used for the second block, intended to bracket the 50% point. Reasonable estimates of 50% thresholds can sometimes be obtained this way with only two blocks of trials.

c. Method of Adjustment

The method of adjustment is similar to the method of limits, except that it is the observer, rather than the experimenter, who controls the stimulus. If discrete steps are used, and these can be heard or felt (for example, by the action of a pawl and detents every 0.1 log unit), then the sequences for the two methods could be the same, assuming that the observer follows a prescribed routine exactly. Ordinarily this is not the case: there are usually no sensory cues other than visual and kinesthetic ones, and usually observers are not given explicit instructions.

This method is best for continuously viewed fields, but it can also be used with discrete flashes. If so, adjustments are mostly made between flashes, and there is no immediate sensory feedback resulting from the movement of the knob. However, delayed feedback does occur, and the method still works. Much is left to the whim of the observer: One person may tend to start high and then reduce intensity slowly until a couple of flashes are unseen and let it go at that; another may rack back and forth for a long time, moving the flashes in and out of the visible range, refusing to settle upon a final position until a stream of flashes has passed that are seen about half the time. After each final setting just as with the method of limits, the experimenter must offset the wedge to start a new trial. Especially when discrete flashes are used, most of the concerns related to the method of limits apply to the method of adjustment.

The method of adjustment is especially likely to be used in Situation 2, in which stimuli are steadily present; here it offers real advantages. Consider brightness-matching as an example. Visualize a divided field with a standard intensity set into the left side. Suppose the observer's knob is connected to the neutral wedge that varies the intensity of the field seen on the right. As the observer rotates the knob, feedback is instantaneous. In the neighborhood of a match this sometimes proves very useful. For example, as one racks back and forth in the neighborhood of a luminance match, a very pronounced reversal of apparent contrast occurs. The left field (though physically constant) appears to darken as the right one brightens, and vice versa. Especially if the fields are perfectly contiguous, and there are no spectral differences, very accurate settings become possible.

For trichromatic color-matching, the observer adjusts three knobs, each of which controls the intensity of one of three additive primaries. When these are mixed in suitable proportions, they cause the mixture field to match a standard. In using this method the observer, in effect, solves three simultaneous equations in three unknowns, using himself as an analog computer. For trichromatic color-matching the method of adjustment is probably the only one that will produce results within a reasonable length of time.

A major problem with the method of adjustment is that it introduces an undesirable motor component into the judgmental process. Consider a brightness-match made under difficult conditions in which a considerable "dead zone" exists within which an acceptable brightness-match holds over a range of intensities of the variable stimulus. Faced with this situation the observer has no alternative other than to bisect the dead zone by turning the knob back and forth, seeking end points at which the variable stimulus first clearly looks brighter on one side, then dimmer on the other, finally splitting the difference. Although the ambiguous dead zone cannot be seen, it is felt kinesthetically, and a skilled observer can learn to bisect it into two equal angular parts. However, the corresponding stimulus intensity change depends on an arbitrary relation between the rotation of the knob and the change in intensity caused by the rotating wedge that it controls. For example, the intensity corresponding to the final setting would differ for a wedge that is linear in transmittance compared with one that is linear in density.

An advantage of the method of adjustment is that it is fast and pleasant for the observer, who has something more interesting to do than just to say yes or no. Making reliable settings requires skill, but this can be learned, just as for any other motor task.

4. Variations on the Single-Stimulus Method

a. The Concept of Criterion

Judging whether or not a flash of light is seen is not a straightforward matter. Nor is it easy to say, when the difference is small, whether a comparison stimulus differs from a standard. After many trials it may seem to an observer to be a toss-up whether to say yes or no. The observer must adopt a criterion by which to decide between the two responses. The methods so far described leave much to be desired, because they give the experimenter no information about the observer's choice of criterion.

b. Blank Trials

In times past it was sometimes customary for the experimenter to introduce some "blank" trials during an experiment. Using all the usual warning signals and shutter noises, the experimenter would silently and surreptitiously slide an opaque screen into the light beam just before the shutter opened. In this earlier era, to be caught saying yes on a blank trial was considered to be rather naughty, and the observer would be reprimanded. Nevertheless, a record of the percentage of yes responses on blank trials did provide the experimenter with an index of the observer's criterion.

c. Forced Choice

The use of forced choice in vision research dates back to World War II, after which Blackwell (1946) introduced the procedure for the collection of visual contrast-threshold data that had been intended for military application. Instead of being asked to say yes or no in response to a stimulus presentation, observers were required instead to state in which of *n* possible locations the stimulus had been presented (spatial forced choice). In a variation of the method, observers were asked to state in which of a number of temporal intervals, denoted by auditory signals, the stimulus had occurred (temporal forced choice). Choices were forced in the sense that observers were required to guess, even when they felt very certain that nothing at all had been seen.

The use of forced-choice procedures provides a stabilizing influence and a handle on the observer's criterion, because they generate responses that can be objectively scored right or wrong. When an observer is uncertain, there is no reason for him not to make a best guess as to where (or when) a stimulus might have appeared. Blackwell (1953) demonstrated that forced choice yields lower and more stable thresholds and increases the "inferred validity" of the measurements. The latter was gauged in part by varying the range and spacing of the stimuli used. Context effects (similar to the range effect previously described) were found to be smaller for forced choice than for the yes–no method. The work also showed that a temporal forced-choice method was superior to a spatial one.

d. The Up-and-Down (Staircase) Method

According to Dixon and Massey (1956) the up-and-down method originated for the testing of impact explosives to determine from what height they should be dropped to yield a 50% chance of detonation. A sample would be dropped from an arbitrary height; if it did not explode, the next one was tossed off one story higher. If that one did not explode, the height was increased another story, and so on until detonation occurred. When this happened, the height was decreased by one story for the next sample to be tested. The general rule: When the explosive detonates, decrease the height of the drop by one step for the next trial; when it does not, increase the height.

Applied to visual testing this method became known as the staircase method. The basic procedure is the same: If the observer says yes, reduce the stimulus intensity for the next trial by one step. If the observer says no, increase it by one step. With this procedure, which is really a variation of the single-stimulus method, the experimenter must decide in advance only the size of the step and at what intensity to present the first stimulus. As the experiment proceeds, the range of stimuli presented becomes determined exclusively by the behavior to the observer. The range effects, so bothersome for the single-stimulus method, are virtually eliminated.

However, this unvarnished form of the method would permit a devious observer to foil the experimenter easily by alternating yes and no responses, and a conscientious observer would also be influenced by knowing the intensity of each stimulus relative to the preceding one. This problem could be minimized by not informing the observer about the true nature of the stimulus–response dependencies, but this deception would be practically impossible to maintain for long. Instead, it is better to break up the stimulus–response dependency somehow. A standard method for doing this, introduced by Cornsweet (1962), is called the *double staircase* procedure. Two separate schedules are prepared; within each one the procedure just described is followed strictly. However, the experimenter alternates randomly between schedules, often causing the stimulus–response dependency to be one or more steps removed from that of the immediately preceding trial, in a manner that the observer cannot predict, comprehend, or remember. An alternative procedure has been called *scrambling* (Boynton, *et al.*, 1961). A small number of *reversal* trials (usually about 10%) are designated randomly beforehand or are determined randomly prior to each trial. On such trials a yes response is followed by a stimulus that is one step more intense (rather than a step dimmer); and after a no response, a stimulus one step dimmer is presented. Like the double-staircase, the scrambling procedure keeps the observer confused about actual stimulus–response dependencies.

In addition to eliminating the range-effect problem, the staircase method is more efficient than the standard single-stimulus procedure, because it tends to concentrate most of the stimuli near the final 50% threshold; as noted above, this is the most sensitive portion of a frequency-of-seeing function.

Many methods can be used to estimate the 50% threshold when the staircase method is used. One is to calculate the percentage of yes responses at each stimulus intensity and then fit curves as with the constant-stimulus method. Because of the likelihood of very few trials at the extremes of the tested range, data for intensities with too few responses (for example <5) should be ignored. Additionally, because the number of stimulus presentations at each intensity is variable, the percent of yes values should be weighted accordingly before thresholds are calculated. Another procedure is to base the thresholds on computations that make use of the location of reversal points on the intensity scale, much as one would do with the method of limits. Other calculational procedures are discussed by Dixon and Massey (1956) and by Wetherill (1963).

The starting intensity turns out to be relatively unimportant, although some time is wasted if it differs by too many steps from the final threshold. A desirable procedure is not to count the first few trials in a series, waiting, for example, for the first reversal to occur.

The staircase method is widely used today, because it seems to combine the best features of the method of limits and the method of single stimuli. It can also be used in conjunction with a forced-choice procedure, in which case the stimulus sequence depends on whether each previous response was right or wrong.

5. Correcting for Chance

In a four-alternative forced-choice experiment, observers can close their eyes and still score 25% correct on the average. Therefore, if it seems important to determine a "true" probability of seeing, some means must be found to translate an uncorrected score of 25% to a corrected one of 0%. At the other performance extreme, however, if a subject scores 100%, he must receive full credit, because it is virtually impossible to be correct on every trial if nothing is actually seen (the probability of doing so is $(1/4)^n$, where n is the number of trials; this probability is less than one in a million for perfect performance after only 10 trials).

The graph of Fig. 3 shows these two points boldly indicated. A chance-correction formula that often has been used is based on a straight line drawn between these two points. However, there is no sound basis for preferring this straight line to some nonlinear function.

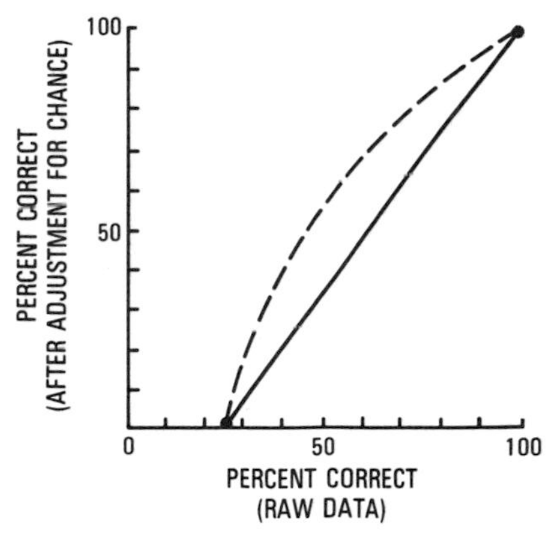

Fig. 3. In a four-alternative forced-choice experiment the two points shown in this graph are determinable from straightforward statistical arguments. The most commonly used formula for chance correction is based on a straight line connecting the points as shown. Other functions such as the dashed curve could be used instead.

6. Difference Thresholds

The sample experiment around which most of the discussion of methods has revolved has been a threshold determination based on single flashes (Situation 1). When such flashes are superimposed on a background, the measured threshold is a kind of a difference measure, known as an *increment threshold*. However, this is not the variety of difference threshold usually considered in the extensive older literature on psychophysical methods, which relates to Situation 2.

It all started with Weber's lifted weights. The experimental question that interested him (and later, Fechner) was the just noticeable difference between successively compared weights. The usual procedure was to lift a reference weight first, followed by the weight to be assessed, after which a judgment would be rendered regarding the relative heaviness of the second weight. The procedure adds some complications to the simpler cases in which only one stimulus is presented or two stimuli are presented simultaneously (for example, one weight in each hand).

For one thing, a new class of time errors is introduced. The impression created by the second stimulus must be compared with a memory of the first one that bridges the interstimulus interval. Another problem that received a great deal of attention was whether to allow subjects to use an "equal" category (Woodworth, 1938), which is similar to allowing an observer to say "I don't know" when a single flash is presented. Some felt that the data of the "equal" category should be used as the prime determinant of the just noticeable difference. Although a consensus eventually arose that its use was undesirable, there are times when an "equal" category makes sense. For example, in developing a method of color scaling (Boynton & Gordon, 1965), we at first allowed observers to use only two responses. The first of these signalled the primary hue that was perceived and the seccnd a secondary hue. Stimuli around 575 nm are always called yellow first and may be called reddish or greenish as well. Although our original plan was to force a second response—which, in effect, disallows an "equal" category—some preliminary work and a dash of introspection led to the opposite decision. The change of plan was based on the observation that some stimuli are seen as pure yellow, that is, there is no trace of any secondary hue. Moreover, when there is a trace of red or green, there is never any doubt about which of them is seen. In such a case little would be gained by forcing a second response based on wholly nonsensory considerations.

More often, when forced to guess, observers do provide additional valid sensory information. Consider for example a two-alternative temporal forced-choice experiment in which an observer is first asked to state yes or no, and then, without regard for the first response, he must indicate in which of two time intervals the stimulus, even if not seen, occurred. Experiment shows that even for the not-seen trials, subjects will pick the correct interval at a rate much better than chance.

An aspect of the difference threshold that merits discussion concerns the attitude that psychophysicists have taken toward it. The earlier view seems to have been that for any particular condition there should exist a particular, specific stimulus difference that elicits an inherently just-noticeable sensory difference. From this point of view the proper methodology is the one that proves capable of measuring the stimulus difference without bias, and one should beware of errors of various kinds that might contaminate the measure. A more reasonable view is that there is no such thing as a just-noticeable difference that transcends method. All that one can expect are consistent measures, and one can hope that despite disagreements about absolute values the difference thresholds measured by one method will correlate well with those measured by some other procedure.

C. THE THEORY OF SIGNAL DETECTION (TSD)

The theory of signal detection originated in the context of communications theory, outside the realm of psychophysics, and was related to studies of hearing before spilling over into the visual domain (Green and Swets, 1966; Egan, 1975). As applied to vision it is imagined that for a given stimulus, repetition results in a continuum of subjective intensities. Therefore, as repetition continues, a distribution of responses (which is usually assumed to be normal) builds up along the subjective intensity scale. Because of "noise" in the visual system, a sensation will sometimes occur that is mistaken for a stimulus when, in fact, no stimulus has been presented. When a physical stimulus is presented, its subjective intensity is assumed to add to that caused by whatever noise is present, leading to a distribution of subjective intensities related to signal-plus-noise. These two distributions, of noise alone and of signal-plus-noise, are depicted in Fig. 4.

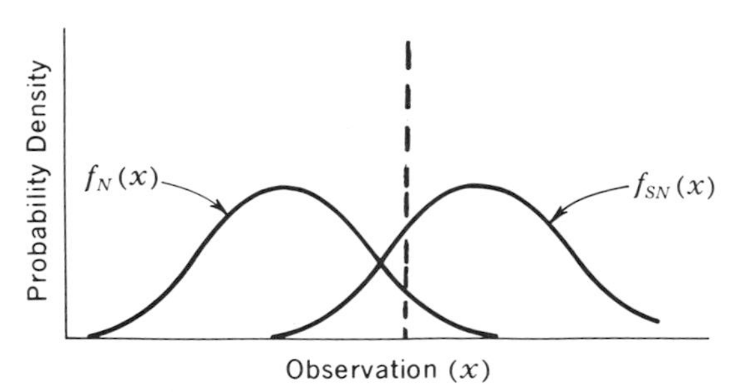

Fig. 4. Illustration of basic concepts of TSD. The abscissa, represents a continuous scale related to sensory magnitude. The distribution at the left represents the probability that a given magnitude of sensation will occur as a result of noise in the absence of a physical stimulus. The distribution at the right represents the probability that a given magnitude of sensation will occur in the presence of a physical stimulus, which adds its signal to that of the noise. The dashed line represents the observer's *criterion*, above which he will say yes and below which, no. For this particular criterion the observer is much more likely to say yes when a signal is present than when it is not, because noise alone seldom exceeds the criterion. (Adapted from Corso, 1967, p. 448; from Swets, Tanner, and Birdsall, 1961.)

The observer's criterion is represented by some point along the sensory continuum below which the response is no and above which it is yes. A low-criterion observer will often say yes when no stimulus has been presented. A high-criterion observer will seldom do this but is less likely to say yes when a stimulus is actually present.

The TSD experimenters usually present equal numbers of blank and stimulus trials, leading to four stimulus–response possibilities: (1) stimulus is presented and response is yes (a hit); (2) stimulus is presented and response is no (a miss); (3) no stimulus presented and response is yes (a false alarm); and (4) no stimulus presented and response is no (a correct rejection). One purpose of using TSD is to separate the observer's criterion from his sensitivity. For a given separation of the signal and the signal-plus-noise distributions, as in Fig. 4, the percentage of hits will increase as the criterion moves to the left, but so will the percentage of false alarms. If the observer becomes more sensitive for some reason, or the stimulus intensity is increased without a change in criterion, the signal-plus-noise distribution shifts to the right. This will increase the percentage of hits but, unlike a criterion shift, not that of false alarms.

The statistic called d' is used to estimate the separation between the peaks of the two distributions. Another called β assesses the location of the criterion point. The calculations depend on an analysis of the distribution of responses in the four possible categories of stimulus–response relations, often displayed on a receiver–operator–characteristic (ROC) curve, which is a plot of hits versus false alarms for a stimulus of constant intensity.

The TSD procedure has most often been used in visual experiments in which very dim stimuli of a single intensity are used, not a common situation for most substantive vision experiments. The observer's criterion is then manipulated to see whether TSD can account for the results. It is rare that a criterion d' will be adopted by the experimenter so that, in the manner of classical psychophysics, combinations of physical variables can be manipulated to achieve it.

To this writer, TSD may make good sense for experiments in audition, but for vision its use often merely substitutes for frequency-of-seeing or serves as a fancy method to correct for chance. In the latter context its use depends, like all chance-correction procedures, on several underlying assumptions that are seldom fully met or even testable. For a visual experiment like that of Situation 1 it seems improbable that there is enough noise in the visual system for the TSD approach to be very meaningful. After all, the observer knows where to look, what to look for, and when to look for it. It seems most unlikely that a random event will occur that meets the appearance criteria of a real stimulus.

The use of TSD has been taken to imply that increment thresholds should increase with the intensity of a background field, because such a manipulation causes an increase in the noise within the visual system. The expected result occurs, of course, but the explanation seems wrong. The effect of a background is not to add noise along the same subjective dimension on which the stimulus appears. Rather, the background lowers the gain of the system, by means of a variety of mechanisms that are reasonably well understood. Indeed, a background is effective even when invisible because of image stabilization (Sparrock, 1969).

Those interested in TSD have often attempted to test the theory by manipulating the observer's criterion while holding constant the stimulus intensity. Although the expected results often occur in reasonable agreement with the theory, they do not prove its validity. The original method that was used to manipulate the observer's criterion used a payoff matrix, the effects of which may be understood in terms of an extreme example. An observer will never produce a false alarm if the penalty for doing so is death. But his number of hits would also be zero because of the very high criterion that he wisely would adopt. Conversely, if an observer were paid $100 for each hit without any penalty for a false alarm, he would be wise to adopt a low criterion and always say yes in order to collect on every hit.

Nachmias and Steinman (1963) showed that rather than use a payoff matrix, one can obtain similar information by allowing the subject to use a rating scale. For example, zero could mean "absolutely no," 5 could mean "certain enough to bet my life on it," and the intervening integers could apply to intermediate estimates of certainty about whether a stimulus had been presented on a given trial. After the fact, the experimenter repeatedly divides the data according to various criterion levels so that, for each integer selected, all responses rated at that level or higher are called yes and lower ones, are no. This type of rating procedure works very well. Whether or not data are analyzed in the context of TSD, there is seldom any reason not to use ratings in what have traditionally been yes–no experiments; doing so provides additional information at no cost.

D. METHODS FOR ESTIMATING SENSORY MAGNITUDES

1. Introduction

Earlier it was indicated that to use the null-instrument approach successfully, one must assume that a monotonic relation exists between stimulus and response. That such relations exist is confirmed by everyday experience and simple experiment. For example, if one turns down the intensity of a light bulb by means of a dimmer, the sensation of brightness clearly decreases monotonically with the diminution of the physical intensity of the light. Consequently, there should be no difficulty in establishing an ordinal scale of brightness as a function of physical intensity. But one would like to do better by establishing interval or ratio scales.

Some of the earliest efforts at subjective visual scaling antedate even the work of Fechner. According to Boring (1942), Plateau instructed an artist to paint a gray "half way between" a white and a black and found evidence that this middle gray had a reflectance independent of the absolute level of illumination. The result implies a linear relation between ratios of sensation and ratios of stimulus intensity, which is contrary to Fechner's logarithmic law. This presaged a controversy that continues to the present.

In more modern times a major effort in scaling sensory magnitudes arose from the practical need to establish scales of loudness (Stevens and Davis, 1938). The decibel (db) scale for sound frequency is logarithmically related to the scale for physical sound intensity relative to threshold; if loudness were to grow linearly with sound level in decibels, Fechner's relationship would be validated. But the issue is complicated by complex sounds containing two or more frequencies. Because the prediction of overall loudness (in db) as the sum of the decibel amounts of the components did not work out well, other forms of scaling were attempted. These scaling methods have subsequently been applied to a wide variety of sensory domains.

Psychometric methods, as opposed to psychophysical ones, are concerned with the scaling of stimuli of unknown physical dimensionality according to the reactions that these stimuli produce on human subjects; that is, psychometrics concerns measurement of response differences. As Guilford (1954) notes in his classic text on psychometric methods, psychophysics and psychometrics stem from very different traditions and are usually treated as being more separate than they deserve to be. There is no rule stating that psychometric methods cannot be applied to stimuli amenable to physical specification, in which case they could be considered as varieties of psychophysical methods. For this reason they will be discussed here briefly.

2. Psychometric Methods

An experimental example will again be useful to provide a context within which to launch a brief discussion of psychometric methods. Consider that the stimulus material consists of ten photographs of human female faces, all taken by the same photographer using controlled conditions of camera settings, lighting, and film processing. The objective of the experiment is to develop a scale along which the photographs can be ordered by male subjects according to a scale of perceived feminine beauty. This is a typical sort of subjective problem in which, because beauty is in the eye of the beholder and thus depends in unknown ways on multidimensional aspects of the stimuli, an independent physical specification of the photographic images is not possible. This is quite unlike the situation for psychophysical experiments of the sort previously discussed, in which a single physical dimension (such as light intensity) is manipulated and can be independently measured.

Using the method of *rank order* an observer is asked to sort the photographs in any way that orders the depicted faces from most to least beautiful. For any particular observer the result, by definition, will be an ordinal scaling obtained by the most direct means imaginable.

Using the method of *paired comparisons* an observer is presented with all possible pairs $[n(n-1)/2]$ of pictures, one pair at a time. (For the example given, each picture is viewed 9 times among a total of 45 pairs.) The use of this method is capable of revealing intransitive relationships that cannot show up using the rank-order method. (For example, A might be rated as more beautiful than B and B as more beautiful than C, and yet C might be judged more beautiful than A.) This kind of result, if it occurs reliably, indicates the need for multidimensional scaling.

Formal procedures are available for treating the data obtained by this method, based on Thurstone's (1927) *law of comparative judgment*. Thurstone's method is based on a conception that sensory dimensions exist (not unlike the notion, introduced years later, that underlies the theory of signal detection). The method makes use of the inherent variability associated with comparisons; this is useful if the spacing of stimuli is sufficiently tight. By analyzing the probabilities of choice among stimuli, it is possible, given certain assumptions, to order the stimuli along a scale having interval properties.

The *method of categories* requires a subject to sort stimuli into a limited number of bins, usually having useful labels. For example, three such labels might be *beautiful*, *neutral*, and *ugly*. The method is more likely to be used when the number of items to be sorted is large, and up to seven categories often prove useful. A common example is rating essay examinations on the familiar ABCDF scale.

These are the three main methods of unidimensional psychometric scaling. The rank-order and paired-comparison methods become cumbersome with large numbers of stimuli, because all of the stimuli must be compared with all of the others, and the number of judgments increases almost as the square of the number of stimuli to be judged.

Methods of dealing quantitatively with the results of these methods are available in Guilford (1954) and in Chapters 8–10 of this volume.

An example of a widely used type of multidimensional scaling method is the *analysis of proximities* developed by Shepard (1966). Rather than imagine stimuli ordered along a single dimension, the optimal degree of dimensionality is left unspecified at the outset. The method proceeds by trial and error and requires a digital computer. The input to the program is a series of difference scores. For the example being used, subjects would first be asked to judge, for all 45 possible pairs, how different the depicted members of each pair are judged to be in terms of beauty. (The sign of the difference is ignored.) The computer program begins by ordering points that represent the stimuli randomly in one-dimensional space, that is, along a line. The program seeks this ordering so that for the 45 pairs there is an optimal agreement among the rank-orders of the 45 difference judgments and of the 45 points. After this has been completed, the procedure is repeated using points randomly positioned in two-dimensional space, with the same eventual goal. The process may be continued for as many dimensions as desired.

A solution in three dimensions is always better than in two, and in general the more dimensions there are, the better the solution will be up to a limit, at which the number of dimensions equals the number of stimuli, and the solution becomes trivial. For the lower orders of dimensionality the geometrical increase in the number of pairs compared, relative to the number of points to be represented, means that the solution is overdetermined. This is true even when the initial input in nonmetric. Consider the following example. A circle is drawn and 10 points are randomly spotted around its circumference. The distances separating the 45 pairs of points are measured and ranked from 45 (longest) to 1 (shortest). If the scaling program used is in two dimensions, it will turn out that the 10 points will fall along an almost perfect circle despite the discarding of the original metric information (interpoint distances). This is a powerful demonstration, one that provides some hope that metric relations can be derived when only ordinal, nonmetric data are available, as is usually the case with psychological measurement.

The meaning of the two or more dimensions in the final plot is not provided by the method. Such meaning must be injected by the investigator on some other basis. Moreover, formal methods do not exist for deciding upon the number of dimensions that provide an adequate result. Again, this decision depends importantly on what meaning, if any, can be given to each added dimension.

3. Rating Scales

The use of rating scales has much in common with the method of categories just discussed. There are many versions, but typically the rater is asked to indicate where a particular stimulus falls along a horizontal line representing a continuum. At least two anchor points must be used to provide a frame of reference, and often verbal indicators appear at points equally spaced along the scale. Potential problems with the design and use of rating scales are too numerous to mention here. Rating scales are seldom used in psychophysical experiments.

4. Fractionation

This method attempts to develop a ratio scale. It is assumed that there is a zero point on the sensory continuum and that observers can make meaningful and reliable judgments of ratios. A common procedure is to ask the observer to set the stimulus intensity so that the sensation (brightness, for example) is half as great as a reference stimulus that has been previously presented. An experiment by Garner (1954) casts considerable doubt on the ability of observers to do this meaningfully. Garner, who was concerned with the development of a loudness scale, elected to use the method of constant stimuli. Each of three groups of observers was given a different range of comparison stimuli. The sound intensity judged for half-loudness, no matter what the range of stimuli, occurred exactly in the middle of the range that was used. This huge context effect strongly suggests that there is no unique sensation corresponding to half-loudness.

5. Magnitude Estimation

There are many variations on this method, which requires the assignment of numbers to represent the intensities of sensory experiences. In one version the observer is presented with a stimulus of some random intensity and is told to assign a number like 10 to it (or, in another version of the method, any number that comes to mind). The reference stimulus might or might not be presented again during the experiment (in one version of the method it is not presented at all). Other stimuli are then judged in terms of the remembered reference stimulus or, if one is not used, in the context of the entire series of stimuli that are seen. After many repetitions geometric means are taken, and the numerical estimates are plotted as a function of stimulus intensity. In general the result is linear on a log-log plot if there are no intrusive induction effects, suggesting a power-law relationship between stimulus and sensation. Results using magnitude estimation generally differ from those obtained using rating scales with the same observers.

A detailed discussion of this method is beyond the bounds of this chapter. Opinions about it vary widely, pro and con. For a favorable view see Stevens (1974). For negative views see Garner (1959) and Anderson (1974).

E. MISCELLANEOUS CONCERNS ABOUT PSYCHOPHYSICAL EXPERIMENTS

The following is a potpourri of concerns about psychophysical experiments, based on the author's experience, that relate to all of the methods discussed.

1. Fixation

To stimulate a known retinal region, exact fixation is important. Fixation is a natural act that requires no special training and minimal instructions. Yet the untrained observer may unwittingly let fixation drift, and in experiments where the fixation is eccentric to the stimulus, there may be a strong tendency to "cheat" by sneaking a look directly at the stimulus. The only way to be certain that an observer is properly fixated is to monitor the position of the eyes. There are a variety of methods available (Alpern, 1971) for doing this, ranging from some that cost thousands of dollars to a simple telescope arrangement, or even a peep-hole through which the observer's eye can be seen. It is no great feat to automate the procedure so that, should the eye wander off target, a signal is given or stimulus presentation is inhibited.

For the experiment of Situation 1 a ring-shaped fixation target was mentioned, because for foveal stimulation a fixation point would superimpose upon the stimulus and affect its visibility. Fortunately, the keen ability of humans to fixate the exact center of a ring is well documented (Steinman, 1965). The center of an imaginary square suggested by four points of light (or four black dots, if a bright background is used) also works satisfactorily. (If the stimuli are delivered to the peripheral visual field, there is no reason not to use an actual fixation point.) In some experiments, for example, those concerned with color-matching, no fixation point is used, and the subject is free to move his eyes within the stimulus area; the stimulus itself, being steadily exposed, provides sufficient reference.

2. Instructions

If an experiment can be carried out with monkeys as subjects, and if it generates the same kind of results as those obtained with humans, this constitutes *prima facie* evidence that something real, basic, and sensory is under assessment and that the experiment is not one whose result depends entirely on the subtleties of verbal instruction. The mere mumbling of words at a human observer does not ensure understanding: Practice trials and feedback are also needed, much as when learning a motor skill such as golf or automobile driving. Macaque monkeys have essentially the same vision as humans, but, of course, they cannot talk or understand human speech. Therefore, the training procedures that are used must be very explicit compared with the shorthand verbal instructions usually given to humans. The difference between training the monkey to observe, compared with instructing the human, is like that between programming a computer and giving much less precise verbal instructions to an assistant.

Rewards are essential for training monkeys, and they are similarly useful for humans. Some professional monkey subjects work 8 hours a day and receive all of their liquid diet in the form of brief squirts of juice for correct responses. Sometimes human and monkeys data are virtually indistinguishable (see, for example, Harwerth and Sperling, 1971). Although verbal instructions can shortcut the training process for humans, they can by no means entirely substitute for it.

3. Rewards

An advantage of the forced-choice methods is that rewards can be given, if desired, for correct responses. We advocate this as a device to maintain the motivation of the observer. Many years ago while preparing to do some tedious search-and-recognition experiments for the Air Force, the author happened to convey to a friend (who was a devout Skinnerian) an intention to post scores and reward subjects for good performance on a weekly basis. He scoffed at this, pointing out that reinforcement is truly effective only when given immediately after a response. On his advice our observers subsequently received all of their rewards for good performance in the form of nickels that rolled down a chute, landing with a satisfying "clank" into an accessible container, always as a reward for correct responses. Not all correct responses were rewarded, although knowledge of results was always given. One of our observers earned $1000 in nickels and, like most of the others, never seemed to tire of the game. We spiked the procedure with occasional super-reward (50¢) trials; these occured unpredictably and were announced after the fact. In our opinion there are hundreds of experiments in the perceptual literature, including many with paid observers, in which immediate rewards could and should have been used to acquire more stable data. Instead, trial-by-trial rewards seem to have been mostly used by those interested in signal-detection theory, for the limited purpose of manipulating the observers' criterion.

4. Tactics (Method of Adjustment)

In a color-matching experiment, in which the final match is often not absolutely perfect, some observers may take 15 min to make a single match. They fiddle, sweat, retreat, and regroup. It is best to make such settings relatively quickly, even if each one proves slightly less reliable than it would be if given more adjustment time. The advantage is that many more matches can be made in an experimental session. Unlike rotten apples, a few bad settings will not spoil the barrel, provided that there are enough settings.

5. Set of the Observer

The best psychophysical judgments of simple stimuli seem to occur when the observer is almost in a trance. The feeling is something like that of the experienced automobile driver who suddenly realizes that for the past several minutes his behavior has been controlled at a wholly unconscious level; nevertheless, a long series of complex discriminations and smoothly coordinated motor responses will have successfully occurred. This state is not to be confused with one of gross inattention, which could produce bad data (or an automobile accident). Perhaps, when driving, it is best to be fully conscious and alert, but when psychophysical judgements are concerned, the more automatic they become the better.

6. Bullheaded Objectivity

It was remarked earlier in conjunction with the method of single stimuli that the typical range-effect causes the threshold to shift somewhat toward the center of the range; one explanation is that observers may be tempted to equalize their yes and no responses over a session. The proper attitude of the skilled perceptual observer, like that of the skilled golfer, is to take each shot as it comes, forget about the past, and not worry about the future. Although this is not wholly possible in either context, the more skilled the practioner, the closer it is approximated. We should like to believe that if the light source in the optical system burned out in mid-session, we could give 50 no responses in a row without being tempted to say yes because a perceptible stimulus seemed due. Especially when experimenters serve as their own observers, a stubbornly objective attitude is very important. No matter how much a cherished theory might depend upon a particular outcome, one should be concerned only to find the truth.

REFERENCES

Alpern, M. (1971). Effector mechanisms in vision. *In* "Experimental Psychology" (J. W. Kling and L. A. Riggs, eds.), 3rd ed. Holt, New York.

Anderson, N. H. (1974). Algebraic Models in perception. *In* "Handbook of Perception" E. C. Cartette and M. P. Friedman, eds.), Vol. 2. Academic Press, New York.

Bartleson, C. J. (1980a). Introduction. *In* "Optical Radiation Measurements: Vol. 2. Color Measurement," (F. Grum and C. J. Bartleson, eds.), Vol. 2, pp. 1–9. Academic Press, New York.

Bartleson, C. J. (1980b). Colorimetry. *In* "Optical Radiation Measurements: Color Measurement," (F. Grum and C. J. Bartleson, eds.), Vol. 2, pp. 33–148. Academic Press, New York.

Blackwell, H. R. (1946). Contrast thresholds of the human eye. *J. Opt. Soc. Am.* **36**, 624–643.

Blackwell, H. R. (1953). "Psychophysical Thresholds," Bulletin No. 36. University of Michigan Engineering Research Institute, Ann Arbor, MI.

Boring, E. G. (1942). "Sensation and perception in the history of experimental psychology." Appleton-Century-Crofts, New York.

Boring, E. G. (1955). Psychology. *In* "What is Science?" (J. R. Newman, ed.). Simon and Schuster, New York.

Boynton, R. M. (1961). Some temporal factors in vision. *In* "Sensory Communication" (W. Rosenblith, ed.), Wiley, New York.

Boynton, R. M. and Gordon, J. (1965). Bezold-Brücke hue shift measured by color-naming technique. *J. Opt. Soc. Am.* **55**, 78–86.

Boynton, R. M., and Kambe, N. (1980). Chromatic difference steps of moderate size measured along theoretically critical axes. *Color Res. Appli.* **5**, 13–23.

Boynton, R. M., Sturr, J. F., and Ikeda, M. (1961). Study of flicker by increment-threshold technique. *J. Opt. Soc. Am.*, **51**, 196–201.

Cobb, P. W. (1932). Weber's law and the Fechnerian muddle. *Psychol. Rev.* **39**, 533–551.

Cohen, M. R., and Nagel, E. (1934). "An Introduction to Logic and Scientific Method." Harcourt, New York.

Cornsweet, T. N. (1962). The staircase-method in psychophysics. *Am. J. Psychol.* **75**, 485–491.

Corso, J. F. (1967). "The Experimental Psychology of Sensory Behavior." Holt, New York.

Dixon, W. F., and Massey, F. J. (1956). "Introduction to Statistical Analysis." McGraw-Hill, New York.

Egan, J. P. (1975). "Signal Detection Theory and ROC Analysis." Academic Press, New York.

Fechner, G. (1966). "Elements of Psychophysics," Vol. I (H. E. Adler, D. H. Howes, and E. G. Boring eds. and translators). Holt, New York.

Ferguson, A. (Chm) (1932). Quantitative estimates of sensory events. Final report of the Committee to consider and report upon the possibility of quantitative estimates of sensory events. *Adv. Sci.* **1**, 331–349.

Finney, D. J. (1947). "Probit Analysis." Cambridge University Press.

Garner, W. R. (1954). Context effects and the validity of loudness scales. *J. Exp. Psychol.* **48**, 218–224.

Garner, W. R. (1959). An argument for the use of discriminability scaling procedures in scaling sensory intensities. *Acta Psychol.* **15**, 94–97.

Green, D. M., and Swets, J. A. (1966). "Signal Detection Theory and Psychophysics." Wiley, New York.

Guilford, J. P. (1954). "Psychometric Methods." McGraw-Hill, New York.

Harwerth, R. S., and Sperling, H. G. (1971). Prolonged color blindness induced by intense spectral lights in rhesus monkeys. *Science* **174**, 520–523.

Johnson, H. M. (1929). Did Fechner measure "introspectional" sensations? *Psychol. Rev.* **36**, 257–284.

Johnson, H. M. (1945). Are psychophysical problems genuine of spurious? *Am. J. Psychol.* **58**, 189–211.

Kling, J. W., and Riggs, L. A. (1971). "Experimental Psychology," 3rd ed., pp. 352–358. Holt, New York.

Metfessel, M. (1947). A proposal for quantitative reporting of comparative judgments. *J. Psychol.* **24**, 229–235.

Nachmias, J., and Steinman, R. M. (1963). Study of absolute visual detection by the rating-scale method. *J. Opt. Soc. Am.* **53**, 1206–1213.

Newman, E. B. (1974). On the origin of "scales of measurement." *In* "Sensation and Measurement" (H. Moskowitz, B. Scharf, and J. C. Stevens, eds.), Reidel, Dordrecht, Netherlands.

Shepard, R. N. (1966). Metric structure in ordinal data. *J. Math. Psychol.* **3**, 287–315.

Sparrock, J. M. B. (1969). Stabilized images: Increment thresholds and subjective brightness. *J. Opt. Soc. Am.* **59**, 872–874.

Steinman, R. M. (1965). Effect of target size, luminance, and color on monocular fixation. *J. Opt. Soc. Am.* **55**, 1158–1165.
Stevens, S. S. (1946). On the theory of scales of measurement. *Science*, **103**, 677–680.
Stevens, S. S. (ed.) (1951). "Handbook of Experimental Psychology." Wiley, New York.
Stevens, S. S. (1974). Perceptual magnitude and its measurement. *In* "Handbook of Perception" (E. C. Carterette and M. P. Friedman, eds.), Vol. 2. Academic Press, New York.
Stevens, S. S., and Davis, H. (1938). "Hearing, its Psychology and Physiology." Wiley, New York.
Swets, J. A., Tanner, W. P. Jr., and Birdsall, T. G. (1961). Decision processes in perception. *Psychol. Rev.* **68**, 301–340.
Thurstone, L. L. (1927). A law of comparative judgment. *Psychol. Rev.* **34**, 273–286.
Wetherill, G. B. (1963). Sequential estimation of quantal response curves, *J. Roy. Stat. Soc. B* **25**, 1–48.
Woodworth, R. S. (1938). "Experimental Psychology. Holt, New York.

7

Thresholds and Matching

C. J. BARTLESON

Research Laboratories
Eastman Kodak Company
Rochester, New York

LIST OF SYMBOLS

ANOVA	analysis of variance
CE	constant error
D_a	average deviation from mean: $D_a = (\sum\lvert x_i - \bar{X}\rvert)n^{-1}$
D'_a	average deviation from median: $D'_a = (\sum\lvert c_i - m\rvert)n^{-1}$
d'	SDT sensitivity: $d' = (\mu_N - \mu_{S+N})(\sigma_{N,S+N})^{-1}$
DL	difference limen
$E(X)$ or E	expectation: $E = \bar{X}$
e_k	an event according to some criterion k
F	variance ratio: $F = (\text{greater } s_x^2)(\text{lesser } s_x^2)^{-1}$
G	geometric mean: $G = \{\prod[S \cdot p(S)]\}^{1/n}$
γ_1	population skewness
γ_2	population kurtosis
g_1	sample skewness
g_2	sample kurtosis

ISBN-0-12-304905-9

H	harmonic mean: $H = n\sum[S \cdot p(S)]^{-1}$
IU	interval of uncertainty
L	luminance or limen
LL	lower limen
ℓ	likelihood ratio: $\ell = [p(e_k \mid S + N)][p(e_k \mid N)]^{-1}$
μ	population mean: $n^{-1}\sum(S_i)$
m	sample median; the value of S such that the area under the probability density distribution is equal above and below m.
M_r	rth moment about the mean of a distribution
m_r	sample midrange: $m_r = \frac{1}{2}(S_{max} - S_{min})$
N	noise in SDT (and sample or origin as specified in discussions of probability and statistics)
N_r	rth moment about the origin of a distribution
n	sample size
p	probability of "yes" or "greater" response: $p = p(\text{yes} \mid \langle S_s, S_i \rangle)$
PE	probable error: $PE \cong 0.6745\sigma$
$p(S)$	normal probability density function: $p(S) = (2\pi\sigma)^{-\frac{1}{2}}\{\exp[-(S - \mu)^2(2\sigma)^{-2}]\}$
$P(S)$	cumulative normal probability function: $P(S) = (2\pi\sigma)^{-\frac{1}{2}}\int\{\exp[-(S - \mu)^2(2\sigma)^{-2}]ds\}$
PSE	point of subjective equality
q	probability of "no" or "smaller" response: $q = (1 - p)$
R	response (and number of successes in discussions of probability)
RL	absolute limen (threshold)
ROC	receiver operating characteristic in SDT
r^2	coefficient of determination: $r^2 = (r_{xy})^2$
r_{xy}	correlation coefficient: $r_{xy} = (s_{xy})(s_x s_y)^{-1}$
S	stimulus or (in SDT) signal
SDT	signal detection theory
σ	population standard deviation: $\sigma = \{(n - 1)^{-1}(\sum[S - \mu]^2)\}^{\frac{1}{2}}$
s_x	sample standard deviation
s_{xy}	covariance
UL	upper limen
$V(X)$ or V	variance: $V = s_x^2$
v	coefficient of variation: $v = (100s_x)\bar{X}^{-1}$
$\bar{X}$	sample mean
x_i	sample i stimulus value
z	standard normal deviate: $z = (S - \mu)\sigma^{-1}$

A. INTRODUCTION

1. General Considerations

Matching and threshold methods of visual measurement are treated together in this chapter, because they represent two sides of the same coin: *discrimination.* Threshold discrimination is the process of finding the minimum stimulus or difference in stimuli that can be distinguished as different in appearance from some standard. Matching is the process of finding that stimulus or difference in stimuli that cannot be discriminated from some standard. The same experimental methods can be used to determine either a match or a threshold. Although some methods are logically more direct for determining matches and others more direct for determining thresholds, it is primarily in the way the experimental data are treated that we distinguish between the two concepts. In short, thresholds and matches comprise two aspects of discrimination; the treatment of data and the form in which they are reported depend on which kind of discrimination information they are to represent.

In this chapter we shall examine what is meant by a match and what is meant by a threshold in the context of visual measurement. The probabilistic basis for both concepts will be discussed, and the various measures of probability relations that are used to specify each of them will be outlined. Psychophysical methods used to measure both matches and thresholds will be described and examples given. Finally a general description of signal-detection theory and its application to threshold measurement will be given. The purpose of the chapter is to introduce the reader to the theory and application of matching and threshold measurement so that the one best suited for a particular task can be chosen from among the various methods available.

It will be convenient to use a number of standard abbreviations throughout this chapter, which are given in the List of Symbols. Each abbreviation and the concept it stands for will be described in the text, but it will be helpful to have a summary of them for easy reference.

a. Absolute and Terminal Thresholds

The concept of the threshold was originally intended to apply to stimuli, in particular, to intensities of stimuli. One can imagine a continuum of stimulus intensities from zero to some extremely high value. At some point on this continuum there will be a stimulus of adequate intensity to arouse a response in the organism. Stimulus values above that point are often called adequate stimuli; those below it are inadequate stimuli. Perception occurs with adequate stimuli and fails with inadequate stimuli. The *absolute threshold* is that stimulus value below which perception fails.

At the other extreme of the stimulus continuum, in the region of extremely high stimulus values, there is a point above which perception fails as a result of saturation or damage to the organism. That stimulus is called the *terminal threshold.* For obvious reasons, little work has been done to explore characteristics of terminal thresholds; that which has been done has been for the most part to establish tolerances for safe practices.

Between the extremes of absolute and terminal thresholds is an infinitude of *difference thresholds*; that is, there are infinitely many adequate stimuli that are just perceptibly different from some standard or reference stimulus. In a sense the absolute threshold is the lower limiting case of difference thresholds. The comparison standard for an absolute threshold may be thought of as zero stimulus intensity. Similarly the terminal threshold may be thought to have as its standard that stimulus that evokes a failure in perception. We will not discuss the terminal threshold in this chapter, because relatively few quantitative data are available and because visual measurement, the subject of this volume, can take place only over the range of intensities for which stimuli evoke visual perceptions. The remainder of this discussion will then center on matching and threshold discrimination from (and including) the absolute threshold to (but excluding) the terminal threshold.

To reiterate, the original sense of the absolute threshold was a point on a continuum of stimulus intensity below which an observer cannot sense the presence of the physical quantity in question and above which he can sense it. Figure 1 illustrates this concept. Stimuli to the left of the vertical dashed line labeled S_t are not adequate to evoke a physiological response; hence, the probability that they will be detected is zero. Stimuli to the right of S_t are adequate to elicit a response, and so the probability that they will be detected is unity. The point of transition from inability to detect a stimulus to ability to detect it was originally called the stimulus threshold. More exactly, because most of the earliest determinations of thresholds were reported by German-speaking workers, it was called the *Reiz Limen*, a combination of the German word for stimulus (der Reiz) and the Latin for threshold (Limen). The abbreviation RL is still used by many non-German workers and will be used throughout this chapter.* This choice of words tends to emphasize the concept of threshold as a stimulus value as opposed to some aspect of response. The response is the *criterion*; the stimulus that meets that criterion is the *specification* of threshold. The absolute threshold is the least intense stimulus that can evoke a response under a given set of viewing conditions. Any stimulus greater than this threshold is thought to be perceptible according to the dichotomous concept of threshold discussed above.

b. Increment Thresholds

The same concept can be applied to *increment thresholds*. An increment threshold is a measure of the increase in stimulus intensity required to detect a change from a background level of intensity. It is represented in Fig. 1 by the vertical dashed line labeled S_b. Stimuli to the left of S_b, when added to the background intensity, are not seen as more intense, whereas addition of those

* The German language incorporates other words for threshold, for example, *der Schwellenreiz*, *der Schwellenwert*, and *die Reizschwelle*, all of which emphasize that the threshold refers to some minimal value of the stimulus. Nonetheless, the earlier mixture of German and Latin is more often used throughout the world. Because German words for response, *die Sinneswahrnehmung*, and sensation, *die Sinnesempfindung*, begin with the letter S, distinctions between stimulus and response were made in German by the abbreviations *R* and *S*, respectively. Unfortunately, the cognate English words begin with S and R, respectively (stimulus and response). In this chapter the original German–Latin forms for absolute and difference thresholds will be used to form abbreviations, but the symbols *S* and *R*, when standing alone, will refer to the English stimulus (or signal) and response terms.

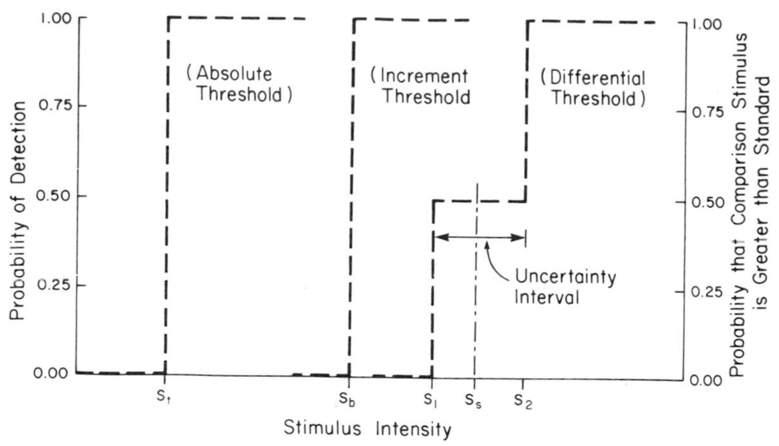

Fig. 1. Simplistic concepts of discontinuous thresholds are illustrated by the dashed lines. The absolute threshold is illustrated on the left, increment thresholds in the middle, and difference thresholds on the right. See text for discussion.

to the right of S_b yields results that are seen to be more intense than the background.

In the general case, this concept can be extended to the process of subtracting a stimulus from the background; this threshold is referred to as a *decrement threshold.* Thus increment thresholds are determined by adding stimuli to a background and decrement thresholds by subtracting stimulus intensity from the background. Together, these two concepts are closely allied to what is commonly referred to as a difference threshold.

c. Difference Thresholds

The difference threshold is more general than increment and decrement thresholds in the sense that the concept of difference threshold includes substitution as well as addition and subtraction. That is, a difference threshold may be determined for single stimuli presented with or without a background and compared explicitly or implicitly with a standard or reference stimulus of constant intensity. Usually the test and standard stimuli are presented side by side (often as halves of a bipartite field), and the question asked of the observer is whether a difference can be seen between them.

Difference thresholds will be abbreviated DL here, after the German *Differenz Limen.* They are also frequently called just-noticeable differences (JNDs). Most often, however, they are reported not as differences but as ratios of stimulus intensities in percentage form. Regardless of how they are reported, they always represent thresholds, so it is appropriate to use the Latin word for threshold (limen) here.

Figure 1 illustrates a simple dichotomous concept of the DL by the series of stepped dashed lines on the right in the graph. The DL can be either positive or negative. The increase in stimulus intensity that can just be detected is sometimes called the upper limen (UL), and the decrease in intensity that can just be detected is called the lower limen (LL), in order to distinguish direction. In some cases the UL and the LL may be unequal, and such a distinction is useful. In Fig. 1, the LL is bounded by stimulus S_1 and the UL stimulus S_2. These two stimulus intensities are unequal, indicating that there is an interval of uncertainty (IU) or range of stimulus intensities that cannot be discriminated as different from the standard stimulus S_s. If the threshold functions for differences are symmetrical about the standard (that is, $S_s - S_1 = S_2 - S_s$), as they are in Fig. 1, then the DL is equal to half the interval of uncertainty (DL = IU/2 and DL = UL = LL).

d. MATCHES

The midpoint of the interval of uncertainty is often taken as the point of subjective equality (PSE). In Fig. 1, PSE = $(S_1 + S_2)/2$. The PSE represents the condition of match.

Determination of the match-point as the midrange of the interval of uncertainty, the same interval that is used in determining difference thresholds, emphasizes the close relationship between thresholds and matches as merely different aspects of the same discrimination process. The PSE is also a necessary ingredient in determining the DL when threshold functions are asymmetrical. Unlike symmetrical distributions, where UL = LL, the two limen cannot be derived from the interval of uncertainty alone in asymmetrical distributions. Instead, the limen are determined separately as the distances along the stimulus continuum from the PSE and the transitions to just perceptibly greater (UL) and just perceptibly less (LL).

e. DISCRIMINATION AS A PROBABILISTIC PROCESS

The graph of Fig. 1 has been helpful in introducing some basic concepts of thresholds and matching, but in practice, determinations of thresholds and matches do not yield discrete, discontinuous results such as those shown in the figure. Instead, the probability of discrimination usually varies continuously over some range of stimulus intensities. In other words, the likelihood of seeing or detecting a stimulus or a change in stimuli gradually increases from certain lack of detection to certainty of detection. In Chapters 4 and 5 it was noted that many factors affect the momentary physiological state of the organism. Its sensitivity varies from moment to moment in an essentially random fashion. Although these fluctuations may not be large compared with the level of physiological activity elicited by an adequate stimulus, they do cause momentary departures from constant sensitivity. The result is that a stimulus that is detected at one moment or on one trial may not be detected at another moment or on a different trial. In addition to physiological fluctuations, there are other variable factors that affect threshold determinations. In any such task the observer must decide when to report "yes, I see it" and when to respond "no, I don't see it." The criteria used to make this decision may vary from moment to moment. Finally, the stimuli themselves may vary. For example, the rate at which photons are delivered to the eye varies with any real source of illumination, and this may be a large or small fraction of the quantum integration required for detecting the light. In short, there are a number of factors that may vary from any one instantaneous determination of a threshold or match to another. Let us examine two hypothetical examples to understand the effect of such variations on the threshold distribution function.

Suppose that a 20° diameter circle of spectral light with wavelength 507 nm is presented to an observer for a 1-sec interval after the observer has become adapted to the dark. Further, assume that the luminance of the light is fixed at a single value somewhere in the region of 0.2 to 2.0×10^{-6} $\text{cd} \cdot \text{m}^{-2}$; this luminance is then near the absolute threshold for these conditions. The observer's task is to say *yes* if he sees the light or *no* if he does not. Let us present the light many times over a series of trials, giving the observer a suitable period between presentations to allow him to readapt to the dark. Typically the observer will say *yes* on some trials and *no* on other trials. If we assume that there is no variation in the photon flow rate from one presentation to another, we may evaluate the observer's responses in terms of *yes* and *no* reactions to that intensity of light under those conditions of viewing. The probability of a *yes* response to the stimulus may be symbolized as $p(\text{yes} \mid S)$, which is to be read as "the probability of a yes given S." Similarly the probability of no given S is symbolized as $q(\text{no} \mid S)$. These probabilities can be computed simply as the relative frequencies of their occurrences. For example, if the light was presented 100 times, and the observer responded yes on 20 trials and no on 80 trials, then $p(\text{yes} \mid S) = 0.2$ and $q(\text{no} \mid S) = 0.8$. Together their sum is 1.0 because, as relative frequencies, $q(\text{no} \mid S) = [1 - p(\text{yes} \mid S)]$.

We could then say that the observer is expected to detect that stimulus intensity 20% of the time under those conditions. However, that is not a very high probability that the light will be detected; it means, after all, that it will not be seen 80% of the time. If we want to find out what stimulus intensity would be required to be more certain, say, 75% certain, we must present lights of other intensities as well. If we were to use several lights with small enough intervals of intensity differences among them and choose enough lights to cover a reasonably large range of intensities, we could then determine the probabilities for yes responses to each of them. From this information we can find the stimulus intensity for which the probability of detection is essentially zero (below which we should not expect the observer ever to see the light) and the stimulus intensity for which the probability of detection is essentially unity (above which we should expect the observer always to see the light). In addition, we should have a measure of the probabilities of detecting the lights of intermediate intensities. When all of these probabilities are plotted against stimulus intensity, we generate a graph of the distribution of probabilities of detection with respect to stimulus intensity.

Figure 2(a) illustrates such a graph. The data points do not follow the simple, vertical, dichotomous threshold functions of Fig. 1 (reproduced in Fig. 2 as the dashed line). Instead, the data describe an approximately S-shaped curve along which the probability of detection gradually increases from near zero to near unity. We can choose any level of probability as a criterion of detection. For example, we might choose a level of 50%, meaning that the observer is likely to detect that intensity 50% of the time under those conditions. The RL is then specified as the stimulus intensity on the abscissa corresponding to a 50% probability on the ordinate. Suppose, however, that the observer's ability to detect the light is a matter of considerable consequence, as it might be in navigation or signalling applications. Then we could use the same data to choose some other criterion level corresponding to higher probability of detection, possibly 75%. The same graph will tell us which higher stimulus intensity would be appropriate to this more conservative criterion. The point is that the data describe probabilities of detection as a function of stimulus intensities in a way that depends only on the conditions of the experiment (the observer, the viewing and adaptation

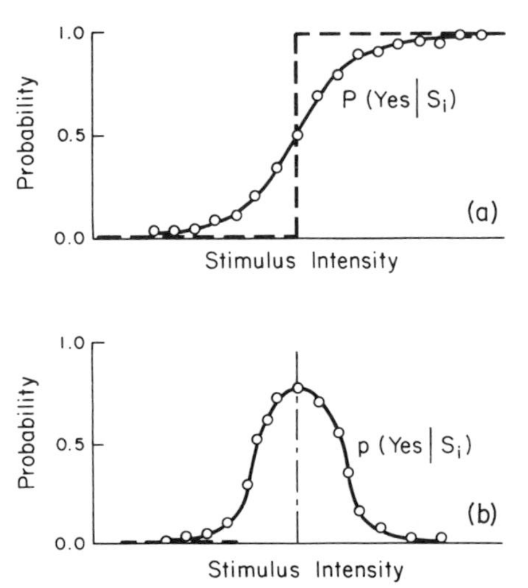

Fig. 2. Hypothetical probability distributions. (a) represents probability of yes responses for determination of absolute threshold. (b) represents probabilities of yes responses for matching.

conditions, and so on), but the criterion may be determined by ulterior considerations. There is, then, no single level of probability that serves to define *the* threshold. Any nonzero probability level may be operationally defined as the threshold. Levels most often chosen are 50%, 60%, about 68%, and 75%. The higher the level the more conservative the detection criterion in the sense that the more certain we are that it will be seen.

A second illustration of the probabilistic nature of responses deals with matching the intensities of two lights. Assume that we have a standard light of known intensity that occupies 2° diameter area of the central visual field. The problem is to determine the intensity of a test light of the same size and chromaticity that provides a brightness match to the standard when they are seen side by side. When the test light is presented for a brief period, we ask the observer whether he is satisfied that it matches the standard. Again, let us have the observer respond to many presentations of the test light. Perhaps in 100 trials he will say yes 65 times and no 35 times. The probability of his being satisfied with the match is reasonably high (65%), but would it have been higher if we had chosen a different intensity for the test light? The way to find out, of course, is to present the test light at many different intensities with many trials for each intensity level. As with the determination of the absolute threshold, we should like to choose sufficiently large numbers of intensities and suitably small intervals of difference in intensity so that we can describe satisfactorily the transition of the observer's responses from a near-zero probability of matching because the light is too dim to a near-zero probability of matching because the light is too bright.

When this is done, and the measured probabilities of yes responses are plotted against stimulus intensity, a graph such as that shown in Fig. 2(b) results. The probability that the observer will call a given stimulus intensity a match to the standard varies continuously from near zero when the intensity is too low, through some maximum value (not necessarily unity), to near zero again when the intensity is too high. In other words, the yes-response probabilities are distributed, or dispersed, over a range of stimulus values. They are not all located at some single value such as that represented by the vertical broken line; their densities are high for some stimulus values and low for others. That being so, our best estimate of the matching stimulus intensity is some measure corresponding to the maximum probability value or to the central tendency of the *probability density distribution* (the name used to describe the kind of distribution shown in Fig. 2(b), where probability versus stimulus intensity is assumed to form a continuum).

The stimulus value corresponding to the peak probability is the *mode* of the distribution (determined by calculus from the probability density function or, more simply, as the peak ordinate point of the probability density distribution). Measures of central tendency include the *midrange*, the average of the smallest and largest stimuli for which the probability is just not zero;

$$S_{\text{midrange}} = m_{\text{r}} = \left[\frac{S_{\max} - S_{\min}}{2}\right]; \tag{1}$$

the *median*, that stimulus value which divides the probability density distribution exactly in half

$$S_{\text{median}} = m; \qquad p(\text{yes} \leqq m) = 0.5 \quad \text{and} \quad p(\text{yes} \geqq m) = 0.5; \tag{2}$$

one or another measure of the mean, the simple *arithmetic mean*, or average,

$$S_{\text{arithmetic mean}} = \mu = \left\{\frac{\sum[S \cdot p(S)]}{n}\right\}; \tag{3}$$

or the *harmonic mean*, generally appropriate to data involving rates,

$$S_{\text{harmonic mean}} = H = n\sum[S \cdot p(S)]^{-1}; \tag{4}$$

or the *geometric mean*, which is sometimes used with skewed distributions,

$$S_{\text{geometric mean}} = G = \sqrt[n]{\prod[S \cdot p(S)]}, \tag{5}$$

where S stands for stimulus intensity, $p(S)$ represents the associated probability or frequency of occurrence of a yes response, $\sum$ means cumulative sums, and $\prod$ means cumulative products.

Notice that in matching, as for threshold detection, there is no single criterion for deciding which stimulus intensity corresponds to a match. By far the most often used estimate is the arithmetic mean, but this may not be the most appropriate value if the probability density distribution is asymmetrical or polymodal.

The same concepts elucidated for the simple, dichotomous model of thresholds illustrated in Fig. 1 apply also to the probabilistic models illustrated in Fig. 2. We have assumed that for each test stimulus S_i compared with a standard stimulus S_s there exists a probability that the test will be seen as greater than the standard. We may express that probability as $p(\text{yes}|\{S_s, S_i\})$, which should be read as "the probability of yes given S_i compared with S_s." The probability that the test will be seen as smaller than the standard is the complementary expression. These probabilities are assumed to satisfy three requirements: monotonicity, continuity, and certain limit constraints. *Monotonicity* means that if two stimuli, S_i and S_j, are unequal in the sense that $S_i < S_j$, then $p(\text{yes}|\{S_s, S_i\}) < p(\text{yes}|\{S_s, S_j\})$ when neither probability is zero or unity. *Continuity* implies that $p(\text{yes}|\{S_s, S_i\})$ can be mathematically differentiated and, therefore, is a continuous function of S_s and S_i. *Limit constraint* simply means that the limits of probability are zero and unity:

$$\lim_{S_i \to \infty} p(\text{yes}\,|\,\{S_s, S_i\}) = 1 \quad \text{and} \quad \lim_{S_i \to 0} p(\text{yes}\,|\,\{S_s, S_i\}) = 0.$$

The probability density distribution is represented by values of $p(\text{yes}|\{S_s, S_i\})$. These can be taken to be the partial derivative of the *cumulative probability distribution*,

$$P(\text{yes}\,|\,\{S_s, S_i\}) = \int_{-\infty}^{+\infty} p(\text{yes}\,|\,\{S_s, S_i\})\cdot dS_i.$$

The cumulative probability distribution is the S-shaped curve of Fig. 2(a). It has often been called the *psychometric function*, a term introduced by Urban (1908) and derived from two Greek words (ψυχή and μέτρον) meaning measurement of the mind. Psychologists often use the work *psychometrics* to mean measurement of differences among people with respect to some specified task, particularly when the stimulus is complex or cannot be characterized; for example, intelligence quotient (IQ) scores are psychometric measures, and Rayleigh ratios, determined with an anomaloscope, are psychometric measures of color vision capability. The key element is a measure of differences among responses of people. In contrast with this is the concept of the cumulative probability distribution of discrimination among stimuli; here, also, $P(\text{yes}\,|\,\{S_s, S_i\})$ may represent data averaged over trials of a single observer or averaged over trials of many observers. Accordingly, the appellation *psychometric* will be eschewed here. Instead, the function will simply be called the cumulative probability distribution and will be symbolically distinguished from the probability density distribution by the use of an uppercase P rather than a lowercase p.

If the stimulus corresponding to a match is to be determined as the median of a unimodal probability density function, then the point of subjective equality is found where

$$\text{PSE} = P(\text{yes}\,|\,\{S_s, S_{1/2}\}) = \int_{0}^{S_{1/2}} p(\text{yes})\,|\,\{S_s, S_i\})\cdot dS_i = 0.5. \tag{6}$$

If the difference limen is taken as one-half the interquartile range, then

$$\text{DL} = \left[\frac{P(\text{yes}\,|\,\{S_s, S_{3/4}\}) - P(\text{yes}\,|\,\{S_s, S_{1/4}\})}{2}\right]. \tag{7}$$

Other examples will be detailed in later sections of this chapter. The foregoing illustrate the point that all of the relations described in connection with Fig. 1 apply as well to the probability distributions of Fig. 2.

2. Weber's law

Over the 20-yr period ending about 1849, Ernst Heinrich Weber, a German anatomist and physiologist at Leipzig University, collected experimental evidence on difference thresholds for a number of sense departments. He studied touch, weight, and length sensitivities to stimulus differences.

Combining his experimental results with previous data gathered by others on sensitivity to changes in musical pitch, he formulated a simple stimulus relation that he believed applied generally to difference thresholds for all sense departments (Weber, 1834). Simply stated, Weber's idea was that the difference between one stimulus and another that is just noticeably different is a constant fraction of the first. Stated mathematically

$$\left(\frac{\Delta S}{S}\right) = k, \tag{8}$$

where ΔS represents the DL, S the original stimulus, and k is a constant. Stated in this form the relationship is called the *Weber fraction*. Multiplying both sides by S, one obtains *Weber's law*,

$$\Delta S = k \cdot S. \tag{9}$$

The implication of Weber's law is illustrated in Fig. 3. The horizontal line represents a constant Weber fraction. The diagonal line represents a "perfect" detection system in which the DL is a constant stimulus value. At moderate intensities the constant Weber-fraction form is found for most experimental determinations of thresholds. Accordingly, the difference in stimulus intensity that can be detected tends to increase in proportion to the increasing stimulus

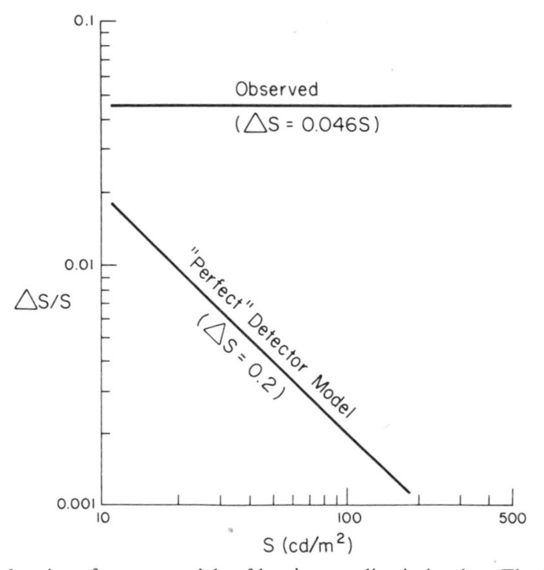

Fig. 3. Weber fractions for two models of luminance discrimination. The "perfect" detector model assumes constant photon flow rate and discontinuous threshold functions; DL is constant with S. The upper function, labeled "observed," is a probabilistic threshold function that follows Weber's law; DL is proportional to S.

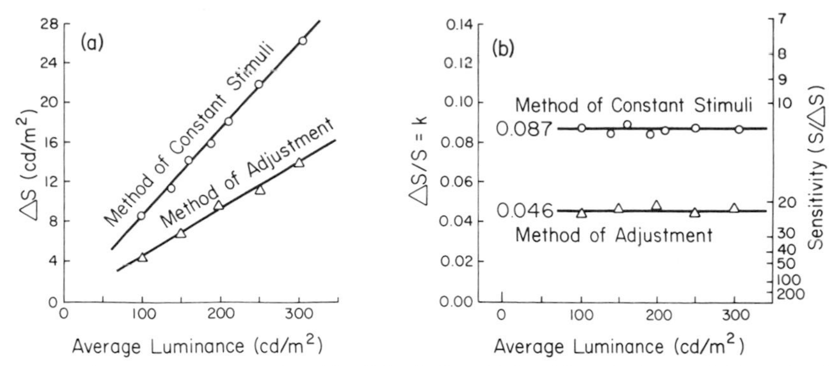

Fig. 4. Functions representing (a) Weber's law ($\Delta S = kS$) and (b) the Weber fraction ($k = \Delta S/S$) determined by two different psychophysical methods: method of constant stimuli (circles) and method of adjustment (triangles). Sensitivity, calculated as the reciprocal of the Weber fraction, is shown on the ordinate farthest to the right.

level. Figure 4(a), in which ΔS is plotted against S, illustrates this last point. Figure 4(b), in which $\Delta S/S$ is plotted against S, depicts Weber fractions.

Two things should be noted about the graphs of Fig. 4. First, there are two lines in each graph. One of these lines represents data determined by the experimental method of constant stimuli and the other by the method of adjustment; these methods will be described in a later section of this chapter. The point to be noted now is that the slope of the function representing Weber's law or the value of the Weber fraction may differ from one kind of experimental method to another. *Relative sensitivity* is conventionally defined as the reciprocal of the Weber fraction; a scale of sensitivity is shown at the right of Fig. 4(b). We may say, then, that experimental methods differ in their sensitivity. In the example illustrated in the figure, the method of adjustment yields sensitivity measures nearly twice as high as those found by the method of constant stimuli.

Earlier in this chapter it was pointed out that any of several measures could be used to determine DL from a given set of experimental data. Because ΔS and DL are one and the same, it follows that ΔS can also be determined according to different criteria. The probability level selected to define the threshold may differ from one data-reduction method to another. Even for a single experimental method and data reduction method, the measure of dispersion used to define the interval of uncertainty and the DL may differ. In turn, these choices influence the size of the Weber fraction. Consider, for example, three common measures of dispersion: the standard deviation σ, the probable error PE, and the average deviation D_a; all of these terms will be defined in a later section, but the important point at this juncture is that each can yield a somewhat different value of dispersion. When a probability of detection of 50% is used, the Weber fraction k is equal to the measure of dispersion divided by the stimulus value of the PSE:

$$k_1 = \frac{\sigma}{\text{PSE}},$$

or

$$k_2 = \frac{PE}{\text{PSE}},$$

or

$$k_3 = \frac{D_a}{\text{PSE}},$$

in the examples. Each k will be different, exactly how different depends on the shape of the probability density distribution, and these differences are noted by the subscripts. If we recall that the PSE may be determined as midrange, mode, median, or one of several kinds of means, then it is clear that there are a number of degrees of freedom in calculating the Weber fraction. The value of k depends on the choices of measures to represent ΔS and S. The choices may be arbitrary (but should not be capricious) and usually are suggested by the form of the probability density distribution. In any event it should be clear that there is no single value of Weber fraction for any one sense department, despite implications found in much of the literature.

The second thing that should be noted about the graphs of Fig. 4 is that the range over which the stimuli were varied is rather small (only about 3:1), and it is centered well within the range of photopic luminances. When a much broader range of intensities is used, the Weber fraction does not remain constant. It tends to increase at very low intensities and sometimes increases at very high intensities as well. Typically, results for very large ranges of intensities resemble those shown in Fig. 5, the data of which are taken from a classic study of luminance discrimination by König and Brodhun (1889). It is obvious that the Weber fraction is far from constant over the approximately eight decades of luminance that are illustrated. Note, however, that the abscissa of the graph displays uniform intervals of the logarithm of luminance. If, instead, luminance were to be arrayed arithmetically, the Weber fraction could more easily be seen to be near its minimum over most of the luminance range.

The lack of constancy of the Weber fraction found in most experiments in which the DL are determined for intensive stimulus continua has been the subject of controversy in discussions about the utility of Weber's law. It has sometimes been suggested that the Weber fraction has little, if any, value as an index of sensitivity, and that is certainly true if only a single value of the fraction is quoted without qualification. Woodworth and Schlosberg (1954, pp. 224–225) offered the sensible view that Weber's law "holds as a rough

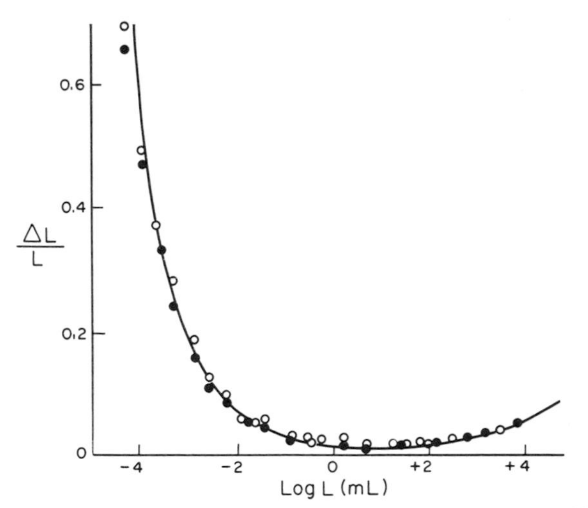

Fig. 5. Weber fractions $\Delta L/L$ for luminance discrimination as a function of log L (in mL) as reported by König and Brodhun (1889).

empirical generalization in the mid-ranges of most senses.... By and large, these mid-ranges are the working areas of the senses... [and] Weber's law furnishes us with a valuable description of the discriminating power of the important sensory ranges."

A number of suggestions have been made for "correcting" Weber's law; Guilford (1954) reviews several. Most of these corrections fall into one of two categories: (a) empirical corrections that may apply well to certain experimental results but not to all results and (b) general forms that are based on certain theoretical considerations but are only fair predictors of experimental results. The first class is not of general interest and need not concern us here. The most common form of the second category involves modification of Weber's law by an additive constant. It is often referred to as the *generalized form of Weber's law* and may be stated as

$$\Delta S = k \cdot S + c. \tag{10a}$$

The constant c may be taken to represent some minimum stimulus value S_0, which is not necessarily the absolute threshold RL but depends on the conditions of viewing,

$$\Delta S = k \cdot S + S_0. \tag{10b}$$

S_0 is, nevertheless, a threshold. Furthermore, for any given viewing condition, it is the lower limit of discrimination. If we symbolize that limit as $\ll$RL$\gg$, then the general form of Weber's law may be expressed in terms of limen as

$$\mathrm{DL} = k \cdot S + \ll \mathrm{RL} \gg, \tag{11}$$

where $\ll$RL$\gg$ does not necessarily equal RL.

Thus far the discussion has been concerned with discrimination along intensive stimulus continua. That is, an increase in stimulus value is taken to evoke an increase in response. However, there are also some stimuli or aspects of stimuli that may be considered qualitative rather than intensive; a change in the stimulus may evoke a change in the quality of response. Figure 6 illustrates such a case. Difference limen $\Delta\lambda$ were determined for changes in wavelength λ that could just be detected; the data are from another classic experiment reported by König and Dieterici (1884). Obviously the DL do not reflect the same form of relation to stimulus values here as in Fig. 4. The reason is that wavelength λ does not correspond to a simple progression of stimulus intensity. These experiments measured sensitivity to change in quality rather than quantity of stimuli. The result is that Weber's law does not apply to the results. The measured quantity ΔS still represents the DL, but the Weber fraction is not constant with stimulus value, and we would not expect it to be.

B. PROBABILITY DISTRIBUTIONS

Thus far we have considered matching and thresholds as statistically determined values of stimuli that satisfy one of two classes of perceptual criteria: a stimulus or difference in stimuli that is detected or a difference in stimuli that is not detected. The concepts of upper and lower limen and of increment and decrement thresholds imply, in addition, the idea of *order*, an awareness of greater or smaller.

The remainder of this chapter will describe psychophysical methods that are useful for determining such thresholds or matches. Because the underlying

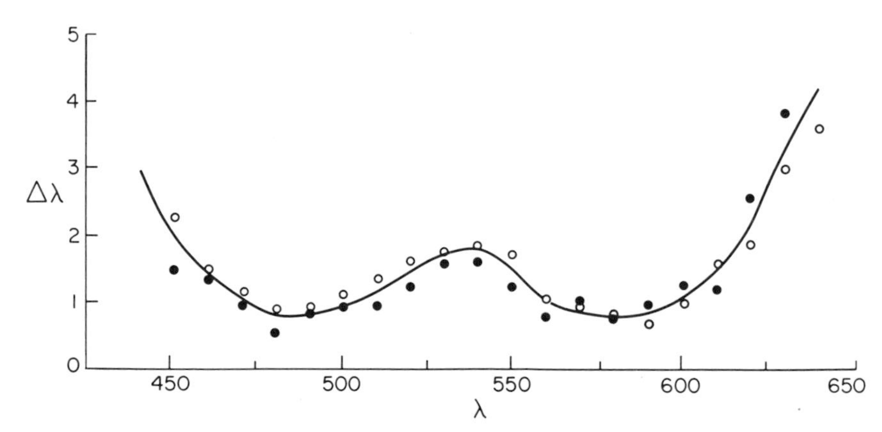

Fig. 6. Difference limen $\Delta\lambda$ for wavelength λ discrimination as reported by König and Dieterici (1884).

processes are probabilistic, it will be useful first to discuss probability distributions and some characteristics of probability relationships.

1. Probability Distributions

All probabilities have five important properties:

1. The probability of an event ω is zero if and only if the event cannot occur; $p(\omega) = 0$ if and only if ω does not exist.
2. Probabilities are always between 0 and 1; $0 \leqq p(\omega) \leqq 1$.
3. If the probability of an event is 1, the event must occur; $p(\omega) = 1$ if and only if ω always exists.
4. If events ω and ζ are mutually exclusive or inconsistent, their probabilities add; $p(\omega \vee \zeta) = p(\omega) + p(\zeta)$.
5. When the probabilities of an event ω and its negation Ω are added, all possible outcomes are included, and, therefore, the summed probability must be 1; $p(\omega) + p(\Omega) = 1$, hence $p(\Omega) = 1 - p(\omega)$.

In each case the term *probability* means the numerical value of the chance occurrence of one or several possible outcomes of an unpredictable event. We may define a *sample space* as the mathematical set of all elementary events that are possible outcomes of some simple experiment. If X represents a function that associates a real number with each elementary event in a sample space S, then X is called a random variable that is defined on the sample space S; in other words, a random variable X is a real-valued function defined on a sample space. X is the *probability function.* In general, any statement of a probability function having a set of mutually exclusive and exhaustive events for its domain is a *probability distribution.* The probability distribution is, then, a set of probabilities for the elementary events in S for the random variable X defined on S.

The purpose of stating these rules in the general, set-theory case is to emphasize that any distribution or function that meets the requirements of the conditions outlined by the statement constitutes a probability distribution or its characterizing function. These distributions or functions may vary from one kind of process to another; the shapes of the graphical representations of the functions may differ. Figure 7 illustrates a number of possible probability distributions. They differ in shape: their peak probability values (modes) occur at different places along the abscissa; their centers of gravity (medians) are located at different places along the abscissa; their spreads or dispersions along the abscissa also vary. Each distribution represents a quite different probability function. In short, there may be many different classes of probability functions. Different classes result from different processes.

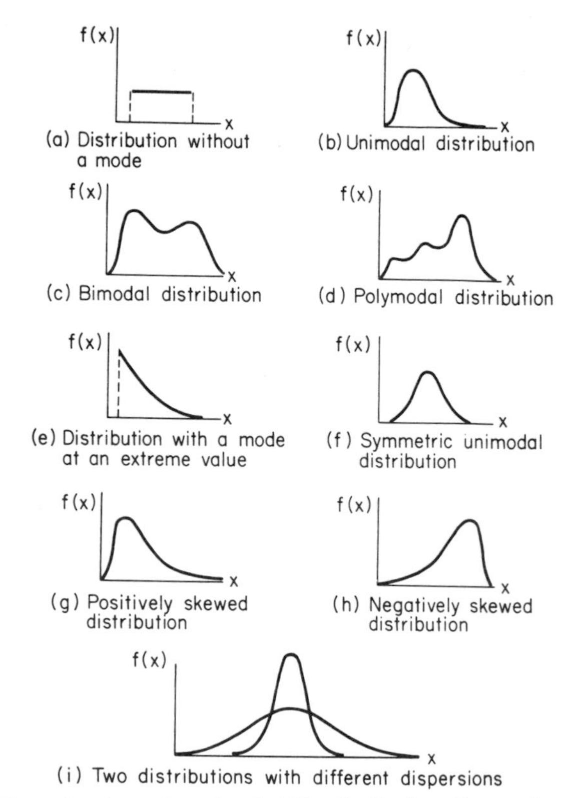

Fig. 7. Examples of a variety of probability distributions. See text for discussion.

Probability theory is based on the outcomes of some well-defined process that leads to a single, well-defined result. There are a number of processes for which the probability of an outcome can be well defined. Some of the best known among these are Bernoulli processes (with the outcome either one of only two events), the binomial process (in which exactly r successes are observed in n independent Bernoulli trials), Pascal distributions (a process in which the number of successes in a fixed number of Bernoulli trials is specified), geometric distributions (a special case of the Pascal process in which sampling continues only until the first success), Poisson distributions (a continuous, rather than finite, analogy to the binomial process), beta distributions (a continuous cognate to the Bernoulli process), and gamma distributions (corresponding conjugate family to the Poisson process). There are others as well. These probability processes are useful for defining probability mass functions (for the finite case) and density functions (for the continuous case) where asymmetrical distributions of matching and threshold data are involved. The beta function, for example, can be made to fit data of virtually all kinds, from flat (no mode) distributions to those with negative or positive extreme modes, and anything in between. We are interested in fitting a function to data to be able to make the best possible estimates of measures, such as mode or mean for PSE, and variance or some other measure of spread for IU, and to compute probability as a function of stimulus intensity for setting criteria for thresholds. Details of such probability processes can be found in standard textbooks on mathematical statistics (for instance, Kendall and Stuart, 1958/1961/1966; Derman, Gleser, and Olkin, 1973; Winkler and Hays, 1975).

When data form a bimodal or, in general, polymodal distribution, no one of these probability functions will adequately fit them. In fact, when the data yield a distribution with more than one mode, there is usually something wrong in the experiment. A bimodal result often indicates that more than one task was addressed or more than one perceptual criterion was used by different observers or by the same observer at different times. In such cases, the best thing to do is to try to analyze the data to determine the nature of the problem. Often such results can be segregated into two or more classes of unimodal distributions, so that each one can be treated by fitting a standard function to it.

Many symmetric distributions can be fitted closely with the *normal distribution*, a continuous (density) distribution based on random incidence of deviations from a centroid. It is sometimes called a *Gaussian distribution*, after the German scientist Carl Friedrich Gauss, but the normal probability function was also discovered independently by Gauss' contemporaries: the French–English mathematician Abraham de Moivre and the French astronomer–mathematician Pierre Simon LaPlace. Here the distribution will simply be called the normal distribution. Use of the normal distribution to characterize data is so common, and so much of statistical methodology and scaling is based on the assumption of normality, that the whole of the next section will be devoted to a summary of that family of distributions.

2. The Normal Distribution

Much of probability theory involves calculating probabilities for various sample outcomes from a defined process. The probabilities are calculated from functions that completely specify the details of the process. Statistical theory is somewhat different. Generally, when we are dealing with a statistical problem, the process is not defined. We may not know everything about the population or the process, but we do have some information about them. In a psychophysical experiment we have at least gathered results from a random sampling of the population or process. That information can be analyzed from descriptive and inferential points of view. *Descriptive statistics* is the term used for a set of procedures to describe or summarize experimental information.

Inferential statistics refers to methods for using the data to draw inferences about a population or process. The normal distribution is by far the most often encountered distribution in descriptive and inferential statistics. There are a number of reasons for this. Of course, there may be a priori knowledge that a process or a population is theoretically normal; therefore, the data represent normally distributed probabilities plus associated measurement errors. The normal distribution also serves as a good approximation to a number of other theoretical distributions that are more difficult to deal with, or about which less is commonly known than for the normal distribution. Often, however, sampling distributions involving large values of n are closely approximated by the normal distribution even when the population from which the samples were drawn is not normal. This is a consequence of what is called the *central limit theorem* which, simply stated, defines the normal distribution as the limiting form for large n of a wide variety of sampling distributions. Because most psychophysical methods yield data that fall into this latter class, the normal distribution is also most often encountered in psychophysics.

a. SHAPE OF THE NORMAL DISTRIBUTION

The normal distribution is completely specified only by its mathematical rule, that is, by the *normal density function.* If we symbolize density as y and set it equal to some function of x, then the equation for the normal density function is

$$y = f(x) = \left(\frac{1}{\sqrt{2\pi\sigma}}\right)e^{-\left(\frac{(x-\mu)^2}{2\sigma^2}\right)}. \tag{12}$$

Because π, e, and 2 are constants, the operative part of Eq. (12) is the exponent. This form of equation dictates that the distribution is symmetrical about its mean μ and that it has two inflection points, at $\pm\sigma$. Figure 8 illustrates these attributes of the normal distribution. In Fig. 8(a), the perfect symmetry of the distribution about μ is illustrated. In addition there is exact coincidence of all measures of central tendency; that is, the mode, the midrange, and all forms of mean lie at the same point $\bar{X}$. Figure 8(b) illustrates the positions of the two inflection points at $\pm\sigma$. The mean score μ or stimulus value on the abscissa is 10, and σ is specified as 1. Therefore, the inflection points are located at scores of 9 and 11. Thus in terms of σ, the position of the mean can be specified as 0σ and those of the inflection points as $+1\sigma$ and -1σ. These intervals may be used to define a new abscissa scale related to the size of σ. This has been done in Fig. 8(c). The new scale is labeled a z scale. The symbol z represents the *standard normal deviate*. It was previously noted that, except for a scale factor that also involves σ, the working part of Eq. (12) is the exponent. The standard normal deviate z is simply

$$z = \left[\frac{x-\mu}{\sigma}\right], \tag{13}$$

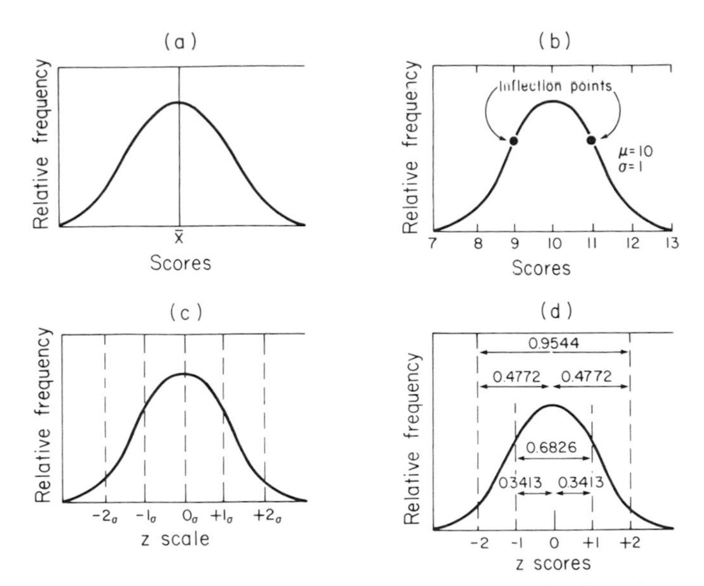

Fig. 8. Properties of the normal distribution. See text for discussion.

which inspection reveals as being related to the square root of the exponent of Eq. (12), up to a constant. Intervals of z are equal to intervals of σ and are often called z scores as in Fig. 8(d). The advantage of z scores is that any normal distribution specified in terms of them has exactly the same shape, with $\mu = 0$ and $\sigma = 1$. In other words, z scores provide a general (relative) description of all normal distributions, so results of different experiments can be compared directly. The density function of such a standardized distribution is

$$f(z) = \left(\frac{1}{\sqrt{2\pi}}\right)e^{-z^2/2}. \tag{14}$$

The shape of the normal distribution is such that most results are found close to the mean; the larger the absolute value of z, the more removed is the result from the mean. This means that results with large absolute values of z are less likely to occur than those with small zs. Probabilities are measured as the area under the curve, symbolized as A, between the mean μ and a given z value. Figure 8(d) illustrates a number of As. In the standard normal density distribution of Fig. 8(d), the area under the curve between the mean ($\mu = 0z$) and $z = +1$ is 0.3414 (or a little more than 34% of the total area). The area between $z = 0$ and $z = -1$ is also 0.3414, because the standard normal distribution is symmetric. Together, the two areas sum to 0.6826 (or about 68% of the total area of the normal curve). Analogously, the areas between the mean and $\pm 2z$ add to 0.9544 (roughly 95% of the total area). Table I lists the relations among values of z, x, y, A, and $1 - A$ for z intervals of 0.05 from 0.00

TABLE I. Normal Distribution[a]

z	x	y	A	$1 - A$	z	x	y	A	$1 - A$
0	μ	0.399	0.0000	1.0000	±1.50	$\mu \pm 1.50\sigma$	0.1295	0.8664	0.1336
±0.05	$\mu \pm 0.05\sigma$	0.398	0.0399	0.9601	±1.55	$\mu \pm 1.55\sigma$	0.1200	0.8789	0.1211
±0.10	$\mu \pm 0.10\sigma$	0.397	0.0797	0.9203	±1.60	$\mu \pm 1.60\sigma$	0.1109	0.8904	0.1096
±0.15	$\mu \pm 0.15\sigma$	0.394	0.1192	0.8808	±1.65	$\mu \pm 1.65\sigma$	0.1023	0.9011	0.0989
±0.20	$\mu \pm 0.20\sigma$	0.391	0.1585	0.8415	±1.70	$\mu \pm 1.70\sigma$	0.0940	0.9109	0.891
±0.25	$\mu \pm 0.25\sigma$	0.387	0.1974	0.8026	±1.75	$\mu \pm 1.75\sigma$	0.0863	0.9199	0.0801
±0.30	$\mu \pm 0.30\sigma$	0.381	0.2358	0.7642	±1.80	$\mu \pm 1.80\sigma$	0.0790	0.9281	0.0719
±0.35	$\mu \pm 0.35\sigma$	0.375	0.2737	0.7263	±1.85	$\mu \pm 1.85\sigma$	0.0721	0.9357	0.0643
±0.40	$\mu \pm 0.40\sigma$	0.368	0.3108	0.6892	±1.90	$\mu \pm 1.90\sigma$	0.0656	0.9426	0.0574
±0.45	$\mu \pm 0.45\sigma$	0.361	0.3473	0.6527	±1.95	$\mu \pm 1.95\sigma$	0.0596	0.9488	0.0512
±0.50	$\mu \pm 0.50\sigma$	0.352	0.3829	0.6171	±2.00	$\mu \pm 2.00\sigma$	0.0540	0.9545	0.0455
±0.55	$\mu \pm 0.55\sigma$	0.343	0.4177	0.5823	±2.05	$\mu \pm 2.05\sigma$	0.0488	0.9596	0.0404
±0.60	$\mu \pm 0.60\sigma$	0.333	0.4515	0.5485	±2.10	$\mu \pm 2.10\sigma$	0.0440	0.9643	0.0357
±0.65	$\mu \pm 0.65\sigma$	0.323	0.4843	0.5157	±2.15	$\mu \pm 2.15\sigma$	0.0396	0.9684	0.0316
±0.70	$\mu \pm 0.70\sigma$	0.323	0.5161	0.4839	±2.20	$\mu \pm 2.20\sigma$	0.0355	0.9722	0.0278
±0.75	$\mu \pm 0.75\sigma$	0.301	0.5467	0.4533	±2.25	$\mu \pm 2.25\sigma$	0.0317	0.9756	0.0244
±0.80	$\mu \pm 0.80\sigma$	0.290	0.5763	0.4237	±2.30	$\mu \pm 2.30\sigma$	0.0283	0.9786	0.0214
±0.85	$\mu \pm 0.85\sigma$	0.278	0.6047	0.3953	±2.35	$\mu \pm 2.35\sigma$	0.0252	0.9812	0.0188
±0.90	$\mu \pm 0.90\sigma$	0.266	0.6319	0.3681	±2.40	$\mu \pm 2.40\sigma$	0.0224	0.9836	0.0164
±0.95	$\mu \pm 0.95\sigma$	0.254	0.6579	0.3421	±2.45	$\mu \pm 2.45\sigma$	0.0198	0.9857	0.0143
±1.00	$\mu \pm 1.00\sigma$	0.242	0.6827	0.3173	±2.50	$\mu \pm 2.50\sigma$	0.0175	0.9876	0.0124
±1.05	$\mu \pm 1.05\sigma$	0.230	0.7063	0.2937	±2.55	$\mu \pm 2.55\sigma$	0.0154	0.9892	0.0108
±1.10	$\mu \pm 1.10\sigma$	0.218	0.7287	0.2713	±2.60	$\mu \pm 2.60\sigma$	0.0136	0.9907	0.0093
±1.15	$\mu \pm 1.15\sigma$	0.206	0.7499	0.2501	±2.65	$\mu \pm 2.65\sigma$	0.0119	0.9920	0.0080

± 1.20	$\mu \pm 1.20\sigma$	0.194	0.7699	0.2301	± 2.70	$\mu \pm 2.70\sigma$	0.0104	0.9931	0.0069
± 1.25	$\mu \pm 1.25\sigma$	0.183	0.7887	0.2113	± 2.75	$\mu \pm 2.75\sigma$	0.0091	0.9940	0.0060
± 1.30	$\mu \pm 1.30\sigma$	0.171	0.8064	0.1930	± 2.80	$\mu \pm 2.80\sigma$	0.0079	0.9949	0.0051
± 1.35	$\mu \pm 1.35\sigma$	0.160	0.8230	0.1770	± 2.85	$\mu \pm 2.85\sigma$	0.0069	0.9956	0.0044
± 1.40	$\mu \pm 1.40\sigma$	0.150	0.8385	0.1615	± 2.90	$\mu \pm 2.90\sigma$	0.0060	0.9963	0.0037
± 1.45	$\mu \pm 1.45\sigma$	0.139	0.8529	0.1471	± 2.95	$\mu \pm 2.95\sigma$	0.0051	0.9968	0.0032
± 1.50	$\mu \pm 1.50\sigma$	0.130	0.8664	0.1330	± 3.00	$\mu \pm 3.00\sigma$	0.0044	0.9973	0.0027
					± 4.00	$\mu \pm 4.00\sigma$	0.0001	0.99994	0.00006
					± 5.00	$\mu \pm 5.00\sigma$	0.000001	0.9999994	0.0000006
± 0.000	μ	0.3989	0.0000	1.0000	± 1.036	$\mu \pm 1.036\sigma$	0.2331	0.7000	0.3000
± 0.126	$\mu \pm 0.126\sigma$	0.3958	0.1000	0.9000	± 1.282	$\mu \pm 1.282\sigma$	0.1755	0.8000	0.2000
± 0.253	$\mu \pm 0.253\sigma$	0.3803	0.2000	0.8000	± 1.645	$\mu \pm 1.645\sigma$	0.1031	0.9000	0.1000
± 0.385	$\mu \pm 0.385\sigma$	0.3704	0.3000	0.7000	± 1.960	$\mu \pm 1.960\sigma$	0.0584	0.9500	0.0500
± 0.524	$\mu \pm 0.524\sigma$	0.3477	0.4000	0.6000	± 2.576	$\mu \pm 2.576\sigma$	0.0145	0.9900	0.0100
± 0.674	$\mu \pm 0.674\sigma$	0.3178	0.5000	0.5000	± 3.291	$\mu \pm 3.291\sigma$	0.0018	0.9990	0.0010
± 0.842	$\mu \pm 0.842\sigma$	0.2800	0.6000	0.4000	± 3.891	$\mu \pm 3.891\sigma$	0.0002	0.9999	0.0001

[a] Ordinates y at $\pm z$ and areas A between $-z$ and $+z$ of the normal distribution

TABLE II. Cumulative Normal Distribution

z	x	Area	z	x	Area	z	x	Area
−3.25	$\mu - 3.25\sigma$	0.0006	−1.00	$\mu - 1.00\sigma$	0.1587	1.05	$\mu + 1.05\sigma$	0.8531
−3.20	$\mu - 3.20\sigma$	0.0007	−0.95	$\mu - 0.95\sigma$	0.1711	1.10	$\mu + 1.10\sigma$	0.8643
−3.15	$\mu - 3.15\sigma$	0.0008	−0.90	$\mu - 0.90\sigma$	0.1841	1.15	$\mu + 1.15\sigma$	0.8749
−3.10	$\mu - 3.10\sigma$	0.0010	−0.85	$\mu - 0.85\sigma$	0.1977	1.20	$\mu + 1.20\sigma$	0.8849
−3.05	$\mu - 3.05\sigma$	0.0011	−0.80	$\mu - 0.80\sigma$	0.2119	1.25	$\mu + 1.25\sigma$	0.8944
−3.00	$\mu - 3.00\sigma$	0.0013	−0.75	$\mu - 0.75\sigma$	0.2266	1.30	$\mu + 1.30\sigma$	0.9032
−2.95	$\mu - 2.95\sigma$	0.0016	−0.70	$\mu - 0.70\sigma$	0.2420	1.35	$\mu + 1.35\sigma$	0.9115
−2.90	$\mu - 2.90\sigma$	0.0019	−0.65	$\mu - 0.65\sigma$	0.2578	1.40	$\mu + 1.40\sigma$	0.9192
−2.85	$\mu - 2.85\sigma$	0.0022	−0.60	$\mu - 0.60\sigma$	0.2743	1.45	$\mu + 1.45\sigma$	0.9265
−2.80	$\mu - 2.80\sigma$	0.0026	−0.55	$\mu - 0.55\sigma$	0.2912	1.50	$\mu + 1.50\sigma$	0.9332
−2.75	$\mu - 2.75\sigma$	0.0030	−0.50	$\mu - 0.50\sigma$	0.3085	1.55	$\mu + 1.55\sigma$	0.9394
−2.70	$\mu - 2.70\sigma$	0.0035	−0.45	$\mu - 0.45\sigma$	0.3264	1.60	$\mu + 1.60\sigma$	0.9452
−2.65	$\mu - 2.65\sigma$	0.0040	−0.40	$\mu - 0.40\sigma$	0.3446	1.65	$\mu + 1.65\sigma$	0.9505
−2.60	$\mu - 2.60\sigma$	0.0047	−0.35	$\mu - 0.35\sigma$	0.3632	1.70	$\mu + 1.70\sigma$	0.9554
−2.55	$\mu - 2.55\sigma$	0.0054	−0.30	$\mu - 0.30\sigma$	0.3821	1.75	$\mu + 1.75\sigma$	0.9599
−2.50	$\mu - 2.50\sigma$	0.0062	−0.25	$\mu - 0.25\sigma$	0.4013	1.80	$\mu + 1.80\sigma$	0.9641
−2.45	$\mu - 2.45\sigma$	0.0071	−0.20	$\mu - 0.20\sigma$	0.4207	1.85	$\mu + 1.85\sigma$	0.9678
−2.40	$\mu - 2.40\sigma$	0.0082	−0.15	$\mu - 0.15\sigma$	0.4404	1.90	$\mu + 1.90\sigma$	0.9713
−2.35	$\mu - 2.35\sigma$	0.0094	−0.10	$\mu - 0.10\sigma$	0.4602	1.95	$\mu + 1.95\sigma$	0.9744
−2.30	$\mu - 2.30\sigma$	0.0107	−0.05	$\mu - 0.05\sigma$	0.4801	2.00	$\mu + 2.00\sigma$	0.9772
−2.25	$\mu - 2.25\sigma$	0.0122	—	—	—	2.05	$\mu + 2.05\sigma$	0.9798
−2.20	$\mu - 2.20\sigma$	0.0139	—	—	—	2.10	$\mu + 2.10\sigma$	0.9821
−2.15	$\mu - 2.15\sigma$	0.0158	0.00	μ	0.5000	2.15	$\mu + 2.15\sigma$	0.9842
−2.10	$\mu - 2.10\sigma$	0.0179	—	—	—	2.20	$\mu + 2.20\sigma$	0.9861
−2.05	$\mu - 2.05\sigma$	0.0202	—	—	—	2.25	$\mu + 2.25\sigma$	0.9878
−2.00	$\mu - 2.00\sigma$	0.0228	0.05	$\mu + 0.05\sigma$	0.5199	2.30	$\mu + 2.30\sigma$	0.9893
−1.95	$\mu - 1.95\sigma$	0.0256	0.10	$\mu + 0.10\sigma$	0.5398	2.35	$\mu + 2.35\sigma$	0.9906
−1.90	$\mu - 1.90\sigma$	0.0287	0.15	$\mu + 0.15\sigma$	0.5596	2.40	$\mu + 2.40\sigma$	0.9918
−1.85	$\mu - 1.85\sigma$	0.0322	0.20	$\mu + 0.20\sigma$	0.5793	2.45	$\mu + 2.45\sigma$	0.9929
−1.80	$\mu - 1.80\sigma$	0.0359	0.25	$\mu + 0.25\sigma$	0.5987	2.50	$\mu + 2.50\sigma$	0.9938
−1.75	$\mu - 1.75\sigma$	0.0401	0.30	$\mu + 0.30\sigma$	0.6179	2.55	$\mu + 2.55\sigma$	0.9946
−1.70	$\mu - 1.70\sigma$	0.0446	0.35	$\mu + 0.35\sigma$	0.6368	2.60	$\mu + 2.60\sigma$	0.9953
−1.65	$\mu - 1.65\sigma$	0.0495	0.40	$\mu + 0.40\sigma$	0.6554	2.65	$\mu + 2.65\sigma$	0.9960
−1.60	$\mu - 1.60\sigma$	0.0548	0.45	$\mu + 0.45\sigma$	0.6736	2.70	$\mu + 2.70\sigma$	0.9965
−1.55	$\mu - 1.55\sigma$	0.0606	0.50	$\mu + 0.50\sigma$	0.6915	2.75	$\mu + 2.75\sigma$	0.9970
−1.50	$\mu - 1.50\sigma$	0.0668	0.55	$\mu + 0.55\sigma$	0.7088	2.80	$\mu + 2.80\sigma$	0.9974
−1.45	$\mu - 1.45\sigma$	0.0735	0.60	$\mu + 0.60\sigma$	0.7257	2.85	$\mu + 2.85\sigma$	0.9978
−1.40	$\mu - 1.40\sigma$	0.0808	0.65	$\mu + 0.65\sigma$	0.7422	2.90	$\mu + 2.90\sigma$	0.9981
−1.35	$\mu - 1.35\sigma$	0.0885	0.70	$\mu + 0.70\sigma$	0.7580	2.95	$\mu + 2.95\sigma$	0.9984
−1.30	$\mu - 1.30\sigma$	0.0968	0.75	$\mu + 0.75\sigma$	0.7734	3.00	$\mu + 3.00\sigma$	0.9987
−1.25	$\mu - 1.25\sigma$	0.1056	0.80	$\mu + 0.80\sigma$	0.7881	3.05	$\mu + 3.05\sigma$	0.9989
−1.20	$\mu - 1.20\sigma$	0.1151	0.85	$\mu + 0.85\sigma$	0.8023	3.10	$\mu + 3.10\sigma$	0.9990
−1.15	$\mu - 1.15\sigma$	0.1251	0.90	$\mu + 0.90\sigma$	0.8159	3.15	$\mu + 3.15\sigma$	0.9992
−1.10	$\mu - 1.10\sigma$	0.1357	0.95	$\mu + 0.95\sigma$	0.8289	3.20	$\mu + 3.20\sigma$	0.9993
−1.05	$\mu - 1.05\sigma$	0.1469	1.00	$\mu + 1.00\sigma$	0.8413	3.25	$\mu + 3.25\sigma$	0.9994

TABLE II (*cont*). Selected Values

Area	x	z
0.50	μ	0
0.60	$\mu + 0.252\sigma$	+0.252
0.75	$\mu + 0.674\sigma$	+0.674
0.84	$\mu + 1.000\sigma$	+1.000
0.90	$\mu + 1.272\sigma$	+1.272
0.95	$\mu + 1.645\sigma$	+1.645

to 5.00. In addition, the bottom of Table I provides the same relations for intervals of 0.1 in A up to $A = 0.9$ and for diminishing intervals up to $A = 0.9999$.

Because the normal probability density function represents a set of continuous random variables, the occurrence of any exact value of x may be regarded as having zero probability. This means simply that we must consider probabilities for intervals of values of x rather than discrete values; the process is continuous, not discrete. Table I and Fig. 8(d) illustrate one kind of interval in x that is of interest: the interval taken relative to the mean position of x. Such an interval of probability tells us how likely (probable) it is that a particular result will differ from the average result by an interval corresponding to $\pm z$ about the mean. A second kind of interval that is of interest is that from a limiting value of x to some other value of x. The limits of x are $\pm\infty$. The cumulative probability corresponds to the interval from $x = -\infty$ to $x = x_i$ and (for $x_i = +\infty$) is represented by the *cumulative normal distribution*:

$$F(x) = \int_{-\infty}^{+\infty} \left\{\frac{1}{\sqrt{2\pi\sigma}}\right\} e^{-(x-\mu)^2/2\sigma^2}\, dx. \tag{15}$$

Table II lists cumulative probabilities (areas under the curve from $x = -\infty$ to $x = x_i$) for 0.05 intervals of z from $z = -3.25$ to $z = +3.25$. It can be seen from this table that the area under the curve from $x = -\infty$ to $x = \mu$ (where $z = 0.00$) is 50% and that to $x = \mu - 2\sigma$ (where $z = -2$) is only 2.28%. The difference between these two values is 47.72%, which is the interval between the mean and $-2z$ shown in Fig. 8(d) and half the value of A corresponding to the interval between $+2z$ and $-2z$ in Table I.

The general relation between intervals of abscissa values of x, a real physical variable, and standard normal deviate z intervals may be stated by the two following equations:

$$P(x_i \leqq x \leqq x_j) = P\left[\frac{x_i - \mu}{\sigma} \leqq z \leqq \frac{x_j - \mu}{\sigma}\right], \tag{16}$$

and

$$P(z_i \leqq z \leqq z_j) = P[(\mu + z_i) \leqq x \leqq (\mu + z_j)]. \tag{17}$$

For example, suppose that the mean μ of a normal distribution of luminances adjusted to match the perceived brightness of a sample to that of a standard is $100\ \text{cd}\cdot\text{m}^{-2}$ and the standard deviation σ of the distribution of matching luminances is $\pm 10\ \text{cd}\cdot\text{m}^{-2}$; what is the probability that the "true" matching luminance is between 92 and $108\ \text{cd}\cdot\text{m}^{-2}$? Substituting $x_i = 92$, $x_j = 108$, and $\mu = 100$ in Eq. (16), we have

$$P(92 \leqq x_{\text{true}} \leqq 108) = \text{P}\left[\frac{92 - 100}{10} \leqq z \leqq \frac{108 - 100}{10}\right]$$
$$= P(-0.8 \leqq z \leqq +0.8)$$

which from Table II corresponds to the interval between 0.7881 and 0.2119; therefore, the probability that the true value of the matching luminance lies between 92 and $108\ \text{cd}\cdot\text{m}^{-2}$ is 0.5762, or just over $57\frac{1}{2}\%$.

b. Statistics of the Normal Distribution

Expectation is formally defined as the integral (or sum in the discrete case) of the products of values of the random variable X and the corresponding values of the probability density function of X (or the mass function for discrete cases). Symbolizing the expectation of X as $E(X)$, we restate it mathematically as

$$E(X) = \int_{-\infty}^{+\infty} x \cdot f(x) \cdot dx. \tag{18}$$

Mathematical characteristics that summarize a distribution are called *moments*. The moments of a distribution are simply the expectations of different powers of the random variable X. They may be taken with respect to variation about the origin of X (that is, where $x = 0$), about the mean of X (where $x = \mu$), or for that matter, about any arbitrary value of $X(x = x_0)$. In the general case the rth moment about the origin N_r is expressed as

$$N_r = E(X^r) = \frac{1}{n}\sum y_i x_i^r, \tag{19}$$

where y is the frequency of x and $n = \sum y$. When the moment is taken about the mean, where $\mu = E(X)$, the mean is subtracted from X before the power is taken,

$$M_r = E[X - E(X)]^r = \frac{1}{n}\sum y_i(x_i - \mu)^r. \tag{20}$$

Although r may assume any nonzero, positive, integer value, we are generally concerned only with the first four moments of a distribution that approximates normality. Measures of these moments are taken as statistics relating to each moment. For the standard normal distribution, such as defined in Eq. (12), illustrated in Figs. 8(c) and (d), and listed in Table I, these statistics are formed from relationships between N_r and M_r, the rth moments about the origin and mean, respectively. They may be expressed in general form as follows:

$$M_1 = 0 = \mu \qquad \text{(called mean)}$$

$$M_2 = N_2 - N_1^2 = \sigma^2 \qquad \text{(called variance)}$$

$$M_3 = N_3 - 3N_2N_1 + 2N_1^2 = \gamma_1 \qquad \text{(called skewness)}$$

$$M_4 = N_4 - 4N_3N_1 + 6N_2N_1^2 - 3N_1^3 = \gamma_2 \qquad \text{(called kurtosis)}$$

$$\vdots$$

$$M_k = \sum \frac{k!}{k!(k-r)!r!} N_{k-r}(-N_1)^r.$$

M_1 (the mean) is a measure of central tendency. M_2 (the variance) is a measure of the spread about the mean. M_3 (the skewness) is a measure of asymmetry about the mean; when γ_1 is positive, there is an elongated tail to the right of the curve, and when it is negative, the tail is to the left of the curve. M_4 (the kurtosis) is a measure of "peakedness" of the distribution; positive values of γ_2 indicate a sharp peak (*leptokurtic* curve), and negative values indicate a flat (*platykurtic*) curve.

In virtually all experimental situations the data only approximate the normal distribution. Usually the tails of the distribution curve are higher than those of a true normal distribution, and they do not stretch out to $\pm\infty$. In addition, even if the data did form an exact match for the shape of the normal distribution, we could not determine the true moments without an infinitely large sample, which, of course, we never have. Accordingly, the usual statistics that must be relied on to characterize any real set of data are only estimates of the statistics relating to the moments of the distribution. They may be calculated from real data derived from a sample of the entire population in the following ways.

The Sample Mean The sample statistic corresponding to the mean μ of a population is called the *sample mean* and is symbolized $\bar{X}$. It is calculated as

$$\bar{X} = \frac{\sum x_i}{n}. \tag{21}$$

$\bar{X}$ is the arithmetic mean of the sample and is identical to other measures of central tendency when the distribution is symmetric. When the distribution is asymmetric, other measures of central tendency such as the midrange m_r, median m, harmonic mean H, or geometric mean G may be more appropriate. They were defined earlier in this chapter; see Section A.1.d

The Sample Variance A measure of the dispersion about the mean of a sample is the *sample variance*, symbolized s_x^2. It may be calculated as

$$s_x^2 = \frac{\sum(x_i - \bar{X})^2}{n - 1}, \tag{22}$$

or

$$s_x^2 = \frac{n\sum x_i^2 - (\sum x_i)^2}{n(n - 1)}. \tag{23}$$

The Sample Standard Deviation The square root of the sample variance is called the *sample standard deviation*, symbolized as s_x,

$$s_x = \sqrt{s_x^2}. \tag{24}$$

The Sample Average Deviation The arithmetic mean of the absolute values of the differences between observed values and the mean or median is called the *sample average deviation*. The two forms of average deviation D_a are

$$D_\mathrm{a} = \frac{\sum|x_i - \bar{X}|}{n}, \tag{25}$$

and

$$D'_\mathrm{a} = \frac{\sum|x_i - m|}{n}. \tag{26}$$

The Coefficient of Variation A measure of relative dispersion is called the *coefficient of variation* v and is equal to the standard deviation s_x divided by the mean $\bar{X}$ and multiplied by 100 to provide percentage values:

$$v = \frac{100\, s_x}{\bar{X}}. \tag{27}$$

The Sample Skewness A measure of the symmetry of the sample distribution is called the *sample skewness*, symbolized g_1. The value of g_1 for a normal (symmetric) distribution is zero. When g_1 is positive, the sample distribution has an elongated tail to the right (positively skewed), and when g_1 is negative, the tail is elongated to the left (negatively skewed). The degree of elongation is related to the absolute value of g_1. The measure may be calculated in either of two ways:

$$g_1 = \left\{ \frac{\left(\frac{\sum x_1^3}{n}\right) - 3\left(\frac{\sum x_1^2}{n}\right)\left(\frac{\sum x_i}{n}\right) + 2\left(\frac{\sum x_i}{n}\right)^3}{\left[\left(\frac{\sum x_i^2}{n}\right) - \left(\frac{\sum x_i}{n}\right)^2\right]^{3/2}} \right\}, \tag{28}$$

or

$$g_1 = \left\{ \frac{n^{1/2} \sum (x_i - \bar{X})^3}{[\sum (x_i - \bar{X})^2]^{3/2}} \right\}. \tag{29}$$

The Sample Kurtosis A measure of the peakedness of the sample distribution is called the *sample kurtosis*, symbolized as g_2. The value of g_2 for the normal distribution (which is *mesokurtic*) is zero. Positive values of g_2 indicate that the distribution is leptokurtic (that is, it has a sharper peak and longer tails than the normal distribution). Negative values of g_2 are found for platykurtic distributions (ones with flatter peaks and shorter tails than the normal distribution). Values of g_2 may be calculated by either of two methods:

$$g_2 = \left\{ \frac{\left(\frac{\sum x_i^4}{n}\right) - 4\left(\frac{\sum x_i^3}{n}\right)\left(\frac{\sum x_i}{n}\right) + 6\left(\frac{\sum x_i^2}{n}\right)\left(\frac{\sum x_i}{n}\right)^2 - 3\left(\frac{\sum x_i}{n}\right)^4}{\left[\left(\frac{\sum x_i^2}{n}\right) - \left(\frac{\sum x_i}{n}\right)^2\right]^2} \right\} - 3, \tag{30}$$

or

$$g_2 = \left\{ \frac{n \sum (x_i - \bar{X})^4}{[\sum (x_i - \bar{X})^2]^2} \right\} - 3. \tag{31}$$

Thus far the measures defined have been descriptive statistics, measures that characterize the location and the shape of a distribution. There are three inferential statistical measures that should also be mentioned here. When two variables are involved (or pairs of observations from two observers are to be compared), it is convenient to calculate the *sample covariance* or two of its derivative statistics: the *correlation coefficient* and *coefficient of determination.* Although not strictly measures of the shapes of distributions, these statistics are based on measures of mutual shapes characterized by their variances.

Covariance The sample covariance is one measure of the relationship between pairs of observations from two variables (or, in the general case, when two or more variables are considered). Measures of covariance may be important when some condition of the experiment cannot be held constant. Covariance analysis allows us to describe and (if desired) remove the effects of this changing condition from the experimental data. A measure of sample covariance, symbolized s_{xy}, may be calculated as either

$$s_{xy} = \frac{n \sum x_i y_i - (\sum x_i)(\sum y_i)}{n(n-1)}, \tag{32}$$

or

$$s_{xy} = \frac{\sum (x_i - \bar{X})(y_i - \bar{Y})}{n-1}. \tag{33}$$

Note that the covariance s_{xy} is not represented by a squared symbol as is variance s_x^2; this is because it corresponds approximately to a product of two different standard deviations denoted by the subscript xy.

The Correlation Coefficient One method of testing hypotheses that assume linear relations between pairs of results is called regression analysis, a special case of variance or covariance analysis. A measure that indicates the degree of linear relationship between two sets of numbers is called the *simple correlation coefficient*, symbolized as r_{xy} or simply r; there are also multiple correlation coefficients that indicate the degree of linearity among the combined relationships of several independent variables and a given dependent variable. The value of r varies between -1 and $+1$; the higher its absolute value the greater the linear correlation. r may be calculated as

$$r_{xy} = \frac{s_{xy}}{s_x s_y}, \tag{34}$$

or

$$r_{xy} = \left[\frac{n \sum x_i y_i - \sum x_i \sum y_i}{([n \sum x_i^2 - (\sum x_i)^2][n \sum y_i^2 - (\sum y_i)^2])^{1/2}} \right]. \tag{35}$$

The Coefficient of Determination A measure of the portion of the variance for one variable that can be explained by its linear relationship with a second variable is called the *coefficient of determination* r_{xy}^2 or simply r^2. As implied by its symbol, the coefficient of determination is the square of the simple correlation coefficient,

$$r_{xy}^2 = (r_{xy})^2. \tag{36}$$

There are, of course, other descriptive statistics that are useful from time to time. However, the ones just summarized are probably those most often used to characterize normal or quasi- normal distributions. There also are many other inferential statistics beyond those few mentioned here. Some will be discussed in later sections of this chapter, but the reader should consult other textbooks that deal with inferential statistical analysis for further details (for example Dixon and Massey, 1957; Winkler and Hays, 1975; Mosteller and Tukey, 1977; Draper and Smith, 1981).

C. REPRESENTING EXPERIMENTAL DATA WITH PROBABILITY DISTRIBUTIONS

1. How to Tell When Experimental Data Fit a Standard Distribution

Frequently, threshold data (both RL and DL determinations) will resemble a cumulative normal density distribution by inspection. It is well, however, to test the "goodness of fit" of the data to the cumulative normal distribution. There are a number of ways to make such a test. We shall describe only two, Pearson's χ-squared test and a graphical analysis. The reader is referred to the textbooks cited at the end of the preceding section for details about other tests of goodness of fit. The kinds of tests described here are not limited to testing agreement between experimental data and the normal distribution; they can also be used (with appropriate modification) to test the fit of differences among data from other kinds of distributions. The normal distribution is used here as a standard of comparison, because many sets of data resemble that distribution, because random differences are often distributed normally, and because it will serve adequately to illustrate the methods of test.

a. Pearson's χ-Squared Test of Goodness of Fit

If the data form a normal distribution, we should be able to predict them well from the probabilities along the normal curve; that is, the expected values from the normal curve should match the experimentally determined values. We wish to test the *null hypothesis* H_0 that there is no difference between our experimental data and the normal expectation. The normal function represents a family of distributions, the members of which all have the same basic shape but differ in their means and variances. The standard normal distribution has a mean of zero ($\mu = 0$) and standard deviation (square root of variance) of unity ($\sigma = 1$). We can calculate the sample mean $\bar{X}$ and sample standard deviation s_x for a distribution of experimental data on the assumption that the distribution is normal. These two values $\bar{X}$ and s_x, representing two points on the distribution, can be normalized to 0 and 1, matching the corresponding points on the standard normal distributions. For example, suppose that we have generated experimental data in which the values of these two sample statistics are $\bar{X} = 50$ and $s_x = 10$. The factor that will normalize the spread of the distribution is σ/s_x, which is 1/10 in our example. The additive constant that will locate the mean properly is $\mu - \bar{X}$, which is 0 − 50 here. Therefore, if we transform all data values by the equation $x_i' = 0.1\,(x_i - 50)$, we will create a new distribution in x_i' with a mean of zero and unit standard deviation. It will be the same shape as the original distribution of x_i but will be normalized for direct comparison with the standard normal distribution.

The method of comparison most often used is one invented by the English mathematician Karl Pearson. It is based on the finding that the multivariate normal distribution approximates the multinomial distribution for large values of n. To the extent that there are differences between x_i' values and the expected values E_i of the standard normal distribution, there will be a distribution of differences generated by comparing the two over values of i. Because we are dealing with intervals in the cumulative normal distribution, we should categorize the data in some way. It will make the comparison simple if we choose categories that are all of equal probability (that is, of equal area under the normal curve). When the observed data for each category j have the expected value subtracted from them, are squared, and then are divided by the expected value for that category, and the quotients are summed over all categories, the resulting distribution is called χ-squared. Stated mathematically,

$$\chi^2 = \sum_j \frac{[(x_i')_j - (E_i)_j]^2}{(E_i)_j}, \tag{37}$$

which is known as *Pearson's χ-squared statistic*. When the null hypothesis is true, that is, when the experimental distribution has the same shape as the normal distribution (or, in general, matches the expectation of any reference distribution), then χ^2 is distributed with $j - 1$ degrees of freedom according to the function

$$f(\chi^2) = h(\eta)\exp(-\chi^2/2)(\chi^2)^{(\eta/2)-1}, \qquad \text{for} \quad \chi^2 \geq 0 \quad \text{and} \quad \eta > 0, \tag{38}$$

where $h(\eta)$ is a constant relating to degrees of freedom η through a gamma function. The χ-squared distribution has statistics relating to its first four moments that are functions of η:

$$\begin{aligned} \text{mean} &= \eta \\ \text{variance} &= 2\eta \\ \text{skewness} &= 2(2/\eta)^{1/2} \\ \text{kurtosis} &= (12/\eta) + 3. \end{aligned}$$

A simple example will illustrate the application of the χ^2 test of goodness of fit. Suppose that we have collected yes responses from an observer in an absolute threshold experiment. The observer has responded to many presentations of many lights during the experiment. We wish to test whether the results form a normal distribution. First, the data are categorized. Let us use ten categories, each representing 10% probability; that is, the area under the normal curve within each category is 10% of the total area. Each category will extend over a different range of stimulus intensities, so the easiest way to categorize the data is to plot the results (as normalized data, x_i') and count the number of yes responses in each of the categories. An example is given in Table III.

TABLE III. Example of Calculating χ^2 for Ten Categories of Hypothetical Threshold Data

$\sum A$	j	O	E	$(O-E)$	$(O-E)^2$	$\frac{(O-E)^2}{E}$
0.1	1	7	10	−3	9	0.9
0.2	2	6	10	−4	16	1.6
0.3	3	10	10	0	0	0.0
0.4	4	13	10	+3	9	0.9
0.5	5	15	10	+5	25	2.5
0.6	6	16	10	+6	36	3.6
0.7	7	14	10	+4	16	1.6
0.8	8	7	10	−3	9	0.9
0.9	9	7	10	−3	9	0.9
1.0	10	5	10	−5	25	2.5
		$\sum = 100$	$\sum = 100$			$\chi^2 = 15.4$
$\eta = (j-1) = 9$						

The first column shows the cumulative area represented by each of the 10 categories listed in column 2. The observed counts are shown in the third column, and the fourth column lists the expectations from the normal distribution (because we chose the categories to represent constant area under the normal curve, they each have the same expectation). Next, in accordance with Eq. (37), we take the differences between pairs of observations and expectations (column 5), square them (column 6), and divide by the expectation (column 7). The sum of column 7 is χ^2 for $\eta = 9\ (= j - 1)$. In this example, $\chi^2 = 15.4$. We can now determine whether the experimental results resemble a normal distribution to some acceptable level of significance.

This can be done computationally, but use of standard χ^2 tables simplifies the process greatly. Table IV provides such information. Values of the χ^2 distribution are shown for various degrees of freedom ($df = \eta = j - 1$) and levels of significance (the corresponding percentiles are also given; these are the complements of significance levels). At the 5% significance level, which is a 95 percentile, a value of about 16.9 is critical for 9 degrees of freedom. The value that we just calculated is 15.4, less than the critical value. Accordingly,

TABLE IV. Values of the χ^2 Distribution[a]

Degrees of freedom (η)	Percentiles 90	95	$97\frac{1}{2}$	99
	Significance 0.10	0.05	0.025	0.01
1	2.7055	3.8415	5.0239	6.6349
2	4.6052	5.9915	7.3778	9.2103
3	6.2514	7.8147	9.3484	11.345
4	7.7794	9.4877	11.143	13.277
5	9.2364	11.071	12.833	15.086
6	10.645	12.592	14.449	16.812
7	12.017	14.067	16.013	18.475
8	13.362	15.507	17.535	20.090
9	14.684	16.919	19.023	21.666
10	15.987	18.307	20.483	23.209
11	17.275	19.675	21.920	24.725
12	18.549	21.026	23.337	26.217
13	19.812	22.362	24.736	27.688
14	21.064	23.685	26.119	29.141
15	22.307	24.996	27.488	30.578
16	23.542	26.296	28.845	32.000
17	24.769	27.587	30.191	33.409
18	25.989	28.869	31.526	34.805
19	27.204	30.144	32.852	36.191
20	28.412	31.410	34.170	37.566
30	40.256	43.773	46.979	50.892
40	51.805	55.759	59.342	63.691
50	63.167	67.505	71.420	76.154
60	74.397	79.082	83.298	88.379
70	85.527	90.531	95.023	100.43
80	96.578	101.88	106.63	112.33
90	107.57	113.15	118.14	124.12
100	118.50	124.34	129.56	135.81

[a] For large values of degrees of freedom ($\eta \gtrsim 20$), χ^2 may be estimated by the formula

$$\chi_\alpha^2 = \eta[1 - (2/9\eta) + z_\alpha(2/9\eta)^{1/2}]^3,$$

where z is the standard normal deviate corresponding to the α percentile of interest, and η is the number of degrees of freedom.

we could accept the null hypothesis with reasonable certainty that our experimental data resemble a normal distribution well enough for our purposes. That being so, we may calculate statistics of the normal distribution to characterize the experimental data. On the other hand, if the calculated χ^2 value were larger than the critical value, we should not accept the null hypothesis at that level of significance.

A few rules of thumb should be mentioned in connection with χ-squared tests of goodness of fit. Usually a significance level of 5% is a reasonable criterion to set for testing; loosely, this means that 95% of the differences between an experimental distribution and the normal one are the result of random variations and are not due to systematic differences between the two distributions. This level of significance corresponds to the 95th percentile of the χ^2 distribution and is represented in Table IV by the column labeled 95. Next, recall that the χ^2 distribution is based on the fact that the multinomial and normal distributions are essentially the same for large sample size n. In general we can be more certain of psychophysical results when they are based on very large numbers of observations. As n is made smaller and smaller, the appropriateness of inferential statistics (such as χ^2 tests) is more tenuous. The number of observations in any one category should generally be at least 10–30, and not more than 20% of the categories should contain five or fewer observations. These considerations help to dictate the number of categories that should be used to represent the data. Ideally, the more categories the better, if each category contains enough observations. To satisfy the demand for a large enough number of observations in each category, it may be necessary to reduce the number of categories. Table IV shows that the critical value of χ^2 diminishes as degrees of freedom are reduced for any one level of significance. It is well, therefore, to make as many observations as practical in the experiment and to select the largest number of categories for χ^2 testing that is consistent with the need to have an arbitrary minimum number of observations in all categories. Finally, the selection of category widths can be according to any arbitrary, consistent rule (for example, constant stimulus intervals, constant z intervals, or constant probability), but the χ^2 test is sensitive to the choice of category intervals. Intervals of equal probability have been recommended here for two reasons: (1) Departures from normality in the central portion of the distribution are more likely to be detected, because the stimulus intervals are smallest in this part of the distribution, where we are most interested in the results (that is, whether it be a match or a threshold determination, we are less interested in the extremes than in the central portion of the distribution); and (2) computations are somewhat simplified by having equal expectations for all categories.

There are a number of other tests of goodness of fit, both parametric and nonparametric, some highly specialized and some quite general. They all provide one or another inferential statistics intended to aid in deciding whether two or more distributions match well enough for some purpose. The reader will find descriptions of these other tests in textbooks such as Mosteller and Tukey (1977), Draper and Smith (1981), and Winkler and Hays (1975). For most purposes, however, the Pearson χ-squared test is well suited to psychophysical data on thresholds and matching.

b. Graphical Analysis

Graphical representation of experimental data is the oldest and, in some ways, still the most attractive way to understand many kinds of data. It has been said by Riggs (1963) that experimenters

> are forever plotting the results of their experiments on graph paper for the very good reason that a graph enables the viewer to see at once the general trend of the results without being distracted by the actual numerical values of the individual observations... [and] he is often not content to let [the data] speak for themselves [so] he draws a smooth line through them.... this commonplace act is a first-rate example of inductive reasoning. [pp. 47–49]

A good representation, often of even quite complex variation, can be obtained by bringing experience and common sense to the analysis of a graphical representation of experimental results.

Graphical analysis offers a simple but imprecise way of determining whether a set of experimental data resembles a normal distribution. The advantages of such a method are that it can be performed quickly and with no need for computations. The disadvantages are that the method is imprecise and incorporates no formal test for goodness of fit. In many situations, however, the method will serve well enough. Experimenters often fail to obtain sufficiently large sample sizes, with the result that the data are imprecise enough to make it difficult or impossible to distinguish between results of a formal analysis, such as the χ-squared test, and the informal graphical analysis.

Graphical analysis of data with respect to the normal distribution evaluates the relationship between z scores and some function of the stimuli, such as their intensities. This kind of graphical analysis is similar to the mathematically more formal *probit analysis* (Finney, 1971). A probit is simply $z + 5$; probit analysis is a method of determining the linear regression of probits on stimulus values by the method of maximum likelihood. If the relative frequencies of yes responses in a threshold experiment are normally distributed with stimulus intensity, they will form a straight-line relation with intensity when plotted as standard normal deviates.

Figure 9 illustrates this assertion. The inset of this figure shows the relative frequencies (probabilities) of yes responses for each of a series of stimulus intensities. The figure could as easily show cumulative frequencies of match responses in a matching experiment. The main graph of Fig. 9 shows the same data plotted as z scores (on the left ordinate) or the equivalent cumulative frequencies arrayed along normal probability intervals (on the right ordinate) against an abscissa of relative stimulus intensity.

The data of the example in Fig. 9 are both normal and "well behaved"; that is, they fit the cumulative normal probability density curve well and have very little scatter about that curve. Real experimental data seldom provide results as clear-cut as those in Fig. 9. Usually data that may be reasonably normal

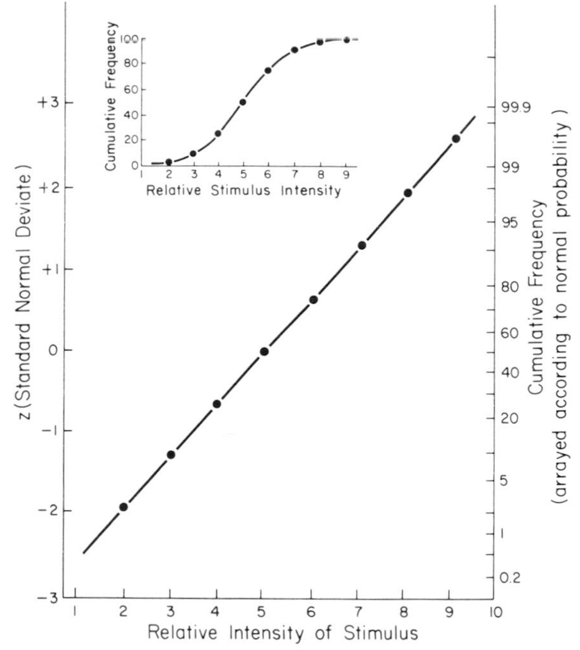

Fig. 9. Results of a hypothetical experiment to determine a difference threshold. Probability values plotted against stimulus values (inset) form an S-shaped curve, but when cumulative probabilities (right ordinate) or z scores (left ordinate) are plotted against stimulus values, the result is a straight line for normally distributed data.

differ from those of the figure in two principal ways: They exhibit scatter, and they depart from the straight line that implies normality. We shall discuss each of these in turn.

The question of scatter or imprecision in the experimental data is one that usually can be addressed through choice of sample size. If the true underlying function is smooth, then the results will tend to be more and more smooth as the number of observations is made larger and larger. This is not an infallible rule, but it is true often enough so that it can be taken as a useful first-order approximation. Figure 10 illustrates the tendency toward smooth results as sample size increases. In this example the stimulus continuum was sampled in only six places. Sampling at more places would provide more points through which to fit a curve. Given a particular number of sampling positions on the stimulus continuum, the more replications that are made for each observer and the larger the number of observers participating in the experiment, the more likely are the data to form a smooth result.

Integration also helps to smooth data. Generally, the more steps of integration involved the smoother the result. For example, plotting the data as a cumulative frequency distribution yields a smoother result than a graph of

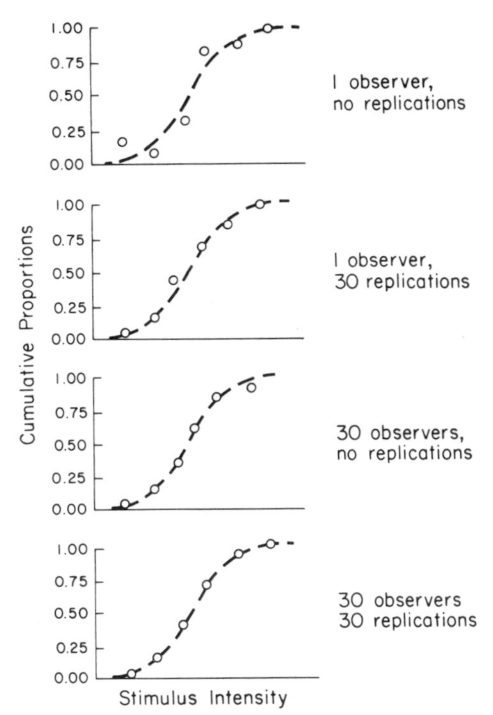

Fig. 10. Examples of precision, or scatter, of data as a function of number of observations. See text for discussion.

the individual frequencies. Plotting the average of several observers' averages over replications yields a smoother result than does a graph of the individual points for all replications and all observers.

Often, experimental data will depart from normality by not forming a straight-line relation between z scores and some function of intensity (see Fig. 9). For a large class of experimental data the departures will be only at, or most obvious at, the extremes of the graph. If these departures are such that they gradually increase with the absolute value of z and are nearly insignificant between about $\pm 2z$, then the data may be accepted as reasonably normal over the central 95% of the distribution. The graph of Fig. 9 offers a simple way to determine how well the data fit a normal distribution over the central range: when they form a linear array between $-2z$ and $+2z$, then the central 95% of the distribution matches a normal curve fairly well. Even the range from $-1.5z$ to $+1.5z$, corresponding to the central 87% of the curve, is often acceptable as an approximate fit. However, in such cases, the sample mean and standard deviation are best taken from the graph rather than calculated from numerical data.

2. What to Do When the Data Are Not Normally Distributed

Sometimes, experimental data may not be linearly related to z scores. The graphical relationship of Fig. 9 may be curved at all locations. When this happens, there are two things to do: Fit the data with some other function, or find some transformation of stimulus intensity, frequency of response, or both that will yield a fit to the normal curve.

a. Fitting Data with a Beta Function

Any of a number of functions may be appropriate for fitting data that are not normally distributed. However, one of the most versatile functions, in the sense of providing empirical fitting over a broad gamut of curve shapes, is the beta function. The density function of the beta distribution of probability of yes responses p for number of yes responses r over n observations, where $n > r > 0$, is of the form

$$B(p) = \left[\frac{(n-1)!}{(r-1)!(n-r-1)!}\right] p^{r-1}(1-p)^{n-r-1}, \quad \text{if } 0 \leqq p \leqq 1, \quad 0 \text{ elsewhere.} \tag{39}$$

The mean E and variance V of a beta distribution are then

$$E = \frac{r}{n}, \tag{40}$$

and

$$V = \frac{r(n-r)}{n^2(n+1)}. \tag{41}$$

The most precise way of determining the parameters of the beta function that fit the data is to carry out a nonlinear regression computation. Several such algorithms have been proposed (for instance, Bard, 1974), and some are included in various statistical packages such as SAS (1979). However, before taking recourse to curve fitting with density functions such as the beta distribution, it is advisable first to make every effort to transform the data in some manner to fit a normal curve, so that advantage can be taken of the copious information available about normal curves.

b. Re-expressing the Data to Fit a Normal Curve

More often than not, data that yield a curved graph of z scores versus stimulus intensity can be transformed or re-expressed in a way that provides a linear relation on such a graph. Either the stimulus axis or the relative frequency axis can be transformed (or both), but it is preferable to operate on the stimulus axis, the independent variable. Often, simply taking the logarithm of stimulus intensity will remedy the problem; possibly, the square root or exponential will suffice. Curve fitting and regression are as much an art as a science. The reader who is not familiar with the techniques used in re-expressing data to produce linear relations should consult a textbook on the subject [for instance, Mosteller and Tukey, 1977; Draper and Smith, 1981]. The advantages that accrue to a re-expression that yields a fit to the normal function are that all the tables and standard equations for the normal distribution may be used in the analysis and that graphical assessment is simple and straightforward. It is well worth some investment in time to try to fit the data to a normal curve.

3. How to Determine Descriptive Statistics from a Probability Graph

When the data fit a normal curve, a graph of z scores plotted against some function of stimulus, such as that shown in Fig. 9, is useful, because the mean and standard deviation can be picked off the graph very simply. For example, the mean is that stimulus intensity at which the line intersects $z = 0$, or in other words, the point of 50% cumulative frequency. Standard deviation, on such a graph, is related to the slope of the line; the lower the slope the greater the standard deviation. This is so because the standard deviation corresponds to that intensity at which the line intersects the point $z = +1$ (or $z = -1$, because the graph is symmetric about the mean). Other statistics that are simply related to the mean or standard deviation may also be determined readily from such a graph. For example, the probable error PE (that interval about the mean containing 50% of the total area), is simply the intensity interval between $z = \pm 0.674$.

Whether or not the data fit a normal curve, some general descriptors can be obtained from graphs of cumulative relative frequency plotted against stimulus intensity. Suppose that a difference threshold experiment had been conducted in which observers were shown a series of stimuli in ascending order of intensities, and observers were asked to indicate when they perceived the brightness of the test stimulus to be greater than that of a reference stimulus of constant intensity. Then, in a second experiment using the same reference stimulus, a series of tests of decreasing intensity was also presented, and the observers were requested to indicate when they perceived the test to be less bright than the reference. Figure 11(a) might represent the results from the first example. The curve labeled *greater* represents the relative frequencies of such responses, but the curve labeled *less* is merely the complement of the first, because it does not represent an independent series of judgments. Accordingly, the point at which the two curves cross is equal to a relative frequency of 50%. By dropping a line from this point to the abscissa, one can specify the PSE (point of subjective equality). The second experiment, in which independ-

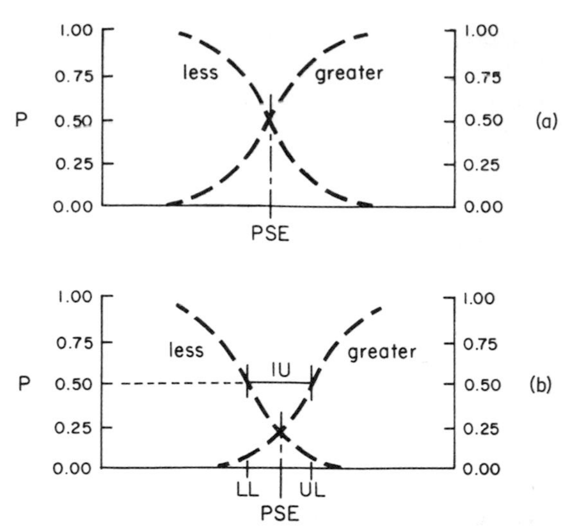

Fig. 11. Results of two hypothetical experiments to determine a difference threshold. In a, only judgments of "brighter" responses (*greater* than the standard) were recorded. The curve labeled *less* is simply the complement of that labeled *greater*. In b, both "brighter" and "darker" responses were recorded from independent trials. The curves labeled *greater* and *less* do not cross at $p = 0.5$. Instead, there is an interval of uncertainty (IU) between their positions at $p = 0.5$.

ent determinations of greater and less were made, might be represented by Fig. 11(b). Here the two cumulative frequency distributions do not necessarily cross at a relative frequency of 50%. Instead, there is an interval between the curves at the 50% ordinate level. The size of this interval, when projected downward onto the intensity axis, defines the IU (interval of uncertainty). If the two curves are symmetric about their cross-over point, then the midrange of the IU will correspond to the same stimulus as does the PSE found at the projected crossover point; if the curves are not symmetric, then the PSE at the crossover is the better indication of the stimulus intensity that is neither greater nor less than the standard. The UL and LL are simply the stimulus intervals between PSE and the projections to the abscissa of the points on the two curves that intersect the 50% probability (relative frequency) level.

D. PSYCHOPHYSICAL METHODS

1. General Considerations

A number of psychophysical methods were described in Chapter 6. These included the method of constant stimuli, the method of adjustment, the method of limits, and several variations of these three methods. The general theories and the methodologies of these psychophysical techniques were summarized in that chapter and will not be repeated here. This section will set forth examples of experiments conducted according to the three methods just listed and will include detailed analyses of the experimental results. The purpose is to provide the reader with a step-by-step guide to data reduction and analysis for typical threshold and matching experiments. To facilitate comparison of the results of the methods, a single experimental situation will be considered.

The experimental situation is illustrated in Fig. 12. The observer was presented with a 2° diameter subtense bipartite field centered in a square surround field that subtended 30° on a side. The chromaticity coordinates of the surround and both halves of the bipartite field were $x = 0.44758$ and $y = 0.40744$ in the CIE 1931 chromaticity specification metric; these coordinates are the same as those for CIE Illuminant A, representing an incandescent lamp with a correlated color temperature of about 2856 K. The luminance of the surround was $8.4\ \mathrm{cd \cdot m^{-1}}$ and was uniform over its entire area. One half of the bipartite field consisted of luminance of $10.7\ \mathrm{cd \cdot m^{-2}}$ and served as an invariant standard. The standard could be presented in either

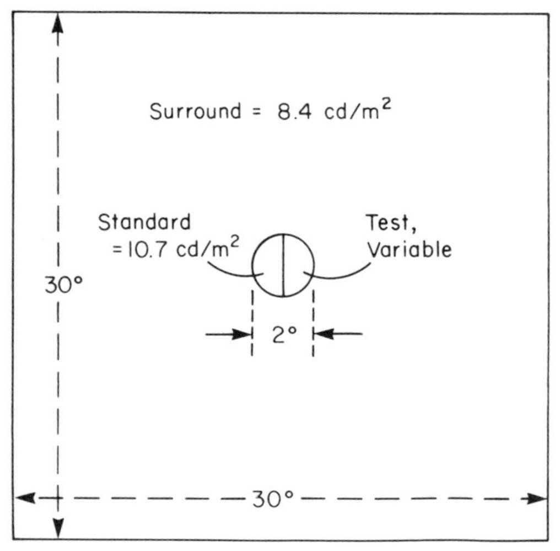

Fig. 12. Experimental array of stimuli for examples of threshold experiments. A bipartite field subtending a visual angle of 2° diameter is surrounded by a uniform field of 30° diameter subtense on each side. A standard stimulus of $10.7 \cdot \mathrm{m^{-2}}$ is presented in one half of the bipartite field; the luminance of the other half is adjusted by the experimenter or by the observer, depending on the experimental method used. CIE chromaticity coordinates of the surround and both halves of the bipartite field are all equal to those of CIE illuminant A: $x = 0.44758$, $y = 0.40744$. The bipartite field is presented for 3 sec and replaced by a field identical to the surround for 10 sec during each cycle of repetitive presentations.

the left or the right side of the bipartite field, according to the plan of a particular experiment. The luminance of the other half of the bipartite field could be varied either by the experimenter or by the observer, according to the type of experiment being conducted. The bipartite field (consisting of both standard and test fields) was presented for a 3-sec period and was replaced by a uniform continuation of the surround field for a 10-sec period during each of many cycles of presentation.

The surround field served to maintain a reasonably steady state of adaptation in the observer. It was large enough to stimulate that central portion of the observer's retina containing most of the cone photoreceptors when the observer fixated in the center of the field. No restraints or fixation fiducial marks were used, so that normal free-viewing resulted. However, head and eye movements tended to concentrate viewing on the center of the target. The central, circular, bipartite field subtended a small enough visual angle (2°) so that it tended to stimulate primarily that area of the observer's retina that contained few rod photoreceptors. Because the luminance of the test and standard fields was greater than that of the surround, a tachistoscopic presentation was used to prevent undue adaptation to the bipartite field. That is, the bipartite field was presented for only about 23% of the duty cycle, and the surround luminance was substituted in the central 2° diameter area during the other 77% of the cycle, to minimize the tendency to adapt to the higher luminance of the standard and matching test fields.

Three experiments were carried out with this viewing arrangement. In all three cases, data were collected for one observer (with normal color vision and corrected acuity) who was experienced in psychophysical experimentation. In the various experiments the observer's task was to respond verbally, indicating whether the test field was brighter or darker than the standard field, or to adjust the luminance of the test field to match the brightness of the standard field.

The three psychophysical methods used were constant stimuli, adjustment and limits. The following sections will set forth the results of each experiment, provide examples of the reduction of the data, and give inferential summaries of the results. Representative examples of the application of the techniques for descriptive and inferential statistics described in the preceding sections will be used.

2. Method of Constant Stimuli

The first experiment was conducted according to the method of constant stimuli (which has also been called the single-stimulus method). Nine test luminances were chosen, ranging from 0.6 to 1.6 log $cd \cdot m^{-2}$. Except for the two extreme intervals the differences between adjacent levels were 0.1 log unit luminance; the two end intervals were each 0.2 log unit. The observer's task, when presented with a test stimulus, was to indicate whether it was brighter or darker than the standard field in the left half of the bipartite field (see Fig. 12). There are many ways in which the order of presentation could be arranged. In this experiment the order chosen was systematic, with stimuli either increasing or decreasing in luminance. The two orders were used at different times (on different occasions but with the same observer) so that the two sets of results were independent. The observer was asked to respond yes if the test stimulus appeared brighter than the standard and no if it appeared darker than the standard. Both ascending and descending orders of presentation were repeated 30 times.

Table V lists the frequency of yes responses for the ascending order and the number of no responses for the descending order of presentation. These are the raw data from the experiment. Probabilities of responses (brighter or darker) can be calculated by dividing the frequency values by 30 (the number of replications). These probabilities were calculated and are plotted in the probability versus luminance coordinates of the graph in Fig. 13; open circles represent the ascending-order responses, and the triangles correspond to the descending-order data.

a. Graphical Regression Method

Recall from the earlier discussion of graphical analysis that normally distributed results should form a straight line when plotted against cumulative normal probability intervals (or *z* scores) such as those used on the ordinate of

TABLE V. Raw Data for Differential Luminance Threshold Experiment[a]

Luminance $\log[\text{cd/m}^2]$	Ascending order (brighter responses)	Descending order (darker responses)	Replications (trials)
0.6	5	23	30
0.8	8	20	30
0.9	11	18	30
1.0	13	16	30
1.1	16	13	30
1.2	19	9	30
1.3	21	8	30
1.4	23	6	30
1.6	26	3	30

[a] Standard at $10.7\ \text{cd}\cdot\text{m}^{-2}$ ($\log L = 1.03$); independent determinations for ascending and descending stimulus presentations of 3 sec each; data for a single observer; bipartite field of 2° diameter; surround illuminated to $8.4\ \text{cd}\cdot\text{m}^{-2}$; chromaticity of test and surround fields equivalent to that of CIE I11. A; psychophysical method of constant stimuli.

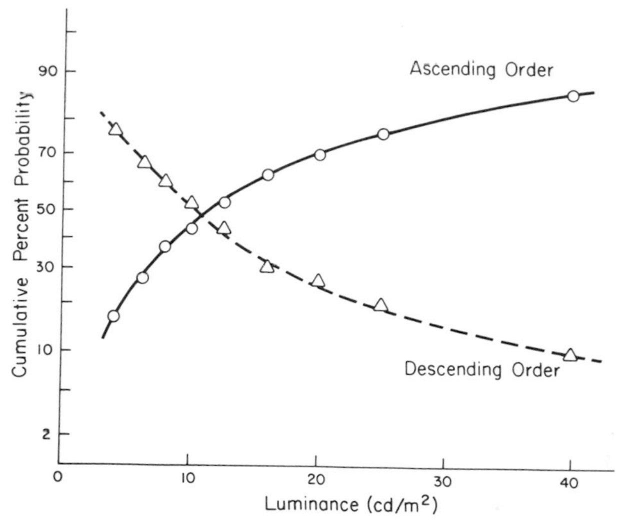

Fig. 13. Cumulative percent probability versus luminance for "brighter" responses to ascending-ordered stimuli (circles) and "darker" responses to descending-ordered stimuli (triangles) in an experimental determination of difference thresholds by the method of constant stimuli.

Fig. 13. The data shown in that figure do not form straight lines; the loci of points for both ascending and descending orders are curved. Rather than attempt to fit the results with some other probability function, a re-expression transformation was sought for the stimulus values. By simply converting luminances ($\text{cd} \cdot \text{m}^{-2}$) to log luminances ($\log \text{cd} \cdot \text{m}^{-2}$), we solved the problem, as illustrated in Fig. 14. The same data plotted as log luminance versus cumulative percent probability intervals now form straight lines. This means that the experimental results are normally distributed in log luminances.

That being so, a simple linear regression of z scores on log luminances should describe the results. The slopes and intercepts of the lines in Fig. 14 can be determined by geometry. However, it is usually more accurate and equally simple to determine them by algebraic calculation. The z scores may be symbolized by y and log luminances by x, and a regression of y on x is determined computationally as

$$y = a + bx, \tag{42}$$

where

$$a = \left[\frac{\sum x^2 \sum y^2 - \sum x \sum xy}{n \sum x^2 - (\sum x)^2}\right] \tag{43}$$

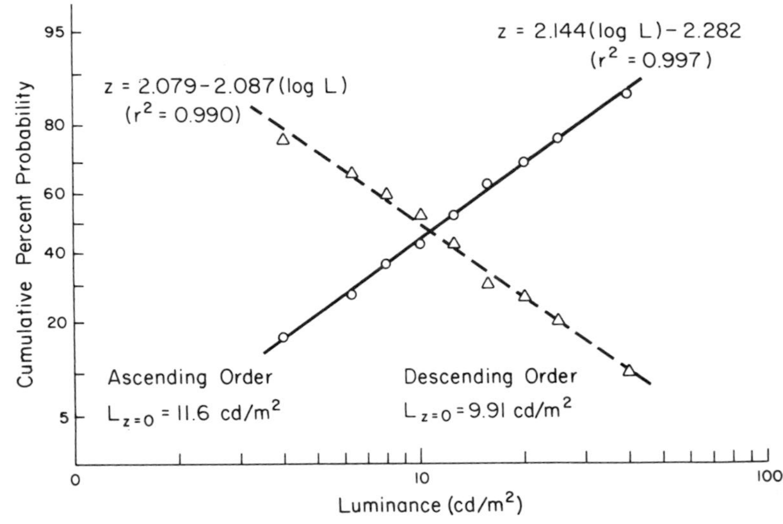

Fig. 14. Same data as shown in Fig. 13. Ordinate intervals are cumulative percent frequency. Abscissa shows luminance arrayed along logarithmic intervals.

and

$$b = \left[\frac{n\sum(xy) - \sum x \sum y}{n\sum x^2 - (\sum x)^2}\right]. \tag{44}$$

These expressions may be solved by any of a number of readily available computer programs or desk-calculator algorithms. The resultant regression equations are shown in Fig. 14 for the two orders of stimulus presentation. Given these equations we can calculate the log luminances corresponding to 50% probability in the two sequences. The $z = 0$ (or 50% probability value) for the ascending order is 1.064 and for the descending order is 0.996; these log luminances correspond to 11.6 and 9.91 $\text{cd} \cdot \text{m}^{-2}$, respectively. In other words, the threshold for a 50% probability was higher with the ascending order than with the descending order of stimulus presentation. This is a common finding in such experiments. It simply means that there is an interval of uncertainty (the IU) in which the observer is unable to detect a difference in brightness between the test stimulus and the standard stimulus (probably resulting from adaptation in this case). In this example the IU is 0.068 log units wide (that is, $1.064 - 0.996 = 0.068$) corresponding to the luminance interval from 9.91 to 11.6 $\text{cd} \cdot \text{m}^{-2}$. In other words, any stimulus with a luminance within that range is equally likely to appear as bright as the 10.7 $\text{cd} \cdot \text{m}^{-2}$ standard on a series of mixed trials when approached from higher or lower luminances (and, of course, for the observer and viewing conditions in question).

The point of subjective equality PSE, or best estimate of the luminance of the stimulus that provides a brightness match for the standard, is taken as the midpoint of the interval of uncertainty. That is $0.068/2 = 0.034$ and $0.996 + 0.034 = 1.030 \log L$, or a luminance of $10.7\ \mathrm{cd \cdot m^{-2}}$. In this example the PSE matches the standard. Sometimes the two are not the same because of time errors, adaptation effects, or other bias factors that intrude into the experiment.

The determination of the linear regression of z scores on log luminance yields a measure of goodness of fit in the value of the coefficient of determination r^2. In the example cited here, $r^2 = 0.997$ for the ascending regression and 0.990 for the descending one. These values are reasonably high. As indicated earlier in this chapter, r^2 is the square of the ratio of covariance to the product of standard deviations of x and y; that is $r^2 = (s_{xy}/s_x s_y)^2$. The standard deviation in log luminance s_x may also be taken as a measure of differential sensitivity to changes in log luminance. It can be shown that the modulus of this standard deviation is the reciprocal of the slope b of Eq. (44); $s_x = 1/b$. In the example, differential sensitivity is equal to 0.466 for the ascending order and 0.479 for the descending order, or about 0.473 on average. In other words, a unit change in z corresponds to $\sim 3\ \mathrm{cd \cdot m^{-2}}$.

We see from this analysis that PSE $= 10.7\ \mathrm{cd \cdot m^{-2}}$ (that is, a log luminance of 1.03) and the IU corresponds to $1.17\ \mathrm{cd \cdot m^{-2}}$ (a log luminance range of 0.068). Although the UL and LL are equal in log luminance ($0.034 \log \mathrm{cd \cdot m^{-2}}$), they correspond to 0.89 and $0.79\ \mathrm{cd \cdot m^{-2}}$, respectively, in luminance units. The best estimate of DL as IU/2 could be expressed as $0.034 \log \mathrm{cd \cdot m^{-2}}$.

b. Manual Least-Squares Method

The regression can also be determined without recourse to a graph of the data, although the graphical display was useful in recognizing that log luminance, rather than arithmetic luminance, was linearly related to the z scale. Table VI will help to illustrate the manual method of regression calculation.

Both ascending (top) and descending (bottom) orders of presentation are shown in Table VI. The first column lists the luminances of the nine stimuli used in the experiment. Column 4, labeled p, gives the probabilities of yes (top) or no (bottom) responses for each of the stimuli over all trials. The other four columns list transformations of the data in columns 1 and 4 that will be used in calculating the simple regression of z values on $\log L$. The regression is based on the relation

$$z = \alpha + \beta S, \tag{45}$$

TABLE VI. Manual Calculation of a Least-Squares Solution to Differential Luminance Threshold Experiment

L (cd·m^{-2})	log L	(log L)^2	p	z	z(log L)
I. Ascending order					
3.98	0.6	0.36	0.167	−0.97	−0.58
6.31	0.8	0.64	0.267	−0.63	−0.50
7.94	0.9	0.81	0.367	−0.34	−0.31
10.0	1.0	1.00	0.433	−0.17	−0.17
12.6	1.1	1.21	0.533	+0.09	+0.10
15.9	1.2	1.44	0.633	+0.33	+0.40
20.0	1.3	1.69	0.700	+0.53	+0.69
25.1	1.4	1.96	0.767	+0.73	+1.02
39.8	1.6	2.56	0.867	+1.11	+1.78
Sums:	9.9	11.67		+0.68	+2.43
II. Descending order					
3.98	0.6	0.36	0.767	+0.73	+0.44
6.31	0.8	0.64	0.667	+0.43	+0.34
7.94	0.9	0.81	0.600	+0.25	+0.23
10.0	1.0	1.00	0.533	+0.09	+0.09
12.6	1.1	1.21	0.433	−0.17	−0.19
15.9	1.2	1.44	0.300	−0.52	−0.62
20.0	1.3	1.69	0.267	−0.63	−0.82
25.1	1.4	1.96	0.200	−0.84	−1.18
39.8	1.6	2.56	0.100	−1.29	−2.06
Sums:	9.9	11.67		−1.95	−3.77

where $S = \log L$, $\beta = s_{xy}/s_x^2$ [in which s_x^2 is the variance of S from Eq. (23), and s_{xy} is the covariance from Eq. (32)], and $\alpha = z - \beta(\bar{S})$. The denominators of Eqs. (23) and (32) cancel so that β may be expressed as

$$\beta = \left[\frac{n\sum(zS) - (\sum S)(\sum z)}{n\sum S^2 - (\sum S)^2}\right]. \tag{46}$$

There were nine stimuli; hence, the value of n is 9. Column 6 of Table VI provides the sum zS (which equals 2.43); column 2 sums S; column 5 sums z; column 3 sums S^2; and the square of the sum of S is simply the square of the sum shown at the bottom of column 2. Using the values for the ascending order of presentation given in Table VI, we can write Eq. (46) explicitly as

$$\beta = \left[\frac{9(2.43) - (9.9)(0.68)}{9(11.67) - (9.9)^2}\right] = 2.16.$$

Having calculated β, we may now compute $\alpha = z - \beta(\bar{S})$ as

$$\alpha = [(0.68)(9^{-1}) - 2.16(9.9)(9^{-1})] = -2.30.$$

Thus, Eq. (45) becomes

$$z = -2.30 + 2.16S,$$

which represents the regression equation of z on $\log L$ for the ascending-order data of Table VI. A similar procedure may be used to calculate the regression for the descending-order data.

The explicit expressions of Eq. (45), when solved for $z = 0$, yield 50% probability values of 1.065 log $\text{cd} \cdot \text{m}^{-2}$ for ascending order and 0.995 log $\text{cd} \cdot \text{m}^{-2}$ for descending order. The mean of these two values is 1.030 log $\text{cd} \cdot \text{m}^{-2}$, which is the PSE estimate from this method of analysis. The difference between the two values is 0.070, which is the IU, and the DL equals half that value (0.035 log $\text{cd} \cdot \text{m}^{-2}$).

The foregoing paragraphs show that the regression of z values on stimulus values can be determined easily from statistics of the normal distribution. The method relies on the normality of the distribution of response probabilities with respect to stimulus values. If that requirement were not satisfied—for example, if luminances rather than log luminances had been used—then the method would yield incorrect results.

c. Probit Analysis

A third method by which the regression of z on $\log L$ can be determined is that called *probit analysis*. This method differs from the regression methods described in Sections a and b in that a *maximum-likelihood* model is used rather than a model that provides a least-squares solution. The least-squares criterion of regression selects constants (a,b or α,β) of the linear equations so that the sum of the squared errors between the prediction line and the data is as small as possible. The maximum-likelihood model selects these constants so that the probability of prediction is highest for all the data. When both variables of the regression are normally distributed, that is, when they form a bivariate normal distribution, the maximum-likelihood estimates and the least-squares regression coefficients are identical. However, when the distributions differ, the maximum-likelihood method, in effect, applies different statistical weights, and the two methods yield different estimates of the regression coefficients.

Weights can also be used with the least-squares regression method. For example, Guilford (1954) describes a method for applying weights to the probability data in order to give greatest weight to those proportions closest to 50% and to observations with smallest mean-square errors. Many of the advantages sought by earlier workers through the use of weights are provided

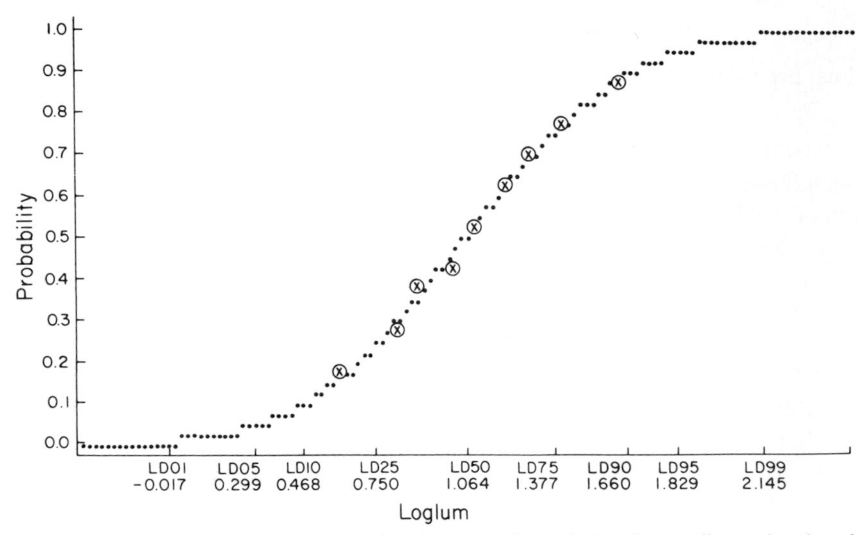

Fig. 15. Maximum-likelihood regression from probit analysis of ascending-order data in experimental example of the method of constant stimuli.

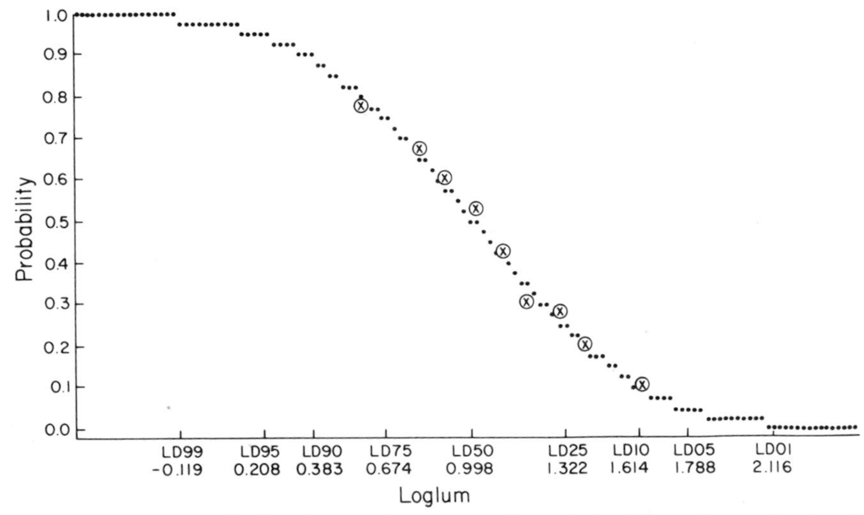

Fig. 16. Maximum-likelihood regression from probit analysis of descending-order data in experimental example of the method of constant stimuli.

by the method of maximum-likelihood regression. When that method of regression is carried out with probits $(z + 5)$ versus stimulus values, the process is called probit analysis. The methodological details are described in textbooks on statistics (for example, Finney, 1971), but today, computer programs for probit analyses are readily available (for example, SAS, 1979).

The data of Table V were used in one such computer program: the SAS Probit Procedure (SAS, 1979). This program tested the fit of the data to a normal distribution and computed a value of $\chi^2 = 0.153$ for ascending order and $\chi^2 = 0.592$ for descending order of presentation, indicating that the data were normally distributed to a high level of confidence. The computer then provided an explicit statement of the linear relation between probits and log L, corresponding to Eq. (45): $z = 2.151S - 2.289$ for ascending order and $z = -2.082S + 2.087$ for descending order, where $S = \log L$.

Figures 15 and 16 show computer-derived graphs of probability versus log L where the best-fitting cumulative normal function is plotted with dots and the circled Xs correspond to the input data. The abscissas of the two graphs combine log L values (labeled Loglum) with percentiles of the cumulative normal distribution (labeled LD50, and so on, standing for "lethal dose," a descriptor that reflects the method's original application to biological assay problems). Figure 15 shows at a glance that the 50% probability for ascending order corresponds to 1.064 log L, and Fig. 16 indicates a 50% value of 0.998 log L for the descending order. Tables VII and VIII provide listings of the log L values corresponding to probabilities from 0.01 to 0.99 in increments of 0.01, together with 95% (1.96σ) limits about each point for both orders of presentation.

Table IX summarizes the regression estimates of all three regression methods discussed in this section. The equations shown in this table have constants that differ for either order of presentation by less than 1%. Although the results are slightly different, each method yields a PSE, rounded to one decimal place, of 10.7 $\text{cd}\cdot\text{m}^{-2}$. It may be inferred from the comparisons in Table IX that when data are normally distributed and well-behaved, as they are in the experiment of this example, any one of the three methods described will provide a satisfactory result. The least-squares regression methods require no distributional assumptions and may be classed as curve-fitting techniques. The method of maximum likelihood does require distributional assumptions to determine the likelihood function and may be classed as a statistical procedure for optimum prediction. When the data are normally distributed, both classes yield essentially the same results, but the methods for arriving at them are quite different. When the distributions are not normal, the two methods will yield different results. As a general rule, the maximum-likelihood regression may be a more reliable indicator of the true relation. With packaged computer programs readily available, the considerable additional complexity of the maximum-likelihood method need not present a problem to those who have access to computers. For those who must make calculations by hand, the manual least-squares regression method is easier to use.

TABLE VII. Probit Analysis on log L (Ascending Order)[a]

		95% fiducial limits	
Probability	log L	Lower	Upper
0.01	−0.01734698	−0.44692618	0.23075067
0.02	0.10935992	−0.27325920	0.33124367
0.03	0.18975139	−0.16322789	0.39515806
0.04	0.25022675	−0.08055806	0.44334082
0.05	0.29941880	−0.01339160	0.48261284
0.06	0.34138896	0.04371156	0.51610555
0.07	0.37800089	0.09372191	0.54553005
0.08	0.41087205	0.13844765	0.57192875
0.09	0.44076706	0.17907506	0.59598617
0.10	0.46828546	0.21642638	0.61817734
0.15	0.58221900	0.37046841	0.71065712
0.20	0.67276976	0.49190912	0.78514415
0.25	0.75045425	0.59499614	0.85014567
0.30	0.82021732	0.68625339	0.90983713
0.35	0.88486321	0.76915936	0.96680763
0.40	0.94620583	0.84569672	1.02299937
0.45	1.00555553	0.91701946	1.08009356
0.50	1.06396426	0.98386303	1.13963092
0.55	1.12237299	1.04691279	1.20296210
0.60	1.18172269	1.10711458	1.27117725
0.65	1.24306531	1.16580151	1.34521941
0.70	1.30771120	1.22465873	1.42623866
0.75	1.37747427	1.28574005	1.51610605
0.80	1.45515876	1.35176410	1.61817055
0.85	1.54570952	1.42701570	1.73884667
0.90	1.65964306	1.52008650	1.89229769
0.91	1.68716146	1.54238130	1.92954538
0.92	1.71705647	1.56653880	1.97007271
0.93	1.74992763	1.59303467	2.01470127
0.94	1.78663956	1.62255419	2.06461660
0.95	1.82850972	1.65614072	2.12162594
0.96	1.87770177	1.69550671	2.18869844
0.97	1.93817713	1.74378571	2.27127203
0.98	2.01856860	1.80780276	2.38120068
0.99	2.14527550	1.90841603	2.55474739

[a] Probit analysis, differential luminance threshold, psychophysical method of constant stimuli, one observer, 30 replications for each stimulus.

TABLE VIII. Probit Analysis on log L (Descending Order)[a]

Probability	log L	95% fiducial limits	
		Lower	Upper
0.01	2.11579709	1.87576476	2.53800496
0.02	1.98486528	1.77238693	2.35662091
0.03	1.90179325	1.70658591	2.24174964
0.04	1.83930140	1.65694292	2.15547991
0.05	1.78846909	1.61644901	2.08541931
0.06	1.74520282	1.58188594	2.02588312
0.07	1.70726676	1.55149478	1.97376764
0.08	1.67329955	1.52420368	1.92718389
0.09	1.64240772	1.49930846	1.88489288
0.10	1.61397175	1.47632020	1.84603611
0.15	1.49623922	1.38017008	1.68613129
0.20	1.40266914	1.30207733	1.56071968
0.25	0.32239434	1.23320461	1.45510379
0.30	1.25030508	1.16869948	1.36272294
0.35	1.18350365	1.10590582	1.28023103
0.40	1.12011562	1.04252393	1.20575112
0.45	1.05878697	0.97693535	1.13795681
0.50	0.99843066	0.90811975	1.07550427
0.55	0.93807435	0.83551584	1.01684002
0.60	0.87674570	0.75866960	0.96030338
0.65	0.81335767	0.67686519	0.90424599
0.70	0.74655624	0.58882891	0.84699670
0.75	0.67446699	0.49238821	0.78665142
0.80	0.59419219	0.38381574	0.72063528
0.85	0.50062211	0.25621887	0.64472780
0.90	0.38288957	0.09464403	0.55024769
0.91	0.35445360	0.05549642	0.52755027
0.92	0.32356178	0.01292513	0.50293533
0.93	0.28959456	−0.03393019	0.47591580
0.94	0.25165850	−0.08631066	0.44578963
0.95	0.20839223	−0.14610794	0.41148765
0.96	0.15755993	−0.21642946	0.37125466
0.97	0.09506807	−0.30296582	0.32187830
0.98	0.01199604	−0.41812083	0.25636102
0.99	−0.11893577	−0.59983635	0.15331466

[a] Probit analysis, differential luminance threshold, psychophysical method of constant stimuli, one observer, 30 replications for each stimulus.

TABLE IX. Comparison of Analyses of Differential Luminance Threshold Experiment (Method of Constant Stimuli)

Graphical regression method	Manual least-squares method	Probit analysis method
I. Ascending order		
$z = 2.144(\log L) - 2.282$	$z = 2.16(\log L) - 2.30$	$z = 2.151(\log L) - 2.289$
$r^2 = 0.997$	—	$\chi^2 = 0.153$
$\log L_{z=0} = 1.064$	$\log L_{z=0} = 1.065$	$\log L_{z=0} = 1.064$
II. Descending order		
$z = -2.087(\log L) + 2.079$	$z = -2.08(\log L) + 2.07$	$z = -2.082(\log L) + 2.087$
$r^2 = 0.990$	—	$\chi^2 = 0.592$
$\log L_{z=0} = 0.996$	$\log L_{z=0} = 0.995$	$\log L_{z=0} = 0.998$
III. Combined summary		
PSE = 1.030 [10.7 $cd \cdot m^{-2}$]	1.030 [10.7 $cd \cdot m^{-2}$]	1.031 [10.7 $cd \cdot m^{-2}$]
IU = 0.068 [1.17 $cd \cdot m^{-2}$]	0.070 [1.18 $cd \cdot m^{-2}$]	0.066 [1.16 $cd \cdot m^{-2}$]
UL = 0.034 [0.89 $cd \cdot m^{-2}$]	0.035 [0.95 $cd \cdot m^{-2}$]	0.033 [0.85 $cd \cdot m^{-2}$]
LL = 0.034 [0.79 $cd \cdot m^{-2}$]	0.035 [0.76 $cd \cdot m^{-2}$]	0.033 [0.79 $cd \cdot m^{-2}$]

3. Method of Adjustment

The same experiment as that illustrated in Fig. 12 was repeated for a single observer using the psychophysical method of adjustment (also called the method of average error, method of reproduction, or method of equivalent stimuli). The observer's task was to match the brightnesses of the two halves of the bipartite field. Sixty trials were made. The test field was presented on the left in half the trials and on the right in the other half. In both cases, 15 trials involved approaching the match from a lower luminance than that of the standard (ascending order), and 15 trials approached a match from a higher luminance (descending order). The four space-order combinations were presented in random sequence. The raw data are listed in Table X as log luminances of the 60 matches. The purpose of using different positions and orders was to determine whether any space or order biases existed in the experiment. The data of Table X may be tested for homogeneity to examine this question. Two such tests will be used: a two-way analysis of variance and a χ-squared test of homogeneity.

a. Analysis of Variance Test of Homogeneity

Analysis of variance is the most well known of a variety of statistical inference tests designed to compare any and all mean differences in terms of their reliability. The analysis of variance (abbreviated ANOVA) is a relatively simple method for simultaneous comparison of several treatments (such as the

TABLE X. Raw Data from Differential Luminance Threshold Experiment (Log Luminances of Matches)

Trial	Test field on right	Test field on left	
1	1.052	1.034	
2	1.016	1.057	
3	1.049	1.014	
4	1.043	1.029	Descending order
5	1.058	1.022	
6	1.026	1.028	
7	1.054	1.031	
8	1.034	1.031	
9	1.049	1.059	
10	1.035	1.050	
11	1.023	1.025	
12	0.998	1.015	
13	1.044	1.016	
14	1.018	1.049	
15	1.010	1.018	
16	1.040	1.031	
17	1.005	1.017	
18	1.005	1.018	
19	1.031	1.006	
20	1.007	1.035	Ascending order
21	1.026	1.029	
22	1.034	1.051	
23	1.035	1.031	
24	1.030	1.045	
25	1.036	1.032	
26	0.999	1.014	
27	1.046	1.048	
28	1.021	1.029	
29	1.043	1.027	
30	0.993	1.061	

space and order treatments used in this example experiment) to draw inferences about the relations among the treatments and the variable under measurement (log luminance of the match in this instance). A two-way ANOVA procedure allows us to address the following questions in the experiment described here:

1. Are there systematic effects resulting from the position of presentation (left or right) of the test stimulus?
2. Are there systematic effects resulting from the order (up or down) of approach to a matching luminance?
3. Are there systematic effects resulting neither from order nor from position alone, but attributable only to the combination of order and position?

The first two questions are said to concern *main effects*, and the third question is related to an *interaction effect*. The main effects will be distinguished here by subscripts j and k and their interaction by jk. Each match y (over all stimuli i) may be considered the sum of systematic effects plus random error e:

$$y_{ijk} = \mu + \Delta_j + \Delta_k + \Delta_{jk} + e_{ijk}, \quad (47)$$

where Δ_j is the effect of position, $\Delta_j = \mu_j - \mu$; Δ_k is the effect of order, $\Delta_k = \mu_k - \mu$; Δ_{jk} is the interaction of position and order, $\Delta_{jk} = \mu_{jk} - \mu_j - \mu_k + \mu$; and μ stands for the true grand mean of x_i.

The foregoing definitions indicate that complete absence of position and order effects leads to absolute equality of all means for the four quadrants of data listed in Table X. The feature of the data tested by ANOVA is, then, absolute equality of mean results with all treatments.

Details of ANOVA procedures are given in standard textbooks on statistics. The availability of computer programs for carrying out ANOVA tests permits anyone with access to a computer to apply the method easily. Table XI shows the results of such a computer analysis of the data from the experiment under discussion here. The key test statistic in this table is labeled F value. It is a variance ratio, sometimes called Snedecor's F test, after the man who computed tables of the distribution of variance ratios and named the ratio F in honor of R. A. Fisher, who provided the original mathematical foundation for the test method. The variance ratio is defined as

$$F = \frac{\text{greater estimate of population variance}}{\text{lesser estimate of population variance}}.$$

When the results of two treatments are compared, one will generally provide a different estimate of the total population variance from the other, although they may, of course, give the same estimates. Those two estimates are the ones used in calculating the F ratio. Tables have been drawn up to list the critical values of F that will be exceeded with a given level of probability for various sample sizes; such tables are found in most standard textbooks on statistics (for instance, Dixon and Massey, 1957; Fisher and Yates, 1963; Pearson and Hartley, 1967/1972; Burington and May, 1969; Winkler and Hays, 1975). All the F ratios in Table XI are well below critical levels, even for the 5% significance levels. With the computer program used here (SAS ANOVA Procedure: SAS, 1979) it is not necessary to go to tables of F values, because the probabilities that the variance ratios are larger than critical are also computed and printed in the report, these are shown in Table XI under the symbols PR > F. The table also gives the computed value of the *coefficient of variation* (labeled C.V.). This coefficient is equal to 100 times the standard deviation (STD DEV) divided by the mean (Response Mean). Its value in

TABLE XI. ANOVA Test of Data Homogeneity[a]

Source	DF	Sum of squares	Mean square	F value	PR > F	R-square	C.V.
Model	3	0.00097373	0.00032458	1.21	0.3149	0.060822	1.5906
Error	56	0.01503587	0.00026850		STD DEV		Response mean
Corrected total	59	0.01600960			0.01638590		1.03020000

Source	DF	ANOVA SS	F value	PR > F
Position	1	0.00014107	0.53	0.4716
Order	1	0.00043740	1.63	0.2071
Position times order	1	0.00039527	1.47	0.2301

Means

Position		N	Response
L		30	1.03173333
R		30	1.02866667
Order		**N**	**Response**
Down		30	1.03290000
Up		30	1.02750000
Position	**Order**	**N**	**Response**
L	Down	15	1.03186667
L	Up	15	1.03160000
R	Down	15	1.03393333
R	Up	15	1.02340000

[a] Differential luminance threshold, psychophysical method of adjustment, analysis of variance procedure, dependent variable: response

Table XI is reasonably low. All of these indicators suggest that the small differences among the computed means shown at the bottom of the table are not statistically significant. In short, the ANOVA test leads us to infer that the data from our sample experiment are homogeneous in the sense that there were no significant order–position biases affecting the data.

b. χ-SQUARED TEST OF HOMOGENEITY

A second kind of test of homogeneity is the χ-squared test discussed earlier in this chapter. Table XII shows the results of three χ-squared tests of homogeneity: comparing the distributions by position, by order, and with the

TABLE XII. χ-Squared Test of Homogeneity

Comparison	Computed χ^2	df	Critical χ^2 ($\alpha = 0.05$)	Inference
Response by position	34.13	29	42.55	No difference
Response by order	38.53	29	42.55	No difference
Total response by normal	52.30	59	77.93	No difference

normal distribution. The computed χ-squared values are all less than the critical values for a 5% significance level. The inferences that may be drawn from this are that the position and order treatments yielded essentially the same results and, further, that they were reasonably normally distributed about the population mean.

The results of these tests of homogeneity suggest that it is appropriate to pool all of the data from the four quadrants of Table X. They are, in fact, all plotted on the same probability graph in Fig. 17. They are scattered closely

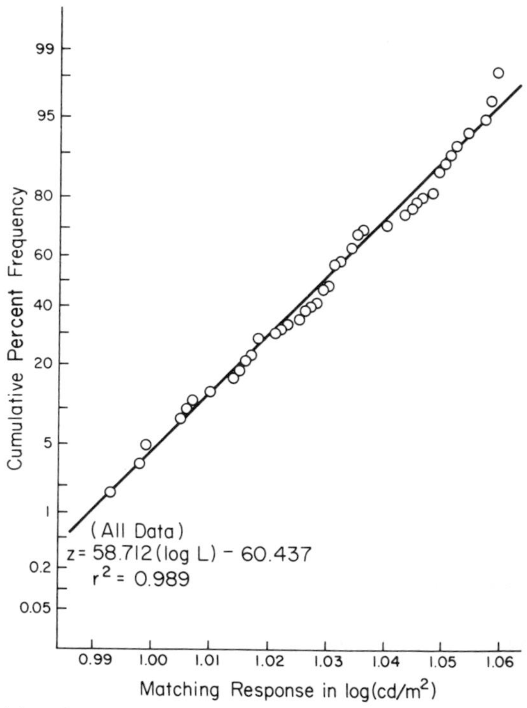

Fig. 17. Pooled data from ascending and descending orders, left and right presentations, of method of adjustment experiment plotted against cumulative frequency.

about a straight line (implying a normal distribution for the coordinates of this kind of graph) and may be represented by a linear function of z on $\log L$ with a coefficient of determination r^2 of 0.989:

$$z = 58.712 \log L - 60.437.$$

c. Frequency Analysis

Table XIII shows the results of a frequency analysis of the pooled data. Matching log luminances (in the column labeled Response) are shown together with their frequencies, cumulative frequencies, proportions in percent, and cumulative percentages.

In the example experiment used here, the data were found to be homogeneous. Sometimes, however, they are inhomogeneous, and in such cases it is helpful to analyze them separately according to treatment. The methods described in this and the foregoing section may be used, as well as other methods. To illustrate another commonly applied method of analysis, the data will be treated separately by the order treatment. The method involves deriving a frequency histogram for each order and determining a regression equation of z on $\log L$ for each of the two treatments. Table XIV and Fig. 18 illustrate this method.

Eight cells of 0.01 log $\text{cd} \cdot \text{m}^{-2}$ intervals were used to create the frequency histogram. The number of matches in each cell was counted (shown in column 4 of Table XIV), and those counts were used to estimate the probabilities of match for each cell. The cumulative probabilities of matches have been plotted in the probability graph of Fig. 18. Linear regressions of z on $\log L$ were determined for each order, ascending and descending. These equations and their associated coefficients of determination are shown in Fig. 18. Note that this frequency histogram method, which is somewhat coarser than the other methods used in this section, provides considerably more scatter (and, hence, lower values of r^2) than was found in the earlier analyses. Nonetheless, a comparison of indicators, listed in Table XV, shows that there is not a very large difference between the PSE estimates determined by the methods that pooled orders and that separated orders; the PSE values differ by only 0.02 $\text{cd} \cdot \text{m}^{-2}$. Proportionately larger differences may be seen between estimates of the values for IU. Still, the difference between intervals of uncertainty of 1.03 $\text{cd} \cdot \text{m}^{-2}$ from the analysis of pooled data and 1.01 $\text{cd} \cdot \text{m}^{-2}$ as an average from the separated order data cannot be considered very large.

The example experiment discussed in this section is a simple matching paradigm along a single, intensive dimension. The method of adjustment is also used for other kinds of experiments. In colorimetry, for example, color-matching is usually done by the method of adjustment. The perceptual

TABLE XIII. Frequency Analysis of Data (Differential Luminance Threshold, Psychophysical Method of Adjustment)

Response	Frequency	Cumulative frequency	Percentage	Cumulative percentage
0.993	1	1	1.667	1.667
0.998	1	2	1.667	3.333
0.999	1	3	1.667	5.000
1.005	2	5	3.333	8.333
1.006	1	6	1.667	10.000
1.007	1	7	1.667	11.667
1.01	1	8	1.667	13.333
1.014	2	10	3.333	16.667
1.015	1	11	1.667	18.333
1.016	2	13	3.333	21.667
1.017	1	14	1.667	23.333
1.018	3	17	5.000	28.333
1.021	1	18	1.667	30.000
1.022	1	19	1.667	31.667
1.023	1	20	1.667	33.333
1.025	1	21	1.667	35.000
1.026	2	23	3.333	38.333
1.027	1	24	1.667	40.000
1.028	1	25	1.667	41.667
1.029	3	28	5.000	46.667
1.03	1	29	1.667	48.333
1.031	5	34	8.333	56.667
1.032	1	35	1.667	58.333
1.034	3	38	5.000	63.333
1.035	3	41	5.000	68.333
1.036	1	42	1.667	70.000
1.04	1	43	1.667	71.667
1.043	2	45	3.333	75.000
1.044	1	46	1.667	76.667
1.045	1	47	1.667	78.333
1.046	1	48	1.667	80.000
1.048	1	49	1.667	81.667
1.049	3	52	5.000	86.667
1.05	1	53	1.667	88.333
1.051	1	54	1.667	90.000
1.052	1	55	1.667	91.667
1.054	1	56	1.667	93.333
1.057	1	57	1.667	95.000
1.058	1	58	1.667	96.667
1.059	1	59	1.667	98.333
1.061	1	60	1.667	100.000

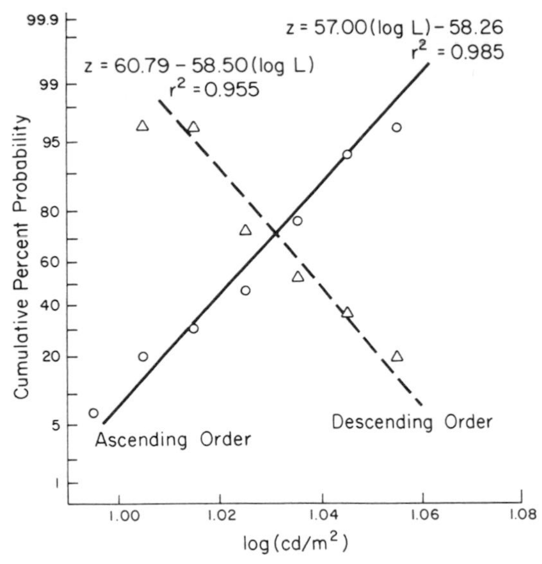

Fig. 18. Separate determinations of cumulative percent probability versus log cd · m^{-2} for ascending- and descending-order data from experiment illustrating the method of adjustment.

TABLE XIV. Frequency Histogram of Data from First Differential Luminance Threshold Experiment

Cell	L^a	Log L^a	Count	Probability	Cumulative probability
I. Ascending order					
1	9.89	0.995	2	0.067	0.067
2	10.12	1.005	4	0.133	0.200
3	10.35	1.015	3	0.100	0.300
4	10.59	1.025	5	0.167	0.467
5	10.84	1.035	9	0.300	0.767
6	11.09	1.045	5	0.167	0.934
7	11.35	1.055	1	0.033	0.967
8	11.61	1.065	1	0.033	1.000
II. Descending order					
1	9.89	0.995	1	0.033	1.000
2	10.12	1.005	0	0.000	0.967
3	10.35	1.015	7	0.233	0.967
4	10.59	1.025	6	0.200	0.734
5	10.84	1.035	5	0.167	0.534
6	11.09	1.045	5	0.167	0.367
7	11.35	1.055	6	0.200	0.200
8	11.61	1.065	0	0.00	0.000

[a] Midpoints of 0.01 log cd · m^{-2} cells

TABLE XV. Comparison of Analyses of Differential Luminance Threshold Experiment, Method of Adjustment (Combined Summary)

Orders pooled	Orders separated
PSE = 1.030 = 10.72 $cd \cdot m^{-2}$	PSE = 1.031 = 10.74 $cd \cdot m^{-2}$
IU = 0.011 = 1.03 $cd \cdot m^{-2}$	IU = 0.005 = 1.01 $cd \cdot m^{-2}$
UL[a] = 0.005 = 0.12 $cd \cdot m^{-2}$	UL = 0.003 = 0.07 $cd \cdot m^{-2}$
LL[a] = 0.005 = 0.13 $cd \cdot m^{-2}$	LL = 0.002 = 0.05 $cd \cdot m^{-2}$

[a] Based on *PE* (probable error)

variation there is qualitative rather than quantitative. Both central tendencies and variabilities must be determined over two dimensions of chromaticity or over three tristimulus dimensions. This may be done by computing two-dimensional ellipses or three-dimensional ellipsoids, as appropriate, according to some level of significance α. Simple methods for computing ellipses are given in Jackson (1956) and for tristimulus ellipsoids in Wyszecki and Stiles (1967).

4. Method of Limits

Six of the stimuli listed in Table V (those ranging in log L from 0.8 to 1.3) were used in an experiment conducted according to the psychophysical method of limits. A double-staircase plan was used (see Chapter 6). That is, there were two plans of presentations, randomly interspersed, in which the same observer used in the other examples in this section was presented with one of the six stimuli to which he responded either yes (meaning brighter than the standard) or no (meaning darker than the standard); he was then presented with the next lower or higher luminance stimulus, according to a given plan of presentation. The process was repeated until he had responded the same way on two successive occasions following a reversal of response; and then the direction of change in test luminance was reversed. Figure 19 illustrates the method. The observations were carried out over 30 trials, and they are identified numerically along the top of the figure to form columns. The two plans, ascending- and descending-luminance orders, are identified as runs A and B. The choice of whether to use run A or B on any given trial was determined by a random-choice computer program. Log L values of 0.9 and 1.2 were selected as starting points. The experiment began with run A, so the observer was shown a test field of 0.9 log L. He responded no (darker than the standard), and, accordingly, a minus sign was entered in column 1 opposite log L = 0.9. The next trial was also from plan A, so the next higher log L

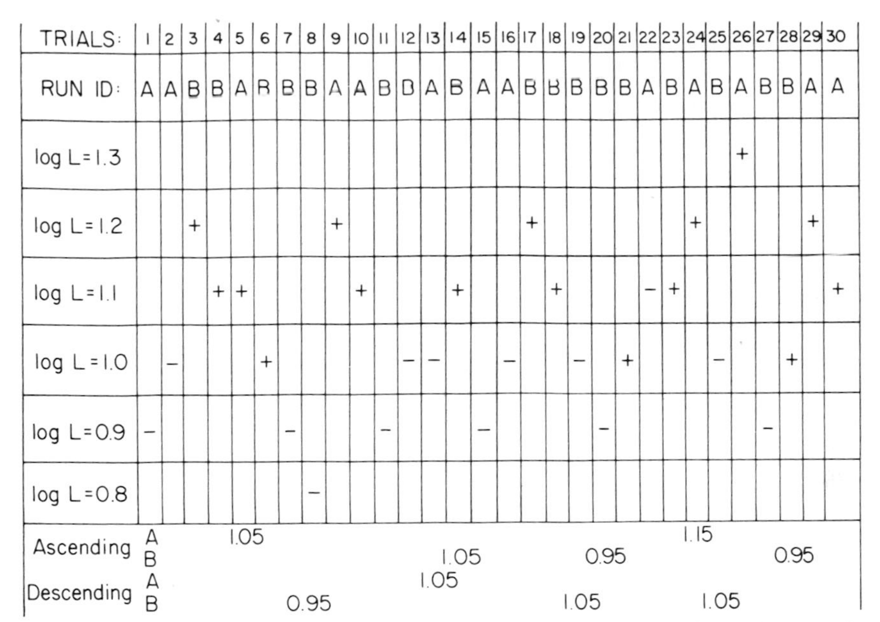

TRIALS:	1	2	3	4	5	6	7	8	9	10	11	12	13	14	15	16	17	18	19	20	21	22	23	24	25	26	27	28	29	30
RUN ID:	A	A	B	B	A	B	B	B	A	A	B	B	A	B	A	A	B	B	B	B	B	A	B	A	B	A	B	B	A	A
log L = 1.3																										+				
log L = 1.2			+						+								+							+					+	
log L = 1.1				+	+					+				+				+				−	+							+
log L = 1.0		−				+						−	−			−			−		+				−			+		
log L = 0.9	−						−				−				−					−							−			
log L = 0.8								−																						
Ascending A					1.05																			1.15						
Ascending B														1.05						0.95								0.95		
Descending A													1.05																	
Descending B								0.95											1.05						1.05					

Fig. 19. Plan and results of double-staircase method of limits experiment. See text for discussion.

stimulus, 1.0, was presented. The observer again responded no; therefore, a minus sign was entered in the column for trial 2 next to log $L = 1.0$. Trial 3 was from run B, the descending order, so a test stimulus of log $L = 1.2$ (the upper starting value) was shown to the observer. He responded yes (brighter), and a plus sign was entered in the column for the third trial opposite the log $L = 1.2$ position. Trial 4 was also from run B; the stimulus elicited another yes response, and that was recorded by the plus sign in the appropriate position of the chart. The fifth trial was from plan A again. The response was yes. After three more trials, the next presentation according to run A was encountered, that is, trial 9. Because the last response elicited in run A (that in trial 5) had been yes, the ascending order of presentation was continued by showing the stimulus with a log L of 1.2. This also evoked a yes response, the second in succession after a reversal of responses in run A (the reversal occurred on trial 5), so the next presentation according to the plan for run A (which occurred randomly on trial 10) used the next lower luminance of log $L = 1.1$.

The experiment continued in this way until 30 trials had been carried out. Figure 19 sets forth the plan of the experiment, its results, and indications of the log luminances corresponding to response reversals. The reason for using two runs, A and B, randomly intermixed, was to prevent the observer from deducing a pattern of the order of presentation. The reason that two successive, identical responses were used as the criterion for reversing direction was to increase the likelihood that the preceding reversal did not occur by chance alone. These are possibilities that can be tested by analyzing the data, but first it will be instructive to describe how the log L data are derived from the experiment.

The entries at the bottom of Fig. 19 list the log Ls resulting from this experiment. They are shown on different rows according to whether they resulted from ascending or descending order and from run A or run B. The first entry occurs below trial 5. Its position along the rows indicates that it occurred in the ascending, run A, sequence. The previous run A response was negative for log $L = 1.0$ (trial 2). The response was positive to log $L = 1.1$ (in trial 5). The estimated log luminance at which the reversal took place is then assumed to be midway between these two values of log L, at log $L = 1.05$. This assumption is the one most frequently used to deduce the reversal point in such experiments. The choice of stimulus intervals will influence the results; when the intervals are too large, the estimates will be too coarse. If the experimenter succumbs to the temptation to alter the plan because of a misguided idea that the data will be more precise if he changes the size of the intervals, then the estimates of PSEs will be based on data with different underlying criteria. For these reasons it is always a good idea to conduct exploratory experiments first, to establish both the range and the intervals of stimuli to be used in a method of limits experiment. In this example the log L interval of 0.1 appeared reasonable from the results of a short, informal, preliminary experiment. Some care was used in designing the experiment, both with regard to establishing stimulus intervals and developing a double-staircase plan. The experiment was not particularly well designed, however, with respect to its length. Notice that in 30 trials only 9 reversals occured. A better plan would have involved about 100 trials for each of runs A and B. In short, a better experiment would have been about six times as long as the one illustrated in Fig. 19. Nonetheless, this one will illustrate techniques for reducing and analyzing data from a method-of-limits experiment.

In Fig. 19, two reversals are listed for ascending order with run A, one reversal for descending order with run A, and three reversals for each of the remaining combinations. Those log L values estimated for points of reversal can be averaged in a number of ways. Table XVI lists seven arithmetic means from the log L data. Means for all data, for those corresponding to ascending order and to descending order, for data from only the first 15 trials and only the last 15 trials, and for data from the two runs, A and B, are shown in the table, together with the associated standard deviations. These means provide the necessary information to examine questions about response biases.

TABLE XVI. Mean PSE and Standard Deviation

Grouping	n	Mean[a]	Standard deviation[a]
All data	9	1.03	0.07
Descending order	5	1.03	0.05
Ascending order	5	1.03	0.08
First 15 trials	4	1.03	0.05
Last 15 trials	5	1.03	0.08
Run A	3	1.08	0.06
Run B	6	1.00	0.05

[a] Expressed as log L

1. Was there a different result for ascending and descending orders?
2. Was there a different result for runs A and B?
3. Was there a time effect, that is, was there a difference in mean response between the first and second halves of the experiment?

These are questions that can be addressed by applying a two-way ANOVA test, as was done in the example of the experiment conducted according to the method of adjustment. Table XVII summarizes the results of such a test for the time (first 15 trials versus second 15 trials) and order (ascending and descending) main effects and their interaction. Although the run effect could have been included in the test, it was not. The results of Table XVII may be taken to imply that there were no statistically significant differences between results from the effects or from their interaction at the 1% significance level. In other words, all the data should be pooled in order to estimate the PSE and limen. When this was done, the PSE was taken as the arithmetic mean of log L (which would correspond to the geometric mean luminance), and the IU was taken to be twice the *probable error*. The *PE* is equal to $\sim 0.6745(s_x)$. The *PE* has often been taken as an index that is proportional to the DL and is assumed by many workers, but not by all, to be an inverse measure of sensitivity (Guilford, 1954; p. 98). Table XVIII shows the values that were determined in the column labeled *orders pooled*.

TABLE XVII. Two-Way Analysis of Variance for Homogeneity of Data

	SS	df	F	Significance
Time (T)	0.0001	1	0.063	$p > 0.01$
Order (O)	0.0001	1	0.063	$p > 0.01$
Interaction ($T \times O$)	0.002	1		
Total	0.0022	3		

TABLE XVIII. Comparison of Analyses of Differential Luminance Threshold Experiment (Method of Limits)

Orders pooled	Orders separated
PSE = 1.028 = 10.72 cd · m^{-2}	PSE = 1.030 = 10.72 cd · m^{-2}
IU[a] = 0.094 = 1.24 cd · m^{-2}	IU = 0.005 = 1.01 cd · m^{-2}
DL[a] = 0.047 = 0.62 cd · m^{-2}	DL = 0.003 = 0.51 cd · m^{-2}

[a] Based on PE (probable error)

Suppose that, for the sake of argument, an ANOVA test had not been used, and it was merely assumed that the ascending and descending orders of presentation would yield significantly different results. Would the results have been different? The answer, in this case, is yes. Figure 20 shows the now familiar linear functions in log probability coordinates. Note, however, that there are only two means for each order that can be plotted. There is no justifiable choice other than a straight line to be drawn through the two points if they must be connected. The data are too sparse from our experiment to provide a robust indication of the underlying probability functions. The crossover point of the two lines was taken as an estimate of PSE, and the average standard deviation was used to estimate the IU; the DL was taken as IU/2. The resulting values are listed in Table XVIII in the column labeled *orders separated*.

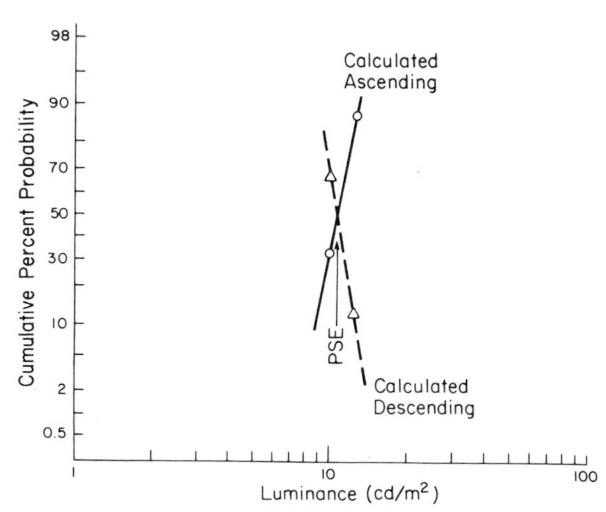

Fig. 20. Separate determination of probability functions for ascending- and descending-order data from experimental example of the method of limits.

The PSE estimated in this manner does not differ much from that estimated when all the data were used (orders pooled column). However, the IU and DL are very different. In describing the experiment, it was pointed out that the design was not very good from the standpoint of the number of trials included. The effect of this shortcoming shows up as the differences between IU or DL estimates in Table XVIII and as the paucity of data and steepness of slopes of the functions of Fig. 20. Not all such experiments would necessarily yield such outcomes, but the important point to be made by this example is that you cannot tell ahead of time what the outcome might be unless the data are tested and found to be robust and homogeneous.

We performed a test of homogeneity by applying the two-way ANOVA test to the data (although this was ignored for the sake of the argument in the last part of the example), and we found no statistically significant differences between orders of presentation. This result indicates that the better way of analyzing the data would be that in which the results from both orders are pooled. Validity, a loose term referring to how much utility attaches to something, generally increases as sample size increases. Inferential statistics are measures intended to aid in deciding whether to accept or reject hypotheses. The null hypothesis in our example would be that there is no difference between the results from the two order treatments. Two kinds of errors can be made: Rejecting the null hypothesis when, in fact, it is true; and accepting it when it is false. These are known simply as *type-I* and *type-II errors*, respectively. The probability of a type-I error (usually symbolized α) is what has been referred to here as significance. In most tests, α is chosen to be reasonably small, often about 5% ($\alpha = 0.05$). The probability of making a type-II error (β) is fixed by the choice of α and the sample size. Often, the probability of making a type-II error (that is, accepting the null hypothesis when it is false) will be ten times as large as the α significance level chosen, unless the sample size is very large. Suppose that our ascending and descending results with a brightness-matching experiment (using a different standard from that in the examples drawn from Fig. 12) were found to be 1.70 and 1.72 log cd $\cdot$ m^{-2}. We might be inclined to think that a difference of only 0.02 log cd $\cdot$ m^{-2} was not significant, so our null hypothesis would be that there is no difference between the two means. If we had a sample size of 10 (one more than the number of reversals in the method of limits example we have been discussing), and we chose a 5% significance level (that is, $\alpha = 0.05$) for testing, then the probability of a type-II error would be 0.903, less than 10% different from certainty. If the same level of significance is used, the probability of a type-II error can be reduced only by increasing sample size: it is 0.83 for $n = 25$, about 0.71 for $n = 50$, 0.484 for $n = 100$, and 0.021 for $n = 400$. In this example we would need a sample size of about 360 for $\beta = \alpha = 0.05$. The necessary sample size would be different for other values of α (the smaller the value of α the larger n needs to be for a given value of β); it also differs according to the disparity between elements being compared in the null hypothesis. Certainty of results, which loosely means their validity, depends strongly upon sample size.

The other characteristic of data referred to previously was robustness (see, Tukey, 1960). Distributions can differ in a number of ways. Two of the most common differences are in the tails and in the central portions of the distributions. In a sense the F test is a measure of the robustness of efficiency. When two estimates of the same population have unequal variances, the ratio of the smaller to the larger variance is a measure of the relative efficiency of the estimate with the larger variance. There are other approximate measures that may be used. For example, the sample average deviation, defined in Eq. (25), divided by the sample standard deviation, Eq. (24), yields a ratio of about 0.88 for a normal distribution, but the ratio differs for other distributions. Tukey (1960) computed that a complex distribution consisting of two normal distributions, each with the same mean but one with a standard deviation three times as large as that of the first, when combined with only 1% of the broader distribution in the mix resulted in a ratio of s_x/D_a of 1.44 rather than 0.88. Thus, small differences in the tails of distributions can greatly alter the calculated variances. Measures of DLs, which are often functions of variance, are particularly dependent on variations in tails of distributions.

The way to determine whether data are homogeneous and robust is to perform tests, such as those used in the earlier part of this section on the method of limits. The way to avoid problems with data that are not sufficiently robust is to increase the sample size. A careful experimenter would then have conducted many more trials than the 30 used in the example experiment of this section. Lacking sufficient number of trials, the prudent analyst would have tested the homogeneity of the data and, finding no justification for analyzing orders separately, would have pooled the data to determine the best estimates of PSE and DL from the existing data. The point to be made is simply that care must be used in both the design and the analysis of a psychophysical experiment. There are no shortcuts to accuracy.

Perhaps the most common failing of experiments from the standpoint of accuracy is the use of sample sizes that are too small. There is a best compromise between the antagonistic requirements of sufficient sample size, with attendant high cost of time and labor, and efficiency of experimentation with minimum commitment of effort on the part of both experimenter and observers. The expenditure of some additional time in planning an experiment to provide a satisfactory level of accuracy, as well as of precision, rewards the experimenter with less equivocal and more satisfactory results. We hope that some of the analytical methods and design considerations discussed in these sections will help those who are not expert in psychophysical experimentation to attain these rewards with greater certitude.

5. Signal-Detection Theory

Signal-detection theory (for example, Green and Swets, 1966) is an approach to threshold determination based on decision theory. Signal-detection theory (SDT) usually dispenses with the concept of a threshold as a probabilistic cutoff point (although it need not do so) and substitutes the concept of a continuum of events e that may occur either in the presence of noise N alone or of combined signal and noise $S + N$. The observer knows only that an event has taken place. It is his job to decide on the likelihood that the event resulted from a signal. Based on the estimated likelihood and the costs and rewards involved in the various possible outcomes, the observer is thought to set his criterion for responding yes or no. In this sense SDT is essentially similar to classical statistical *hypothesis testing*. This similarity is illustrated in Fig. 21.

Figure 21(a) shows a matrix of possible SDT outcomes. The observer may respond either yes or no to the presence of either signal or noise. The probability that he responds yes when the signal is present is referred to as a *hit* in *SDT*. By comparing top and bottom matrix arrays of the figure, it can be seen that an SDT hit corresponds to a type-I success $(1 - \alpha)$ in hypothesis testing, in which the null hypothesis H_o is accepted when it is true. When the observer responds yes to noise alone, SDT describes this as a *false alarm*. The false alarm corresponds to a type-II error (β) in hypothesis testing. When the

(a)

Stimulus	RESPONSE: Yes	RESPONSE: No
Signal	p (Yes | S) hit	p (No | S) miss
Noise	p (Yes | N) false alarm	p (No | N) correct rejection

(b)

True Situation	DECISION: Accept H_0	DECISION: Reject H_0
H_0	p (Yes | H_0) $1 - \alpha$ type I success	p (No | H_0) α type I error
H_1	p (Yes | H_1) β type II error	p (No | H_1) $1 - \beta$ type II success

α = p(type I error) = p(rejecting $H_0 | H_0$ is true)

β = p (type II error) = p (accepting $H_0 | H_0$ is false)

Fig. 21. Information matrices comparing SDT probabilities in (a) with those of classical statistical hypothesis testing in (b).

observer responds no to a signal, SDT labels the response a *miss*; and this corresponds to a type-I error (α) on the bottom matrix. Finally, a *correct rejection* in SDT is simply the equivalent of a type-II success $(1 - \beta)$ or *power* measure in classical hypothesis testing. The concept of SDT is, then, basically the same as that of hypothesis testing but is applied to the observer's decision process rather than to that of the data analyst.

Figure 22 may help to illustrate the assumed relationships between responses and $S + N$ combinations in SDT. Given the distribution of noise N

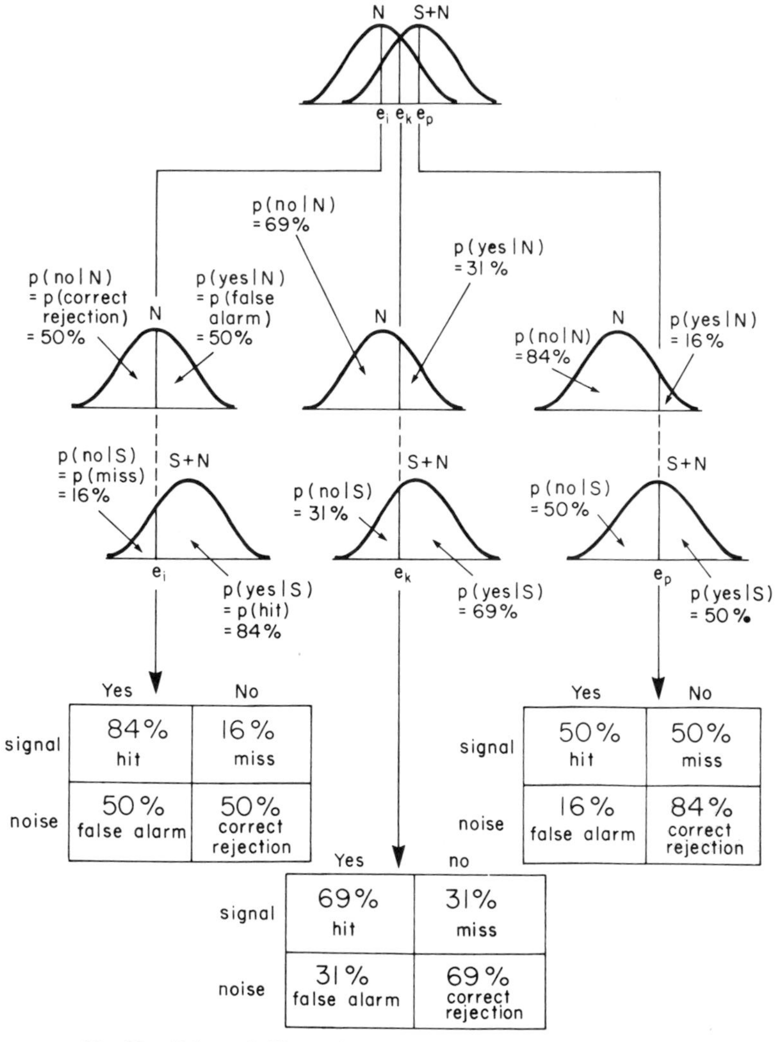

Fig. 22. Schematic illustration of decision-theory relations in SDT.

and signal plus noice S + N shown at the top of Fig. 22, the observer may base his response criterion on any of three events: e_i, e_k, or e_p. For e_i the probabilities of correct rejection and false alarms are equalized, both equal 50%. Because of the position of the noise distribution with respect to the signal distribution, this criterion results in an 84% probability of a hit, that is, of responding yes in the presence of S. Suppose, however, that the observer is penalized as much for misses and false alarms as he is rewarded for hits and correct rejections. It would then pay him to use a different criterion, one in which there were equally small probabilities of misses and false alarms. This criterion corresponds to event e_k. Finally, when the observer is penalized for false alarms and rewarded for hits and correct rejections, he might adopt a criterion that equalizes hits and misses such as that corresponding to e_p. The point of these examples is that different decisions can be made according to a variety of criteria for the same distributions of $S + N$ and N.

The *likelihood ratio* ℓ_{e_k} is the probability that the event occurs, given that the signal is present, divided by the probability of that event given noise alone:

$$\ell_{e_k} = \left[\frac{p(e_k \mid S + N)}{p(e_k \mid N)}\right]. \tag{48}$$

In Fig. 22, the likelihood according to Eq. (48) is 1.0 at event e_k. It is less than unity for e_i and greater than unity for e_p. If the observer based his decision entirely on likelihood, then he would use event e_k to the best of his ability, because it yields the correct likelihood ratio. However, there may be considerations other than likelihood that enter into the decision. In the general case the observer needs to know the joint probability rather than the conditional probability of Eq. (48). That is, what is the probability of an event *and* signal relative to the probability of the event *and* noise? This joint probability is given by

$$\left[\frac{p(e_k \cdot \{S + N\})}{p(e_k \cdot N)}\right] = \left[\frac{p(s)}{p(n)}\right]\ell_{e_k}, \tag{49}$$

which is the likelihood ratio of Eq. (48) multiplied by the prior probability of the signal relative to the noise. Ordinarily the subject should adopt a criterion for which the joint probability of Eq. (49) is unity; that is, he should say yes whenever it is at least an event bet that the event was produced by a signal.

The criterion that the observer actually adopts can be inferred from a graph of the kind illustrated in Fig. 23, which is usually referred to as a *receiver operating characteristic* (*ROC*) in signal-detection theory. It is similar to the *operating characteristic* (OC) of conventional hypothesis testing; the OC is simply the complement of the power curve that describes the probability of rejecting H_o as a function of the parameter of interest (for example, Winkler

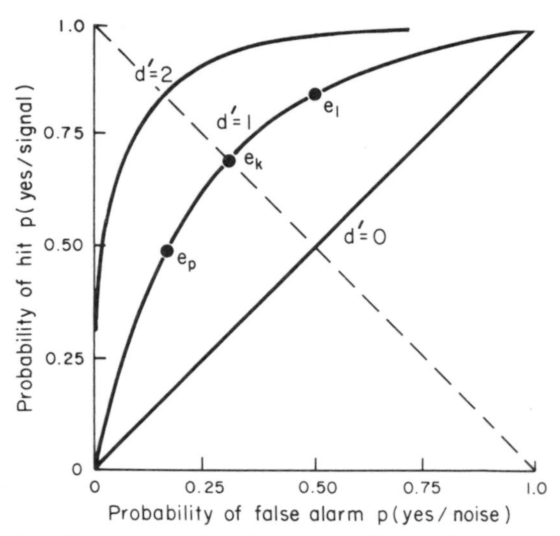

Fig. 23. Receiver operating characteristic. See text for explanation.

and Hays, 1975). The ordinate of Fig. 23 shows probabilities of hits (yes, given a signal), and the abscissa is a scale of probabilities of false alarms (yes, given noise). Liberal criteria produce points toward the lower part of the area defined by the dashed diagonal line passing through the unity coordinates, and conservative criteria yield points in the upper part of the diagonally defined area. The dashed diagonal defines the locus of all conditions for which the likelihood ratio of Eq. (48) is unity, that is, where the probability distributions for signal and noise cross (e_k in Fig. 22). The degree of curvature is related to the differences along the stimulus continuum between the means for noise and signal. The stronger the signal, in the sense of its mean being located at a higher stimulus value than is the mean of the noise, the more steeply curved is the result; in Fig. 23, curve $d' = 2$ represents a stronger stimulus than does curve $d' = 1$. The consequence of steeper curves is simply that the probability of a hit is greater than that for a false alarm. When both probabilities are equal, as they are in the straight line $d' = 0$, there is no difference between the means of the signal and noise distributions. These differences represent, inversely, the degree of overlap between signal and noise distributions. They are measured by the parameter d', which is simply equal to the stimulus difference between the means of the two distributions divided by their (assumed to be common) standard deviations:

$$d' = (S_{e_p} - S_{e_i})(\sigma)^{-1}, \tag{50}$$

where S stands for stimulus value and the other symbols are as in Fig. 22. The measure d' is said to represent the observer's sensitivity. This sensitivity varies along the direction of the dashed diagonal line. The criteria that the observer uses will vary along the curves that go from the origin to the coordinate position 1,1. This ability to separate measures of sensitivity and criterion severity is often said to be the chief advantage of SDT.

There are some disadvantages to SDT. A very large number of trials should be used for stable results. This arises from two factors. First, the measures derive as much from probability distributions of noise as from signal, and therefore, as many measurements of the noise distribution should be made as of the signal distribution. Second, because sensitivity is probabilistic, enough measurements need to be made to determine d' to some reasonable degree of stability. Usually these factors mean that several hundred trials should be run for a careful SDT experiment. Rating-scale and forced-choice procedures (see Chapter 6) are often used in SDT.

Once the SDT experiment has been conducted, data reduction is not unlike that described for classical psychophysical methods in earlier sections of this chapter. The ROC can be plotted as z scores of hits versus z scores of false alarms rather than in the linear probability coordinates of Fig. 23. This method generally yields linear relations. The standard regression methods can then be used to determine the relationship between the two z scales. The differences in intercepts from the origin (where $z = 0$) are then related to d' sensitivity through Δz.

This very brief discussion of SDT has not covered all the advantages claimed for it by its supporters. However, the discussion has sketched the major attributes of SDT in the context of psychophysical methodology, of which SDT is simply one more method.

REFERENCES

Bard, Y. (1974). "Nonlinear Parameter Estimation." Academic Press, New York.

Burington, R. S., and May, D. C. (1969). "Handbook of Probability and Statistics with Tables," 2nd ed. McGraw-Hill, New York.

Derman, C., Gleser, L. J., and Olkin, I. (1973). "A Guide to Probability Theory and Application." Holt, New York.

Dixon, W. J., and Massey, F. J., Jr. (1957). "Introduction to Statistical Analysis." McGraw-Hill, New York.

Draper, N., and Smith, H. (1981). "Applied Regression Analysis," 2nd ed. Wiley, New York.

Finney, D. J. (1971). "Statistical Methods in Biological Assay," 2nd ed. Griffin Press, London.

Fisher, R. A., and Yates, F. (1963). "Statistical Tables for Biological, Agricultural and Medical Research," 6th ed. Oliver and Boyd, Edinburgh.

Green, D. M., and Swets, J. A. (1966). "Signal Detection Theory and Psychophysics." Wiley, New York.

Guilford, J. P. (1954). "Psychometric Methods," 2nd ed. McGraw-Hill, New York.

Jackson, J. E. (1956). Quality control methods for two related variables. *Ind. Qual. Control* **12**, 2–6.

Kendall, M. G., and Stuart, A. (1958/1961/1966). "The Advanced Theory of Statistics," Vol. I–III. Griffin, London.

König, A., and Brodhun, E. (1889). Experimentelle Untersuchungen ueber die psychophysische Fundamentalformel in Bezug auf den Gesichtssinn. *Sitzungsber. Preuss Akad. Wiss. Berlin* **27**, 641–644.

König, A., and Dieterici, C. (1884). Ueber die Empfindlichkeit des normalen Auges für Wellenlängenunterschiede des Lichtes, *Ann. Phys. Chem.* **22**, 579–589.

Mosteller, F., and Turkey, J. W. (1977). "Data Analysis and Regression." Addison-Wesley, Reading, Mass.

Pearson, E. H., and Hartley, H. O. (1967/1972). "Biometrika Tables Vol. I, 3rd ed., and Vol. II, 2nd ed. Cambridge University Press, Cambridge.

Riggs, D. S. (1963). "The Mathematical Approach to Physiological Problems." Massachusetts Institute of Technology Press, Cambridge.

SAS [Statistical Analysis System] (1979). "SAS User's Guide." SAS Institute, Cary, N.C.

Tukey, J. W. (1960). A survey of sampling from contaminated distributions. *In* "Contributions to Probability and Statistics," I. Olkin, S. G. Ghurye, W. Hoeffding, W. G. Madow, and H. B. Mann, eds.), pp. 448–485. Stanford University Press, Stanford, California.

Urban, F. M. (1908). "The Application of Statistical Methods to Problems of Psychophysics." Psychological Clinic Press, Philadelphia.

Weber, E. S. (1834). "De pulsu, resorptione, auditu et tactu: annotationes anatomicae at physiologicae." H. Köhler, Leipzig.

Winkler, R. L., and Hays, W. L. (1975). "Statistics: Probability, Inference, and Decision," 2nd ed., Holt, New York.

Woodworth, R. S., and Schlosberg, H. (1954). "Experimental Psychology." Holt, New York.

Wyszecki, G., and Stiles, W. S. (1967). "Color Science." Wiley, New York.

8

Measuring Differences

C. J. BARTLESON

Research Laboratories
Eastman Kodak Company
Rochester, New York

ISBN-0-12-304905-9

A. INTRODUCTION

1. General

The purpose of this chapter is to provide an introduction to methods for measuring suprathreshold sensory differences (intervals or distances) among stimuli. The next chapter will deal with methods for measuring sensory magnitudes (ratios) of suprathreshold stimuli, and Chapter 10 will discuss techniques for multidimensional scaling, techniques that are useful when stimuli can be ordered according to more than one sensory attribute that may be of interest.The emphasis throughout these three chapters will be on simple methods that can be applied to problems involving measurement of optical radiations by human observers relying on their visual mechanisms. In short, these chapters will present a view of elementary psychophysics. Little space will be devoted to discussions of theory or to detailed variations on methodology, subjects treated well elsewhere; useful works will be cited as appropriate throughout the text. Instead, the bulk of these chapters will be an exposition of basic concepts, straightforward methods of experimentation, and reasonably simple algorithms for constructing measurement scales from raw experimental data. I hope to provide enough information so that the novice can understand the basic concepts of psychophysical suprathreshold scaling and can design and conduct a measurement experiment without feeling dismay at the complexity of the undertaking.

I shall assume that the reader has progressed through this volume in an orderly fashion and that some understanding of the anatomical, optical, and physiological workings and limitations of the visual mechanism have been gained from the first five chapters. I shall also assume that some familiarity with the concepts, purpose, and background of psychophysics has been acquired from Chapter 6 and that Chapter 7 has provided an understanding of the distinction between threshold and suprathreshold conditions as well as an appreciation of the fundamental probabalistic nature of visual measurement, in which statistical methods play such an important analytical role.

2. Variables

The last chapter was concerned with determining when two stimuli matched or just did not match: the threshold condition. This and the two following chapters concern determinations of the extent to which stimuli do not match: the suprathreshold condition. Such measurements involve (1) manipulation or selection of stimuli by the experimeter, (2) controlled presentation of the stimuli to observers, (3) collection of data representing observations of the stimuli, and (4) reduction of the data to form a measurement scale that is homomorphic with the observations. To form a measurement scale, certain assumptions are invoked about how people perceive the differences among stimuli and how people use words or numbers to express their judgments. A few preliminary comments on these matters will be helpful.

The ways in which stimuli and responses differ are called *variables*. A variable is some property or characteristic of stimuli or responses that may take on different values at different times or under different conditions. Stimulus variables, referred to as *independent variables*, are systematically varied by the experimenter. Response variables, known as *dependent variables*, are the variables on which the observer's judgment or performance is measured. The measurement task of psychophysics is to relate dependent variables to independent variables, that is, to express variations in response quantitatively as a function of variations in stimulation.

3. Samples

The first task of the experimenter who wishes to determine a psychophysical relationship is to select a plan for manipulating the stimuli to provide a range and distribution of stimulus differences that are of interest in the measurement experiment. In some cases the experimenter can change stimulus intensities or qualities by direct adjustment. For example, the irradiance provided by a light source can be adjusted continuously by varying the distance between the light and the object it illuminates. The irradiance could also be adjusted in increments by placing neutral (spectrally nonselective) optical filters of different transmittances in the illuminating beam between the source and the object.

Often the experimenter cannot adjust the stimulus directly; instead, a number of fixed stimuli must be selected for presentation sequentially to the observer. For example, an experiment may involve an attempt to determine which of several fluorescent brightening agents produces the "whitest" white tablecloth. Each brightening agent may result in a slightly different chromaticity, and each may be present in a different concentration, which will affect the relative radiance of the cloth. The experimenter must decide which chromaticities and which concentrations are of interest. In short, the population or domain of all possible variations must be sampled under the conditions of the experiment and be presented to the observer with the stimuli contained in the selected sample. (It is generally good practice to select a sample that has a distribution or gamut of variations somewhat larger than that of immediate interest to allow for differences of opinion among observers and to help define the psychophysical relationship more unequivocally within the gamut of immediate interest.) The sample then, is a portion of a population and is chosen by a clearly defined set of procedures. In general there are five ways of choosing a sample: the random independent sample, the stratified sample, the contrast sample, the purposeful sample, and the incidental sample. Each of these is appropriate to a different question that may be of interest.

The *random independent sample* has two major properties. First, every item in the population has an equal chance of being selected; this is the random property of the sample. Second, the selection of any one item in no way affects the selection of any other item; this is what makes the sample independent. Random independent samples are not usually the most efficient way of sampling a population; too many items must be selected to provide a reasonable range of variation and to achieve randomness. Sometimes, however, the random independent sample is the most appropriate one to use. For example, in an attempt to measure the influence of camera exposure on the overall lightness of photographic transparencies produced by amateur photographers, a random independent sample is the only kind that will satisfactorily include both the effects of the extent and frequency of errors in camera exposure that result from differences in photographic skills and the effects of different kinds of scenes.

The *stratified sample* has two properties in addition to those just mentioned. First, the population is classified according to some scheme that represents distinctions of interest among its members. To continue with the photographic example, we may be interested in determining separately the influence of camera exposure on overall transparency lightness for pictures taken with natural daylight and with artificial (photoflash and photoflood) lighting. The kind of lighting used to take the pictures then constitutes a class distinction. The second additional property of stratified samples is that the number of items included in each class conforms to the proportion of its incidence in the population as a whole. If 75% of all transparencies result from daylight exposure and 25% from artificial light, then those are the proportions in which the two classes should be present in the stratified sample. Finally, the items chosen within each class, according to the constraints imposed, should form a random independent subsample.

The *contrast sample* has the same properties as a stratified sample but also intentionally includes an excess of items in certain categories that are of particular interest. Suppose that the illuminants used to take pictures were also to be classified according to their irradiances and color temperatures. In most natural situations, higher levels of irradiance are associated with sources of higher color temperature. Therefore, we may be more interested in what happens with high-irradiance, high-color-temperature and low-irradiance, low-color-temperature conditions than what happens with their complements. We would then "oversample" the domain of greater interest to form a contrast sample. This would allow us to determine the influence of these conditions with somewhat greater certainty, but we must be aware that the results no longer apply directly to the population as a whole; we would not be justified in averaging over all classes, for example.

The *purposeful sample* is one in which either a particular selection is made of a relatively small number of items that represent the population as a whole or a selection is made of items that vary systematically in some attribute. In the first instance the purpose is to provide a small sample that can be taken to represent the population without random independent sampling. This approach is often used when it is necessary to reduce the number of stimuli to as few as possible, but one still hopes that something can be deduced about the entire population. The second approach is used when the influence of a particular variable is of interest, without regard to how frequently it may occur. This kind of sample is probably the one most often used in visual psychophysics.

The *incidental sample* may also be of two varieties: a random independent sample of one class or subgroup that is of overriding interest or a collection of items that already exist but cannot be added to. Special interest, convenience, and availability are the criteria that most often determine when an incidental sample should be used. It is generally better to avoid such a choice whenever possible, however.

Proper selection of a sample is not independent of the rest of the design of an experiment. Unfortunately, many experimenters approach a measurement problem as if this were not so. A sample of some kind is generated, and then the question is asked: "How shall we measure the (attribute) of these?" Of course, this is sometimes unavoidable with incidental samples, but it is poor practice to select a sampling plan without first considering the experimental method that is to be used. The reason is that both choices depend on the purpose of the experiment, the reason for making the measurements. What is to be learned? Sometimes the purpose of an experiment is to describe the intensity of visual perceptions associated with changes in the level of the independent variable over a range that may be either broad or restricted. Sometimes small differences are of interest, sometimes large ones are. We may want to determine the dependence of the response on one independent variable or on more than one at the same time. We may want to distinguish as precisely as possible among similar stimulus differences, or we may wish to map a global relationship with as much accuracy as possible. Sometimes it is sufficient merely to know the order in which stimuli stand with respect to one another. All these different considerations require different experimental plans involving choices of experimental method and sampling method.

The choices are so numerous and depend on so many particular considerations that it is impossible to provide a useful discussion of experimental design in the space available here. Some comments on the classes of conditions that are appropriate to the experimental methods that are presented in this and the next two chapters will be made in conjunction with descriptions of those methods. However, the reader who wishes to explore the subject of experimental design should consult one or more of the many textbooks that have been published on that subject (for instance, Brownlee, 1949; Fisher, 1951; Kempthorne, 1952; Lindquist, 1953; Woodworth and Schlosberg, 1954; Cochran and Cox, 1957; Cox, 1958; Chapanis, 1959; Edwards, 1960; Myers, 1966; Sidowski, 1966; Kling and Riggs, 1971; Winer, 1971).

4. Scales

The choice of experimental method largely determines the kind of measurement scale that can be constructed from the data. Stevens (1946, 1951) has classified scales according to their mathematical power or, more exactly, their *transformation invariance*. The term transformation invariance refers to the mathematical transformations that can be made of the scale and still retain unchanged the property that is uniquely determined by the scale. Mathematical power is inversely related to the number of such transformations. The least powerful scale has been called the *nominal scale*. It consists simply of unique labels; each item has a different label, as with names of flower varieties or numbers on the jerseys of football players. Many people do not even consider such labeling to form a scale.

The *ordinal scale* consists of an ordered progression of integers that is isomorphic (no ties are permitted) with the order of the intensities or attributes of the items scaled. When six horses run a race, the order in which they cross the finish line comprises an ordinal scale of position with respect to the winner. The integrity of such a scale is maintained over any monotone transformation (linear or nonlinear) because the scale tells us only the relative order of the items in question with respect to a specified attribute.

An *interval scale* determines order as well, but it also specifies ratios of differences uniquely. That is, the relative sizes of intervals are specified by an interval scale. The information that is unique is the ratio of one interval to another; a minimum of three stimuli are involved. Such ratios remain invariant over any linear transformation, a more restrictive condition than for an ordinal scale; hence, the result is a more powerful scale.

A *ratio scale* defines, in addition, ratios of magnitudes uniquely; hence, a minimum of two stimuli are required to express the information of a ratio scale. Such a scale remains invariant only over transformations that are linear and pass through the origin. That is, a ratio scale of measurement is determined uniquely up to a multiplicative constant.

Finally, an *absolute scale* is one that defines magnitudes. This can be done by counting or by convention. The number of apples in a barrel represents a point on an absolute scale of numerosity. If there are 347 apples in the barrel, no mathematical transformation will yield an identical (base ten) number. The Kelvin scale of thermodynamic temperature is another kind of absolute scale, one that was declared to be absolute by agreement that certain theoretical structures and measurement conventions were true (however, they cannot be immutably true in the sense that numerosity is).

Of the foregoing scale types, all but the nominal and ordinal share the property that they are linear and additive. Each is based on the summation of some unit. The information provided by a less powerful scale can also be obtained from more powerful scales. There are some other kinds of scales for which this is not true, because they are not linear and additive without transformation (one is the scale that Stevens has called the *logarithmic interval scale*), but we shall not discuss these in this chapter. Here we shall deal with methods for deriving ordinal scales and, especially, interval scales that are linear and additive up to transformations of the type $y = ax + b$.

5. Examples

To illustrate experimental methods and construction of scales from raw data, one example will be used throughout this chapter. Other data will also be used where appropriate, but a single experimental example will facilitate comparison of results obtained by applying the different methods to a common problem.

The example is one in which colorfulness, or chromatic content, of nine colored papers was scaled by different methods. Figure 1 shows a graph of the

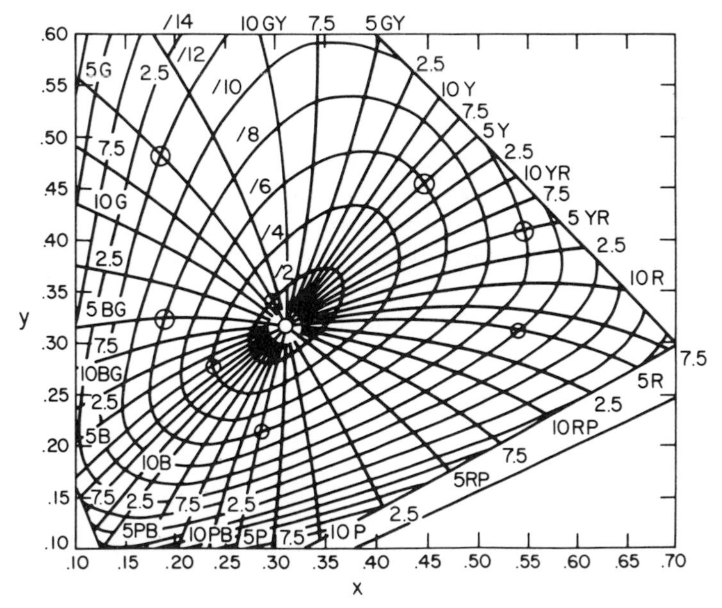

Fig. 1. CIE 1931 chromaticity coordinates of chroma samples, shown as circles, compared with Munsell hue, value, and chroma contours.

TABLE I. Specifications of Samples Used in Illustrative Example

	CIE 1931			Munsell		
Stimulus number	x	y	β_v	H	V	C
1	0.3100	0.3308	0.1202	N	4	0
2	0.5385	0.3130	0.1210	5R	4	12
3	0.2859	0.2149	0.1197	5P	4	8
4	0.2360	0.2781	0.1200	5B	4	4
5	0.2960	0.3413	0.1206	5G	4	2
6	0.1887	0.3235	0.1202	5BG	4	8
7	0.1843	0.4805	0.1202	5G	4	12
8	0.4448	0.4551	0.1208	5Y	4	6
9	0.5430	0.4099	0.1203	5YR	4	10

CIE 1931 chromaticity coordinates of the nine samples. Table I lists the CIE chromaticity coordinates and luminance factors of the samples together with their approximate Munsell hue H, value V, and chroma C notations.

As shown in the table the samples were all of approximately the same luminance factor (or Munsell value), but they varied in Munsell hue and chroma. Each sample was 38 × 50 mm. They were viewed at a distance of 355 mm by each observer (the observers all had normal color vision). The samples were illuminated with 2200 lux of light that had a relative spectral power distribution very similar to that of CIE illuminant D_{65}. The samples were presented in a 1 × 2.5 m viewing booth painted with a spectrally nonselective, matte surface paint of about 40% reflectance. The observers were allowed to adapt to the surround for 5 min before each judging session began.

Different judgments were made in each of the various experiments. Altogether, judgments were made by rank order, paired comparison, successive category, magnitude estimation, and multidimensional difference. This chapter will discuss the rank-order, paired-comparison, and successive-category scaling methods and will show how interval scales of colorfulness can be formed from the experimental data in each case.

B. THE RANK-ORDER METHOD

1. Determining Ranks

When the number of stimuli is small, as in the example here, it is a simple matter to determine rank order. The experimenter places all the stimuli in front of the observer at once. The experimenter is careful to arrange the stimuli in a different random order, without replacement, for each observer. Only one observer at a time is present in any one session. The observer is instructed to place the stimuli in order according to the attribute being scaled. A representative set of instructions would be as follows:

> Here are 9 colored papers. I would like you to place them in order of how colorful they are. Put the most colorful paper at the right and the least colorful one at the left. You may move the papers around and rearrange them until you are satisfied that they are ordered according to colorfulness from left to right. Please tell me when you have finished. You may begin when you are ready.

These instructions were given to each of 100 observers, and their responses were recorded. Table II summarizes all the responses. Each entry in the table represents the number of times a rank (shown on the left as rows) was assigned to each colored paper (shown at the top as columns). These are the raw data resulting from the experiment. We will calculate three things from these data: the simple rank order, an interval scale determined according to the method of comparative judgments, and an interval scale determined according to the method of normalized ranks. The rationale underlying derivations of interval scales from rank-order data will be discussed when the two methods for estimating interval scales are described. First, however, we shall determine the rank order.

The simplest way to determine a rank order from such data is to calculate and order the mean ranks (symbolized M_R). This is done by multiplying the ranks and frequencies, summing each column, and dividing by the number of stimuli; values of M_R are shown at the bottom of the columns in Table II. The M_R values are then ordered, and successive integers are assigned to form a rank order. In the example of Table II there are no ties, but this is not always the case. When there are ties, that is, when two values of M_R are the same, any arbitrary rule may be used to break the ties in forming a rank order. For example, one may go back to the raw frequency data to see whether a distinction can be made between a pair of stimuli with tied M_Rs (in other words, does the tie result from rounding in calculating the mean rank?). Another rule might be to give precedence to the nominal numbering of the stimuli. In any case, whatever rule is chosen, it should be used consistently in a single experiment. Because there were no ties in the example here, it is a simple matter to rank the M_Rs to form a scale of ranks (symbolized R), which is shown below the values for M_R in Table II.

The size of the rank number increases as the stimulus has less of the attribute scaled. For this reason a mean choice M_C value is often computed so that the sizes of the numbers increase as the attribute scaled increases. The mean choice is defined as $M_C = n - M_R$, where n is the number of stimuli. Table II also shows values of M_C. As with ranks, it is possible to specify the choices as integers; for example, $C = R - (n - 1)$.

TABLE II. Frequency with Which Stimuli Were Assigned Ranks

J_R (judged rank)	Stimuli									C (choice)
	1	2	3	4	5	6	7	8	9	
1	0	73	0	0	0	0	24	0	3	9
2	0	25	0	0	0	10	60	0	5	8
3	0	2	2	0	0	10	5	0	81	7
4	0	0	33	0	0	65	0	1	1	6
5	0	0	52	17	0	10	10	10	1	5
6	0	0	11	12	0	2	0	75	0	4
7	0	0	2	68	10	0	0	12	8	3
8	13	0	0	3	77	3	1	2	1	2
9	87	0	0	0	13	0	0	0	0	1
M_R	8.87	1.29	4.78	6.57	7.13	3.96	2.17	6.04	3.29	
R	9	1	5	7	8	4	2	6	3	
M_C	0.13	7.71	4.22	2.43	1.87	5.04	6.83	2.96	5.71	
p	0.016	0.964	0.528	0.304	0.234	0.630	0.854	0.370	0.714	
z	−2.144	+1.799	+0.070	−0.513	−0.726	+0.332	+1.054	−0.332	+0.565	
Estimated chroma	−0.8	13.3	7.1	5.0	4.2	8.0	10.6	5.7	8.9	
Chroma	0	12	8	4	2	8	12	6	10	

2. Estimating Intervals from Rank-Order Data

There are two basic ways to estimate an interval scale from rank-order data, the *comparative-judgment* method and the *normalized-rank* method. The first draws upon the logical argument that to order the stimuli the observer must compare each stimulus with every other stimulus, directly or indirectly. Were this not so, at least to a good approximation, it is difficult to know how it would be possible to choose an order by other than chance. Accordingly, we may consider the entries in Table II as proportions of choice. One way to calculate a proportion p for each stimulus is to divide the mean choice by one less than the number of stimuli; that is, $p = M_C/(n - 1)$. Values of p are also shown in the table.

a. The Comparative-Judgment Method of Data Reduction

As with proportions, or probabilities, for threshold values in Chapter 7, we may now convert values of p to standard normal deviates z either from Table II in Chapter 7 or by computing them directly from the equation for the normal curve. The latter has been done to derive the z values shown in Table II of this chapter.

The z scale is taken as a scale of intervals or differences. The underlying rationale for this is based on what Thurstone (1927) referred to as a *law of comparative judgments.* As noted in earlier chapters there are many factors that may cause the momentary response of an observer to differ from his response to the same stimulus at a different time. Over many trials the responses to a stimulus will usually be normally distributed. When two stimuli are compared, what is actually being compared is the difference between two points on two such distributions. Again, over many trials, the differences between two distributions are compared. Just as the underlying distributions are normal, so is the distribution of their differences. Therefore, a scale with equal differences will be one along which intervals correspond to equal probabilities or areas under the normal curve. That is precisely what is represented by a scale of z values, as was shown in Chapter 7. Thurstone's law of comparative judgments will be discussed in somewhat more detail in the next section where the method of paired comparisons is dealt with. It will be sufficient here to point out that the comparative-judgment method of reducing rank-order data assumes that all pairs have been compared and, therefore, that the proportions represent probabilities corresponding to differences that are normally distributed. Thus, a scale of equal intervals may be represented by standard normal deviates (zs).

In our example we have ulterior information about the actual chromas of the samples that were scaled. We can, then, compare our z-scaled results with

TABLE III. Frequencies Times z Values of Centile Positions for Rank-Order Data

	Stimuli										
R	1	2	3	4	5	6	7	8	9	P	z
1	0	+11.61	0	0	0	0	+3.82	0	+0.48	0.944	+1.59
2	0	+2.43	0	0	0	+0.97	+5.82	0	+0.49	0.833	+0.97
3	0	+0.12	+0.12	0	0	+0.59	+0.30	0	+4.78	0.722	+0.59
4	0	0	+0.92	0	0	+1.82	0	+0.03	+0.03	0.611	+0.28
5	0	0	0	0	0	0	0	0	0	0.500	0.00
6	0	0	−0.31	−0.34	0	−0.06	0	−2.10	0	0.389	−0.28
7	0	0	−0.21	−4.01	−0.59	0	0	−0.71	−0.47	0.278	−0.59
8	−1.26	0	0	−0.29	−7.47	−0.29	−0.10	−0.19	−0.10	0.167	−0.97
9	−13.83	0	0	0	−2.07	0	0	0	0	0.056	−1.59
Means	−1.677	+1.573	+0.068	−0.516	−1.126	+0.337	+1.093	−0.330	+0.579		
Estimated chroma	0.1	13.3	7.2	4.7	2.3	8.3	11.3	5.6	9.2		
Chroma	0	12	8	4	2	8	12	6	10		

the "true" answer to see how well they agree. Figure 2 shows a graph of the z values from Table II plotted again Munsell chroma. The equation shown in the graph is the linear regression of z-scale values on chroma; it has a coefficient of determination of 0.914, which is reasonable but not exceptionally high for predictive purposes. The inverse regression, of chroma on z-scale value, is chroma $= 3.5886z + 6.8470$, with the same coefficient of determination. Estimated chromas are shown for each of the stimuli at the bottom of Table II, where they are compared with the chromas listed for the stimuli in Table I. The latter data and the graph of Fig. 2 indicate that this simple rank-order experiment has produced data that can be treated as comparative judgments to form an interval scale having a relationship to the original chromas that is, at least, not unreasonable.

b. The Normalized-Rank Method of Data Reduction

Another kind of distribution assumption can be made about the rank-order data. It also invokes a normal distribution but does so in a different way from that of the comparative-judgment method. If the stimuli form a random independent sample, or if they come from a population that is normally distributed with respect to the attribute in question and are sampled in any way that might reflect that normality, then it is not unreasonable to assume that the ranks themselves might form a normal distribution with respect to the scaled attribute. This is more apt to be true in biological samples or attributes that have to do with living organisms than it is with our example of colored papers. However, it may be that the responses to the papers somehow form a distribution that tends toward normality. In any case it will be instructive to analyze the data from Table II according to the *method of normalized ranks* to demonstrate how the method is used.

The frequencies of Table II are first converted to what Guilford (1954) called *centile positions* P. These correspond to the middles of the rank areas and are defined as $P = (R - 0.5)/n$. Table III lists the centile positions for 9 ranks in the penultimate column. The last column lists the corresponding z values. The entries in the table are the products of these zs and one-tenth the frequencies of Table II. The averages of each of the columns (also shown in Table III) constitute the mean centile scale values.

They are plotted against the original chromas in Fig. 3, in which the linear regression equation is again shown. The coefficient of determination for that relationship is 0.9669. The inverse regression was used to calculate the estimated chroma values shown at the bottom of Table III. The equation is chroma $= 4.048$ (scale value) $+ 6.888$.

Estimating an interval scale relationship from rank ordering of stimuli is not unlike practicing alchemy. We gathered no information about intervals or

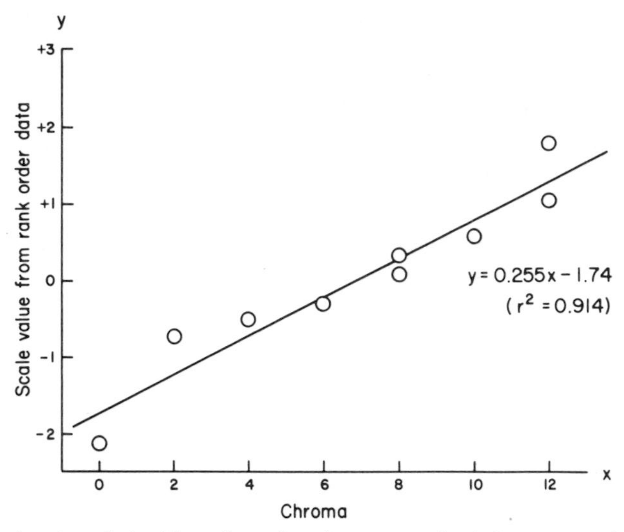

Fig. 2. Scale values, derived from the rank-order comparative-judgements method, compared with chroma values.

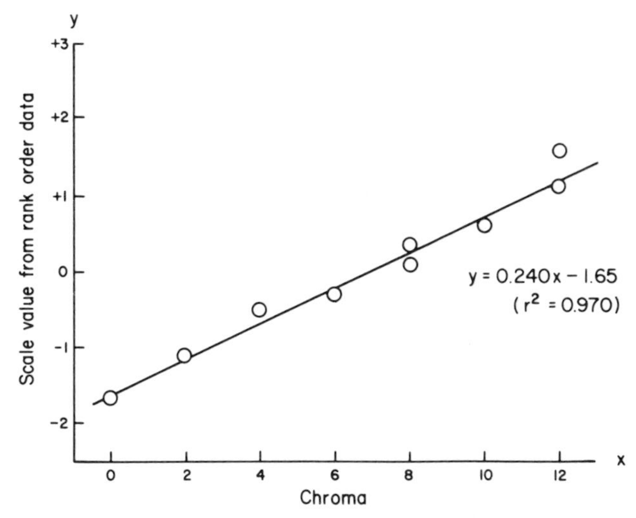

Fig. 3. Scale values, derived from the rank-order normalized-rank method, compared with chroma values.

differences; all that was done was to have observers place the stimuli in a rank order. However, 100 observers made such judgments, and they did not all agree about the order. The fact that there was less than complete unanimity is what provides the information (that is, the uncertainty) that allows us to make such estimates. When the number of observations is large enough, particularly when the number of observers is large, there will often be disagreements or uncertainties in rankings. When there are few observations, the chances are that there may not be enough uncertainty to form a reasonable estimate of an interval scale from rank-order data. As we shall see, however, even the paired-comparison method of determining an interval scale involves only judgments of order, but it does so by pairs. That is, every pair is judged explicitly, and this opens the way for more uncertainty with which to estimate an interval by making assumptions about the distributions of differences among stimuli.

C. THE PAIRED-COMPARISON METHOD

1. The Law of Comparative Judgments

The *law of comparative judgments* was set down formally by Louis Leon Thurstone (1927) as an equation relating the proportion of times any stimulus i is judged greater, according to some attribute, than any other stimulus j in terms of the discriminal differences of the two stimuli on a judgment continuum. This equation postulates three things: (1) that each stimulus gives rise to a discriminal process that may be characterized by a value on the continuum, (2) that over time the discriminal process gives rise to a normal distribution of values, and (3) that the average of the distribution and its standard deviation relate to the value and discriminal dispersion, respectively, of the stimulus. Stated as the difference in scale values R between two stimuli i and j, the equation of the law of comparative judgements is

$$R_i - R_j = z_{ij}(\sigma_i^2 + \sigma_j^2 - 2r_{ij}\sigma_i\sigma_j)^{\frac{1}{2}}, \quad (1)$$

where R_i and R_j represent the scale values of stimuli i and j, σ_i and σ_j represent the discriminal dispersions of stimuli i and j, r_{ij} is the correlation between the discriminal processes, and z_{ij} is the normal deviate corresponding to the proportion of times stimulus j is judged greater than stimulus i.

Equation (1) represents the complete law of comparative judgments. Thurstone stated five cases for that law, and Mosteller (1951a) has since added a sixth case (usually referred to as Case Va). Each of these cases invokes certain assumptions, and the form of expression for scale values varies for simplifying assumptions. Table IV sets forth the expressions for the six cases. Case I assumes replications over the judgments of one observer. For Case II, replications are over observers. In both cases the expression is the same as in Eq. (1). Case III makes the simplifying assumption that the discriminal processes are independent, and thus their correlations are zero. Case IV assumes that, in addition, the discriminal dispersions are nearly equal, whereas Case V assumes that they are actually equal. Case Va relaxes the assumption of Case V in that the correlation of the discriminal processes may simply be

TABLE IV. Cases of the Law of Comparative Judgment

Case	σ Constraints	r Constraints	Expression
I	σ_i unrelated to σ_j	$0 < r_{ij} \leq 1$	Equation 1
II	σ_i unrelated to σ_j	$0 < r_{ij} \leq 1$	Equation 1
III	σ_i unrelated to σ_j	$r_{ij} = 0$	$R_i - R_j = z_{ij}(\sigma_i^2 + \sigma_j^2)^{1/2}$
IV	$\sigma_i \cong \sigma_j$	$r_{ij} = 0$	$R_i - R_j = (2^{1/2}/2)z_{ij}(\sigma_i^2 + \sigma_j^2)^{1/2}$
V	$\sigma_i = \sigma_j$	$r_{ij} = 0$	$R_i - R_j = z_{ij}\sigma(2)^{1/2}$
Va	$\sigma_i = \sigma_j$	$r_{ij} = k$	$R_i - R_j = z_{ij}\sigma(2[1 - r_{ij}])^{1/2}$

constant rather than zero. It is common practice to set the scale unit equal to $\sigma(2)^{\frac{1}{2}}$ in Case V and equal to $\sigma(2[1 - r_{ij}])^{\frac{1}{2}}$ in Case Va, so that the scale unit is simply equal to z_{ij} in both cases.

In practice, we seldom know the values of σ and r (although they can be estimated). Instead, experimenters assume one of the cases. Occasionally, Case III is assumed to apply, but more often Case V or Va is used so that the scale values are simply values of z. In practice, then, we usually need only to determine the proportion of times one stimulus is preferred over another and calculate the standard normal deviate corresponding to that proportion to determine the interval-scale value of a stimulus. This can be done by comparing each stimulus with every other stimulus in turn to decide which stimulus shows more of the attribute under study. The experimental method by which this is done is called the *method of paired comparisons.*

2. The Experimental Method of Paired Comparisons

In this method data are gathered by asking observers to respond with an ordinal judgment between two stimuli. In its complete form the method involves showing all possible pairs of stimuli in the collection to each observer (replications may be over presentations or over observers or both, but replications are required to ascertain the proportion of times that each stimulus is ranked above each other stimulus in the collection). The total number of permutations of n things taken two at a time is $N = n(n - 1)$. Because the total number of permutations includes each pair twice (that is, i is compared with j on one occasion, and j is compared with i on another), the usual practice is to deal only with combinations in which $N = (n/2)(n - 1)$. With 10 stimuli there would then be 45 pairs for each observer to judge; with 20 stimuli there would be 190 pairs; 435 with 30 stimuli; and so on. The number of pairs increases rapidly with the number of stimuli; N increases as $(n^2 - n)/2$, in fact. It is desirable, therefore, either to keep n as small as possible or to reduce N by leaving out some of the pairs.

Sometimes it is neither possible nor desirable from ulterior considerations to use a small number n of stimuli. In such cases the experimental labor can be reduced by decreasing the number of pairs judged. There have been many plans suggested for effecting such reductions. The best of these operate on the premise that most information is gained from comparisons of similar stimuli. When obviously dissimilar stimuli are compared, the judgments are likely to be unanimous: proportions of 0 or 1. The corresponding normal deviates are $-\infty$ and $+\infty$, respectively, and cannot be manipulated mathematically to determine scale values. Obviously we would prefer to have proportions greater than 0 and less than 1; Bock and Jones (1968) recommend proportions between 0.01 and 0.99, and Guilford (1954) suggests they be between 0.023 and 0.977 (corresponding to ± 2 on the z scale). In essence the reasoning applied here is that the central tendency of the distribution of discriminal differences is more well determined than are its tails. Hence, most reliable information is to be found for conditions in which the proportions are near 0.50, and least reliable results are those near unity or zero proportions. The problem then reduces to one of determining which pairs will result in unanimous judgments and which will provide uncertainty.

The experimenter can determine this during the experiment by trial and error, or he can arbitrarily decide ahead of time which pairs to leave out. The latter approach has the longer history. When the stimuli differ along some intensive physical continuum, this method is appropriate. The experimenter can preview the stimuli and decide which pairs are likely to be judged the same way by all observers; he then eliminates these pairs from the experiment. The problem with this approach is that the experimenter's opinion may not be the same as those of all the observers. The result may be either inclusion of extreme proportions or exclusion of (useful) proportions that are not extreme. The second approach allows observers to apply their own sorting process during the experiment. That is, as the data are gathered, pairs that differ by more than an amount that the observer assigns the same judgment to on the second replication are eliminated from further consideration. This is reminiscent of the staircase approach to the method of limits described in Chapter 7. This kind of sorting algorithm can result in as few as $n \log_2 n$ pairs rather than $(n/2)(n-1)$ pairs. With 32 stimuli, for example, this means that as few as 160 pairs (32 times 5) may be required rather than the full 496 pairs (16 times 31). These sorting algorithms have been published (for example, Shell, 1959) and are available as interactive computer programs (for example, Whaley, 1979) that can be used during experimentation.

For all pairs that are presented, the observer is asked to indicate which member of the pair has more of the attribute in question. A typical set of instructions (used for an experiment in which the brightness of cloths containing fluorescent brighteners was determined) might be as follows:

> You will be shown a number of pairs of cloth swatches. Please indicate which member of the pair is brighter.

The experimenter records each response and presents the observer with the next pair in the series. In an example experiment with fluorescent brighteners, there were 8 stimuli which were presented as random-ordered pairs to each of 22 observers. All 28 combinations were used to form the pairs. The data were recorded as the number of the stimulus preferred in each pair. Table V shows the data that were collected. Each row represents the responses of a single observer. The entry in each column is the number of the brighter stimulus. The columns are ordered by the sequence of pairs: 1,2 then 1,3 then 1,4 through 1,8, and then 2,3 and 2,4 and so on, up to 7,8. Table V represents the raw experimental data. The following section will use those data to derive an interval scale under the assumption of Thurstone's Case V.

TABLE V. Case V Pair-Comparison Whiteness Data

OBS	PAIR1	PAIR2	PAIR3	PAIR4	PAIR5	PAIR6	PAIR7	PAIR8	PAIR9	PAIR10	PAIR11	PAIR12	PAIR13	PAIR14	PAIR15	PAIR16	PAIR17	PAIR18	PAIR19	PAIR20	PAIR21	PAIR22	PAIR23	PAIR24	PAIR25	PAIR26	PAIR27	PAIR28	SUBJECT
1	2	1	4	1	6	7	1	3	2	2	2	7	2	3	3	3	3	3	4	4	4	4	5	7	8	6	8	7	A
2	1	1	1	1	1	1	1	3	2	2	2	7	2	3	3	3	7	3	4	4	7	4	5	7	5	7	6	7	B
3	1	1	1	5	1	7	1	2	2	2	2	7	2	3	3	3	7	3	5	4	7	4	5	7	5	7	6	7	C
4	2	3	4	5	6	1	8	2	4	2	2	2	2	4	5	6	7	8	4	4	4	4	6	5	8	7	6	8	D
5	1	1	4	1	1	1	1	3	4	2	2	7	8	3	3	3	3	3	4	4	7	4	6	7	5	7	6	7	E
6	1	3	1	1	1	7	8	3	2	2	2	7	8	3	3	3	3	3	4	4	7	4	6	7	8	7	6	7	F
7	1	1	1	1	6	7	1	3	4	2	6	7	8	4	3	3	7	3	4	6	7	4	6	7	8	7	6	7	G
8	1	3	1	1	1	7	1	3	2	2	2	7	2	3	3	3	7	3	4	4	7	4	6	7	8	7	8	7	H
9	2	3	4	1	1	7	8	3	4	2	6	7	2	3	3	3	7	3	4	4	7	4	6	7	8	7	8	8	I
10	1	3	4	1	1	1	1	3	4	2	6	7	8	4	3	6	3	3	4	4	4	4	6	7	8	7	8	7	J
11	1	1	4	1	6	7	1	3	2	2	6	7	8	4	3	6	3	3	4	4	7	4	6	7	8	7	6	7	K
12	1	3	1	1	1	7	1	3	2	2	6	7	2	3	3	3	3	3	4	4	7	4	6	7	5	7	6	7	L
13	2	1	1	5	6	7	8	2	4	5	2	2	8	4	5	6	3	8	4	4	4	4	5	5	8	6	8	8	M
14	1	1	1	1	1	7	1	3	2	2	6	7	2	3	3	3	7	3	4	4	7	4	6	7	5	7	6	7	N
15	2	1	1	5	1	1	1	3	4	5	6	2	2	4	3	3	3	3	4	4	4	4	5	7	5	7	6	7	O
16	2	3	4	1	6	7	1	2	4	5	6	7	2	4	3	3	7	8	4	4	4	4	6	7	8	7	8	7	P
17	2	3	4	5	6	7	8	3	4	2	6	2	8	4	3	6	7	8	4	4	4	4	6	5	8	6	8	8	Q
18	1	1	1	1	1	7	1	3	4	2	6	7	2	3	3	3	3	3	5	4	7	4	6	7	8	7	6	7	R
19	1	1	4	1	1	1	1	3	4	5	6	7	2	3	3	6	7	8	4	4	4	4	6	5	8	6	6	8	S
20	1	3	4	1	1	7	8	3	4	2	2	2	2	4	3	6	7	3	4	4	4	8	6	7	5	7	8	8	T
21	2	3	1	1	1	7	1	3	4	5	6	7	2	4	3	3	3	3	4	6	7	4	6	7	8	7	8	7	U
22	1	3	1	1	1	1	1	2	4	2	2	2	2	4	5	3	7	8	4	4	4	4	5	7	5	7	6	8	V

a. Estimation of Interstimulus Differences

The raw data of Table V can easily be converted into a *matrix of frequencies*, such as that shown in the SUMMAT at the top of Table VI. This matrix shows the frequency or number of times that a stimulus represented by a column was found to be brighter than a stimulus represented by a row. For example, in column 1, row 2, the entry is 14, which means that stimulus 1 was found to be brighter than stimulus 2 by 14 of the 22 observers. Note that the diagonal entries in the matrix are 0s, because the observers did not compare each stimulus with itself. The frequency matrix can be converted to a *proportion matrix* simply by dividing each entry by the number of observations it represents. Thus the 14 represents a proportion of $14/22 \cong 0.636$, as shown in the PMAT in Table VI. In this case a complete paired-comparison experiment was conducted; therefore, the proportion p is determined as $p_{ij} = f_{ij}/N$, where N is the number of observations and f is frequency. In the general case, in which the frequency matrix may be incomplete, the equation is $p_{ji} = f_j/(f_i + f_j)$ and $p_{ij} = f_i/(f_i + f_j)$. Because we shall assume that when a stimulus is compared with itself and a choice is forced, it would be represented by a proportion of one-half; the diagonal entries of the proportion matrix have, therefore, been made 0.5.

Each of the entries in the proportion matrix represents the result of a direct comparison of one stimulus with another. However, the data also include many indirect comparisons. For example, the difference between stimuli 1 and 2 was observed not only directly but also indirectly, if we consider the difference between 1 and 3 compared with 2 and 3, between 1 and 4 compared with 2 and 4, and so on. Without defining exactly what is meant by a response as yet, we can represent this process for our 8 stimuli as differences among responses R as follows:

$$
\begin{aligned}
R_2 - R_1 &= (R_2 - R_1) - (R_1 - R_1) = \Delta_1 \\
R_2 - R_1 &= (R_2 - R_2) - (R_1 - R_2) = \Delta_2 \\
R_2 - R_1 &= (R_2 - R_3) - (R_1 - R_3) = \Delta_3 \\
R_2 - R_1 &= (R_2 - R_4) - (R_1 - R_4) = \Delta_4 \\
R_2 - R_1 &= (R_2 - R_5) - (R_1 - R_5) = \Delta_5 \\
R_2 - R_1 &= (R_2 - R_6) - (R_1 - R_6) = \Delta_6 \\
R_2 - R_1 &= (R_2 - R_7) - (R_1 - R_7) = \Delta_7 \\
R_2 - R_1 &= (R_2 - R_8) - (R_1 - R_8) = \Delta_8 \\
&\vdots \\
R_8 - R_7 &= (R_8 - R_7) - (R_7 - R_7) = \Delta_{34} \\
R_8 - R_7 &= (R_8 - R_8) - (R_7 - R_8) = \Delta_{35}
\end{aligned}
$$

The mean difference between stimuli 1 and 2 is the average of Δs 1 through 8; the mean difference between stimuli 2 and 3 is the average of Δs 9 through 15; and so on for each of the adjacent interstimulus differences. Both Rs and Δs are generally expressed as normal deviates z. There are also proportions p' corresponding to each of these average differences, and they may not be the same as the proportions for direct comparisons p shown in the PMAT of Table VI. The extent to which p and p' values differ relates to how well the scale is determined. The two proportion matrices may differ, because the standard deviations are unequal (that is, failure of Case V to apply), because the data represent a multi-dimensional rather than a unidimensional problem, or because the data are not normal. It is well to test the data to see whether the p and p' matrices agree sufficiently well for a particular purpose. There are a number of ways to do this.

If the problem is a multidimensional one, there will generally be evidence of what are called *triangle inequalities* among scale values. For example, R_2 may be greater than R_3, and R_1 greater than R_2, but R_3 is found to be greater than R_1. These inconsistencies usually are obvious from inspection of the z matrices. When they occur, the best course of action is to repeat the experiment using a multidimensional scaling method (*vide infra*). If that cannot be done, the problem can sometimes be eliminated by excluding from the data base those stimuli that cause the problem; however, this procedure throws away useful and sometimes critically important information.

A statistical *Student's t test* can be made to compare every p and p' value (for example, Dixon and Massey, 1957) to assess the differences between the two matrices; but this is a cumbersome method, because it requires a test for every entry in the matrices. Mosteller (1951b,c) has provided a way of performing a χ-squared test (see Chapter 7) on the entire matrices. Because the proportions p and p' are not necessarily normally distributed, particularly towards the extremes, the χ-squared test is performed on matrices whose elements are transformed to expressions that are assumed to be normally distributed. In particular, the elements of the transformed matrices comprise the angle whose sine is the square root of the proportion; that is, $\theta = \arcsin(p)^{\frac{1}{2}}$ and $\theta' = \arcsin(p')^{\frac{1}{2}}$. The value of χ-squared is then

$$\chi^2 = (N/821)\sum(\theta - \theta')^2, \tag{2}$$

where N is the number of observations for each pair of stimuli. The significance of a computed χ-squared value is determined (as in Chapter 7) by reference to the critical value for the degrees of freedom df involved. In this instance df is determined as

$$df = [(n - 1)(n - 2)]/2, \tag{3}$$

where n is the number of stimuli in the experiment.

TABLE VI. *Z*-Scale Positions of Case V Pair-Comparison Whiteness Data

	Col 1	Col 2	Col 3	Col 4	Col 5	Col 6	Col 7	Col 8
SUMMAT								
Row 1	0	8	11	10	5	7	15	6
Row 2	14	0	17	14	5	12	16	7
Row 3	11	5	0	11	3	7	12	6
Row 4	12	8	11	0	2	2	12	1
Row 5	17	17	19	20	0	16	18	14
Row 6	15	10	15	20	6	0	18	9
Row 7	7	6	10	10	4	4	0	7
Row 8	16	15	16	21	8	13	15	0
PMAT								
Row 1	0.5	0.363636	0.5	0.454545	0.227273	0.318182	0.681818	0.272727
Row 2	0.636364	0.5	0.772727	0.636364	0.227273	0.545455	0.727273	0.318182
Row 3	0.5	0.227273	0.5	0.5	0.136364	0.318182	0.545455	0.272727
Row 4	0.545455	0.363636	0.5	0.5	0.0909091	0.0909091	0.545455	0.0454545
Row 5	0.772727	0.772727	0.863636	0.909091	0.5	0.727273	0.818182	0.636364
Row 6	0.681818	0.454545	0.681818	0.9090901	0.272727	0.5	0.818182	0.409091
Row 7	0.318182	0.272727	0.454545	0.454545	0.181818	0.181818	0.5	0.318182
Row 8	0.727273	0.681818	0.727273	0.954545	0.363636	0.590909	0.681818	0.5
ZMAT								
Row 1	0	−0.348756	0	−0.114185	−0.747859	−0.472789	0.472789	−0.604585
Row 2	0.348756	0	0.747859	0.348756	−0.747859	0.114185	0.604585	−0.472789
Row 3	0	−0.747859	0	0	−1.0968	−0.472789	0.114185	−0.604585
Row 4	0.114185	−0.348756	0	0	−1.33518	−1.33518	0.114185	−1.69062
Row 5	0.747859	0.747859	1.0968	1.33518	0	0.604585	0.908458	0.348756
Row 6	0.472789	−0.114185	0.472789	1.33518	−0.604585	0	0.908458	−0.229884
Row 7	−0.472789	−0.604585	−0.114185	−0.114185	−0.908458	−0.908458	0	−0.472789
Row 8	0.604585	0.472789	0.604858	1.69062	−0.348756	0.220884	0.472789	0
SCLMAT								
Row 1	1.81538	−0.943493	2.80785	4.48136	−5.7895	−2.24056	3.59545	−3.7265

The corresponding equation for Case III (in which $z_{ij} = [R_i - R_j]/[\sigma_i^2 + \sigma_j^2]^{\frac{1}{2}}$ and $\sigma_i^2 \neq \sigma_j^2$) is given by

$$df = [(n-1)(n-4)]/2. \tag{4}$$

In the foregoing paragraphs it has been assumed that the proportions p_{ij} relate to a normal model; that is

$$P_{ij} = (2\pi)^{-\frac{1}{2}} \int \exp[-\tfrac{1}{2}(R_j - R_i)^2]\, d(R_j - R_i). \tag{5}$$

Other distribution functions have also been proposed from time to time. The *arcsine transformation* of Mosteller (1951b) has already been mentioned. In addition, Bradley and Terry (1952) have proposed the *squared hyperbolic secant* function as a replacement for the normal ogive; it is given by

$$P_{ij} = \tfrac{1}{4} \int \operatorname{sech}^2[(R_j - R_i)/2]\, d(R_j - R_i). \tag{6}$$

The *logistic function* (for example, Berkson, 1953) is also used rather frequently. It defines the proportions as

$$P_{ij} = \frac{1}{1 + \exp\{-(a + b[R_j - R_i])\}} \tag{7}$$

where a and b are constants of the linear equation up to which p_{ij} is determined. Each of these proposals has advantages and disadvantages. We shall examine the use of the normal model [Eq. (5)] and the logistic model [Eq. (7)] in the next section, in which estimation of scale values is discussed.

b. Estimation of Scale Values

Scale values for each of the stimuli are determined from the proportion matrix by transformation through one of the models discussed above. By far the most frequently used model is the normal one. Values of p are converted to standard normal deviates z by computation or through reference to tables (see Chapter 7). A matrix of z values is constructed. The matrix in Table VI that is labeled ZMAT is a z matrix. We wish to project the z matrix onto a single continuum to form a scale of response values. It is possible to do this by least-squares regression techniques (for instance, Torgerson, 1958; Mosteller and Tukey, 1977; Draper and Smith, 1981), but Mosteller (1951c) has shown that essentially the same result can be obtained by the established technique of taking the sum of each of the columns in the z matrix. That is what has been done to form the SCLMAT of scale values in Table VI. Each stimulus, labeled COL1 through COL8, occupies the relative position on the scale indicated by the numbers listed (in ROW1) in the SCLMAT. This represents the classical result of a paired-comparison experiment.

As noted in the beginning of this chapter, there are difficulties with the classical method when pairs are missing or when there is a unanimity of judgment. The standard normal deviates for 0 and 1 are $-\infty$ and $+\infty$, and these cannot be manipulated mathematically to form a response scale. The logistic function has been used in a modified method (Maxwell, 1974) that circumvents these problems, however. Without listing the derivation, it is possible to define a logistic scale value as

$$V = \log_e[(f_{ij} + 0.5)/(N - f_{ij} + 0.5)], \tag{8}$$

where N stands for the number of times the pair was judged, and f is frequency. The arbitrary additive constant, 0.5, merely prevents there being any zeros in the frequency matrix. Obviously the greater the number of observations the smaller the proportionate contribution of this additive constant; hence, it is a good idea to include as many judgments as possible when using this method. Another approach would be to use a smaller additive constant but it is better practice to use many judgments.

Table VII shows how the method works with the data on fluorescent brighteners of Table V. The SUMMAT again provides a matrix of raw frequencies. These are transformed according to Eq. (8) to produce the LGMAT. As was done with z scores, the columns of the LGMAT are summed, and the values labeled SCALE are taken as the scale values for the stimuli.

In this case there were only 22 observations of each of the 28 pairs of stimuli. However, the correspondence between the scales of Tables VI and VII is very high. Table VIII shows the ZSCL values from Table VI (together with the LGSCL values of Table VII) and the estimations of the ZSCL values, labeled ZHAT; the residual error for each sample is shown in the column labeled ZRES. The predicted values of ZHAT were determined by the simple linear regression equation ZHAT = 0.641(LGSCL) − 2.0520. This equation had a coefficient of determination r^2 of 0.99996, so it can be seen that the two scales agree very well.

There are a number of other methods for dealing with missing or extreme values in the proportion matrix. These are discussed in detail in textbooks on scaling (for instance, Guilford, 1954; Torgerson, 1958). Most involve estimating interpolation and extrapolation functions for the rows and columns of the p matrix that contain missing elements. All require assumptions to be made that are difficult to prove or refute. I find the Maxwell method of Eq. (8) is satisfactory for most problems. It has been incorporated into a BASIC language computer program (called COMPJUDGE) that also provides a number of pertinent inferential statistics (Buyhoff and Hull, 1980) and which is available from Gregory J. Buyhoff, Department of Forestry, Virginia Polytechnic Institute and State University, Blacksburg, Virginia 24961. A useful companion BASIC program for a shell-sort algorithm has been published by Whaley (1979).

TABLE VII. LG-Scale Positions of Case V Pair-Comparison Whiteness Data

	Col 1	Col 2	Col 3	Col 4	Col 5	Col 6	Col 7	Col 8
SUMMAT								
Row 1	0	8	11	10	5	7	15	6
Row 2	14	0	17	14	5	12	16	7
Row 3	11	5	0	11	3	7	12	6
Row 4	12	8	11	0	2	2	12	1
Row 5	17	17	19	20	0	16	18	14
Row 6	15	10	15	20	6	0	18	9
Row 7	7	6	10	10	4	4	0	7
Row 8	16	15	16	21	8	13	15	0
LGMAT								
Row 1	0	−0.534082	0	−0.174353	−1.15745	−0.725937	0.725937	−0.931558
Row 2	0.534082	0	1.15745	0.534082	−1.15745	0.174353	0.931558	−0.725937
Row 3	0	−1.15745	0	0	−1.71765	−0.725937	0.174353	−0.931558
Row 4	0.174353	−0.534082	0	0	−2.10413	−2.10413	0.174353	−2.66259
Row 5	1.15745	1.15745	1.71765	2.10413	0	0.931558	1.41369	0.534082
Row 6	0.725937	−0.174353	0.725937	2.10413	−0.931558	0	1.41369	−0.351398
Row 7	−0.725937	−0.931558	−0.174353	−0.174353	−1.41369	−1.41369	0	−0.725937
Row 8	0.931558	0.725937	0.931558	2.66259	−0.534082	0.351398	0.725937	0
SCALE								
Row 1	2.79745	−1.348756	4.35825	7.05623	−9.01603	−3.51239	5.55953	−5.79489

TABLE VIII. Comparison of Scale Values, Z-Scale versus LG-Scale

Sample	ZSCL	LGSCL	ZHAT	ZRES
1	1.8154	2.7974	1.7950	0.002420
2	−0.9435	−1.4481	−0.9292	−0.014300
3	2.8078	4.3582	2.7964	0.011414
4	4.4814	7.0562	4.5276	−0.046214
5	−5.7895	−9.0160	−5.7851	−0.004431
6	−2.2406	−3.5124	−2.2537	0.013140
7	3.5954	5.5595	3.5672	0.028222
8	−3.7265	−5.7949	−3.7182	−0.008251

3. An Application of the Paired-Comparison Method to Chroma Scaling

We may now return to the problem of scaling chroma that was used as an example in the section on rank ordering. It has not yet been dealt with in this section, because the stimulus differences were generally so obvious that observers had little trouble in determining colorfulness orders. Similarly, there were many unanimous cases in the paired-comparison scaling of the samples listed in Table I. Whenever that is the case, the paired-comparison method is not best suited to the scaling problem; paired-comparison methods work best when the stimuli are similar enough to be confused often. However, the method can be applied even to our chroma samples, and it will be instructive to see how the scale values so derived compare with those based on the rank-order method.

The same 100 observers were shown all possible pairs (36) of the 9 colored papers. They were asked to indicate which member of each pair was most colorful. The data were recorded in a matrix similar to that of Table V, and a frequency matrix was determined from the raw data. That matrix is shown at the top of Table IX. Note that there are many entries with values of 100 and 0. The logistic model of Eq. (8) was used to generate the LGMAT shown in Table IX, and its columns were summed to form the SCALE.

If the scale values so derived are appropriate, we might expect them to be linearly related to the original chroma values (analogous to what was done to assess the utility of the scales derived from the rank-order method). Figure 4 shows that the scale derived from the paired-comparison experiment is, in fact, very nearly linearly related to chroma. The equation of the line that passes through or near the points in that figure is $y = 6.62x - 45.61$, where y is the scale of values from Table IX and x stands for chroma. The coefficient of determination is 0.993.

TABLE IX. Paired-Comparison Logistic Scale for Chroma Samples (100 Observers)

	Col 1	Col 2	Col 3	Col 4	Col 5	Col 6	Col 7	Col 8	Col 9
SUMMAT									
Row 1	50	100	100	100	100	100	100	100	100
Row 2	0	50	1	0	0	0	51	0	2
Row 3	0	99	50	0	0	49	99	1	100
Row 4	0	100	100	50	0	100	100	100	100
Row 5	0	100	100	100	50	100	100	100	100
Row 6	0	100	51	0	0	50	100	1	0
Row 7	0	49	1	0	0	0	50	0	0
Row 8	0	100	99	0	0	99	100	50	100
Row 9	0	98	0	0	0	0	100	0	50
LGMAT									
Row 1	0	5.3033	5.3033	5.3033	5.3033	5.3033	5.3033	5.3033	5.3033
Row 2	−5.3033	0	−4.19469	−5.3033	−5.3033	−5.3033	0.0396091	−5.3033	−3.67377
Row 3	−5.3033	4.19469	0	−5.3033	−5.3033	−0.0396091	4.19469	−4.19469	5.3033
Row 4	−5.3033	5.3033	5.3033	0	−5.3033	5.3033	5.3033	5.3033	5.3033
Row 5	−5.3033	5.3033	5.3033	5.3033	0	5.3033	5.3033	5.3033	5.3033
Row 6	−5.3033	5.3033	0.0396091	−5.3033	−5.3033	0	5.3033	−4.19469	5.3033
Row 7	−5.3033	−0.0396091	−4.19469	−5.3033	−5.3033	−5.3033	0	−5.3033	−5.3033
Row 8	−5.3033	5.3033	4.19469	−5.3033	−5.3033	4.19469	5.3033	0	5.3033
Row 9	−5.3033	3.67377	−5.3033	−5.3033	−5.3033	−5.3033	5.3033	−5.3033	0
SCALE									
Row 1	−42.4264	34.3454	6.45153	−21.2132	−31.8198	4.15508	36.0541	−8.38939	22.8428

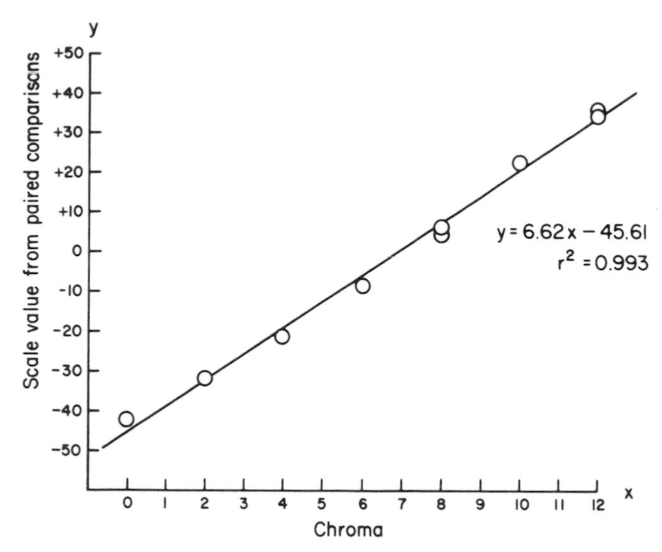

Fig. 4. Scale values, derived from the paired-comparison logit method, compared with chroma values.

Even here, then, the paired-comparison method provides a reasonable interval scale representation of the original chroma values, but a better way to treat stimuli with obvious differences would be to use a scaling method such as the category or rating method.

D. THE RATING-SCALE METHOD

1. Forms of Rating Scales

Rating scales are perhaps the most commonly used methods for estimating relationships among stimuli, and at the same time they are perhaps the most misunderstood form of scaling. Nearly everyone has said or heard said at one time or another: "I'd rate (it) 7 on a scale of 1 to 10." This is an example of the application of one form of numerical rating scale. Many people believe that the estimate tells us something about the magnitude of the attribute being rated. It does not. What it does tell us is the rater's estimate of the proportion or ratio of the difference between what is being rated and the anchor that represents the bottom of the scale and the difference between what is being rated and the anchor that represents the top of the scale. The estimate is a proportion of two distances. It is, in other words, a relative statement.

The anchors may be tangible or intangible. They may be absolute extremes, or they may be arbitrary, intermediate positions along some continuum. In either case the anchors are assigned the same *moduli* or scale values: for example, 0 and 10, 0 and 100, least and most, and lowest and highest. The proportion represented by the rater's estimate relates only to the overall interval represented by the two anchors. We should not be misled by the fact that anchors have the same moduli; there is no basis for intercomparison of estimates among rating scales without some explicit knowledge of the ratio of overall intervals between the anchors on different scales.

There are many forms of rating scales. Generally the classes of rating scales may be taken as three: numerical, adjectival, and graphical. There are many variations within each class, and classes may also be mixed.

The *numerical rating scale*, in its pure form, consists of a range of numbers. These numbers are almost always integers, but they may be either all positive or both negative and positive distributed symmetrically about 0, which is taken to represent an indifference point. An attribute to be scaled is associated with the numbers. The direction of increase in the attribute is also made explicit. For example, an observer may be asked to estimate how pleasant he finds a painting by using a scale of 0 to 100 for which 0 represents least pleasant imaginable and 100 represents most pleasant imaginable.

An *adjectival rating scale* does the same thing without recourse to numbers by introducing intermediate adjectives that are intended to imply a series of equal intervals. In the example of pleasantness of paintings, we might define the rating scale as most unpleasant imaginable, very unpleasant, moderately unpleasant, mildly unpleasant, neither pleasant nor unpleasant, mildly pleasant, moderately pleasant, very pleasant, and most pleasant imaginable. The key is to make the scale symmetrical about an indifference point. Most people prefer to state the indifference point explicitly. Therefore, the number of adjectives is most often an odd one. There is both empirical and experimental evidence that 5, 7, or 9 moduli are most satisfactory for many purposes (for example, Miller, 1956; Attneave, 1959; Norwich, 1981).

The *graphical rating scale* uses only the two extreme anchors and connects them with a line. The observer's task is to place a mark along the length of the line, the position of which is to represent the relationship of the estimate to the anchors. This method makes the relative nature of rating scales most obvious. The experimenter obtains data by measuring the length of the line segment (from either end) and dividing it by the measured length of the complete line. The graphical rating scale is an exact analog of Comrey's (1950) method of constant-sum direct scaling, the first supplying the observer with a picture of the distance to be partitioned and the Comrey technique requiring the observer to carry out the numerical division in his head.

There are many potential problems with rating-scale methods. The observers may ascribe different meanings to the adjectives used. People may use numbers in different ways. Sometimes both adjectives and numbers are presented together; and not only may there be differences among observers in the use of numbers and adjectives, but also a given observer may use the number scale in a different way from the way in which he uses the adjectival scale. The rating-scale methods are particularly prone to the effects of adaptation to the distribution of stimuli by the observers. Usually there is a central-tendency effect in which observers are loath to use the ends of the scale; I find it helpful in trying to avoid serious problems of this kind to use a 9-point scale when I am really interested only in a 7-point one. A related problem is that some observers may use the end points and then find that some stimuli are greater or less than those that have been assigned to the end points, so that it is impossible to discriminate among near-extreme stimuli. There are many other kinds of problems that can arise with rating scales; Guilford (1954) discusses a number of them.

Generally, data are analyzed from rating-scale experiments by simple arithmetic methods. Numerical weights may be assigned to adjectives, and the products of frequencies of choice and weights are summed over observations for a given stimulus to determine its scale position. When numerical ratings are used, the assigned numbers are summed over observations for each stimulus. In the graphical method, the proportions are summed. There are other methods of reducing rating-scale data (Guilford, 1954; Torgerson, 1958), but generally the methods used are crude by the standards of comparative-judgment data-reduction methods. This is perhaps the major distinction between rating-scale methods and categorical scaling methods, which are essentially a special case of rating scales. Of the rating-scale methods, I prefer the graphical technique used by Newhall (1939) in which adjectives are assigned to the two end points, and the end points are connected by a line along which the observer indicates the relative position of his estimate of stimulus magnitude [Newhall used a sliding pointer or a bar, but a line drawn on paper can be used just as effectively (Freyd, 1923).] Schiffman, Reynolds, and Young (1981, p. 32) suggest a 5-in. line as the best length. An example of the application of this method to our chroma samples will be given in the next section.

2. An Example of a Graphical Rating Scale

The samples of Table I were presented, one at a time, to each of 22 observers. A single sheet of paper was presented along with each sample. The 210 × 298 mm paper had a 135-mm line drawn along the longer dimension, parallel to the edge and centered. At the left of the line was printed "No Colorfulness," and at the right appeared "Very Colorful"; it looked like this:

No Colorfulness ______________________________ Very Colorful

The observer was asked to place a pencil mark on the line to indicate how colorful he found the sample. The instructions were

> You will be shown each, in turn, of a series of colored samples. We would like to know how colorful each of these samples is. There will be a paper accompanying each sample. The paper contains a line running from "No Colorfulness" to "Very Colorful." Please use the pencil that you will be given to place a mark on the line at the position where you think the colorfulness of the sample belongs. You may place marks at the ends of the line but not off the line beyond the ends.

The data collected in this way were analyzed by measuring the number of millimetres from the left-hand end of the line to the pencil mark and then dividing that number by 135 (the number of millimetres along the entire line). Data for the 22 observers are listed in Table X. The data are shown as percentages of total line length. The mean positions for each of the 9 stimuli are shown at the bottom of the table. These are simply arithmetic averages of each column. The bottom row of Table X lists the original chroma values for each of the stimuli.

Figure 5 shows how these simple averages relate to the chromas. The linear regression (shown in the figure) has a coefficient of determination of 0.995, which suggests that this relatively crude method provided results that are highly correlated with the original chroma values of the samples.

The experiment required about 5 min for each observer, less than 2 hr altogether for the total of 22 observers, and took only about 15 min to analyze. Therein lies the great appeal of rating-scale methods. They are simple, fast, and easy to analyze. In many cases, including the example given here, they also provide about as much (and as good) information as that resulting from a more laborious, more complex experiment. However, that is not always the case. The problem is that we seldom know when a rating-scale result will be as good as that from some other scaling technique. Usually this is because we do not know precisely what it is that observers do; we only know what we ask them to do. The most common unknown is whether observers partition the rating scale (or ascribe meanings to the rating adjectives) that represent equal underlying intervals of perceptual differences. Largely for these reasons, a special case of rating scales has been developed in which assumptions are made about the distributions of judgments provided by observers. This case is referred to as the *successive-category* method of scaling.

TABLE X. Chroma: Rating Scale from No Colorfulness to Very Colorful

	Stimuli								
Observers	1	2	3	4	5	6	7	8	9
1	0	90	60	30	15	60	90	45	75
2	0	100	50	21	10	63	100	51	80
3	0	90	65	35	17	55	88	40	81
4	2	95	62	28	19	59	90	44	70
5	0	98	60	29	14	62	90	44	75
6	8	80	58	33	11	60	100	48	74
7	0	90	54	30	20	61	87	46	75
8	1	85	59	25	18	60	85	44	71
9	0	90	62	32	18	57	91	47	65
10	0	80	55	20	15	61	93	35	75
11	0	85	65	25	13	60	90	40	72
12	0	87	70	35	16	59	87	37	85
13	0	90	50	30	15	59	89	43	74
14	0	88	63	37	17	51	91	45	76
15	0	78	65	19	12	56	82	45	72
16	0	90	58	30	14	63	86	33	79
17	0	99	60	39	15	67	89	49	75
18	5	90	55	30	15	57	90	45	75
19	0	85	57	29	17	61	86	43	73
20	0	90	59	27	14	63	90	40	68
21	0	98	61	31	18	59	90	42	77
22	1	90	63	30	15	57	87	39	76
Mean position:									
	0.77	88.95	59.59	29.32	15.36	59.55	89.59	42.95	84.68
Chroma:	0	12	8	4	2	8	12	6	10

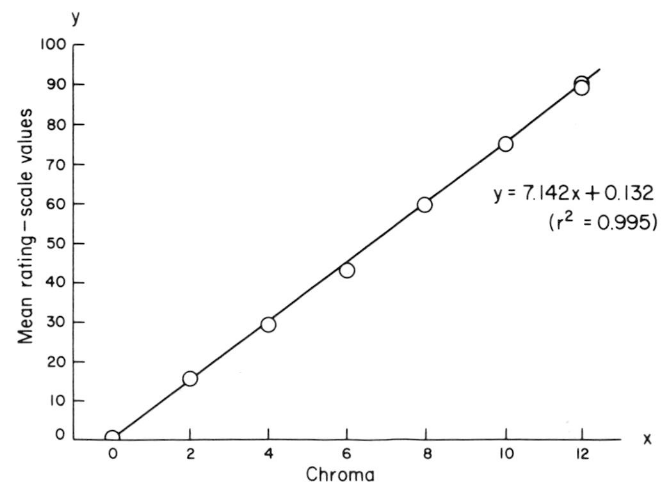

Fig. 5. Scale values, derived from the mean rating-scale method, compared with chroma values.

E. THE CATEGORY METHOD

1. The Law of Categorical Judgments

The *law of categorical judgments* (Torgerson, 1954) is an extension of Thurstone's (1927) law of comparative judgements. The law of categorical judgements may be expressed mathematically as

$$B_k - R_j = z_{jk}(\sigma_j^2 + \sigma_k^2 - 2r_{jk}\sigma_j\sigma_k)^{1/2}, \tag{9}$$

where

B_k = the mean location of the kth category boundary,
R_j = the mean response to stimulus j,
σ_k = the dispersion of the kth category boundary,
σ_j = the dispersion of stimulus j,
r_{jk} = the correlation between momentary positions of stimulus k and category boundary k on the scale, and
z_{jk} = the normal deviate corresponding to the proportion of times stimulus j is placed below boundary k.

Comparison of Eq. (9) with Eq. (1) for the law of comparative judgments will show that the two expressions are of the same form. The difference between them is simply that the law of categorical judgements relates to the relative positions of stimuli with respect to category boundaries rather than with respect to one another. Essentially the same conditions apply to Eq. (9) as those that make up the six cases of Eq. (1).

The underlying assumptions of the law of categorical judgments have been stated by Torgerson (1958):

> 1. The psychological continuum of the subject can be divided into a specified number of ordered categories or steps.
> 2. Owing to various and sundry factors, a given category boundary is not necessarily always located at a particular point on the continuum. Rather, it also projects a normal distribution of positions on the continuum. Again, different category boundaries may have different mean locations and different dispersions.
> 3. The subject judges a given stimulus to be below a given category boundary whenever the value of the stimulus on the continuum is less than that of the category boundary. [p. 206]

The law of comparative judgments is merely a formal statement of a rating-scale and data-analysis method that has a long history of use in scaling (Titchner, 1905; Boring, 1942, 1950; Guilford, 1954). There are many forms of category scaling and a wide variety of experimental techniques and data-reduction algorithms that have been used in category scaling (for instance, Guilford, 1954; Woodworth and Schlosberg, 1954; Torgerson, 1958; Bock and Jones, 1968; Kling and Riggs, 1971). The purpose of this chapter is not to present a definitive review but, rather, to attempt to convey some appreciation of the underlying central concept of categorical methodology. I have, therefore, selected an example of a single, representative experimental method to provide data that will be analyzed by each of two techniques. The data relate to the common example of this chapter: the scaling of colorfulness of nine Munsell chroma samples. The two analytical techniques represent nearly opposite ends of the spectrum of sophistication in such data analysis.

2. An Experimental Method for Category Scaling

A common experimental method of category scaling was used to gather data about the colorfulness of nine Munsell samples that differed in Munsell hue and chroma. Twenty-two observers participated in the experiment. They were asked to rate the colorfulness of each sample on a 9-point scale on which numbers at both ends were associated with adjectives. The instructions were as follows:

> You will be shown each of a series of colored papers. We would like to know how colorful you think each paper is. Please express your opinion on a scale of numbers from 1 to 9 where *1 represents a complete lack of colorfulness and 9 represents the most colorful thing you can imagine.* Use numbers between 1 and 9 to represent equal intervals of colorfulness. For example, you might think of the scale in the following way:
>
> 1. Least imaginable colorfulness
> 2. Very little colorfulness
> 3. Mildly colorful
> 4. Moderately colorful
> 5. Colorful
> 6. Moderately highly colorful
> 7. Highly colorful
> 8. Very highly colorful
> 9. Highest imaginable colorfulness
>
> where the difference in colorfulness between categories 3 and 4 is the same as the difference in colorfulness between categories 7 and 8, and so on. You may not use fractions or decimals; you must use integers. The integers should be from 1 to 9; no larger or smaller integers may be used.

As with most rating-scale methods, this experiment was very efficient of time and labor. Each observer completed all 9 observations in about 2–3 min. The data were recorded as the category number assigned by each observer to each stimulus. The raw data are tabulated in Table XI.

TABLE XI. Raw Data from Category Scaling; Entries Are Category Assignments

OBS	Subject	Sample 1	Sample 2	Sample 3	Sample 4	Sample 5	Sample 6	Sample 7	Sample 8	Sample 9
1	1	1	7	5	3	2	5	7	3	6
2	2	1	7	5	3	2	4	7	4	6
3	3	1	6	4	3	2	5	7	4	7
4	4	1	6	6	3	2	5	8	4	6
5	5	2	7	5	3	1	5	7	4	5
6	6	1	7	6	3	2	5	7	4	5
7	7	1	7	6	3	2	5	8	4	6
8	8	1	7	4	3	2	6	7	4	6
9	9	1	8	5	2	3	5	7	3	6
10	10	1	7	5	3	3	5	8	4	6
11	11	1	7	5	3	3	5	8	4	6
12	12	1	7	5	3	2	5	7	4	5
13	13	1	8	5	4	2	5	7	5	6
14	14	1	8	5	3	3	6	7	5	6
15	15	1	7	6	3	2	6	7	4	6
16	16	2	6	5	3	2	5	7	4	6
17	17	1	6	5	3	2	5	7	3	6
18	18	2	7	6	4	2	5	7	4	6
19	19	2	7	5	3	2	5	6	4	6
20	20	1	7	5	3	2	5	7	4	6
21	21	1	7	5	3	2	5	7	4	6
22	22	1	7	5	3	2	6	7	4	6
Category means		1.18	6.95	5.14	3.05	2.14	5.14	7.14	3.95	5.86
Chromas		0	12	8	4	2	8	12	6	10

3. The Mean-Category-Value Method

The simplest way of determining scale values from the raw data of Table XI is to follow the most common practice for rating-scale data and sum each of the stimulus columns. Underlying such a practice is the tacit assumption that the observers were both capable of and inclined to follow the instructions exactly with respect to the equal-interval properties of the category scale. Virtually all scaling theorists take this to be an unwarranted assumption. Certainly, it is a great deal to expect of observers. However, the method has been included here to illustrate the extent to which scales derived on the basis of such simplistic assumptions can resemble scales derived by other techniques.

The category means are shown at the bottom of each column in Table XI, along with the original Munsell chroma values for each of the stimuli. These data have been plotted in Fig. 6, where the category means are shown against chroma. The regression equation in the figure has an r^2 value of 0.9976, reasonably high.

Frequently it is found that simple category means yield scale values that are very close to those determined by other methods. In the example of colorfulness scaling we might take the Munsell chroma values as the "true" solution, and when the category means are compared with them, we find that they agree to a high level of correlation. In other words, the simple method of determining scale values can be both accurate and consistent with results determined by more sophisticated techniques.

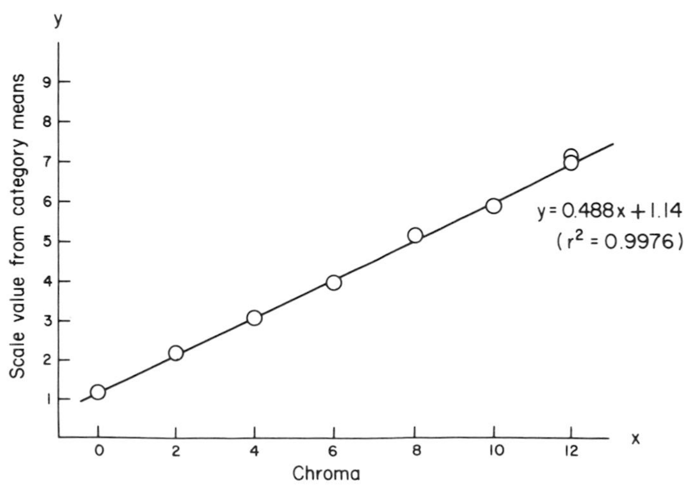

Fig. 6. Scale values, derived from the successive-category means method, compared with chroma values.

4. The Categorical-Judgment Method

It is seldom that we know the "true" solution. Most often we seek a solution without benefit of such foreknowledge. Unless the results of the experiment are of little consequence, the prudent course is to derive an interval scale by invoking the law of categorical judgments. There are a number of mathematical techniques for doing this. All of them involve similar algorithms and deal with four basic matrices.

The first of these matrices is a frequency matrix derived directly from the raw data. The top matrix in Table XII shows a frequency matrix from the data

TABLE XII. Calculation Matrices Used in Determining an Interval Scale

Observed frequencies

	1	2	3	4	5	6	7	8	9
1	18.000	4.000	0.0	0.0	0.0	0.0	0.0	0.0	0.0
2	0.0	0.0	0.0	0.0	0.0	4.000	15.000	3.000	0.0
3	0.0	0.0	0.0	2.000	15.000	5.000	0.0	0.0	0.0
4	0.0	1.000	19.000	2.000	0.0	0.0	0.0	0.0	0.0
5	1.000	17.000	4.000	0.0	0.0	0.0	0.0	0.0	0.0
6	0.0	0.0	0.0	1.000	17.000	4.000	0.0	0.0	0.0
7	0.0	0.0	0.0	0.0	0.0	1.000	17.000	4.000	0.0
8	0.0	0.0	3.000	17.000	2.000	0.0	0.0	0.0	0.0
9	0.0	0.0	0.0	0.0	3.000	18.000	1.000	0.0	0.0

Observed cumulative frequencies

	1	2	3	4	5	6	7	8
1	18.000	22.000	22.000	22.000	22.000	22.000	22.000	22.000
2	0.0	0.0	0.0	0.0	0.0	4.000	19.000	22.000
3	0.0	0.0	0.0	2.000	17.000	22.000	22.000	22.000
4	0.0	1.000	20.000	22.000	22.000	22.000	22.000	22.000
5	1.000	18.000	22.000	22.000	22.000	22.000	22.000	22.000
6	0.0	0.0	0.0	1.000	18.000	22.000	22.000	22.000
7	0.0	0.0	0.0	0.0	0.0	1.000	18.000	22.000
8	0.0	0.0	3.000	20.000	22.000	22.000	22.000	22.000
9	0.0	0.0	0.0	0.0	3.000	21.000	22.000	22.000

Observed cumulative proportions

	1	2	3	4	5	6	7	8
1	0.818	1.000	1.000	1.000	1.000	1.000	1.000	1.000
2	0.0	0.0	0.0	0.0	0.0	0.182	0.864	1.000
3	0.0	0.0	0.0	0.091	0.773	1.000	1.000	1.000
4	0.0	0.045	0.909	1.000	1.000	1.000	1.000	1.000
5	0.045	0.818	1.000	1.000	1.000	1.000	1.000	1.000
6	0.0	0.0	0.0	0.045	0.818	1.000	1.000	1.000
7	0.0	0.0	0.0	0.0	0.0	0.045	0.818	1.000
8	0.0	0.0	0.136	0.909	1.000	1.000	1.000	1.000
9	0.0	0.0	0.0	0.0	0.136	0.955	1.000	1.000

TABLE XII. (*cont.*)

Normal deviates

	1	2	3	4	5	6	7	8
1	0.908	1.515	2.246	2.844	3.686	4.521	5.230	5.634
2	−4.521	−3.915	−3.184	−2.585	−1.744	−0.908	1.097	1.501
3	−3.271	−2.665	−1.934	−1.335	0.748	1.583	2.292	2.696
4	−2.297	−1.691	1.335	1.934	2.775	3.611	4.320	4.724
5	−1.691	0.908	1.639	2.238	3.079	3.915	4.624	5.028
6	−3.627	−3.020	−2.290	−1.691	0.908	1.744	2.453	2.857
7	−5.304	−4.697	−3.967	−3.368	−2.527	−1.691	0.908	1.312
8	−2.434	−1.828	−1.097	1.335	2.177	3.012	3.721	4.125
9	−3.874	−3.268	−2.537	−1.938	−1.097	1.691	2.400	2.804

Category boundary	estimate	Scale code	Scale value	Standard deviation
1	−1.498	1	−2.049	0.617
2	−1.108	5	−1.153	0.467
3	−0.664	4	−0.721	0.392
4	−0.299	8	−0.457	0.405
5	0.240	3	0.107	0.452
6	0.718	6	0.135	0.408
7	1.211	9	0.287	0.395
8	1.399	2	0.845	0.474
		7	1.022	0.423

of Table XI. The number of times each stimulus was assigned to each category makes up the elements of the matrix. The n rows represent the stimuli, and the columns correspond to the m categories; $m = 9$ in our example. The frequency matrix is the basic one from which the other three matrices are derived in successive steps. Recall that the law of categorical judgments deals with proportions of times that a given stimulus is assigned to a position below a given category boundary. We are, in other words, interested in cumulative frequencies rather than raw frequencies. Accordingly, a second (cumulative-frequency) matrix is constructed by changing the elements of the frequency matrix to running sums across each of the rows. This is shown in the second matrix of Table XII. Notice that we now have only $m - 1$ columns, corresponding to the k category boundaries. That is, column 1 now corresponds to the sum of 0 (the starting point for accumulation of frequencies) and the frequency for the element in the first column of the initial frequency matrix. Column 2 of the cumulative-frequency matrix adds to that figure the next element in the row, and so on until all $k = m - 1$ category boundaries have been included in the accumulation. The third of the four matrices common to this method is then constructed by simply converting the cumulative frequencies to cumulative proportions. The elements of the cumulative-proportion matrix are now available for use in constructing the final, transformation matrix. A nonlinear transformation is used. The form of the transformation is dictated by the particular model chosen to represent the assumed form of discriminal dispersions between stimuli and category boundaries.

It is at this point that the various methods of data analysis reach their first point of divergence. As with analyses of paired-comparison proportion matrices, any of a number of different dispersion models can be assumed for transforming the cumulative proportion matrix of category-scaling data. The most common choice is the normal model, for which the cumulative proportions are converted to z values. Other options include the logistic transform (Berkson, 1953), the similar squared-hyperbolic–secant transform (Bradley and Terry, 1952; Luce, 1959), the arcsine transform (Mosteller, 1951c) and transforms based on Laplace functions (Dawkins, 1969). Basically these different models yield cumulative distributions that are all roughly S-shaped. The Dawkins (1969) model differs most from the others, because its frequency-distribution shape is concave about the mean, consisting of two Poisson exponentials for which the difference function is a Laplace function. However, there is generally little to choose among the different models *on the basis of real experimental data*; that is, the data from real experiments are fallible and are never precise enough to permit a clear choice about which model best fits the results. Any distinctions that are made must be based on theoretical considerations. The transformation model used in Table XII is the normal one; it was assumed that $r_{jk} = 0$ and both σ_j^2 and σ_k^2 were constant in Eq. (9), the expression for the law of categorical judgments. The matrix labeled "normal deviates" then has elements that are the z values corresponding to the cumulative proportions of the matrix shown just above it.

Comparison of the cumulative-proportion and normal-deviate matrices reveals something that represents a second point of divergence among methods for analyzing data from such experiments. In particular, note that the top row ($n = 1$) of the cumulative-proportion matrix displays elements that reach unity proportions by the second category boundary. That is, stimulus 1 was only assigned to categories 1 and 2. This means that the entries in the cumulative-proportion matrix must be 1.0 for all but column (boundary) 1. Look at the top row of the normal-deviate matrix, however. It displays z values that increase progressively up to 5.634 in column 8. It is in the performance of this bit of legerdemain that different analytical methods also diverge. They use different treatments for unanimous cases (or extreme proportions).

The oldest method for filling in missing (or 0 and 1) elements of the cumulative-proportion matrix is an algebraic one that takes advantage of the fact that each row of the proprotion matrix (when arranged in increasing order of column sums) provides an approximate solution to Eq. (9). In terms of z values this approximation can be represented (graphically or computationally) by a straight line. The missing values are simply interpolated or extrapolated from that line. The line can be determined from each row, or it can be computed from the average of all rows. Another method involves performing a least-squares or maximum-likelihood regression for the entire matrix. Guilford (1954), Torgerson (1958), Bock and Jones (1968), and Edwards (1951) summarize these methods in detail. The method used in Table XII was proposed by Gulliksen (1954, 1956). It is a least-squares solution that also yields the scale values for category boundaries and individual stimuli, as shown at the bottom of the table.

The standard deviation for each of the scale values is also shown at the bottom of Table XII. The average of these is about 0.45, which may be divided by the number of categories (9) to obtain a rough indication of the precision of the scale. The figure is 5%. My experience has been that the scale precision obtained with carefully conducted successive-categories experiments ranges from about 2 to 7%. When we consider that many commercially available photometers and colorimeters yield similar values for scale precision, the 5% figure for this scaling experiment suggests that suprathreshold visual measurements need not be considered inferior in precision to other common measurements of optical radiations.

Precision is not the same as accuracy, however. Two kinds of accuracy are required here. One relates to how well the results correspond to the original chroma ("true") values. The other is the degree to which the least-squares regression reconstitutes the input proportion matrix. The latter question influences the former, of course, and so it will be examined first.

Table XIII shows a back substitution of proportions computed from the matrix-wide regression in the column labeled CALC. The computed proportions are compared with the observed ones (OBSV) shown in the third column. The first two columns identify the stimuli and categories, respectively (because nothing was assigned to category 9, it does not appear on the table). Notice that the back-substituted values are not particularly good matches to the original proportions except where the highest proportion occurs in an extreme category. The group for stimulus 6 is a good example of the problem; although the mode of the computed distribution of proportions also occurs at category 5, the original distribution is much more leptokurtic than that of the back substitution. The computer routine that was used also computed a χ-squared value, which is shown at the bottom of Table XIII. In this case the degrees of freedom number only 10, because each unanimous proportion in the matrix uses up one degree of freedom. The critical χ-squared value for $\alpha = 0.05$ and $df = 10$ is 18.307, much lower than the calculated 107.07, so we

TABLE XIII. Back Substitution to Predict the Proportion Matrix

Indices				
Stimulus	category	OBSV	CALC	DIFF
1	1	0.818	0.814	0.004
1	2	0.182	0.122	0.060
1	3	0.0	0.051	−0.051
1	4	0.0	0.010	−0.010
1	5	0.0	0.002	−0.002
1	6	0.0	0.000	−0.000
1	7	0.0	0.000	−0.000
1	8	0.0	0.000	−0.000
2	1	0.0	0.000	−0.000
2	2	0.0	0.000	−0.000
2	3	0.0	0.001	−0.001
2	4	0.0	0.007	−0.007
2	5	0.0	0.093	−0.093
2	6	0.182	0.293	−0.111
2	7	0.682	0.386	0.296
2	8	0.136	0.099	0.038
3	1	0.0	0.000	−0.000
3	2	0.0	0.003	−0.003
3	3	0.0	0.041	−0.041
3	4	0.091	0.141	−0.050
3	5	0.682	0.431	0.251
3	6	0.227	0.296	−0.069
3	7	0.0	0.081	−0.081
3	8	0.0	0.005	−0.005
4	1	0.0	0.024	−0.024
4	2	0.045	0.138	−0.093
4	3	0.0864	0.396	0.467
4	4	0.091	0.301	−0.210
4	5	0.0	0.134	−0.134
4	6	0.0	0.007	−0.007
4	7	0.0	0.000	−0.000
4	8	0.0	0.000	−0.000
5	1	0.045	0.230	−0.185
5	2	0.773	0.308	0.465
5	3	0.182	0.314	−0.132
5	4	0.0	0.114	−0.114
5	5	0.0	0.032	−0.032
5	6	0.0	0.001	−0.001
5	7	0.0	0.000	−0.000
5	8	0.0	0.000	−0.000
6	1	0.0	0.000	−0.000
6	2	0.0	0.001	−0.001
6	3	0.0	0.023	−0.023
6	4	0.045	0.118	−0.072

6	5	0.772	0.460	0.312
6	6	0.182	0.323	−0.141
6	7	0.0	0.071	−0.071
6	8	0.0	0.003	−0.003
7	1	0.0	0.000	−0.000
7	2	0.0	0.000	−0.000
7	3	0.0	0.000	−0.000
7	4	0.0	0.001	−0.001
7	5	0.0	0.031	−0.031
7	6	0.045	0.204	−0.158
7	7	0.773	0.437	0.336
7	8	0.182	0.141	0.041
8	1	0.0	0.005	−0.005
8	2	0.0	0.049	−0.049
8	3	0.136	0.251	−0.114
8	4	0.773	0.346	0.426
8	5	0.091	0.306	−0.215
8	6	0.0	0.041	−0.041
8	7	0.0	0.002	−0.002
8	8	0.0	0.000	−0.000
9	1	0.0	0.000	−0.000
9	2	0.0	0.000	−0.000
9	3	0.0	0.008	−0.008
9	4	0.0	0.061	−0.061
9	5	0.136	0.384	−0.247
9	6	0.818	0.410	0.408
9	7	0.045	0.128	−0.082
9	8	0.0	0.007	−0.007
Largest deviation				0.47
Average deviation				0.00
Standard error of deviations				0.1462

1	1.76
2	8.29
3	5.73
4	19.66
5	22.40
6	8.55
7	8.65
8	17.09
9	14.94
Total χ-square	107.07[a]
Degrees of freedom	10

[a] Critical χ-square ($\alpha = 0.05, df = 10$) = 18.307
($\alpha = 0.05, df = 36$) = 50.964

must conclude that the reconstituted matrix is a poor match to the input matrix. Even if the degrees of freedom were equal to the total number of implied pairs $[n(n-1)/2 = 36]$, the critical value for χ-squared at $\alpha = 0.05$ would be only 50.964. Obviously there is not a good fit between the original and reconstituted matrices. This is a result of having too many unanimous decisions in the data; it is for this reason that the chroma experiment was not used as the first example in the section on paired comparisons. In short, there is simply not enough confusion in the data from these stimuli for a comparative-judgment model to work well.

The problem is manifested in the comparison of scale values determined in Table XII with the original chroma values. The two sets of values are plotted in Fig. 7. That the accuracy is not very high is evidenced by the relatively low coefficient of determination (0.9742) shown for the simple regression of the data in this figure. Here, then, is a case for which the simple scaling methods provide a more accurate solution than does the more sophisticated one. The reader is warned, however, that this is not always the case. Were there more confusion in the data, the categorical-judgment method of data reduction would likely provide the better answer. In general the more sophisticated method is to be preferred. Where computers are available, there is no reason not to use the categorical method rather than the category-means approach, particularly when a test can be made to determine how well the model fits the experimental data. If the fit is not good with the categorical-judgment model,

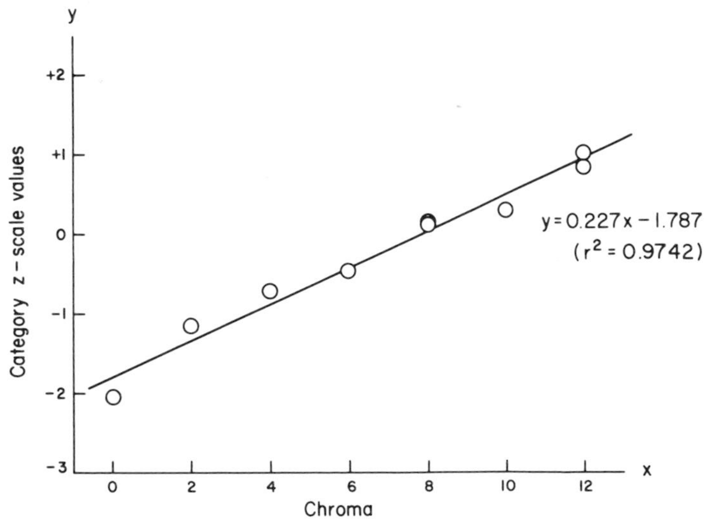

Fig. 7. Scale values, derived from normal deviates of the successive-category boundaries method, compared with chroma values.

as in the present example, the category means and rank order can always be determined from the same data. We can see that it would be misleading and even inaccurate to use the scale values of the categorical-judgment solution in Table XII in preference to those of the category-means solution given in Table XI. However, we would not know this without having obtained both solutions and performed the back substitution and χ-squared test. Obtaining enough information allows us to back down to a less elegant solution or even to a less powerful method (such as rank order) when appropriate.

F. SOME REFLECTIONS ON INTERVAL SCALING

1. Summary of Methods

We have seen that even the simple rank-ordering method (which provides only ordinal information directly) can be used to construct a scale of equal perceptual intervals by invoking certain assumptions about the observers or the stimuli. The paired-comparsion experiment also asks only questions of order of the observers. It does so explicitly for all possible pairings of the stimuli. Again, assumptions are made about the discriminal dispersions of responses and stimuli. Even the category-scaling method, which might appear to address suprathreshold differences directly, relies on assumptions about discriminal dispersions to form a scale. All of these methods have in common the need to manipulate uncertainty distributions. Only the simple rating-scale and category-means methods of analyzing categorical data address differences directly. They do so by assuming that observers preserve the equal-interval properties of the scale when their judgments are made. This assumption would appear to be theoretically untenable (Boring, 1920; Bartlett, 1940; Cattell, 1944; Gulliksen, 1950; Burke, 1953; Coombs *et al.*, 1954; Guilford, 1954; Torgerson, 1958; Coombs, 1964; Savage, 1970), yet experience indicates that it is often well enough grounded that useful results can be obtained. In general, however, it is uncertainty or confusion that is the kernel of interval scaling methods. Stevens (1975, p. 10) has dubbed these methods *poikilitic* (from a Greek word ΠΟΙΚΟΣ meaning a noisy sound). He classes them with limen (JNDs) and says they are "a measure of variability or 'noise'—those nasty perturbations that set limits in our sensory resolving power." This is exactly what lies at the bottom of threshold scaling; Fechner's limen is determined from uncertainty of judgments. One classification of scaling methodology groups the interval methods under the rubric "generalized Fechnerian scaling" (Baird and Noma, 1978). In Chapter 7 (Section A.1.e) discrimination was discussed as a probabilistic process. These same concepts of threshold discrimination begat the law of comparative judgments (which concerns discrimination among suprathreshold stimuli), which in turn begat the law of categorical judgments (which concerns discrimination among suprathreshold stimuli and category boundaries). The discriminal dispersion is the root from which these scaling methods grow.

The key implication of all this is that without discriminal dispersion, there is no basis for deriving a scale of intervals according to the laws of comparative and categorical judgments. No uncertainty equals no scale. We have seen that the major problem in analyzing comparative-judgment matrices (from either paired-comparison or categorical experiments) is what to do about unanimous decisions, that is, how to treat certainty when uncertainty is what is needed. This problem is intimately bound up with the more general question of when to use a particular scaling method.

2. Which Method Should Be Used When?

Interval scales can be determined psychophysically by two general methods: (1) those that invoke either Thurstone's law of comparative judgments or Torgerson's law of categorical judgments as theories for inferring perceptual differences from discriminability (these include the paired-comparison method and the rank-ordering, rating-scale, and category methods when the data are reduced by drawing on one of these theories that postulates a relationship between discriminability and perceived difference); and (2) those that make the assumption that the observer is able to respond directly in a manner that does not violate the required properties of an interval scale (rating-scale and category methods that do not rely upon discriminability for data analysis belong in this category). Categorized in this way, it is clear that the distinction resides with the matter of discriminability. That property, in turn, depends on the variability of observations.

When there is no variability, a theory for relating discriminability to perceived difference is of no practical use, no matter how elaborate, logical, and appealing the theory may be. It simply does not apply. Accordingly, the alternative is to use a method of the second category elucidated above, one that does not depend on discriminability for establishing an interval scale. The category-means method of categorical scaling or the simple rating-scale approach will suffice in such cases. One might ask, with some justification, that if there is no doubt about the outcome (there is no uncertainty and, hence, no discriminal dispersion of responses), what is to be learned beyond simple order? It may be that rank ordering will supply all that need be known. Possibly, we do not even need an experiment at all (remember that observations require labor and take time, so we should be reasonably certain that the cost of the experiment is worth the additional information that results from it). On the other hand, there may be a genuine need to derive a quantitative psychophysical relationship. Perhaps the number of stimuli is fixed; we cannot add more to increase the likelihood of creating uncertainty. Perhaps the range of variation in stimulus attributes is so large that the labor that would result from increasing the number of stimuli would be prohibitively costly. There are, in fact, many reasons why we might want to attempt to derive a quantitative scale even though there is no prospect of encountering uncertainty in the data. When this is the case, our recourse for an interval scale is to one of the more simplistic techniques; if a ratio scale (which will not necessarily yield interval-scale properties) is desired, we can turn to ratio-scaling techniques.

When there is ample uncertainty, then the scaling methods and data analysis techniques that rely on a theoretical relationship between discriminability and perceived difference are better choices. Some of the same experimental methods used when there is no uncertainty can also be used when uncertainty abounds: rank ordering, rating scales, and category methods are examples. The difference is only in the way the data are treated. In addition, the paired-comparison method can be used. The major advantage of the paired-comparison method is that it provides greatest "leverage". Using the paired-comparison experimental method is somewhat like placing stimulus differences under a magnifying glass in that very small differences can be discriminated (in fact, when the differences are not small enough, the method does not work very well). The price of this extra magnification is experimental labor. Paired comparisons require significantly more time and labor than the other methods discussed in this chapter. However, with modern sorting algorithms, it is possible to minimize the additional cost of a paired-comparison experiment. When the discriminal dispersions are ample and ubiquitous, there is no better method than paired comparisons (I have consistently obtained scale precisions of 1 to 2% with paired-comparison assessment of "well-confused" data). It requires little of observers (only ordinal judgments between members of pairs of stimuli) and yields precise results with high discrimination capability.

In practice, discriminability is neither completely lacking nor totally ubiquitous. Choice of method would be simple if that were the case, but it is not often so. Instead, we usually find that there is confusion throughout much of the data matrix but not all of it. The choices here are more numerous and the reasons for them not always so obvious as in the two cases just discussed. Perhaps only one or two stimuli cause the certainty. We might want to discard them and treat the remaining, confused stimuli. We might choose to use one of the several techniques for dealing with missing elements in the data matrix. We might decide that the best course of action is to treat the experiment as a pilot study and design a more appropriate experiment, which we hope will yield less equivocal results. In short, we must bring to bear some understanding of interval-scaling methods and associated techniques for data analysis together with imagination in an attempt to produce results that are most useful. Such considerations should also take account of the cost of experimentation. Needs must be balanced against cost.

3. Some Further Considerations about Good Experimental Practice

The best way to get your money's worth out of a psychophysical experiment is to design it carefully and conduct it carefully. The question that precipitates the experiment should be clear and as simple as possible; it usually is not a good idea to try to answer several questions with a single experiment if, in the process of doing so, the best method for addressing any one of the questions must be compromised. Given a clear and succinct statement of the problem, the experimental method that addresses the question most directly and simply should be carefully detailed. Don't go hunting for a mouse with an elephant gun; if rank ordering will provide the needed information, do not use paired-comparison or successive-category methods. In general choose the simpler method whenever there is a choice. As we have seen, simple methods may yield as much or almost as much information as more complicated ones. Finally, be meticulous about the conduct of the experiment. Select observers carefully according to a sampling plan just as the stimuli are selected by the most appropriate sampling plan. Select enough observers. One prestigious journal used to reject arbitrarily any paper submitted to it if the experiment reported did not include at least 30 observers. The more observations the better, from the standpoint of statistical considerations alone, but there is also a matter of sampling a wide enough variety of opinions for the results to be reasonably representative, so both the number of observers and the number of replications should be as large as practically possible. Compose instructions with care to make sure that they ask the question you want to ask and nothing more. Keep them simple, clear, and concise. Try not to impose on the observers to discount extraneous qualities of the stimuli (it is better to eliminate them yourself if it is possible to do so) or to abstract more than one thing at a time from a welter of dimensions of variability. Be casual rather than explicit about the definition of the attribute that you want to scale; the observer may have an entirely different concept of what certain words mean. Use as few anchors and moduli as possible. In short, let the observer tell *you*, with his response, what *he* thinks. Your job as experimenter is to unravel the meaning of the results, not to impose your preconceived notions on the observers.

G. CONCLUSIONS

When properly practiced, the methods outlined here are capable of providing useful interval scales of suprathreshold perceptual differences. These scaling methods are probably the most widely used techniques for determining psychophysical interval scales. A carefully designed and conducted experiment yields information about the ratios of differences among stimuli. The results do not tell us anything about how much greater one thing is than another; we need a ratio scale for that. Also, the results represent a projection onto a single continuum. Sometimes this is wholly appropriate, as when we wish to determine "quality" of photographs or television images where many factors affect quality, but we seek a unidimensional characterization of only the affective value of the images. Sometimes both stimulus and response dimensionality are singular, as when we wish to scale the perceived brightnesses of lights of the same correlated color temperatures and spatial extents, presented to the same areas of the observer's retinae under constant viewing conditions. At other times, unidimensional scales may not be appropriate. There may be more than one response dimension involved, and observers may differ in the ways they weight the various dimensions, or the same observer may change the way he weights them at different times. Often these problems show up as inconsistencies in the data matrix that have been called triangle inequalities. Often they may be hidden in the apparent imprecision of the data. We need to use multidimensional scaling methods to treat such problems adequately. However, wherever a unidimensional result is appropriate, interval-scaling techniques offer a rich variety of choices for determining a psychophysical scale.

REFERENCES

Attneave, F. (1959). "Applications of Information Theory To Psychology." Holt, Rinehart and Winston, New York.

Baird, J. C., and Noma, E. (1978). "Fundamentals of Scaling and Psychophysics." John Wiley and Sons, New York.

Bartlett, R. J. (1940). Measurement in psychology. *Rep. Br. Assoc. Adv. Sci.* **1**, 422–441.

Berkson, J. (1953). A statistically precise and relatively simple method of estimating the bio-assay with quantal response, based on the logistic function, *J. Am. Stat. Assoc.* **48**, 565–600.

Bock, R. D., and Jones, L. V. (1968). "The Measurement and Prediction of Judgment and Choice." Holden-Day, New York.

Boring, E. G. (1920). The logic of the normal law of error in mental measurement. *Am. J. Psychol.* **31**, 1–33.

Boring, E. G. (1942). "Sensation and Perception in the History of Experimental Psychology." Irvington, New York.

Boring, E. G. (1950). "A History of Experimental Psychology." 2nd ed. Appleton-Century-Crofts, New York.

Bradley, R. A., and Terry, M. E. (1952). The rank analysis of incomplete block designs, I. The method of paired comparisons. *Biometrika* **39**, 324–345.

Brownlee, K. A. (1949). “Industrial Experimentation.” Chemical Publishing, New York.

Burke, C. J. (1953). Additive scales and statistics. *Psychol. Rev.* **60**, 73–75.

Buyhoff, G. J., and Hull, R. B., IV (1980). Computation of law of comparative judgment scales on a microcomputer. *Behav. Res. Method Instrum.* **12**, 465.

Cattell, R. B. (1944). Psychological measurement: Normative, ipsative, interactive, *Psychol. Rev.* **51**, 292–303.

Chapanis, A. (1959). “Research Techniques in Human Engineering.” Johns Hopkins University Press, Baltimore.

Cochran, W. G., and Cox, G. M. (1957). “Experimental Designs,” 2nd ed. John Wiley and Sons, New York.

Comrey, A. L. A. (1950). Proposed method for absolute ratio scaling. *Psychometrika* **15**, 317–325.

Coombs, C. H. (1964). “A Theory of Data.” John Wiley and Sons, New York.

Coombs, C. H., Raiffa, H., and Thrall, R. M. (1954). Some views on mathematical models and measurement. *Psychol. Rev.* **61**, 132–144.

Cox, D. R. (1958). “Planning Experiments.” John Wiley and Sons, New York.

Dawkins, R. (1969). A threshold model of choice behavior. *Anim. Behav.* **17**, 120–133.

Dixon, W. J., and Massey, F. J., Jr., (1957). “Introduction to Statistical Analysis.” McGraw-Hill, New York.

Draper, N., and Smith H. (1981). “Applied Regression Analysis,” 2nd ed. John Wiley and Sons, New York.

Edwards, A. L. (1951). “Psychological Scaling by Means of Successive Intervals.” University of Chicago Press, Chicago.

Edwards, A. L. (1960). “Experimental Designs in Psychological Research.” Rinehart, New York.

Fisher, R. A. (1951). “The Design of Experiments.” Oliver and Boyd, Edinburgh and London.

Freyd, M. (1923). The graphic rating scale. *J. Educ. Psychol.* **14**, 83–102.

Guilford, J. P. (1954). “Psychometric Methods,” 2nd ed. McGraw-Hill, New York.

Gulliksen, H. (1950). “Theory of Mental Tests.” John Willey and Sons, New York.

Gulliksen, H. (1954). A least squares solution for successive intervals assuming unequal standard deviations. *Psychometrika* **19**, 117–139.

Gulliksen, H. (1956). A least squares solution for paired comparisons with incomplete data. *Psychometrika* **21**, 125–134.

Kempthorne, O. (1952). “The Design and Analysis of Experiments.” John Wiley and Sons, New York.

Kling, J. W., and Riggs, L. A. (eds.) (1971). “Woodworth and Schlosberg's Experimental Psychology,” 3rd ed. Holt, Rinehart and Winston, New York.

Lindquist, E. F. (1953). “Design and Analysis of Experiments in Psychology and Education.” Houghton-Mifflin, Boston.

Luce, R. D. (1959). “Individual Choice Behavior.” Van Nostrand, New York.

Maxwell, A. E. (1974). The logistic transformation in the analysis of paired-comparison data. *Br. J. Math. Stat. Psychol.* **27**, 62–71.

Miller, G. A. (1956). The magical number seven, plus or minus two: Some limits on capacity for processing information. *Psychol. Rev.* **63**, 81–97.

Mosteller, F. (1951a). Remarks on the method of paired comparisons: I. The least squares solution, assuming equal standard deviations and equal correlations. *Psychometrika* **16**, 3–9.

Mosteller, F. (1951b). Remarks on the method of paired comparisons: II. The effect of an aberrant standard deviation when equal standard deviations and equal correlations are assumed. *Psychometrika* **16**, 203–206.

Mosteller, F. (1951c). Remarks on the method of paired comparisons: III. A test of significance when equal standard deviations and equal correlations are assumed. *Psychometrika* **16**, 207–218.

Mosteller, F., and Tukey, J. W. (1977). "Data Analysis and Regression." Addison-Wesley, Reading, Mass.

Myers, J. L. (1966). "Fundamentals of Experimental Design." Allyn and Bacon, Boston.

Newhall, S. M. (1939). The ratio method in the review of the Munsell colors. *Am. J. Psychol.* **52**, 394–405.

Norwich, K. H. (1981). The magical number seven: Making a "bit" of "sense". *Percept. Psychophys.* **29**, 409–422.

Savage, C. W. (1970). "The Measurement of Sensation." University of California Press, Berkeley.

Schiffman, S. S., Reynolds, M. L., and Young, F. W. (1981). "Introduction to Multidimensional Scaling: Theory, Methods, and Applications." Academic Press, New York.

Shell, D. L. (1959). A high-speed sorting procedure. *Commun. ACM*, **July 1959**, 30–32.

Sidowski, J. B. (ed.) (1966). "Experimental Methods and Instrumentation in Psychology." McGraw-Hill, New York.

Stevens, S. S. (1946). On the theory of scales of measurement. *Science* **103**, 677–680.

Stevens, S. S. (1951). Mathematics, measurement, and psychophysics. *In* "Handbook of Experimental Psychology" (S. S. Stevens, ed.), pp. 1–49. John Wiley and Sons, New York.

Stevens, S. S. (1975). "Psychophysics: Introduction to its Perceptual, Neural, and Social Prospects." Wiley, New York.

Thurstone, L. L. (1927). A law of comparative judgment, *Psychol. Rev.* **34**, 273–286.

Titchner, E. B. (1905). "Experimental Psychology." McMillan, New York.

Torgerson, W. S. (1954). A law of categorical judgment. *In* "Consumer Behavior," L. H., Clark, (ed.), pp. 92–93. New York University Press, New York.

Torgerson, W. S. (1958). "Theory and Methods of Scaling." John Wiley and Sons, New York.

Whaley, C. P. (1979). Collecting paired-comparison data with a sorting algorithm. *Behav. Res. Method Instrum.* **11**, 147–150.

Winer, B. J. (1971). "Statistical Principles in Experimental Design." McGraw-Hill, New York.

Woodworth, R. S., and Schlosberg, H. (1954). "Experimental Psychology, 2nd ed. Holt, Rinehart and Winston, New York.

9

Direct Ratio Scaling

C. J. BARTLESON
Eastman Kodak Company
Research Laboratories
Rochester, New York

A. INTRODUCTION

1. General Considerations

Chapters 7 and 8 have dealt primarily with measurement methods that are based on discrimination. Only the simple rank order and those rating and category scaling methods that rely on average responses fall within the province of what might be called *direct scaling*, the kind of scaling that does not invoke assumptions about discriminal dispersions. Direct scaling takes the observer's estimate of a magnitude* (or his adjustment of stimulus intensity to

* The usual way in which such scales are described, and the very name of the class of scaling methods to which they belong, involves the term *magnitude*. I will continue that common practice throughout this chapter. However, it is not magnitudes that are specified, although this has been claimed by some (see, Stevens, 1975; Zwislocki and Goodman, 1980). What is specified is a ratio, the ratio in which one sensation stands with respect to some other implicit or explicit sensation. Shepard (1981, p. 40) has stated that magnitude-estimation data show that in the method "it is equal ratios of physical magnitude that are psychologically equivalent," or expressed in functional notation: $R(S_i, S_j) = R(kS_i, kS_j)$. In the general case this expression amounts to a functional relation between response and ratios of stimulus intensities: $R(S_i, S_j) = f(S_j/S_i)$. In short, a minimum of two stimuli must be involved (even if only by implication) and what is determined is the *perceived ratio* in which they stand; the magnitude of neither member of the pair is determined directly.

ISBN-0-12-304905-9

match a magnitude number) as the indication of the size of his response. This assumption that observers can make proper evaluations of the magnitudes of their sensory experiences is a fundamental point of difference between direct scaling and indirect scaling based on discrimination. It is also a controversial point. These are matters that are discussed in detail in Luce, Bush, and Galanter's *Handbook of Mathematical Psychology* (1963–1965).

Direct scaling has a long history in psychophysics. Plateau (1872) used a method of bisection in which an observer was assumed to be able to recognize a sensation that was half as large as another through the simple expedient of matching two differences. The reciprocal process was introduced a few years later (Merkel, 1888, 1889) under the name "die Methode der doppelten Reize" (the method of doubling stimuli), but it is clear that what was meant was doubling sensations, not stimuli. In this method, matching sensations of the same kind was not involved. Fullerton and Cattell (1892) used a method in which observers were instructed to produce stimuli that elicited both multiples and fractions of the sensation for a standard stimulus (although the method has usually been referred to as the *fractionation method* of scaling). All of these methods involve the assumption that sensations can be addressed directly, unlike Fechner's (1860) belief that sensations *per se* could not be measured; the unit of sensation could be determined only from discriminal dispersions.

This dichotomy of attitudes is one that has generated considerable argument over the years. In 1932 a committee was established by the British Association for the Advancement of Science to determine whether it is possible to make useful quantitative estimates of sensory events. The committee's final report (Ferguson, 1940) indicates the committee's inability to reach a consensus. A number of minority reports were also generated (for instance, Gage, 1934; Campbell, 1938), some of them harshly polemic. The central question to the arguments about direct and indirect measurements of sensory events can be reduced to a question of how measurement is defined. Generally the various definitions of measurement fall into or closer to one of two broad categories; they have been called narrow and broad views of measurement. In the narrow view, measurement is the assignment of numbers to objects or events along some dimension by comparing them to units along some other dimension through the operation of adding these units together. Additivity plays a paramount role in this definition. The broad view defines measurement as the assignment of numbers to objects or events according to any consistent rule or set of rules. Additivity does not play a crucial role; instead, the requirement is only that measurement scales be isomorphic or homomorphic with the objects or events under consideration.

It is no accident that the broad or liberal definition of measurement was proposed by Stanley Smith Stevens (1946). He was the leading proponent of direct scaling during the 20th century. Stevens (1975) credits the rapid development of direct ratio scaling in the 1930s to the invention of the telephone:

> It was the invention of the telephone, with its attendent development of electrical instruments, that set in motion a revival of Merkel's procedure of ratio production. Merkel had produced a sound intensity by dropping a small metal ball from a given height onto a block of ebony. He controlled the sound intensity by varying the height of the fall—a control that was uncertain at best. The invention of electrical oscillating circuits and telephone receivers injected a new order of precision and convenience into auditory experiments. [p. 20]

By 1938 Stevens was able to report a fundamentally significant finding from his and others' work on ratio scaling of audition: that observers can judge the ratio of two sensations and that they can do so with reasonable consistency (Stevens and Davis, 1938). He spent the remaining 37 years of his life developing a schema of direct scaling that is widely accepted throughout the field of psychophysics today.

2. Direct Ratio Scaling

There are two broad classes of ratio scaling: Magnitude estimation and magnitude production. The two methods are simply complementary ways of addressing the same question: How many times greater is one thing than another?

a. Magnitude Estimation

The scaling procedure called *magnitude estimation* involves asking an observer to match a number to the perceived magnitude of the attribute under test when the stimulus is presented by the experimenter. One early application of the method presented two sounds in rapid succession and asked observers to tell how much louder or softer the second sound was by stating the ratio in which the two sounds stood (Richardson and Ross, 1930); another presented sounds that decreased in loudness by successive steps and asked observers to state what percentage each was of the original sound (Ham and Parkinson, 1932). Stevens (for example, 1934, 1936, 1955a) developed the method to the point where sounds were presented one at a time in random order; the observer's task was to assign a number to each sound to match its loudness. Not only numerosity but also other sensory attributes were matched with sounds (Stevens, 1966). Although in his early work Stevens used anchors and moduli, he later found that the results were not only as consistent but less biased, for technical reasons, when the observer was allowed to use any numbers he felt comfortable with and was not forced to refer to an anchor (for example, Stevens, 1975).

A typical set of instructions would then be of the following kind:

> You will be presented with each, in turn, of a series of circular lights. We would like you to tell us how bright each light is by assigning a number to its brightness. Call the first light any number that seems appropriate to you. Then assign numbers to successive lights in such a way that they reflect your impression of how bright they are. There is no limit to the range of numbers that you may use. You may use whole numbers, fractions, or decimals. But try to make each number match the brightness that you see.

The raw data are recorded as number-magnitude. They are averaged by taking their geometric means, regardless of whether replications are over observations or subjects (or both). The reason for this is that the dispersions tend to be normally distributed over the logarithms of responses rather than their arithmetic values (Stevens, 1955b). This is what we might expect if observers were dealing with ratios; equal intervals of logarithms correspond to equal ratios. The general finding for psychophysical scales relating response to stimulation that are determined in this way is that either individual or pooled responses tend to form a power function of some general kind. That is, we may characterize the results of magnitude scaling as

$$R = \alpha S^{\beta} + \gamma, \tag{1}$$

where R = response, S = stimulus intensity, and α, β, and γ are constants that depend upon several factors.

The general form of Eq. (1) can be used to characterize a number of scale forms. Some of them are shown in Table I. Stevens (for example, 1975) and others (for example, Marks, 1974a) generally have referred to the results of direct-magnitude experiments as following the psychophysical power law. However, almost all such results can be represented best by one or the other expressions that are referred to in Table I as logarithmic interval scales. Curiously, Stevens (1957, p. 176) called logarithmic interval scales "empirically useless," although essentially all his magnitude-scaling results fell into the class I or II log interval scale group. He referred to them as "ratio scales," treating the log interval scale as more or less a theoretical curiosity. Shepard (1981) has pointed out that Stevens' original classification of scale types (Stevens, 1946) would require that a ratio scale (as distinct from a judgment) must have the property of invariance for both ratios of differences and ratios of magnitudes. The log interval scale does not. It has invariance only for ratios of magnitudes (for instance, $R_1/R_2 = R_3/R_4$ = constant which corresponds to ratios of *logarithmic* intervals [$(\log R_1 - \log R_2)/(\log R_3 - \log R_4)$ = constant]).

Stevens himself (for example, 1975; Stevens and Galanter, 1957) and many others have found that for certain kinds of sensory attributes, interval scales and log interval scales (alias "ratio scales" in Stevens' terminology) are nonlinearly related. In many instances the first is approximately the square

TABLE I. Scale Types of the Form $y = \alpha x^{\beta} + \gamma$

Scale Name	Degrees of freedom	Parameter constraints	Parameters to be determined	Scale invariance	Permissible transformations
Absolute	0	—	—	$\frac{y}{x} = 1$	$y = x$
Difference	1	$\alpha = \beta = 1$	γ	$y - x = \gamma$	$y = x + \gamma$
Ratio	1	$\beta = 1$ $\gamma = 0$	α	$\frac{y}{x} = \alpha$	$y = \alpha x$
Interval	2	$\beta = 1$	α, γ	$\frac{\Delta y}{\Delta x} = \alpha$	$y = \alpha x + \gamma$
Power I	1	$\alpha = 1$ $\gamma = 0$	β	$\frac{\log y}{\log x} = \beta$	$y = x^{\beta}$
Power II	2	$\alpha = 1$	β, γ	$\frac{\log(y - \gamma)}{\log x} = \beta$	$y = x^{\beta} + \gamma$
Log interval I	2	$\gamma = 0$	α, β	$\frac{\Delta \log y}{\Delta \log x} = \beta$	$y = \alpha x^{\beta}$
Log interval II	3	—	α, β, γ	$\frac{\Delta \log(y - \gamma)}{\Delta \log x} = \beta$	$y = \alpha x^{\beta} + \gamma$

root of the second (for instance, Marks, 1974b). This seems to be true only for those classes of continua that have been called *prothetic*. These are basically intensive continua for which the standard deviations of responses tend to be proportional to response magnitude. In such cases complementary attributes are related as reciprocals; for example, lightness is found to be the reciprocal of darkness, roughness the reciprocal of smoothness, and loudness of softness, and so on for other opposites (Torgerson, 1961; Stevens and Guirao, 1962; Stevens and Harris, 1962; Stevens, 1975). Expressed differently, the logarithms of lightness and darkness, are complements. This is not so with other sensory attributes classed as *metathetic*, which are basically those for which qualitative or substitutive changes occur and for which the standard deviations of responses tend to be constant regardless of response level. In these, opposites are complementary, and log interval scales are linearly related to interval scales. The situation is not quite as clear-cut as this because of differences found among observers (for instance, Bartleson and Breneman, 1973) and concepts represented by different kinds of experimental instructions (for instance, Indow, 1974).

The distinction between type I and type II log interval scales is one of induction. The type I form generally applies when there is no induction and type II when there is induction. Figure 1 shows an example of magnitude-estimation results for scaling the brightnesses of small areas under conditions of dark surround, without induction, and of the same areas presented against

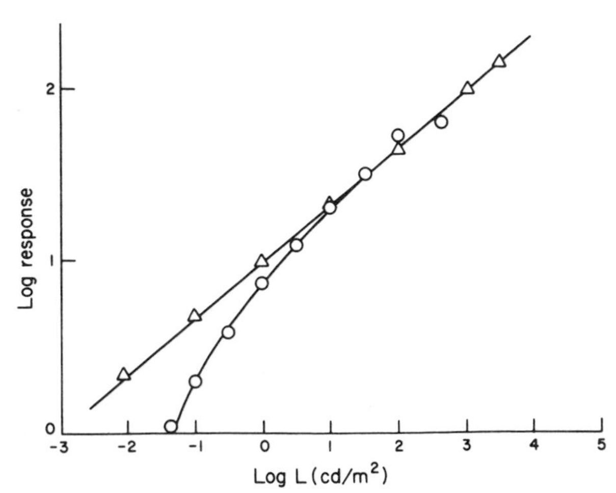

Fig. 1. Scaled lightness (log response) as a function of log luminance (log L in cd · m^{-2}). Triangles represent data obtained with dark surround. Circles represent data obtained with illuminated surround providing induction.

an illuminated surround, with induction (Bartleson, 1980). The induction (caused by the illuminated surround, in this case) depresses log responses more at the low end of the scale than at the high end. This effect has variously been called contrast, inhibition, induction, masking, and adaptation (Stevens, 1975). Stevens has represented its effect on the psychophysical function essentially as a threshold phenomenon, in which the stimulus increment above threshold is used to characterize intensity of stimulation rather than the absolute stimulus level. His expression is

$$R = \alpha(S - S_o)^\beta, \tag{2}$$

where S_o is somehow related to the threshold stimulus for the conditions under consideration. This "correction" or adjustment technique, introduced by Ekman (1956), has been called the *stimulus correction*, for which the size of β varies with induction. Jameson and Hurvich (1964) prefer a mathematically equivalent expression which they call the *response correction* method, in which β remains constant (at the threshold value) and induction is accounted for by different values of γ, following the form of Eq. (1).

Numerosity is most often used as the response data of magnitude estimation. Several experiments have provided information that suggests that our perceptions of numerosity are not linear with number (for example, Attneave, 1962; Schneider and Lane, 1963; Marks, 1968, 1974b; Banks and Coleman, 1981); but if we consider the assignment of number to represent response attributes on a continuum corresponding to some other mode of perception as merely another form of cross-modality matching (Stevens, 1966), then we can identify perceived numerosity as an arbitrary standard for scales of response ratios, so that intercomparisons of such scales have relative utility. There are many ways in which observers can misuse numbers, but the compelling aspect of the nearly 50-year history of experiences with modern magnitude-estimation experiments is how very consistent observers are in their replicate use of numbers. Although standard deviations of 50% are sometimes encountered, my experience has been that carefully controlled magnitude-estimation experiments commonly yield standard deviations in the range of 8 to 18%, only about twice that of interval scaling or roughly four times that of threshold scaling methods. Generally the precision of the results of an experiment will be dramatically improved by a small investment in time to train observers to use numbers to represent ratios. So much has been written about the theory of direct scaling that there sometimes seems to be a danger that the overlay of theoretical complexity may mask the underlying simplicity of the experimental method. It is a simple process, so straightforward that even five year old children have been able to perform cross-modality matching of loudness and brightness with results about as precise as those of adults (Bond and Stevens, 1969). Similar results have been obtained in a number of other experiments (Anderson, 1979). In my own work with the method, I have found little or no difference between results from expert (psychophysically experienced) and inexpert (psychophysically inexperienced) observers; if there has been any trend, it is a tendency for nonscientists to form a more consistent group then scientists. The casual observer performs better than the one who studies and agonizes over every decision.

b. Magnitude Production

Whereas magnitude estimation requires the observer to assign a number to a stimulus, *magnitude production* requires that the observer adjust the intensity of a stimulus to match his perception to a number. One method is the logical complement of the other. The experimeter calls out a number, and the observer adjusts stimulus intensity to provide a match. Generally the process requires some preliminary experimentation to determine an appropriate range of numbers. The numbers are presented in random order, but they should comprise a geometric series when ordered. The problem with this method is that the numbers may not be natural to the observer. That is, left to his own devices, he might choose to use lower or higher numbers, and the range of numbers used by the experimenter may be greater or smaller than that which the observer would choose himself.

Usually the results differ between the two methods of ratio scaling. When psychophysical functions from each method are plotted in double logarithmic coordinates, the slope of the magnitude-production results tends to be slightly higher than that for the estimation method, a result that has been called a *regression effect*. Stevens (1975) suggests that both methods be used and the data combined to obtain the best estimate of the "true" psychophysical function. Although that is generally good practice, it is not always possible to perform a magnitude-production experiment; for example, the stimuli may be fixed in intensity and limited in number. There may also be a question about whether the numbers used by the experimenter are congruent with the observer's natural desires; if not, there is a possibility of biasing or distorting the results from magnitude production.

c. Cross-Modality Matching

Although assignment of numbers to perceived stimulus intensities may be considered a common form of cross-modality matching, there are other forms of the method that do not require the observer to respond verbally or in any other way that specifies numerosity. These other methods involve making direct matches between perceptions along two qualitatively different sensory continua. The observer is provided with some means for adjusting the intensity of the stimulus on the second continuum. He is presented with a stimulus to be measured and adjusts the intensity of the matching stimulus until he is satisfied that the two perceived intensities are equal. The data are recorded either as stimulus intensities of the (standard) matching stimulus or preferably as the perceived magnitude corresponding to the matching stimulus intensity determined by ulterior means.

Stevens and his colleagues have used force of handgrip and loudness of sounds for many cross-modality matching experiments (Stevens, 1975). Sound intensity of a random ("white") noise signal makes a good standard for cross-modality matching in visual experiments. There are several reasons for this. First, the perceptual scale of loudness, called *sones*, where perceived loudness increases as approximately the $\frac{2}{3}$ power of sound pressure, is well established so that reasonable approximations to perceptual ratios can be made without necessarily having to establish the psychophysical function relating loudness to sound pressure [for reviews of many experiments see Stevens (1975) or Marks (1974a)]. Experimental methods in which loudness is used as the matching attribute are also convenient. The observer can wear earphones without being distracted or causing distraction. Audio generators are readily available and easy to adjust for intensity of their output. Either a white-noise generator or a sine-wave generator set to about 400 or 1000 Hz serves well as a stimulus-generating device. It is important, however, that problems resulting from hearing losses be avoided (if the observer has diminished sensitivity at the sound frequency used, he may run out of adjustment range on the audio generator, and his data will differ in level from those of others, although his hearing impairment should not affect the ratios of his matches if the range of adjustment is adequate). For this reason there is a marginal advantage to the use of white-noise generators. Cross-modality matches between loudness and other intensive attributes such as brightness and lightness are easy to make, but matching loudness to colorfulness and, especially, to hue is considerably more difficult.

3. Some Problems with Direct Scaling

Interval-scaling methods are very sensitive to adaptation. Different results are often obtained for different distributions of stimuli. Different results are also obtained when stimuli are presented systematically in different orders, which has been called the *hysteresis effect*. However, when interval scales of prothetic and metathetic continua (for example, loudness and pitch or colorfulness and hue) are compared, they tend to be linearly related. This is not generally true for the results of magnitude estimation or production experiments. That is perhaps the single most significant problem with the direct-scaling methods, for it raises questions of validity.

There are a number of ways in which direct magnitude scales are distinguished from interval scales with regard to their outcomes for prothetic and metathetic continua. These are bound up with the distinctions between the two kinds of continua. Prothetic continua involve quantity or intensity; metathetic continua concern quality or location. The relative errors of scales for prothetic continua tend to be constant, whereas the absolute errors are constant for metathetic continua. Errors are distributed normally as logarithms in prothetic cases and arithmetically in metathetic cases. There is a strong hysteresis effect for prothetic continua and little or none for metathetic continua. Prothetic and metathetic continua are nonlinearly related in a magnitude-scaling results and linearly related in interval-scaling results. There have been many essays on possible theoretical differences between the two kinds of scaling (for example, Savage, 1970; Krantz, 1972; Marks, 1974a; Stevens, 1961, 1975; Torgerson, 1961; Shepard, 1981; Banks and Coleman, 1981) together with claims and counterclaims for which method provides the "true" scale, but experimental data do not yield a means for choosing between them on the basis of validity. The theoretical issues remain unresolved.

We are left with the empirical fact that interval scales provide measures of differences but do not yield ratios that agree with those found by direct ratio estimation, and scales of the magnitude-estimation variety provide indications of ratios but do not yield differences that agree with those found by interval-scaling methods. Except for the empirical observation that the square roots of magnitude estimates tend to agree with differences from interval-scaling experiments, there seems to be no practical basis for relating the results of one kind of experiment to those of the other. There is no straightforward way to examine the validity of either method.

Some workers, myself among them, take the position that both methods can provide useful results, both being merely indications of underlying perceptual relationships, and the fact that they do not permit us to use all the rules of arithmetic and algebra does not detract greatly from their utility as indicators of response relationships. I have frequently compared the square roots of magnitude estimations to interval-scale differences to see whether there is reasonable correspondence. When there is (which is what I have found for most well-controlled experiments), I ascribe no special significance to it, and certainly no theoretical implications, but I am satisfied that two such different methods tend to converge on the same or a similar solution. In fact, convergence of results from a variety of experimental approaches to the same problem is possibly the most comfort that can be drawn from psychophysical experimentation. When there is convergence toward a common solution, we tend to accept the results as valid; when there is not such convergence, we suspect the results.

B. AN EXAMPLE OF MAGNITUDE ESTIMATION OF COLORFULNESS

1. Design of the Experiment

The nine colored papers that were used in the common experimental examples in Chapter 8 were used again in a magnitude-estimation experiment. They were scaled individually by each of 25 observers with 4 replications each (a total of 100 observations for each paper). The instructions were as follows:

> You will be presented with each, in turn, of a series of colored papers. Each paper is a different color. We would like you to tell us how colorful each paper is by assigning a number to its colorfulness. Call the colorfulness of the first paper any number that seems appropriate to you. Then assign numbers to successive papers in such a way that they respresent your impression of how colorful they are. If the second paper is twice as colorful as the first, the number you assign to it should be twice as large; if it is only one-fourth as colorful, the number should be only one-fourth as large; and so on. There is no limit to the range of numbers that you may use; the scale does not have an upper limit. You may use integers, fractions, or decimals. If you see no colorfulness, then assign the number zero. Just try to make the numbers match the colorfulness that you see in each paper.

The viewing conditions were the same as those used in Chapter 8. The observers were some of the same ones who performed the rank ordering of the nine colored papers.

The experimental data were recorded, and geometric means were computed for each stimulus:

$$\bar{R}_i = \left(\prod R_i\right)^{1/n}, \tag{3}$$

where n represents the number of observations, i stands for a particular stimulus, and $\prod$ designates cumulative products. Equation (3) can be expressed in log form as a summation,

$$\overline{\log R_i} = (1/n)\left(\sum \log R_i\right), \tag{4}$$

to compute the mean log response.

The mean responses over all 100 observations are listed in Table II. One of the entries in the table is not a geometric mean: that for stimulus number 1, which was a nominally neutral (gray) sample. Some of the 100 observations of that stimulus resulted in zero responses. Geometric means cannot be computed for any sample where even one response is zero; it takes only one zero to make the geometric mean equal zero, so Eq. (3) cannot yield a true measure of the central tendency in such cases.

A different procedure must be used when there are zeros. Usually the best choice is to determine the mode (the most frequent) of the distribution of responses. If the raw data do not exhibit a singular mode, a log normal

TABLE II. Summary of Results of Magnitude-Estimation Scaling of Colorfulness

Sample	$\bar{R}$ Geometric mean response	$\bar{R}^{1/2}$	$\log \bar{R}$	Standard deviation of $\log \bar{R}$
1	[1.0][a]	1.0	0.000	—
2	127.5	11.3	2.106	0.148
3	60.1	7.8	1.779	0.104
4	12.5	3.5	1.097	0.073
5	3.5	1.9	0.544	0.023
6	50.4	7.1	1.702	0.121
7	121.6	11.0	2.085	0.139
8	33.6	5.8	1.526	0.103
9	101.8	10.1	2.008	0.114

[a] mode

distribution may be calculated to fit the raw data and its mode determined. In the case of stimulus 1 it was possible to determine the mode from the raw data.

The table also lists the square roots of the mean responses, the logarithms of responses, and associated standard deviations. This method has been shown to provide good estimates of the confidence limits and dispersion of the geometric mean (Alf and Grossberg, 1979) even when the data do not exactly form a log normal distribution (for instance, Luce and Mo, 1965). The latter question can be evaluated by a variation on the analysis of variance for testing goodness of fit (Coleman *et al.*, 1981). The standard deviations of the mean log responses vary from about 1.1 to 30.1% of response when converted to arithmetic values. On average, they are about 6.5% of response value. If the logarithmic standard deviations were all constant (say, at about their average of 0.103), then the scale errors would be a constant proportion of response level. That is not quite the case; their values tend to be somewhat higher for the lower response values. Despite this apparent failure of the observers to use ratios the same way near the zero end of the scale as they did at higher response levels [a situation encountered in a number of experiments (Bartleson, 1970)], the data of Table II are generally quite satisfactory representations of typical magnitude-estimation results.

2. Results of Magnitude-Estimation Scaling of Colorfulness

Munsell chroma (Munsell, 1909) is now specified according to data that were collected by an interval-scaling technique (Newhall, 1940; Newhall *et al.*,

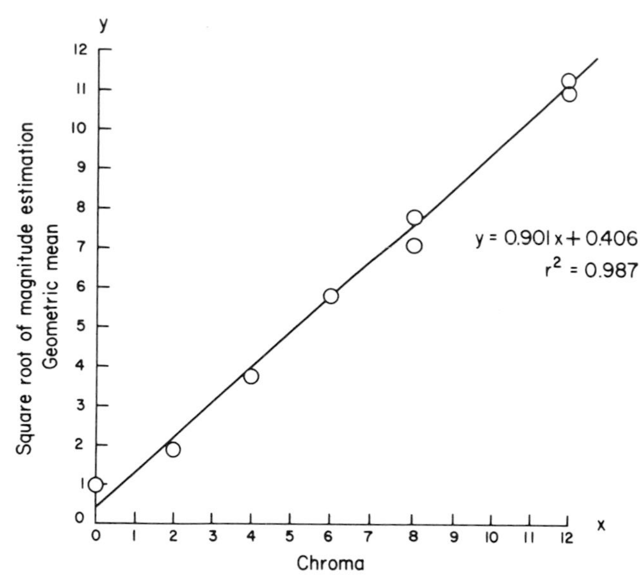

Fig. 2. Comparison of the square roots of magnitude estimation geometric mean estimates of colorfulness with nominal chroma of samples.

1943). We should, therefore, not expect the geometric-mean magnitude estimations to be linearly related to Munsell chroma. However, if the approximate square-root relationship between interval and direct ratio estimation results holds true for chroma, we might expect the square roots of the geometric means to be linearly related to chroma.

Figure 2 shows that this was the case. The square roots of $\bar{R}$ values have been plotted against original chroma, and the result is essentially a linear relationship. The coefficient of determination is reasonably high: $r^2 = 0.987$. By chance the raw square-root values are themselves very close to the actual chroma values; that is, the slope of the regression equation shown in Fig. 2 is near unity, and the intercept is near zero.

We can compare the estimated chromas from this direct ratio scaling experiment with the chroma values estimated from the various interval-scaling methods described in Chapter 8. This has been done in Table III, where each of the estimations is shown together with the nominal chroma values for each sample. The square roots of the sums of squares of deviations of the estimates from the nominal chroma are shown at the bottom of Table III. That for the magnitude estimation results is just midway along the rank order of these deviations for all seven methods listed. Accordingly, it should be reasonable to conclude that the magnitude estimation data (expressed as square roots) are about as accurate as the interval-scaling methods described in Chapter 8.

TABLE III. Comparison of Estimated Chroma Values

Sample	Nominal chroma	Normalized ranks	Comparative ranks	Paired comp. logit	Rating means	Category means	Categorical judgment	Magnitude estimation
1	0	0.1	−0.8	0.5	0.1	0.1	−0.9	0
2	12	13.3	13.3	12.0	12.0	11.9	11.4	12.0
3	8	7.2	7.1	7.9	8.0	8.2	8.3	7.9
4	4	4.8	5.0	3.7	4.0	3.9	4.7	4.0
5	2	2.3	4.2	2.1	2.1	2.1	2.9	1.7
6	8	8.3	8.0	7.5	8.0	8.2	8.4	7.4
7	12	11.3	10.6	12.3	12.1	12.3	12.2	11.8
8	6	5.6	5.7	5.6	5.8	5.7	5.9	6.0
9	10	9.2	8.9	10.3	10.1	9.7	9.1	10.8
Square root of sums of squares of deviations		2.10	3.50	0.97	0.28	0.62	1.89	1.07

What the magnitude-estimation data tell us that the interval-scale data do not tell us is the ratios in which the colorfulnesses of the samples stand with respect to one another. For example, sample 2 is about 4 times as colorful as sample 8. The interval-scale data tell us that the difference between samples 2 and 8 is about 1.8 times as great as the difference between samples 8 and 9. To the extent that we are satisfied to rely on the square-root relationship, we can estimate either kind of information from both scales by invoking the proper transformation. However, that relationship should not be blindly relied on to provide data; there is some evidence that the nonlinear relationship between category and magnitude results may obtain only when small numbers of categories (around 7) are used but not when large numbers of categories are involved or when scales are open-ended (for example, Banks and Coleman, 1981). The approximate square-root relationship is useful for comparing results of one kind of experiment with those of another when the same samples have been scaled for the same attribute both ways. It allows us to search for convergence, but there is as yet no sound theoretical basis for the nonlinear association of direct-ratio and interval-scaling results.

REFERENCES

Alf, E. F., and Grossberg, J. M. (1979). The geometric mean: Confidence limits and significance test. *Percept. Psychophys.* **26**, 419–421.

Anderson, N. H. (1979). Algebraic rules in psychological measurement, *Am. Sci.* **67**, 555–563.

Attneave, F. (1962). Perception and related areas. *In* "Psychology: A Study of a Science" (S. Koch, ed.), Vol. 4, pp. 619–659. McGraw-Hill, New York.

Banks, W. P., and Coleman, M. J. (1981). Two subjective scales of number. *Percept. Psychophys.* **29**, 95–105.

Bartleson, C. J. (1970). Differences and ratios in color scaling. *In* "Tagungsbericht Internationale Farbtagung COLOR 69" (M. Richter, ed.), pp. 386–390. Stockholm, Musterschmidt-Verlag, Göttingen.

Bartleson, C. J. (1980). Measures of brightness and lightness. *Farbe* **28**, 132–148.

Bartleson, C. J., and Breneman, E. J. (1973). Differences among observers in scaling brightness. *Farbe* **22**, 200–212.

Bond, B., and Stevens, S. S. (1969). Cross-modality matching of brightness to loudness by 5-year-olds. *Percept. Pscychopys* **6**, 337–339.

Campbell, N. R. (1938). "Symposium: Measurement and its Importance for Philosophy" (Aristotelian Soc. Supplement, Vol. 17). Harrison, London.

Coleman, B. J., Graf, R. G., and Alf, E. F. (1981). Assessing power function relationships in magnitude estimation. *Percept. Psychophys.* **29**, 178–180.

Ekman, G. (1956). "Subjective Power Functions and the Method of Fractionation" (Report No. 34 of the Psychological Laboratory). University of Stockholm, Stockholm.

Fechner, G. T. (1860). "Elemente der Psychophysik." Breitkopf and Härtel, Leipzig.

Ferguson, A. (1940). Final report on quantitative estimates of sensory events. *Adv. Sci.* **2**, 331–349.

Fullerton, G. S., and Cattell, J. M. (1892). "On the Perception of Small Differences." University of Pennsylvania Press, Philadelphia.

Gage, F. H. (1934). An experimental investigation of the measurability of visual sensation. *Proc. R. Soc. London B* **116**, 123–128.

Ham, L. B., and Parkinson, J. S. (1932). Loudness and intensity relations. *J. Acoust. Soc. Am.* **3**, 511–534.

Indow, T. (1974). Scaling of saturation and hue shift: Summary of results and implications. *In* "Sensation and Measurements" (H. R. Moskowitz, ed.), pp. 351–362. Reidel, Dordrecht, The Netherlands.

Jameson, D., and Hurvich, L. M. (1964). Theory of brightness and color contrast in human vision. *Vision Res.* **4**, 135–154.

Krantz, D. H. (1972). Visual scaling. *In* "Handbook of Sensory Physiology" (D. Jameson and L. M. Hurvich, eds.), Vol. VII/4. Visual Psychophysics, pp. 660–689. Springer, Berlin.

Luce, R. D., Bush, R. R., and Galanter, E. H. (1963–1965). "Handbook of Mathematical Psychology," Vols. I–III. J. Wiley and Sons, New York.

Luce, R. D., and Mo, S. S. (1965). Magnitude estimation of heaviness and loudness by individual subjects: A test of a probabilistic response theory. *Br. J. Math. Statis. Psychol.* **18**, 159–174.

Marks, L. E. (1968). Stimulus range, number of categories, and form of the category scale. *Am. J. Psychol.* **81**, 467–479.

Marks, L. E. (1974a). "Sensory Processes, The New Psychophysics." Academic Press, New York.

Marks, L. E. (1974b). On scales of sensation: Prolegomena to any future psychophysics that will be able to come forth as a science. *Percept. Psychophys.* **16**, 358–376.

Merkel, J. (1888). Die Abhängigkeit zwischen Reiz und Empfinddung, I. *Philos. Stud.* **4**, 541–594.

Merkel, J. (1889). Die Abhängigkeit zwischen Reiz und Empfinddung, II. *Philos. Stud.* **5**, 245–291.

Munsell, A. H. (1909). On the relation of the intensity of chromatic stimulus (physical saturation) to chromatic saturation. *Psychol. Bull.* **6**, 238–239.

Newhall, S. M. (1940). Preliminary report of the O. S. A. subcommittee on the spacing of the Munsell colors. *J. Opt. Soc. Am.* **30**, 617–645.

Newhall, S. M., Nickerson, D., and Judd, D. B. (1943). Final report of the O. S. A. subcommittee on the spacing of the Munsell colors. *J. Opt. Soc. Am.* **33**, 385–418.

Plateau, J. A. F. (1872). Sur la mesure des sensations physiques, et sur la loi qui lie l'intensité de ces sensations à l'intensité de la cause excitante. *Bull. Acad. R. Belg.* **33**, 376–388.

Richardson, L. F., and Ross, J. S. (1930). Loudness and telephone current. *J. Gen. Psychol.* **3**, 288–306.

Savage, C. W. (1970). "The Measurement of Sensation." University of California Press, Berkeley, California.

Schneider, B., and Lane, J. (1963). Ratio scales, category scales, and variability in the production of loudness and softness. *J. Acoust. Soc. Am.* **35**, 1953–1961.

Shepard, R. N. (1981). Psychological relations and psychophysical scales: On the status of "direct" psychophysical measurement. *J. Math. Psychol.* **24**, 21–57.

Stevens, S. S. (1934). The volume and intensity of tones. *Am. J. Psychol.* **46**, 397–408.

Stevens, S. S. (1936). A scale of measurment of psychological magnitude: Loudness. *Psychol. Rev.* **43**, 405–416.

Stevens, S. S. (1946). On the theory of scales of measurement. *Science* **103**, 677–680.

Stevens, S. S. (1955a). Decibels of light and sound. *Phys. Today* **8**, 12–17.

Stevens, S. S. (1955b). On the averaging of data. *Science* **121**, 113–116.

Stevens, S. S. (1957). On the psychophysical law. *Psychol. Rev.* **64**, 153–181.

Stevens, S. S. (1961). The psychophysics of sensory function. "Sensory Communication" (W. A. Rosenblith, ed.), pp. 1–33. M.I.T. Press, Cambridge, Massachusetts.

Stevens, S. S. (1966). Matching functions between lightness and ten other continua. *Percept. Psychophys.* **1**, 5–8.

Stevens, S. S. (1975). "Psychophysics. Introduction to its Perceptual, Neural, and Social Prospects." J. Wiley and Sons, New York.

Stevens, S. S., and Davis, H. (1938). "Hearing: Its Psychology and Physiology." J. Wiley and Sons, New York.

Stevens, S. S., and Galanter, E. H. (1957). Ratio scales and category scales for a dozen perceptual continua. *J. Exp. Psychol.* **54**, 377–411.

Stevens, S. S., and Guirao, M. (1962). Loudness, reciprocality, and partition scales. *J. Acoust. Soc. Am*. **34**, 1466–1471.

Stevens, S. S., and Harris, J. R. (1962). The scaling of subjective roughness and smoothness. *J. Exp. Psychol.* **64**, 489–494.

Torgerson, W. S. (1961). Distances and ratios in psychological scaling. *Acta Psychol.* **19**, 201–205.

Zwislocki, J. J., and Goodman, D. A. (1980). Absolute scaling of sensory magnitudes: A validation. *Percept. Psychophys.* **28**, 28–38.

10

Multidimensional Scaling

C. J. BARTLESON

Research Laboratories
Eastman Kodak Company
Rochester, New York

ISBN-0-12-304905-9

A. INTRODUCTION

1. Dealing with More Than One Dimension

Many bodies of data involve observations associated with stimuli that vary in ways that cause changes in more than one attribute of perception. When either dependent (response) or independent (stimulus) variables number more than one, the problem is referred to as *multivariate*. Often when there are several response variables, the situation is called *multidimensional*. Unfortunately, there are two common definitions of *dimension* that are used in connection with perceptions. Titchner (1910) proposed a definition that corresponds to the usual mathematical usage of the term, that is, as equivalent to degrees of freedom. He proposed that to be a dimension, a perceptual attribute must be capable of independent variation while all other dimensions are held constant. Some people have argued that Titchner's definition of dimension is too strict. They prefer a more liberal definition offered by Stevens (1934). His is the logical complement to Titchner's definition. According to Stevens, an attribute must be capable of being held constant while all other attributes are varied. This criterion yields phenomenologically distinct response characteristics but not necessarily unique ones. I prefer to use the neutral phrase *response attributes* when referring to perceptual phenomena and to speak of multivariate situations when stimulus variables are subject to covariation or when it is not important to distinguish between dependent and independent variables. However, the terms multivariate and multidimensional are so firmly embedded in the lexicons of statistics and psychophysics that it would be confusing to use different terms here. Accordingly, the term multivariate will be used in this chapter to refer to situations in which we are concerned with the analysis of m points in n space, that is, when each of m observers or samples has associated with it an n-dimensional vector of responses. The term multidimensional will be reserved for situations in which interest centers on the coexistence of a number of response attributes (regardless of the nature of the stimuli or even whether they are known), particularly when we are concerned with differences among a plurality of coexistent responses. Somewhat oversimplified, then, the distinction is one of vector relationships on the one hand and interpoint distances on the other. Unfortunately, even this simplification does not ensure clarity, because *multiple regression* techniques can be thought of in either vector or distance terms. These techniques properly belong in this chapter as well; they deal with descriptive or predictive models that relate a dependent variable to covariation in two or more independent variables. All of these methods, concepts, and models have one thing in common: They deal with covariation in *spaces* (or, more generally, *manifolds*) of orders two or greater. The scaling methods described in Chapters 8 and 9 deal with variation along a single continuum. They are referred to as *unidimensional*. The logical semantic complement of unidimensional is *multidimensional*; for this reason this chapter has been entitled "Multidimensional Scaling." The reader should be aware, however, that what will be discussed is that class of scaling situations in which more than one dimension is dealt with at the same time. The chapter will first describe multiple regression methods, because they are frequently useful when the dependent variables are known with reasonable certainty. Multivariate techniques will be sketched briefly; they are most often useful in studying behavioral problems, usually those in which the independent variables are thought to be understood. Considerable space will be devoted to multidimensional scaling methods in which the object is to define the dimensionality of responses without necessarily knowing what either the dependent or independent variables are.

B. MULTIPLE REGRESSION

1. Linear Regression

Multiple regression techniques are often used to derive a psychophysical function when a single response continuum is thought to be related to more than one stimulus continuum. If it is assumed that stimulus attributes combine linearly or that some transformation of them can be found that provides linear combination, then it is possible to determine a *multiple linear regression* equation in several independent variables. In Chapters 8 and 9 we have considered first-order linear regression models in which a dependent variable y is related to a single independent variable x:

$$y = \beta_0 + \beta_1 x + \varepsilon, \tag{1}$$

where β_i are "true" parameter values sought by the regression process, and ε is the error term representing the extent to which the regression fails to determine exactly the true parameters. Equation (1) determines a line. The object of multiple regression is, instead, to determine a surface. Stated differently, the conditional expectation of Eq. (1) is $E(Y|x)$ and corresponds to a line (or a curve in the general case). When k independent variables are involved, the conditional expectation is $E(Y|x_1, x_2, \ldots, x_k)$, which corresponds to the theoretical regression surface of Y on the k variables. Recall from Chapter 7 that the expectation is equal to the mean of a distribution. The conditional mean may vary for different values of x, and this function of x is called the regression of Y on X. For the simplest case of two independent variables, we may then rewrite Eq. (1) as a multiple regression equation:

$$y = \beta_0 + \beta_1 x_1 + \beta_2 x_2 + \varepsilon. \tag{2}$$

The least-squares criterion of fitting such a surface then is

$$\text{minimize} \quad \sum \varepsilon^2 = \sum \{y - (\beta_0 + \beta_1 x_1 + \beta_2 x_2)\}^2 \tag{3}$$

Estimates b_i of the parameters β_i can be calculated for Eq. (2) as follows:

$$b_2 = \left[\frac{\begin{array}{c}[\{n \sum x_{1i}^2 - (\sum x_{1i})^2\}\{n \sum x_{2i} y_i - (\sum x_{2i})(\sum y_i)\}] \\ -[\{n \sum x_{1i} x_{2i} - (\sum x_{1i})(\sum x_{2i})\}\{n \sum x_{1i} y_i - (\sum x_{1i})(\sum y_i)\}]\end{array}}{[\{n \sum x_{1i}^2 - (\sum x_{1i})^2\}\{n \sum x_{2i}^2 - (\sum x_{2i})^2\}][n \sum x_{1i} x_{2i} - (\sum x_{1i})(\sum x_{2i})]^2} \right], \tag{4}$$

and

$$b_1 = \left[\frac{\{n \sum x_{1i} y_i - (\sum x_{1i})(\sum y_i)\} - b_2 \{n \sum x_{1i} x_{2i} - (\sum x_{1i})(\sum x_{2i})\}}{n \sum x_{1i}^2 - (\sum x_{1i})^2} \right], \tag{5}$$

and

$$b_0 = \left[\frac{\sum y_i - b_2 \sum x_{2i} - b_1 \sum x_{1i}}{n} \right], \tag{6}$$

where

$$\begin{aligned} \sum y_i &= b_0 n + b_1 \sum x_{1i} + b_2 \sum x_{2i}, \\ \sum x_{1i} y_i &= b_0 \sum x_{1i} + b_1 \sum x_{1i}^2 + b_2 \sum x_{1i} x_{2i}, \\ \sum x_{2i} y_i &= b_0 \sum x_{2i} + b_1 \sum x_{1i}^2 x_{2i} + b_2 \sum x_{2i}^2, \end{aligned}$$

and n is the number of observations. The equations for $\sum y_i$, $\sum x_{1i} y_i$, and $\sum x_{2i} y_i$ are called the *normal equations*; they are solved simultaneously after inserting the sums calculated from the data in order to estimate b_0, b_1, and b_2 according to Eqs. (4)–(6). The coefficient of determination r^2 may be determined as follows:

$$r^2 = \left[\frac{b_0 \sum y_i + b_1 \sum x_{1i} x_{2i} + b_2 \sum x_{2i} y_i - (1/n)(\sum y_i)^2}{(\sum y_i^2) - (1/n)(\sum y_i)^2} \right]. \tag{7}$$

Even with only three parameters, for the bivariate case of Eq. (2), the equations for estimating parameter values are cumbersome to calculate by hand. With more variables the equations become even more unwieldy. There are as many normal equations as there are parameters, so that in the general case

$$\begin{aligned}
\sum y_i &= nb_0 + b_1 \sum x_{1i} + b_2 \sum x_{2i} + \cdots + b_k \sum x_{ki} \\
\sum x_{1i} y_i &= b_0 \sum x_{1i} + b_1 \sum x_{1i}^2 + b_2 \sum x_{1i} x_{2i} + \cdots + b_k \sum x_{1i} x_{ki} \\
&\vdots \\
\sum x_{ki} y_i &= b_0 \sum x_{ki} + b_1 \sum x_{ki} x_{1i} + b_2 \sum x_{ki} x_{21} + \cdots + b_k \sum x_{ki}^2.
\end{aligned}$$

When the number of parameters is large, it is simpler to express the data in matrix algebra form and to take advantage of computer algorithms for solving the regression. Suppose that we wish to determine k parameters from a sample size of n observations (generally n should be no smaller than between 5 and 10 times k). We then have n equations of data, of the form of Eq. (2), with k unknown parameter values:

$$\begin{aligned}
y_1 &= \beta_0 + \beta_1 x_{1,1} + \beta_2 x_{2,1} + \cdots + \beta_k x_{k,1} + \varepsilon_1, \\
y_2 &= \beta_0 + \beta_1 x_{1,2} + \beta_2 x_{2,2} + \cdots + \beta_k x_{k,2} + \varepsilon_2, \\
y_3 &= \beta_0 + \beta_1 x_{1,3} + \beta_2 x_{2,3} + \cdots + \beta_k x_{k,3} + \varepsilon_3, \qquad (8) \\
&\vdots \\
y_n &= \beta_0 + \beta_1 x_{1,n} + \beta_2 x_{2,n} + \cdots + \beta_k x_{k,n} + \varepsilon_n.
\end{aligned}$$

We may define the matrices as

$$\mathbf{Y} = \begin{bmatrix} y_1 \\ y_2 \\ y_3 \\ \cdot \\ \cdot \\ \cdot \\ y_n \end{bmatrix}, \quad \mathbf{X} = \begin{bmatrix} 1 & x_{1,1} & x_{2,1} & \cdots & x_{k,1} \\ 1 & x_{1,2} & x_{2,2} & \cdots & x_{k,2} \\ 1 & x_{1,3} & x_{2,3} & \cdots & x_{k,3} \\ \cdot & & & & \\ \cdot & & & & \\ \cdot & & & & \\ 1 & x_{1,n} & x_{2,n} & \cdots & x_{k,n} \end{bmatrix}, \quad \mathbf{B} = \begin{bmatrix} \beta_1 \\ \beta_2 \\ \beta_3 \\ \cdot \\ \cdot \\ \cdot \\ \beta_n \end{bmatrix}, \quad \mathbf{E} = \begin{bmatrix} \varepsilon_1 \\ \varepsilon_2 \\ \varepsilon_3 \\ \cdot \\ \cdot \\ \cdot \\ \varepsilon_n \end{bmatrix},$$

so that Eqs. (8) can be expressed in matrix form as

$$\mathbf{Y} = \mathbf{BX} + \mathbf{E}. \qquad (9)$$

The least-squares criterion is then expressed as

$$\text{minimize} \sum \varepsilon^2 = (\mathbf{E}'\mathbf{E}) = (\mathbf{Y} - \mathbf{XB})'(\mathbf{Y} - \mathbf{XB}), \qquad (10)$$

where the prime (′) stands for transpose; that is, $\mathbf{E}'$ is the transpose of matrix $\mathbf{E}$. The solution is then simply expressed as

$$\mathbf{B} = (\mathbf{X}'\mathbf{X})^{-1}(\mathbf{X}'\mathbf{Y}), \qquad (11)$$

where $(\mathbf{X}'\mathbf{X})^{-1}$ is the inverse of matrix $(\mathbf{X}'\mathbf{X})$.

The coefficient of determination, expressed in matrix alegbra, is

$$r^2 = \left[\frac{\mathbf{B'X'Y} - (1/n)(\sum y_i)^2}{\mathbf{Y'Y} - (1/n)(\sum y_i)^2}\right]. \tag{12}$$

Details of the regression method can be found in textbooks on regression analysis (for example, Winkler and Hays, 1975; Draper and Smith, 1981). Readers not familiar with matrix algebra should consult one or more of the many textbooks on that subject (such as Aiken, 1942; Campbell, 1977; Davis, 1973; Munakata, 1979).

There are several kinds of regression schemes available in various computer programs. Program packages such as MINITAB (Ryan *et al.*, 1976), SPSS (Nie *et al.*, 1975), BMDP-79 (Dixon, 1979), and SAS (Helwig and Council, 1979) all offer a choice of regression models. The question of which model to use for a given problem is not one that can be answered by any simple rule of thumb. Usually we want to gather as many data as possible in order to define as many parameters as are necessary to provide high predictive power, but at the same time we want to gather as few data as possible in order to minimize the time and expense of experimentation. Draper and Smith (1981) and Daniel and Wood (1980) discuss in some detail the various considerations that might go into a choice of method. I have found that for most of the visual psychophysical work that I have done, a least-squares criterion standard multiple-regression program such as the GLM (General Linear Models) procedure of SAS works well.

Whichever program is selected, it is well to examine correlations, deviations, and residuals to decide how many parameters are adequate. The most common such measures are r^2, s^2, and the C_p statistic (all of which are supplied in the SAS GLM procedure, incidentally). The coefficient of determination achieved by the least-squares fit is r^2 (the square of the multiple correlation coefficient); it generally increases as the number of relevant parameters increases but does so at an ever-diminishing rate. The value of s^2 is the residual mean square, the estimates of Eqs. (3) or (10), in effect; its value generally decreases as the number of relevant parameters increases but does so at an ever-diminishing rate. Mallows' (1973) statistic labeled C_p is defined as

$$C_p = \frac{\text{(residual sum of squares)}}{s^2 - (n - 2p)},$$

where p stands for the number of parameters (including β_0), and n is the number of observations. The expectation of C_p is approximately p (that is, $E(C_p) \cong p$). Therefore, when calculated values of C_p are plotted against values of p, the result should be a straight line of unit slope if the model is appropriate to the data. Plotted points that lie at some distance from the 45° line in such a graph may be taken to suggest that the number of parameters is wrong.

a. EQUIVOCAL AND IMMUTABLE NONLINEARITIES

Linear regression algorithms can be applied to a wide variety of problems, many of which appear to be nonlinear. Not all nonlinear functions are inherently or immutably nonlinear. Many are what I have referred to here as equivocally nonlinear. I mean by this that some transformation can be found that will lend itself to expression in the form of Eqs. (8).

A transformed variable ξ can often be found and substituted in the equation. For example, $\xi = \ln x_i$ could be expressed in either of two ways:

$$y_1 = \beta_0 + \beta_1 \ln x_1 + \beta_2 \ln x_2 + \cdots + \beta_k \ln x_k,$$

or

$$y_1 = \beta_0 + \beta_1 \xi_1 + \beta_2 \xi_2 + \cdots + \beta_k \xi_k,$$

when the raw data are of the form $e^y = f(x)$. Similarly, a multiplicative form such as $y = \beta_0 x_1{}^{\beta_1} \cdot \beta_0 x_2{}^{\beta_2} \cdot \beta_0 x_3{}^{\beta_3} \cdot \varepsilon$ transforms to

$$\ln y = \ln \beta_0 + \beta_1 \ln x_1 + \beta_2 \ln x_2 + \beta_3 \ln x_3 + \ln \varepsilon$$

and so on. Even apparently quite complex nonlinear functions will often yield to re-expression so that linear regression methods can be applied. When such re-expression is used, the residuals and statistics that should be evaluated are those of the transformed variables.

Some expressions are inherently nonlinear, however. Draper and Smith (1981, p. 224) give two examples:

$$y = \beta_0 + \beta_1 e^{-\beta_2 x} + \varepsilon,$$

and

$$y = \beta_0 + \beta_1 x + \beta_2 \beta_3^x + \varepsilon.$$

There simply is no ξ re-expression that will yield a multiple linear form for equations such as these. When this is the case, there is no recourse but to apply nonlinear regression techniques.

2. Nonlinear Regression

Nonlinear regression is essentially no different from linear regression. If we know what form of nonlinear function might be appropriate to our data, and if we can compute $f(y \mid x)$ and $E(Y \mid x)$, then we can determine an exact regression surface. We do this by minimizing the differences between the observed data and the expectations based on whatever function we select as a model for the nonlinearity involved. The trick is to select a nonlinear function that is appropriate to the data. It is as much an art as a result of experience; certainly it is not a science. With unidimensional and bidimensional data, it is possible to plot the data, examine them, and decide what kind of nonlinear function would provide the same shape as the curve we see on the graph paper. This is not so easy with data in three or more dimensions; in fact, the greater the dimensionality the more difficult the task. It is possible to generate graphs of all possible combinations of pairs of dimensions and examine them, but when there are more than three dimensions involved, even this becomes a monumental task for all but skilled geometricians and topologists.

All is not hopeless, however. Fortunately, psychophysical data from experiments in which attempts are made to measure or specify responses to various aspects of light do not often involve more than three dimensions. Usually the experimenter can arrange such an outcome by designing the experiment in such a way as to limit the number of independent variables. Regardless of the outcome, proper choice of mathematical model is the key to success in nonlinear regression. If a poor choice is made, the regression likely will not converge on a solution that has high correlation with the experimental data. It is well worthwhile to try several models that might be appropriate to find the one that works best.

The model may be as simple or as complex as necessary, but there is benefit in selecting the simplest model that will do the job. One reason is that nature has a way of using Occam's razor; there is often underlying simplicity in natural processes if we can see through the haze of apparent complexity that is largely a result of our ignorance of the process. Therefore, simpler is often "truer." Another reason to seek simplicity over complexity is that calculations are not as difficult. In addition, some computer programs require estimation of differential values for each of the parameters to be determined—it is much easier to find the partial of a simple expression than of an unnecessarily complex one—and virtually all programs need estimates of starting values. If the function chosen as a nonlinear model is too complex, the computer may get struck in some local minimum of the response surface and will think it has reached a solution (computers are not intelligent; it is always up to the experimenter to guide things in the right direction). There are a few textbooks devoted to the problem of nonlinear parameter estimation (for example, Goldfeld and Quandt, 1972; Bard, 1974; Beck and Arnold, 1977).

The normal equations for the nonlinear regression differ from those of Eq. (9) for the linear case. Here we have the dependent variable as some nonlinear function of both parameters and independent variables, not just their products:

$$y = f(\beta_0, \beta_1, \ldots, x_1, x_2, \ldots) + \varepsilon,$$

so that the nonlinear normal equations are of the form

$$\mathbf{X}'\mathbf{F}(\beta) = \mathbf{X}'\mathbf{Y},$$

where $\mathbf{X} = \partial\mathbf{F}/\partial\beta$. The iterative process for minimizing the sums of the squares of errors begins with some initial value of β, then computes a value Δ such that $\varepsilon(\beta + \Delta) < \varepsilon(\beta)$. There are several ways of doing this, but the end result is to achieve the same object as with linear regression: to minimize the differences between the observations and the function being fitted to them.

As with linear regression methods, there are several computational packages available that offer methods for solving nonlinear regression problems. BMDP-79, SPSS, and SAS are among them. The SAS NLIN procedure will be used in an example in the next section.

3. An Example Multiple-Regression Problem

To illustrate an application of both linear and nonlinear multiple regression analysis, I shall draw upon some recently reported data (Bartleson, 1982) that were gathered to study the combined influence of sharpness and graininess of photographic prints on their perceived quality. The spatial frequency response and elemental granular structure of image-forming components of photographic films are finite. Therefore, the corresponding light-modulation characteristics of prints made from film negatives will vary according to the degree to which the image is enlarged. In addition, different films have different frequency response and noise characteristics. These characteristics can be measured by microphotometry of suitable targets, but a question of interest to designers and users of photographic materials is: To what extent do these variations impair the quality of prints?

The data listed in the columns of Table I identified as Q, S, and G are mean category values determined from the observations of 30 subjects who used a 9-point successive-category scaling method to assess quality, sharpness, and graininess, respectively, of 43 90 × 125 mm prints. Figure 1 illustrates the mean results. Scaled sharpness S values are plotted against the complement of graininess N. The numeric figures in the graph represent the independently scaled quality values. The problem is to determine the concatenation of S and N on Q; that is, how do covariations in S and N combine to influence Q?

One approach to this question is to determine the multiple linear regression of S and N on Q. This approach assumes that the two factors combine linearly to determine quality (all other factors that could vary in the experiment were carefully controlled and invariant, so dealing with combinations of only sharpness and graininess should be sufficient to characterize all differences among the prints). In fact, a multiple linear regression was carried out on the data of Table I, using the SAS GLM procedure. The regression equation that resulted from this analysis was

$$Q = 0.661S + 0.383N - 0.921, \tag{13}$$

for which the value of r^2 is 0.9148. That coefficient of determination

TABLE I. Multiple Regression Quality versus Sharpness versus Graininess

PRINT	Q	S	G	N	QQ	$QQRES$	$QHAT$	$QRES$
1	8.1	7.9	1.5	8.5	8.09263	−0.00737	7.56228	0.5377
2	7.6	7.6	1.8	8.2	7.77533	0.17533	7.24886	0.3511
3	6.6	7.0	3.5	6.5	6.56142	−0.03858	6.20023	0.3998
4	5.6	6.5	5.0	5.0	5.30373	−0.29627	5.29442	0.3056
5	4.4	6.1	6.1	3.9	4.21600	−0.18400	4.60811	−0.2081
6	3.4	6.2	6.7	3.3	3.58755	0.18755	4.44417	−1.0442
7	2.3	4.8	7.7	2.3	2.36583	0.06853	3.13492	−0.8349
8	1.8	4.5	8.1	1.9	1.88045	0.08045	2.78316	−0.9832
9	7.4	7.1	1.6	8.4	7.52248	0.12248	6.99490	0.4051
10	5.9	6.7	3.8	6.2	6.24409	0.34409	5.88681	0.0132
11	5.2	6.1	5.1	4.9	5.11279	−0.08721	4.99156	0.2084
12	4.3	5.9	6.1	3.9	4.18687	−0.11313	4.47585	−0.1759
13	3.5	5.3	6.6	3.4	3.60481	0.10481	3.88736	−0.3874
14	2.5	4.7	7.4	2.6	2.70007	0.20007	3.18383	−0.6838
15	1.8	4.3	8.2	1.8	1.75445	−0.04555	2.61256	−0.8126
16	5.5	4.7	1.8	8.2	5.31264	−0.18736	5.33112	0.1689
17	5.2	5.3	4.1	5.9	5.33930	0.13930	4.84597	0.3540
18	4.5	4.9	5.1	4.9	4.63489	0.13489	4.19801	0.3020
19	3.9	4.7	6.3	3.7	3.75713	−0.14287	3.60562	0.2944
20	3.5	4.6	6.5	3.5	3.56564	0.06564	3.46280	0.0372
21	2.3	4.0	7.6	2.4	2.41992	0.11992	2.64424	−0.3442
22	1.7	3.2	8.2	1.8	1.70116	0.00116	1.88514	−0.1851
23	4.2	3.8	1.8	8.2	4.29337	0.09337	4.73596	−0.5360
24	4.6	4.3	4.2	5.8	4.55273	−0.04727	4.14634	0.4537
25	4.1	4.0	5.1	4.9	4.06171	−0.03829	3.60285	0.4972
26	3.2	3.7	6.5	3.5	3.25652	0.05652	2.86764	0.3324
27	2.8	3.5	6.9	3.1	2.91760	0.11760	2.58200	0.2180
28	2.0	3.2	7.8	2.2	2.10869	0.10869	2.03852	−0.0385
29	1.7	2.6	8.2	1.8	1.62530	−0.07470	1.48837	0.2116
30	3.0	2.7	1.7	8.3	2.94724	−0.05276	4.04689	−1.0469
31	3.1	2.9	4.1	5.9	3.13398	0.03398	3.25888	−0.1589
32	3.3	3.1	5.4	4.6	3.21844	−0.08156	2.89266	0.4073
33	2.6	2.6	6.5	3.5	2.54083	−0.05917	2.14022	0.4598
34	2.0	2.4	7.5	2.5	2.04965	0.04965	1.62452	0.3755
35	2.2	2.5	7.1	2.9	2.28327	0.08327	1.84403	0.3560
36	1.6	2.1	8.2	1.8	1.49610	−0.10390	1.15772	0.4423
37	2.4	2.3	1.9	8.1	2.43892	0.03892	3.70568	−1.3057
38	2.6	2.6	4.6	5.4	2.76025	0.16025	2.86877	−0.2688
39	2.5	2.4	5.5	4.5	2.47843	−0.02157	2.3941	0.1086
40	2.3	2.4	6.5	3.5	2.36072	0.06072	2.00796	0.2920
41	2.1	2.2	6.9	3.1	2.09870	−0.00130	1.72233	0.3777
42	1.7	1.9	8.0	2.0	1.52181	−0.17819	1.10215	0.5978
43	1.4	1.7	8.2	1.8	1.31058	−0.08942	0.89321	0.5068

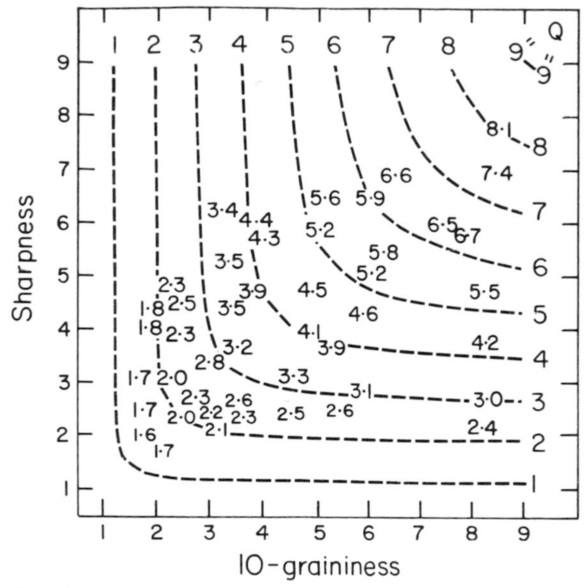

Fig. 1. Scaled quality (numerics) of photographic prints as functions of scaled sharpness and the complement of scaled graininess. Dashed curves are computed from Eq. (14) of the text. Reprinted with permission from Bartleson, 1982, J. Phot. Sci.

corresponds to a multiple correlation coefficient of 0.9565, which may seem reasonably high. It might be satisfactorily high in a behavioral survey, but it is not good enough for predictive purposes in an experiment such as the one cited here.

To see why this is so, we shall examine the data and the results more closely. The two far-right columns of Table I list the quality estimates predicted by Eq. (13) (labeled *QHAT* in the table) and the residual errors (labeled *QRES*) for each of the 43 sample prints. Values of *QHAT* have been plotted against the scaled quality values in Fig. 2. It is obvious from the graph that although the correlation is reasonably high, there is something wrong. Virtually all the departures from a 45° line (representing perfect agreement) are in the same direction. This peculiarity is even more obvious in Fig. 3, in which the residual errors have been plotted against scaled quality. This graph is an excellent illustration of why it is important to examine residuals in any analysis. If the analysis is appropriate, the residual errors will be randomly distributed about the horizontal line that represents their mean. When this is not so, something is wrong. Usually the model is inappropriate. If there is an obvious trend or pattern to the residuals, or if they are biased in one direction (as in Fig. 3), the best thing to do is to try another model; often the pattern of the residuals will suggest something about how the model should differ from that first used.

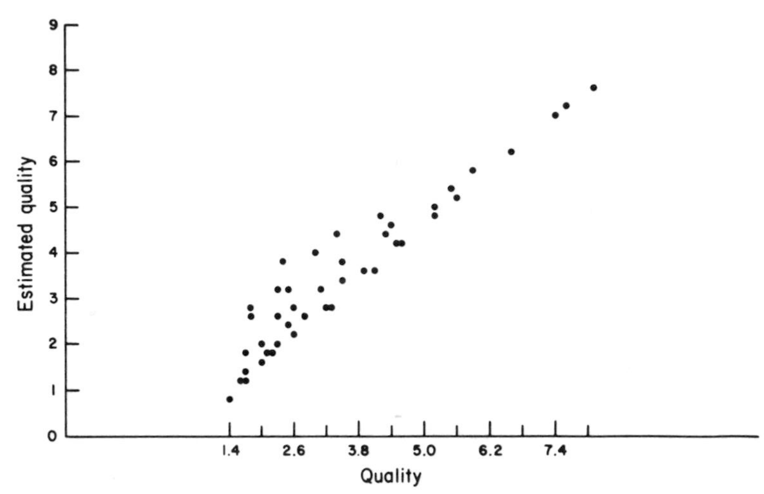

Fig. 2. Estimated quality, from Eq. (13) of the text, plotted against scaled quality for results of Table I in the text.

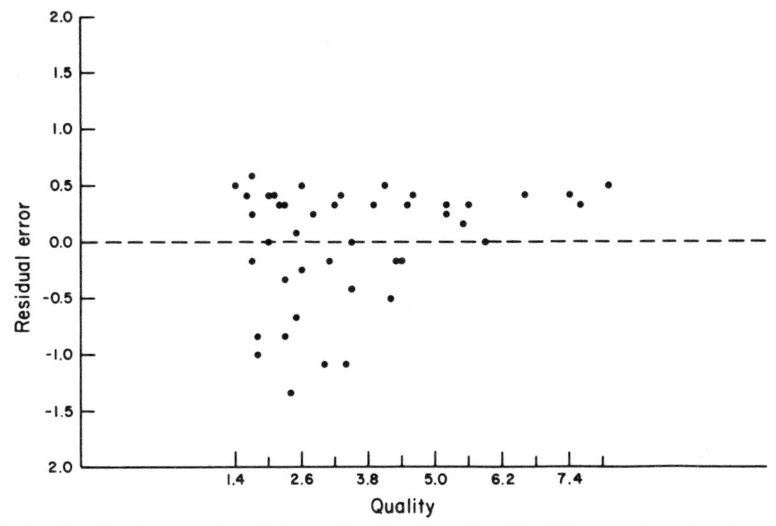

Fig. 3. Residual errors of predictions of quality, from Eq. (13) of the text, plotted against scaled quality.

Since the first model tested was a simple multiple linear combination of S and N, it seems reasonable to select a nonlinear model for a second attempt to find an appropriate regression. If the linear combination model had been appropriate, contours of iso-quality would be diagonal lines in Fig. 1 whose (negative) slopes should be related to the relative contributions of S and N to the level of Q. Careful inspection of Fig. 1 will show that constant-slope straight lines cannot be passed through or near all points having similar Q values. For example, there is a series of points about midway up the graph that have Qs (from right to left) of 4.2, 3.9, 4.1, 3.9, and 4.3. There is no way in which a straight line could be drawn in these coordinates that would pass suitably near these similar-quality points. They almost form the corner of a rectangle. This is not true of the points for Q values of 6.7, 6.5, and 6.6, however. They are almost on a straight diagonal line. This difference provides a valuable clue to the choice of nonlinear model from among the many nonlinear functions available. In particular, it suggests that the nonlinear function we must seek is one in which the degree of nonlinearity varies over different levels of Q.

Without detailing all of the logical and empirical steps that led to a satisfactory model, it is sufficient here to say that after many hours of practicing the art of curve-fitting, a suitable model was found. When it was used in the SAS NLIN procedure for nonlinear regression, the best regression was found to be

$$Q = (0.413S^{-3.4} + 0.422N^{-3.4})^{-1/3.4} - 0.532. \tag{14}$$

The value of r^2 for this regression is 0.9963, which corresponds to a multiple correlation coefficient of 0.9981, a significant improvement over the fit of Eq. (13).

Predictions made from Eq. (14) are listed in Table I in the column labeled QQ. The corresponding residual errors are shown in column $QQRES$. Figure 4

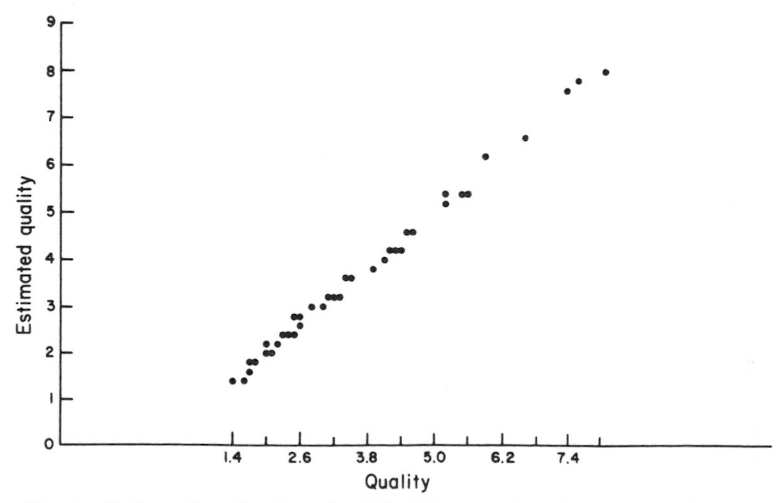

Fig. 4. Estimated quality, from Eq. (14) of the text, plotted against scaled quality.

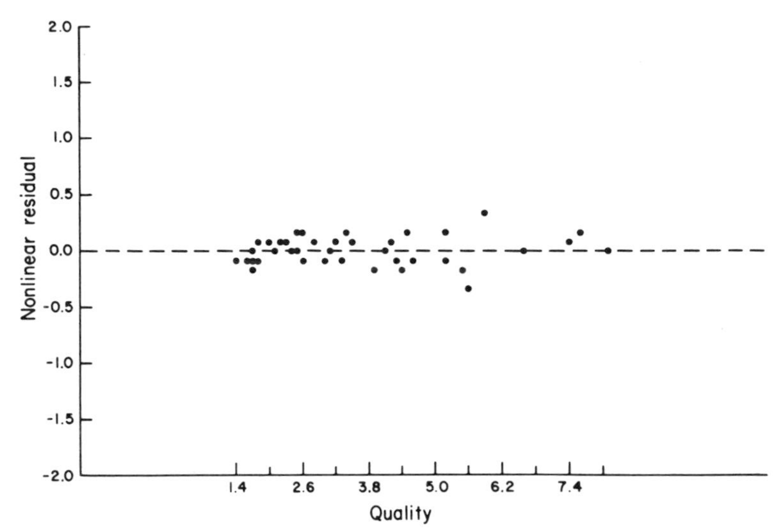

Fig. 5. Residual errors of predictions of quality, from Eq. (14) of the text, plotted against scaled quality.

shows the predicted quality values plotted against scaled quality, and Fig. 5 shows the residuals plotted against quality. Now the scaled quality values are well predicted, and the residual errors are evenly and randomly distributed across quality levels. In short, the model appears to fit the data well. The form of the model is such that increasingly more weight is given to an independent variable as its value decreases. This kind of relationship, in which the poorest element tends to determine quality, is just what is needed to mimic the different kinds of curvature exhibited by the results discussed above. In fact, the dashed curves shown in Fig. 1 are iso-quality contours computed from Eq. (14). Inspection will show that they are reasonably congruent with the results.

This kind of problem would be difficult to analyze in other ways, because it involves a highly nonlinear (and even noneuclidean) concatenation of covarying factors. Even though it is a multidimensional problem, the example has shown that it can be handled well by an orderly progression of multiple-regression techniques, starting with simple methods and building to complex, nonlinear ones. In short, there are some multivariate problems for which multiple regression may offer the most direct or least confusing path toward a useful solution. As this example demonstrated, however, it is best not to be satisfied with simple solutions without first finding out how much improvement can be gained by proceeding to a higher level of analytical and model complexity.

C. MULTIVARIATE ANALYSIS

1. Forms of Multivariate Analysis

There are a number of well-established techniques for analyzing multivariate data other than multiple-regression methods. They include principal-components analysis, factor analysis, cluster analysis, correspondence analysis, and canonical correlation analysis. Some of these are metric techniques, involving quantitative measures of distance, and others are nonmetric metric methods that deal with nominal and ordinal data. Although this chapter will concentrate on multidimensional scaling methods, it will be helpful to understand how this class of methods developed and how it differs from other multivariate statistical techniques. For that reason, each of the analytical methods mentioned above will be described briefly.

2. Principal-Components Analysis

The technique of *principal-components analysis* was described by Pearson (1901) and developed by Hotelling (1933). The basic idea of principal-components analysis is to describe a dispersion of m points in an n-dimensional space by introducing a new set of orthogonal coordinates that order the sample variances. The method is sometimes known as *eigenvector analysis*, or *analysis of latent roots*. The first principal component (or eigenvector) accounts for the largest fraction of the total variance in the sense that the projections of data points onto it have maximum variance among all possible linear coordinates. The second principal component has maximum variance subject to its being orthogonal to the first. The same holds true for each succeeding principal component. Computer programs that are used in determining principal components frequently print out 10 to 15 eigenvectors. They show the proportion of total variance accounted for by each and the cumulative sum over the eigenvectors ordered by magnitude. This kind of result is possibly the most valuable information that is supplied by the principal-components analysis. It allows the experimenter to determine how many dimensions may be necessary, in a practical sense, to characterize a given set of results. The cumulative proportion of variance accounted for by the first few vectors usually approaches 100% quite rapidly; the remaining vectors represent diminishingly important factors or simply noise.

Principal-components analysis can be performed on the correlation matrix or on the covariance matrix, but the components so derived are not invariant over linear transformation. Two of the methods used are called *QR* (for instance, Businger, 1965) and *singular value decomposition* (for instance,

Businger and Golub, 1969). The major disadvantage of the method of principal-components analysis is the requirement that all components be orthogonal. Hence, there may be more parsimonious ways (that is, with fewer dimensions) to characterize the data.

3. Factor Analysis

Factor analysis differs from principal-components analysis in that a matrix $\mathbf{\Lambda}$ of unknown parameters called *factor loadings* is determined for a set $\mathbf{F}$ of hypothetical variables called *common factors* in addition to an n-dimensional vector $\mathbf{U}$ of hypothetical variables called *unique factors*:

$$\mathbf{Y} = \mathbf{\Lambda} \cdot \mathbf{F} + \mathbf{U}. \tag{15}$$

The factor analytic model of Eq. (15) yields a set of factor-loading vectors that may be rotated to derive the most easily interpreted solution. There are a number of ways to perform the computations, but the two most well-known are the *principal-factor method* (Thurston, 1931; Thomson, 1934; Harman, 1967) and the *maximum-likelihood method* (Lawley, 1940). As with the method of principal components, the method of factor analysis requires data that are at least interval-scale level of mathematical power. The major advantage of factor analysis is that a minimum dimensionality can be deduced from the data by combining dimensions with multiple correlations. Three other methods for examining correlations and deducing dimensionality should also be mentioned before discussing multidimensional scaling. These methods may be applied to ordinal and, in some instances, even nominal level data. They are called cluster analysis, canonical correlation, and correspondence analysis.

4. Cluster Analysis

The growth of mainframe computers in recent years has made it possible to perform iterative computations that would have been prohibitive in time and cost in years past. One of the benefits of this phenomenon is the development of a class of iterative search algorithms called cluster analysis in which the basic scheme is to find groupings of data for which the units within each group are more similar than the units across groups. There are basically two kinds of clustering schemes: *hierarchical* (for example, Hartigan, 1967; Johnson, 1967) and *nonhierarchical* (for example, Ball and Hall, 1965; Friedman and Rubin, 1967). In a sense, one method is the complement of the other. Hierarchical clustering forms similarities by clustering data that are merged from previous stages. Nonhierarchical forms begin with one grand cluster and divide the data into more and more clusters. Although both metric and nonmetric data can be used, similarity or distance figures for intercluster differences are derived so that the major clusters can be identified and related. This information often provides a clue to the underlying dimensionality of the problem.

5. Canonical Correlation Analysis

Hotelling (1936) developed a method for studying associations between two sets of variables that is called *canonical correlation analysis*. The basic idea is to find two linear combinations of each set of variables, by iteration, that have maximal correlation. Then from among them one finds two sets that are orthogonal and have maximal correlation, and so on, until n (or some arbitrary limit $< n$) combinations have been found. The method may also be used with more than two variables (for instance, Horst, 1965; Kettenring, 1971). The derived linear functions are called *canonical variates*. Graphs of the original data transformed according to the canonical variates often provide useful clues to dimensionality and sometimes to aberrant data (uncommon observations called "flyers," unexpected relationships, and so on).

6. Correspondence Analysis

A special case of canonical correlation analysis has been developed in France in recent years under the name *L'analyse des correspondances* (Benzécri, 1973; Greenacre, 1981). It is basically a way to display data from two-way contingency tables (two-way classification tables that show frequencies or ratings for subdivisions of the two major classes under consideration) as points in corresponding spaces of minimal dimensionality. In common with many other multivariate techniques, correspondence analysis derives an estimate of low-rank matrices from singular value decomposition, or canonical form, of the matrix represented by the contingency table of data (for example, Marshall and Olkin, 1979). The low-dimensional, normalized euclidean vector spaces thus derived can be superimposed to obtain joint displays of data representing the different classes in the contingency table. As with factor analysis and most other multivariate techniques mentioned here, a major disadvantage of correspondence analysis is that attributes, or factors, or dimensions must be identified a priori.

7. Further Information

Multivariate analysis concerns the study of dependencies, associations, and relationships among responses and stimuli. Many of the techniques for such analysis were developed to seek basic factors that influence intelligence-test scores (Spearman, 1904). A large body of theory and methodology has been developed during the twentieth century. This section has given a superficial sketch of only some of those multivariate analytical techniques that are related to multidimensional scaling. They share certain similarities but differ in many ways, often subtly. The reader who wishes to pursue details of theory and methodology of these analytical techniques should consult some of the many textbooks that have been written on the subject (for instance, Anderson, 1958; Green and Carroll, 1976; Morrison, 1976; Gnanadesikan, 1977; Mardia *et al.*, 1979). Experimental methods used to gather data for multivariate analysis have also been developed. They are summarized in textbooks dealing with behavioral research and experimental psychology (for example, Kerlinger, 1964; Cattell, 1966; Myers, 1966; Winer, 1971).

Multidimensional scaling springs from the well of multivariate analysis, but it differs from multivariate techniques in one important respect. The multivariate methods collect data essentially by categorization, classification, or assumptions about the nature and number of dimensions involved in a problem. The experimenter makes certain decisions ahead of time. In some cases these are merely to list the items or factors to be surveyed or measured. In other instances, the experimenter may decide how many dimensions should be included in the study and even what they are.

D. MULTIDIMENSIONAL SCALING

1. How Multidimensional Scaling Differs from Multivariate Analysis

Modern multidimensional-scaling techniques require essentially no a priori decisions by experimenters about the form of results. The data of the scaling experiment are used to determine dimensionality directly. In addition, the power metric (*vide infra*) by which differences may best be measured within the dimensional manifold can also be obtained directly from the data. These are perhaps the major advantages of multidimensional scaling. They allows us to measure and understand multivariate relationships even when the underlying dimensions are not known a priori. Modern multidimensional scaling (MDS) techniques can be applied even to nonmetric situations for which only ordinal information exists. MDS permits us to represent distances, orders, or similarities perceived among stimuli as spatial maps. Similar objects are located near one another on the map. Dissimilar stimuli occupy distant positions on the map. Maps may be of any dimensionality, from straight lines to 5- or 6-dimensional (or even more) representations of the least-dimensional manifold that best represents the data. The major disadvantage of MDS, effectively, the price to be paid for such sweeping versatility, is that it is relatively expensive and time-consuming to carry out.

There are basically three ways in which MDS differs from multivariate analytical techniques such as factor analysis. The first difference may seem small or even trivial at first glance: MDS deals with interpoint distances in space, whereas factor analysis, for example, concerns differences among vector angles in space. Even in common euclidean space, distance is an easier concept with which to deal than vector angle; in noneuclidean space, vector angles are confusing to all but practiced mathematicians and geometricians. Related to this fact is the generally larger number of dimensions associated with methods such as factor analysis when compared with the parsimony of dimensions typical of MDS results. Most multivariate analytical methods are based on an assumption of linearity among variables, and this is one of the principal reasons why more dimensions result from such techniques. MDS does not require assumptions of linearity or orthogonality. Neither does the method require that spaces be euclidean. This is particularly useful when dealing with data from experiments on visual perception, because the underlying relationships are almost never linear and frequently are noneuclidean as well. Thus requirements for linearity of relationships constitute a severe limitation on the utility of multivariate techniques when they are applied to visual perception, and at the same time the lack of such requirements for MDS methods provides a strong advantage for those techniques.

The third basic difference between the two classes of techniques for data gathering and analysis is that the experimenter is not obliged to impose his preconceived notions on the data or their analysis in MDS. In factor analysis, for example, the experimenter must bring a "shopping list" (Schiffman *et al.*, 1981, p. 14) of attributes to the experiment. The problem is that some of the items on the shopping list may be irrelevant, and some important attributes may not even appear on the list. When the data are analyzed according to the list of items, experimental noise is generated, the results are not always easy to interpret, and important (even primary) relationships may go unnoticed. MDS calls for no such preconceived notions about the outcome of experiments (composing a list of important attributes before the experiment is conducted surely represents a preconception of what its outcome should be in the mind of the experimenter); it merely gathers data about perceived similarities or differences and analyzes them in such a way as to yield the most parsimonious dimensional manifold that is consistent with the observations. MDS does not tell what those dimensions represent. This is the other side of the coin; attributes are not specified ahead of time, but neither are they identified in terms of variables in the outcome. That is a job for the experimenter, one in which he must bring to bear whatever intuition and creativity he has. The real challenge of interpretation in MDS experiments is to figure out what the derived dimensions might represent in terms of perceptual and physical attributes rather than as abstruse mathematical relations.

2. What MDS Does

MDS programs start with a set of data that represents scaled similarities (that is, measures of how alike things are) or dissimilarities (distances or differences, that is, measures of how unlike things are). These data are arranged in a starting set of coordinates, which are assumed or calculated by the computer. *Distances* between all pairs of data points are then computed and compared with the data; the differences are called *disparities*. A particular method of calculating distances must be used to measure distances and disparities. The experimenter can specify the method to be used. He can also specify the number of dimensions. The object is to determine disparities that are monotone transformations of the data that are as nearly as possible the same as the distances between original pairs of data points (usually according to a criterion not unlike least squares). Typically, the first arrangement will not yield sufficiently small residual differences. A measure of the size of the residual, called *stress*, is used to evaluate the effectiveness of the arrangement. The computer then rearranges the coordinates and recomputes distances. The result is again evaluated in terms of the stress measure. This process is repeated until a sufficiently low value of stress is obtained or, more frequently, until the reduction in stress with further rearrangement yields only a small, criterion decrease in the stress value. This entire process is usually carried out for a number of methods for calculating distance (called *power metrics*) and for each of several orders of dimensional manifold. Analysis of final stress values as a function of number of dimensions for each of the forms of power metric studied will generally indicate the optimal manifold that describes the original data. The computed organization of the data in an n-dimensional space comprises an n-dimensional map that is called a *configuration*. The final configuration represents the solution to a multidimensional scaling problem.

We will examine these basic concepts in more detail, but first we should examine the ways in which multidimensional scaling data can be gathered.

3. Gathering Data

Any of the scaling methods described in Chapters 8 and 9 can be used to gather MDS data. That is, ordinal, interval, or ratio scaling methods can be used.

a. Ordinal Methods: The Method of Triads

The object of ordinal MDS scaling is to determine the isomorphic ranking of all possible interpoint differences among the stimuli. There are a number of ways of doing this, but the most often used is called *the method of triads.* In this method the observer is presented with three stimuli and asked two questions about the order of their similarities: Which pair is most similar? Which pair is least similar? The process is repeated for all possible triads. Usually only all possible combinations of three stimuli are used (permutations are of little interest unless a left-right or top-bottom bias effect is suspected); thus the total number of triads to be shown is

$$N = \frac{n(n-1)(n-2)}{6}. \tag{16}$$

N increases rapidly with the number of stimuli n used. With 10 stimuli, for example, $N = 120$; with $n = 20$, $N = 1140$; with $n = 30$, $N = 4060$; and with $n = 100$, $N = 161{,}700$.

There have been several computer algorithms that treat ordinal (nonmetric) data of the kind obtained by a triad experiment to determine interpoint orders (for example, Torgerson, 1952; Shepard, 1962a,b; Kruskal, 1964a,b). Since the interpoint data are not distances but merely ordinal relations, the term *proximity* is most often used to describe interpoint relationships, thereby avoiding distinctions between metric and nonmetric data. A problem with many of the computer algorithms that treat ordinal data is that they require a complete *proximity matrix.* In other words, Eq. (16) applies to triadic experiments in which a complete proximity matrix is required, which poses problems when the number of stimuli is large.

How Many Stimuli Are Required? It is almost always desirable to have as many stimuli as possible from the standpoint of determining results as thoroughly as possible. At the same time, labor, time, and cost of experimentation should be kept at a minimum, which militates against large numbers of stimuli. This is particularly true when the method of triads is used, because as Eq. (16) shows, the total number of triads to be presented to each observer quickly becomes unwieldy when the number of stimuli increases beyond about 10. There is also another consideration that must go into the choice of number of stimuli, and this concerns a more fundamental problem. The problem is the number of stimuli that are required to perform the mathematical computations necessary to determine a given number of dimensions. More dimensions require more stimuli. I have found a simple rule of thumb to be useful: the minimum number of stimuli should be between 5 and 10 times the number of dimensions. Table II shows how this simple rule compares with other recommendations made by people who have worked

TABLE II. Recommended Numbers of Stimuli for Number of Dimensions

Number of dimensions	Minimum number of stimuli	Source
1	5–10	This text
1	6	Spence & Domoney (1974)
1	6	Young (1970)
2	10–20	This text
2	9	Kruskal & Wish (1978)
2	11	Spence & Domoney (1974)
2	11	Young (1970)
2	12	Schiffman, Reynolds, & Young (1981)
3	15–30	This text
3	13	Kruskal & Wish (1978)
3	17	Spence & Domoney (1974)
3	17	Young (1970)
3	18	Schiffman, Reynolds, & Young (1981)
4	20–40	This text
4	17	Kruskal & Wish (1978)

with multidimensional scaling. In addition, most computer programs also set a minimum for the number of stimuli (or at least total number of observations) that may be used in determining a specified number of dimensions.

b. Interval Methods

In large part because Eq. (16) yields such large numbers of stimulus combinations, many multidimensional scaling experiments are carried out with all combinations of pairs, where

$$N = \frac{n(n-1)}{2}. \tag{17}$$

Here, however, ordinal scaling in inappropriate. Remember that it is the proximities among stimuli, not the stimuli themselves, that are being scaled. It is not enough to say that one stimulus has more or less of something than does another. What we need to know is how relatively similar are the stimuli, that is, what are the differences between members of a pair compared with all possible differences among all pairs? To rank-order all possible differences would require that they all be presented at once or that some efficient scheme be devised for presenting and combining data from some subset of all possible differences. The most efficient such subset is that of the method of triads. Therefore, at least interval-level data must be gathered about the differences between members of all possible pairs if only one pair is presented at a time.

Either of the direct interval scaling methods described in Chapter 8 can be used to obtain such data: rating scales or successive categories methods. My preference for an interval method for use in multidimensional scaling is a 9-point, numerical, category scale with raw data entered for each observer (or a simple category mean for an average proximity matrix). The proximity matrix can be treated as either interval or ordinal level data. That is, even though interval data were gathered, the proximity matrix may be assumed to represent nothing more mathematically powerful than ordinal information. The advantages of adopting such a nonmetric approach are that one need not make assumptions about the power of the data (did the observers really respond according to uniform intervals or did they do something else?), a broader choice of computer algorithms is available, and the computer algorithm is permitted to determine distance functions without imposing the experimenter's hopes or notions on the raw data. This kind of treatment is possible today only because multidimensional-scaling programs are available that allow an experimenter to select the level of mathematical power at which the data are to be treated; in the early days of MDS, only ordinal, nonmetric options could be used.

A representative set of instructions for an 11-point rating scaling of pairs in which a specific attribute is of interest would be:

> You will be presented with each of a number of pairs of samples. We would like to know how much they differ in (attribute). If they are extremely different, call their difference 10; if they are the same, call the difference 0. Use numbers ranging from 0 to 10 to indicate how different you think they are.

If a specific attribute is not sought, but merely an overall difference, then the words "in (attribute)" would be deleted. The instructions illustrated scale dissimilarities. The same kind of instructions could be used to scale similarities. The same numerical scale could be used. Either similarities or dissimilarities can be retrieved from the same scale, because one is considered to be the complement of the other.

A few words about zeros are called for at this point. In Chapter 8 we saw that zeros create problems in interval scaling, because their standard normal deviates and their logarithms are indeterminate. In Chapter 9 it was pointed out that zeros create problems in magnitude estimation, because geometric means and their logarithms are indeterminate. Zeros also cause problems in multidimensional scaling, but for a different reason. Many computer programs treat zeros as missing values. In some programs, counterinstructions can be given. If not, the presence of a zero may bias the results. Obviously, if we instruct observers to assign a value of zero to the difference between identical members of a pair, we must expect to encounter the odd zero from time to time. For this reason, some workers prefer not to allow observers to use zeros, but this imposes a rather unnatural situation on the observers. If someone sees no difference, he should be entitled to say so in a natural way and that means saying that he sees zero difference. It might seem perfectly reasonable to add a constant to all scale values, because the interval scale is invariant over transforms of the form $y = ax + b$. However, adding a constant biases the multidimensional intercept position; it distorts the multidimensional map according to the size of the added constant. What are we to do then? The best approach is to use a computer program that does not treat zeros as missing values. If that is not possible, then a combination of two things can be done. First, add a very small constant, say, one-tenth to one-hundredth of the maximum scale value. Second, analyze the data according to the next lowest level of power (this may not always be necessary, but it is a prudent step). If the data gathered are of interval level, treat them as ordinal data. In this way problems with zeros can be avoided.

c. Ratio Methods

Magnitude estimation can also be used to scale similarities or dissimilarities between members of all possible pairs (although scaling the magnitudes of dissimilarities or differences seems the logically more straightforward alternative). Again, Eq. (17) defines the number of pairs for any given number of stimuli. There are, however, some sorting algorithms that may be used to reduce the labor of experimentation, just as with paired-comparison interval-scaling techniques described in Chapter 8. Spence (1977), Young and Cliff (1972), and Young *et al.* (1978a) have proposed algorithms for interactive similarity scaling that reduce the number of pairs to between a fourth and a half that required for a complete proximity matrix. Some workers have applied sorting techniques to the scaling itself rather than to determine interstimulus proximities (for example, Boorman and Arabie, 1972; Rosenberg and Sedlak, 1972), a method somewhat similar to cluster analysis discussed in Section D.4 of this chapter.

When magnitude estimations are gathered, precautions must be taken to avoid treating the data as if they formed an interval metric. In Chapter 9 it was pointed out that, for reasons that remain inexplicable, scales determined by magnitude estimation and by interval scaling are nonlinearly related. Accordingly, if dissimilarities are scaled by magnitude estimation, similarities cannot be assumed to be the complement; instead they would be represented by reciprocals of the dissimilarities. In addition, multidimensional manifolds of log intervals, or ratios, are difficult to interpret. Therefore, some method should be found to convert the magnitude estimations to something that approximates intervals or interstimulus distances. The square-root approximation discussed in Chapter 9 may be helpful in this regard. I find that magnitude estimations of dissimilarities are quickly and easily obtained from trained observers. When the square roots of the geometric-mean proximity matrix are taken, and these transformed proximities are entered as interval or ordinal level data, multidimensional programs such as SAS ALSCAL and KYST (*vide infra*) yield results that are useful, and when I have compared them with results from interval (category) scaled data analyzed by the same programs, I have found only small differences between the two sets of results.

A representative set of instructions for scaling dissimilarities by magnitude estimation would be as follows:

> You will be shown each of a series of pairs of samples. We are interested in how different the members of each pair are. They may differ in a number of ways. We are not interested in any one kind of difference any more than any other. That is, what we want to know is your impression of the *overall* difference between samples. Please assign a number to the size of the overall difference that you see between the samples in each pair. If the difference is large, you should use a relatively large number; if it is small, use a small number. The numbers should reflect your impression of how large or small the differences are. You may use any numbers you wish, whole numbers, fractions, decimals, but please use numbers that you think are the most appropriate to the sizes of differences that you see.

These instructions call for direct estimation of the *sizes* of differences. A more direct determination of the ratios in which members of a pair stand with respect to a specified attribute can be obtained by changing the instructions slightly:

> You will be shown each of a series of pairs of samples. We are interested in how different the samples are in (attribute). We would like you to tell us how many times greater in (attribute) the greater sample is. If it is twice as (attribute), then you should assign the number 2; if only one-and-a-half times as (attribute), then assign a ratio of 1.5. When the two members of each pair have the same (attribute), then they stand in a ratio of 1. In short, simply tell us which sample is greater and the ratio of the (attribute) of that sample to the other by using numbers equal to or larger than 1. The ratios should reflect your impression of how many times more (attribute) you see in the greater member of each pair.

Here the instructions set the stage for the purest estimates of ratios that it is probably possible to obtain. In addition, there are no difficulties with zeros, because the scale runs between unity and some higher number. Square roots of the ratio responses can be taken, and the exact similarities will still be represented by ones on the transformed scale.

4. Kinds of MDS Analysis

Several kinds of MDS analysis have been developed. For this exposition they can be classified according to three quite general categories: classical MDS methods, multidimensional unfolding methods, and weighted MDS methods. The following sections will describe each of these methods briefly.

a. Classical MDS

Modern MDS methods developed from attempts to overcome the problems resulting from requirements for linearity, interval level power, and larger-than-necessary dimensionality associated with factor analysis multivariate methods. Attneave (1950) analyzed similarity judgments according to fixed, specified dimensions which he inferred from theoretical and empirical considerations. Other early attempts at multidimensional scaling include Richardson's (1938) method of triads, Klingberg's (1941) multidimensional rank order, Torgerson's (1952, 1958) method of complete triads, and work by Young and Householder (1938) on interpoint distance estimates. It was not until the 1960s, with the work of Shepard (1962a,b) and Kruskal (1964a,b), that MDS became practical and reasonably popular, however. The key to the popularity of the Shepard–Kruskal approach lies in the assumption that input data represent only ordinal information. This means a nonmetric approach to spatial mapping can be used in which proximities derived by any of a variety of scaling techniques are represented in space according to the order of interpoint distances. Earlier methods scaled proximities as distances directly. The result of the nonmetric assumption is that minimum dimensionality and monotonicity are more easily determined regardless of the scale power of the input data. Kruskal's major contribution was to determine a measure of goodness of fit called stress.

Basically, classical MDS represents stimuli as points in a manifold such that their interrelations are properly ordered. Once the positions of the points are determined in the manifold, their coordinates may be specified and interpoint distances computed from them. The procedure requires a square input matrix in which the number of columns equals that of the rows (representing all interpoint proximities), a choice of dimensionality (that is, how many mathematical dimensions comprise the manifold), a model for a power metric by which distances are measured in the manifold, and a measure of stress (goodness of fit) which compares the residual differences between distances computed in the manifold with those deduced from the input data. Each of these features will be discussed in the following sections, but first it will be helpful to describe two classes of variations on classical MDS methods: multidimensional unfolding and weighted MDS.

b. Multidimensional Unfolding

Coombs (1964) developed an approach to preference scaling called *multidimensional unfolding*. It is used today to treat rectangular matrices of data (in which the numbers of rows and columns are unequal). For example, six attributes of ten stimuli may be evaluated on rating scales, forming a 10×6 rectangular matrix. Multidimensional unfolding (MDU) permits us to represent both modes (that is, the ten rating scales and the six stimuli) in a joint manifold. In such a joint manifold, each stimulus is located in relation to its position on the rating scales. In the simplest case, where preference for chromaticity of near-white cloths is rated, for example, the stimulus specifications of the cloths will be located near the mean preferences expressed for them. MDU can be performed for each of the observers as well as for their mean. In this way, differences among the individual's preferences can be displayed in the joint manifold.

For technical reasons, there are often severe computational difficulties in performing MDU analyses, but similar results can be obtained by replicated MDS techniques in which different weights are assigned to the matrices representing data from different observers. More generally, different weights can be assigned to different matrices. When the matrices are square and represent proximities, the distances in the manifold correspond to similarities among members of pairs of stimuli. When the matrices are rectangular, distances in the manifold represent the degree of relation between stimuli and preference ratings (or any other set of things). Often the matrices (either square or rectangular) represent data from different observers, and the analyses are carried out in such a way as to indicate individual differences in scaling. Because this is a form of weighted MDS, it will be discussed in the next section, which deals with that subject.

c. Weighted MDS

Carroll and Chang (1970) developed a computer program called INDSCAL in which a matrix of weights is used in conjuction with each observer's data matrix. Thus, in *weighted multidimensional scaling* (WMDS), each observer has a personal solution-space. Usually the observer weights are the square roots of the proportion of variance in the transformation for each subject, and these can be plotted as vectors of the dimensions to examine both how well each observer performs and how much agreement exists among observers. Vector angle is taken to represent degree of agreement, and vector magnitude is taken as an index of performance.

These weights may be used to form a single group stimulus manifold as well as a manifold for each observer. The scaled data are averaged over observers for the group constellation, and the product of stimulus data and individual weights for the dimensions are plotted to represent individual differences in the same solution-space.

A number of computational variations have been proposed for WMDS (for example, McGee, 1966, 1968; Green and Rao, 1972; Shepard *et al.*, 1972; Green and Carroll, 1976; Kruskal and Wish, 1978; Borg, 1981; Schiffman *et al.*, 1981), and a number of computer packages include one or more of them. In addition, some computer programs will accept metric as well as nonmetric data.

5. MDS Input and Output

It will be helpful for the reader to understand a few characteristics of MDS data before proceeding to a description of some MDS programs. These include data matrices, levels of measurement of data, dimensions, power metrics, and stress.

a. Data Matrices

The form in which data are input to an MDS program is in the arrangement of a table, that is, a matrix of numbers that displays the experimental measurements in some orderly way. Basically there are two forms of data matrices: square and rectangular. Square matrices represent proximities, in which interpoint distances between all pairs of stimuli are represented. For example, if there are 4 stimuli and all 16 permutations of pairs were judged, the *square proximity matrix* would be of the form:

Stimuli		Stimuli: *A*	*B*	*C*	*D*
	A	δ_{11}	δ_{12}	δ_{13}	δ_{14}
	B	δ_{21}	δ_{22}	δ_{23}	δ_{24}
	C	δ_{31}	δ_{32}	δ_{33}	δ_{34}
	D	δ_{41}	δ_{42}	δ_{43}	δ_{44}

where the δs represent the data determined experimentally. It is more common to obtain data for the combination of pairs, rather than for all permutations, and to assume that any one stimulus, when compared with itself, would yield a zero proximity measure. Accordingly, most square data matrices are represented only by the *lower-half, triangular matrix*:

Stimuli		Stimuli: *A*	*B*	*C*	*D*
	A	—	—	—	—
	B	δ_{21}	—	—	—
	C	δ_{31}	δ_{32}	—	—
	D	δ_{41}	δ_{42}	δ_{43}	—

The lower-half matrix represents a square matrix. The complete square matrix is said to be symmetric, because corresponding elements are assumed to be equal; that is, $\delta_{41} = \delta_{14}$. However, a complete square matrix in which all elements (or all except the diagonal elements) are determined experimentally need not be symmetric; it may be that the experimental results are such that

$\delta_{41} \neq \delta_{14}$. Most MDS computer programs accept lower-half matrices and assume that they represent symmetric square matrices.

The second form of data matrix is rectangular. A rectangular matrix may represent proximities among pairs of stimuli but more often is used to array ratings against stimuli. An example of such a matrix is:

		Stimuli			
		A	B	C	D
Attributes	1	ρ_{11}	ρ_{12}	ρ_{13}	ρ_{14}
	2	ρ_{21}	ρ_{22}	ρ_{23}	ρ_{24}
	3	ρ_{31}	ρ_{32}	ρ_{33}	ρ_{34}
	4	ρ_{41}	ρ_{42}	ρ_{43}	ρ_{44}
	5	ρ_{51}	ρ_{52}	ρ_{53}	ρ_{54}
	6	ρ_{61}	ρ_{62}	ρ_{63}	ρ_{64}

,

where the ρs represent ratings according to designated attributes. A rectangular matrix cannot be symmetric, of course, but it can be either complete or incomplete. Some MDS computer programs can treat incomplete rectangular matrices, generally by procedures for making assumptions about what to do with missing values.

b. Measurement of Levels of Data

Chapters 8 and 9 described properties of data scales according to Stevens' (1951) classification of measurement level. The four levels discussed were nominal, ordinal, interval, and ratio. MDS experimental data can be gathered according to any of these four levels. Some MDS computer programs treat any measurement level of input data as ordinal, but more often there is a choice that may be made in selecting the algorithm level to be used by a particular program. It is sometimes desirable to select an algorithm measurement level that is lower in power than that of the experimental data, but the level selected should not be higher than that corresponding to the experimental data. Often, choosing an ordinal level for the computer algorithm will provide results that are not much different from those determined for the measurement level of the data, and such a choice sometimes helps to avoid imposing the experimenter's preconceived notions about the nature of the data on the results. It should be stressed again that although the experimenter instructs observers to do something, there is no guarantee that they actually do it.

c. Dimensions

Determination of the dimensionality of the manifold that best represents the solution to an MDS problem involves a judgment and a decision by the experimenter. What is judged is whether additional dimensions improve the goodness of fit of the solution. They almost always do so, but the extent to which they do tends to decrease as the number of dimensions increases. The decision that must be made, then, is whether the additional complexity of added dimensions significantly improves the utility of the results.

There are two inferential statistics that can aid in assessing the extent of improvement in goodness of fit. One is the coefficient of determination r^2, and the other is stress S. Figure 6 illustrates a typical form of the changes in r^2 and S as the number of dimensions of a solution manifold increases. The data relate to an MDS experiment in which the colorfulnesses of nine chroma samples of constant luminance factor were determined (one of the examples of the application of MDS that will be discussed later in this chapter). The graph shows that r^2 increases rapidly up to two dimensions but does so only very slowly beyond two dimensions. Conversely, S decreases rapidly up to two dimensions, but the decrements are smaller beyond that point. The relationship between S and number of dimensions typically has a "bend at the elbow," and this is usually a good indication of the minimum dimensionality of an appropriate manifold. In the example of Fig. 6, the bend occurs at an abscissa value corresponding to two dimensions.

The r^2 relationship also shows an abrupt change at two dimensions. However, the S values indicate somewhat more separation between adjacent pairs of values for higher dimensions (when both ordinate scales are

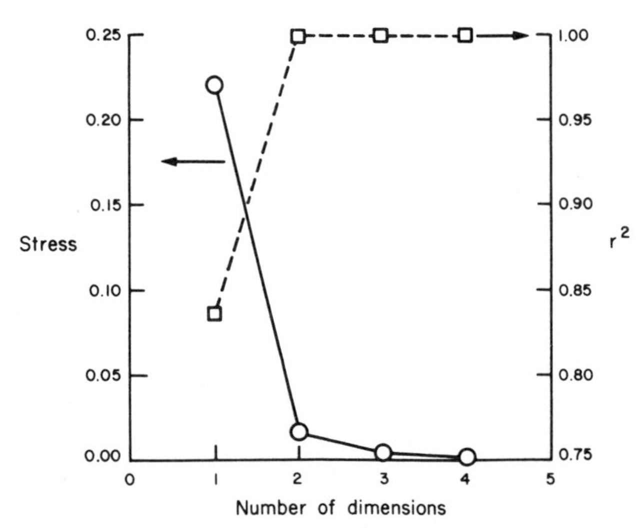

Fig. 6. Stress (formula 1), shown on the left, and r^2, shown on the right, plotted against number of dimensions for configurations of 1, 2, 3, and 4 dimensions for data from color scaling. See text for discussion.

represented by intervals of the same size). Sometimes the bend is not as sharp as that shown in Fig. 6. The decision is not as easy to make in such cases; so plotting both S and r^2 may be helpful. However, S values are generally to be preferred, because they are measures specifically designed for evaluating goodness of fit of MDS configurations.

d. STRESS

The purpose of classical MDS is to match the order of the input proximity measures δ_{ij} with a set of distances d_{ij} in the solution configuration. Shepard (1962a,b) plotted proximity ranks against derived distances to provide an indication of goodness of fit that is somewhat similar to the correlogram of standard regression analysis. Kruskal (1964a,b) carried this analogy with regression analysis one step further by deriving a quantitative measure of fit based on an evaluation of residual differences. When the Shepard diagram relation (δ_{ij} vs d_{ij}) is forced to be monotone, the disparity distances $\hat{d}_{ij}$ may be measured along the abscissa (of d_{ij} values) to represent the departures from δ_{ij} along the monotonic line. The squared deviations of d_{ij} from $\hat{d}_{ij}$ are summed for all pairs of points in the manifold to provide a measure S^*, called *raw stress*.

$$S^* = \sum_{i=1}^{N} \sum_{j=i+1}^{N} (d_{ij} - \hat{d}_{ij})^2. \tag{18}$$

To normalize the value of stress, a scaling factor T_1^* is used,

$$T_1^* = \sum_{i=1}^{N} \sum_{j=i+1}^{N} d_{ij}^2. \tag{19}$$

The first formula for *stress* is then expressed as

$$S_1 = (S^*/T_1^*)^{1/2}. \tag{20}$$

An alternative scaling factor T_2^* was proposed by Kruskal and Carroll (1969), which is based on the deviations from the average interpoint distances ($\bar{d}_{ij}$).

$$T_2^* = \sum_{i=1}^{N} \sum_{j=1+1}^{N} (d_{ij} - \bar{d}_{ij})^2, \tag{21}$$

which provides a second measure of stress,

$$S_2 = (S^*/T_2^*)^{1/2}. \tag{22}$$

There are a number of other measures of goodness of fit (for instance, Schiffman, *et al.*, 1981), but those of Eqs. (20) and (22) are probably the most commonly used. They are often referred to as stress formula 1 and formula 2, respectively. Each is really an inverse measure of how well the solution fits the data, and for that reason they are sometimes called badness of fit indices: the larger the value of S_1 or S_2, the poorer the fit. That is why we seek a practical minimal value for stress (see Fig. 6).

e. POWER METRICS

The distance between any two sets of points i and j has been symbolized in the preceding paragraphs as d_{ij}. Nothing has yet been said about how d_{ij} is measured. In other words, the metric for distance has not been specified, but it is necessary to do so when attempting to solve any real problem. There are many kinds of metrics, but the class of metrics that is most commonly used in MDS is referred to as the *power metric* or often, in general form, as the *Minkowski metric* (after the Russian-born mathematician Hermann Minkowski from whose work the form derives).

The concept of metric distance requires that a valid measure is one that can be described by a mathematical function having all four of the following properties:

1. The distance between any two points is never negative: $d_{ij} \geqq 0$.
2. The distance between two identical points is always zero: $d_{ii} = 0$.
3. Distance is symmetric: $d_{ij} = d_{ji}$.
4. The sum of the distances between two points by way of a third point is always equal to or greater than the direct distance between the two original points: $d_{ij} + d_{jk} \geqq d_{ik}$.

The Minkowski metric meets these requirements and because of its generality includes many forms of geometrical relationships. It may be expressed as

$$d_{ij} = \left(\sum_{\kappa=1}^{N} [x_{i\kappa} - x_{j\kappa}]^r \right)^{1/r}, \qquad 1 \leqq r \leqq \infty, \tag{23}$$

where x is the projected distance of points i and j along κ coordinates, N is the number of dimensions, and r is the power of the metric.

When $r = 1$, distance d_{ij} is simply the sum of the projected distances along the κ coordinates; this particular geometry is often called the city block model, because it resembles the distance that must be traversed when walking from one location to another along city streets that are laid out in a rectangular pattern. When $r = 2$, the model is the familiar euclidean one in which distance is the square root of the sum of the squares of the coordinate distances (from the Pythagorean theorem). Although Minkowski metrics of all powers r may be translated, only the euclidean model (where $r = 2$) may also be rotated. When $r = \infty$, distance is equal to the largest coordinate distance, that is, the maximum coordinate difference dominates all other coordinate differences. In a two-dimensional graph (for which $\kappa = 2$), distances of equal size in the all-positive quadrant would be arrayed along a diagonal line of slope $= -1$ when $r = 1$; when $r = 2$, such isodistances would be described by a circle whose center is located at the origin; and when $r = \infty$, constant distance is described by a pair of lines, parallel to the abscissa and to the ordinate, that meet at a point diagonally opposite to the origin. Powers of r between 2 and ∞ form isodistance contours that are increasingly boxlike as the value of r approaches ∞.

Some computer programs permit a choice of integer values for r; in other words it is not always necessary to assume that the data can be described only in an ($r = 2$) euclidean metric. Note, however, that rotation of the manifold applies only to the case of $r = 2$. As with the number of dimensions deduced from measures of stress, it is possible to determine the value of r that is most appropriate for the data from stress measures. This is done by plotting the values of stress (computed for a selected number of dimensions) against values of r. The graph that results is usually U-shaped with minimum stress corresponding to the optimum value of r. By determining solutions for several values of r for each of several numbers of dimensions, it is usually possible to find both the optimum dimensionality and power metric for a given set of data. The ability to make such determinations directly from the experimental data is one of the most important advantages of MDS.

6. Some MDS Computer Programs

There are a dozen or more MDS computer programs available at present. This section will describe briefly only six programs, four of which are general-purpose MDS programs and two that are designed specifically to analyze individual differences. In each case the available options and details of the programs are much more numerous than those that are set forth here. The reader who wishes to pursue further details of one or more of the programs will find references cited in connection with those that are discussed here. In addition, comparisons of these and other MDS programs have been published recently by Schiffman *et al.* (1981), a practical text that is highly recommended for understanding and using currently popular programs.

a. ALSCAL

ALSCAL, an acronym for Alternating Least-Squares sCALing, is an MDS routine developed by Takane *et al.* (1977) and Young *et al.* (1978b). The program and documentation entitled "ALSCAL-4 User's Guide" is available from Forrest W. Young, L. L. Thurstone Psychometric Laboratory, University of North Carolina, Chapel Hill, North Carolina 27514. A version called PROC ALSCAL, with similar input and identical output features, has been incorporated into the SAS package of programs (Helwig and Council, 1979; Reinhardt, 1980) and is available from The SAS Institute, P.O. Box 10066 Raleigh, North Carolina 27605.

ALSCAL (and SAS PROC ALSCAL, which will not be further distinguished in this section) is perhaps the most easily used and powerful MDS program available today. It provides great flexibility with simple instructions and robust computational characteristics. Observations from a number of subjects may be treated in a variety of ways: as replications, as individual differences, and with analysis according to ordinal, interval, or ratio levels to derive configurations of 1 to 6 dimensions. Its major drawback is that it deals only with euclidean power metrics (where $r = 2$). This is because the program operates in alternating least-squares minimization of a measure called SSTRESS that involves squared distances. Table III summarizes some of the features of ALSCAL and compares them with those of five other programs.

b. KYST

KYST is an acronym for the names of those workers whose programs were modified, rearranged, and combined to provide the refinement: Kruskal, Young, Shepard, and Torgerson. KYST-2A (the latest version at the time of this writing), together with documentation and user's guide (Kruskal *et al.*, 1977), is available from Bell Laboratories, Computing Information Library, Irma Biren, Supervisor, 600 Mountain Avenue Murray Hill, New Jersey 07974.

KYST provides everything that ALSCAL does except individual differences. However, it does permit use of general Minkowski metrics (which ALSCAL does not). In addition, KYST permits adjustment of the threshold value for missing data with a minimum (and default) level of -1.23×10^{20}; therefore, zeros can be dealt with easily. Table III gives a simplified summary of KYST's features.

c. MINISSA

MINISSA, an acronym for Michigan–Israel–Nijmegen-Integrated-Smallest-Space-Analysis (Roskam and Lingoes, 1970; Lingoes *et al.*, 1979), is a useful program for analyzing a single proximity matrix representing results for one observer or the mean over several observers. It is considerably less flexible than those programs referred to above. Balanced against this, however, is its appealing simplicity and two useful options, one for dealing with zeros and missing values and a second for adding more points to a fixed manifold.

The program is available from James C. Lingoes, Computing Center Station, 1005 North University Building, University of Michigan, Ann Arbor, Michigan 48109.

d. POLYCON

POLYCON is quite similar to KYST, differing mostly in the degree of internal complexity of the source program, a maximum of 5 rather than 6 dimensions, and the fact that it is constrained to Minkowski powers of only 1 and 2. It is a model for conjoint analysis of n-way data, where $n = 1$, 2, or 3 (Young, 1972, 1973).

The program is available from Forrest W. Young, L. L. Thurstone Psychometric Laboratory, University of North Carolina, Chapel Hill, North Carolina 27514.

TABLE III. Simplified Summary of MDS Program Characteristics

	Input data									Power metric			Stress Minimization		
Program name	Metric	Nonmetric	Square matrix	Rectangular matrix	1 Matrix	>1 Matrix	WMDS	MDU	Maximum number of dimensions	City-block model	Euclidean	General Minkowksi	Least-squares distance	Least-squares $(distance)^2$	Maximum likelihood
ALSCAL	X	X	X	X	X	X	X	X	6		X			X	
INDSCAL	X		X			X	X		10		X			X	
KYST	X	X	X	X	X	X		X	6	X	X	X	X		
MINISSA		X			X				10	X	X		X		
MULTISCALE	X		X		X	X	X		10		X				X
POLYCON	X	X	X	X	X	X		X	5	X	X		X		

e. INDSCAL

INDSCAL (INdividual Differences multidimensional SCALing) is a program developed by Carroll and Chang (1970) for analyzing spatial differences among the solution manifolds of individual observers. It is available from the Bell Laboratories Computing Information Center (see KYST). Basically, it is a model for multidimensional three-way analysis, assessing distances among stimuli and observers' responses to their proximities. The program, therefore, requires a minimum of two data matrices (and has a maximum limit of about 100 matrices). It can provide solutions in euclidean spaces of 1 to 10 dimensions. A simplified summary of INDSCAL features is given in Table III. The program is easy to use and is probably the best known of those designed for analysis of individuals' data.

f. MULTISCALE

MULTISCALE (Ramsay, 1977, 1978) also provides analysis of individuals' data. In addition, it has certain other features that make it unusual among MDS programs. As indicated in Table III, MULTISCALE derives a solution by the method of maximum likelihood rather than least-squares techniques. For technical reasons, this and other features of MULTISCALE permit application of a number of statistical inference tests, such as χ^2, so that the program can be used to confirm whether the data and results satisfy hypotheses about the experimenter's expectations for them. Most other MDS programs merely provide us with pictures of the relationships among the data; MULTISCALE permits us to test the inferences that are drawn from such pictures by applying statistical testing techniques. As with INDSCAL, MULTISCALE is often considered to complement some other MDS routine. It may stand alone as an MDS program, however. As such, its features are similar to those of INDSCAL, as indicated in Table III, but its methods are very different.

MULTISCALE is available from International Educational Services, 1525 East 53rd Street, Room 829 Chicago, Illinois 60615 together with program documentation (Ramsay, 1978).

E. EXAMPLES OF MDS

This section will give two examples of multidimensional scaling. One deals with multidimensional unfolding to derive a configuration that represents certain aspects of preferences for illumination in a lounge area. The second addresses the problem of scaling chromatic samples that is common to Chapters 8 and 9. In both cases the SAS ALSCAL routines will be used.

1. Differences in Preferences among Luminaires

Some years ago I had occasion to investigate the preferences of 25 observers for illumination conditions in a waiting room. Six different sets of luminaires were installed at different times. The luminaires differed in correlated color temperature from "warm" to "cold" in appearance, in geometry of illumination from specular to diffuse, and in their light output from bright to dim. The observers rated each of these characteristics and their preferences on a graphical rating scale. The adjectives set forth above represented the end points of the rating scales (with pleasant and unpleasant representing the preference end points). These data have been used here as input to the ALSCAL procedure of the SAS package of programs (Reinhardt, 1980; Young and Lewyckyj, 1980). The data should not be taken to imply anything about general preferences for illumination, even in conjunction with similar applications; the task and questions at issue when the data were gathered were quite specific to the particular conditions studied. However, the data do provide means for demonstrating how an MDS program such as ALSCAL can be used to derive information of the kind commonly obtained by MDU methods.

Because it was desirable to determine how the warm–cold, specular–diffuse, and bright–dim aspects of the situation related to one another as well as how they related to pleasantness–unpleasantness, two numbers were determined from each rating. One represented the proportion (expressed in percent of total line length) of bright, and the other (its complement) was taken to represent dim, for example. The same procedure was used for each of the other ratings. The data matrix, averaged over all 25 observers, is the 8×6 element matrix shown in Table IV. The purpose of using ratings for both directions on the scale was to determine two points in the configuration that could be taken to define the direction of variation for each attribute scaled.

TABLE IV. Dissimilarity of Luminaire Systems, Data for Mean of 25 Observers

	OBS	L1	L2	L3	L4	L5	L6
Bright	1	14	26	45	63	85	96
Cold	2	18	40	42	90	90	19
Specular	3	25	36	48	60	79	88
Pleasant	4	35	82	80	38	40	34
Dim	5	86	74	55	37	15	4
Warm	6	82	60	58	10	10	81
Diffuse	7	75	64	52	40	21	12
Unpleasant	8	65	18	20	62	60	66

TABLE V. File: ALSCAL SAS

```
(1) Options Center;
(2) CMS FILEDEF INPUT DISK LAMP DATA A1;
(3) TITLE DISSIMILARITY OF LUMINAIRE SYSTEMS;
(4) TITLE2 DATA FOR MEAN OF 25 OBSERVERS;
(5) DATA;
(6) INFILE INPUT;
(7) INPUT (L1-L6) (6*3.);
(8) PROC PRINT;
(9) PROC ALSCAL PLOT SHAPE = RECTANGU ROWS = 8 LEVEL = INTERVAL
    DIMENS = 2.
```

Table V illustrates the simplicity of program commands required to perform the SAS ALSCAL analysis on an IBM terminal with batch processing in an IBM 4310 computer. Lines 1–7 are merely commands for locating the data file, titles, and read format. Line 8 calls for a printout of the data (as shown in Table IV). Line 9 invokes the ALSCAL routine and indicates that a rectangular matrix with eight rows is used as input, that the data level is interval, and that the solution is to be determined for two dimensions. With these simple statements the program provides the output shown in Table VI and Fig. 7.

Table VI lists coordinates for the columns (luminaires) and rows (attributes). In addition, values for r^2 and stress (formula 2) are shown in the table. The symbols used to plot each set of coordinates in Fig. 7 are also given in Table VI. That figure shows that although the stress value is not particularly low, the configuration is useful in illustrating the relationships sought.

Lines have been drawn through the coordinates representing the complementary points on each rating scale. The lines indicate that (for the conditions of this particular experiment) dim–bright and diffuse–specular variations are arrayed in a manner similar to those of the pleasant–unpleasant axis. Warm–cold varies along a line that is nearly at right angles to the pleasant–unpleasant axis. Therefore, we might conclude that the correlated color temperature of the lighting made little or no difference to how pleasant the installation was judged to be by these observers. Specularity and brightness did make a difference in pleasantness, however, because those axes are nearly coincident with that of pleasantness.

The position of each of the six lamps with respect to these axes is also shown in the figure. To determine the rating position of any one lamp with respect to a particular attribute it is only necessary to construct a line from the lamp position perpendicular to the axis in question. Lamp 6, for example, was found to provide the most pleasant conditions. It was judged to be a cold lamp. The next most pleasant situation was that provided by Lamp 5, a warm lamp. In

TABLE VI. Dissimilarity of Luminaire Systems Data for Mean of 25 Observers, Configuration Derived in 2 Dimensions, Stimulus Coordinates

Stimulus number	Plot symbol	Dimension 1	Dimension 2
Column			
1	1	1.2807	−0.8323
2	2	1.3535	−0.1056
3	3	1.0388	−0.4094
4	4	−0.9911	1.1716
5	5	−1.4827	0.8213
6	6	−0.8855	−1.3831
Row			
1	7	1.4143	0.8110
2	8	0.3825	−1.5500
3	9	1.2168	0.8148
4	A	−1.1365	−0.6066
5	B	−1.2989	−0.2969
6	C	−0.4865	1.2636
7	D	−1.0592	−0.3169
8	E	0.6539	0.6184

Note: Stress and squared correlation RSQ in distances; Kruskal's stress formula 2 is used; stress = 0.203; RSQ = 0.961.

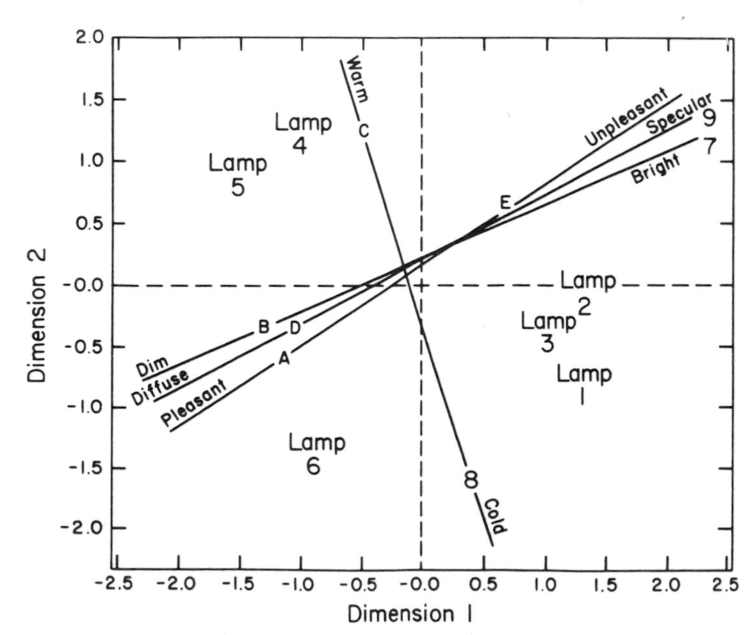

Fig. 7. Multidimensional unfolding configuration showing axes of attributes, as labeled, and coordinate positions of samples. See text for discussion.

this way, by inspection of the configuration, it is possible to draw a number of inferences about the relationships among the lamps tested and the attributes evaluated. A single experiment, requiring only four rating-scale decisions for each of six luminaires, yielded data that, when input to the MDS program with only those commands shown in Table V, resulted in the configuration of Fig. 7. This configuration provides a picture that conveys a wealth of information to the experimeter. It would be difficult to obtain as much information with so little investment in experimental labor and analysis by techniques other than those offered by such multidimensional scaling techniques.

2. Color Scaling of Nine Chromatic Samples

The second example of an MDS problem is one that involves the same colored papers that were scaled for chromaticness in Chapters 8 and 9. Each of ten observers made magnitude estimations of the differences in color appearance among all pairs of the nine colored papers. Ten times the square roots of those magnitude estimations were entered as proximity measures in the data matrices of Table VII. These individual, lower-half, triangular matrices were input to SAS ALSCAL using commands as simple as those illustrated in Table V for the previous example. In this case, however, the complete matrix shape was square, and only an ordinal level was called for in the analysis. Whether one chooses to accept the validity of the square-root relationship between magnitude- and interval-scaling results, there is no question that the data must represent at least ordinal relations. The program was then commanded to provide configurations for the individual observers and for their mean.

Figure 8 shows the coordinate positions for each of the nine stimuli, plotted for each observer. The numbers represent the stimuli. Contours have been drawn around the scatter of results for each of the stimuli. The "+" represents the mean observer's coordinates within each of the contours. There is some scatter, that is, the results differed somewhat among observers. Figure 9 illustrates this scatter in a graph of distances versus disparities. The amount of scatter shown in that figure is reasonably representative of what should be expected in a successful experiment of this kind.

Table VIII lists the stress and r^2 values found for each of the ten observers' configurations. They are all satisfactory although they do vary in stress over a ratio of about 6:1 (from 0.016 to 0.099). Similar information can be obtained by examining the individuals' weights along the two dimensions. Because the weights are approximately the square roots of the variances of the transformations, we would expect a situation in which there are relatively high r^2

TABLE VII. Example of Colorfulness Scaling, Two-Dimensional Analysis

D1	D2	D3	D4	D5	D6	D7	D8	D9	Subject
									A
39									A
26	38								A
18	59	26							A
7	45	37	13						A
26	65	43	18	22					A
38	74	62	37	33	24				A
20	38	43	31	17	37	38			A
32	23	46	45	35	56	57	20		A
									B
4									B
3	4								B
2	6	3							B
1	4	3	2						B
3	7	4	2	2					B
4	7	5	3	3	2				B
2	4	4	3	2	4	3			B
3	3	5	4	4	6	6	2		B
									C
65									C
45	70								C
35	100	50							C
15	90	65	25						C
50	125	80	35	45					C
70	150	120	75	65	50				C
40	75	85	60	35	70	80			C
65	45	90	90	70	110	110	40		C
									D
46									D
34	35								D
20	50	20							D
5	50	35	20						D
20	60	40	20	25					D
42	70	65	34	30	20				D
24	40	46	28	15	35	35			D
30	25	40	45	33	55	55	20		D
									E
17									E
13	17								E
10	20	13							E
5	19	15	8						E
13	24	18	10	11					E
17	26	23	16	15	12				E
11	17	18	14	9	16	17			E
15	12	19	30	16	22	22	10		E

TABLE VII. (*cont.*)

D1	D2	D3	D4	D5	D6	D7	D8	D9	Subject
									F
55									F
34	53								F
22	73	34							F
8	65	44	13						F
34	100	62	22	28					F
53	115	94	52	45	30				F
25	55	62	44	20	55	53			F
44	30	67	65	49	85	85	25		F
									G
26									G
18	25								G
12	34	16							G
6	30	22	10						G
15	40	30	15	15					G
25	45	40	25	22	16				G
15	25	30	20	12	25	25			G
22	16	30	30	25	35	35	15		G
									H
43									H
29	42								H
20	55	29							H
8	50	35	15						H
28	72	48	20	24					H
42	82	69	41	36	26				H
22	42	48	34	41	42	44			H
35	25	51	50	39	62	65	22		H
									I
16									I
13	16								I
11	19	13							I
6	18	15	9						I
13	22	17	11	12					I
16	24	21	16	15	12				I
11	16	17	14	10	16	17			I
15	12	18	18	15	20	19	12		I
									J
40									J
30	35								J
15	50	25							J
5	40	30	15						J
25	60	50	20	20					J
35	75	60	40	25	30				J
20	35	45	30	20	30	40			J
25	25	50	45	35	60	55	20		J

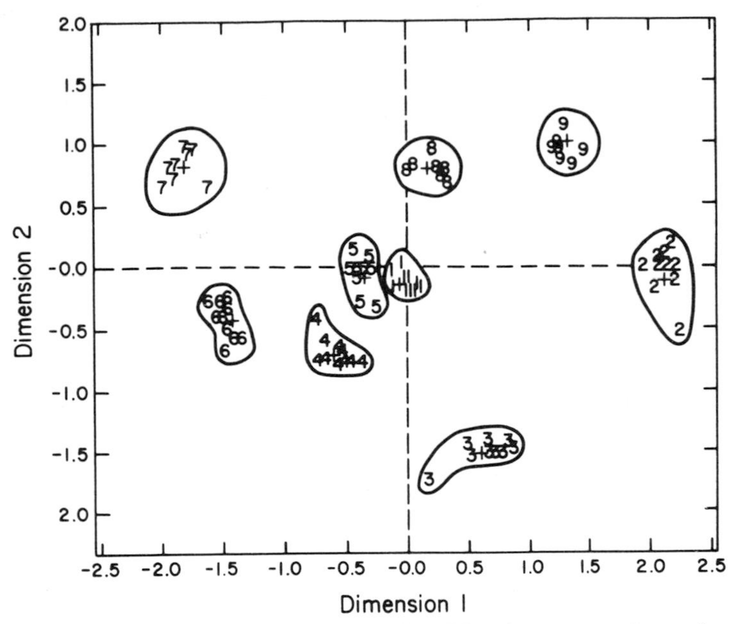

Fig. 8. Two-dimensional configuration of color differences among chromatic samples. Individual results for each of samples 1–9 are shown; they are enclosed in contours for clarity. The + symbols represent the mean coordinates over all ten observers.

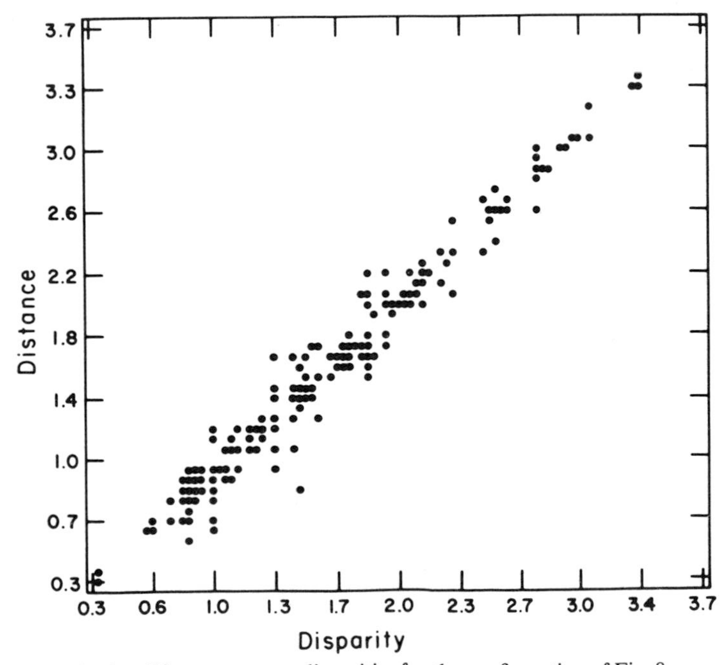

Fig. 9. Distances versus disparities for the configuration of Fig. 8.

TABLE VIII. Stress and r^2 Values for Individual Observers

Observer	Stress*	r^2
1	0.018	0.998
2	0.099	0.940
3	0.044	0.988
4	0.085	0.953
5	0.028	0.995
6	0.016	0.998
7	0.042	0.989
8	0.061	0.976
9	0.041	0.990
10	0.070	0.969

* Equation (20) in text

values (such as those listed in Table VIII) to result in vector magnitudes of weights that are approximately equal, which is the case in Fig. 10 in which the weight vectors are plotted. That figure tells us two things that Table VIII does not, however. The first is that there was not much difference among observers with respect to the relative weighting applied to the two dimensions; all ten vectors have similar angles. Second, they are all reasonably close to 45°, which means that both dimensions were approximately equally weighted. The

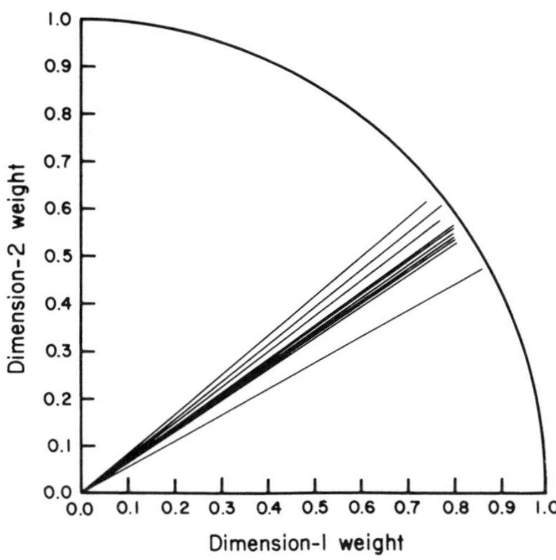

Fig. 10. Dimension-1 weights versus dimension-2 weights for the individual data of Fig. 8. See text for discussion.

consistency of vector angles means that it would be justified to pool all the data, as averages over all ten observers and determine an average configuration.

Figure 11 shows the configuration derived for the mean observer. The associated value of stress (formula 1) is 0.022, and $r^2 = 0.997$. The stimuli are specified in Table I and Fig. 1 of Chapter 8. It will be seen from that table that stimulus 1 was a gray patch of Munsell value 4; stimulus 8 was a 5Y (yellow) 6 chroma of the same value. This information can be used to try to reconstruct the color relationships among the nine stimuli.

By assuming that the Munsell notation represents a reasonably uniform visual metric of color, we may take stimulus 1 (the gray patch) to be the center of a series of concentric circles with radial extents corresponding to chroma levels. A radial line from stimulus 1 that passes through the coordinate position for stimulus 8 should represent a line of Munsell 5Y (see Table 1 in Chapter 8). Four other radii may be constructed to divide the circle into five equal portions. Together, these five radii should represent the principal hues (5Y, 5R, 5P, 5B, and 5G) of the Munsell diagram. The intermediate hue lines would, of course, be located at intermediate angular extents. The radial

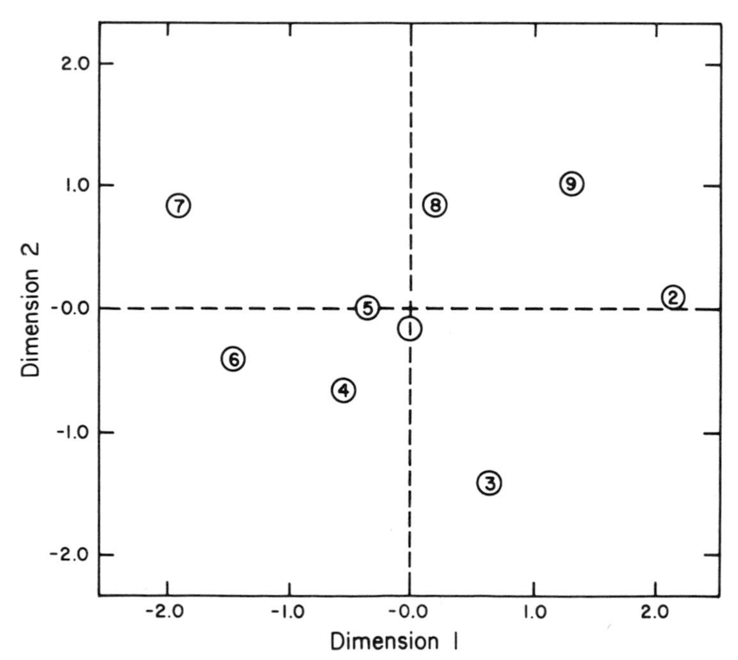

Fig. 11. Two-dimensional configuration for the average observer of the data illustrated in Fig. 8.

distance from the center of the circle to the point for stimulus 8 may be taken as a circle of 6 units in radius. Other concentric circles can then be constructed in proportion. The result, illustrated in Fig. 12, represents an overlay of one plane (of constant value) of the Munsell notation color solid. If the ten observers who took part in this experiment agree on average with the Munsell spacing, then the hues and chromas of all nine stimuli should lie on the overlayed diagram at positions corresponding to the Munsell notations for the stimuli as shown in Table I of Chapter 8.

As Fig. 12 shows, the stimulus coordinates of the configuration agree very well with the original Munsell spacing. The MDS experiment has thus determined both hues and chromas of the samples. If we estimate the chromas from the overlay, we can then plot them against the original chroma values to see how well this MDS technique has provided a solution to the chromaticness scaling problem that was used as examples in Chapters 8 and 9. Figure 13 illustrates that relationship. The MDS-estimated chromas have high correlation with the original values; the relationship has a coefficient of determination r^2 of 0.9997. In this way, a unidimensional aspect of the two-dimensional problem can be abstracted from the MDS configuration.

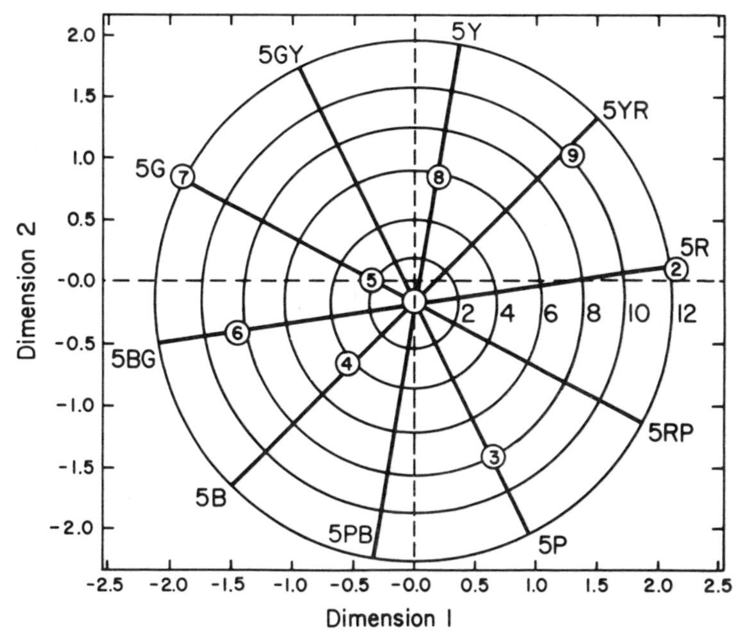

Fig. 12. Same as Fig. 11 but with a Munsell notation overlay constructed on the configuration. See text for details.

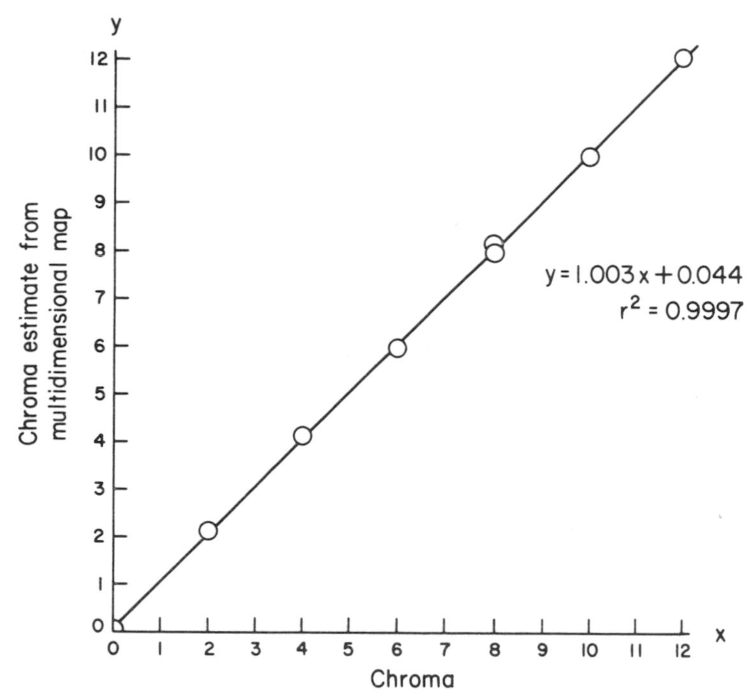

Fig. 13. Chroma estimated from Fig. 12 plotted against original chroma of nine samples. This graph may be compared with similar ones in Chapters 8 and 9.

F. CONCLUSIONS

Although this chapter has discussed only a few methods for scaling responses to complex stimuli and discussed them only superficially, I hope that it will convey the idea that there are a number of techniques for systematically classifying data in ways that aid the experimenter to understand relationships among data. Multiple regressions, multivariate analyses, and multidimensional scaling methods are only tools; they are tools designed to aid the experimenter to understand his data. Just as with statistical methods in general, we must not assume that the tool unfailingly serves up the correct answer, any more than a shovel or a bulldozer can construct a building. The shovel and bulldozer are tools that differ in their power to move earth, but they are only two of the tools that must be used to construct a building, and behind the construction process must be an architect who conceives and guides the whole process. It is the same in psychophysics. The experimenter must guide the process, using whichever tools are most helpful wherever they are appropriate.

Chapters 7–10 have all stressed the need for an experimenter to define his task clearly and to choose from among a broad battery of tools those that seem best to fit the needs of the task at hand. It has been recommended here that the simplest tool that can do the job be used. This is an efficient way to work, a way to avoid unnecessary expense and wasted effort. However, the tool cannot be too simple to cope with the job at hand, because that also wastes efforts (leading either to improper inferences or to the necessity to repeat experiments). Just as we would not use a bulldozer to turn over one square yard of soil in a garden, neither would we choose to use a hand shovel to dig the basement for a skyscraper building. When a light-measurement problem is obviously unidimensional, we should choose a unidimensional scaling technique, and it should be the simplest one that we think will yield the information we seek. If there is a question in our minds, we should try more than one approach to understand the problem better and enable ourselves to choose the solution that seems most valid. When a problem involves more than one dimension, particularly more than one response dimension, then we should eschew unidimensional methods for one or another of the multidimensional scaling tools that are at our disposal. The purpose of the whole process of scaling is, after all, to understand what is implied by experimental data. We can do this effectively, and sometimes only, if we use the tools that are appropriate to the job at hand.

REFERENCES

Aiken, L. R. (1942). "Determinants and Matrices," 2nd ed. Oliver and Boyd, Edinburgh.

Anderson, T. W. (1958). "An Introduction to Multivariate Statistical Analysis." John Wiley & Sons, New York.

Attneave, F. (1950). Dimensions of similarity. *Am. J. Physchol.* **63**, 516–556.

Ball, G. H., and Hall, D. J. (1965). "ISODATA, A Novel Method of Data Analysis and Pattern Classification." Stanford Research Institute, Standford, California.

Bard, Y. (1974). "Nonlinear Parameter Estimation." Academic Press, New York.

Bartleson, C. J. (1982). The combined influence of sharpness and graininess on the quality of color prints. *J. Photogr. Sci.* **30**, 33–38.

Beck, J. V., and Arnold, K. J. (1977). "Parameter Estimation in Engineering and Science." John Wiley & Sons, New York.

Benzécri, J. P. (1973). "L'analyse de donness; Tome 2: Analyse des correspondances." Dunod, Paris.

Boorman, S. A., and Arabie, P. (1972). Structural measures and the method of sorting. *In* "Multidimensional Scaling: Theory and Applications in the Behavioral Sciences" (R. N. Shepard, A. K. Romney, and S. C. Nerlove, eds.), Vol. I, pp. 225–249. Seminar Press, New York.

Borg, I. (ed.) (1981). "Multidimensional Data Representations: When and Why?" Mathesis Press, Ann Arbor, Michigan.

Businger, P. A. (1965). Algorithm 254. Eigenvalues and eigenvectors of a real symmetric matrix by the QR method. *Commun. ACM* **8**, 218–219.

Businger, P. A., and Golub, G. H. (1969). Algorithm 358. Singular value decomposition of a complex matrix. *Commun. ACM* **12**, 654–565.

Campbell, H. G. (1977). "An Introduction to Matrices, Vectors, and Linear Programming," 2nd ed. Prentice-Hall, Englewood Cliffs, New Jersey.

Carroll, J. D., and Chang, J. J. (1970). Analysis of individual differences in multidimensional scaling via an N-way generalization of Ekart-Young decomposition. *Psychometrika* **35**, 283–319.

Cattell, R. B. (1966). "Handbook of Multivariate Experimental Psychology." Rand McNally, Chicago.

Coombs, C. H. (1964). "A Theory of Data." John Wiley & Sons, New York.

Daniel, C., and Wood, S. F. (1980). "Fitting Equations to Data," 2nd ed. John Wiley & Sons, New York.

Davis, P. J. (1973). "The Mathematics of Matrices," 2nd ed. John Wiley & Sons, New York.

Dixon, W. J. (ed.) (1979). "BMD Biomedical Computer Programs." University of California Press, Berkeley.

Draper, N., and Smith, H. (1981). "Applied Regression Analysis," 2nd ed. John Wiley & Sons, New York.

Friedman, H. P., and Rubin, J. (1967). On some invariant criteria for grouping data. *J. Am. Stat. Assoc.* **62**, 1159–1178.

Gnanadesikan, R. (1977). "Methods for Statistical Data Analysis of Multivariate Observations." Wiley, New York.

Goldfeld, S. M., and Quandt, R. E. (1972). "Nonlinear Methods in Economics." North-Holland, Amsterdam.

Green, P. E., and Carroll, J. D. (1976). "Mathematical Tools for Applied Multivariate Analysis." Academic Press, New York.

Green, P. E., and Rao, V. R. (1972). "Applied Multidimensional Scaling: A comparison of Approaches and Algorithms." Holt, Rinehart, and Winston, New York.

Greenacre, M. J. (1981). Practical correspondence analysis. *In* "Interpreting Multivariate Data" (V. Barnett, ed.), pp. 119–145. John Wiley & Sons, New York.

Harman, H. H. (1967). "Modern Factor Analysis," 2nd ed. University of Chicago Press, Chicago.

Hartigan, J. A. (1967). Representation of similarity matrices by trees. *J. Am. Stat. Assoc.* **62**, 1140 1158.

Helwig, J. T., and Council, K. A. (eds.) (1979). "SAS User's Guide." SAS Institute, Cary, North Carolina.

Horst, P. (1965). "Factor Analysis of Data Matrices." Holt, Rinehart, and Winston, New York.

Hotelling, H. (1933). Analysis of a complex of statistical variables into principal components. *J. Educ. Psychol.* **24**, 417–441, 498–520.

Hotelling, H. (1936). Relations between two sets of variates. *Biometrika* **28**, 321–377.

Johnson, S. C. (1967). Hierarchical clustering schemes. *Psychometrika* **32**, 241–254.

Kerlinger, F. N. (1964). "Foundations of Behavioral Research." University of Chicago Press, Chicago.

Kettenring, J. R. (1971). Canonical analysis of several sets of variables. *Biometrika* **58**, 433–451.

Klingberg, F. L. (1941). Studies in measurement of the relations among sovereign states. *Psychometrika* **6**, 335–352.

Kruskal, J. B. (1964a). Multidimensional scaling by optimizing goodness of fit to a nonmetric hypothesis. *Psychometrika* **29**, 1–27.

Kruskal, J. B. (1964b). Nonmetric multidimensional scaling: A numerical method. *Psychometrika* **29**, 115–129.

Kruskal, J. B., and Carroll, J. D. (1969). Geometrical models and badness-of-fit functions. *In* "Symposium on Multivariate Analysis," Vol. II, pp. 639–671. Academic Press, New York.

Kruskal, J. B., and Wish, M. (1978). "Multidimensional Scaling." Sage Press, Beverly Hills, California.

Kruskal, J. B., Young, F. W., and Seery, J. B. (1977). "How to Use KYST-2A, A Very Flexible Program to do Multidimensional Scaling and Unfolding." Bell Laboratories, Murray Hills, New Jersey.

Lawley, D. N. (1940). The estimation of factor loadings by the method of maximum likelihood. *Proc. R. Soc. Edinburgh A* **60**, 68–82.

Lingoes, J. C., Roskam, E. E., and Borg, I. (eds.) (1979). "Geometric Representations of Relational Data." Mathesis Press, Ann Arbor, Michigan.

McGee, V. E. (1966). The multidimensional analysis of "elastic" distances. *Br. J. Math. Stat. Psychol.* **19**, 181–196.

McGee, V. E. (1968). Multidimensional scaling of N sets of similarity measures: A nonmetric individual differences approach. *Multi. Behav. Res.* **3**, 233–248.

Mallows, C. L. (1973). Some comments on C_p. *Technometrics* **15**, 661–675.

Mardia, K. V., Kent, J. T., and Bibby, J. M. (1979). "Multivariate Analysis." Academic Press, New York.

Marshall, A. W., and Olkin, I. (1979). "Inequalities: Theory of Majorization and its Applications." Academic Press, New York.

Morrison, D. F. (1976). "Multivariate Statistical Methods." McGraw-Hill, New York.

Munakata, T. (1979). "Matrices and Linear Programming." Holden-Day, San Francisco.

Myers, J. L. (1966). "Fundamentals of Experimental Design." Allyn and Bacon, Boston.

Nie, N. H., Hull, C. H., Jenkins, J. G., Steinbrenner, K., and Bent, D. H. (1975). "SPSS Statistical Package for the Social Sciences," 2nd ed. McGraw-Hill, New York.

Pearson, J. (1901). On lines and planes of closest fit to systems of points in space. *Philos. Mag.* **2**, 559–572.

Ramsay, J. O. (1977). Maximum likelihood estimation in multidimensional scaling. *Psychometrika* **42**, 241–266.

Ramsay, J. O. (1978). "MULTISCALE: Four Programs for Multidimensional Scaling by the Method of Maximum likelihood." National Educational Resources, Chicago.

Reinhardt, P. S. (ed.) (1980). "SAS Supplemental Library User's Guide." SAS Institute, Cary, North Carolina.

Richardson, M. W. (1938). Multidimensional psychophysics. *Psychol. Bull.* **35**, 659.

Rosenberg, S., and Sedlak, A. (1972). Structural representations of perceived personality trait relationships. *In* "Multidimensional Scaling: Theory and Applications in the Behavioral Sciences" (R. N. Shepard, A. K. Romney, and S. B. Nerlove, eds.), pp. 133–162. Seminar Press, New York.

Roskam, E. E., and Lingoes, J. C. (1970). MINISSA-I: A FORTRAN IV(G) program for the smallest space analysis of square symmetric matricesm. *Behav. Sci.* **15**, 204–205.

Ryan, T. A., Joiner, B. L., and Ryan, B. F. (1976). "MINITAB Student Handbook." Duxbury Press, Duxbury, Massachusetts.

Schiffman, S. S., Reynold, M. L., and Young, F. W. (1981). "Introduction to Multidimensional Scaling: Theory, Methods, and Applications." Academic Press, New York.

Shepard, R. N. (1962a). The analysis of proximities: Multidimensional scaling with an unknown distance function I. *Psychometrika* **27**, 125–140.

Shepard, R. N. (1962b). The analysis of proximities: Multidimensional scaling with an unknown distance function II. *Psychometrika* **27**, 219–246.

Shepard, R. N., Romney, A. K., and Nerlove, S. B. (eds.) (1972). "Multidimensional Scaling: Theory and Applications in the Behavioral Sciences." Seminar Press, New York.

Spearman, C. (1904). 'General intelligence': Objectively determined and measured. *Am.* J. Psychol. **15**, 201–293.

Spence, I. (1977). Incomplete experimental designs for multidimensional scaling. *In* "Multidimensional Data Analysis of Large Data Sets" (R. Colledge and J. N. Rayer, eds.). Ohio State University Press, Columbus, Ohio.

Spence, I., and Domoney, D. W. (1974). Single subject incomplete designs for nonmetric multidimensional scaling. *Psychometrika* **39**, 469–470.

Stevens, S. S. (1934). The attributes of tone. *Proc. Natl. Acad. Sci. U.S.A.* **20**, 457–459.

Stevens, S. S. (1951). Mathematics, measurement and psychophysics. *In* "Handbook of Experimental Psychology" (S. S. Stevens, ed.), pp. 1–49. John Wiley & Sons, New York.

Takane, Y., Young, F. W., and de Leeuw, J. (1977). Nonmetric individual differences multidimensional scaling: An alternating least-squares method with optimal scaling features. *Psychometrika* **42**, 7–67.

Thomson, G. H. (1934). Hotelling's method modified to give Spearman *g*. *J. Educ. Psychol.* **25**, 366–374.

Thurstone, L. L. (1931). Multiple factor analysis. *Psychol. Rev.* **38**, 406–427.

Titchner, E. B. (1910). "A Text-Book of Psychology." Macmillan, New York.

Torgerson, W. S. (1952). Multidimensional scaling: I. Theory and method. *Psychometrika* **17**, 401–419.

Torgerson, W. S. (1958). "Theory and Methods of Scaling." John Wiley & Sons, New York.

Winer, B. J. (1971). "Statistical Principles in Experimental Design." McGraw-Hill, New York.

Winkler, R. L., and Hays, W. L. (1975). "Statistics," 2nd ed. Holt, Rinehart, and Winston, New York.

Young, F. W. (1970). Nonmetric multidimensional scaling: Recovery of metric information. *Psychometrika* **35**, 455–474.

Young, F. W. (1972). A model for conjoint analysis algorithms. *In* "Multidimensional Scaling: Theory and Applications in the Behavioral Sciences," Vol. I., (R. N. Shepard, A. K. Romney, and S. Nerlove, eds.). Seminar Press, New York.

Young, F. W. (1973). POLYCON: A program for multidimensionally scaling one-, two-, or three-way data in additive, difference, or multiplicative spaces. *Behav. Sci.* **18**, 152–155.

Young, F. W., and Cliff, N. (1972). Interactive scaling with individual subjects. *Psychometrika* **37**, 385–415.

Young, F. W. and Lewyckyj, R. (1980). "ALSCAL-4 User's Guide." University of North Carolina, Chapel Hill, North Carolina.

Young, F. W., Null, C. H., and Sarle, W. (1978a). Interactive similarity ordering. *Behav. Res. Method Instrum.* **10**, 273–280.

Young, F. W., Takane, Y., and Lewckyj, R. (1978b). Three notes on ALSCAL. *Psychometrika* **43**, 433–435.

Young, G., and Householder, A. S. (1938). Discussion of a set of points in terms of their mutual distances. *Psychometrika* **3**, 19–22.

PART

Advanced Methods of Photometry and Colorimetry

11

Photometric Measurements

PETER K. KAISER

Department of Psychology
York University
Downsview, Ontario

A. INTRODUCTION

The metrology of radiant energy falls into two broad classes: radiometry and photometry. Radiometry is concerned with the measurement of radiant energy. Photometry, though involved with measuring radiant energy, places

ISBN-0-12-304905-9

certain constraints on such measures. These constraints include measuring energy contained within the visible portion of the electromagnetic spectrum and weighting these measures as a function of the spectral sensitivity of the visual system. The resultant measure is intended to correspond to the perception of brightness for normally sighted human beings. This chapter is concerned with photometry and its relation to the perception of brightness.

B. PHYSICAL PHOTOMETRY

There are two classes of photometry: *Visual photometry* and *physical photometry*. I shall deal with the latter first because it is conceptually the simpler of the two. Physical photometry involves the measurement of the visible portion of the electromagnetic spectrum (light) by means of a physical radiation detector. The spectral sensitivity of this detector is adjusted to correspond to that of the *CIE 1924 Standard Photopic Photometric Observer* [$V(\lambda)$] or the *CIE 1951 Scotopic Photometric Observer* [$V'(\lambda)$]. The CIE Standard Photopic, $V(\lambda)$, and Scotopic, $V'(\lambda)$, functions are tabulated in Table I and illustrated in Fig. 1. More complete tables describing these standard observers can be found in CIE publications 18 (1970) and 41 (1978) and in Wyszecki and Stiles' *Color Science* (1982). Very few commercial physical photometers are designed to make light measurements in the scotopic (nighttime) range of vision. They are usually designed to have spectral response characteristics that approximate the CIE $V(\lambda)$ function. The key word here is approximate. As can be seen in Table I, a reasonably precise description can be made of the $V(\lambda)$ and $V'(\lambda)$ functions. However, different manufacturers have varying success in mimicking these standards. The major deviations between the spectral response of the detector system and that of the standard observer's usually occur at the spectral extremes. Here the discrepancy can be quite large. However, for many applied purposes the approximations are quite good, and consistency among photometers made by the same manufacturer and by different manufacturers is usually within several percent when measuring broadband radiation.

The principle upon which these photometers operate is the same as that which defines the basic photometric quantity: *luminance*. The defining equation for photopic luminance given by the CIE is

$$L_m = K_m \int_{360}^{830} L_{e,\lambda} V(\lambda)\, d\lambda, \tag{1}$$

where L is the luminance in $\mathrm{cd.m^{-2}}$, $L_{e,\lambda}$ is the spectral radiance in $\mathrm{Wm^{-2}.sr^{-1}.nm^{-1}}$, $V(\lambda)$ is the spectral luminous efficiency for photopic vision, and K_m is the maximum spectral luminous efficiency (683 lm/W). The defining

TABLE I. Spectral Luminous Efficiency for the Standard CIE Observer

λ mm	Photopic vision $V(\lambda)$	Scotopic vision $V'(\lambda)$
380	0.000039	0.000589
390	0.000120	0.00221
400	0.000396	0.00929
410	0.00121	0.0348
420	0.00400	0.0966
430	0.0116	0.1998
440	0.0230	0.328
450	0.0380	0.455
460	0.0600	0.567
470	0.0910	0.676
480	0.129	0.793
490	0.208	0.904
500	0.323	0.982
510	0.503	0.997
520	0.710	0.935
530	0.862	0.811
540	0.954	0.650
550	0.995	0.481
560	0.995	0.329
570	0.952	0.208
580	0.870	0.121
590	0.757	0.0655
600	0.631	0.0332
610	0.503	0.0159
620	0.381	0.00737
630	0.265	0.00334
640	0.175	0.00150
650	0.107	0.000677
660	0.0610	0.000313
670	0.0320	0.000148
680	0.0170	0.0001715
690	0.00821	0.0000353
700	0.00410	0.0000178

Note: The table lists luminous efficiency values only from 380 to 700 nm in 10-nm steps. Complete values in 1-nm steps are available in Wyszecki & Stiles, (1982) and CIE Publication No. 18 (E-1.2) for photopic vision (from 380 to 830 nm) and in CIE Compte Rendu, 12th Session, Stockholm, 1951, Vol. III, pp. 32–40, for scotopic vision (from 380 to 780 nm).

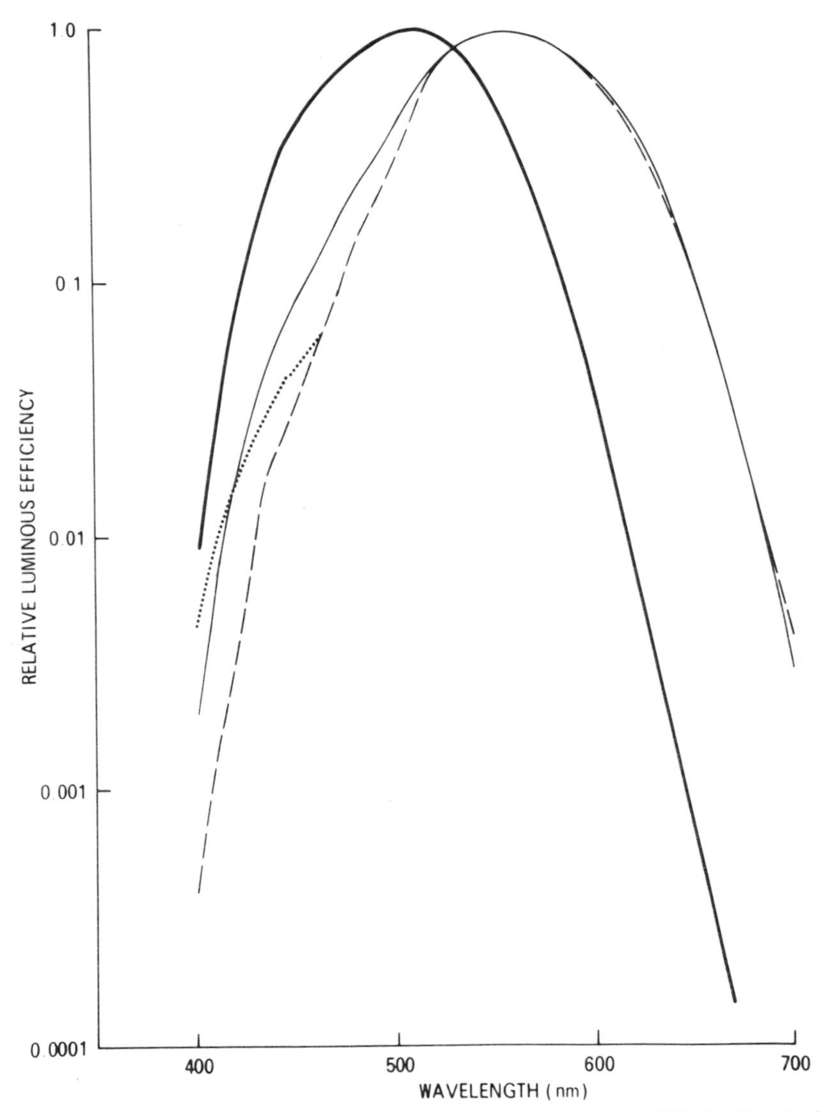

Fig. 1. CIE relative luminous efficiency functions. Heavy line: scotopic, $V'(\lambda)$ function; broken line: photopic, 2°, V(λ) function; dotted segment: Judd's modification of $V(\lambda)$; thin line: photopic, 10° function.

equation for *scotopic luminance* is

$$L = K'_m \int_{360}^{830} L_{e,\lambda}\, V'(\lambda) d\lambda, \tag{2}$$

where L and $L_{e,\lambda}$ are the same as Eq. (1), $V'(\lambda)$ is the spectral luminous efficiency for scotopic vision, and K'_m is the maximum spectral luminous efficiency (1700 scotopic lm/W).

The purposes of the constants K_m and K'_m are to fix the values of L at 60 photopic and 60 scotopic cd/m^2, respectively, when measuring a light generated by a blackbody radiator at the temperature of freezing platinum.

Equation (1) and photopic luminance have an inherent field-size limitation. It is applicable to the measurement of light that subtends a retinal visual angle of about 2° diameter. It is still uncertain whether field sizes of substantially less than 1° diameter are adequately served by 2° photopic luminances measures (CIE, 1978). However, it is clear that sensitivity functions obtained by fields of at least 10° diameter are broader and more sensitive at shorter wavelengths than the 2° function (Fig. 1). Although I know of no photometers that are calibrated for 10° luminous efficiency functions, for most applied purposes this would probably be the more suitable calibration. The higher short-wave sensitivity is probably due to rod receptor involvement, which does not occur to an appreciable degree in the 2° function.

Whereas physical photometry involves light measurement by a physical detector, visual photometry uses a human observer to make the measurement. That is, the observer makes some sort of a null visual judgment. This will be described more fully below. The advantages of physical photometry are greater reliability by the person making the measurement and among people making the measurements than is obtained by visual photometry; also physical photometry more precisely measures light consistent with the CIE definition of luminance. The major disadvantage of light measures derived from physical photometry is that they frequently do not correspond to the perception of brightness. This problem is more fully discussed in Section H. Brightness is usually the quantity in which we are interested. This chapter will not elaborate further on physical photometry. The interested reader can obtain more information from CIE Publication No. 18 (E-1,2) 1970, *Principles of Light Measurement*; Wyszecki and Stiles' *Color Science*; and Chapter 3 of Volume 2 of this treatise.

As can be seen in Eqs. (1) and (2), the photometric measures of light really make light "an unusual quantity, something not completely physical, not psychological, but psychophysical" (CIE, 1978). $L_{e,\lambda}$ represents a radiometric physical quantity that is then modified by a "physiological" quantity $V(\lambda)$ that is obtained from human observers by some appropriate psychophysical procedure.

The shape or form of the $V(\lambda)$ function one obtaines depends critically on the procedure used to measure it. The CIE $V(\lambda)$ function was derived primarily by means of heterochromatic flicker photometry and by the step-by-step (or cascade) brightness-matching method (Gibson and Tyndall, 1923; Kaiser, 1981). These and other methods for evaluating spectral sensitivity will be discussed.

C. VISUAL PHOTOMETRY

Visual photometry can be described as being performed either with a visual photometer or with a specialized optical system and associated instrumentation. Visual photometers used to be common but have given way to physical photometers. The reader is referred to Walsh's (1958) book, *Photometry*, for a review of many of these devices. Two of the more common visual photometers include the Macbeth Illuminometer and the SEI Photometer. Neither of these is now commercially available. However, they still exist in some laboratories and are occasionally used. Figures 2 and 3 illustrate the functional designs of both of these instruments.

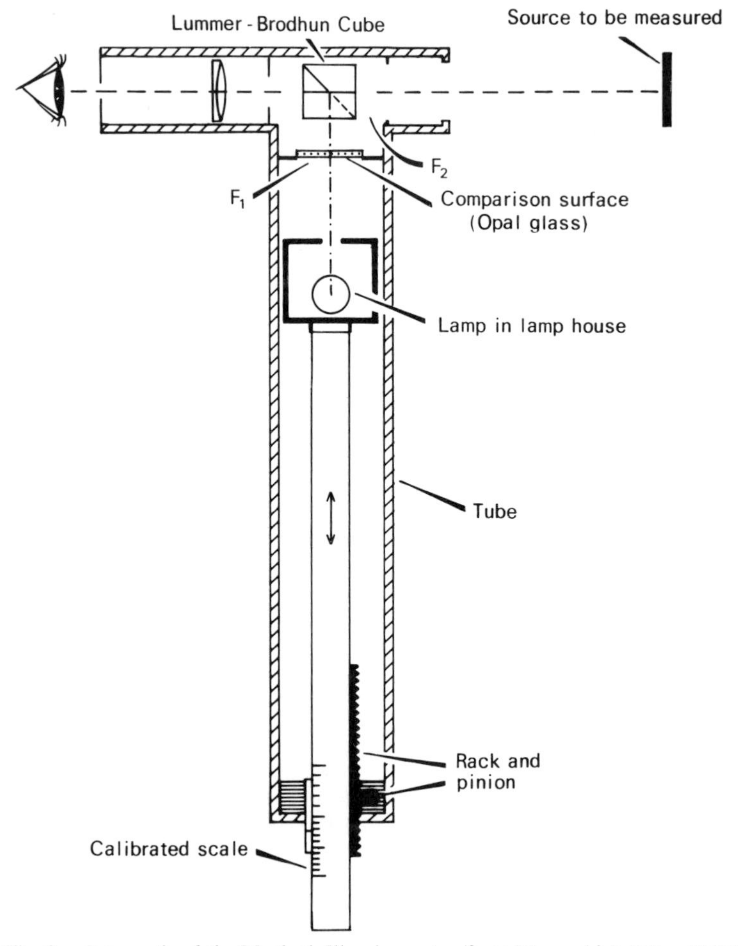

Fig. 2. Schematic of the Macbeth Illuminometer (from Wyszecki & Stiles, 1967).

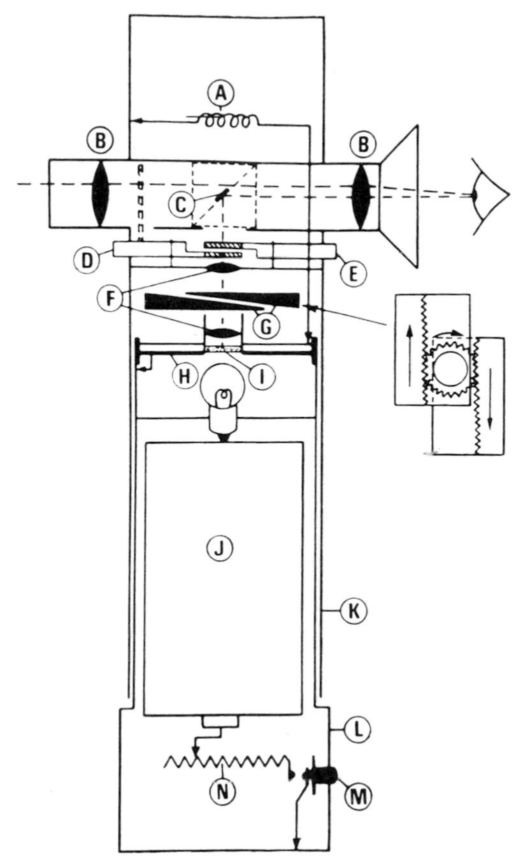

Fig. 3. Schematic of the SEI Exposure Photometer: A, microammeter coil; B, telescope lenses; C, mirror spot; D, range shift disc; E, color-matching disc; F, collecting lenses; G, optical wedges; H, photoelectric cell; I, diffusing screen; J, dry battery; K, exposure, density, and brightness scales; L, stop and film-speed scales; M, lamp switch; N, rheostat (Salford Electrical Instruments).

D. VISUAL PHOTOMETERS

These devices operate on the principle that when the operator looks through the eyepiece of the photometer two fields are seen: a reference field and the field or area whose light is to be measured. The operator is instructed to adjust the reference field until it is equal in brightness to the test field. The luminance reading is then taken from the appropriate scale on the device. These instruments require calibration against a reference source of known luminance.

In those rare instances in which the color of the test field is exactly the same as that of the reference field, the brightness match is easily, reliably, and accurately accomplished. However, as the color difference between these two fields increases, the ease of measurement and the reliability decrease. The intra observer reliability during repeated measures in a brief period of time is better than the intraobserver reliability made across days. The interobserver reliability is usually considerably worse than the intraobserver reliability. Such observations are contrasted with the highly reliable measurements made with physical photometers regardless of which competent operator is making the measurements.

E. RESEARCH PHOTOMETRY

There is another class of photometric assessments that are made, but these generally are research related methods. It is frequently required that one measure the amount of light presented to observers in experiments. A particular experiment may require a series of differently colored lights to be presented to an observer so that they are perceived as equally bright. If a physical photometer is used that purports to make brightness measurements (and many do so purport), the probability is that lights of equal luminance (which is what the photometer will measure) will not be equally bright. If the investigator intends that the light stimuli be equally bright or have some other constant effect on the visual system, then some version of visual photometry would have to be performed. I noted previously that visual photometry usually requires the observer to make a null judgment. In the discussion that follows I include under research photometry those methods in which light is directed to the eye and the observer makes some response. This response may involve adjusting a dial, pushing a button, making verbal responses, or making some involuntary physiological response. The following is a list of various procedures that allow one to evaluate the intensive quality of light:

Brightness matching
Flicker photometry
Critical flicker frequency
Absolute threshold
Increment threshold
Minimally distinct border
Visual acuity
Pupilometry
Electrophysiology
Preferential looking
Behavioral responses

In each of the methods listed, light is presented to the visual system of the observer (human or animal), and some response is observed or measured. If the task is merely to measure the amount of light on some visually relevant dimension, the amount of light is varied until the criterion response is obtained. If a spectral sensitivity function is desired, then it is also necessary to make radiometric measurements of the light to determine the energy used to obtain the criterion response. This will become more clear as each of the above methods is described.

1. Preliminary Considerations

Before performing some psychophysical procedure to evaluate the amount of light, one must decide the purposes for which the measurement is to be performed. For example, if the main purpose is to evaluate the ability of the photopic (cone, daylight) system to respond to light, then it is important to have a stimulus test-field of no more than 2° diameter visual angle, and it should be centrally fixated so that only the fovea is used. On the other hand, if one is concerned only with the scotopic system (rod, nighttime), then it is important not to confine the stimulus to the central fovea. If the measurements are concerned with more general or applied considerations, then it might be desirable to use a stimulus field of approximately 10° diameter visual angle. This angular subtence, if not viewed too far eccentrically, will include both the foveal cone and rod regions of the retina.

If one is measuring light of very short durations (100 msec or less), for example, strobe lights, then it is important to measure these lights under the temporal conditions that will be used. If this is impractical, then the amount of light should be reduced by the same factor that the duration is increased. This procedure relies on *Bloch's law*, which states that under about 100 msec, duration × intensity = constant. Such considerations may be important when measuring the luminance of flashing signal lights.

Signal lights frequently can subtend exceedingly small visual angles, especially when viewed from great distances. *Ricco's law* states that there is an inverse relationship between the size of a light stimulus and the intensity. This reciprocal relationship holds up to about 10 min of arc diameter. Therefore, when measuring lights that under the conditions in which they will be used subtend 10 min diameter of visual angle or less, it would be important to use the same visual angle as under the conditions of interest or make the necessary intensive adjustments.

If one wishes to equate two or more broadband spectral lights for brightness, the best procedure would be to do heterochromatic brightness-matching. It has been clearly shown that measurements made with a physical photometer, heterochromatic flicker photometry, and minimally distinct border are almost surely not going to yield equally bright lights (Kaiser, 1971; Wagner and Boynton, 1972). The reasons will be discussed in Section H.

For precise vision research one should not equate spectral stimuli by means of a physical photometer. Physical photometers, as noted previously, approximate the spectral sensitivity of the CIE photopic observer. For precise experimentation one should equate the stimuli based on the spectral sensitivity of one's own observers. This is analogous to the difference between making sound-level measurements and sensation-level measurements in audition. This caveat is especially critical in nonhuman studies in which the organism's spectral sensitivity can be drastically different from that of the CIE $V(\lambda)$ function. Even when working with humans, caution must be exercised. Though the average human has a spectral sensitivity similar to CIE $V(\lambda)$, Fig. 1 shows that there is a significant deviation at the short-wavelength end of the spectrum (Judd, 1951b), and Fig. 27 of Chapter 5 indicates significant variability among observers over the entire visible spectrum.

Investigators usually equate the amount of light on some behavioral criterion. When an observer's spectral sensitivity and the spectral radiance of the stimuli are known, then a luminance measure of the stimulus can be computed using Eq. (1) provided that the additivity of luminances is obeyed (Boynton & Kaiser, 1967). The computational procedure can be easily accomplished with the aid of a computer. The ways in which one determines individual spectral sensitivites and how colored stimuli are equated are discussed below.

One of the most common mistakes is to measure scotopic luminance levels (below about $0.003\,cd/m^2$) with a physical $V(\lambda)$ corrected photometer. This incorrect procedure can easily occur, because most physical photometers are not calibrated for the standard scotopic luminous efficiency function $V'(\lambda)$. Therefore, such measures are frequently made using the photopic filter and thus yield incorrect results. There is a range of light levels that stimulates both the photopic and scotopic systems: the *mesopic* region. Light levels in this range will be incorrectly measured with either photopic or scotopic physical photometers. The CIE $V(\lambda)$ function seems to be appropriate for point sources such as signal lights viewed at great distances or LEDs (light emitting diodes) viewed at normal distances. The data that are available on this point, suggest that $V(\lambda)$ probably does not yield large errors for point sources, even for narrowband spectral stimuli (CIE, 1978; Ikeda *et al*., 1982; Nusinowitz and Kaiser, 1983). The major deviation between brightness measures and luminance occurs at short wavelengths. For point sources, however, the insensitivity of the fovea to short wavelengths brings the spectral sensitivity function to approximate coincidence with the $V(\lambda)$ function. It has been shown that the human visual system is insensitive to short-wavelength stimuli when viewed foveally and when the visual angle of the stimulus is less than about 12 min of arc diameter (Ingling, *et al*., 1970). The reduced short-wave spectral sensitivity for near point sources may, just coincidentally, match the short-wave sensitivity of the CIE $V(\lambda)$.

F. METHODS FOR EQUATING COLORED LIGHTS FOR BRIGHTNESS AND FOR DETERMINING SPECTRAL SENSITIVITY FUNCTIONS

The difference between equating lights on some intensive dimension and determining a spectral sensitivity function is that the latter involves the extra step of measuring the radiance of the lights that were equated to some common reference criterion. Typically, one then plots the reciprocal of the radiance required to satisfy the measurement criterion as a function of wavelength to represent the spectral sensitivity function.

1. General Considerations

When one is interested in measuring the spectral sensitivity function of an observer, the spectral distribution of the variable chromatic field is usually controlled by a monochromator. A monochromator is any device that isolates narrow bands of the spectrum, each of which evokes a hue appearance that is not significantly altered by making the spectral band still narrower. Typically one uses a grating or a prism monochromator so that the resultant wavelength has a narrow bandwidth (usually less than 10 nm). Other means of controlling wavelength can work equally well. For instance, it is possible to use a series of narrowband filters such as interference filters. This method requires that each filter be placed in the optical system as required, either manually or by some automatic procedure such as a mechanized filter wheel that can be rotated. Closely related to the interference filter is the continuous interference filter. The difference between these two types of filters is that the former transmits light only in one restricted wavelength range. Continuous interference filters have the capability of changing wavelength continuously over the entire visible spectrum. Continuous interference filters are more difficult to use because the required optical system is more complex. Figure 4 shows typical simplified arrangements using these two types of spectral filters. For a more complete discussion of optical systems used for color vision research, the reader is referred to Boynton (1966).

An interference filter should be placed in a collimated (parallel) beam of light. If the light rays do not hit the interference filter orthogonally, the wavelength emerging from the filter will be slightly different from the calibrated value. A continuous interference filter must be placed in a part of the light beam that is very small. This is most effectively accomplished by placing it in a part of the optical system where a minified image of the source is projected. One must be careful that the angle at which the light rays converge onto the continuous interference filter are not too sharp, or wavelength distortion will occur. Therefore, a rather long focal-length lens to image the light onto the filter should be used.

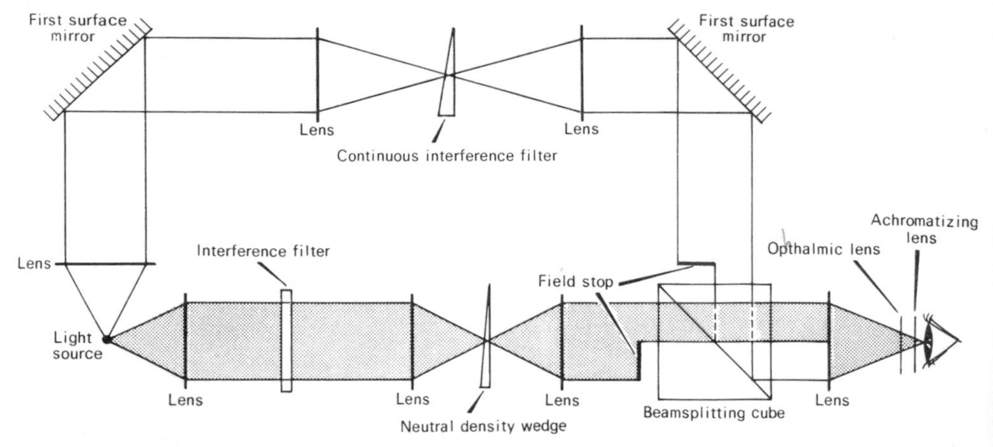

Fig. 4. Schematic of hypothetical two-channel Maxwellian view optical system. Note that the interference filter is placed in the collimated beam, whereas the continuous interference filter is placed in a focal plane conjugate with the source.

If one prefers to use a grating monochromator, it can be placed with its entrance slit at the same position at which the continuous interference filter is located in Fig. 4.

2. Brightness Evaluation

Brightness can be evaluated by one of several matching procedures or by estimating brightness magnitude. In heterochromatic brightness-matching there are always two fields: the reference field and the variable field. The variable field changes in spectral distribution and brightness. The spectral distribution is adjusted by the experimenter, and the brightness is adjusted by the observer, or by the experimenter on the observer's instructions, or is determined by the psychophysical procedure. The reference field is typically (although not always) a broad-spectrum field (white-appearing). Often investigators control the spectral distribution of the reference field to match one of the CIE Illuminants: for example, A, B, C, or D_{65} (Wyszecki and Stiles, 1982). A recent paper has suggested that 570 nm would make an effective reference field (Kaiser, 1978).

a. Bipartite Fields

In basic experiments the use of simultaneously presented bipartite fields is the most frequent approach to heterochromatic brightness-matching. In this method the idea is to have a field (frequently a circular one) that is divided in two parts. The division can be either horizontal or vertical (another variant is a disc-annulus). The dividing line between the fields should be clear but very small, not more than several seconds of arc. One half of the field is the reference and the other the variable field. The task of the observer is to adjust the brightness of the variable field until it is equal in brightness to the reference field. The word adjust is to be taken in its most general sense. Indeed, the observer may vary an appropriate control to change the brightness of the variable field directly. Or the experimenter may employ one of several other psychophysical procedures to vary the chromatic field systematically and obtain responses from the observer as to whether or not the reference and chromatic fields are equally bright. Discussion of these psychophysical procedures is beyond the scope of this chapter. However, the interested reader can find a good discussion of available methods in Chapter 6 of this volume and in Volume 1 of *Woodworth and Schlosberg's Experimental Psychology* (Kling and Riggs, 1972).

b. Slow Flicker

This method, developed by Ikeda and Shimozono (1978), bears a similarity to flicker photometry, except that the flicker frequency is very slow, clearly below color fusion, that is, 2 Hz or less. Two alternating (reference and variable) fields are employed. The quality of these fields must be such that when they are equal in spectral distribution and luminance (that is, metameric), it is possible to substitute one field for the other without a visual indication that a change took place. It is important that the two fields be spatially identical and that the temporal substitution of the fields be without artifact. That is, there should be no temporal overlapping of the fields nor should there be an interval when neither of the two fields is present.

Once two fields satisfy the above criteria, the following procedure is followed. The observer adjusts the brightness of the chromatic field until both fields are equally bright. The colors will be different, but the intensive quality of the two fields will be the same. It is also possible to pay attention to the transition point when going between the reference and variable chromatic field. This point, according to Ikeda and Shimozono (1978), can be adjusted for minimum perceptual flicker. However, this criterion does not yield equal brightness but rather something akin to equal luminance. This concept will be further discussed in the flicker photometry section.

c. Successive Brightness-Matching with Separated Fields

This method is probably the closest to normal, everyday kinds of brightness judgments. In point of fact, very few scientific experiments are conducted with this approach. With this method one has a reference located in one position and the variable chromatic field located at some distance from the reference. This distance may be large or small. To distinguish it from the bipartite field condition, however, the fields have to be sufficiently separated to require at least some short-term memory for brightness when comparing the brightness of the reference with that of the variable chromatic field. This is, of course, what one does in many practical situations.

Recent work by Uchikawa and Ikeda (1981) on wavelength discrimination has shown that as the time difference between viewing a reference and a variable field increases, the size of the just noticeable wavelength difference increases. The farther apart two fields are, the longer it takes to view one, then the other. Therefore, it is not unreasonable that the variability in brightness-matching would increase as the separation between fields and, hence, the time between viewing successive fields increases.

d. Step-by-Step Brightness-Matching

Step-by-step brightness-matching (also called cascade brightness-matching) is a method, in addition to flicker photometry, that comprised the data for the 1924 CIE Standard Photopic Photometric Observer. The step-by-step procedure avoids the problem of making brightness matches with large color differences found in heterochromatic brightness-matching. This is accomplished by making brightness matches between spectral lights that do not differ much in wavelength. It is difficult to define exactly what the size of the wavelength difference should be in practice. The idea is that the chromatic difference between the two colors to be matched should be small or nearly imperceptible.

The procedure for this method operates as follows. One adjusts a 400-nm stimulus to be equally bright to a 410-nm stimulus. Then a brightness-match is made between a 420-nm stimulus and the 410-nm, and so on. One thus moves across the spectrum in a step-by-step fashion until the desired spectral range is covered. The problem with this method is the systematic accumulation of errors. If, for example, brightness-matching errors are made in the process of stepping through the spectrum, they will accumulate. This will result in the red end of the spectrum not appearing equally bright to the blue end. Errors in judgment are bound to be made; it is called variability. These errors will not have a systematic effect if they are random. A good procedure is to traverse the spectrum at least twice, once in each direction, and average the data to help counteract systematic errors due to the direction of stepping through the spectrum.*

* The step by step procedure described in this chapter provides a good representation of the principle behind the method. Gibson and Tyndall (1923) describe a procedure where it is not necessary to use as many different wavelengths as is implied by stepping through the spectrum in sufficiently small steps so that very little if any color difference will be perceived. In essence the more practical procedure involves determining the slope of the sensitivity function at about 20 locations throughout the spectrum and then drawing a smooth best-fitting curve through these slopes. I would like to thank Dr. Uchikawa for calling this procedure to my attention.

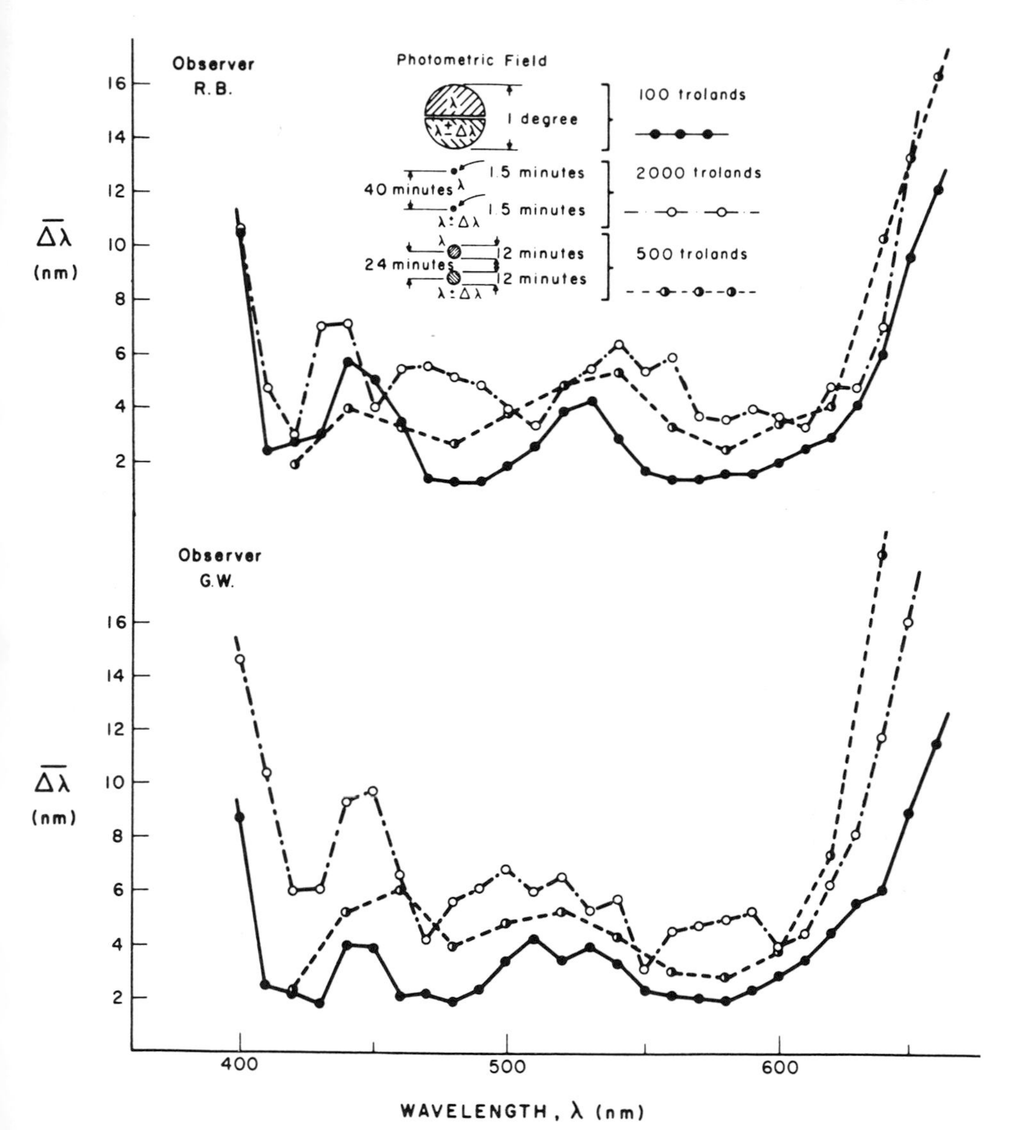

Fig. 5. Wavelength discrimination curves for different field sizes and retinal illuminances (Wyszecki & Stiles, 1967).

An interesting problem arises with the choice of step size. As can be seen in Fig. 5, the size of the just noticeable wavelength difference as a function of wavelength is not constant. At the ends of the spectrum the just noticeable difference can be as great as 5 to 6 nm and in the middle as small as 1 nm or less. Fig. 6 shows the amount of white light required to just noticeably desaturate a monochromatic light of purity equal to 1.0, as a function of

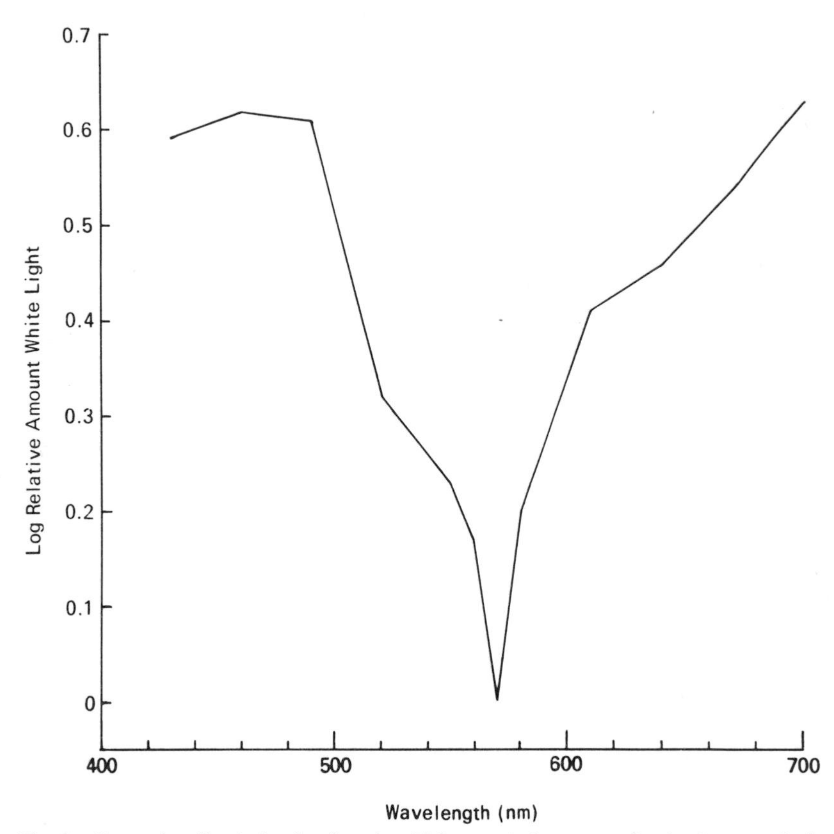

Fig. 6. Saturation discrimination function. This curve is the average for six observers. It shows how much white light had to be added to narrowband monochromatic light to obtain a just noticeable difference in saturation while maintaining equal brightness. The slow flicker method was used. (Adapted from Kaiser, Comerford, and Bodinger, 1976.)

wavelength. The point is that if one keeps the step size constant, the perceived magnitude of the hue and saturation (that is, the color) difference changes with wavelength. With constant step size, therefore, brightness matches are made with little or no color difference in some parts of the spectrum, though in other parts there clearly is a perceptible color difference.

One of the best and most complete descriptions of the step-by-step brightness-matching method is that by Gibson and Tyndall (1923). The reader who plans to use this technique is encouraged to read that paper. It provides a complete description of the step-by-step method, the basis for the recommendation for the standard spectral sensitivity function that the CIE adopted in 1924. A more recent application of the step-by-step method can be found in Wagner ad Boynton (1972).

e. Magnitude Estimation

The methods discussed, so far, have one feature in common. They have an explicit reference field against which the variable chromatic field is compared. The magnitude-estimation method can have such a reference field, but it is not necessary. Further, in the methods already discussed, the object was to equate the brightnesses of a variable chromatic field and a reference field. The magnitude-estimation procedure requires the observer to estimate the brightness of the variable field. If a reference (anchor) is used, then the observer would be told that it should be accorded some arbitrary value (a modulus), and the magnitude of the variable chromatic field should be estimated relative to the modulus of the anchor.

The magnitude-estimation method has been promulgated primarily by Stevens (1975). He pointed out that one need not necessarily employ a reference. The observer can use whatever number system he wants as long as it is consistently used as a ratio scale. Another approach employs cross-modality matching. For example, an observer would be asked to squeeze a handgrip dynamometer to represent the brightness of a light. The brighter the light the harder the observer would squeeze. A more complete discussion on the magnitude-estimation method is beyond the scope of this chapter. The reader is referred to Chapter 10 of this volume and to the *Woodworth and Schlosberg's Experimental Psychology* (Kling and Riggs, 1972) for a fuller description and additional useful references.

Cavonius and Hilz (1973) used the magnitude-estimation method to determine spectral sensitivity functions. They determined magnitude as a function of radiance for a number of different wavelengths. Then using a constant-magnitude criterion they determined the associated radiance. The radiance was then plotted as a function of wavelength. The method is not very practical because it takes a long time to collect the required data. Brightness-matching information can also be derived from the original raw magnitude estimates. The assumption to be made here is that all chromatic stimuli that are assigned the same magnitude would be perceived as equally bright. I know of no research to support or refute this assumption for the case of heterochromatic comparisons.

3. Heterochromatic Flicker Photometry

Heterochromatic flicker photometry is probably the most frequently used method for equating chromatic stimuli on the intensive dimension and for deriving spectral sensitivity functions. This procedure was the more prevalent one used with the more than 200 observers whose data constitute the basis for the CIE $V(\lambda)$ function (Kaiser, 1981; LeGrand, 1968). Heterochromatic flicker photometry requires two fields that are identical in perceived size and shape. They can be any shape; however, circular fields are usually the easiest to obtain. It can be difficult to make the two fields perceptually identical in size. I shall discuss this point in more detail later. The reference and variable chromatic field occupy the same spatial location but are presented one at a time by alternating them.

The manner in which they are alternated is rather important. If the two fields are spectrally identical and of equal brightness, the observer should see little or no change when one field is substituted for the other. There are basically two ways to alternate the fields: square-wave and sinusoidal-temporal functions. In both cases the wave forms of the reference and the chromatic field are usually presented 180° out of phase. When sinusoidal modulation is being used, a slight phase shift may be desirable; see Walraven and Leebeek (1964) for more information on this point.

The perceptual criterion in flicker photometry is a threshold amount of flicker when flicker frequency and the radiance of the chromatic field are properly set. When the radiance of the chromatic field is either increased or decreased from the minimum point, the perceived flicker increases. If the flicker rate is too high, there will be an extended interval where no flicker is perceived. Consider the case for which the flicker rate is very high, for example, 30 Hz. Under this condition, especially in the middle of the visible spectrum (when white is the reference light), the observer will have a rather wide range of chromatic radiance variation and still maintain zero perceived flicker. On the other hand, if the flicker frequency is set so that a very small amount of flicker is perceived, then it is easy to set the radiance of the chromatic light reliably within a few percent. When the two fields are alternated slowly, for example, 2 Hz, one sees first one field then the other. As the rate of alternation is increased, a point is reached at which the colors of the two fields just fuse. For example, when a red light is alternated with a white one, the colors fuse, and a pulsing pink is observed. When this occurs, it is not yet possible to get good minimum flicker, because the perceived flicker of pink light cannot be sufficiently reduced with a radiance adjustment. However, if the flicker frequency is increased, a point can be reached at which the radiance of the red light can be adjusted to obtain a threshold amount of perceived flicker.

Thus far, I have described the major principle associated with the criterion for heterochromatic flicker photometry. In fact, the setting of the flicker frequency is somewhat more involved. It can be done in at least two ways. It is possible to use one flicker frequency for all wave-lengths, or one can choose the optimum flicker frequency for each wavelength. By optimum, one means the frequency that yields the narrowest range of minimum flicker for each wavelength as the radiance of the chromatic field is varied. Arguments can be made for both choices. The argument for a constant flicker frequency is that one is not using different temporal conditions for different wavelengths.

However, when the flicker frequency is constant, one will encounter varying degrees of perceived flicker at the minimum flicker point, which will be wavelength dependent. For example, if one adjusts the flicker frequency so that a threshold amount of flicker is perceived at the short wavelengths, then zero flicker over a certain range of radiances will be observed in the middle of the spectrum. On the other hand, if the flicker frequency is set for minimum flicker with a wavelength near 570 nm, then it may be difficult to achieve a reliable setting of minimum flicker at the spectral extremes. I prefer to adjust the flicker frequency at the shortest wavelength to be used and put up with a small range of zero flicker near 570 nm. It is not too difficult to determine the mid range of this zero flicker interval.

Some people prefer to adjust the flicker frequency for an optimal setting at each wavelength. The result with this procedure would be a reasonably homogeneous variability of radiance adjustments across the spectrum. I know of no study that has compared results of these two approaches. I know of no study that has compared results of these two approaches. Bornstein and Marks (1972) in their work on critical flicker frequency have shown that between 20 and 35 Hz comparable sensitivity functions are obtained. Above 35 Hz a change occurs with long wavelength stimuli. Because they were working with critical flicker frequency, they did not use frequencies below 20 Hz. To what extent we can generalize their results below 20 Hz remains an open question. Determining the optimum flicker frequency for wavelength adds to the length of time needed to complete the heterochromatic flicker photometry procedure. In the varying flicker-frequency case the physiological conditions are held constant, and the stimulus conditions vary. In the constant-frequency case the stimulus is held constant, but the physiological conditions vary.

It was mentioned previously that it is important for the reference and variable field to be identical in size and shape. At first glance this might seem an easy requirement to meet. All one would have to do is to use a common limiting aperture or field stop. However, chromatic aberration of the visual system does not permit identical imaging of different wavelengths onto the retina. Therefore, the spectral extremes project images of different sizes from each other as well as from wavelengths from the middle of the spectrum. An achromatizing lens (Bedford and Wyszecki, 1957) placed just before the eye will correct for the axial chromatic aberration.

If the chromatic aberration problem is not handled, one sees the following. The center of the field will appear homogeneous, and minimum flicker can be achieved. But there will be a thin ring about the outside edge of the fused fields, and it will be constantly flickering. Although this does not prevent one from doing flicker photometry, it does make it more difficult. The solution for perfect temporal superpositioning is to use an optical system that allows one to adjust the size of the field stop individually for each channel.

4. Critical Flicker Frequency

The critical flicker frequency (cff) method is similar to that of heterochromatic flicker photometry. The major difference is that with the cff method the variable chromatic field is alternated with a reference field of zero radiance, that is, it is alternated with a dark field. Bornstein and Marks (1972) asked their observers to start with a dim chromatic field that produced zero flicker and to increase the chromatic radiance until a just noticeable flicker was apparent. Starting with zero flicker might be preferable, because one has greater control over chromatic and brightness adaptation. Starting with a bright chromatic light that produced clear flicker could cause the visual system to become at least partially adapted to this level of chromatic light, and the results could thereby be affected. As it turns out, Marks and Bornstein (1972) report that decreasing chromatic light to make the flicker disappear yielded the same luminous efficiency functions as did increasing the radiance. The cff method uses one flicker frequency at a time for all wavelengths. Marks and Bornstein used flicker frequencies ranging from 20 to 45 Hz. They found that between 20 and 35 Hz the sensitivity functions derived by the cff method are nearly the same as Judd's (1951b) modification of the CIE $V(\lambda)$ function. However, at 40 to 45 Hz there was a clear attenuation in sensitivity of wavelengths longer than 560 nm.

Although the cff method has a long history, it is not frequently used. One of the earliest proponents of the method was Ives (1912), who gave up on it as a practical method because of the uncertainty of the results. An extensive review of cff is beyond the scope of this chapter. However, interested readers are referred to Brown's (1965) chapter on "Flicker and Intermittent Stimulation" in Graham's *Vision and Visual Perception.* Other reviews on the subject of flicker can be found in Kelly's (1971) chapter entitled "Flicker" in the *Handbook of Sensory Physiology, Visual Psychophysics.* A chapter by Boynton (1979) entitled "Some Temporal and Spatial Factors in Color Vision" nicely summarizes and discusses the complex interactions that occur when manipulating the temporal and spatial properties of chromatic visual stimuli.

5. Absolute Threshold

The detectability of chromatic lights has been a concern primarily in basic color-vision research, but it is also of applied interest. The question here is: How much energy as a function of wavelength is required to say "yes I saw something"? There are two other threshold questions that can be asked: How much energy is required to see color, and how much energy is required to just detect the color correctly? Walraven (1962) determined the names given to spectral colors as a function of their radiance but not the radiance required for correct color recognition. These approaches, when discussed under the aegis of absolute threshold, assume that the chromatic stimuli are presented in an otherwise dark field. When one presents the stimulus superimposed on a nonzero background, the concept of increment threshold pertains (see Section F.6).

One can think of the absolute detectability of chromatic lights as one end of a continuum of brightness-matching. It will be recalled that in the brightness-matching case the task was to set the brightnesses of the chromatic lights equal to that of a common reference light. In the absolute-threshold method, one effectively sets brightness equal to an internal reference corresponding to the minimum radiance for detection of light (that is, absolute threshold).

In the absolute-threshold method the selection of field size, stimulus duration, and retinal location are of paramount importance. When one wants to assess the spectral sensitivity of the cone system by this method, it is necessary to guarantee that the stimulus falls on the rod-free portion of the retina, the fovea centralis. When assessing the scotopic system it is necessary to project the stimulus outside of the fovea centralis.

There are two considerations that need to be taken into account when keeping the stimulus in the fovea centralis. The size of the stimulus must subtend no more than about 1° diameter in visual angle. Even if one has the appropriate field size to insure stimulating the rod-free part of the retina, one must guarantee that the observer is looking in the right direction so that the stimulus will fall on the fovea centralis. This requires a fixation target.

Fixation targets for absolute-threshold measurements have to be devised carefully. One cannot, for example, use a fixation spot that is looked at directly by the observer, because it would light-adapt precisely that part of the retina on which the test stimulus will be projected, causing an elevation in threshold. One might wish to argue that such a fixation target would represent a constant and thereby merely shift the spectral sensitivity function vertically without altering its shape. It would be very difficult to provide a fixation target, viewed by the fovea, that is capable of raising the threshold equally for all wavelengths. Boynton, Kandel, and Onley (1959) have shown, for example, how the shape of spectral sensitivity functions change under different chromatic adaptation conditions (see Fig. 28 in Chapter 5 of this volume).

One can devise a pattern of fixation lights that is seen by the rods and not by the cones. Such a stimulus consists of four fixation lights located in the corners of an imaginary diamond. The distance separating these four lights should be at least 5°, and the spectral distribution should come from the short-wavelength end of the spectrum, where the rods are considerably more sensitive than the cones. The intensity of the blue lights must be adjusted so that when one looks at the center of the diamond, the lights are visible, but when one looks directly at any one of them, it is invisible. Under these conditions one has only to assume that an insignificant amount of rod–cone interaction (Drum, 1981) is occurring and thereby not affecting the results of the cone thresholds.

Assuming now that some appropriate means has been devised to insure foveal (or another desired location) presentation of the stimulus, one now procedes to determine the minimum amount of light required for obtaining the response "yes I see something" or "the color of the stimulus is. . . ." The presentation of the various radiances to obtain one of these responses is done by some appropriate psychophysical method. Discussion of these methods is beyond the scope of this chapter. However, see Chapter 6 in this volume and the signal-detection theory discussions in *Woodworth and Schlosberg's Experimental Psychology* (Kling and Riggs, 1972). In recent years some use of the signal-detection approach has been implemented in preference to classical psychophysics in vision research (for instance, Olzak & Thomas, 1981).

A major use of the threshold method in photometry is for the purpose of determining spectral sensitivity functions. To accomplish this, one merely plots the reciprocal of the radiance of each wavelength required for threshold as a function of wavelengths. A number of investigators have done this and found that the results are similar to those obtained by heterochromatic brightness-matching (Hsia and Graham, 1952; Sperling and Lewis, 1959; Guth and Lodge, 1973). Figure 7 shows a comparison between a brightness-matching function and the average threshold data of several investigators (CIE, 1978).

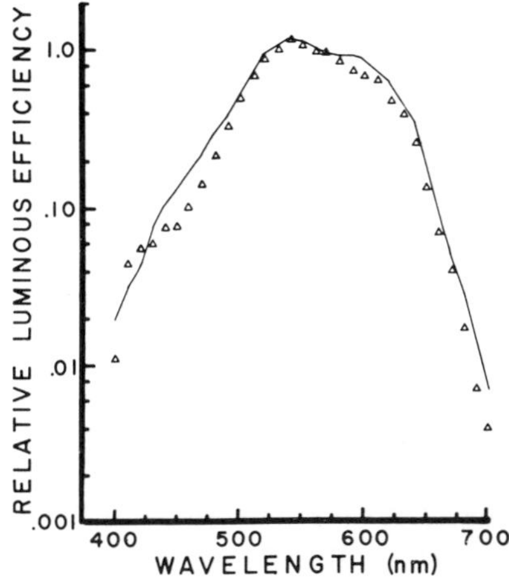

Fig. 7. Comparison of a luminous efficiency function obtained with the method of absolute-threshold light detection (triangles) and a heterochromatic brightness-matching (solid line). (CIE, 1978).

6. Increment Threshold

In basic vision research it is of interest to know how the visual system operates under different brightness and chromatic adaptation conditions. From a practical point of view it is also of considerable interest to know the spectral sensitivity properties under different adaptation conditions. For example, does one require the same radiance for a particular signal light for correct recognition when viewing it against a blue sky as when viewing it against the green of foliage (assuming equivalent light-adaptation)? When one chooses to proceed with some visual photometric procedure, it is important that the use to which this information will be put is clearly understood, because it can dictate whether absolute or increment thresholds should be measured.

The procedures to be followed with the increment-threshold method are similar to those of the absolute threshold. The problems associated with insuring that the test stimuli are projected onto the fovea are more easily dealt with, because fixation targets can be placed on the background field.

Increment-threshold techniques have been used in several important research areas. Stiles (1978) used a two-color increment-threshold method to study what he has called pi-mechanisms, which may represent some functional combinations of the spectral sensitivities of basic color mechanisms. The most authoritative source of this work is a recent compilation of his works (Stiles, 1978). Enoch (1972) presents an excellent review and synopsis of Stiles' work. The most recent work on pi-mechanisms using the Stiles' two-color increment-threshold technique can be found in the work of Pugh and his colleagues (for example, Pugh & Mollon, 1979; Sigel & Pugh, 1980). The references to Stiles' work have concentrated on the chromatic mechanisms by employing chromatic adaptation and increment thresholds. Sperling and Harwerth (1971) used very bright neutral background fields and the increment-threshold technique to determine photopic spectral increment-threshold sensitivity functions. They presented chromatic stimuli on a bright background and determined the radiance of the chromatic stimuli required for them to be just visible.

7. Minimally Distinct Border

The minimally distinct border method is a procedure introduced by Boynton and Kaiser (1968). This method uses a bipartite field with its components sufficiently juxtaposed so that if the radiance and spectral distribution in each half are identical, the two halves appear as one spatially homogeneous field. This criterion is critical to the method.

Assume that half of a bipartite field is white, and the other half is red. In the minimally distinct border method the observer is asked to adjust the radiance of the chromatic field until the saliency of the border between the two fields is minimized. One might expect that the border is minimized when the two fields are equally bright. As it turns out, this occurs only when the colors of the two halves of the bipartite field are the same or very nearly so.

Use of the minimally distinct border yields spectral sensitivity functions and light measures similar to those obtained with flicker photometry (Kaiser, 1971; Wagner and Boynton, 1972). Additivity of luminances is obtained with the minimally distinct border method as it is with flicker (Boynton and Kaiser, 1968). Thus in those experimental situations in which the temporal modulation of the light is unwarranted or impractical, the minimally distinct border method is a reasonable alternative to flicker photometry.

As noted, the precise juxtapositioning of the reference and test fields is critical. There are reports in which this method was used with Maxwellian view (Sommers and Fry, 1974; Ingling *et al.*, 1978). My experience is confined to projecting the fields onto a magnesium carbonate block and having the observers look at the block as they would a projection screen. This procedure reduces some of the problems encountered with the Maxwellian view system and allows the observer to make easily the rather critical final adjustments required for precise juxtapositioning. Figure 8 shows a schematic of a system that has been successfully used. The reader will note that the use of an achromatizing lens just before the eye. It is used to compensate for the chromatic aberration of the eye, a procedure that is critical for the successful use of the minimally distinct border method.

Deriving a spectral sensitivity function with the minimally distinct border method is similar to the procedures just described. The observer adjusts the radiance of a number of lights of different wavelengths to satisfy the criterion of minimally distinct border between the sample and the constant reference field; then radiometric measurements are made of the amount of chromatic light that was used. When the reciprocal of the measured radiance is plotted as a function of wavelength, a spectral sensitivity function is obtained. For observers with normal vision this function should be similar to that obtained by heterochromatic flicker photometry and should bear a marked resemblance to Judd's (1951b) modification of the CIE $V(\lambda)$.

In addition to the papers cited above, a number of other studies have been conducted using this criterion. Among these are Kaiser and Greenspon (1971); Tansley and Boynton (1976); Kaiser *et al.* (1971).

8. Visual Acuity

Although visual acuity is an infrequently used criterion for evaluating the spectral sensitivity of the visual system or for measuring light, its use dates back at least to the beginning of this century (Ives, 1912). In this method one

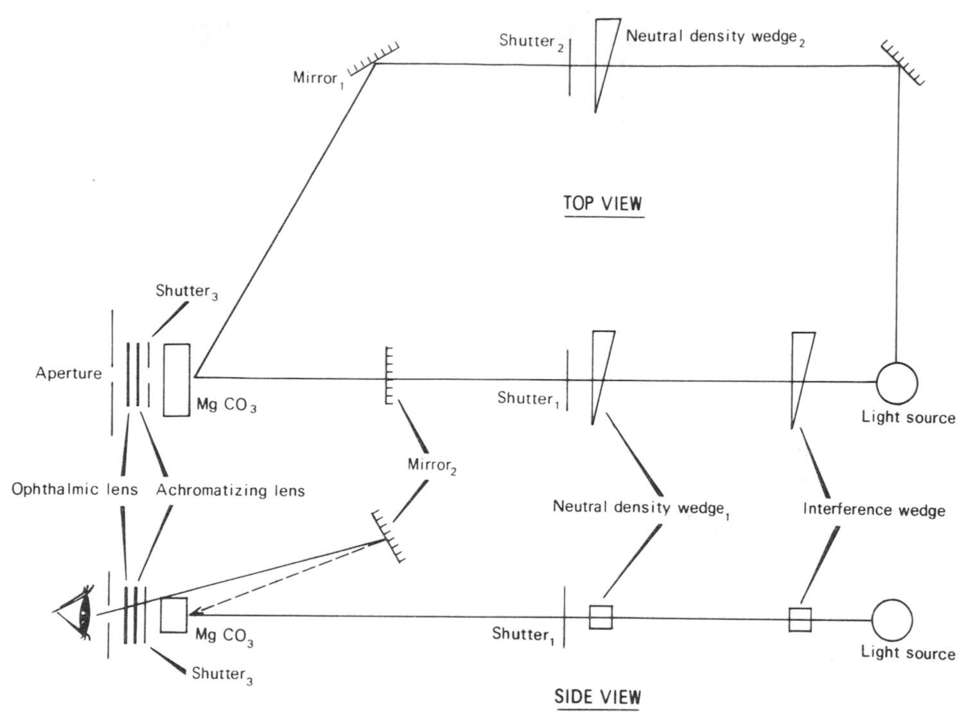

Fig. 8. Schematic of an optical system used for the minimally distinct border method. As seen in the side view the observer views the fields that are projected onto the magnesium carbonate block by looking at mirror 2, which is located above the optical axis.

chooses an acuity target with spatial separation of a constant size and determines the radiance of chromatic light required for the separation to be just detectable. The target may contain the illuminant with the background black, or the target may be black with the background containing the illuminant. The spatial separation can be generated in a number of ways, all relatively common in visual acuity measurements. However, for critical work the most common test targets are *Landolt Cs* and *square-wave gratings* (Fig. 9). One of the desirable features of these particular test targets is that the observer does not have to be asked whether the spatial separation is perceived. Rather, the observer is asked the direction that the gap in the C faces or the orientation of the bars in the grating. Obviously, to answer these questions correctly the observer must be able to detect the spatial separation in these targets.

The visual acuity method, especially when using rather fine gratings (for example, visual angle of about 3 min of arc, which corresponds to about 10 cycles per deg), requires that great care be taken to avoid the artifacts associated with the plane of focus and the eye's spherical and chromatic

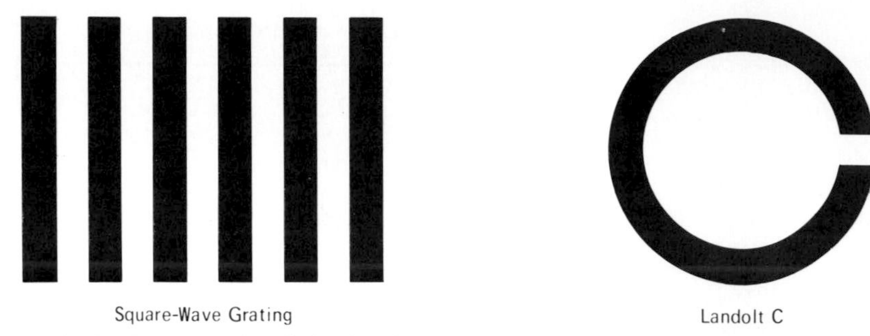

Fig. 9. Two examples of visual acuity targets: square-wave grating and Landolt C.

aberrations. The procedures for avoiding these problems are tricky and extensive discussions are beyond the scope of this chapter. We have already discussed the use of an achromatizing lens to avoid axial chromatic aberration. The reader is referred to an excellent discussion of how to minimize axial chromatic aberration and spherical aberration by Meyers *et al.* (1973); the methods section of this paper provides a complete description of how the authors handled these problems.

The minimum visual angle that can just be resolved is a function of several variables. The most important for our purposes is luminance. Over a luminance range of approximately six log units, visual acuity will change approximately 80-fold (Riggs, 1965). Because the eye is differentially sensitive as a function of wavelength, this dependence of visual acuity on luminance provides the mechanism by which visual acuity can be used as a criterion for the assessment of chromatic light.

The procedure operates as follows: A suitably chosen square-wave grating, for example, with bars separated by 3 min of arc is selected. One set of the alternating bars constitutes a constant, the absence of light or black bars being convenient. The alternate bars are variable in wavelength and radiance. The object of this procedure is to adjust the radiance of the chromatic bars to the minimum level at which the orientation of the grating can be correctly identified. This is done for the wavelengths of interest, and the resulting radiance is measured. As with previous procedures, the reciprocal of these radiances as a function of wavelength provides the spectral sensitivity function. Another example of the visual acuity criterion for determining spectral sensitivity can be found in Brown *et al.* (1960).

Ives (1912) indicated that the main problem with visual acuity as a criterion is its "extreme lack of definiteness." He further noted that unless it can be shown that this criterion gives results "very different from that given by other methods, and much more important, visual acuity would not appear to deserve much attention" (p. 154). In this same paper Ives noted that to his knowledge there were no measurements on record regarding whether or not additivity holds using visual acuity as a criterion. Studies by Meyers *et al.* (1973) and by Guth and Graham (1975) show that when the experiment is properly conducted, additivity of luminances does hold with this criterion. Graham and Guth conducted the additivity experiment using Landont Cs, whereas Meyers *et al.* used a grating. Meyers *et al.* provide physiological explanations for the additivity of brightnesses using flicker photometry, minimally distinct border and visual acuity. However, they conclude that the additivity found with visual acuity probably has a different basis than that proposed to explain the additivity found with flicker and minimally distinct border.

9. Pupilometry

It is well known that the size of the pupil varies as a function of the amount of light entering the eye. Furthermore, the pupils of both eyes react consensually even when light is put into only one eye. This is quite useful, because it allows one to stimulate one eye while recording the pupillary response from the other.

A recent and complete review describing the methods and results of pupilometry is provided by Hedin (1978). Hedin compares pupilometry photometry with other visual photometric methods. The method used by him to measure pupil diameter involved illuminating the eyes with near infrared energy (920 nm) and monitoring them with an infrared-sensitive TV camera. In this way an image of the eye was presented on the TV monitor, but the observer did not see the illuminating light. With the aid of calipers Hedin measured the pupil diameter to the nearest half millimeter, as a function of radiance and wavelength. He then used a criterion pupil diameter and determined the associated radiance at each wavelength. The reciprocal of this radiance was plotted as a function of wavelength to derive the spectral sensitivity function. He found that the spectral sensitivity derived from pupilometry showed agreement with the sensitivity functions obtained of absolute threshold. That is, there was an elevated sensitivity at the long and short wavelengths compared with the sensitivity of the CIE $V(\lambda)$.

Other methods have been used to measure pupil diameter. For example, one can take advantage of the fact that the pupil has a rather long reaction time to flashes of light. This allows one, theoretically, to stimulate the eye with a given illuminant and then quickly photograph the pupil using a photoflash. This procedure would cause problems for the repeated determinations of pupil diameter frequently required in photometric assessments. Rather long periods of time would have to be taken between measurements to insure that the adaptive condition of the eye recovered from the high-intensity photoflash. This problem could be avoided by means of infrared photography.

The method described above requires the direct measurement of the pupil from the photographed or televised image. Another method takes advantage of the fact that the pupil reflects less light than the iris. When a calibrated area of light is imaged on the iris near the edge of the pupil, it is possible to record the amount of reflected light from the eye as a function of pupil diameter. The smaller the pupil the more the reflected light. Although one obtains continuous measures of pupil diameter, this method lacks the same precision as obtained when discrete measures are made with calipers. For one thing, the moving eye causes considerable difficulty in maintaining proper calibration.

One of the major problems using pupil diameter for photopic light measurements is the influence of rod activity or intrusion. The rods play a major role in determining the pupil diameter. Therefore, when attempting to evaluate cone spectral sensitivity or making photopic measurements, it is important that the rod activity be suppressed. This can be accomplished by means of a large bright adapting field to suppress the activity of the rods. The cones are considerably less sensitive (except beyond 650 nm) than the rods. When a homogeneous field is used, the stimulus is incremented on the adapting field. Another alternative uses a large adapting field with a hole in the center. The large field contains the adapting light that suppresses the rod sensitivity, and the test stimulus is presented through the central hole. The difficulty with both of these adapting field methods is that they restrict the already rather narrow range over which the pupil diameter varies.

Another method frequently used to insure that only the cones are being activated is to light adapt the whole eye and then dark adapt it. The stimulus is presented when dark adaptation progresses to the point at which constant cone thresholds are reached. Also one can take advantage of the fact that the cone system has higher temporal resolving power than the rods. Thus intermittent stimulation at a frequency resolved by the cones but not the rods could be used. A recent paper by Young and Alpern (1980) discusses pupil responses to chromatic lights and the rather stringent method they used to insure rod-free pupillary responses. This paper is suggested to the reader for the use they make of an infrared television pupilometer originally described by Green and Maaseidvaag (1967).

10. Electrophysiological Recording Methods

There are three major methods involving electrophysiological techniques: single-unit recordings, visually evoked cortical potentials (VECP), and electroretinograms (ERG).

Electrophysiological recordings seem to have an intuitive and mystical appeal to the uninitiated in this area. Frequently when a phenomenon is demonstrated electrophysiologically, it somehow takes on a greater value than the psychophysical demonstration. Although the author is not an electrophysiologist, some of the magic and mystery has been removed for him by an all-too-brief sojourn involving extracellular recordings at the Center for Visual Science, University of Rochester. A similar experience is recommended to all psychophysicists. Unfortunately, the brief discussion that follows will not allow the uninitiated to engage in these techniques, but perhaps it will pique the interest of some readers to seek out laboratories where work is being done and to learn firsthand for themselves. I shall provide explanations and, more importantly, critical references for more detailed explanations so that the reader will, at least, find it easier to read the literature on these techniques.

The last ten years or so have been quite exciting in the field of electrophysiology. The advent of microelectronics and computers has been to current advances what the vacuum-tube amplifiers were to the initiation of electrophysiology in the first place. The reader may wish to start a study of these methods by reading Thompson and Patterson's (1973) edited book, *Bioelectronic Recording Techniques, Part A, Cellular Processes and Brain Potentials.*

a. Single-Unit Recording

There seems to be at least one great mystery in the technology of single-unit recordings. This has to do with the art of making good electrodes. I know of no really good references on the subject. Descriptions, when they do exist in journal articles, are woefully inadequate for the novice. The making of electrodes seems to be best learned by doing, preferably under the guidance of an expert. Consequently, I will not touch further on this subject despite its importance in electrophysiological recordings.

There are two major classes of single-unit recordings: extracellular and intracellular. In the former, the electrode does not penetrate the cell from which the recordings are being made; in the latter, the cell is penetrated, hopefully without damaging it. The technology required for these two types of recordings is somewhat different and discussion of them is beyond the scope of this chapter.

Single-unit recordings are not obtained from human subjects. Animals that are frequently used include monkeys, cats, squirrels, fish, and various amphibia. When considering the topic of photometry, the most relevant animal is the monkey, primarily the macaques. This species has been shown to possess a visual system close to that of humans (De Valois and Jacobs, 1968).

Whereas humans require extensive instructions and training to perform many of the psychophysical procedures, animals require considerable physiological preparation. This preparation includes sedation, paralyzation, insertion of intravenous tubes to deliver nutrients, liquids, drugs, and so on. Because the animals are frequently paralyzed so that eye position can be carefully controlled, they must also be artificially respirated. Artificial respiration requires the insertion of an intratracheal tube. Various vital signs must be monitored, such as expired carbon dioxide, core body temperature, and heart rate. It is also common practice to maintain the animal under light sedation by means of an appropriate nitrous oxide–oxygen mixture. This description is clearly not intended as a primer for those attempting single-unit recordings. The intent is merely to give a flavor of the procedures and skills required to engage in this type of research.

When one is recording from only one cell, it is not possible to measure the spectral sensitivity of the entire visual system. Indeed, all one can do is to determine the spectral characteristics of the particular cell from which one is recording. When such recording is obtained from a receptor cell, it will be descriptive of the neural processing within that receptor. However, when records are obtained from a horizontal cell, ganglion cell, lateral geniculate nucleus (LGN) cell, or cortical cell, then one learns about the processing of neural information in these particular cells as modified by the preceding stages. This may seem an almost trivial comment. However, it is possible to find statements, for example, that color vision is processed in the LGN. This statement stems undoubtedly from the important work of DeValois and his colleagues, who concentrated their early efforts on investigating the spectrally opponent and nonopponent processing properties of the LGN. However, the color-vision processing begins at the most distal part of the neural system, the outer segments of the receptors (Marks *et al.*, 1964). Many other investigators have helped to understand color-vision processing at other neurophysiological stages, for example, the horizontal cells (Svaetichin, 1956) and the cortex (Gouras, 1974).

When evaluating the spectral sensitivity of a cell, one uses a procedure similar to that in psychophysics: A criterion is chosen, and the radiance required to reach that criterion is determined. The criterion is frequently a given firing rate of action potentials (in spikes per second). Using this procedure, DeValois *et al.*, (1966) were able to convolute the data of various types of LGN cells to compute the spectral sensitivity of the spectrally opponent as well as the spectrally nonopponent systems. The spectral sensitivity of the spectrally nonopponent cells bears a similarity to that of the CIE 1924 Standard Photopic Observer. Thus the nonopponent system is frequently referred to as the luminance system. The spectral sensitivity of the monkey's opponent system as recorded from the LGN bears a similarity to human spectral sensitivity determined under very bright neutral adaptation as measured by Sperling and Harwerth (1971). A review of color vision relevant to electrophysiological investigations in the central nervous system can be found in a chapter by DeValois (1973).

DeValois and colleagues used rather large stimulus fields in their early work on the LGN. Using a greater variety of stimulus configurations, Gouras (1968) and Wiesel and Hubel (1966) have shown that a number of spatial and temporal subtleties were missed by the pioneering DeValois research. The work of DeMonasterio *et al.*, (1975) shows how selective chromatic adaptation helps to unravel additional complexities and temporal factors in the LGN. These comments are not meant as a criticism of the early work but rather as a clue to the reader that one cannot merely stick an electrode somewhere in the visual system, irradiate the retina with light, and suppose that the desired information will come forth. Indeed, today many more intriguing complexities are being revealed, including the importance of cone–cone and rod–cone interactions, the importance of chromatic adaptation, sizes of receptive fields, and so on.

An interesting paper by Gouras and Zrener (1979) provides a concise description of the preparation required for recording ganglion cell activity in the rhesus monkey in response to flickering colored lights. In this paper they show how luminance flicker and color flicker can both be explained by activity within the spectrally opponent cells. Most color-vision models assume that luminance flicker occurs only from the non-opponent system (Ingling and Tsou, 1977; Boynton, 1979; Guth *et al.*, 1980; and others). Thus it would seem that our understanding of the mechanisms behind heterochromatic flicker photometry may not yet be firmly established.

b. Visually Evoked Cortical Potentials

When a surface electrode is placed over the occipital cortex with a reference electrode suitably placed elsewhere, one can record a rather complicated appearing wave form. This wave form is the result of the spontaneous electroencephalitic activity (EEG), electromyographic activity (EMG), and visually induced cortical activity. The latter activity will be very small relative to all the other ongoing cortical responses. However, by means of computer averaging techniques, it is possible to ferret out these cortical activities associated with visually presented stimulation.

Visually evoked cortical potentials (VECP) have been called by various other names such as evoked potentials, cortical responses, visually evoked responses, occipitograms, and occipital response. However, Riggs and Wooten (1972) have suggested *visually evoked cortical potentials* (VECP), because it more aptly describes the electrical changes that occur in the cortex in response to visual stimulation.

As noted, the amplitudes of VECPs are very small, especially when recorded from the human scalp using gross surface electrodes. The VECP seems virtually buried in the other electrical activity present in the cortex. Fortunately, most of this activity is random with respect to the temporal relationships of specific visual inputs. Therefore, when repeatedly obtained gross critical potentials, near the time of a visual stimulus, are arithmetically averaged, the random fluctuations will sum to near zero, whereas the nonrandom potential associated with the visual stimulus sums to a larger value.

For a more complete description of the VECP as it relates to visual psychophysics, the reader is referred to the excellent chapter by Riggs and Wooten (1972). More general discussions regarding the use of cortically evoked potentials can be found in Regan's (1972) book *Evoked Potentials in Psychology, Sensory Physiology and Clinical Medicine*. For more recent information the reader is referred to Regan (1975).

Regan (1970) reported that spectral sensitivity functions determined by measuring the amplitudes of the evoked potentials do not yield precise results. However, if a Fourier analysis is performed on the wave form, and one measures the amplitudes of the second harmonic, results comparable in precision to psychophysics are obtained. Regan measured the spectral sensitivity of observers using flicker photometry and this second harmonic component of the evoked scalp potential. Both methods yielded the same results within the accurarcy permitted by the luminance steps.

Dobson (1976) used the VECP to measure the spectral sensitivity of two-month-old infants. She used a technique somewhat different from Regan's. She also recognized the variability exhibited between response amplitude and energy. Therefore, she employed a criterion of implicit time. This is the time between the stimulus onset and the peak of the first major positive deflection that occurred between 100 and 260 msec after stimulus onset. The implicit time versus energy functions for each wavelength were used to calculate the spectral sensitivity functions. When Dobson compared these sensitivity functions with psychophysically determined functions from other observers, the correspondences were quite good. She tested five infants, and three of the comparisons are remarkable when one considers all the differences among observers and experimental conditions. Dobson also compared the VECP data from two adults with the psychophysically determined functions, and the fits are much better.

This method of measuring spectral sensitivity has great potential in obtaining such data from naive observers and from those who are for various reasons unable to interact sufficiently to participate in psychophysical experiments (for example, very young infants). Consider, for example, using the evoked potential for measuring spectral sensitivity functions for clinical purposes on stroke victims, people with multiple sclerosis, and so on.

c. ELECTRORETINOGRAM (ERG)

The gross electrophysiological activity of the retina can be recorded by embedding a corneal metallic electrode in a scleral contact lens and placing a reference electrode at some other convenient location (for instance, the forehead). Such recordings are possible because of the good contact made with the retina by media surrounding the retinal tissue and the electrical contact these media have with the cornea. Readers wishing a more detailed description of the logic and methods behind ERG recordings are referred to the Riggs and Wooten chapter "Electrical Measures and Psychophysical Data on Human Vision" (1972). In addition to providing an excellent introductory description of ERG methods and logic, it also provides many useful references for the reader who wishes to delve further into the subject matter.

An excellent paper by Aiba *et al.* (1967) provides a description and data on the use of ERG for photometric purposes. These authors show how ERG varies with radiance and how this information can be converted to a spectral sensitivity function.

The major component of the ERG comes from the activity of the rods. Thus if one wants, for example, to study the cone contribution to the ERG, it is necessary somehow to suppress this rod component. Johnson and Cornsweet (1954) attempted to do it by taking advantage of the higher temporal resolution of the cone system over that of the rods. They obtained spectral sensitivity functions that tended to be shifted towards short wavelengths more than would be expected by the photopic system. This would suggest that they were not fully successful in suppressing rod activity. Aiba *et al.* (1967) superimposed foveally fixated, narrowband test stimuli on a bright blue background. This background would saturate the rods and thus suppress their contribution to the ERG. For each of eight wave bands, Aiba *et al.* determined the amplitude of the ERG as a function of the test stimulus intensity. Plotted on log–log coordinates these amplitude versus intensity functions yielded straight lines denoting underlying power functions. In order to obtain an action spectrum, Aiba *et al.* merely took a criterion ERG amplitude and determined the associated stimulus intensity. Actually, the observer was presented with a continuous train of square-wave pulses of chromatic light, and the ERG was summed with a computer of average transients for 200 accumulations. The measurements of the ERG amplitude were made from these computer-averaged ERGs. Plotting this intensity as a function of wavelength yielded a spectral sensitivity function for one observer that was virtually identical to Judd's modification of CIE $V(\lambda)$. The second observer for which they also present data follows this modification of the CIE function except for the two shortest wavelengths, for which the observer's sensitivity was somewhat greater. Aiba *et al.* also psychophysically measured the sensitivity functions using a cascade brightness-matching method. These psychophysically determined functions are reasonable fits to the ERG and CIE functions. Aiba *et al.* made similar measurements under other conditions including nonfoveal stimulation and lower luminances levels. This paper is a must for the novice considering spectral sensitivity measurements using the ERG.

Thus it is seen that the ERG can also be used to evaluate the physiological responses to light as a function of wavelength. This technique would be used primarily for research purposes due to the difficulty in implementing it. For example, it would be much less invasive to attach scalp electrodes to measure VECPs than to attach scleral lenses to the eye for ERG measurements. Whereas VECPs give information about the processing of information at the cortical level as modified by the preceding stages of the visual mechanism, the ERG provides information about the processing of information at the retinal level.

The ERG is suitable for use with human and animal observers. Because of the distance of the recording electrode from the retina, gross potential is recorded. However, techniques have been developed to record local ERG's in animals using procedures that do not require sacrificing the animal or the animal's eye (Boynton & Baron, 1975; Baron *et al.*, 1979). This procedure requires most of the same preparations briefly noted previously for single-unit recordings. The implementation of the local ERG technique involves inserting a cannula into the eye and through it an electrode that is placed into the foveal region to such a depth as to maximize the late receptor potential. Baron *et al.* (1979) report that the receptor potentials they record by this technique are "most likely monitoring cone receptor activity" (p. 109). The reader is referred to the Baron *et al.* paper for a fuller description of the method and references to several other papers that Boynton and his colleagues have published since 1970 using this technique to study the role that the receptors play in color vision. For example, Boynton and Baron (1975) used this method to study the temporal characteristics of the receptors, which have clear relevance on the visual systems response to flicker photometry.

11. Preferential Looking

Although electrophysiological methods (VECP) have been used to measure the spectral sensitivity of infants (Dobson, 1976), one can understand the hesitency parents may have in allowing their babies to be subjected to such experimental procedures. Clearly, two-month-old infants are not going to cooperate verbally in a psychophysical experiment. Teller and her colleagues have developed a procedure that requires merely observing where an infant's gaze is directed (Teller *et al.*, 1978; Peeples and Teller, 1978).

Infants are comfortably held and positioned in front of a stimulus panel. This panel is homogeneous in color so that if a stimulus is presented on it the question raised is "does the infant look at it?" It has been previously determined that if stimuli are visible and attractive, infants will stare at them. Teller and her colleagues present two stimuli to the infant, one to the infant's right and one to the left, and determine in which direction the infant's attention is drawn. An observer hidden from view watches the infant through a peephole in the stimulus board. This observer does not know at the time of the stimulus presentation where the different stimuli are placed. But "trial by trial feedback to the observer is used to optimize the observer's use of all possible relevant cues provided by the infant" (Teller *et al.*, 1978).

Teller *et al.* (1978) report that color vision in infants has been studied many times. However, in discrimination experiments it is necessary to know that the infant's discrimination is based on color information not brightness information. Previous investigators have used adult spectral sensitivity functions to equate stimuli for luminance. Although this is a good approximation (Dobson, 1976), a need does exist (Teller *et al.* argued) to determine wavelength discrimination properly for stimuli equated on the brightness dimension. This need leads quite naturally to using the preferential-looking technique for measuring infant spectral sensitivity. Peeples and Teller (1978) did this research and found that though adults seemed to be somewhat more sensitive than infants in absolute terms, the shape of their sensitivity functions were similar. These results would support the VECP-based functions reported by Dobson and justify using adult sensitivity functions as first approximations to infant values. However, just as one would prefer to measure an adult's spectral sensitivity function rather than use CIE $V(\lambda)$, the same would be true for infants. Teller's forced-choice preferential-looking technique would seem to provide a workable method.

12. Behavioral Responses

Due to ethical considerations the single-unit method discussed above requires that animals be used rather than humans. However, it is a rather large conceptual leap from electrophysiological recordings to perceptual responses. Thus it is of interest to evaluate the effect that light stimuli have on animals in some other way to provide a converging operation for the conclusions one may wish to draw from single-unit recordings. Because animals cannot talk or respond to complex verbal instructions, other methods must be used to obtain more behaviorally oriented responses to light stimuli.

Behavioral methods have been used to study the color-vision capabilities of goldfish (Beauchamp *et al.* 1979), cats (LaMotte & Brown, 1970), monkeys (DeValois and Jacobs, 1968) and pigeons (Blough, 1955). An excellent account of visual psychophysics in animals is given by Blough and Yeager (1972). Many animals can be conditioned to respond in a particular way to a specific stimulus. Such responses can include pushing a lever and moving in a given direction. It is also possible to monitor certain autonomic responses. For example, the heart rate of goldfish has been conditioned in response to spectral stimuli (Beauchamp *et al.*, 1979).

As noted previously, research is frequently conducted on animals, because ethical considerations preclude the investigation from being conducted on humans. However, because the main concern frequently is human not animal vision, the behavioral animal research provides an important link. This link becomes even stronger when the behavioral experiment conducted on the animal is repeated using the same apparatus and procedure with humans. DeValois and Jacob's (1968) brightness sensitivity tests are an excellent example of this approach.

DeValois and Jacob presented their subjects with four colored stimuli. One of them was flashing. The task of the observers (human and monkey) was to detect which of the four stimuli was flashing. The brightness of the stimuli was varied and thresholds determined. If the brightness was below threshold, the observers would respond randomly in this four-alternative forced-choice method.* When the monkey chose the correct light, it received a reward. When testing the scotoptic sensitivity these investigators used low flicker rates and when testing the photopic system, high flicker rates. They found that macaque monkeys had virtually identical photopic and scotoptic sensitivity properties as those of humans. Squirrel monkeys had the same scotoptic sensitivity as normal humans, but the photopic sensitivity was reduced in the red end of the spectrum much like that of red blind or red weak (protanopic, protoanomalous) humans.

It is frequently believed that cats are color blind. Indeed, such a statement appeared in a brochure that my veterinarian was planning to make available to his clients. After I gave him the information contained in Sechzer and Brown (1964) and LaMotte and Brown (1970) he decided to change the pamphlet. Both of these studies employed behavioral methods of testing cats' visual capabilities, which is perhaps the only way to determine conclusively the color-vision capabilities of an animal.

G. BRIGHTNESS AND LIGHTNESS

The visually meaningful measurement of light, thus far, has only been concerned with the problem that luminance measures are frequently inadequate. We saw that lights of equal luminance, especially chromatic lights,

* The apparent contradiction in terms is commonly used by psycho-physicists to mean that the observer is provided with four (or *n*) options, one of which must be chosen.

usually were not equally bright. The reader will have noticed that we concentrated on the measurement of lights. We have not considered two additional matters of equal importance: the measurement of reflected light from spatially complex reflecting surfaces (photographs, for example,) or the relationships between brightness and luminance. Brightness is not a linear function of luminance.

Earlier in this chapter we made reference to the following variables that could influence the brightness of a light: duration, area, chromaticity, and, in a loose way, adaptation. However, we have not considered the influence of various induction effects, such as, chromatic, light and dark adaptation. Further, we were implicitly assuming that if the source of the light we were evaluating was related to anything, it was to a light bulb, LED, laser, beacon, signal light, and so on. Thus we were considering unrelated colors. If instead of measuring the light from an incandescent bulb, we measured the light being reflected from an apple in a fruit bowl, the concept of related colors would pertain. Light reflected from a color photograph is another example of a situation in which the concept of related colors applies. The reason for introducing the concept of related and unrelated colors is that, with unrelated colors, one usually refers to the brightness of that which is being measured. However, with related colors the concept of lightness is introduced.

In the fourth edition of the CIE International Lighting Vocabulary lightness is defined as: "The brightness of an area judged relative to the brightness of a similarly illuminated area that appears white or highly transmitting."

A magnitude-estimation procedure is described in Section F.7. With this procedure one obtains brightness estimates as a function of luminance. Only if the function relating brightness to luminance of unrelated colors is plotted on log–log coordinates is a straight line of a positive slope obtained. Slopes of approximately one-third are frequently obtained for unrelated colors, and this has been used by the CIE in an equation that relates lightness to luminance:

$$\mathrm{L} = 116(Y/Y_N)^{1/3} - 16; \qquad Y/Y_N > 0.008856, \tag{3}$$

where Y = luminance of the stimulus, Y_N = luminance of a reference stimulus.

The specification of the reference stimulus is arbitrary. Recall that in the definition, reference is made to "a similarly illuminated area." Because lightness is a relative measure, its precise value will depend on the value of the reference stimulus Y_N. In Eq. (3), L does not take into account the effects of chromatic adaptation, light or dark adaptation, or aspects of related colors such as chromatic contrast and brightness contrast. Y is one of three tristimulus values used to measure colors and is equivalent to the luminance of the measured area. Because Y is dependent on $V(\lambda)$, the luminous efficiency of the CIE standard observer is taken into account. We have already seen that this luminous efficiency function is not fully representative of most observers;

it does not represent total sensitivity of the visual mechanism, and it does not account for induction among related colors. Therefore, CIE luminance measures (Y) are not fully in agreement with subjective brightness or lightness measures.

For further information on this fascinating and complicated topic of how the perception of lightness varies with luminance, color, and so on, the reader is referred to Bartleson and Breneman (1967), Bartleson (1980), Jameson (1969), Stevens (1969), and Hurvich and Jameson (1963). These references by no means exhaust the sources to which one can go, but they do represent some leaders in this field and thus provide a good starting point for further reading.

1. Induction Effects

Consider two white circular areas of light of equal luminance. They will appear equally bright. If the amount of light falling on these areas is held constant while a much brighter surround field is placed around one of them,

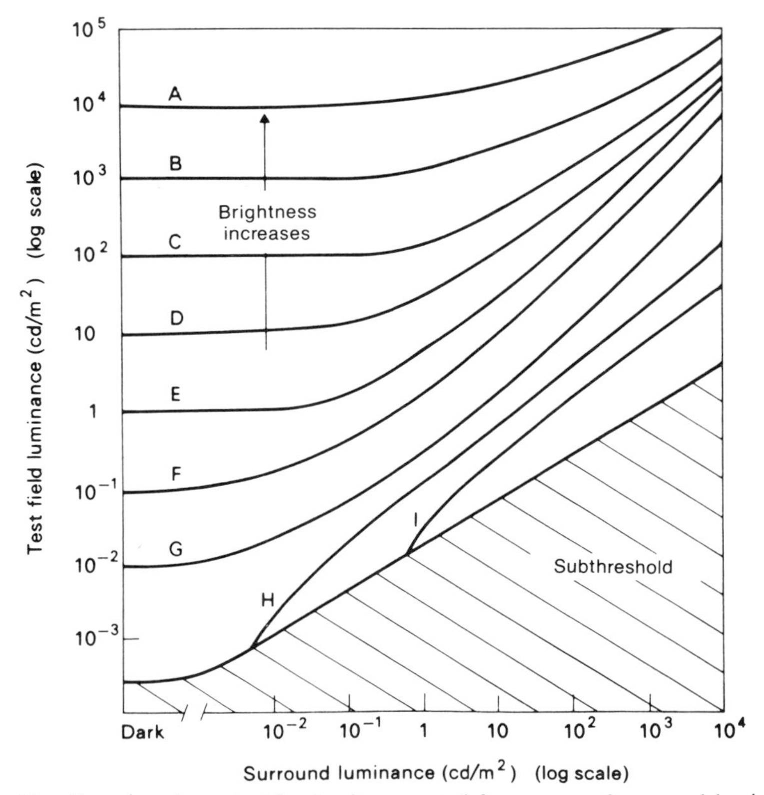

Fig. 10. Contours of constant luminosity measured for a range of surround luminances (Saunders, 1968).

the original circular area will no longer appear equally bright. The one with the surround field will appear distinctly darker, even though the amount of light being reflected from it is exactly the same as from the field without the surround field. Figure 10 shows how the subjective impression of light changes as a function of surround illumination (after Saunders, 1968).

2. Light Adaptation

Recall when you were a child and went to a Saturday afternoon movie. You waited in the bright sunshine to gain entrance to the theater. When you went inside, you felt as though you were almost blind. It was very difficult to see the seats or the people sitting in them. Then after some minutes you were able to see reasonably well. You were even able to recognize faces though before you could not even see that people were there. The amount of light in the theater did not increase; you adapted to the dark. In so doing, your impression of how much light there was in the theater increased. Figure 11 illustrates how our

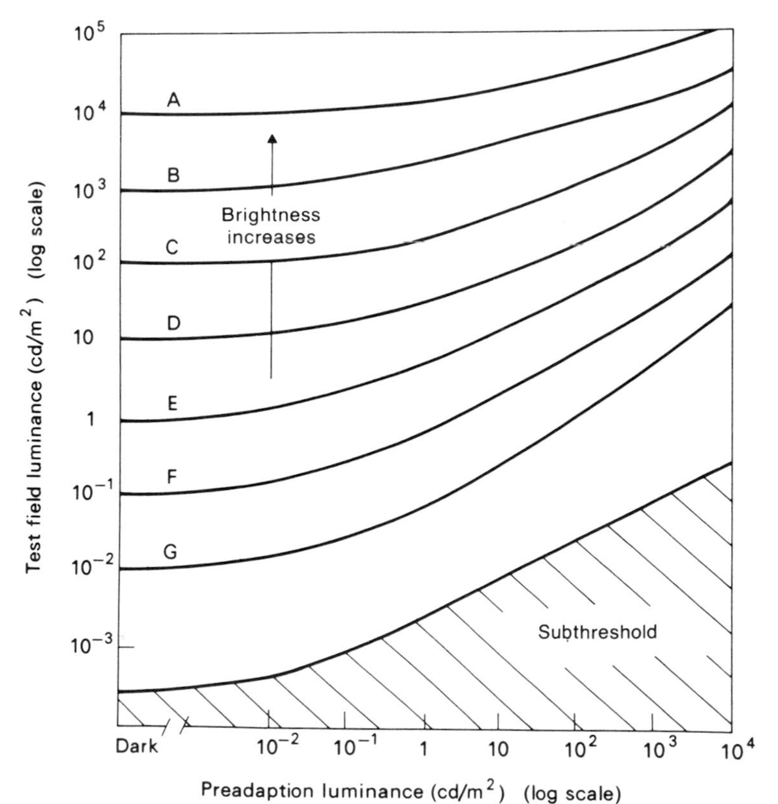

Fig. 11. Contours of constant luminosity measured with a dark background when observer is preadapted to a range of luminance levels (Saunders, 1968).

subjective impression of light changes as a function of the adaptive state of the visual system.

Bartleson (1980) attempted to handle many of the variables that effect our subjective impression of light. His basic equation is also a function of the ratio of two luminances raised to the one-third power. However, he has derived a number of empirically based coefficients and constants to account for chromatic adaptation, induction effects by surrounds of various luminances, and the purity of the stimulus. His formulae are extensive and complex, and it is not possible to elaborate on them here. The reader is referred to Bartleson's *Die Farbe* paper and to the summary in Chapter 5 of this volume for a fuller description of his approach and many useful references.

H. THEORETICAL CONSIDERATIONS

Visual scientists from many disciplines are working to unravel the mysteries of the visual system. Our understanding of how the visual system works is imperfect. However, significant advances have been made. For the purposes of this chapter, our main concern is related to the mechanisms involved in the perception of brightness and lightness. Specifically we are concerned with the way in which the physical properties of wavelength, radiance, spatial, and temporal distributions affect our perception of light (direct or reflected) quantity.

Because a critical factor in visual perception includes the processing of color information, color-vision models are perhaps the most useful for our purposes. Today, the two-stage color-vision model (initial *trichromatic* followed by *opponent-responses*) is the most widely accepted. The first stage involves the quantum catch by the visual photopigments that are located in the outer segments of the visual receptors. There are two classes of photoreceptors: the rods and the cones (Fig. 12). The rods contain only one photopigment, rhodopsin, and are specialized for low light level vision. These are the receptors we use at night when there is no artificial ambient illumination. The reader may have noticed that under these conditions all objects appear to be various shades of gray or nearly so. The rod system, also called the scotoptic system, is color blind. Our inability to distinguish among various hues except by brightness is due to the fact that the rods contain only one photopigment.

The receptors referred to as cones, which make up the photopic system, are used primarily for high light level vision, normally operative under daylight conditions. The cones contain three types of photopigments: One photopigment is mainly responsible for catching short wavelength quanta (blue-catching), another for catching middle wavelength quanta (green-catching), and the last for catching longer wavelength quanta (red-catching). The

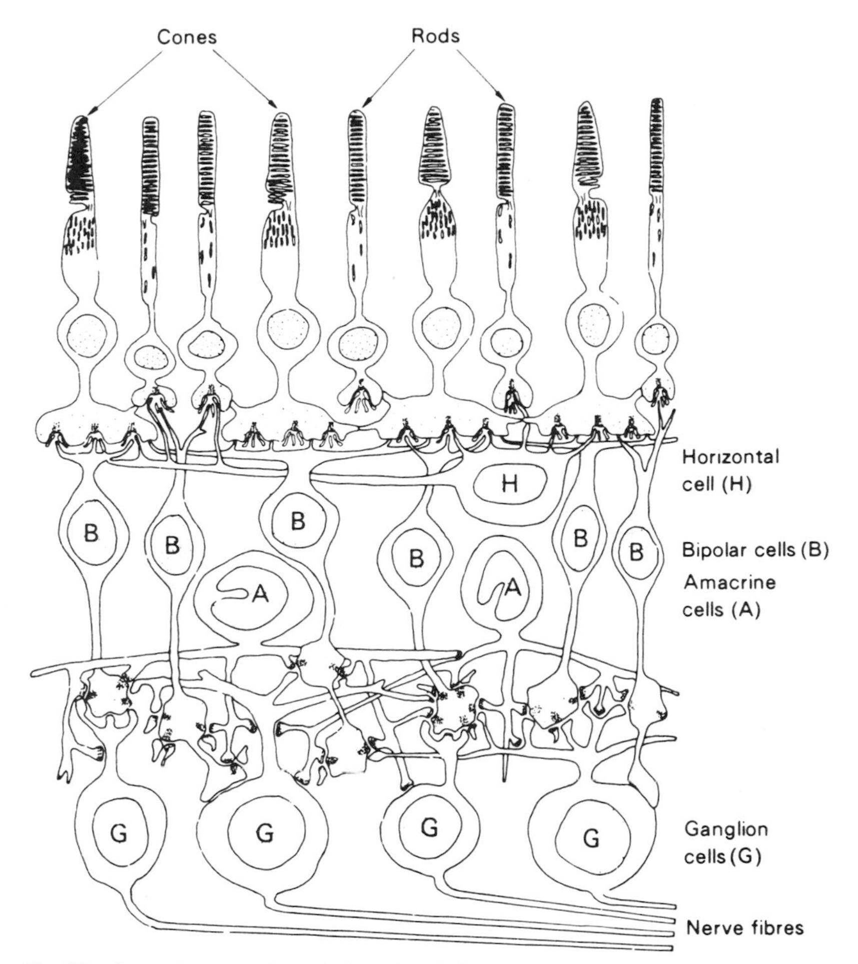

Fig. 12. Synaptic connections of the retina similar to Fig. 5 of Chapter 5.

descriptors in parentheses are to be taken as shorthand jargon for identifying these photopigments. In point of fact, not only are energy quanta not colored, but all of these photopigments are capable of absorbing quanta from a very broad range of the visible spectrum. Their peak absorbences are at short, middle, and long wavelengths.

The second stage involves neuronal processing that begins in the inner segments of the receptors and travels through the layers of the retina to the lateral geniculate nucleus (LGN) in the middle of the brain and up to the cortex (see Fig. 13). Information about this activity is obtained inferentially by psychophysical experimentation or directly by electrophysiological research.

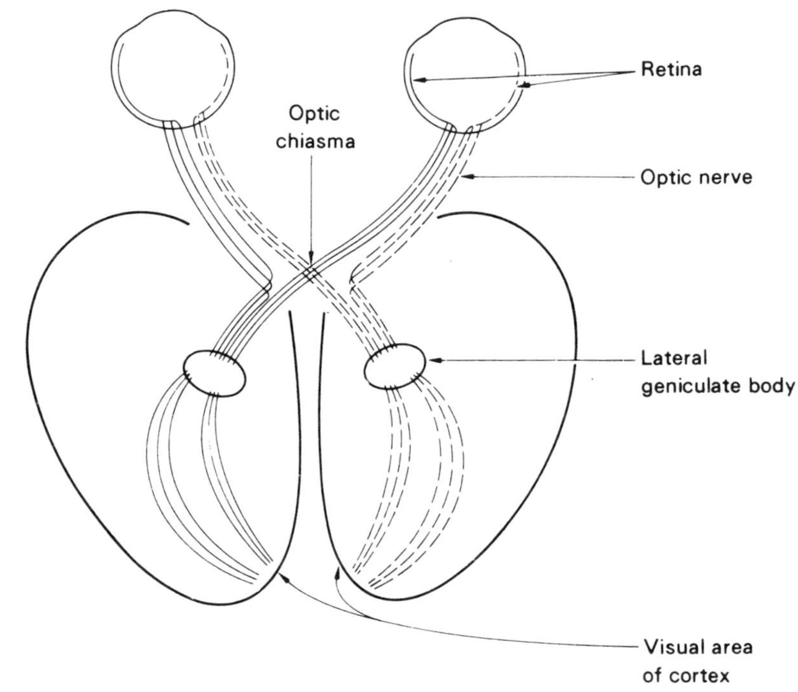

Fig. 13. Schematic diagram of the binocular visual pathways from the retina through the lateral geniculate body to the visual cortex.

There are several examples of models that operate on this two-stage model framework. The reader who is interested in descriptions of these models is referred to Chapter 5 of this volume and to Vos and Walraven (1971, 1972), Ingling (1977), Ingling and Tsou (1977), Guth *et al.* (1980), and Boynton (1979). Judd (1951b) provides an historical perspective of such models.

For the purposes of this chapter some common features of these models are worth exploring. The photopigment stage is sometimes referred to as the trichromatic stage because of the three classes of photopigments in the cones. In the post-receptor level of retinal processing, the visual information is transformed from trichromatic processing to an opponent processing. At this level there are two major channels: an additive and subtractive. The red and green quanta catching receptors feed into the additive achromatic system. This system is variously called the black–white opponent system, the spectrally nonopponent system, the luminance system, or the achromatic system. This system does not process color information but only achromatic intensive information.

The second class of neural information, processed by the spectrally opponent system, is the system that receives input from all three types of cone receptors. The spectrally opponent system consists of two opponent channels: a longwave versus middlewave, and a shortwave versus middlewave plus longwave. These chromatically opponent channels are frequently referred to as the red versus green channel and the blue versus yellow channel. The precise description of these channels has not met with universal agreement. However, the general way in which they contribute to visual information processing is agreed upon. Figure 14 shows a schematic used by Boynton (1979) to illustrate these channels.

The output of the achromatic, nonopponent channel combines with the outputs of the two chromatic, opponent channels nonlinearly. Guth *et al.* (1980) and Ingling and Tsou (1977) present models of the visual system in which the outputs of the achromatic and the two chromatic channels combine according to a vector sum; for further details see Chapter 5. The vector addition in terms of the model of Guth *et al.* is

$$L = (A^2 + T^2 + D^2)^{1/2}, \tag{4}$$

where L = vector luminance, A = activity associated with the achromatic or nonopponent system, T = activity associated with the tritanopic or r-g spectrally opponent chromatic system, D = activity associated with the deuteranopic or y-b spectrally opponent chromatic system.

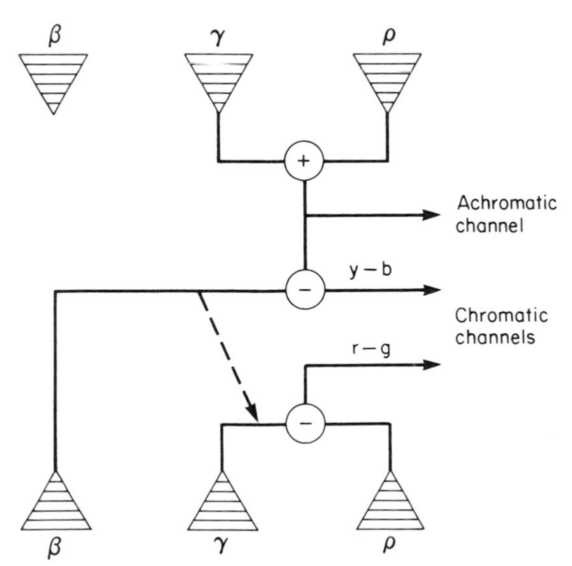

Fig. 14. Schematic of the opponent-color model of human color vision. The ρ, γ, and β cones at the top represent the same three cones that are shown at the bottom. The luminance channel is activated by the ρ and γ cones, the chromatic y–b channel is activated by the ρ, γ, and β cones, and the r–g chromatic channel is activated by the ρ and γ cones (Boynton, 1979).

The opponency of the chromatic channels has been demonstrated electrophysiologically and can be described as follows. With a micro-electrode placed in an optic nerve fiber, it is possible to monitor the electro-physiological activity of the optic nerve as a function of stimulus input. Because there are no synapses between the ganglion cell axons of the retina and the LGN, similar recordings can be made at the LGN (DeValois and Jacobs, 1968; Wiesel and Hubel, 1966; DeMonasterio *et al.*, 1975). These recordings show that there are two major classes of cells: those that respond by increased excitability to all wavelengths or by decreased excitability to all wavelengths and those cells that exhibit an increase in excitability to some wavelengths and a decrease in excitability to other wavelengths. Those cells whose excitability is either increased or decreased to all wavelengths are called the spectrally nonopponent, luminance, or achromatic cells. Those cells whose activity increases to one range of wavelengths and decreases to another range are called spectrally opponent cells. These are the cells that comprise the chromatic channels and yield information about the hue and brightness of the stimulus.

It can be seen in Fig. 14 that a subtractivelike process is represented in the chromatic channels. When an appropriate amount of red and green light is projected into the eye, the signal from the r-g opponent chromatic channel will be zero. However, there will be outputs from the achromatic and y-b channel. The (photopic) output of the achromatic channel will be the sum of ρ and γ cone input; these inputs also contribute to the y-b channel. Because there is no β cone activity in the example, the output of the y-b channel will be yellow. Perceived brightness is hypothesized to be a function of the chromatic and achromatic channels; therefore, these two will combine according to Eq. (4), and the result will be less than the linear sum of the chromatic plus the achromatic activity.

According to current color-vision models, these chromatic and achromatic channels relate to human psychophysical responses in a number of ways. Consider the following additivity of luminances and additivity of brightness experiments. A blue light is adjusted so that it is equal in brightness to a reference white field. Now a yellow light is adjusted so that it also evokes the same brightness reponse as this reference white field. One now takes these amounts of yellow and blue light, reduces them by 50%, and mixes them together. If additivity of brightness held, the result of mixing this blue and yellow light should be a white (blue and yellow are complementary colors) that is equally bright to the reference white. In fact, the mixture will be clearly less bright than the reference.

The reason for this additivity failure is that the subjective impression of brightness is due to the combined outputs of the chromatic and achromatic channels. Because the chromatic channels are differencing channels, a cancellation of chromatic output occurs. Yellow light is obtained by equal activation of the ρ and γ cones. Thus the chromatic output of the r-g channel will be zero. Yellow light is mixed with blue until the mixture appears white and equal to the reference. Thus the chromatic output of the y-b channel is zero. Because there is no chromatic signal coming from either opponent channel, the only channel that is responding to the yellow-plus-blue light would be the achromatic channel. It is seen that under these conditions, whatever brightness signal would be derived from the chromatic channels is lost, which would explain the failure of brightness additivity.

If this experiment is repeated with the criterion of flicker photometry (see Section F.8) additivity will be obtained. That is, when the amount of yellow light that was matched to the reference white on the basis of minimum flicker is

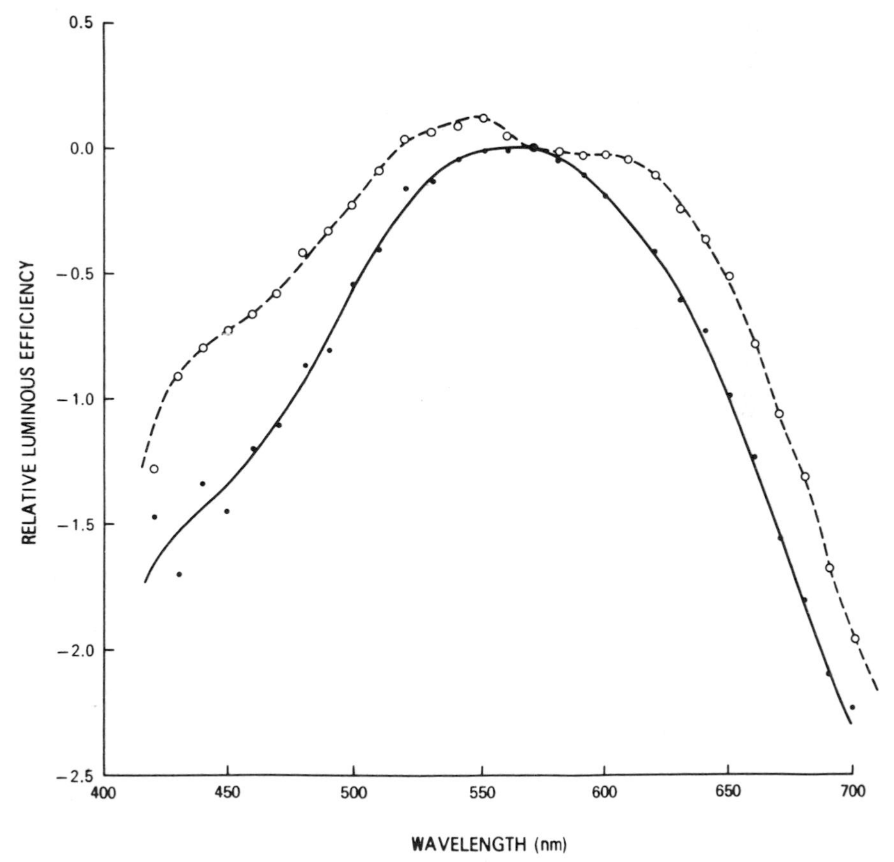

Fig. 15. A comparison of luminous efficiency functions determined by heterochromatic brightness-matching (dashed curve) and heterochromatic flicker photometry (solid curve). These are the average data for five observers (Kaiser, 1981).

halved, and the same is done for the blue, and the resultant quantities are then mixed together, the sum will be equally bright to the reference. The reason this occurs is because the time constant of the chromatic system is less than that of the achromatic system. Thus when the flicker criterion is employed, only the achromatic system is mediated (for example, Boynton and Kaiser, 1968; and Chapter 5).

We now have the necessary background to understand why lights equal in luminance usually are not equally bright. A very large proportion of the data that were used to define the CIE $V(\lambda)$ function came from flicker photometric research. Thus luminance measures reflect primarily the activity of the achromatic channel. Brightness-matching, on the other hand, reflects the combined activity of the chromatic channels and the achromatic channel. Therefore, because brightness-matching reflects the activity of the chromatic and achromatic channels, and luminance corresponds only to the activity of the achromatic channel, less energy is required for brightness matches to a reference white than when flicker is used. The comparison of brightness-matching spectral sensitivity functions with those derived from flicker photometry show this differential in required energy by the brightness-matching function exhibiting a greater sensitivity (Fig. 15).

I. CHOOSING THE BEST METHOD

In this chapter many different methods have been discussed for evaluating light that is directed or reflected into observers' eyes. The purpose is to provide some information to help the reader decide which method is the best for a particular task. To choose the best method the investigator must first decide the purpose of the measurement.

For many applied purposes one can safely use a well calibrated physical photometer. This instrument will provide luminance measures that have the following characteristics: the measures will be based on an internationally agreed upon system of light measurement, the procedure for obtaining this measure is easy to describe, and the measurement will be reasonably accurate and easily repeatable.

If this route is taken, the investigator should insure that if the light levels are reasonably high (above several $cd \cdot m^2$), then the photometer should have a relative spectral response function that suitably matches the CIE $V(\lambda)$ function to provide photopic luminance measures. If the light levels to be measured are below about $0.0001\ cd \cdot m^2$, then the photometer should have a relative spectral response similar enough to the CIE $V'(\lambda)$ function to insure scotopic luminance measures. These relative spectral response functions results from the response characteristics of the instrument's photodetector, the spectral transmittances of optical filters in the beam, and spectral efficiencies of lenses, heat absorbers, and mirrors through which the light must pass to be detected. For light levels between photopic and scotopic levels (that is, mesopic) the investigator can do one of several things. The light measure can be taken with either a scotopic or photopic photometer, and the one used must be explicitly specified. Though this approach will yield light measures that may be very unrepresentative of brightness perception, other people will be able to repeat what was done. This approach has limited usefulness but can be better than nothing.

The CIE Vision Committee (TC-1.4) is working on the problem of measuring light under mesopic conditions. Currently there are two proposed methods for handling this problem. Palmer (1966, 1967, 1968) has proposed a weighted combination of photopic and scotopic luminance measures according to

$$L = (MS + P^2)/(M + P), \tag{5}$$

where L, S, and P are in $\mathrm{cd \cdot m^{-2}}$ and $M = 0.06\ \mathrm{cd \cdot m^{-2}}$. Kokoschka and Bodmann (1976) have proposed a somewhat more complicated approach involving the concept of equivalent luminance (Kowaliski, 1969).

At present there is little basis for choosing one approach over the other. Palmer's empirical formula is easier to use but is only an approximation to an adequate solution to the problem. The Kokoschka and Bodmann equation is much more complicated but has the virtue of being more adequately based on basic vision data under a wide range of light level conditions. For a complete description of their approach to mesopic photometry the reader is referred to their 1976 publication. These approaches to light measurement can be used when attempting to set light levels according to some specifications or when one is attempting to set specifications.

In those instances when light measurements are made of complex reflecting or transmitting sources (that is, photographs or transparencies), the lightness measures as described in Section G are recommended.

In those cases in which one cannot assume the luminous efficiency function of a CIE standard observer, it becomes necessary to use one of the other approaches to light measurement. Examples of these cases include light measurements made in which these measures are to relate either to non-human vision or to some special classes of human beings. When visually meaningful light measures for animals are needed, one of the behavioral methods is most appropriate.

Suppose one wanted to adjust the light level of signal lights so that color blind or color defective observers would clearly see these lights as different in brightness, and normally sighted people would perceive both chromatic and brightness differences. Theoretically, it is possible to make the brightness difference such that it is irrelevant for the normally sighted but is an important distinguishing feature for the color defective. If this was the desired goal, a CIE-based photometer would not serve well; some other light measurement approach would be required. Some behavioral method as described above can be used. However, because humans have the ability to communicate, it would be easier to use one of the brightness-evaluation methods.

For basic-research purposes one must first decide why the light measurements are being made. If they are used for the purpose of reporting the light levels that you happened to use, then the appropriate physical photometer can be employed together with a colorimeter to specify chromaticity. If, on the other hand, one wishes to set the light at some physiologically meaningful level, then some form of visual photometry must be made. Suppose one wanted to set a series of chromatic lights to equal brightness for reasons that are relevant to the experimental design. Clearly, setting them for equal luminance will not do the job. If one really means equal brightness, then some brightness-evaluation procedure is appropriate.

If one wanted a luminance like measurement, but one that depended on the spectral sensitivity of the observers, then the lights can be evaluated by means of flicker photometry.

Suppose you were attempting to determine the optimum level of light for extended close work. Perhaps a procedure of measuring the light using a visual acuity criterion would make the most sense. In close work one usually is not making brightness judgments or flicker judgments. However, the ability of the observer to see clearly over an extended period of time is often what is at issue.

Suppose the problem is to specify the minimum amount of light required for signal lights to be just visible and for their colors be correctly identified as well. In a case like this, one would want to use the threshold criterion of correct color recognition. If the minimum amount of light required to detect a light were needed, then an absolute-threshold technique is appropriate.

It is difficult to identify all of the various reasons for making light measurements. However, if the intent is to make visually meaningful light measurements, then it is best to use a method that makes sense with regard to the use to which the measurements will be put. This discussion of the various ways in which visually meaningful light measurements can be made, and the limitations of each of the methods, should aid the reader in deciding how best to measure and specify the amount of light for his own purposes.

ACKNOWLEDGMENT

This chapter was prepared with assistance from the Natural Sciences and Engineering Research Council of Canada grant APA 295.

REFERENCES

Aiba, T. S., Alpern, M., and Maaseidvaag, F. (1967). The electro-retinogram evoked by the excitation of human foveal cones. *J. Physiol.* **189**, 43–62.

Baron, W., Boynton, R. M., and Norren, D. (1979). Primate cone sensitivity to flicker during light and dark adaptation as indicated by the foveal electroretinogram. *Vision Res.* **19**, 109–116.

Bartleson, C. J. (1980). Measurements of brightness and lightness. *Die Farbe* **28**, 132–148.

Bartleson, C. J., and Breneman, E. J. (1967). Brightness perception in complex fields. *J. Opt. Soc. Am.* **57**, 953–957.

Beauchamp, R., Rowe, J. S., and O'Reilly, L. A. (1979). Goldfish spectral sensitivity: Identification of the three cone mechanisms in heart-rate conditioned fish using colored adapting backgrounds. *Vision Res.* **19**, 1295–1302.

Bedford, R. E., and Wyszecki, G. (1957). Axial chromatic aberration of the human eye. *J. Opt. Soc. Am.* **47**, 564–565.

Blough, D. S. (1955). Method for tracing dark adaptation in the pigeon. *Science* **121**, 703–704.

Blough, D. S., and Yeager, D. (1972). Visual psychophysics in animals. *In* "Handbook of Sensory Psychophysics" (D. Jameson, and L. M. Hurvich, eds.), pp. 732–763, Springer-Verlag, N.Y.

Bornstein, H. M., and Marks, L. E. (1972). Photopic luminosity measured by the method of critical frequency. *Vision Res.* **12**, 2023–2033.

Boynton, R. M. (1966). Vision. *In* "Experimental Methods and Instrumentation in Psychology" (J. B. Sidowski, ed.). McGraw-Hill, N.Y.

Boynton, R. M., and Baron, W. S. (1975). Sinusoidal flicker characteristics of primate cones in response to heterochromatic stimuli. *J. Opt. Soc. Am.* **65**, 1091–1100.

Boynton, R. M., and Kaiser, P. K. (1968). Vision: The additivity law made to work for heterochromatic photometry with bipartite fields. *Science* **161**, 366–368.

Boynton, R. M., Kandel, G., and Onley, J. W. (1959). Rapid chromatic adaptation of normal and dichromatic observers. *J. Opt. Soc. Am.* **49**, 654–666.

Boynton, R. M. (1979) *Human Color Vision*, Holt, Rinehart and Winston, Toronto.

Brown, J. L. (1965). Flicker and intermittent stimulation. *In* "Vision and Visual Perception" (C. H. Graham, ed.), pp. 251–320. Wiley, N.Y.

Brown, J. L., Phares, L., and Fletcher, D. E. (1960). Spectral energy thresholds for the resolution of acuity targets. *J. Opt. Soc. Am.* **50**, 950–960.

Cavonius, C. R., and Hilz, R. (1973). Brightness of isolated colored lights. *J. Opt. Soc. Am.* **63**, 884–888.

CIE (1970). "Principles of Light Measurements" Publication CIE No. 18. CIE Bureau Central, Paris. CIE (1971). "Colorimetry" Publication CIE No. 15. CIE Bureau Central, Paris.

CIE (1978). "Light as a True Visual Quantity: Principles of Measurement" Publication CIE No. 41. CIE Bureau Central, Paris.

DeMonasterio, F. M., Gouras, P., and Tolhurst, D. J. (1975). Trichromatic colour opponency in ganglion cells of the rhesus monkey retina. *J. Physiol.* **252**, 197–216.

DeValois, R. L. (1973). Central mechanisms of color vision. *In* "Handbook of Sensory Physiology," (R. Jung, ed.), VII/3, pp. 209–253. Springer-Verlag, N.Y.

DeValois, R. L., and Jacobs, G. H. (1968). Primate color vision. *Science* **162**, 533–540.

DeValois, R. L., Abramov, I., and Jacobs, G. H. (1966). Analysis of response patterns of LGN cells. *J. Opt. Soc. Am.* **56**, 966–977.

Dobson, V. (1976). Spectral sensitivity of the 2-month old infant as measured by the visually evoked cortical potential. *Vision Res.* **16**, 367–374.

Drum, B. (1981). Rod-cone interaction in the dark adapted fovea. *J. Opt. Soc. Am.* **71**, 71–74.

Enoch, J. M. (1972). The two color threshold technique of Stiles and derived component color mechanisms. *In* "Handbook of Sensory Physiology," VII/4 (D. Jameson, and L. M. Hurvich, eds.), pp. 537–567. Springer-Verlag, N.Y.

Gibson, K. S., and Tyndall, E. P. T. (1923). Visibility of radiant energy. *Sci. Papers Bur. Stand.* **19**, 131–191.

Green, D. G., and Maaseidvaag, F. (1967). Closed circuit television pupilometer. *J. Opt. Soc. Am.* **57**, 830–833.

Gouras, P. (1968). Identification of cone mechanisms in monkey ganglion cells. *J. Physiol.* **199**, 533–547.

Gouras, P. (1974). Opponent-colour cells in different layers of the foveal striate cortex. *J. Physiol.* **238**, 583–602.

Gouras, P., and Zrener, E. (1979). Enhancement of luminance by color opponent mechanisms. *Science* **205**, 587–589.

Guth, S. L., and Graham, B. V. (1975). Heterochromatic additivity and the acuity response. *Vision Res.* **15**, 317–319.

Guth, S. L., and Lodge, H. R. (1973). Heterochromatic additivity, foveal spectral sensitivity and a new color model. *J. Opt. Soc. Am.* **63**, 450–462.

Guth, S. L., Massof, R. W., and Benzschawel, T. (1980). Vector model for normal and dichromatic color vision. *J. Opt. Soc. Am.* **70**, 197–212.

Hedin, A. (1978). Pupillomotor spectral sensitivity in normals and colour defectives. *Acta Ophthamol. Suppl.* **137**, 3–83.

Hsia, Y., and Graham, C. H. (1952). Spectral sensitivity of the cones in the dark adapted human eye. *Proc. Natl. Acad. of Sci.* **38**, 80–85.

Hurvich, L. M., and Jameson, D. (1963). "The Perception of Brightness and Darkness." Allyn and Bacon, Boston.

Ikeda, M., and Shimozono, H. (1978). Luminous efficiency functions determined by successive brightness matching. *J. Opt. Soc. Am.* **68**, 1767–1771.

Ikeda, M., Yaguchi, N., Yoshimatsu, N., and Ohmi, M. (1982) *Journal Optical Society America*, **72**, 68–73.

Ingling, C. R. (1977). The spectral sensitivity of the opponent-color channels. *Vision Res.* **17**, 1083–1089.

Ingling, C. R., and Tsou, B. H. (1977). Orthogonal combinations of three visual channels. *Vision Res.* **17**, 1075–1082.

Ingling, C. R., Scheibner, H. M. O., and Boynton, R. M. (1970). Color naming of small foveal fields. *Vision Res.* **10**, 501–511.

Ingling, C. R., Tsou, B. H.-P., Gast, T. J., Burns, S. A., Emerick; J., and Riesenberg, L. (1978). The achromatic channel: 1. The nonlinearity of minimum border and flicker matches. *Vision Res.* **18**, 379–390.

Ives, H. E. (1912). Studies in the photometry of lights of different colors-V, The spectral luminosity curve of the average eye. *Philos. Mag.* **24**, 853–863.

Jameson, D. (1970). Brightness scales and their interpretation. *In* "AIC Proceedings Color 69," pp. 377–385. Musterschmidt-Verlag, Gottingen.

Johnson, E. P., and Cornsweet, T. N. (1954). Electroretinal photopic sensitivity curves. *Nature* **174**, 614–615.

Judd, D. B. (1951a). Basic correlates of the visual stimulus. *In* "Handbook of Experimental Psychology" (S. S. Stevens, ed.), pp. 811–867. Wiley, N.Y.

Judd, D. B. (1951b). Report of U.S. Secretariate, Committee on Colorimetry and Artificial Daylight. "Proceedings CIE I," part 7, p. 11. Paris: Bureau Central CIE.

Kaiser, P. K. (1971). Minimum border as a prefered psychophysical criterion in visual heterochromatic photometry. *J. Opt. Soc. Am.* **61**, 966–971.

Kaiser, P. K. (1978). Request for brightness-matching data and mathematical colour-vision models. *Color Res. Appl.* **3**, 148.

Kaiser, P. K. (1981). Photopic and mesopic photometry: yesterday, today and tomorrow. "Proceedings, Golden Jubilee of Colour in the CIE." The Society of Dyers and Colourists, Bradford, England.

Kaiser, P. K., and Greenspon, T. (1971). Brightness difference and its relation to the distinctness of a border. *J. Opt. Soc. Am.* **61**, 962–965.

Kaiser, P. K., Herzberg, P. A., and Boynton, R. M. (1971). Chromatic border contrast and its relation to saturation. *Vision Res.* **11**, 953–968.

Kaiser, P. K., Comerford, J. P., and Bodinger, D. M. (1976). Saturation of spectral lights. *J. Opt. Soc. Am.* **66**, 818–826.

Kelly, D. H. (1971). Flicker. *In* "Handbook of Sensory Physiology," VII/4 (D. Jameson, and L. M. Hurvich, eds.). Springer-Verlag, N.Y.

Kling, J. W., and Riggs, L. A. (1972). "Woodworth and Schlosberg's Experimental Psychology," 3rd ed. Holt, N.Y.

Kokoschka, S., and Bodmann, H. W. (1976). Ein knosistentes system zur photometrischen strahlungsbewertung im gesampten adaptationsbereich. *In* "Publication CIE No. 36," pp. 217–225. Bureau Central de la CIE Paris.

Kowaliski (1969). Equivalent luminances of colors. *J. Opt. Soc. Am.* **59**, 125–130.

LaMotte R. H., and Brown, J. L. (1970). Dark adaptation and spectral sensitivity in the cat. *Vision Res.* **10**, 703–716.

LeGrand, Y. (1968). "Light Color and Vision" (R. W. G. Hunt, T. Walsh, and F. R. W. Hunt, transl.), 2nd ed. Halstead Press, Somerset, N.J.

Marks, L. E., and Bornstein, M. H. (1973). Spectral sensitivity by constant CFF: Effect of chromatic adaptation. *J. Opt. Soc. Am.* **63**, 220–226.

Marks, W. B., Dobell, W. H., and MacNichol, E. F. (1964). Visual pigments of single primate cones. *Science* N.Y. **143**, 1181–1183.

Meyers, K. J., Ingling, C. R., and Drum, B. A. (1973). Brightness additivity for a grating target. *Vision Res.* **13**, 1165–1173.

Nusinowitz, S., and Kaiser, P. K. (1983). Small field spectral sensitivity, 20th Session, CIE, Amsterdam, Vol. 1: papers, Bureau Central de la CIE, 52 Boulevard, Malesherbes 75008 Paris, France, D101/1–4.

Olzak, L. A., and Thomas, J. P. (1981). Gratings: Why frequency discrimination is sometimes better than detection. *J. Opt. Soc. Am.* **71**, 64–70.

Palmer, D. A. (1966). A system of mesopic photometry. *Nature* **209**, 276–281.

Palmer, D. A. (1967). The definition of a standard observer for mesopic photometry. *Vision Res.* **7**, 619–628.

Palmer, D. A. (1968). Standard observer for large field photometry at any level. *J. Opt. Soc. Am.* **58**, 1296–1299.

Peeples, D. R., and Teller, D. Y. (1978). White adapted photopic spectral sensitivity in human infants. *Vision Res.* **18**, 49–53.

Pugh, E. N., and Mollon, J. D. (1979). A theory of the Pi 1 and Pi 3 color mechanisms of Stiles. *Vision Res.* **19**, 293–312.

Regan, D. (1970). Objective method of measuring the relative spectral luminosity curve in man. *J. Opt. Soc. Am.* **60**, 856–859.

Regan, D. (1972). "Evoked Potentials in Psychology, Sensory Physiology and Clinical Medicine." Chapman and Hall, London.

Regan, D. (1975). Color coding of pattern responses in man investigated by evoked potentials feedback and direct plot techniques. *Vision Res.* **15**, 175–183.

Riggs, L. A. (1965). Visual acuity. *In* "Vision and Visual Perception" (C. H. Graham, ed.), pp. 321–349. Wiley, N.Y.

Riggs, L. A., and Wooten, B. R. (1972). Electrical measures and psychophysical data on human vision. *In* "Handbook of Sensory Physiology, VII/4" (D. Jameson, and L. M. Hurvich, eds.), pp. 690–731. Springer-Verlag, N.Y.

Saunders, J. E. (1968). Adaptation, its effect on apparent brightness and contribution to the phenomenon of brightness constancy, *Vision Research*, **8**, 451–468.

Sechzer, J. A., and Brown, J. L. (1964). Color discrimination in the cat. *Science* **144**, 427–429.

Siegel, C., and Pugh, E. N. (1980). Stiles's Pi 5 color mechanism: Tests of field displacement and field additivity properties. *J. Opt. Soc. Am.* **70**, 71–81.

Sommers, W. W., and Fry, G. A. (1974). Relation of macular pigment and photoreceptor distribution to the perception of a brightness difference. *Am. J. Optometry Physiol. Opt.* **51**, 241–251.

Sperling, H. G., and Harwerth, R. S. (1971). Red-green cone interactions in the increment-threshold spectral sensitivity of primates. *Science* **172**, 180–184.

Sperling, H. G., and Lewis, W. G. (1959). Some comparisons between foveal spectral sensitivity data obtained at high brightness and absolute threshold. *J. Opt. Soc. Am.* **49**, 983–989.

Stevens, S. S. (1969). On predicting exponents for cross modality matches. *Percept. Psychophys.* **6**, 251–256.

Stevens, S. S. (1975). "Psychophysics." Wiley, New York.

Stiles, W. S. (1978). "Mechanisms of Colour Vision." Academic Press, N.Y.

Svaetichin, G. (1956). Spectral response curves from single cones. *Acta Physiol. Scandanavica* **39** (Suppl. 134), 17–46.

Tansley, B. W., and Boynton, R. M. (1976). A line, not a space, represents visual distinctness of borders formed by different colors. *Science* **191**, 954–957.

Teller, D. Y., Peoples, D. R., and Sekel, M. (1978). Discrimination of chromatic from white light by two-month old human infants. *Vision Res.* **18**, 41–48.

Thompson, R. F., and Patterson, M. M. (1973). "Bioelectric Recording Techniques, Part A. Cellular Processes and Brain Potentials," Vol. 1-A. Academic Press, N.Y.

Uchikawa, K., and Ikeda, M. (1981). Temporal deterioration of wavelength discrimination with successive comparison method. *Vision Res.* **21**, 591–596,

Vos, J. J., and Walraven, P. L. (1971). On the derivation of the foveal receptor primarics. *Vision Res.* **11**, 799–818.

Vos, J. J., and Walraven, P. L. (1972). An analytical description of the line element in the zone-fluctuation model of colour vision, Vol. I, II. *Vision Res.* **12**, 1327–1344.

Wagner, G., and Boynton, R. M. (1972). Comparison of four methods of heterochromatic photometry. *J. Opt. Soc. Am.* **62**, 1508–1515.

Walraven, P. L. (1962). "On Mechanisms of Color Vision." Thesis, Institute for Perception RVO-TNO, Soesterberg, Netherlands.

Walraven, P. L., and Leebeek, H. J. (1964). Phase shift of alternating colored stimuli. *Doc. Ophthalmol.* **18**, 56–71.

Walsh, J. W. T. *Photometry* (1958). (3rd ed.) Constable and Co. Ltd., London.

Wiesel, T., and Hubel, D. H. (1966). Spatial and chromatic interactions in the lateral geniculate body of the rhesus monkey. *J. Neurophysiol.* **29**, 1115–1156.

Wyszecki, G., and Stiles, W. S. (1982). "Color Science." (2nd Ed) Wiley, N.Y.

Young, R. S. L., and Alpern, M. (1980). Pupil responses to foveal exchange of monochromatic lights. *J. Opt. Soc. Am.* **70**, 697–706.

OPTICAL RADIATION MEASUREMENTS, VOL. 5

12

Colorimetric Measurement

C. J. BARTLESON
Research Laboratories
Eastman Kodak Company
Rochester, New York

A. INTRODUCTION

Conventional methods of colorimetry are described in Chapter 3 of Volume 2 of this treatise (Bartleson, 1980). These methods are based on matching and threshold evaluations such as those described in Chapter 7 of this volume. Conventional colorimetry provides empirical methodologies for specifying combinations of standard stimuli that match test stimuli according to the perceptions of a standard observer and, in addition, offers equations for transforming such data to estimate the relative sizes of small differences in color. These colorimetric methods will be summarized only briefly; the major part of this chapter will describe certain advanced methods for estimating differences in color and predicting the magnitudes and qualities of color appearances, methods that draw upon existing knowledge of the mechanism of color vision (see Chapter 5 of this volume) to devise structures that are consistent with what is known of the physiology of vision and the observational data of color measurement. In short, this chapter will illustrate the way in which visual science and colorimetric practice are being brought together to form a general schema for the measurement and assessment of color.

Three proposed methods will be described. One deals with threshold (and, by extension, with suprathreshold) color differences; it incorporates a model

ISBN-0-12-304905-9

of threshold sensitivity devised by Guth and coworkers (for example, Guth and Lodge, 1973; Guth *et al.*, 1980; Benzschawel and Guth, 1981). The method not only predicts relative sizes of color differences observed by direct experimentation, but also it predicts a number of other threshold and near-threshold color relationships. The second method to be described deals with suprathreshold color relationships. It involves a model developed by Hunt (1982) that incorporates empirical expressions for relationships implied by considerations of the mechanism of color vision. It provides estimates of the appearances of color stimuli under a variety of viewing conditions. The third method that will be discussed is a model for estimating the influence of chromatic adaptation on the appearance of color stimuli, such as changes that may ensue from differences among the spectral qualities of illuminants. It involves largely empirical relationships among visual sensitivities observed experimentally and implied by considerations of the mechanism of vision; the model has been proposed by Nayatani and co-workers (for example, Nayatani *et al.*, 1980, 1981a,b).

Other models have been proposed for the same or similar purposes; several are referred to in Chapter 5. These three have been selected to illustrate the current status of continuing work to develop theoretically based structures for assessing color. Unquestionably, the models described here will be superseded by others that will provide more exact predictions and, at the same time, will incorporate less empiricism and more theoretically direct structures. The methods discussed should, however, give the reader some appreciation of the considerations that enter into a more advanced structure of color measurement than that exemplified by conventional colorimetry.

B. SUMMARY OF CONVENTIONAL COLORIMETRY

The background, methods, and industrial applications of conventional colorimetry are described in detail elsewhere (for instance, CIE, 1971; Judd and Wyszecki, 1975; Bartleson, 1980; Billmeyer and Saltzman, 1981; Wyszecki and Stiles, 1982). This section will set forth only a brief summary of those methods of colorimetry that have been developed and recommended by the CIE over the past fifty years.

A conventional specification of a color match is expressed as a triad of numbers representing the amounts of three (reference or standard) *primary stimuli* that may be combined by addition to produce a composite stimulus that appears to a standard observer to match the color of the test stimulus. The triad of numbers consists of three *tristimulus values.* Tristimulus values (such as X, Y, Z) may be computed by integration over wavelength λ of the products of functions describing the spectral radiances of the test stimulus $\varphi(\lambda)$ and three color-mixture functions, such as $\bar{x}(\lambda)$, $\bar{y}(\lambda)$, $\bar{z}(\lambda)$, that describe the tristimulus values corresponding to color matches of the spectral components of an equi-energy spectrum:

$$X = k \int \varphi(\lambda)\bar{x}(\lambda)d\lambda,$$
$$Y = k \int \varphi(\lambda)\bar{y}(\lambda)d\lambda, \tag{1}$$
$$Z = k \int \varphi(\lambda)\bar{z}(\lambda)d\lambda.$$

The limits of integration usually are from λ = 360–400 to 700–760 nm. The stimulus function $\varphi(\lambda)$ for objects that are not luminous may consist of reflectance functions $\rho(\lambda)$, transmittance functions $\tau(\lambda)$, or radiance factor functions $\beta(\lambda)$, multiplied by the spectral power of the illuminant irradiating the object $S(\lambda)$. The scale factor k may assume any convenient value; the most often used values are $k = 680\ \ell \cdot W^{-1}$, which yields tristimulus values consistent with luminance (see Chapters 5 and 11), and $k = 100[\int S(\lambda)\bar{y}(\lambda)\,d\lambda]^{-1}$ (which yields tristimulus values consistent with luminance factor). The CIE has defined two standard sets of color-mixture functions: one corresponding to a standard observer who views an approximately 2° diameter subtense field $[\bar{x}(\lambda), \bar{y}(\lambda), \bar{z}(\lambda)]$ and one corresponding to an observer who views an approximately 10° diameter subtense field $[\bar{x}_{10}(\lambda), \bar{y}_{10}(\lambda), \bar{z}_{10}(\lambda)]$. The color-mixture functions for the 2° observer (known as the *CIE 1931 Standard Colorimetric Observer*) are given at 10-nm intervals in Table IV of Chapter 5 here. Data for both observers are given for 1-nm intervals in Chapter 3 of Volume 2 in this treatise (Bartleson, 1980, Tables I and II, pp. 48–68) and in CIE Publication No. 15 (CIE, 1971, Tables 2.1 and 2.2, pp. 93–112).

Chromaticity coordinates represent the proportions of tristimulus values in a color match. They are calculated from the tristimulus values as follows:

$$x = (X)(X + Y + Z)^{-1},$$
$$y = (Y)(X + Y + Z)^{-1}, \tag{2}$$
$$z = (Z)(X + Y + Z)^{-1}.$$

Similar equations apply to the 10° *CIE 1964 Supplementary Standard Colorimetric Observer*. Because $x + y + z = 1$, only two chromaticity coordinates are needed to specify the relative tristimulus characteristic called chromaticity; x and y are used to do so. A graph in which y is plotted against x comprises a *chromaticity diagram*; such a diagram is shown in Fig. 1. The locus of chromaticities corresponding to spectral components of an equi-energy spectrum is usually plotted in such a diagram, and the chromaticities of the

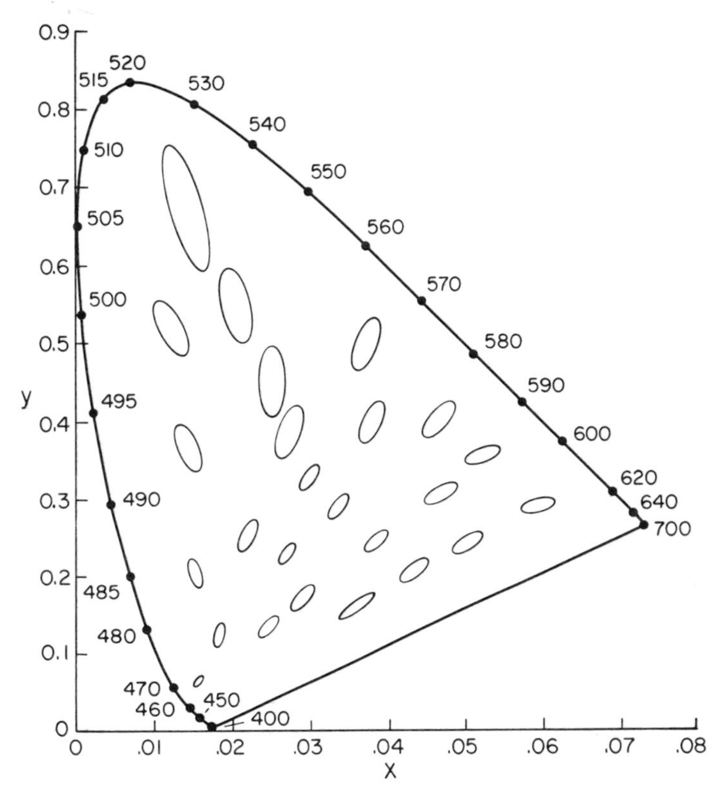

Fig. 1. CIE 1931 chromaticity diagram showing loci of chromaticities corresponding to spectral components of an equi-energy spectrum, joined by a straight line at the spectral extremes. Also shown are discrimination ellipses for 25 chromaticities as reported by MacAdam (1942).

spectral end points are joined by a straight line to depict the area of the diagram containing chromaticities for all possible real color stimuli.

When bivariate normal ellipses representing the intervals of uncertainty about color matches (see Chapter 7) are plotted in a CIE 1931 chromaticity diagram, the sizes and orientations of the ellipses differ throughout the diagram. Figure 1 illustrates such ellipses for data determined by MacAdam (1942); each ellipse is drawn with its principal axis approximately 10/3 the size of a difference limen. The differences among these ellipses imply that the diagram is not perceptually uniform; a perceptually uniform diagram would depict bivariate normal ellipses as circles of constant radius everywhere in the graph. Numerous mathematical transformations of tristimulus values have been proposed to provide a metric system in which such ellipses are rendered as constant-radius circles. None have been completely successful. The one currently recommended by the CIE is the CIE 1976 UCS diagram (where UCS stands for "uniform chromaticity scale"), which has chromaticity coordinates u' and v' defined as

$$u' = (4X)(X + 15Y + 3Z)^{-1},$$
$$v' = (9Y)(X + 15Y + 3Z)^{-1}. \tag{3}$$

Figure 2 illustrates MacAdam's (1942) ellipses plotted in the diagram defined by Eq. (3). The sizes and orientations of the ellipses in Fig. 2 do not vary throughout the diagram as much as those for the same data shown in Fig. 1. The CIE 1976 UCS diagram is, therefore, said to be "more nearly uniform" than the CIE 1931 chromaticity diagram.

The underlying, transformed, tristimulus dimensions of the CIE 1976 UCS diagram are symbolized L^*, u^*, v^* and are defined as

$$L^* = 116(Y/Y_n)^{1/3} - 16,$$
$$u^* = 13L^*(u' - u'_n), \tag{4}$$
$$v^* = 13L^*(v' - v'_n),$$

where the subscript n refers to a neutral ("white") reference. Color differences

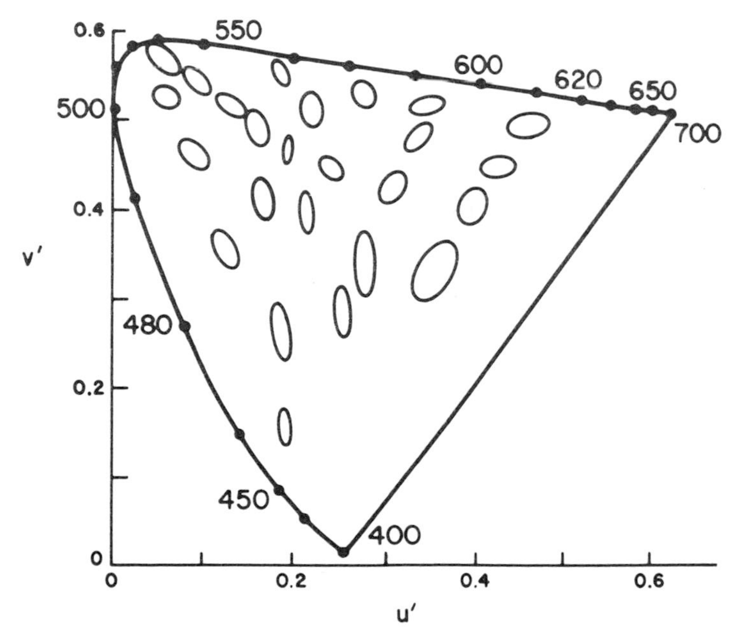

Fig. 2. MacAdam's (1942) discrimination ellipses plotted in the CIE 1976 u', v' chromaticity diagram.

ΔE are then calculated in the space defined by Eqs. (4) as follows:

$$\Delta E_{uv} = [(\Delta L^*)^2 + (\Delta u^*)^2 + (\Delta v^*)^2]^{1/2}. \tag{5}$$

The subscript uv distinguishes such CIE color differences from those similarly calculated in the CIE 1976 L^*, a^*, b^* space.† Components of such color differences may be derived from Eq. (5) and its CIE 1976 L^*, a^*, b^* equivalent; these are described in detail in Chapters 2 and 3 of Volume 2 in this treatise (Hunt, 1980; Bartleson, 1980).

These CIE color-difference equations have limited utility, because they are only approximately uniform in visual extent (see Fig. 2), they relate only to threshold or near-threshold differences in color, and they are appropriate only for adaptation to illumination substantially the same as that of CIE Illuminant D_{65}, representing one phase of natural daylight. They fail to characterize suprathreshold color differences satisfactorily in most applications and fail to characterize properly even threshold color differences in many applications. For these reasons, efforts continue to derive other methods for predicting the sizes of color differences. The next section of this chapter will describe one method that has been developed for estimating the sizes of threshold and near-threshold color differences.

C. THRESHOLD COLOR DIFFERENCES

Guth and Lodge (1973) proposed a color-vision model that subsequently was modified by Guth *et al.* (1980). This model has been used to predict a wide range of experimental results for observers with normal trichromatic vision and for dichromatic observers as well, including: (1) color-matching, (2) hue and chromaticness appearances at or near threshold, (3) foveal spectral sensitivities obtained by flicker photometry, (4) heterochromatic additivity failures, (5) differences between luminance and brightness sensitivities, (6) wavelength discrimination, and (7) discrimination of near-threshold color differences. This section will deal primarily with item 7, color discrimination.

Figure 3 shows a schematic representation of the model. This figure is similar to Fig. 20 in Chapter 5 of this volume, although Fig. 3 is greatly simplified. In the Guth *et al.* model, signals from the cone mechanisms (ρ, γ, β) feed one achromatic (A) and two chromatic (T, D) systems which together

† L^* is as shown in Eq. (4), and a^* and b^* are as follows:

$$a^* = 500[(X/X_n)^{1/3} - (Y/Y_n)^{1/3}],$$
$$b^* = 200[(Y/Y_n)^{1/3} - (Z/Z_n)^{1/3}];$$

this space does not have a chromaticity diagram, because it does not have planes of constant luminance.

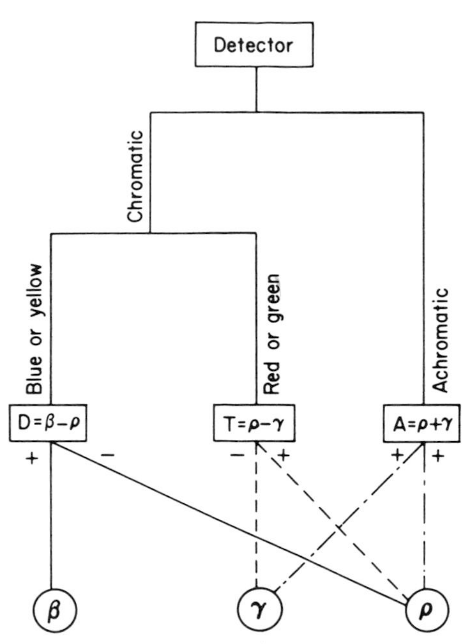

Fig. 3. Schematic diagram of a color-vision model proposed by Guth and co-workers (Guth and Lodge, 1973; Guth, Massof, and Benzschawel, 1980; Benzschawel and Guth, 1981). See text for explanation.

form the second stage of the mechanism. The symbol A stands for a non-opponent, achromatic system, whereas T and D represent opponent, chromatic, tritanopic and deuteranopic systems, respectively. The T system is equivalent to a red–green process, and D is the same as a blue–yellow process. Response characteristics of the second stage are expressed as linear transformations of the cone mechanisms. The linearity assumption is most nearly valid for threshold and near-threshold conditions (Massof and Bird, 1978; Bird and Massof, 1978; Guth *et al.*, 1980). The responses of the second stage are combined by vector addition in euclidean three-dimensional space to predict discrimination. Vectors of equal length represent equally detectable differences, and discrimination is taken to be proportional to differences between the vector angles corresponding to pairs of stimuli.

The *fundamental primaries* used by Guth *et al.* are those of Smith and Pokorny (Boynton, 1979) (see Table V of Chapter 5). The explicit transformations from Judd's (1951a) $\bar{x}'$, $\bar{y}'$, $\bar{z}'$ modified CIE color-mixture coefficients are

$$\begin{aligned} \rho &= 0.2435\bar{x}' + 0.8524\bar{y}' - 0.0516\bar{z}', \\ \gamma &= -0.3954\bar{x}' + 1.1642\bar{y}' + 0.0837\bar{z}', \\ \beta &= 0.6225\bar{z}'. \end{aligned} \tag{6}$$

This equation differs from the Smith and Pokorny equation of Table V in Chapter 5 only in that Eq. (6) normalizes the assumed cone sensitivities at their peaks. In common with most modern models, the fundamental primaries are not linear functions of the CIE 1931 color-mixture functions but are linearly related to one or another set of color-mixture functions that are thought to be more accurate representations of normal trichromatic visual responses.

The *opponent* and *nonopponent* systems are, in turn, linear transforms of the fundamental primaries of Eq. (6):

$$\begin{aligned} A &= 1.0(0.5967\rho + 0.3654\gamma), \\ T &= 1.0(0.9553\rho - 1.2836\gamma), \\ D &= 1.0(-0.0248\rho + 0.0483\beta), \end{aligned} \tag{7}$$

and the relative amounts of A; T, D are

$$\begin{aligned} a &= (A)(A + T + D)^{-1}, \\ t &= (T)(A + T + D)^{-1}, \\ d &= (D)(A + T + D)^{-1}. \end{aligned} \tag{8}$$

At the neutral threshold, $A = T = D$ and, therefore, $a = t = d = 1/3$. The relative distribution coefficients $(\bar{a}, \bar{t}, \bar{d})$ of this second stage of the Guth *et al.* model for the spectral components of an equi-energy spectrum are then formed from the combination of Eqs. (6), (7), and (8) as follows:

$$\begin{aligned} \bar{a} &= 1.0(0.9341\bar{y}'), \\ \bar{t} &= 1.0(0.7401\bar{x}' - 0.6801\bar{y}' - 0.1567\bar{z}'), \\ \bar{d} &= 1.0(-0.0061\bar{x}', - 0.0212\bar{y}' + 0.0314\bar{z}'). \end{aligned} \tag{9}$$

Equation (9) is normalized so that $(A^2 + T^2 + D^2)^{1/2} = 1.0$ at its peak of 545 nm. The sensitivity functions represented by Eq. (9) are shown in Fig. 4. When the expression is normalized for equal threshold detectability at each wavelength of stimulation (that is, when unit vector magnitudes are computed for each spectral stimulus), then the weighting functions are as shown in Fig. 5. Those weights are expressed relative to the threshold condition. In general, in the A, T, D vector space, the criterion for a limen or threshold difference corresponds to the condition in which the vector sum of the changes in A, T, D produced by a stimulus equals a constant.

For example, to predict discriminations arising from changes in chromaticity on a plane of constant luminance or luminance factor, T and D values are

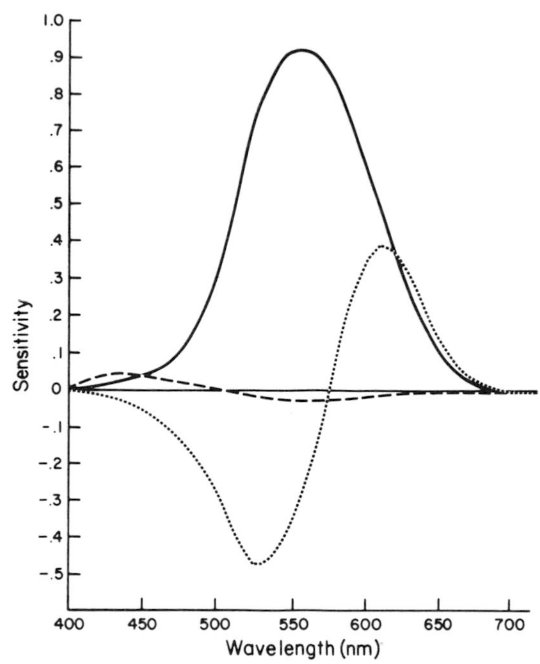

Fig. 4. Threshold-level equal-radiance response functions for A, the achromatic system (solid curves); T, the tritanopic system (short dashes); and D, the deuteranopic system (long dashes).

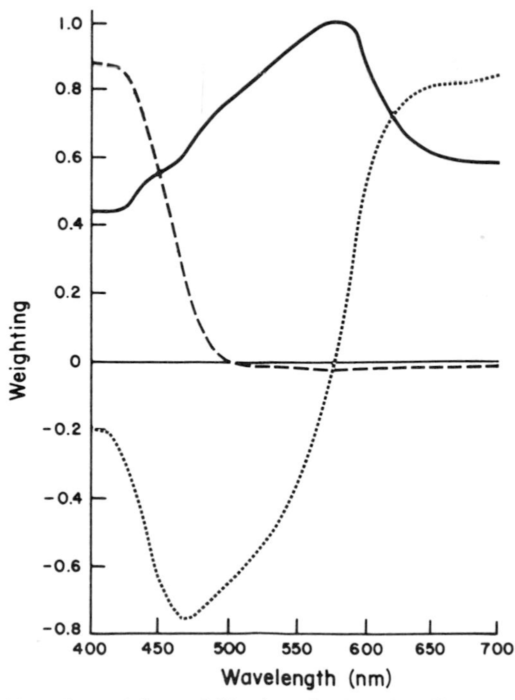

Fig. 5. Threshold-level equal-detectability (equal vector length) response functions for A (curve with maximum value), T (curve with minimum value), and D (remaining curve).

divided by those of A:

$$t_A = T/A,$$
$$d_A = D/A. \qquad (10)$$

When d_A and t_A are plotted orthogonally, the equivalent of a chromaticity diagram results in which distances among points corresponding to equal luminance stimuli are proportional to discriminal differences. Figure 6 illustrates three such diagrams. The loci of chromaticities corresponding to

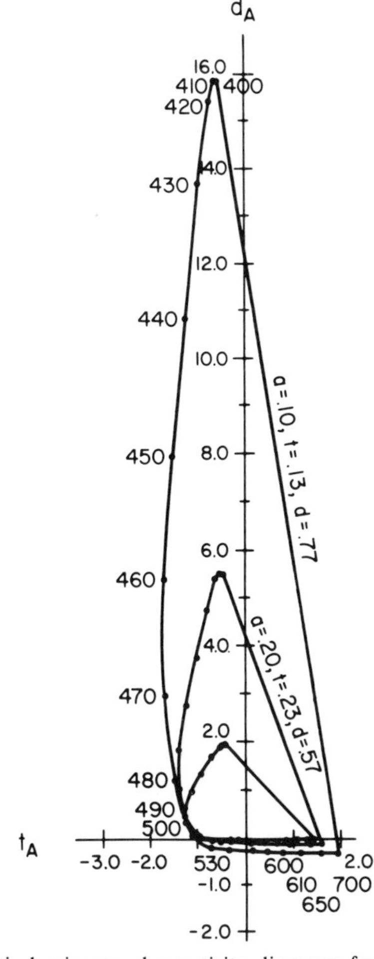

Fig. 6. Unit achromatic luminance chromaticity diagrams for three levels of luminance according to the Guth *et al.* model.

spectral stimuli are shown for three levels of luminance that yield the values of a, t, and d indicated in the graph. At threshold, where $a = t = d = 1/3$, the area contained by the spectrum locus is relatively small. The area defined by the suprathreshold condition where $a = 0.10$, $t = 0.13$, and $d = 0.77$ is much larger and tends to be expanded most in the region of short-wavelength stimuli. The model predicts that a single diagram is inappropriate for stimuli of different luminances. Figure 7 illustrates difference limen for centers of

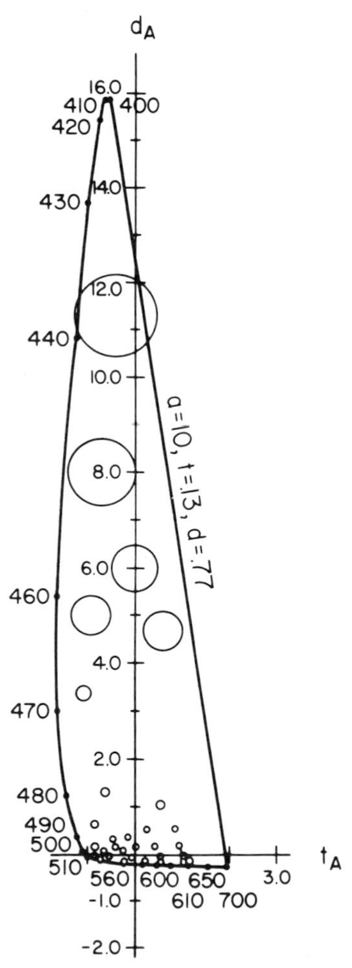

Fig. 7. Circles representing discrimination limen for the high-luminance condition of Fig. 6. Centers of solid circles correspond to MacAdam's (1942) ellipses.

MacAdam's (1942) discrimination data at the highest of the three luminance levels shown in Fig. 6. The circles are computed from the euclidean power metric:

$$\Delta_r = [(t_{A1} - t_{A2})^2 + (d_{A1} - d_{A2})^2]^{1/2}. \quad (11)$$

Note that the absolute sizes of the limen increase with the distance from the origin (representing the neutral or white point at $t_A = d_A = 0.0$); so the radius of Δ_r can be expressed at any point in the plane in terms of the limen corresponding to the neutral point Δ_{r_0} as

$$\Delta_r = \Delta_{r_0}(1^2 + t_A^2 + d_A^2)^{1/2}. \quad (12)$$

In the graph of Fig. 7, $\Delta_{r_0} = 0.004$. This represents the value of the Weber fraction (see Chapters 6 and 7) for the conditions of MacAdam's experiment. To facilitate comparison with MacAdam's data, the solid circles of Fig. 7 have been transformed to the Judd x', y' chromaticity diagram (which closely resembles the CIE 1931 x, y diagram) in Fig. 8 where the sizes and orientations of the ellipses can be compared directly with those shown in Fig. 1. The close correspondence between the 25 pairs of ellipses implies that the Guth *et al.*

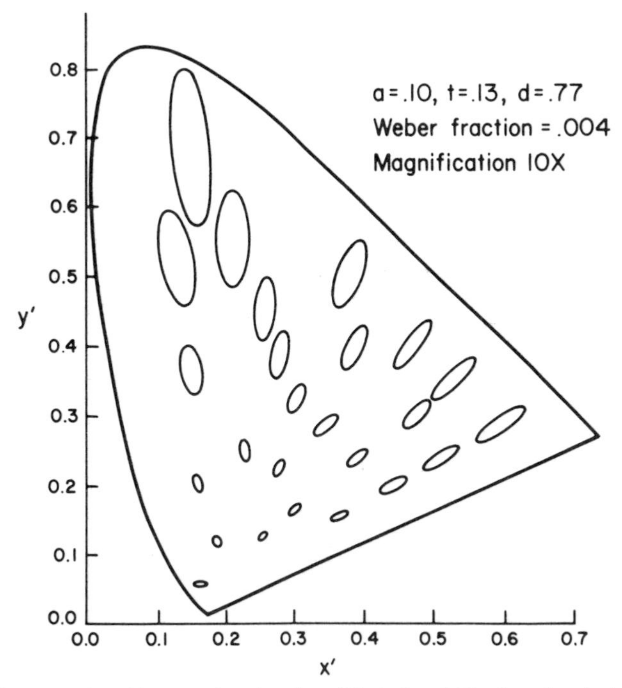

Fig. 8. Ellipses produced by transforming the solid circles of Fig. 7 to the x', y' chromaticity diagram.

model is locally uniform at or near threshold; this assumes, of course, that the MacAdam data are correct representations of threshold discrimination of color differences at constant luminance, an assumption that is, at least, not unreasonable, because other discrimination data tend to resemble those of MacAdam (see Chapter 3 in Volume 2 of this treatise).

Although the linear model may be locally uniform, it is not globally uniform. That is, equally discriminable differences are not represented everywhere in the diagram of Fig. 7 by circles of equal radii. If the t_A vs d_A plane is scaled in terms of the constant radius Δ_r, the resulting space (with coordinates t_{Ar} and d_{Ar}) should be both locally and globally uniform if the underlying mechanism that leads to discrimination ellipses is, in fact, linear. That is, when we plot the MacAdam data as

$$t_{Ar} = \left[\frac{t_A}{(1^2 + t_A^2 + d_A^2)^{1/2}}\right], \qquad d_{Ar} = \left[\frac{d_A}{(1^2 + t_A^2 + d_A^2)^{1/2}}\right], \tag{13}$$

then the ellipses should plot as circles of constant radii at all points if the linear model results in a perceptually uniform threshold color space.

Figure 9 shows that this is not the result when the MacAadm data are plotted according to the coordinate definitions of Eq. (13). Instead, there are systematic departures from uniformity such that distance is especially exaggerated for increasing "blue" responses (that is, when $D > 0$). The discrimination contours generally become increasingly elliptical as positive values of d_{Ar} increase in Fig. 9.

In view of the many possible sources of nonlinearity discussed in connection with the mechanism of vision in Chapter 5, it may not be too surprising that the linear model is not globally uniform even at threshold. (Note that several cascaded, linear processes may yield a net nonlinear transfer function when they contain additive constants, so that the term "nonlinear" should not be taken necessarily to imply that the underlying component mechanisms are nonlinear, only that they result in a combined nonlinear transfer from input to output.) It was pointed out in Chapter 5 that response compression functions, such as that proposed by Naka and Rushton (1966), have been widely useful in characterizing both physiological and psychophysical results in vision science. Benzschawel and Guth (1981, and personal communication) have recently adopted such a compression function in order to develop a nonlinear version of their model. They used compression functions of the following general form:

$$R = \frac{I^n}{\sigma + I^n}, \tag{14}$$

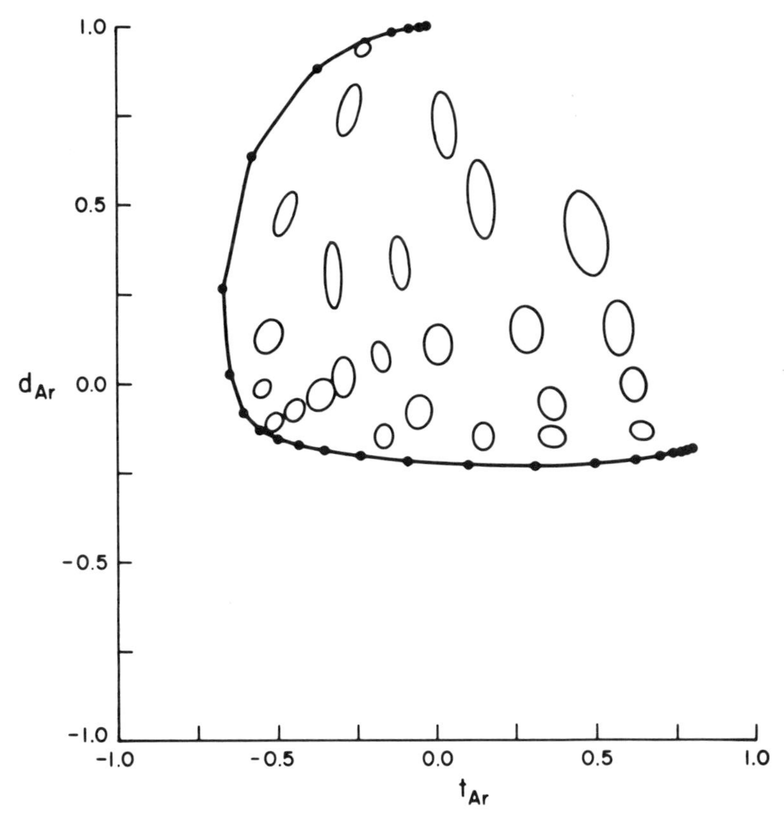

Fig. 9. Discrimination contours for MacAdam (1942) chromaticity centers computed according to Eq. (13) of the text.

where R is the resulting response magnitude, I is response magnitude assuming the linear version of the model, and σ and n are the half-saturation constants and exponents, respectively, which would be expected to differ for the T and D mechanism responses. Note that Eq. (14) is the expression developed by Naka and Rushton (1966) or Boynton (1979; Boynton and Whitton, 1970) to describe intensity response functions of single nerve cells. Benzschawel and Guth analyzed data on discrimination published by Pointer (1974) to derive explicit values for the constants n_t, n_d, and σ_t, σ_d of Eq. (14) but found that satisfactory predictions of those data could be obtained only if different values of σ and n were used for positive and negative values of t_A and d_A That is, different half-saturation constants and exponents were required for each direction of response: "red", "green", "blue", and "yellow". When suitable account was taken of the differences in luminances of the stimuli used in MacAdam's and Pointer's experiments, both sets of data were fit best using $\sigma_r = 1.275$, $n_r = 0.925$; $\sigma_g = 0.825$, $n_g = 1.125$; $\sigma_b = 2.975$, $n_b = 0.750$; and

$\sigma_y = 0.675, n_y = 1.225$. These values may be used to rewrite Eq. (14) as follows:

$$d^* = \frac{d_A^{0.925}}{1.275 + d_A^{0.925}}, \qquad \text{where} \quad d_A > 0,$$

or

$$d^* = \frac{d_A^{1.125}}{0.825 + d_A^{1.125}}, \qquad \text{where} \quad d_A < 0,$$

and

$$t^* = \frac{t_A^{0.75}}{2.975 + t_A^{0.75}}, \qquad \text{where} \quad t_A > 0,$$

or

$$t^* = \frac{t_A^{1.125}}{0.675 + t_A^{1.225}}, \qquad \text{where} \quad t_A < 0.$$

Further, for MacAdam's data, the model was improved by adding a small input from β receptors to the T system. That is, the second line of Eqs. (7) was changed to

$$T = 1.0(0.9553\rho - 1.2836\gamma + 0.05\beta),$$

to calculate the JND contours in Fig. 10. The d/t ratio was 6.0 (approximately

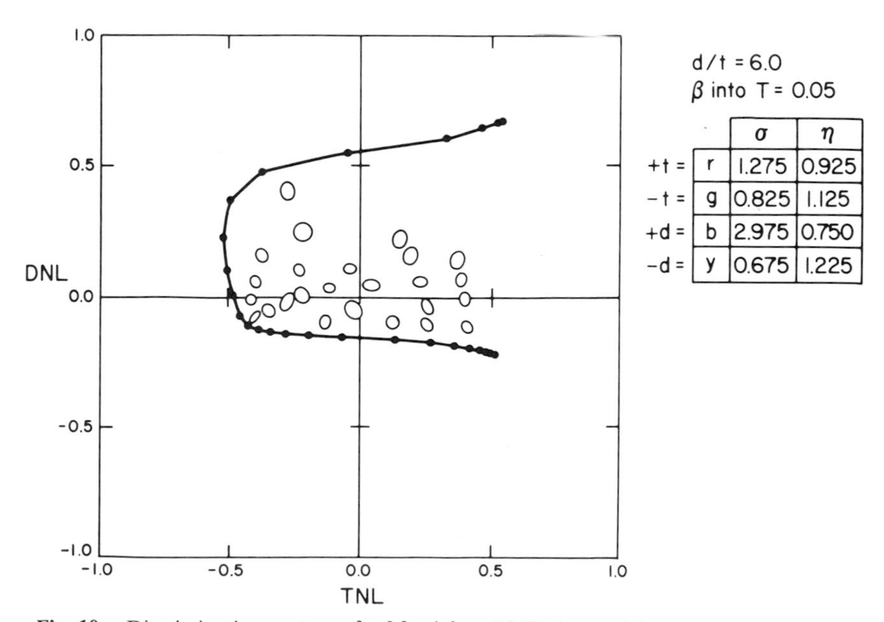

Fig. 10. Discrimination contours for MacAdam (1942) chromaticity centers.

as shown in Fig. 7). For Pointer's data, the amount of β into T was fixed at 0.13, and the d/t ratio was 11.0.

When t^* and d^* are scaled in terms of the constant radius Δr to form the radius-normalized coordinates $t^*_{\Delta r}$ and $d^*_{\Delta r}$ (as in Fig. 10, which may be compared with Fig. 9), the MacAdam data are arrayed more nearly as uniform circles. A quantitative analysis of results in Fig. 10 and Pointer's data revealed that the nonlinear version of the vector model is more uniform both locally and globally than the linear version and the 1931 CIE x, y space, the 1960 u, v diagram, and the L^*, a^*, b^* space.

It can be seen from the foregoing that a metric for color discrimination based on a model of the visual mechanism can be derived to provide

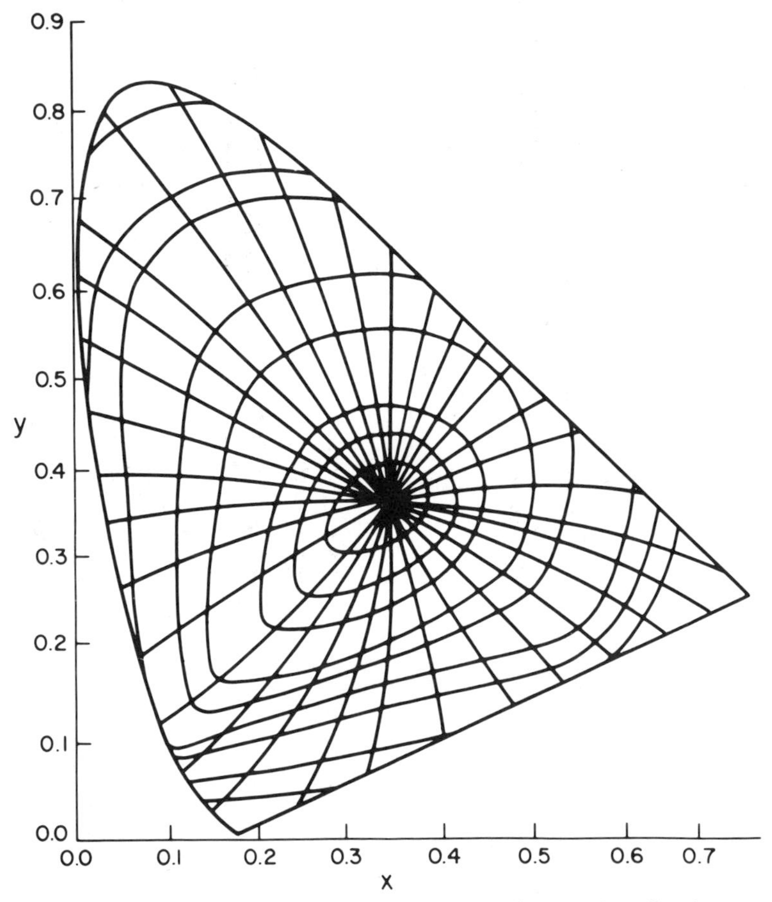

Fig. 11. Contours of isochroma and isohue corresponding to Munsell color system as predicted by Benzschawel and Guth nonlinear model. See text for details.

reasonable predictions of experimental data on color discrimination and, at least in its nonlinear form, is both locally and globally uniform to a reasonable approximation. When optimum parameter values are selected for suprathreshold color differences such as those exemplified by the Munsell color system, the model predicts contours of constant hue and chroma as shown in Fig. 11. Although the contours resemble those of the Munsell color system when plotted in the CIE 1931 chromaticity diagram, the congruence is far from perfect. This kind of paradox, in which threshold and suprathreshold predictions cannot be made equally well with a single model, is a common finding. The Guth–Benzschawel model has many degrees of freedom, allowing it to be applied to both threshold and suprathreshold conditions. Nevertheless, it appears to work best for threshold data, the kind from which it was derived. The next section will discuss a model that was derived from suprathreshold data.

D. SUPRATHRESHOLD COLOR RELATIONS

The preceding section concerned color differences at or near threshold. Those kinds of differences are of interest in sorting materials according to "shades" and in routine quality-control applications. Many other colorimetric applications concern attempts to measure and evaluate large (suprathreshold) differences in colors of samples, often involving a need to characterize the color appearances of samples. In general, threshold techniques do not predict suprathreshold color relations well. For this reason a number of attempts have been made to develop systems for assessing color appearances and large color differences. Many of these attempts have been empirical. Some, however, have tried to draw on existing knowledge of the mechanism of color vision. Because our knowledge of that mechanism is far from complete, the models developed for suprathreshold color assessment are necessarily partly empirical. One of the most recent proposals for a partly empirical suprathreshold model is that offered by Hunt (1982). It will be described in this section. Although it differs in a number of significant ways from other models, it can be described in a manner that permits comparison with threshold models such as the one described in the previous section.

Figure 12 shows a schematic diagram of the Hunt (1982) model. He postulates an overall system consisting of three zones (a concept that has been used before, for example, Müller, 1930; Judd, 1951b; Friele, 1965). The first zone represents reception and processing in the cone mechanisms and is characterized by linear processing. The second zone involves neural combinations of the outputs of the first zone. This second zone has nonlinear processing; in particular, signals are combined according to their square roots.

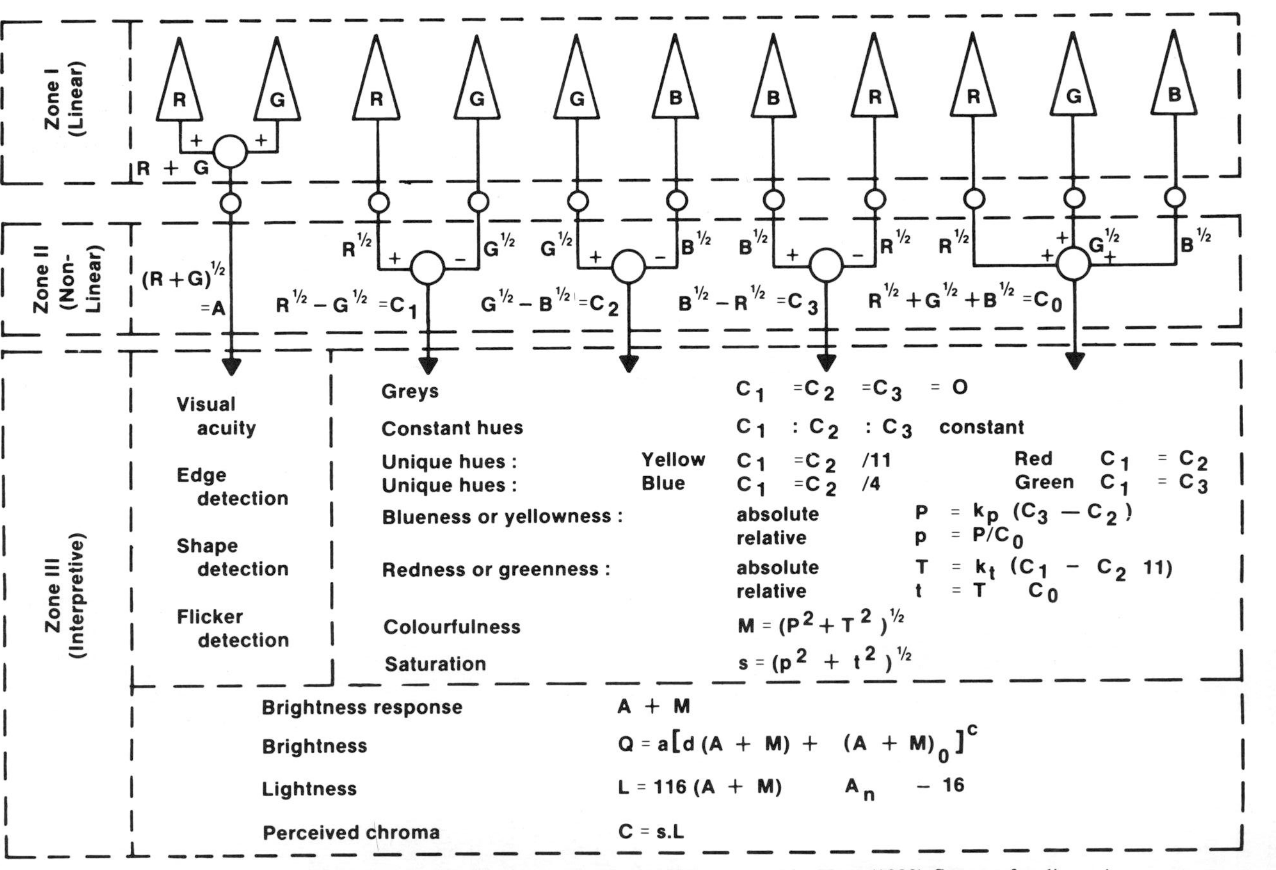

Fig. 12. Schematic diagram of a suprathreshold color-vision model proposed by Hunt (1982). See text for discussion.

The third zone, which Hunt labels "interpretive", conceptually combines the outputs of the second zone to form analogs of various attributes of color appearances.

The cone sensitivity functions of the first zone are based on Estevez's (1979) smoothing of the 2° color-mixture functions of Stiles and Burch (1959); these are listed at 10-nm intervals as $\bar{r}, \bar{g}, \bar{b}$ in Table IV of Chapter 5 in this volume. A set of fundamental primaries ρ, γ, and β were determined from $\bar{r}$, $\bar{g}$, $\bar{b}$ approximately as follows:

$$\begin{aligned}\rho &= 23.8\bar{r} + 62.3\bar{g} + 3.17\bar{b},\\ \gamma &= 8.34\bar{r} + 90.0\bar{g} + 6.94\bar{b},\\ \beta &= 0.00\bar{r} + 3.21\bar{g} + 164.0\bar{b}.\end{aligned} \tag{15}$$

Equations (15) are normalized so that $\rho = \gamma = \beta$ when adaptation is to CIE Illuminant *C*. The ρ, γ, and β signals are labeled *R*, *G*, and *B* by Hunt. His symbols will be used in the remainder of this section, although they should not be taken to imply anything directly about appearances of colors, about receptors, or about receptor sensitivities.

The *R* and *G* signals are additively combined in the first zone to provide the input to the achromatic system in zone 2. Because the second zone compresses incoming signals by taking their square roots, the achromatic signal *A* is

$$A = (R + G)^{1/2}. \tag{16}$$

The underlying spectral sensitivity of *A*, which consists of $R + G$ or simply A^2, closely resembles the Judd (1951a) $V(\lambda)$ function. Although $A^2(\lambda)$ differs somewhat from the modified $V(\lambda)$, Hunt argues that the deviations are well within the range of differences found among results from different (normal) observers when individual $V(\lambda)$ functions are determined by flicker photometry. Accordingly, *A* is taken to relate to contrast sensitivity and other achromatic or luminance related aspects of vision.

Hunt identifies four chromatic signals in the second zone (seven if account is taken of the complements of three difference signals). These are labeled C_i and defined as

$$\begin{aligned}C_1 &= R^{1/2} - G^{1/2},\\ C_2 &= G^{1/2} - B^{1/2},\\ C_3 &= B^{1/2} - R^{1/2},\\ C_0 &= R^{1/2} + G^{1/2} + B^{1/2}.\end{aligned} \tag{17}$$

The input signals are the first-zone output signals: *R*, *G*, *B*. Each of these is raised to the one-half power and is then combined by addition or subtraction to form the zone 2 outputs: C_0, C_1, C_2, C_3, or the complements of C_1, C_2, C_3. The zone 2 output signals form the input to zone 3.

The equivalent of red versus green and blue versus yellow opponent chromatic signals are formed in the third zone and are symbolized T and P, respectively (P is the equivalent of the Guth *et al.* D system, and T is the same in both instances). These chromatic opponent systems are defined as

$$P = k_p[C'_3 - C_2],$$
$$T = k_t[C_1 - C_2/11]. \tag{18}$$

The functions of wavelength defined by A, T, and P may be termed *responsiveness* functions; they are plotted in Fig. 13.

The advantage offered by Hunt's model in having the intervening step (zone 2) between the receptor stage (zone 1) and the stage at which opponent chromatic signals of the red versus green and yellow versus blue kind are formed (zone 3) is that certain definitions of color-appearance anchors can be made easily. These will be discussed before proceeding further with the development of the opponent mechanism. *Neutral color appearances* (white, gray, or black) are defined by the model as satisfying the following condition:

$$C_1 = C_2 = C_3 = 0, \qquad \text{for} \quad \text{neutral colors.} \tag{19}$$

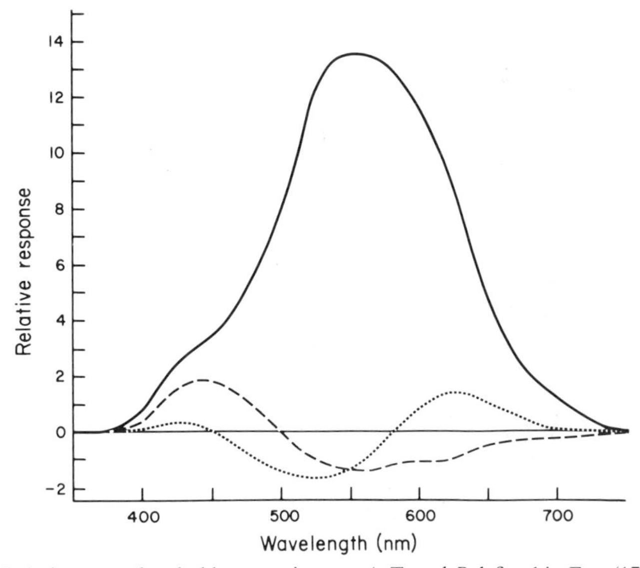

Fig. 13. Relative suprathreshold responsiveness A, T, and P defined in Eqs. (17) and (18) of the text for Hunt's (1982) model normalized for adaptation to CIE Illuminant C.

Constant hue is defined by the condition:

$$C_1/C_2/C_3 = \text{a constant.} \tag{20}$$

Unitary hues are defined by the conditions:

$$\begin{aligned} C_1 &= C_2, && \text{for unitary red,} \\ C_1 &= C_3, && \text{for unitary green,} \\ C_1 &= C_2/11, && \text{for unitary yellow,} \\ C_1 &= C_2/4, && \text{for unitary blue.} \end{aligned} \tag{21}$$

These relationships are simple and, for the most part, obvious from relationships in zone 2 of the schematic representation of the model in Fig. 12. The expressions of Eq. (21) for unitary yellow and blue are the only ones that require discussion. In examining data on interactions among responses and psychophysical determinations of constant-hue loci in chromaticity diagrams, Hunt found that before comparison with the R response, the G response is partially inhibited by the B response. A factor of $1/11$ represents the effect of this inhibition; so relative sensitivity at middle wavelengths is expressed as $(12G/11) - (B/11)$, the $12/11$ being merely a normalizing factor. Therefore, although unitary red and unitary green are simply

$$R^{1/2} - G^{1/2} = G^{1/2} - B^{1/2}, \quad \text{or} \quad C_1 = C_2, \quad \text{for red,}$$

and

$$R^{1/2} - G^{1/2} = B^{1/2} - R^{1/2}, \quad \text{or} \quad C_1 = C_3, \quad \text{for green,}$$

the corresponding expressions for unitary yellow and blue must be

$$R^{1/2} - G^{1/2} = (G^{1/2} - B^{1/2})/11, \quad \text{or} \quad C_1 = C_2/11, \quad \text{for yellow,}$$

and

$$R^{1/2} - G^{1/2} = (G^{1/2} - B^{1/2})/4, \quad \text{or} \quad C_1 = C_2/4, \quad \text{for blue,}$$

when the arithmetic is carried through the calculations.

For binary hue appearances in general, the blueness or yellowness and redness or greenness are related to the extents to which the preceding unitary hue expressions are unequal. Certain other considerations must also enter into the determination, however. Because with binary hues the system model must deal with unequal suprathreshold conditions, account must be taken of the dynamic characteristics of the mechanism of color vision. That is, the fact that for differences in color, responses increase and decrease differently for different hues must also be represented by the model.

For example, it is likely that the retinal responses are less effective in signalling differences and appearance changes in blue–yellow directions than in red–green ones. Possibly this arises because of the relative paucity of β cones in the fovea. Vos and Walraven (1971) estimated that the ratio of ρ to γ to β cones may be about 40:20:1, which would mean that the least populous cone type (β) in the yellow–blue signal is present in only one-twentieth the numbers of the least populous cone type (γ) in the red–green signal. In terms of signal-to-noise considerations the blueness–yellowness correlate should then be divided by $\sqrt{20}$, or about 4.5. Reasoning that the blueness–yellowness of red-appearing colors must be correlated with the difference between C_1 and C_2 and that the blueness–yellowness of green-appearing binary hues must be related to the difference between C_3 and C_1, Hunt has assumed a simple average of the two expressions; that is, yellowness–blueness of binary hues is related to $(C_1 - C_2 + C_3 - C_1)/2$ which simplifies to $(C_3 - C_2)/2$. This expression is part of that for P in Eq. (18). The $\sqrt{20}$ factor relating to relative population densities of cone types then becomes part of the constant k_p in Eq. (18). The redness–greenness of binary hues might similarly have been taken as the average of $C_1 - C_2/11$ and $C_1 - C_2/4$, but on the grounds that unitary blue colors are less well defined in chromaticity than are unitary yellow colors, Hunt chose to use $C_1 - C_2/11$ alone, and this expression is part of that for T in Eq. (18).

In addition, both constants, k_p and k_t, include an empirical factor e_s that varies with dominant wavelength to position binary hue loci properly with respect to the unitary hue loci and an arbitrary factor to represent crosstalk among receptor and neural signal pathways. The factor e_s can be approximated by the following expression:

$$e_s = \frac{0.95 - [(C_1 + C_3/12) + 0.22(C_3 - C_2)]}{[7.30(C_1 + C_3/12)^2 + 0.48(C_3 - C_2)^2]^{1/2}}. \tag{22}$$

The crosstalk factor was deduced by Hunt to be one-hundredth (1%) of the first-stage signal strength and combined from all three cone signals (which corresponds to one-tenth the strength of C_0). That is, the combined signals R_c, G_c, and B_c are represented as

$$\begin{aligned} R_c &= (10/13)R^{1/2} + (1/13)C_0, \\ G_c &= (10/13)G^{1/2} + (1/13)C_0, \\ B_c &= (10/13)B^{1/2} + (1/13)C_0. \end{aligned} \tag{23}$$

The normalizing factor 13 is included merely to ensure that

$$R_c + G_c + B_c = R^{1/2} + G^{1/2} + B^{1/2}.$$

For simplicity the crosstalk is added equally to all three channels.

Thus we have the expressions for the opponent chromatic mechanisms P and T as shown in Eq. (18) [$P = k_p(C'_3 - C_2)$ and $T = k_t(C_1 - C_2/11)$], where

$$k_p = (1/2)(10/13)e_s/4.5,$$

and

$$k_t = (10/13)e_s. \tag{24}$$

The foregoing relations are then used to specify color appearance attributes. Unitary hue specifications are set forth in Eqs. (21), those for constant hue are given in Eq. (20), and the condition for achromaticness is defined by Eq. (19). Chromaticness or colorfulness M is the vector sum of P and T opponent chromatic responses:

$$M = (P^2 + T^2)^{1/2}. \tag{25}$$

Whereas P and T provide specifications of the absolute amounts of blueness–yellowness and redness–greenness, respectively, their relative amounts may be determined simply by reference to the absolute total chromatic response C_0 given in Eq. (17) such that

$$p = P/C_0, \quad \text{for} \quad \text{relative blueness or yellowness,}$$

and

$$t = T/C_0, \quad \text{for} \quad \text{relative redness or greenness.} \tag{26}$$

The relative amount of chromaticness, called *saturation*, s, is then the vector sum of p and t:

$$s = (p^2 + t^2)^{1/2}. \tag{27}$$

A graph of p plotted against t provides a color diagram (as in Fig. 14) with the properties that constant hue is arrayed by straight radial lines and constant saturation is arrayed as concentric circles.

The model proposed by Hunt (1982) then provides a globally uniform metric of *model parameters.* He has shown the model to predict reasonably well a selection of experimental results where suprathreshold color differences or color appearances have been scaled. Thus in Fig. 15, for example, the solid curves show the predictions given by the model for loci of constant hue and saturation (plotted in an arbitrary chromaticity diagram that has approximately the same geometric properties as the CIE 1976 u', v' diagram); the broken curves show the corresponding loci as determined experimentally in the Swedish Natural Color system (plotted in CIE 1976 u', v' and superimposed). It remains to be seen how well this suprathreshold model will stand the test of time and use. In any case it represents a creative attempt to combine results of color-scaling data with known facts of physiology, something that has been difficult to do until now.

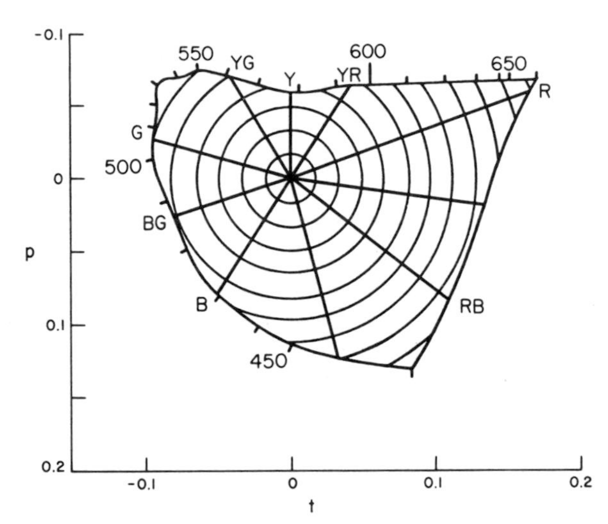

Fig. 14. Diagram of p vs t showing the locus of coordinates for spectral components of an equi-energy spectrum (with wavelengths noted at 50-nm intervals) together with loci of constant saturation s and constant hue. Unitary R, G, B, and Y hue loci are labeled, and binary hue loci RB, BG, YG, and YR are also labeled. Saturation is represented by concentric circles.

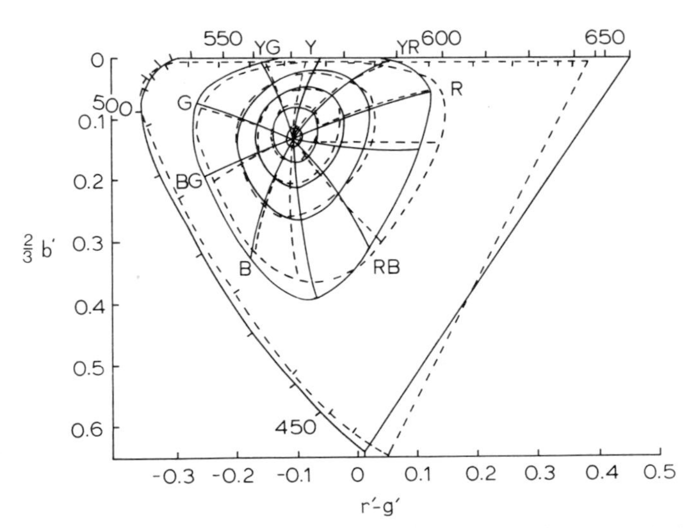

Fig. 15. Loci of constant hue and saturation predicted by part of the grid of Fig. 14 (solid curves), compared with the experimentally determined loci of the Swedish Natural color system (broken curves).

E. CHROMATIC ADAPTATION

The model described in the preceding section incorporates a set of fundamental primaries that are normalized for adaptation to CIE Illuminant C, a relative spectral power distribution adopted in 1931, which was thought to simulate the spectral quality of average daylight. Frequently, however, questions are asked and problems addressed that involve adaptation to illuminants with other spectral qualities. Visual sensitivities differ for different conditions of adaptation. In physiology the term *adaptation* refers to the adjustment of the sensitivity of sensory processes in response to differential excitation of specific receptor systems. That is, the process by which the organism adjusts itself to changes in the environment to maintain physiological stability is called adaptation. Many studies have been conducted to determine the action of the visual mechanism in adapting to different spectral qualities of illumination and to establish engineering data that can be used to predict changes in color appearance of different stimuli that elicit color matches when adaptation varies; these are reviewed elsewhere (for instance, Terstiege, 1972; Jameson and Hurvich, 1972; Bartleson, 1978; Wyszecki and Stiles, 1982).

It has been pointed out that not only color appearances but also large and small color differences vary with chromatic adaptation (Bartleson, 1981). For the threshold and suprathreshold color models described in the preceding two sections to be generally useful, it is necessary to account for differences occasioned by changes in chromatic adaptation. This can be done by using a transformation equation. For example, if Hunt's (1982) model is to be used to analyze data that were gathered in an experiment in which observers were adapted to incandescent illumination, it is necessary first to transform the data to the equivalent specifications that would obtain if the observers had been adapted to CIE Illuminant *C*. This is a process of predicting *corresponding colors*. A corresponding color is defined as the stimulus that, under some different condition of adaptation, evokes the same color appearance as another stimulus when it was seen under the original state of adaptation. Once these corresponding colors are determined for CIE Illuminant *C*, the model may be used to estimate suprathreshold color relations as described in the preceding section.

Chromatic adaptation transforms intended to provide such predictions may be classed in a number of ways (for example, Bartleson, 1978; Wright, 1981a). Most are empirical in the sense that they represent a mathematical method for reconstructing the results of one or more experiments in which adaptation shifts have been studied. Some are based on theory, in the sense that cognizance is taken of existing knowledge of the mechanism of vision or of colorimetric theory. Within each of these two classes, the forms of adaptation transformations may be linear or nonlinear. Fundamental sensitivities may be altered in inverse proportion to the extent of stimulation of each of the three cone types.

Johannes Adolph von Kries (1877, 1904, 1905, 1911) was the first to offer a complete quantitative expression for such a process. Others have found that the linear adjustment of fundamental sensitivities is inadequate to characterize most experimental results properly. Accordingly, nonlinear adaptation transforms have been suggested, particularly in recent years. Some, such as those suggested by Jameson and Hurvich (1972), involve two stages of adjustment: an initial linear coefficient modification of the von Kries type at the receptor level and a subsequent neural response-additivity adjustment. The net effect is a nonlinear transformation. Others, such as Steffen (1955), MacAdam (1961), Bartleson (1979), or Takahama *et al.* (1976), have described nonlinear empirical equations for predicting corresponding colors. Probably the most completely developed quantitative transformation at present is one offered by Nayatani *et al.* (1981a,b). Although it is largely empirical in form, it takes some cognizance of what is known about the visual mechanism and addresses a diversity of experimentally observed phenomena. The remainder of this section will describe this adaptation transformation model in order to illustrate how chromatic adaptation can be addressed to extend the gamut of conditions over which threshold and suprathreshold models of color difference can be used.

Nayatani *et al.* (1981a,b) have developed a nonlinear model that is capable of predicting a variety of different adaptation effects. It is a multistage model with a single transformation expression that involves a power function with a variable exponent. Theoretical bases are given for the chromaticity and luminance dependence of the exponent. The parameters of the model were determined by analyzing the results of many different experiments. The model predicts changes in saturation of chromatic samples when adaptation varies from light to dark (Hunt, 1950, 1952, 1953), changes in corresponding colors when the spectral quality of illumination varies (Aubert, 1865; Exner, 1868), and at least two additional adaptation effects. One relates to the change in contrast among stimuli of different luminance factors when illuminance varies (for example, Hess and Pretori, 1894; Stevens, 1961; Jameson and Hurvich, 1961; Bartleson and Breneman, 1967). The other relates to the perception of achromatic objects when they are illuminated with chromatic light: samples of high luminance factor tend to appear in the hue of the illuminant, whereas samples of low luminance factor appear in the complementary hue; only one level of luminance factor tends to appear achromatic (Helson and Judd, 1932, Helson, 1938; Judd, 1940; Helson, 1964).

The model considers three response mechanisms beginning with a set of fundamental primaries. The fundamental primaries used by Nayatani *et al.* are those suggested by Pitt (1935). Using the R, G, B symbols of Pitt's and the Nayatani *et al.* publications, one obtains

$$\begin{aligned} R &= 0.0711X + 0.9494Y - 0.0156Z, \\ G &= -0.4462X + 1.3173Y + 0.0979Z, \\ B &= 0.9188Z, \end{aligned} \tag{28}$$

where X, Y, Z are CIE 1931 tristimulus values, and R, G, B are the same as tristimulus values determined from ρ, γ, β fundamental primaries. Equation (28) can be rewritten in matrix notation for convenience:

$$\begin{bmatrix} R \\ G \\ B \end{bmatrix} = \mathbf{M} \begin{bmatrix} X \\ Y \\ Z \end{bmatrix}, \tag{29}$$

where

$$\mathbf{M} = \begin{bmatrix} 0.0711 & 0.9494 & -0.0156 \\ -0.4462 & 1.3173 & 0.0979 \\ 0.0000 & 0.0000 & 0.9188 \end{bmatrix}.$$

The same matrix $\mathbf{M}$ may be used to determine the fundamental tristimulus values R_b, G_b, B_b of the background or adapting stimulus.

The second stage of the model determines three responses, R^*, G^*, B^*, as linear functions of the fundamental tristimulus values of the test stimulus R, G, B and the noise components of the visual mechanism R_n, G_n, B_n:

$$\begin{bmatrix} R^* \\ G^* \\ B^* \end{bmatrix} = \begin{bmatrix} k_r & 0 & 0 \\ 0 & k_g & 0 \\ 0 & 0 & k_b \end{bmatrix} \begin{bmatrix} R + R_n \\ G + G_n \\ B + B_n \end{bmatrix}, \tag{30}$$

where

$$\begin{aligned} k_r &= (R_b + R_n)^{-1}, \\ k_g &= (G_b + G_n)^{-1}, \\ k_b &= (B_b + B_n)^{-1}, \end{aligned}$$

and R_n, G_n, B_n are the Weber fractions for the experimental condition (see Chapter 7).

The third stage performs a nonlinear modification of the responses R^*, G^*, and B^* to form operational output responses $\mathscr{R}$, $\mathscr{G}$, $\mathscr{B}$:

$$\begin{bmatrix} \mathscr{R} \\ \mathscr{G} \\ \mathscr{B} \end{bmatrix} = \begin{bmatrix} \mu_r & 0 & 0 \\ 0 & \mu_g & 0 \\ 0 & 0 & \mu_b \end{bmatrix} \begin{bmatrix} (R^*)^{\nu_r} \\ (G^*)^{\nu_g} \\ (B^*)^{\nu_b} \end{bmatrix}. \tag{31}$$

The coefficients μ_i (where $i = r, g$, or b) of Eq. (31) are chosen to satisfy two conditions. The first is that regardless of the absolute chromaticity and luminance of the spectrally nonselective background and the illuminant, brightness constancy holds for a spectrally nonselective sample of medium lightness, for example, a sample with a luminance factor of 0.2. The second condition is that when the luminance factor of a spectrally nonselective sample is equal to that of the adaptation level $[L_0 = (\alpha_0 E)/\pi]$, the sample will appear achromatic regardless of the absolute chromaticity of the illuminant (absolute chromaticity, as used above, refers to the chromaticity with respect to an equi-energy illuminant, α_0 is the luminous reflectance of the background, and the symbol E refers to illuminance in lux). Thus color constancy holds for a spectrally nonselective sample when its luminous reflectance is equal to that of the background to which the observer is completely adapted. In the case of the spectrally nonselective background with a luminance factor of 0.2, brightness and color constancy obtain, and in this case μ_i can be arbitrarily set to unity: $\mu_r = \mu_g = \mu_b = 1$.

The exponents v_r, v_g, v_b are each compression functions of the fundamental tristimulus values of the adapting background R_b, G_b, B_b as follows:

$$v_r = \kappa_1 \left[\frac{6.469 + 6.362(R_b)^{0.4495}}{6.469 + (R_b)^{0.4495}} \right],$$
$$v_g = \kappa_1 \left[\frac{6.469 + 6.362(G_b)^{0.4495}}{6.469 + (G_b)^{0.4495}} \right], \qquad (32)$$
$$v_b = \kappa_2 \left[\frac{8.414 + 8.091(B_b)^{0.5128}}{8.414 + (B_b)^{0.5128}} \right].$$

Figure 16 illustrates the shapes of the functions of Eq. (32) by normalizing to unity for the condition $\theta = R_b = G_b = B_b = 63.66$, which corresponds to an illuminance of 1000 lux for an equi-energy illuminant. The functions each describe an S-shaped compression curve with respect to log θ. The gain of the B system is different from that of the R and G systems. The relationship of these functions is such that v_i in Eq. (31) approaches a constant value at high adapting luminances. This predicts the result found by Breneman (1980) at high adapting luminances, that is, the transformation that best predicts experimental results varies in its degree of nonlinearity with the illuminance to which the observer is adapted. Figure 17 shows the effect of this variation. The three graphs represent predicted contours of Munsell hue and chroma samples at value 5 (0.1977 luminance factor) viewed with adaptation to CIE illuminant A (transformed from CIE Illuminant C). Graph (a) is for adaptation to 1000 lux; graph (b) for adaptation to a level of 30,000 lux; and graph (c) is a linear von Kries prediction, which is the same for all levels of illuminance. The

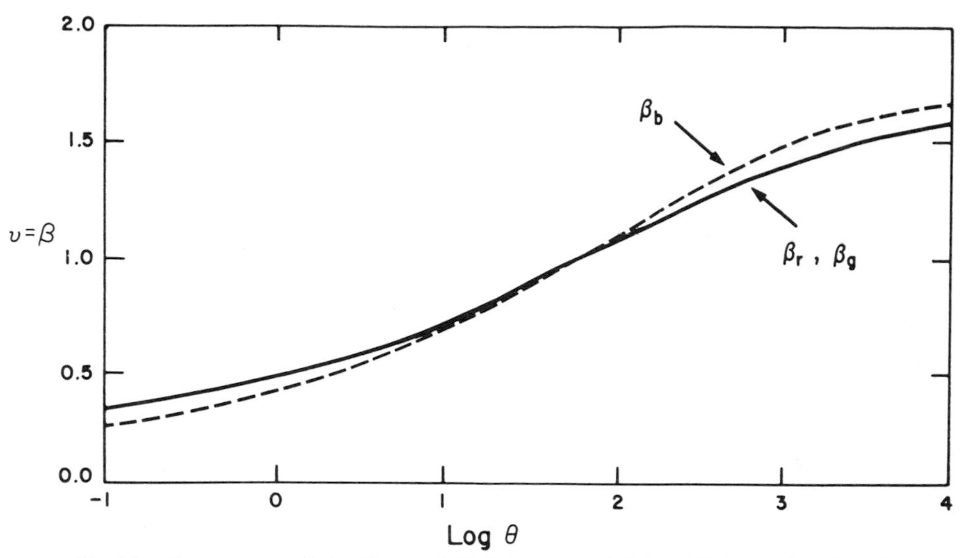

Fig. 16. Exponents v of the Nayatani, Takahama, and Sobagaki chromatic adaptation transformation as a function of fundamental tristimulus values θ. See text for discussion.

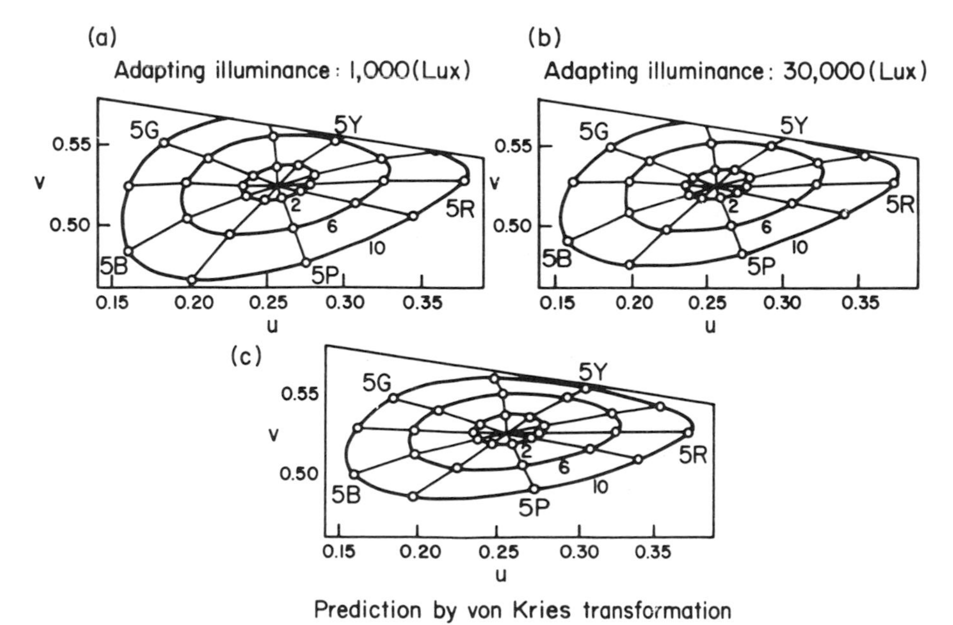

Fig. 17. Predictions of corresponding chromaticities for Munsell samples of value 5. See text for discussion.

shape of the contours in the 30,000-lux graph (b) is more similar to that of the von Kries prediction contours than is that of the 1000-lux graph.

The model predicts a similar kind of nonlinear variation with differences in the luminances of test samples (together with their backgrounds) relative to a reference sample (and background) luminance. Hunt (1950, 1952, 1953) found that the purities required for matching the color appearances of chromatic stimuli varied as their luminances relative to that of the reference change. This is illustrated in Fig. 18 for eight chromatic stimuli each of which was presented with illuminances (for itself and its background) of 12.5, 42, 127, 1020, and 16,900 lux and always compared haploscopically with stimuli (and background) at 127 lux illuminance. The large dot on each curve represents 127 lux illuminances for both eyes. The outermost dot on each curve (labeled A on the curve for sample R) represents the matching chromaticity for the 16,900-lux condition, and the innermost point on each curve (labeled F) corresponds to the matching chromaticity for the 12.5-lux condition. The predictions are for adaptation to CIE Illuminant B; all conditions match those of Hunt's experiment. As with Hunt's experimental results, the model's predictions indicate that for samples viewed at a moderate illuminance the purities must be

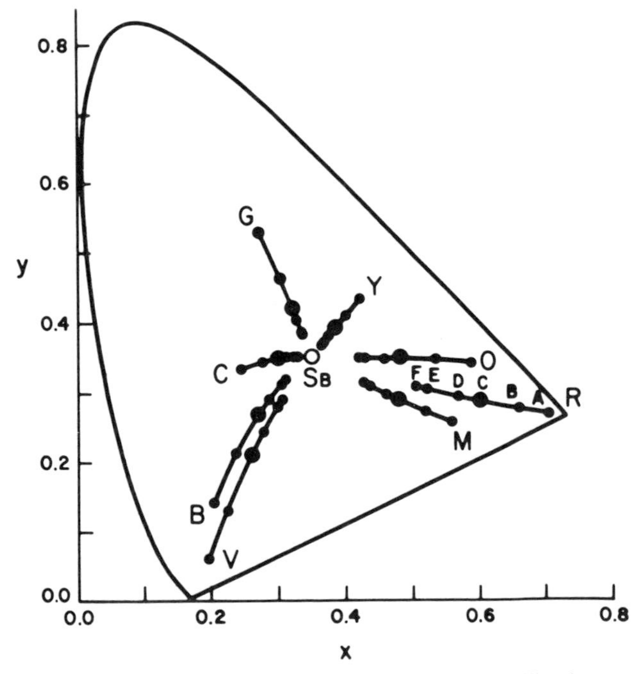

Fig. 18. Predictions of corresponding chromaticities for various illuminances. See text for discussion.

increased to match the color appearances of colors of the same chromaticity viewed at high illuminances, and the purities must be reduced to match samples viewed at low illuminances.

Figure 19 shows the predictions of the model of Nayatani *et al.* for matching lightnesses as a function of illuminance. Each of seven samples was used, and the luminance Y'_{rel} under 1000 lux of CIE Illuminant C was adjusted to match the lightness perceived in the same sample illuminated with the same source at levels from 1 to 10,000 lux. The increased range of log luminances required to effect matches as illuminance increases is illustrated by the spreading of the curves in Fig. 19. The dashed horizontal line at $Y'_{\text{rel}} = 20$ corresponds to the condition of lightness constancy implied by the model.

Figure 20 illustrates the changes in chromaticity required to match the appearances of spectrally nonselective samples of several luminance factors viewed under chromatic illumination. The predictions accord well with experimental results (for instance, Helson, 1964). The open circles in the graphs of Fig. 20 indicate the chromaticity coordinates of the adapting illuminants. The solid circles indicate the chromaticity coordinates of the corresponding colors under CIE Illuminant C for samples of different luminance factors; the points are identified by Munsell value notations V, which relate approximately to luminance factor Y_{rel} in the following way: $Y_{\text{rel}} \cong [(V + 16)/116]^{1/3}[10]^{-1}$. Samples with high luminance factors (or high Munsell value specifications) have appearances that are matched under CIE Illuminant C by samples with chromaticity coordinates that tend toward

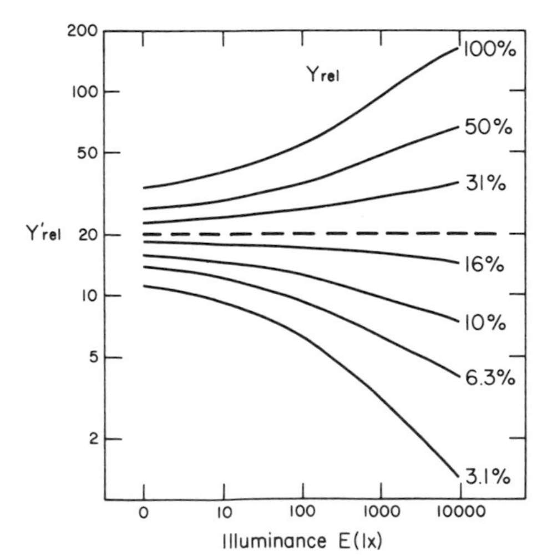

Fig. 19. Predictions of increased lightness contrast as a function of illuminance. See text for discussion.

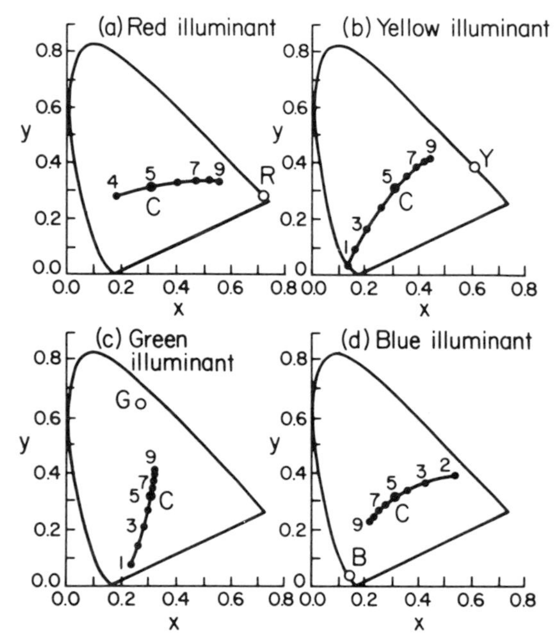

Fig. 20. Predictions of Helson–Judd effect. See text for discussion.

those of the illuminant, whereas samples with low luminance factors are matched when the corresponding chromaticity coordinates tend away from those of the light source. Those samples of Munsell value 5 have the same corresponding chromaticity coordinates as for adaptation to CIE Illuminant C (the large points labeled C). The results are exactly those that have been called the Helson–Judd effect (for example, Burnham *et al.*, 1963).

The foregoing examples illustrate the fact that the Nayatani *et al.* model can predict a variety of adaptation phenomena. Although the model draws heavily on the structure of trichromatic theory rather than opponent-colors theory, it can be used to extend the utility of threshold and suprathreshold models such as those discussed in earlier sections of this chapter. The adaptation model can be restructured in terms of opponent-color theory, and Nayatani and coworkers plan to do so (Nayatani, 1982, private communication).

F. THE FUTURE OF COLOR MEASUREMENT

The three models discussed in the preceding sections are all different. They have in common some degree of empiricism in their structures (the Guth *et al.* model is the least empirical and the Nayatani *et al.* model the most), but the implied or underlying attributes of the mechanisms of vision that they incorporate differ significantly in their details. This is a consequence of the fact that the mechanism of vision is not yet completely understood. In time, when our understanding of the visual mechanism is more complete, a single model undoubtedly will serve to address all three predictive problems that stimulated the development of the separate models. All three phenomena are the result of a single mechanism of vision, so logically all three can be predicted by a single model if it satisfactorily represents the functioning of that mechanism. In fact, an adequate model should predict satisfactorily all significant visual phenomena. The science of vision is still a long way from achieving that degree of utility in vision models. When this goal is attained, we shall be able to measure color appearances and all the relationships implied by that term, including color matches, large and small differences in color, and magnitudes of the separate attributes of color under all conditions of adaptation.

For centuries there have been attempts to derive quantitative methods for estimating color appearances. The first scale of perceived brightness, for example, was described in about 150 B.C. by the Greek protoscientist Hipparchus (Jastrow, 1887). Aristotle, Ptolemy, Alhazen, and Leonardo DaVinci are among those who have attempted to describe color appearances directly (Boring, 1942). Sir Isaac Newton's famous experiments with a prism, carried out at Woolsthorpe in 1666, relied on direct estimates of color appearances, experiments with which he established the correct schema for spectral energies, showed that every stimulus has a complementary with which a mixture can be created to form neutral-appearing light, and, in fact, elucidated two of the three rules that have come to be known as Grassman's laws of color mixture (Newton, 1704; Herivel, 1965; Manuel, 1980). Young (1807a,b). Palmer (Walls, 1956), Helmholtz (1852), Maxwell (1855), and less well-known figures such as Brewster, Herschel, Wollaston, and Grassman (Sherman, 1971) all strove to reduce the estimation of color appearances to a process of quantitative prediction based on some model of color vision. As in psychophysics in general, dynamic problems of color appearance were avoided by confining quantitative structures to the condition of color matches. Colorimetry developed from this milieu. Despite the hopes of many, the conventions of colorimetry adopted by the CIE in 1931 (and those of photometry adopted in 1924) were based entirely on methodologies that *did not* measure color appearance (for example, Wright, 1981b); they measured only stimulus conditions for matching the colors of single objects.

Largely independently, the fields of physiology and sensory psychophysics developed an extensive body of experimental data that related to the functioning of the organism under dynamic conditions of color perception, that is, the assessment of color appearances rather than simply color matches. The physiologist Ewald Hering was one of the most important workers in this area (for example, Hering, 1872, 1874a,b,c,d, 1875, 1878, 1964), although many others concerned themselves with physiological and phenomenological matters of suprathreshold color vision (see, for example, Katz, 1911, 1930, 1935; Parsons, 1924; Boring, 1942, 1950; Wasserman, 1978). Quantitative development of models took a giant step forward beginning in the period of 1951 to 1955 with the work of Leo Hurvich and Dorothea Jameson (for reviews of the subjects of their many publications, see Hurvich, 1978, 1981). They provided a quantitative structure to Hering's opponent-colors theory of vision by supplying experimental measurements of opponent sensitivities under various conditions of stimulation. Physiologists were quick to see the implications to their own work of the Hurvich and Jameson results. The structure of these concepts has increasingly found acceptance throughout most of vision science during the past three decades.

Hurvich and Jameson, aware that much more is unknown about the mechanism of vision than is known, have consistently avoided developing detailed, explicit models. Instead, they have described general, often even qualitative, concepts that provide the bases for research to generate the information necessary to develop quantitative models. In terms of psychophysical data, the relations of receptor activities to variables exhibiting singular and opposite modes of response throughout different parts of the spectrum are expressed generally as

$$\begin{aligned} V_1 &= f_1\{\textstyle\sum[a_{11}\varphi(\lambda)\rho(\lambda) + a_{12}\varphi(\lambda)\gamma(\lambda) + a_{13}\varphi(\lambda)\beta(\lambda)]\}, \\ V_2 &= f_2\{\textstyle\sum[a_{21}\varphi(\lambda)\rho(\lambda) - a_{22}\varphi(\lambda)\gamma(\lambda) + a_{23}\varphi(\lambda)\beta(\lambda)]\}, \\ V_3 &= f_3\{\textstyle\sum[-a_{32}\varphi(\lambda)\rho(\lambda) + a_{32}\varphi(\lambda)\gamma(\lambda) + a_{33}\varphi(\lambda)\beta(\lambda)]\}, \end{aligned} \tag{33}$$

where the symbols are the same as have been used throughout this chapter except that V_i stands for a variable and f_i represents an unspecified function, linear or nonlinear. Equations (33) relate only to part of the process, however. It would be appropriate for an isolated color (seen with a dark surround) under conditions of physiological neutral (for instance, dark) adaptation. In any other situation the variables V_i are formed for every stimulus element in the field of view, and neural responses (which we may continue to symbolize A, C_1, C_2) are formed from their interactions weighted in space, time, and energy terms. Thus for even a simple field consisting of a focal element (c) and a visually completely extensive surround s, we must write

$$\begin{aligned} A &= g_1(V_{1c} - V_{1s}), \\ C_1 &= g_2(V_{2c} - V_{2s}), \\ C_2 &= g_3(V_{3c} - V_{3s}). \end{aligned} \tag{34}$$

The functions f_i and g_i together with the signs of combination and choice of constants permit many degrees of freedom in the development of a model. Hurvich and Jameson have tended to take the view that their structure can be made explicit only as illustrative approximations until such time as sufficient data exist to define the implied relationships unequivocally. That is, we presently do not know enough about the mechanism of vision to specify all the implied relationships correctly, but in the meantime (until we do have sufficiently accurate information), we can use the general structure as a guide to useful experimentation and analysis by starting with simple combinations and approximate values for the parameters in the general model.

Equation (34) becomes more complicated as the stimulating field increases in spatial and energy complexity. In the most general case, for what are usually referred to as complex field conditions, the general expression would be approximately as follows:

$$
\begin{aligned}
A &= g_1(V_{1c} \pm V_{1s1} \pm V_{1s2} \pm \cdots \pm V_{1sn}),\\
C_1 &= g_2(V_{2c} \pm V_{2s1} \pm V_{2s2} \pm \cdots \pm V_{2sn}),\\
C_2 &= g_3(V_{3c} \pm V_{3s1} \pm V_{3s3} \pm \cdots \pm V_{3sn}).
\end{aligned} \tag{35}
$$

Equation (35) specifies neural responses. Still another step is required to specify the conscious responses that we call color perceptions. They are formed from the neural responses by integration and elaboration in the brain. Without accounting for the influence of apperception, the influence of past experiences on current perceptions, we should then have to specify a conscious color response Ξ as

$$\Xi = \Omega(A, C_1, C_2), \tag{36}$$

and the component attributes of Ξ as

$$
\begin{aligned}
\xi_1 &= \zeta_1(\Xi)\\
\xi_2 &= \zeta_2(\Xi),\\
&\vdots\\
\xi_n &= \zeta_n(\Xi)
\end{aligned} \tag{37}
$$

where ξ_i may stand for any attribute of color perception, and ζ_i may be any kind of function.

The future of colorimetry, and color measurement in general, lies in the solutions to Eqs. (36) and (37). The models described in this chapter represent attempts toward attainment of that goal. It should be obvious that they fall considerably short of reaching the goal. Had we, in fact, reached it, this entire volume would take the form of a simple "cookbook" or guide to calculation. Instead, the volume has necessarily involved only a description of those parts of the presumed solution that have been established thus far, together with some discussion of the considerations that should go into attempts to perform visual measurements. We hope that the reader will have gained some appreciation of how to make visual measurements and, in particular, how they might be interpreted.

The conventions of colorimetry have been useful over the past-half-century since the time of their international standardization by the CIE. However, there has been a growing swell of dissatisfaction with standard colorimetric conventions, paralleling the increasing growth of understanding of the limitations of conventional colorimetry. This is recognized by the CIE. What was a single committee on colorimetry in 1931 has fostered additional committees on color rendering (to evaluate appearances of objects illuminated by different light sources) and on vision (to consider the perception of brightness, lightness, and other attributes of color perception). In addition, the colorimetry committee has itself generated a number of subcommittees to consider differences in color, color appearances, the influence of chromatic adaptation, and other matters that go beyond the consideration of simple color matches. In short, the CIE is now directing more and more of its energies to questions of measuring and specifying color appearances. These are questions that are not easily answered at this time; much of the information that is required is still missing. For that reason there is an enhanced awareness among colorimetrists of the value of research in physiology and experimental psychology to questions that colorimetric methodologies of the future must address. There is, then, a blending together of the interests and efforts of people in these various fields of scientific enquiry. The future of colorimetry will necessarily develop from such tandem efforts. Eventually, satisfactory solutions will be at hand for the relationships represented by Eqs. (36) and (37). Until that time, care should be exercised to understand and interpret correctly the meaning and limitations of visual measurements. Much can be learned from such measurements if they are carefully performed and properly interpreted.

ACKNOWLEDGMENTS

The originators of the models discussed in this chapter were most helpful in correcting the first manuscript, supplying additional data and figures, and in graciously granting permission to discuss unpublished work that helps to provide a broader picture of the development and applications of their models. I am grateful to Drs. S. Lee Guth and Terry Benzschawel for their help with the description of the models in Section C, to Dr. Robert W. G Hunt for assistance in describing his suprathreshold model in Section D, and to Dr. Yoshi Nayatani for his guidance in the preparation of Section E on chromatic adaptation. However, I take full responsibility for the descriptions that appear here; if there are any errors, they are mine alone.

REFERENCES

Aubert, H. (1865). "Physiologie der Netzhaut." Morganstern, Breslau.

Bartleson, C. J. (1978). A review of chromatic adaptation. *In* "AIC Color 77" (F. W. Billmeyer, Jr. and G. Wyszecki, eds.), pp. 63–96. Adam Hilger, Bristol.

Bartleson, C. J. (1979). Changes in color appearance with variations in chromatic adaptation. *Color Res. Appl.* **4**, 119–139.

Bartleson, C. J. (1980). Colorimetry. *In* "Optical Radiation Measurements: Color Measurement" (F. Grum and C. J. Bartleson, eds.), Vol. 2, pp. 33–148. Academic Press, New York.

Bartleson, C. J. (1981). On chromatic adaptation and persistence. *Color Res. Appl.* **6**, 153–160.

Bartleson, C. J., and Breneman, E. J. (1967). Brightness perception in complex fields. *J. Opt. Soc. Am.* **57**, 953–957.

Benzschawel, T., and Guth, S. L. (1981). Preliminary report on the development of a uniform chromaticity space. *J. Opt. Soc. Am.* **71**, 1608 (Abstract).

Billmeyer, F. W., Jr., and Saltzman, M. (1981). "Principles of Color Technology," 2nd ed. Wiley (Interscience), New York.

Bird, J. F., and Massof, R. W. (1978). A general zone theory of color and brightness vision. II. The space-time field. *J. Opt. Soc. Am.* **68**, 1471–1481.

Boring, E. G. (1942). "Sensation and Perception in the History of Experimental Psychology." Appleton-Century-Crofts, New York.

Boring, E. G. (1950). "A History of Experimental Psychology," 2nd ed. Appleton-Century-Crofts, New York.

Boynton, R. M. (1979). "Human Color Vision." Holt, Rinehart and Winston, New York.

Boynton, R. M., and Whitton, D. N. (1970). Visual adaptation in monkey cones: Recordings of late receptor potentials. *Science* **170**, 1423–1426.

Breneman, E. J. (1980). Corresponding chromaticities for different states of adaptation with complex fields. Presented at the *ISCC Helson Memorial Symposium on Chromatic Adaptation*, Williamsburg, Virginia, February 1980.

Burnham, R. W., Hanes, R. M., and Bartleson, C. J. (1963). "Color: A Guide to Basic Facts and Concepts." John Wiley and Sons, New York.

CIE [Commission Internationale de l'Eclairage] (1971). "Colorimetry." Publication CIE No. 15, Bureau Central de la CIE, Paris.

Estévez, O. (1979). "On the Fundamental Data-Base of Normal and Dichromatic Color Vision." PhD Thesis, University of Amsterdam, The Netherlands.

Exner, S. (1868). Über einige neue subjektive Gesichtseinungen. *Arch. Physiol.* **1**, 375–394.

Friele, L. F. C. (1965). Further analysis of colour discrimination data. pp. 302–314, *In* "Internationale Farbtagung Luzern 1965, Tagungsbericht." (M. Richter, ed.), pp. 302–314. Musterschmidt-Verlag, Göttingen.

Guth, S. L., and Lodge, H. R. (1973). Heterochromatic additivity, foveal spectral sensitivity, and a new color model. *J. Opt. Soc. Am.* **63**, 450–462.

Guth, S. L., Massof, R. W., and Benzschawel, T. (1980). A vector model for normal and dichromatic color vision. *J. Opt. Soc. Am.* **70**, 197–212.

Helmholtz, H. L. F. von (1852). Über die Theorie der zusammengesetzen Farben. *Ann. Phys. Chem.* **163**, 45–66.

Helson, H. (1938). Fundamental problems in color vision. I. The principle governing changes in hue, saturation and lightness of nonselective samples in chromatic illumination. *J. Exp. Psychol.* **23**, 439–476.

Helson, H. (1964). "Adaptation-Level Theory." Harper and Row, New York.

Helson, H., and Judd, D. B. (1932). A study on photopic adaptation. *J. Exp. Psychol.* **15**, 380–398.

Hering, E. (1872). Zur Lehre vom Lichtsinne. I. Über successiven Lichtinduction. *S.-B. Akad. Wiss. Wien Math.-Nat. Kl. Part III* **66**, 5–24.

Hering, E. (1874a). Zur Lehre vom Lichtsinne. II. Über simultanen Lichtcontrast. *S.-B. Akad. Wiss. Wien Math.-Nat. Kl. Part III* **68**, 186–201.

Hering, E. (1874b). Zur Lehre vom Lichtsinne. III. Über simultanen Lichtinduction und über successiven Contrast. *S.-B. Akad. Wiss. Wien Math.-Nat. Kl. Part III* **68**, 229–244.

Hering, E. (1874c). Zur Lehre vom Lichtsinne. IV. Über die sogenannte Intensität der Lichtempfindung and über die Empfindung des Schwarzen. *S.-B. Akad. Wiss. Wien Math.-Nat. Kl. Part III* **69**, 85–104.

Hering, E. (1874d). Zur Lehre vom Lichtsinne. V. Grundzüge einer Theorie des Lichtsinnes, *S.-B. Akad. Wiss. Wien. Math.-Nat. Kl. Part III* **69**, 179–217.

Hering, E. (1875). Zur Lehre vom Lichtsinne. VI. Grundzüge einer Theorie des Farbensinnes, *S.-B. Akad. Wiss. Wien Math.-Nat. Kl. Part III* **70**, 169–204.

Hering, E. (1878). "Zur Lehre vom Lichtsinne." Carl Gerold's Sohn, Vienna.

Hering, E. (1964). "Outlines of a Theory of the Light Sense" (L. M. Hurvich and D. Jameson, transl.), Harvard University Press, Cambridge, Massachusetts.

Herivel, J. (1965). "The Background to Newton's Principia. A Study of Newton's Dynamical Researches in the Years 1664–1684." Clarendon Press, Oxford.

Hess, C., and Pretori, H. (1894). Messende Untersuchungen über die Gesetzmäsigkeit des simultanen Helligkeits-Contrastes. *Arch. Opthalmol.* **40**, 1–27.

Hunt, R. W. G. (1950). The effect of daylight and tungsten light-adaptation on color perception. *J. Opt. Soc. Am.* **40**, 363–371.

Hunt, R. W. G. (1952). Light and dark adaptation and the perception of color. *J. Opt. Soc. Am.* **42**, 190–199.

Hunt, R. W. G. (1953). The perception of color in 1° fields for different states of adaptation. *J. Opt. Soc. Am.* **43**, 479–484.

Hunt, R. W. G. (1980). Color terms, symbols, and their usage. *In* "Optical Radiation Measurements: "Color Measurement" (F. Grum, and C. J. Bartleson, eds.), Vol. 2. pp. 11–32. Academic Press, New York.

Hunt, R. W. G. (1982). A model of colour vision for predicting colour appearance. *Color Res. Appl.* **7**, 95–112.

Hurvich, L. M. (1978). Two decades of opponent processes. *In* "AIC Color 77" (F. W., Billmeyer, Jr. and G., Wyszecki, eds.), pp. 33–61. Adam Hilger, Bristol.

Hurvich, L. M. (1981). "Color Vision." Sinauer Associates, Sunderland (Massachusetts).

Jameson, D., and Hurvich, L. M. (1961). Complexities of perceived brightness. *Science* **133**, 174–179.

Jameson, D., and Hurvich, L. M. (1972). Color adaptation: Sensitivity, contrast, after-images. pp. 568–581, *In* "Handbook of Sensory Physiology: Visual Psychophysics" (D. Jameson and L. M., Hurvich, eds.), Vol. VII/4. Springer-Verlag, Berlin.

Jastrow, J. (1887). The psycho-physic law and star magnitude. *Am. J. Psychol.* **1**, 112–127.

Judd, D. B. (1940). Hue, saturation and lightness of surface colors with chromatic illumination. *J. Opt. Soc. Am.* **30**, 2–32.

Judd, D. B. (1951a). International Commission on Illumination, Technical Committee No. 7, colorimetry and artificial daylight. Report of Secretariat, United States Committee. *In* "CIE Compute rendu douzième session, Stockholm." CIE, New York.

Judd, D. B. (1951b). Basic correlates of the visual stimulus. *In* S. S., Stevens, (ed.) "Handbook of Experimental Psychology," pp. 811–867. John Wiley and Sons, New York.

Judd, D. B., and Wyszecki, G. (1975). "Color in Business, Science and Industry," 3rd ed. John Wiley and Sons, New York.

Katz, D. (1911). "Die Erscheinungsweisen der Farben und ihre Beeinflussung durch die individuelle Erfahrung." Barth, Leipzig.

Katz, D. (1930). "Der Aufbau der Farbwelt." Barth, Leipzig. (Rev. 2nd ed. of Katz, 1911.)

Katz, D. (1935). "The World of Colour," (R. B. MacLeod and C. W. Fox, transl.). Kegan Paul, Trench, Trubner, London.

Kries, J. A., von (1877). Über die Ermüdung des Sehnerven. *Arch. Opththalmol.* **23**, 17–19.

Kries, J. A., von (1904). Die Gesichtsempfindungen. *In* "Die Physiologie der Sinne III" (W. von Nagel, ed.), pp. 109–282. Vieweg, Braunschweig.

Kries, J. A., von (1905). Die Gesichtsempfindungen. *In* "Handbuch der Physiologie der Menschen" (W. von Nagel, ed.), Vol. 3, pp. 109–282. Vieweg, Braunschweig.

Kries, J. A. von (1911). Die Theorien des Licht- und Farbensinnes. *In* "Handbuch der Physiologische Optik II" (W. von Nagel, ed.), pp. 366–369. Leopold Voss, Hamburg.

MacAdam, D. L. (1942). Visual sensitivities to color differences in daylight. *J. Opt. Soc. Am.* **32**, 247–274.

MacAdam, D. L. (1961). A nonlinear hypothesis for chromatic adaptation. *Vision Res.* **1**, 9–41.

Manuel, F. E. (1980). "A Portrait of Isaac Newton." Frederick Muller, London.

Massof, R. W., and Bird, J. F. (1978). A general zone theory of color and brightness vision. I. Basic formulation. *J. Opt. Soc. Am.* **68**, 1465–1471.

Maxwell, J. C. (1855). Experiments on colour as perceived by the eye, with remarks on colour-blindness. *Trans. R. Soc. Edinburgh* **21**, 275–298.

Müller, G. E. (1930). "Über die Farbenempfindungen." Barth, Leipzig.

Naka, K. I., and Rushton, W. A. H. (1966). S-potentials from colour units in the retina of fish (*Cyprinidae*). *J. Physiol. London* **185**, 536–555.

Nayatani, Y., Takahama, K., and Sobagaki, H. (1980). Estimation of adaptation effects by use of a theoretical nonlinear model. *In* "CIE Proceedings of the 19th Session, Kyoto 1979," pp. 490–494. CIE Publication No. 50, CIE, Paris.

Nayatani, Y., Takahama K., and Sobagaki, H. (1981a). A nonlinear model of chromatic adaptation. *Farbe* **29**, 109–126.

Nayatani, Y., Takahama, K., and Sobagaki, H. (1981b). Formulation of a nonlinear model of chromatic adaptation. *Color Res. Appl.* **6**, 161–171.

Newton, I. (1704). "Opticks." Innys, London.

Parsons, J. H. (1924). "An Introduction to the Study of Colour Vision." Cambridge University Press, Cambridge.

Pitt, F. H. G. (1935). "Characteristics of Dichromatic Vision" (Medical Research Council, Report of the Committee on Physiology of Vision, No. XIV). His Majesty's Stationery Office, London.

Pointer, M. R. (1974). Color discrimination as a function of observer adaptation. *J. Opt. Soc. Am.* **64**, 750–759.

Sherman, P. D. (1971). "Problems in the Theory and Perception of Colour: 1800–1860." Ph.D. Thesis, University of London, London (Cited in Wright 1981b.)

Steffen, D. (1955). Untersuchungen zur Theorie des Farbensehens, *Z. Biol.* **108**, 161–167.

Stevens, S. S. (1961). To honor Fechner and repeal his law. *Science* **133**, 80–86.

Stiles, W. S., and Burch, J. M. (1959). N.P.L. colour-matching investigation: Final report. *Opt. Acta* **6**, 1–26.

Takahama, K., Sobagaki, H., and Nayatani, Y. (1976). Analysis of chromatic-adaptation effect by a linkage model. *J. Opt. Soc. Am.* **67**, 651–656.

Terstiege, H. (1972). Chromatic adaptation. A state-of-the-art report. *J. Color Appear.* **1**, 19–24.

Vos, J. J., and Walraven, P. L. (1971). On the derivation of the foveal receptor primaries. *Vision Res.* **11**, 799–818.

Walls, G. L. (1956). The G. Palmer story. *J. Hist. Med.* **11**, 66–96. (Excerpts from G. Palmer, 1777, "Theory of Colors and Vision." Leacroft, London. *In* "Sources of Color Science." (D. L. MacAdam, ed.). MIT Press, Cambridge, Massachusetts, 1970.)

Wasserman, G. S. (1978). "Color Vision: An Historical Introduction." John Wiley and Sons, New York.

Wright, W. D. (1981a). Why and how chromatic adaptation has been studied. *Color Res. Appl.* **6**, 147–152.

Wright, W. D. (1981b). Historical and experimental background to the 1931 CIE system of colorimetry. *In* "Golden Jubilee of Colour in the CIE," pp. 3–18. The Colour Group (Great Britain). The Society of Dyers and Colourists, Bradford, Yorkshire.

Wyszecki, G., and Stiles, W. S. (1982). "Color Science," 2nd ed. Wiley (Interscience), New York.

Young, T. (1807a). On the theory of light and colours. *In* "Lectures in Natural Philosophy," (T. Young, ed.), Vol. 2. Privately printed for Joseph Johnson by William Savage, St. Paul's Churchyard, London.

Young, T. (1807b). An account of some cases of the production of colours. *In* "Lectures in Natural Philosophy" (T. Young, ed.), Vol. 2. Privately printed for Joseph Johnson by William Savage, St. Paul's Churchyard, London.

Index

D

K

L

M

N

O

P

T

U

V

W

X

Y

Z